Microbiology

A CENTENARY PERSPECTIVE

Microbiology

A CENTENARY PERSPECTIVE

EDITED BY

**Wolfgang K. Joklik • Lars G. Ljungdahl
Alison D. O'Brien • Alexander von Graevenitz
Charles Yanofsky**

WITH A FOREWORD BY
Joshua Lederberg

ASM PRESS

WASHINGTON, D.C.

Copyright © 1999 American Society for Microbiology
1325 Massachusetts Avenue NW
Washington, DC 20005-4171

Library of Congress Cataloging-in-Publication Data

Microbiology : a centenary perspective / edited by Wolfgang K. Joklik…[et al.].
    p.   cm.
    ISBN 1-55581-162-0 (casebound).
    ISBN 1-55581-169-8 (pbk.)
    1. Microbiology. 2. Microbiology—History—20th century Sources.
    I. Joklik, Wolfgang K.
    QR41.2.M57 1999
    579—dc21   99-20173
                                                              CIP

CONTENTS

CONTRIBUTORS

JOHN J. MEKALANOS
Department of Microbiology and Molecular Genetics, Harvard Medical School,
Boston, Massachusetts

ALISON D. O'BRIEN (COMMITTEE CHAIR)
Department of Microbiology and Immunology, Uniformed Services University of
the Health Sciences, Bethesda, Maryland

SECTION III. *General and Applied Microbiology*

HAROLD DRAKE
Lehrstuhl Fur Okologische Mikrobiologie Bitok, Universitat Bayreuth, Bayreuth,
Germany

ROBERT P. GUNSALUS
Department of Microbiology and Molecular Genetics, University of California, Los
Angeles, California

LARS G. LJUNGDAHL (COMMITTEE CHAIR)
Center for Biological Resource Recovery, University of Georgia, Athens, Georgia

JOHN R. SOKATCH
Department of Biochemistry and Molecular Biology, University of Oklahoma
Health Science Center, Oklahoma City, Oklahoma

JUDY D. WALL
Department of Biochemistry, University of Missouri, Columbia, Missouri

WILLIAM B. WHITMAN
Department of Microbiology, University of Georgia, Athens, Georgia

SECTION IV: *Molecular Biology and Physiology*

JONATHAN R. BECKWITH
Department of Microbiology and Molecular Genetics, Harvard Medical School,
Boston, Massachusetts

JEFFREY H. MILLER
Department of Microbiology and Molecular Genetics, University of California, Los
Angeles, California

LUCY SHAPIRO
Department of Developmental Biology, Stanford University, Stanford, California

THOMAS J. SILHAVY
Department of Molecular Biology, Princeton University, Princeton, New Jersey

CHARLES YANOFSKY (COMMITTEE CHAIR)
Department of Biological Sciences, Stanford University, Stanford, California

A. DALE KAISER (INTRODUCTION)
Division of Molecular Microbiology, Stanford University, Stanford, California

NORMAN R. PACE (INTRODUCTION)
Department of Plant and Microbial Biology, University of California, Berkeley,
California

Section V. Virology

ROBERT M. CHANOCK
Laboratory of Infectious Diseases, National Institute for Allergy and Infectious Diseases, National Institutes of Health, Bethesda, Maryland

WOLFGANG K. JOKLIK (COMMITTEE CHAIR)
Department of Microbiology, Duke University Medical Center, Durham, North Carolina

KENNETH N. KREUZER
Department of Microbiology, Duke University Medical Center, Durham, North Carolina

BERNARD MOSS
Laboratory of Viral Diseases, National Institute for Allergy and Infectious Diseases, National Institutes of Health, Bethesda, Maryland

PETER K. VOGT
Department of Molecular and Experimental Medicine, The Scripps Research Institute, La Jolla, California

MILTON B. ZAITLIN
Department of Plant Pathology, College of Agriculture and Life Sciences, Cornell University, Ithaca, New York

ACKNOWLEDGMENTS

As is true with most worthwhile projects, it took many generous people to bring *Microbiology: A Centenary Perspective* to fruition. We are grateful to all. We especially appreciate the devoted work of the committees who donated their valuable time to consider the many worthy papers that might have been included in this volume. Making a limited number of choices from so many magnificent possibilities could not have been an easy task. The committee chairs gave special effort in the final selections and in the preparation of the introductions. In some cases, important noncommittee contributors donated their time to write introductions. Dr. Joshua Lederberg found time amidst his many important projects, including many for ASM, to prepare a special foreword for the book.

To all the scientists whose work was chosen for this volume and to all those whose work built the bridges and connections to allow important scientific discoveries to be made, we are eternally grateful.

Thanks to Dennis Burke, Ric Clark, April Rivers, and the staff at The Publishing Management Connection, L.L.C. (PMC), for all facets of production from permissions gathering, photographing articles, and budget control to the final book. We thank Susan Schmidler for her sensitive design and tasteful cover, William Bell & Eastern Photographic Associates for their expert photography, and Ellie Tupper at ASM Press for her skillful editing and valuable advice.

Eve-Marie Lacroix, Kenneth Niles, and Karen Sinkule at the National Library of Medicine (NLM) generously allowed PMC to use the NLM facilities to photograph the valuable and often fragile journals in their archives. Jayne Campbell, Barbara Todd, and Lawrence Blake at Welch Medical Library at Johns Hopkins University graciously allowed PMC to photograph valuable articles in their archives. Dr. Noel Rose negotiated the Welch visit for PMC and has contributed to ASM and ASM Press in ways too numerous to mention. We would also like to express our gratitude to the services of the medical library of Trinity College in Dublin. We are grateful to all the publishers who granted permission to ASM Press to use the research articles from their journals. We regret the necessity to reproduce some of these historic documents in less than their original pristine condition. Decades of use have taken their toll. Sincere thanks to the scientists who gave advice as we searched for people to chair the committees. Your advice was excellent. We appreciate the help of everyone who took this project seriously and allowed *Microbiology: A Centenary Perspective* to help celebrate ASM's centennial.

ASM Press

Microbiology Past, Present, and Future

JOSHUA LEDERBERG

The advisory committees charged with compiling this collection have concluded a yeoman's task despite many potential frustrations, above all the draconian limits of space. What has been distilled is a set of exciting episodes, with each paper ably introduced to provide the local context of discovery. The introductions are necessarily brief, each merely opening a window to a larger vista of historical and personal biographical portrayals and stories from which many more lessons can be learned.

Other works are available to offer further detail on these seminal stories in microbiology, although none as far as I know concentrates on biographies of microbiologists; an annotated bibliography of such works could be valuable in bringing us forward from Paul de Kruif's *Microbe Hunters* and René Dubos' *Louis Pasteur—Free Lance of Science*. Not to be overlooked are less hagiographic studies, like Gerald Geison's *The Private Science of Louis Pasteur*. Sensitive, well-informed autobiographical works such as François Jacob's *The Statue Within* or Arthur Kornberg's *For the Love of Enzymes* are treasures, but few and far between. But the task assigned to the committees was to represent more the work of microbiologists in history than their lives, and this volume presents the opportunity to see the works essentially as they were initially published.

Because students will often exploit any excuse not to read, particularly not to read works more than 5 years old, not to mention those that predate their own lives, the ready reaccessibility of these historic documents will be of some assistance in connecting 21st-century researchers with their 20th-century roots. This may be less a problem in future, as more of the literature becomes available (and, one hopes, is well archived) in electronic media. Today, science libraries are in a state of transition; there is just no longer room for masses of old paper, and the "right stuff" embedded therein becomes harder to find. It is hard to foresee how all the back issues of print journals could be translated into byte-lingo, although without that the ravages of time and acid paper are bound to diminish our heritage. A valuable task for scientific societies and organizations would be to attend to the long-term preservation of this literature, perhaps on CD or via the Internet.

The American Society for Microbiology maintains extensive and valuable archives at the Albin O. Kuhn Library on the campus of the University of Maryland, Baltimore County. The National Library of Medicine (NLM) also is stepping into the breach with its excellent MEDLINE bibliography. (The delimiters "biography" and "famous persons" tend to point to journal articles of interest; books are harder to find.) In a new program, the NLM is opening a web site for archival material: its prototype collection (which I had the fortune to compile) relates to the same 1944 paper on DNA by O. T. Avery (with Colin, MacLeod, and McCarty) selected for the current volume. Because this paper presents the recognition of the genetic function of DNA, there is little controversy about its seminal role in the dawning of modern biology. See also http://www.profiles.nlm.nih.gov to browse relevant publications, including several books, and personal correspondence pertaining to that discovery. If the NLM archival experiment can be extended—and future barriers are more likely to be institutional and connected with intellectual rather than technical or economic property—there will be a fuller exploitation of the newest media to better understand our historical past.

My own commentary here intentionally minimizes reference to primary sources—several encyclopedic reviews are readily available for detail, and the stories are too complex to be encompassed by single references.

Turning now from process to concept, from trees to forest, can one extract some enduring themes from all the pickaxe work of factual discovery? I believe so, but with some relief that the detail is also beautifully presented in the selections. History is rarely wrapped in tidy, non-orthogonal packages; so my categories admittedly overlap, nor are they all at the same level of abstraction.

Integration of Microbes with Mainstream Biology

As late as 1925, Edmund B. Wilson's magisterial *The Cell in Development and Heredity*, in all its 1232 pages, indexed just three oblique references to bacteria: p. 84, "nucleus" (described as controversial); p. 209, "division," amitotic; and p. 580, "sex" ("In the Bacteria, Cyanophyceae, and certain other low forms no sexual process has thus far been made known . . ."). For their part, textbooks of bacteriology were no more revealing about the relationship of bacterial cells to the rest of the biological world. With the outstanding exception of Topley and Wilson's *Principles of Bacteriology and Immunity* in its numerous editions since 1929, they tended to be even more obfuscating well into the 1950s.

A major conceptual turning point was the publication of René Dubos' *The Bacterial Cell* in 1945, just in the midst of the wave of discovery of spontaneous mutation, of genetic transformation, of (conjugal) genetic recombination—i.e., sex—in bacteria, and soon after of virus-mediated transduction.

These discoveries bolstered the idea that findings in bacteria could be correlated with genes, linkage maps, chromosomes, mutation, and hence Darwinian evolution, as had been worked out for most of the rest of the plant and animal kingdoms. It was particularly important to dispel the confusion between the bacterial culture (or colony) and the single plant or animal organism, by understanding that the culture had to be regarded as a population of potentially disparate units, each capable of clonal propagation. The concept of the "clone" was all-important in understanding, for example, selection for drug resistance; eventually it fed back into macrobiological thinking, encouraging theories such as the origin of cancer in somatic mutation. These axioms are so thoroughly interwoven in today's cell biology that it is hard to recall how many sermons had to be preached in days of yore.

The biochemists were far in the lead, and the very earliest studies of metabolic pathways and enzymes spoke to the underlying unity of biochemistry, almost to a fault. (As Seymour Cohen has pointed out, we do ultimately rely on biochemical disparity for the effectiveness of chemotherapy.) The realization that the same nutritional building blocks—amino acids, vitamins, purines, and pyrimidines—were found in bacteria and other species spoke strongly for that unity. The initial discovery of the amino acid methionine by J. H. Mueller in 1922 as a bacterial growth factor was a particular triumph.

We should understand, however, that bacteriology was originally founded on the idea of a martial struggle: disinfection—ridding the human environment of parasitic germs, or, failing that, immunization to counter them—took priority over fundamental curiosity. The idea that we could discover more of our own nature by dispassionate study of the microscopic "bugs" was beyond the ken of the hygienic enthusiasts. Never mind that we could hope to match the immense reproductivity of microbes, their germinal potential, only by the use of our own wits—that is a lesson we are only now assimilating while being assaulted by HIV, malaria, and tuberculosis as ongoing scourges. By contrast, the "Delft School" of general microbiologists, represented by Martinus W. Beijerinck, Albert J. Kluyver, C. B. van Niel, and R. Y. Stanier, taught that those who "loved" the microbes would learn better how to deal with them than those who hated them.

These movements led to the displacement of the "medical" with a "biological" perspective in microbiological studies, from about the middle of the 20th century on. Now that we are further along with our fundamental concepts and tools, there is a reconvergence, and studies of pathways of microbial pathogenesis and of the dynamics of evolution of virulence are among the most exciting challenges in molecular physiology. Just this decade, when so many pathogens are being DNA sequenced almost by the month, we have these same challenges as prime motivations for functional genomics.

Nevertheless, paradoxes abound. It was a chemist, Louis Pasteur, who taught the doctors about infectious germs. It was a physician, O. T. Avery, immersed in the immunology of pneumonia, who taught the geneticists what genes were made of.

Applied Microbiology and New Models

The iron curtain between micro- and macro-biology having been breached about 50 years ago, many of the most exciting methodological and conceptual breakthroughs in *biology* have used bacteria (and their viruses) for basic tools. Much of the DNA revolution—high points including the Avery work, then the findings of Kornberg and his associates on DNA polymerases and ligases, the initial demonstrations of DNA splicing, and hundreds of other items—falls in this category, many of the seminal papers being represented in this volume.

Lysogenic viruses and their integration into chromosomes were first shown in bacteria; analogous phenomena are paramount in our understanding of retroviruses, of oncogenesis, and of gene therapy in somatic cells of higher organisms right up to humans.

The basic principles having been worked out, including methodological and conceptual analogies between bacterial cultures and animal cell cultures, it has become feasible to move from model microbial systems to targets closer to the animal in development and disease. Of course, eucaryotic life differs from that of the bacteria in many important details. Nevertheless, we can regard the past half-century as a triumph of unified studies. The ASM journal *Molecular and Cellular Biology*, and the renaming of ASM's *Microbiology and Molecular Biology Reviews*, are representative of this exciting trend.

Taxonomy

Meanwhile, we have seen drastic revision in our phylogenetic taxonomy of the little creatures, starting with R. Stanier and C. B. van Niel's separation of procaryotes from eucaryotes and followed by the iconoclastic split off of the Archaea. We have a long way to go in organizing a system that now relegates all multicellular organisms, with humans somewhere between corn and mushrooms, to a smudge on the wall map. And of the cellular organelles, the provenance at least of mitochondria and of chloroplasts from primeval bacteria has strong evidentiary support. In addition, hundreds of retroviral genomes are integrated into our own. We are indeed an evolutionary melting pot.

Viruses and Smaller?

The ultimate origins of viruses remain enigmatic. All viruses are presumably fragments of DNA (or RNA) escaped from some host genome and reshaped by extensive further evolution to invade and proliferate in cells of the same or vastly different species. In their current incarnations, many viruses are episomes capable of cyclical entry and exit from chromosomal havens. Some are conceptually unified with other plasmids and a menagerie of transposable elements in their mutualistic versus parasitic role in the economy of the host cell. With the recent recognition of prions, we have to cope with the prospect of new kinds of self-propagating units, perhaps dependent on shape (versus sequence)-oriented nucleation of protein conformers.

The Future: Evolving Boundaries

In sum, as biological science becomes ever more molecularized, we face a joyous riot of confusion about the definition of "microbiology," with its roles and missions in biomedical research, education, and services. The common denominator of experimental biology is functional genomics; this is elegantly applied to core questions of microbial identity and phylogeny and equally to the flanking fields of metabolism, infectious disease, virology, parasitology, immunology, ecology, and burgeoning applications in biotechnology and pharmaceutics.

The dilemma is in educational design: given that time is finite, what is the core curriculum to produce a "microbiologist"; besides cell biology and genomics, what should be required by way of familiarity with lifestyles of diverse microbes, with their natural history? Will the "general microbiologist" survive? We still have much to learn from comparative insights about ideal experimental laboratory objects, overlooked possibilities of disease etiology, or challenges to our generally accepted physiological and evolutionary models. Most of 20th century biology focused on a few standardized models, such as fruit flies and sea urchin eggs, for diverse and often conflicting purposes. It was strenuous labor to work out the care, feeding, and intangible lore of a novel biological system like *Arabidopsis* or *Caenorhabditis*. But if we had stuck simply with *Escherichia coli* B, of wondrous T-even phage fame, our eyes would have been closed to the marvels of conjugation and lysogeny. Now, genomics offers an easily replicable approach to any new organism, and sequence reports are tumbling out of the chute. Even uncultivable species are succumbing to that sophisticated attack—which, nevertheless, must still be informed by a McClintockian "feeling for the organism."

These tensions have riven other parts of academic biology, creating a universal trend toward the dissolution of phyletic boundaries; botany and zoology have merged into biology, then refissured into molecular, cell, developmental, organismic, ecological, and evolutionary compartments. Will microbiology continue to be defined by taxonomic lines? How will yeasts be related to *E. coli,* on the one hand; to nematodes or to human cells in culture, on the other? Will there be any logic to defining "microbiological" studies as those entailing the use of a microscope and culture media? It is testimony to the success of the microbiologist's perspective that this mode of thinking, embodied by cell culture methodology, pervades all of biology today.

The fissions and fusions will doubtless continue, accompanied by energy releases testifying to the intense dynamism of scientific progress in ways that blur all the boundaries.

GENERAL REFERENCES

Biography
For biographical data, see the biographical memoir series of the U.S. National Academy of Sciences and of the Royal Society (London). For biographical data over historic times, see the *Dictionary* (really encyclopedia) *of Scientific Biography* (Scribners, New York, N.Y., 1981).

History of Microbiology
A broad-ranging history of microbiology up to current times would be a daunting task, and it is difficult to find such works more recent than Patrick Collard's *The Development of Microbiology* (Cambridge University Press, Cambridge, U.K., 1976). Most major textbooks will have introductory chapters on the history of the field. There are many other specialized works, especially on molecular genetics and DNA.

Internet Resources
The World Wide Web has much to offer, including hot links to many external sources, e.g., http://www.profiles.nlm.nih.gov and the American Society for Microbiology site, http://www.asmusa.org/.

The Excitement and Fascination of Science, vol. 1-5
Published by Annual Reviews, Inc., Palo Alto, Calif., these volumes embrace several hundred short memoirs, usually autobiographical, including those of many microbiologists.

Overall Perspectives on Microbiology
For overall perspectives on microbiology, the following monographs and treatises are among the most comprehensive:

Annual Reviews, Inc. *Annual Review of Microbiology.* Annual Reviews, Inc., Palo Alto, Calif.(Volume 1 was published in 1947, and this series has been published annually to date.)

Balows, A. 1992. *The Prokaryotes: A Handbook on the Biology of Bacteria: Ecophysiology, Isolation, Identification, Applications,* 2nd ed. Springer-Verlag, New York, N.Y.

Brock, T. D. 1998. *Milestones in Microbiology*. ASM Press, Washington, D.C.

Brock, T. D. 1998. *Robert Koch: A Life in Medicine and Bacteriology*. ASM Press, Washington, D.C.

Burnet, F. M., and W. M. Stanley (ed.). 1959. *The Viruses: Biochemical, Biological, and Biophysical Properties*. Academic Press, Inc., New York, N.Y.

Dubos, R. 1998. *Pasteur and Modern Science*. ASM Press, Washington, D.C.

Gunsalus, I. C., and R. Y. Stanier (ed.). 1960. *The Bacteria: A Treatise on Structure and Function. Structure*. Academic Press, Inc., New York, N.Y.

Lederberg, J. (ed. in chief). 1992. *Encyclopedia of Microbiology*. Academic Press, Inc., San Diego, Calif.

Neidhardt, F. C., R. Curtiss III, John L. Ingraham, E. C. C. Lin, K. Brooks Low, Jr., B. Magasanik, W. S. Reznikoff, M. Riley, M. Schaechter, and H. E. Umbarger (ed.). 1996. *Escherichia coli and Salmonella: Cellular and Molecular Biology*. ASM Press, Washington, D.C.

Peruski, L. F., and A. H. Peruski. 1997. *The Internet and the New Biology: Tools for Genomic and Molecular Research*. ASM Press, Washington, D.C.

Webster, R. G., and A. Granoff (ed.). 1994. *Encyclopedia of Virology*. Academic Press Ltd., London, U.K.

Diagnostic Microbiology and Epidemiology

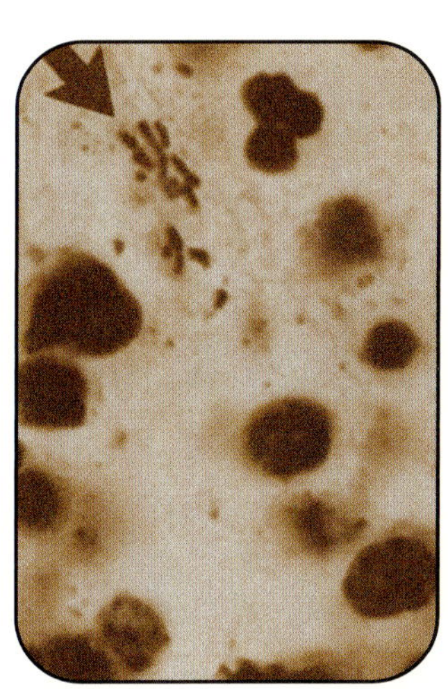

Diagnostic Microbiology and Epidemiology

The Division I committee members who selected the following 11 papers were Mitchell Cohen, Ph.D., Dieter Gröschel, M.D., Edwin H. Lennette, M.D., Ph.D., Josephine Morello, Ph.D., and Alexander von Graevenitz, M.D. (Committee Chair), who also wrote the introductions.

Because we were under a strict page limit, some of our initial selections could not be included because they were too long, e.g., J. H. Brown's monograph, *The Use of Blood Agar for the Study of Streptococci* (The Rockefeller Institute for Medical Research, New York, N.Y., 1919), A. Whitmore's paper on melioidosis (*J. Hyg.* **13**:1–34, 1913), that of E. G. D. Murray et al. on *"Bacterium monocytogenes"* (*J. Pathol. Bacteriol.* **29**:407–439, 1926), that of Rebecca Lancefield on the serological differentiation of streptococci (*J. Exp. Med.* **57**:571–595, 1933), and that of M. D. Eaton et al. on the etiology of primary atypical pneumonia (*J. Exp. Med.* **79**:649–668, 1944). Also, first-rate translations of articles not originally written in English would have taken too much time to complete during the period available to us.

We realize that our selections reflect, at least in part, our particular viewpoints at this juncture and apologize to those whose contributions, although seminal, were not included.

A Protozoön General Infection Producing Pseudotubercles in the Lungs and Focal Necroses in the Liver, Spleen and Lymphnodes

S. T. DARLING

In this paper, Samuel T. A. Darling (1872–1925), at the time (1906) a pathologist in the Panama Canal Zone, described for the first time a case of histoplasmosis on autopsy in a patient thought to have died from tuberculosis. He considered the agent, whose microscopic morphology he described very thoroughly, to be a parasite with a predilection for endothelial and epithelial cells, apparently because it showed what he interpreted as flagella. This "parasite," however, appeared morphologically quite different from ". . . those found by Leishman, Donovan, Marchand, Ledingham, and Wright." The scattered focal necroses in liver, spleen, and lymph nodes also differed microscopically from tuberculous lesions. It remained for H. da Rocha-Lima (*Arch. Schiffs-Tropenhyg.* **16**:79–85, 1912) to conclude, based on comparative tinctorial and histological studies, that Darling's "protozoön" was a fungus. The disease was soon found to occur in Minnesota, but all diagnoses were made post mortem until Dodd and Tompkins reported the first intra vitam diagnosis in 1934 (see C. W. Emmons et al., *Medical Microbiology*, 3rd ed., Lea & Febiger, Philadelphia, Pa., 1977). In the same year, W. A. De Monbreun (*Am. J. Trop. Med. Hyg.* **14**:93–125, 1934) was able to culture the organism in both yeast and mycelial phases. Its soil saprophytism was detected much later (C. W. Emmons, *Public Health Rep.* **64**:892–896, 1949) and the disease in bats and the role of avian habitats were ascertained a few years later still (C. W. Emmons, *Public Health Rep.* **73**:590–595, 1958; L. W. Ajello, *Public Health Rep.* **79**:266–270, 1964). Histoplasmosis is still a disease of considerable importance in the Western Hemisphere and has recently also been observed as an opportunistic infection in patients with AIDS.

ALEXANDER VON GRAEVENITZ

Clinical Notes, New Instruments, Etc.

A PROTOZOÖN GENERAL INFECTION PRODUCING PSEUDOTUBERCLES IN THE LUNGS AND FOCAL NECROSES IN THE LIVER, SPLEEN AND LYMPHNODES.

SAMUEL T. DARLING, M.D.
Pathologist, Ancon Hospital.
ANCON, CANAL ZONE, ISTHMUS OF PANAMA.

On Dec. 7, 1905, while examining smears from the lungs, spleen and bone marrow in a case that appeared to be miliary tuberculosis of the lungs, I found enormous numbers of small bodies generally oval or round. Most of them were intracellular in alveolar epithelial cells, while others appeared to be free in the plasma of the spleen and rib marrow. Tubercle bacilli were absent. The following is an account of the case:

Patient.—C. D., negro from Martinique, aged 27, occupation carpenter; address, Paraiso, a village in the Canal Zone.

History.—The patient had been a resident of the zone three months. While in Martinique he had suffered from some mental disturbance. His present illness dates from Sept. 15, 1905, when he complained of fever and vomiting.

Condition on Admission to Hospital.—On entering Ancon Hospital Dec. 5, 1905, he was mildly delirious and incoherent. Lungs were clear; abdomen was scaphoid; spleen was enlarged.

Blood: Negative for malarial parasites. leucocytosis, 2200. Hemoglobin: 60 per cent. (Dare's).

Feces: Negative.

Temperature: On admission, Dec. 5, 12:30 p. m., 101, pulse 120; Dec. 6, 8 a. m., 95, pulse 96; 4 p. m., 98, pulse 100. The patient died Dec. 6 at 11:30 p. m.

AUTOPSY.

December 7, 8:30 a. m.

Macroscopic and Microscopic Examination.—Body of negro, moderately emaciated; length, 5 feet 8¾ inches; inter nipple distance, 7 3/16 inches; rigor mortis was plus.

The odor on opening thorax was suggestive of pulmonary tuberculosis. The right and left pleuræ were free. There were numerous red blotches (ecchymoses) beneath the visceral pleura of both lungs 8 mm. in diameter. Many small nodules could be felt under the visceral pleura.

The lungs on section were found studded with pale gray hyaline miliary tubercles from 2 to 3 mm. in diameter. The lungs were heavier and more voluminous than normal. The tubercles were not as closely packed or so numerous as is often found in miliary tuberculosis, and the general color of the lungs was bright red.

The peribronchial lymphnodes contained a few small soft recently caseated tubercles. The nodes were enlarged and pigmented .

Heart: This organ was small and normal.

Liver: The liver was enlarged and pale, and there was slight atrophic cirrhosis.

Spleen: This was enlarged to three times the normal in size; the pulp was very firm. The malpighian bodies were distinct. Here and there were a number of small yellow nodules resembling tubercles.

Kidneys: There were a few depressions in a cortex diminished to 8 mm. in depth.

Pancreas: Normal.

Bladder: Normal.

Rib bone marrow: Normal and dry.

Brain: The pia-arachnoid was slightly edematous and more generally adherent to the cortex than normal. The calvarium was very thick.

Intestines: Several specimens of *Tricocephalus dispar* were found in the cecum. There were a few small superficial circular ulcers from 2 to 4 mm. in diameter in the cecum and ileum.

The mesenteric lymphnodes and those at the hilum of spleen were enlarged and pale.

Bacteriologic Examination.—Spleen smears were negative for malarial parasites or pigment. Oval and round bodies were free in the plasma.

In rib bone marrow smears there were traces of intracellular malarial pigment. A number of bodies similar to those in the spleen were seen.

In lung smears tubercle bacilli were absent.

There were myriads of intracellular and extracellular bodies similar to those found in the spleen and the marrow.

A moist coverslip preparation from intestinal ulcers showed motile amebæ.

Anatomic Diagnosis.—Acute miliary tuberculosis, pulmonary type. Tubercul us lymphadenitis, peribronchial. Chronic interstitial splenitis. Atrophic cirrhosis. Chronic interstitial nephritis, slight. Lymphadenitis, mesenteric. Chronic leptomeningitis. Edema of pia-arachnoid. Ulcerative enterocolitis. Amebiasis. General infection by protozoön.

APPEARANCE OF THE PARASITE IN SMEARS.

Lung: This specimen was stained by carbolfuchsin and Gabbet's methylen blue, overstained with polychrome methylen blue, and washed with eosin.

The polychrome blue was prepared as follows:

    Methylen blue, pure medic, Grübl.........g.   1.
    Sodium carbonate, pure.................g.   .5
    Distilled water ....g. 100.

This was placed in thermostat one week, and kept at room temperature for six months.

The excess of blue was removed by washing the smear alternately with alcoholic solution of eosin (.5 per cent in 60 per cent. ethyl alcohol) one second and distilled water a few seconds, until the internal structure of the parasite showed plainly.

The parasite is oviform or round, and is surrounded by a clear refractile non-staining rim, in thickness about 1/5 the diameter of the parasite. This refractile rim is present in all smears, whether previously treated with acid blue or not. The structure is not homogeneous, but consists of a faintly staining substance and a deeply staining one; a clear space or spaces; and chromatin granules. The chromatin granules are generally single, sometimes two or more are counted. One large parasite appeared to have six such dots of chromatin. The granules are often situated in a clear non-staining zone at one side of the darker staining substance; at other times they are situated on the margin or within this substance; and also frequently appearing in the clear refractile capsule. The chromatin granules are generally dot shaped, very rarely elongated. Occasionally two chromatin dots placed together simulated a rod form.

The clear space or spaces resemble vacuoles; at times they resemble the clear non-staining spaces seen in filaria embryos and trypanosomes. The staining substance almost entirely fills the capsule or refractile rim of the parasite. The circular contour of the staining substance is at times broken on one side or place by the clear non-staining zone.

This zone varies in shape, size, and in its relation to the staining substance; being circular, oval, or irregular in form; being three-fourths the size of the entire parasite, or at times barely perceptible on account of its minuteness; being centrally located or excentric; and being single or multiple—two or three.

In size the parasites are from 1 to 4 microns through their greatest diameter; commonly this diameter is 3 microns.

The parasite appears to divide by fission into two equal or unequal elements. One parasite appeared to be dividing into four equal elements. Several parasites with chromatin dots scattered through their substance appeared as pre-segmenting bodies—ready to divide into five or six elements. Occasionally a smaller parasite may be seen close beside a larger one, as though separating from it, the smaller one being about 1 micron in diameter.

Although oval or round in outline, the staining substance,

together with the clear non-staining zone and chromatin granules, give a varying picture, depending on the point of view. Forms suggesting the appearance of familiar objects, such as the eye, a shield, a conch shell, a bullet, or a shuttle are seen. The resemblance of certain parasites to a mammalian embryo in "fetal attitude" is very striking.

In the lung smears the parasite is apparently always intracellular, and the cells contain from 10 to 100 or more parasites. The appearance of free parasites is probably due to the squeezing and breaking up of infected epithelial cells by pressure in making the smear. One unbroken alveolar epithelial cell occupied one-third the diameter of the field, 1/12 oil im. No. 1 oc. B. & L. Parasites had invaded the cell nucleus as well as the cytoplasm, and it was estimated that this cell contained more than 300!

Spleen and rib marrow smears showed fewer parasites, two or three to a field, and they appeared to be extracellular. The

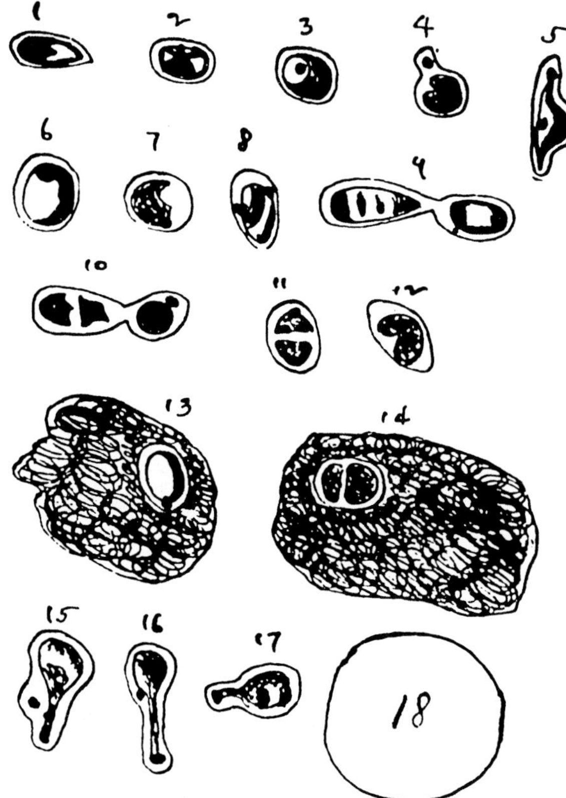

Fig. 1.—(× 2000), 1 to 8, forms of parasite; 9, 10, 11, 12 manner of subdivision; 13, 14, parasites within nuclei of spleen cells; 15, 16, 17, flagellate forms; 18, alveolar epithelial cell containing parasites.

nucleus of a spleen or marow cell appeared now and then to have been invaded. Each parasite had a definite refractile rim, as in the lung smears, and its internal structure could be well made out.

Red blood corpuscles were never invaded.

Three flagellated forms were seen in a lung smear. The distal extremity of one of the flagella contained a rod of chromatin placed at right angles to the flagellum, simulating the relation of centrosome to chromatin filament in *Trypanosoma Lewisi.* The flagella were single, short and thick, without chromatin filaments, and were enclosed by the refractile capsule, continuous with that of the body of the parasite.

EXAMINATION OF SECTIONS.

Sections were fixed in Zenker's solution and stained by eosin and hematoxylin, Van Giesen's method, and polychrome-methylen blue.

Lung: The alveolar capillaries are tortuous and engorged with red blood corpuscles. In places the walls are broken down, stretched, or greatly distended. No leucocytes were

seen within the capillaries. The epithelial cells of the alveolar walls are desquamating or completely shed. In places there appears to be merely a single layer of endothelial cells separating the circulating blood from the alveolar contents. The alveoli are seen to be filled with red blood corpuscles, generally having a washed out appearance; red blood corpuscles and serum; or red blood corpuscles, serum, and large swollen alveolar epithelial cells containing many parasites.

Polymorphonuclear leucocytes are rarely observed in the alveolar contents; a few mononuclear elements are noted. There are no tubercles. The pseudo tubercular areas are made up of alveoli with broken, distorted, or collapsed walls, containing many alveolar epithelial cells distended by parasites. Small vessels or capillaries are seen to pass through the pseudotubercles, but there are no evidences of the hemorrhages seen in other alveoli. Within these areas there are enormous numbers of parasites generally contained within epithelial cells—rarely free. The nuclei of invaded cells stain well, though often more faintly than normal. The cytoplasm of badly infected cells is wanting, and there are numerous distended epithelial cells devoid of cytoplasm and parasites. The infected cells have a distinctly staining rim of cytoplasm, even when their nucleus and cytoplasm are gone.

Liver: There are numerous faintly staining areas ranging in size from that of a single glandular epithelial cell to those one-third the size of a lobule; in which the liver cells and endothelial cells of the portal capillaries are completely transformed by invading parasites. In the larger areas the cytoplasm and nuclei of the invaded cells have disappeared or do not stain. There is a mass of debris, imbedded in which are myriads of parasites. In places the liver cells are normal, in others they have suffered cloudy change. In these latter localities there appears to be a stasis of blood in the portal capillaries due to occlusion of capillaries by enormously distended endothelial cells filled with parasites. The red blood corpuscles are here "washed out."

There is a distinct primary invasion of liver cells in places, although oftener it would seem that many liver cells become invaded after they have had their nutrition cut off by infected overlying endothelial cells.

Around the portal spaces the connective tissue is increased in amount and there is a recent round cell infiltration. The bile capillaries and their epithelium are normal.

Spleen: The splenic spaces are greatly engorged with red blood corpuscles. The connective tissue is moderately increased, its cells are swollen, cloudy, and at times contain parasites. There is cloudy swelling of cells in small areas here and there, and many of these cells contain parasites. There are also numerous free parasites.

Lymphnode from hilum of spleen: The cortical follicles and medullary cords of the dense lymphoid tissue are, with the few exceptions noted, below normal. The capsule and reticulum throughout the node are the seat of degenerative changes. The reticulum of the loose lymphoid tissue encloses many large mononuclear cells possessed of distinctly staining nuclei, and containing many parasites.

There are two cortical follicles, and portions of a medullary cord which have undergone cloudy swelling and necrosis, amid the débris of which are cells containing parasites. The margins of these areas show beginning degenerative changes; many fragmented nuclei are seen, as well as mononuclear cells distended by parasites.

Peribronchial Lymphnode: This node contains several old fibro-caseous tubercles, and one giant cell. The reticulum and capsule of the node are greatly thickened in places. A lymph vessel beneath the capsule contains mononuclear cells infected by parasites.

There is seen to be a general infection by a parasite having a predilection for endothelial and epithelial cells.

The lesions are those of scattered focal necroses of liver, spleen and lymphnodes, with foci of catarrhal pneumonia and hemorrhages in the lungs, in which the lungs play a very passive part, there being absolutely no leucocytic infiltration of the miliary pneumonic nodules.

The infection was a fatal one, there being no other

lesions sufficiently grave to have caused death. The anatomic diagnosis of tuberculosis not being confirmed on examination of sections, save in peribronchial lymph-nodes.

The parasite, as studied from smears, presents certain resemblances to those found by Leishman, Donovan, Marchand, Ledingham and Wright, but the differences are so marked and the lesions so unusual that I feel the case is a unique one.[1]

For the parasite the name *Histoplasma capsulata* is proposed.

I wish to thank Acting Chief Sanitary Officer Dr. H. R. Carter for his kind permission to publish this report.

---

1. Since writing this article I have found the parasite in a second case.

A Micro-Organism Which Apparently Has a Specific Relationship to Rocky Mountain Spotted Fever

H. T. RICKETTS

This paper by Howard T. Ricketts (1871–1910), at the time of publication (1909) a pathologist at the University of Chicago, described for the first time an organism that we now call a rickettsia in his honor. In an earlier publication, Ricketts had already pointed out the importance of rodent ticks *(Dermacentor andersoni)* in the transmission of Rocky Mountain spotted fever. He was careful enough to claim that there was "apparently . . . a specific relationship" between the short bacilli staining with Giemsa and those found in the blood of animals and humans with Rocky Mountain spotted fever, although he was also able to agglutinate the bacilli with serum of infected guinea pigs. The final proof of the causative role of *Rickettsia rickettsii* was provided by S. B. Wolbach a decade later (*J. Med. Res.* **41**:2–197, 1919).

In 1910, Ricketts went to Mexico to investigate epidemics of typhus. He recognized the similarity of the agents of this disease and Rocky Mountain spotted fever, but unfortunately he infected himself and died within 2 weeks. On a proposal of da Rocha-Lima, the agent of typhus was named *Rickettsia prowazekii*, recognizing the role of two microbiologists who died from the disease. Today we know of quite a few other rickettsial diseases and species, but the main threat in the United States remains that from Rocky Mountain spotted fever, which has, however, shifted its geographical center from the Rocky Mountain states to the South Atlantic and western South-Central regions.

ALEXANDER VON GRAEVENITZ

probably their nuclei are subject to unusual disintegration: (2) because of my inability to cultivate a micro-organism of this character from infected blood by the use of ordinary and some unusual culture media, under various conditions of cultivation, or by other means to obtain it in satisfactory concentration.[1]

THE BACILLUS IN THE TICK

Although infected ticks had been examined previously in a more or less cursory manner, their systematic study was not undertaken until recently. In the pursuit of this work advantage was taken of the fact that the disease is transmitted by the infected female to her young through the eggs, as described in a previous report. A repetition of these experiments in the winter of 1907-8, with the help of Dr. Maria B. Maver, resulted in such transmission in 50 per cent. of the ticks used, the fact being determined by allowing the larvæ to feed on normal guinea-pigs. This second series has not been published heretofore.

Female tick No. 40, a *Dermacentor,* from Montana, had produced fatal infections of spotted fever in guinea-pigs 1740 and 1764. A number of eggs from the first day's laying were crushed individually on cover glasses, fixed in absolute alcohol, and stained with Giemsa's stain. Each egg was found to be laden with astonishing numbers of an organism which appears typically as a bipolar staining bacillus of minute size, approximating that of the influenza bacillus, although definite measurements have not yet been made. Various forms are seen depending on the stage of development and the arrangement in which two or more may be found. It is very common to find two organisms end to end, with their poles stained deeply and the intermediate substance a faint blue, resembling a chain of four cocci. When the chromatin is not yet sharply limited to the poles, the somewhat lanceolate forms so often recognized in the blood are seen. Not infrequently delicate bacilli with a uniform distribution of the chromatin are found. These are all interpreted as stages in the evolution of a bipolar organism. They are present in varying numbers in different eggs, but as a rule they are surprisingly numerous, and in some instances they would certainly count into the thousands. Many faintly staining, apparently degenerate forms are encountered.

Examination of the eggs of three dermacentors from Idaho (different specifically from the Montana dermacentor). which were infected from the guinea-pig, showed the presence of the same forms (Ticks 5, 7 and 9).

Conference with zoologic scientists who have made a particular study of the structure of eggs brought out the fact that such bodies are not known as a constituent of the egg of any species of animal.

Although it has not yet been possible to examine the eggs of ticks which are known to be free from spotted fever, the equivalent control has been made through a comparison of the visceral organs of infected and uninfected ticks. The salivary glands, alimentary sac and ovaries of infected females are literally swarming with exactly similar micro-organisms. On the contrary, they appear to be entirely absent from the viscera of the uninfected tick, both male and female.

AGGLUTINATION REACTIONS

The most striking evidence of the probable etiologic relationship of this organism to spotted fever is found

---

A MICRO-ORGANISM WHICH APPARENTLY HAS A SPECIFIC RELATIONSHIP TO ROCKY MOUNTAIN SPOTTED FEVER

A PRELIMINARY REPORT *

H. T. RICKETTS, M.D.
CHICAGO

Since the spring and summer of 1906, bodies which I have referred to in my notes as "diplococcoid bodies," and sometimes short bacillary forms, have been found with considerable constancy in the blood of guinea-pigs and monkeys which were infected with Rocky Mountain spotted fever. They have also been seen in the blood of man but not so frequently. Much more time has been spent on the blood of the experimental animals than on that of man in view of the fact that it could always be obtained in fresh condition.

The form most commonly found is that of two somewhat lanceolate chromatin-staining bodies, separated by a slight amount of eosin-staining substance. The preparations of Giemsa, as furnished by Grübler, has been used almost exclusively, and with variations in the technic the intermediate substance may stain faintly blue.

In spite of the constancy with which these bodies were found, it did not seem justifiable to claim that they represent the microparasite of the disease, for two reasons: (1) the very complex morphology of the blood, especially in febrile states, when various cells and

---

* From the Pathological Laboratory of the University of Chicago

1. Mr. P. G. Heinemann assisted in an extensive series of culture experiments in the spring of 1907.

in the positive outcome of agglutination tests. Fortunately it is so numerous in the eggs that a bacterial emulsion of reasonable concentration for agglutination tests can be made by crushing forty to fifty eggs in about 0.05 c.c. of salt solution. The material is so scant that only the microscopic method could be used. The preparations were made as hanging drops, incubated for two hours, dried, fixed with absolute alcohol and stained.

The serum of the normal guinea-pig either causes no agglutination at all, or at the most produces only slight agglutination in proportions of 1 to 1 and 1 to 9. Dilutions higher than this cause no agglutination. In testing the agglutinating powers of immune serums, three animals which had been infected from different sources and had recovered were used. One (1751) had been infected with a dermacentor from Idaho; another (1692), with a strain handed down direct from guinea-pig to guinea-pig for nearly three years without the intervention of ticks, the original infection having been obtained from the blood of man; the third (1757) with a strain kept in the same way since last spring. Graded dilutions, beginning with 1 to 1 and going as high as 1 to 200, were used in the different series, with the striking result that a complete agglutinating power was present in the three immune serums in dilutions up to 1 to 160. It was somewhat less in a dilution of 1 to 200. The highest dilution which will cause clumping has not been ascertained.

No fresh immune serum from man is at hand, but tests were made with three specimens which are about five, seven and nine months old respectively. They have been preserved in the ice chest with the addition of 0.3 per cent. of chloroform. The peculiar phenomenon of failure to agglutinate in concentrated solution was noted with all three. With the oldest serum no agglutination occurred until the dilutions of 1 to 160 and 1 to 200 were reached, when incomplete clumping was produced. With the second there was no agglutination in the dilution of 1 to 1, distinct clumping in 1 to 10, 1 to 20 and 1 to 40, with little or none in higher dilutions. In the serum of five months standing the reaction was absent in the dilutions of 1 to 1, 1 to 10 and 1 to 20, positive but not complete in 1 to 40, 1 to 80 and 1 to 120, with very little clumping in 1 to 160 and 1 to 200. Normal human serum caused clumping in a dilution of 1 to 1, a very slight amount in 1 to 10, and none at all in the higher dilutions.

The failure of the immune serums from man to agglutinate in concentrated solutions, whereas they did so in higher dilutions, is taken as an example of the action of the so-called proagglutinoids. As explained by Ehrlich's theory, this consists in the occupation of the bacterial receptors by inactive agglutinin which exceeds the active agglutinin in its affinity for the bacteria. With higher dilutions of the serum the agglutinoids are so diluted that they do not completely occupy the bacterial receptors, thus affording a point of attack for the active agglutinin, unless the latter has been eliminated by extreme dilution. This is well known as a property of old agglutinating serums.

BACILLI IN INFECTED SERUM

As a means of concentrating the organisms in the serum of the infected guinea-pig the following experiment was performed: Three cubic centimeters of fresh infected serum were diluted with an equal amount of salt solution, and to this was added 0.3 c.c. of an agglutinating serum from the guinea-pig. The mixture was placed in the incubator for two hours and then centrifugated for about ten hours at a speed of 1800 revolutions. All but the merest drop was then pipetted off and stained preparations were made of the sediment. Examination showed the presence of a moderate number of forms which are identical in appearance and size with those often recognized in ordinary smears of infected blood, and also with the "diplococcoid" forms seen in the egg of the infected tick. No such bodies were found in a control tube of normal serum.

The evidence pointing to this organism as the causative agent in spotted fever, though not complete, is of a striking character. In so far as I know it would be an unheard-of circumstance to obtain such strong agglutination with an immune serum, in the presence of negative controls, unless there were a specific relationship between the organism and the disease. In favor of the specific relationship in this case are also the presence of the organism in large numbers in infected ticks and in their eggs, its absence from uninfected ticks, and the presence of similar forms in the blood and serum of the infected guinea-pig.

Morphologically the organism is a bacillus and somewhat pleomorphic as described. Its resemblance to the bacilli of the hemorrhagic septicemias is striking, and in this connection it is important to note that spotted fever is a hemorrhagic septicemia. It has not been cultivated, although work with this end in view is in progress.

I have devised no formal name for the organism discussed, but it may be referred to tentatively as the bacillus of Rocky Mountain spotted fever. A further study of its characteristics may suggest a suitable name.

That a bacillus may be the causative agent of a disease in which an insect carrier plays an obligate role under natural conditions may be looked at with suspicion in some quarters. Yet, even without the evidence in this case, it would seem to be unscientific to be tied to the more or less prevailing belief that all such diseases must, on the basis of several analogies, be caused by parasites which are protozoon in character.

Further study of the relationship of the bacillus to the disease is being carried on and will be reported at a future date, together with illustrations and a more detailed account of its characteristics.

The Demonstration of Pneumococcal Antigen in Tissues by the Use of Fluorescent Antibody

A. H. COONS, H. J. CREECH, R. N. JONES, AND E. BERLINER

This paper by Albert H. Coons (1912–1978) and his collaborators at Harvard described, for the first time, a method that would be widely used in later years in bacteriology, virology, and parasitology, the so-called direct fluorescent antibody technique. Later developments included the detection of antibody by the indirect fluorescent antibody technique, the use of monoclonal antibodies to improve specificity, and the replacement of the fluorescence microscope by a fluorometer, leading to the solid-phase immunoassays.

The authors used this technique for the detection of pneumococcal antigen in tissues, for which it would rarely be used today. They also provided rudimentary data on sensitivity and specificity of the method. Today, the technique is still used for the detection of leptospirae, treponemata, legionellae, chlamydiae, cryptosporidia, and giardiae; for respiratory viruses like influenza, parainfluenza, and respiratory syncytial virus; and for skin lesions caused by varicella-zoster or herpes simplex virus. The indirect technique is used for serologic tests for rickettsiae, syphilis, and other microorganisms. Finally, there are clinical applications in immunologically mediated kidney and skin diseases. Lately, however, molecular techniques such as PCR have replaced fluorescent antibody techniques in certain areas (e.g., for detecting chlamydiae and certain viruses) because of their higher sensitivity. There is no question, however, that without fluorescent antibody methodology, clinical microbiology would not have progressed the way it has.

ALEXANDER VON GRAEVENITZ

Reprinted with permission from *Journal of Immunology* 45:159–170. Copyright © 1942. The American Association of Immunologists

THE DEMONSTRATION OF PNEUMOCOCCAL ANTIGEN IN TISSUES BY THE USE OF FLUORESCENT ANTIBODY[1]

ALBERT H. COONS,[2] HUGH J. CREECH, R. NORMAN JONES AND ERNST BERLINER

From the Department of Bacteriology and Immunology, Harvard Medical School and School of Public Health, the Department of Anatomy, Harvard Medical School, Boston, and the Chemical Laboratory, Harvard University, Cambridge, Mass.

Received for publication July 13, 1942

The object of this investigation has been to determine the feasibility of using chemically labelled antibodies as reagents for the detection and orientation of antigenic material in mammalian tissue. Such a method requires the retention of specificity by the antibody-molecule during and after the necessary chemical manipulation, a stable chemical linkage between the antibody and its label, and a label that can be detected when present in minute quantities. A further important requirement demands the separation of the labelled-antibody solution from unconjugated tracer-material. In addition a method of this sort would obviously be more useful for many studies if it were possible to determine not only those organs, but those cells which contain the antigen in question. Accordingly we investigated the possibility of employing materials for labelling that could be detected by optical rather than by analytic or radiographic methods.

These studies have been interrupted by the war, but the initial results and the great utility of the method, should it be perfected, urge us to record the experimental data so far accumulated.

Previous investigations with marked antibodies have established the principle that it is possible to introduce chemical groups into the antibody-molecule without destroying its specific reactivity (1–4). They have stimulated, however, very little further study, probably because the labels were themselves difficult to detect. Reiner (1) coupled diazotized atoxyl with concentrated pneumococcal 1 and 2 horse-antibody solutions, and found that the agglutinative titer was unchanged, and mouse-protection still successful. The shift in the isoelectric point made it likely that the antibodies themselves were conjugated. He suggested that this material might be useful for studying the quantitative aspects of the antigen-antibody reaction. Heidelberger, Kendall, and Soo Hoo (5) introduced color as an immunological label by making R-salt-azo-benzidine-azo-egg-albumin, thereby producing an antigen which could be colorimetrically determined by quantitative methods. Marrack (3) applied this procedure to the labelling of antibodies. He prepared R-salt-azo-benzidine-azo-antityphoid and anticholera sera, and found that the organisms were specifically agglutinated

[1] Aided in part by a grant from the International Cancer Research Foundation. The construction of the fluorescence-microscope was made possible by grants to Dr. Allan L. Grafflin by the Ella Plotz Sachs Foundation and the Wellington Fund.

[2] Fellow in the Medical Sciences of the National Research Council.

and colored red. He furthermore proved in the following manner that the anti-body-molecules themselves were dyed. He mixed typhoid-antibody dye with cholera antiserum (unaltered), divided the mixture and added typhoid bacilli to one half and cholera vibrios to the other. Both species were agglutinated, but whereas the clumps of typhoid bacilli were red, the cholera organisms remained unstained. This experiment eliminates, as possible causes of the coloration, non-specific adsorption and mechanical occlusion of colored molecules during the process of agglutination.

Since a repetition of Marrack's experiments with antipneumococcal 2 and 3 sera demonstrated that the red coloration of the organism-antibody complex was barely detectable microscopically, we prepared a fluorescent Pn 3 serum by conjugation with anthracene through the carbamido-linkage—a procedure used by Hopkins and Wormall (6) and by Creech and Jones (7–9). The characteristics of this conjugate have been previously described (10). Its immunological specificity and potency appeared to be unaltered. With this material Marrack's mixing-experiment could be regularly repeated. Pn 3 after exposure to this conjugate assumed a bright blue fluorescence in ultraviolet light. The immunological specificity of the fluorescent antibodies was thus established. In tissues, however, the normal blue fluorescence of many elements make this blue-fluorescing antibody difficult to discern.

EXPERIMENTAL

The preparation of fluorescein-4-isocyanate

Accordingly, attempts were made to synthesize fluorescein-4-isocyanate by the interaction of phosgene and 4-aminofluorescein according to the method previously described (11). The required compound could not be isolated in a completely pure condition because of its instability and tendency to undergo side reactions. But enough crude compound was obtained to demonstrate the existence of the isocyanate which underwent conjugation with Pn 3 antiserum to form chemically modified antibodies possessing a green fluorescence.

The method of Bogert and Wright (12) was used for the preparation of 4-nitrofluorescein from resorcinol and 4-nitrophthalic acid. By the reduction of the nitro compound with stannous chloride and hydrochloric acid, these authors obtained only traces of a substance (m p 281 C) which they believed to be the amino derivative of fluorescein. Analytical confirmation was not obtained. Our attempts to synthesize the amine by this procedure or by procedures involving various metal combinations in the reduction process were unsuccessful. Catalytic hydrogenation of the nitro compound, however, gave a compound (m p 279–281 C) which had the correct analysis for the amine. Treatment of this product with phosgene gave a mixture containing a component which we believe to be the isocyanate.

The nitro compound (800 mg) in 35 ml of absolute alcohol was shaken with Raney nickel in an atmosphere of hydrogen for five hours at which time the theoretical amount of hydrogen had been absorbed. The amine was precipitated by the addition of water and obtained in a crystalline condition (m p

279–281 C) from ethanol-water in a 67 per cent yield. Analysis: Calculated for $C_{20}H_{13}O_5N$: N, 4.03; found: N, 3.88. Platinum may also be used to catalyze the hydrogenation but isolation of the amine is more difficult.

Passage of phosgene into a solution of the amine (500 mg) in acetone caused the formation of a precipitate which disappeared during refluxing for a period of 3 hours. A solution of phosgene in acetone was added during this period to aid in the conversion to the isocyanate. Concentration of the solution and the addition of petroleum ether (70–90 C) produced a brown precipitate which was discarded. From the filtrate, there was obtained a light yellow amorphous powder which was precipitated repeatedly from its solution in acetone by the addition of petroleum ether. The melting point of this material was indefinite and was accompanied by darkening and decomposition. Analyses of several preparations indicated that the completely pure isocyanate had not been isolated. The material undergoes decomposition rapidly to form acetone-insoluble substances, one of which is undoubtedly di-substituted urea.

Preparation of the conjugates

The preparation of fluorescein-4-carbamido antipneumococcal 3 rabbit serum will serve as a specific example of the method utilized for the process of conjugation and purification of the conjugate. A solution of fluorescein-4-isocyanate (25 mg) in 6 ml of dioxane and 3 ml of acetone was added over a 15-minute period to a cold solution, which was continuously stirred, containing 3 ml of Lederle's concentrated antipneumococcal 3 rabbit serum (protein-content, 300 mg) in 20 ml of physiological saline and 3 ml of half-normal carbonate-bicarbonate buffer of pH 9. The temperature was maintained at 0–2 C. After one half-hour, the green-fluorescing solution was dialyzed for 20 hours at 5 C against several changes of physiological saline. Enough unconjugated fluorescein derivative (probably the amine formed by hydrolysis of the isocyanate) passed through the cellophane-membrane to produce a definite fluorescence in the dialysate. The filtered solution of conjugated protein was cooled in ice and enough ammonium sulfate was added by the rotating-membrane technic to bring the concentration to half-saturation. The precipitated protein was collected by centrifugation and filtration and the yellow filter-cake was taken into solution in water and dialyzed against running cold tap-water and then against physiological saline at 5 C. This process was repeated. The solution of conjugate within the membrane had a pronounced fluorescence whereas the dialysate showed only a trace of fluorescence at this stage. The conjugate was precipitated with ammonium sulfate and stored in this medium. The yield was about 70 per cent. When required for use, a portion was dialyzed and the protein-content determined by micro-Kjeldahl analyses.

Several of the preparations were purified further by precipitation of the conjugate from solution by the addition of acetone cooled to −15 C. The precipitate retained its yellow color and gave an aqueous solution with a pronounced green fluorescence. This material was dialyzed and then salted out with ammonium sulfate.

Chemical properties of the conjugates

The proteins in the various antisera were chemically modified by the introduction of the fluorescein prosthetic group and developed a green fluorescence. Undoubtedly, conjugation also occurred with any immunologically inert proteins in the antisera but the immunological investigations demonstrated that the actual antibodies contained chemically combined fluorescein. Studies were conducted by means of ultraviolet spectrophotometry in an effort to ascertain the degree of conjugation. Interpretation of the results was made difficult by the lack of suitable derivatives of fluorescein for calibrative purposes but it seems that in the case of the antipneumococcal serum the fluorescein-content was only about two groups per molecule of protein. The determinations were made according to the general procedure of Creech and Jones (7, 8). The conjugates exhibited maxima at 4900 Å (fluorescein-nucleus) and at 2750 Å (aromatic amino acids of the protein). Conjugates prepared from antistreptococcal serum contained a somewhat higher content of fluorescein than that of the antipneumococcal serum. The slight change in the intensity of the absorption after acetone-treatment of the conjugates indicated that the purification was reasonably complete at this stage.

The use of the isocyanate of fluorescein in the preparation of chemically modified antisera is attended with difficulties not encountered in the preparation and purification of the previously described conjugates from polycyclic aromatic hydrocarbons and proteins. Because of the solubility in water of uncombined fluorescein-derivative, the elimination of the last traces of this material as an adsorbed contaminant on the conjugate is difficult and somewhat uncertain. The instability of the isocyanate of fluorescein and the lack of suitable calibrative standards for ultraviolet spectrophotometry are complicating factors which render uncertain any estimations of the fluorescein-content of the conjugates. The use of fluorescein-4-isocyanate for this research has been reasonably satisfactory for preliminary studies but from the chemical point of view, it would be advisable to use an isocyanate of a more stable green-fluorescing organic molecule for the introduction of the prosthetic group.

The more intense the fluorescence of the prosthetic group, the less is the quantity necessary to introduce into the proteic molecule, and as a consequence there is less danger of destruction of the specificity of the antibody by the experimental conditions required for the conjugation. We have therefore prepared conjugates which fluoresce bright green,—a fluorescence-color which occurs very rarely in human tissue (13). In the tissues of the mouse we have not encountered it at all.

Immunological proprties of the conjugates

The fluorescein-carbamido-antipneumococcal 3 serum-conjugate ("3F") gives a yellow, optically clear solution with a pronounced green fluorescence in visible light and a brilliant bright green fluorescence in ultraviolet light. The agglutinative titer, with dilutions based on the known protein-content before and after conjugation, was unchanged by the chemical manipulation and remained at 1:800. The fluorescence imparted by the antibody to the agglutinated, washed organisms was not visible in a dilution of 1:800 but was of good intensity in that of 1:400. These effects were apparent both to the naked eye and under the fluorescence-microscope. They indicate that the minimal amount of antibody necessary to cause agglutination did not carry enough fluorescent groups to be visible. Improvement in chemical procedures may in the future overcome this relative lack of sensitivity. Marrack's mixing-experiment was easily carried out with this modified antiserum thus demonstrating that the conjugation with the fluorescein-derivative involved the antibody-molecules themselves (table 1).

This reagent was now applied to tissues containing pneumococci 3 and their products with the purpose of determining whether it would specifically reveal these materials *in situ*.

A six-hour 0.1 per cent dextrose, 0.5 per cent rabbit-serum broth-culture of a laboratory strain of Pn 3 was centrifuged, the supernatant broth decanted, and the organisms resuspended in saline ($\frac{1}{20}$ of the original volume). One-half ml of this suspension (equivalent to about 1×10^{10} organisms) was in-

TABLE 1

The effect of mixing type 3 pneumococcal fluorescein-antibody (1:50) with unconjugated type 2 pneumococcal antibody (1:50) of equal agglutinating titer, and adding aliquots of this mixture to suspensions of organisms

| | | FLUORESCENCE | |
PNEUMOCOCCUS TYPE	AGGLUTINATION	Macro	Micro
2	++++	0	0
3	++++	++++	++++

jected intravenously into each of three mice. One was killed at the end of 30 minutes, one at the end of 1 hour, and the third died at the end of 4 hours. A normal mouse was sacrificed as a control. The liver, spleen, kidneys, lungs, and heart of each animal were cut into small pieces and fixed *overnight* in 10 per cent formalin.

Frozen sections of the various tissues about 10 micra thick were washed in water for 10 to 20 minutes, and then placed, with care to avoid folding, in fluorescein-antibody solution ("3F") for from 10 to 20 minutes. They were then washed rapidly with gentle motion in saline, dipped into distilled water to remove the salt, and mounted in neutral reagent-glycerol on new clean slides of ordinary glass under clean coverslips of ordinary glass. Quartz is not necessary. They were examined as soon as convenient under the fluorescence-microscope. The results are summarized in table 2.

No green fluorescence was observed in any of the organs from the normal mouse either before or after staining with 3F. There was no green fluorescence present in unstained sections of the infected mice.

The findings in mouse 3, which died of the infection, deserve a brief description: With the exception of an occasional microscopical abscess in the liver, the fluorescence after staining was limited to the blood vessels of all the organs studied. The antibody-conjugate was deposited in impressive amounts, especially in the liver and spleen. Individual organisms, which were plentiful in these organs as shown by Wright's stain, could not be distinguished as the specific precipitate obscured their outlines. The Kupffer cells were heavily stained and many of the endothelial cells lining the blood vessels were outlined by a fine bright green line which appears under the oil-immersion lens to be less than the diameter of the pneumococcus. This was particularly striking in the kidney where, because of this staining, the vascular bundles running in the pyramids were distinctly visible.

Since the completion of this experiment, about 30 mice have been studied in this way and each showed the same distributions of the antigen. Figure 1 is from a photograph of the liver of one of these mice.

Remarks on technic. Fixation: Ten per cent formalin has been exclusively employed as the fixative. After staining with fluorescein-antibody, fluorescence

TABLE 2

The amount of green fluorescence in the livers of mice that were injected intravenously with 1 × 10^{10} pneumococcus 3 and killed at various intervals, as determined by the examination of frozen sections of the formalin-fixed livers stained with type 3 fluorescein-antibody solution

MOUSE	DURATION OF INFECTION	ORGANISMS BY WRIGHT'S STAIN	FLUORESCENCE
	hours		
1	$\frac{1}{2}$	0	0
2	1	Rare	Rare speck
3	4 (dead)	Numerous	$++++$
4*	0	0	0

* Normal control.

is intense for the first few days after fixation, but prolonged storage of tissue in formalin apparently slowly destroys the antigen so that only slight staining is obtained after about one week, and is impossible after one month. The lability in formalin of any antigen studied must thus be determined.

Staining: In all manipulations with fluorescein-antibody, or other antibody derivatives, one must bear in mind that as with antibodies in general, any protein-precipitant will result in non-specific precipitation.

Fluorescein is a poor dye which fades rapidly, and sections prepared as described above noticeably decreased in intensity of fluorescence after a few days even when stored in the dark. Fluorescein is also an indicator, the wavelength of its emitted light changing from yellow to fluorescence as the pH shifts from 3.6 to 5.6 and its fluorescence ceases at pH 3.8. Accordingly we have maintained a pH of about 6 to 7.

The antibody-solution itself will retain its specific properties for months when kept in the dark in the ice-box. Sterile precautions may be omitted as the dye itself appears to have bacteriostatic properties.

The microscope: Haitinger's monograph (14) should be consulted for a detailed discussion of the principles and methods of fluorescence-microscopy.

The microscope which we have used was one assembled by Dr. Allan L. Grafflin and kindly made available to us. It consists of a Zeiss-Weule electromagnetic-feed carbon arc of the size carrying carbons 6 mm and 8 mm in diameter. The carbons are cored but not impregnated, as so-called "therapeutic" carbons do not burn smoothly enough for optical purposes. The rays from the arc pass through a quartz collecting lens, and quartz-walled cuvettes containing respectively aqueous solutions of p-nitroso-dimethyl aniline (to remove violet

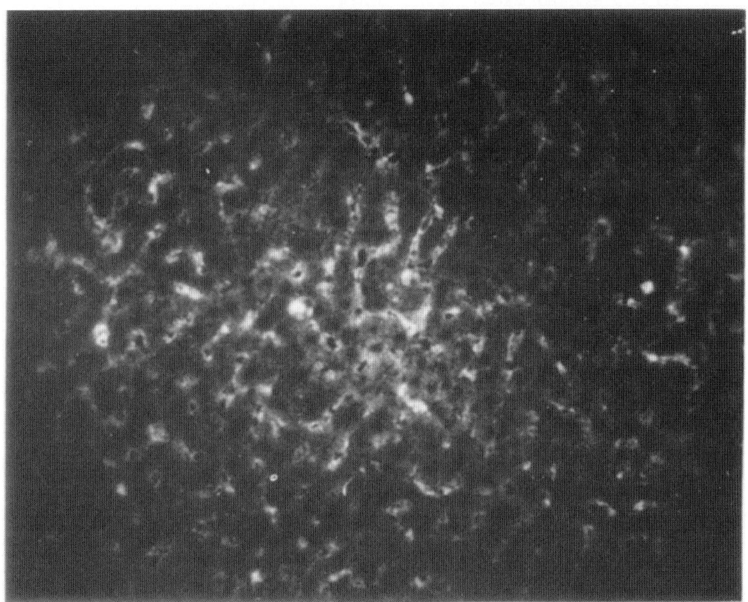

FIG. 1. PHOTOMICROGRAPH THROUGH FLUORESCENCE-MICROSCOPE OF THE LIVER OF A MOUSE MORIBUND WITH PNEUMOCOCCAL 3 INFECTION

The antigen in the liver stained with fluorescein-carbamido-antipneumococcal 3 rabbit serum. Black areas are blood vessels; grey areas are the blue-grey fluorescence of formalin-fixed hepatic cells; white areas and fine white lines represent the green fluorescence of fluorescein-antibody and indicate the location of pneumococcal 3 antigenic material.

20 × apochromatic Zeiss objective, 10 × ocular; in the ocular Wratten filters ⚹2A and ⚹8 (K2.).* Leitz 1 × microphotographic apparatus. Eastman film Super Ortho Press. Exposure: 20 minutes.

rays) and cupric sulfate (to remove red rays). The concentrations of these solutions must be adjusted to the intensity of the light-source, the longitudinal dimension of the cuvette, and the opacity of the glass filter. In our apparatus we used about 2 mg/100 ml of p-nitroso-dimethyl aniline, and 1.5 per cent of cupric sulfate. The solutions should be as dilute as possible without allowing the background of the microscopic field to be more than just detectably red-purple. The middle range of the visible light is removed by Corning light-filter ⚹986.[3] The light-beam then falls upon a totally-reflecting quartz prism

---

* Filter ⚹8 suppresses some of the blue from the liver-cells and thus heightens the contrast. It must be placed above filter ⚹2A because it fluoresces in ultra-violet light.

[3] Corning Glass Company, Corning, New York.

and is brought to a focus on the object by a quartz condensor. Any good microscope may be used providing the lenses do not fluoresce in ultra-violet, since this makes a bright base. Considerable quantities of ultra-violet pass through the optical system of a microscope, and it is therefore necessary to protect the eyes of the observer. For this purpose we have found Wratten filter # 2A[4] very satisfactory as it is colorless, has a sharp cut-off at 0.42 micron, and does not itself fluoresce. It can be put unmounted between the lenses of the ocular. It is so opaque to ultra-violet that in our apparatus it was possible to take photomicrographs requiring 20-minute exposures without fogging the film.

Since the output of light from the object is very low, the main problem is intensity. The use of a 20 × Zeiss apochromat (NA 0.65) greatly increased the intensity over that with a 20 × achromat (NA 0.40). This latter lens was very useful, especially for photographic purposes.

Paraffin-sections may be used in the same way (see below) but they should be left in the aqueous phase and mounted in glycerol. If they are dehydrated after staining and mounted in mineral oil (balsam fluoresces and cannot be used), the fluorescence of fluorescein is completely quenched.

In conjunction with the oil-immersion lens, mineral oil should be used because cedar oil also is fluorescent.

Specificity of the reaction

To establish the specificity of this staining, an attempt was made to prevent it by preliminary saturation of the antigen with unconjugated homologous antibody.

Frozen sections from the liver of a mouse heavily infected with Pn 3 were placed in 1:50 dilutions of type 2 or type 3 unconjugated antibody for 20 minutes. They were then washed in saline for about 3 minutes, and all placed in a 1:50 dilution of 3F for 10 minutes, then washed in saline and water, mounted and examined. The following results were obtained:

Preliminary exposure to unconjugated pneumococcal antibody Pn type	Fluorescence
2	++++
3	0

Thus the staining would seem to be immunologically specific, and we must conclude that the presence of a green deposit in tissues stained with this conjugate indicates the presence of the homologous organism or its specific soluble substance. The occurrence, therefore, of the staining of endothelial cells in the moribund mice described above suggests that free carbohydrate is present in or on those cells. It is to be noted that in the foregoing procedure the conjugated antibody is not brought in contact with living cells, and in consequence the problem of penetration of protein-(antibody) molecules through the normal cell-wall is not at issue. The appearance of the Kupffer cells, however, suggests strongly that cells containing antigen are stained by this method.

   [4] Eastman Kodak Company, Rochester, New York.

Reversibility of the reaction

The experiments described above have been repeated many times, and have invariably indicated sharp and reliable immunologically specific staining. Sometimes, however, with high concentrations of fluorescein-antibody (1:10, 1:5 in terms of original serum) and long exposure, staining of sections pre-exposed to homologous unconjugated antibody occurred. This was rarely as intense as that which was noted after pre-exposure to heterologous antibody, and always appeared first in areas containing the greatest quantity of antigen. The phenomenon suggested that the antigen-antibody reaction might be slowly reversible. To test this hypothesis, the following experiment was undertaken:

Material for testing. Paraffin-sections of the liver of a mouse dead of infection with Pn 3 were fixed in 10 per cent formalin for 18 hours, washed for 24 hours, treated with 70 per cent ethanol, dehydrated, and embedded in paraffin. The sections were mounted on slides prepared with gelatin-adhesive.[5]

Reagents. 1. Antipneumococcal 3 fluorescein-antibody solution freshly dialyzed from ammonium-sulfate precipitate. The agglutinative titer of this solution was 1:80 (1:800 in terms of original serum).
   2. Antipneumococcal 3 rabbit serum (Lederle). Agglutinative titer: 1:1600.
   3. Normal rabbit serum, 1 week old.
   4. 0.9 per cent sodium-chloride solution.
   5. Aluminium ammonium sulfate ($Al_2(SO_4)_3 \cdot (NH_4)_2SO_4 \cdot 24H_2O$) 2 g in 100 ml, 60 ml Aceton C.P., 15 ml.
   6. 0.1 M phosphate-buffer solution, pH 7.4.

Procedure. The slides were run through xylene, 100 per cent, 95 per cent, and 80 per cent ethanol in the usual way, and then put in saline. After about 5 minutes they were removed from the saline, wiped dry except for the tissue itself, and then the tissue-section was covered with the unconjugated-antibody solution or normal rabbit-serum solution. Evaporation was prevented by covering each slide with a petri dish containing a piece of wet filter-paper.

After 10 minutes, the solution was removed, the slide washed for about 30 seconds in saline with gentle motion, wiped dry except for the tissue itself, and then covered with the appropriate dilution of fluorescein-antibody (3F). Alternately a slide was put for 5 minutes in the acetone-alum solution, washed in water for 30 seconds, placed in the buffer-solution for about 1 minute, rinsed in saline, and then exposed to the appropriate dilution of fluorescein-antibody as above. All exposures to fluorescein-antibody were for 10 minutes. All timing was done as accurately as possible with a maximal deviation of 10 per cent from the time stated.

After exposure to fluorescein-antibody all slides were washed for 30 seconds in saline, followed by an exposure of 1 minute in phosphate-buffer solution. They were then rinsed in distilled water and mounted in reagent-glycerol.

The alum-acetone solution caused immediate precipitation of rabbit serum

---

[5] The usual egg-albumin-glycerol adhesive has a blue fluorescence and cannot be used. Slides are dipped in 0.5 per cent gelatin and allowed to dry in a nearly vertical position. They are then dipped in 10 per cent formalin and allowed to dry.

diluted 1:10. This precipitate was insoluble in saline and in weak alkali. The experiment was carried out at room temperature, i.e., about 23 C. The results are presented in table 3.

In several respects they are of interest. A comparison of the findings in slides 1 and 5 demonstrates the immunological specificity of the staining, since it was inhibited by preliminary exposure to homologous antibody but not by exposure to normal rabbit serum. The staining in slide 3 suggests that a 1:100 dilution of the serum used for preliminary exposure does not provide enough antibody-molecules to saturate the available antigen, but slide 4 disproves this by demonstrating that inhibition is practically complete if the deposited

TABLE 3

The reversibility of fluorescein-antibody-antigen reactions

CONJUGATED ANTIPN 3 SERUM 3F	TISSUE TREATED FIRST WITH					
	Antipn 3 serum				Normal rabbit serum	
	1:10		1:100		1:10	
	Saline	Alum-acetone	Saline	Alum-acetone	Saline	Alum-acetone
1:10	1* 0	2 0	3 +++†	4 ±	5 ++++	6 +++
1:100	7 0	8 0	9 0	10 0	11 0	12 0

* Numbers refer to preparations treated in manner indicated.
† Plus marks indicate degree of fluorescence.
Controls (conjugated antibody only):
  A 3F 10 minutes.. ++++
  B 3F 40 minutes.. ++++ (no more
  C 3F 10 minutes followed by washing for 30 seconds in saline... ++++ than A)
  D 3F 10 minutes followed by washing for 30 minutes in saline... ++++
  E 3F 10 minutes followed by washing for 18 hours in saline
      at 4 C.. 0

antibody is fixed to the tissue. The fixation was accomplished by a reagent (alum-acetone) which destroys the solubility of protein but does not itself prevent staining even when applied after a nonspecific-protein solution (slide 6). This result furnishes more direct evidence that there is in fact an equilibrium between deposited and supernatant antibody and the control slide E points in this same direction since it indicates a slow dissociation of the antigen-antibody complex in the presence of merely the reagents used in the experiment.

It should be pointed out that the reactions taking place in the tissue-sections probably involve only the so-called "first stage" of the antigen-antibody reaction, since the antigen is fixed in space and organisms or antigen-antibody aggregates presumably cannot agglutinate or flocculate.

The foregoing experiments serve as examples of the technics employed and suggest the utility of such antibody-derivatives in the investigation of the antigen-antibody reaction.

In addition to the chemical difficulties already mentioned, there have been two failures which must be termed immunological rather than chemical. A preparation of fluorescein-antistreptococcal A (anticarbohydrate) which was very fluorescent failed to cause fluorescence of suspended organisms and to stain organs of mice moribund from infection with a virulent type 6 (group A). This failure may possibly be attributed to the relatively low titer of the serum.

A preparation of fluorescein-anti-rabbit hen-serum markedly lost titer during the chemical manipulations. It failed to stain the liver or spleen of a mouse injected 24 hours before with 1 ml of 2 per cent crude rabbit-globulin solution. This experiment was attempted both after formalin-fixation and with no fixation. In neither instance could staining be demonstrated. Again it would seem that when the concentration of antibody is low, definite staining does not occur.

DISCUSSION

It is obvious that a technic for the demonstration of antigens *in situ* in the animal body is highly desirable for the investigation of many conditions which have been attributed to tissue-damage resulting from the localized union of antigen and antibody. Definitely to establish that a mechanism of this sort does produce a given disease, it is necessary to show that antigens of various sorts are actually present in certain areas that regularly are involved and are absent or in lower concentration in unaffected situations. It was with this general objective in view that the method described above was developed. Before it can be employed effectively, however, it is clear that further modifications are desirable in the chemistry of labeling the antibody with a view to increasing its sensitivity. Moreover, it is probable that only labeled-antibody preparations of high titer will give satisfactory results.

In addition to the principal application of the method which we have in mind, should it prove of general practicability with many antigen-antibody systems, it is evident that the technic could be used in a variety of ways for the study of various aspects of antigen-antibody union both *in vitro* and *in vivo*, including possibly the demonstration of viruses in tissues and cells.

CONCLUSIONS

1. Impure fluorescein isocyanate has been prepared and successfully conjugated through the ureide linkage to pneumococcal 3 antibody.

2. This fluorescein-carbamido-pneumococcal 3 antibody-solution possessed the same agglutinin-titer as the original serum in terms of protein-content, and rendered pneumococcus 3 fluorescent in ultra-violet light.

3. The tissues of mice heavily infected with pneumococcus 3 could be specifically stained in localized areas with this antibody-conjugate, and it is proven that this staining is immunologically specific.

4. Incomplete data are presented which suggest that the antigen-antibody reaction is reversible.

It is a great pleasure to express our great debt for the constant encouragement and counsel of Dr. John F. Enders, the advice of Dr. Allan L. Grafflin, and the kind interest of Professor Louis F. Fieser, and Professor George B. Wislocki.

REFERENCES

(1) REINER, L.   1930   On the chemical alteration of purified antibody-proteins.   Science, **72**, 483–484.

(2) BRONFENBRENNER, J., HETLER, D. M., AND EAGLE, I. O.   1931   Modification of therapeutic sera with a view of avoiding complications of allergic nature.   Science, **73**, 455–457.

(3) MARRACK, J.   1934   Nature of antibodies.   Nature, **133**, 292–293.

(4) EAGLE, H., SMITH, D., AND VICKERS, P.   1936   The effect of combination with diazo compounds on the immunological reactivity of antibodies.   J. Exp. Med., **63**, 617–643.

(5) HEIDELBERGER, M., KENDALL, F. E., AND SOO HOO, C. M.   1933   Quantitative studies on the precipitin reaction.   Antibody production in rabbits injected with an azoprotein.   J. Exp. Med., **58**, 137–152.

(6) HOPKINS, S. J., AND WORMALL, A.   1933   Phenyl isocyanate protein compounds and their immunological reactions.   Biochem. J., **27**, 740–753.

(7) CREECH, H. J., AND JONES, R. N.   1940   The conjugation of horse serum albumin with 1,2-benzanthryl isocyanates.   J. Amer. Chem. Soc., **62**, 1970–1975.

(8) CREECH, H. J., AND JONES, R. N.   1941   The conjugation of horse serum albumin with isocyanates of certain polynuclear aromatic hydrocarbons.   J. Amer. Chem. Soc., **63**, 1661–1669.

(9) CREECH, H. J., AND JONES, R. N.   1941   Conjugates synthesized from various proteins and the isocyanates of certain aromatic polynuclear hydrocarbons.   J. Amer. Chem. Soc., **63**, 1670–1673.

(10) COONS, A. H., CREECH, H. J., AND JONES, R. N.   1941   Immunological properties of an antibody containing a fluorescent group.   Proc. Soc. Exp. Biol. & Med., **47**, 200–202.

(11) FIESER, L. F., AND CREECH, H. J.   1939   The conjugation of amino acids with isocyanates of the anthracene and 1,2-benzanthracene series.   J. Amer. Chem. Soc., **61**, 3502–3506.

(12) BOGERT, M. T., AND WRIGHT, R. G.   1905   Some experiments on the nitro derivatives of fluorescein.   J. Amer. Chem. Soc., **27**, 1310–1316.

(13) HAMPERL, H.   1934   Die Fluorescenzmikroskopie menschlicher Gewebe.   *Virchows Arch.*, **292**, 1–51.

(14) HALTINGER, M.   1938   Fluorescenzmikroskopie: Ihre Anwendung in der Histologie und Chemie.   Akademische Verlagsgesellschaft.   Leipsig.

Aerial Dissemination of Pulmonary Tuberculosis. A Two-Year Study of Contagion in a Tuberculosis Ward

R. L. RILEY, C. C. MILLS, W. NYKA, N. WEINSTOCK, P. B. STOREY, L. U. SULTAN, M. C. RILEY, AND W. F. WELLS

Although airborne transmission of tuberculosis had been assumed since the late 19th century, studies to quantify the risk had been missing. In the 1950s, R. L. Riley et al. started a series of investigations on "air hygiene in tuberculosis." In a cooperative study in Baltimore, Md., the authors exposed guinea pigs continuously for 2 years to air from a ward occupied by patients with active pulmonary tuberculosis, assuming that infectious doses for humans and guinea pigs were approximately the same. With painstaking accuracy and the help of the most recent methodologies available, the study was carried out to conclude that even small numbers of tubercle bacilli can account for the spread of human tuberculosis. Even typing with the help of susceptibility patterns (unfortunately, the technique is not described) was attempted to trace the infections in guinea pigs to a particular human source. This study revealed that patients with the highest counts of tubercle bacilli in their sputa were also the most infectious ones.

This study and the companion ones were pivotal for subsequent air control measures recommended for rooms with tuberculous patients and have assumed new importance with the increase in tuberculosis in general and drug-resistant *Mycobacterium tuberculosis* strains in particular (cf. S. Segal-Maurer and G. E. Kalkut, *Clin. Infect. Dis.* **19**:299–308, 1994).

ALEXANDER VON GRAEVENITZ

Reprinted with permission from *American Journal of Hygiene* (now *Journal of Epidemiology*) 70:185–196. Copyright © 1959. Johns Hopkins University.

AERIAL DISSEMINATION OF PULMONARY TUBERCULOSIS

A TWO-YEAR STUDY OF CONTAGION IN A TUBERCULOSIS WARD [1]

BY

R. L. RILEY, C. C. MILLS, W. NYKA, N. WEINSTOCK, P. B. STOREY,
L. U. SULTAN, M. C. RILEY AND W. F. WELLS [2]

(Received for publication March 26, 1959)

The first report of this series, entitled "Air hygiene in tuberculosis," dealt with the preparation of a pilot ward for the performance of quantitative studies of the infectiousness of the air (1). In the second, basic theoretical concepts were elaborated, upon and the results of the first few months of operation of the pilot ward with human patients were presented (2). The present paper includes data which greatly amplify and, in minor respects, modify the interpretations and inferences presented in the second paper. Studies of the control of contagion by disinfection of the air have been postponed in order to demonstrate beyond question the fact of aerial dissemination and the probability of its predominant importance in the transmission of pulmonary tuberculosis.

EXPERIMENTAL DESIGN

The experimental unit in which these studies were carried out includes a tuberculosis ward with 6 single rooms, a carefully controlled and calibrated closed circuit ventilating system, and a large animal exposure chamber located in the exhaust duct of this system (1, 2). Preliminary experiments demonstrated that the air passing through the exposure chamber was virtually equal in infectivity to the air in the ward itself (1). For the past 2 years the ward has been occupied by tuberculous patients and a number of the guinea pigs breathing air vented from the ward have caught tuberculosis. In this report we shall describe in detail the circumstances surrounding these infections.

RESULTS

Between November, 1956, and November, 1958, 71 guinea pigs breathing air from the pilot ward became infected with tuberculosis. The distribution of these infections in time is shown by the vertical bars at the bottom of figure 1 and also in table 1. The total number of animals exposed is shown at the top of the figure. All animals were tuberculin tested at monthly intervals and positive reactors sacrificed. The diagnosis of tuberculosis was accepted only after pathological demonstration. The number of infections acquired during the course of a single month varied from 0 to 10, the average being approximately 3. The number of guinea pigs exposed at any one time varied between 74 and 187, the average being 156. The number of sputum-positive patients oc-

[1] A cooperative study between the Veterans Administration, The Johns Hopkins University School of Hygiene and Public Health, and the Maryland Tuberculosis Association, Baltimore, Md.

[2] This work would have been impossible without help from every department of the Baltimore Veterans Administration Hospital, and we wish to thank all concerned. Special gratitude is due Dr. Francis O'Grady, Chief of Microbiology and Air Hygiene, for his critical review of the manuscript, and Dr. James D. Murphy, Director of the Hospital, for his patient support at the administrative level.

AM.J.HYG. 1959, VOL. 70: 185–196

Table 1

Occupancy of exposure chamber and fate of animals exposed

Year	1956					1957												1958										
Month	8	9	10	11	12	1	2	3	4	5	6	7	8	9	10	11	12	1	2	3	4	5	6	7	8	9	10	11	12
Sacrificed																													
Tuberculosis					2	3	1	7	4	5	5					1	1		2	8	2	1	3	5	2	2	10	6	1
Tuberculin reaction																													
≤10 mm diam.																													
<3+ induration																													
No central necrosis								4	3		1		2		1	1						1	2			2	1		
Pasteurella																													
Multiple abscesses					2		2	3	7		9		6								3	4					34		
Well animals																													
Natural deaths																													
Pneumonia																													
Acute											1													3		25	5		
Chronic, mostly due to *Pasteurella*		6	11	3	4	3		1	3	10	2	2	2	3	4				2	1	2	6	1	6	5	2			
Other infections, mostly peritonitis					1	2		2			2																1	1	
Trauma	3				2			1	1	1			1	1	1		1			3		2	2		2	2			
Removed alive for observations			7													4		5								4			
Unaccounted for	2	3	5						1	1			1	1	1		2		2	1	2	1	2	2	3	1			
Total removed	5	9	23	3	11	8	3	18	19	17	20	2	12	5	7	6	4	5	6	13	9	15	10	16	12	38	51	7	1
New animals added	37	35			81	3	41	19	10	2	22	14	1	35			11		2	19		28	2		7			4	
Change in occupancy	32	26	−23	−3	70	−5	38	1	−9	−15	2	12	−11	30	−7	−6	7	−5	−4	6	−9	13	−8	−16	−5	−38	−51	−3	
Total occupancy at end of indicated month	73	99	76	73	143	138	176	177	168	153	155	167	156	186	179	173	180	175	171	177	168	181	173	157	152	114	63	60	
Average occupancy for indicated month	68	70	89	74	115	138	165	175	168	158	150	161	159	179	181	171	177	175	172	164	171	187	174	163	154	131	86	61	

cupying the ward during any 1 month varied between 3.5 and 6.

In table 1 the fate of animals occupying the exposure chamber is shown in detail. In general, the health of the animals was good even though they were crowded into a cylindrical space only 5 feet high and 5 feet in diameter. The guinea pigs were distributed throughout this space in 36 cages. In spite of the crowding, their surroundings included the essentials of good hygiene: air of constant temperature from the ward, continuously running tap water, adequate provision of nutritional requirements and sanitary disposal of excreta (1). The floors of the cage baskets were of wire mesh and no bedding material was used. There was one small epidemic of pneumonia which reached its peak in September, 1958, of which the cause remains in doubt (see table 1). The guinea pigs were purchased from a professional supplier and were delivered to us when 5 or 6 weeks of age. Most of the animals that were sick on arrival died during the month's quarantine period and were never put into the exposure chamber. All of the animals were tuberculin tested during the quarantine period and none was found to be positive. Many of the guinea pigs occupied the exposure chamber during most of the 2-year period and grew larger and fatter during this period. The average weight of animals in the exposure chamber increased from 0.8 pound at the beginning to 1.8 pounds at the end of the 2-year period. A total of 373 new animals was used between November 1, 1956, and November 1, 1958.

There were 4 consecutive months in the fall of 1957 when no animals were infected. This unexpected lapse was a mystery at the time. In retrospect it can be seen to coincide with the presence

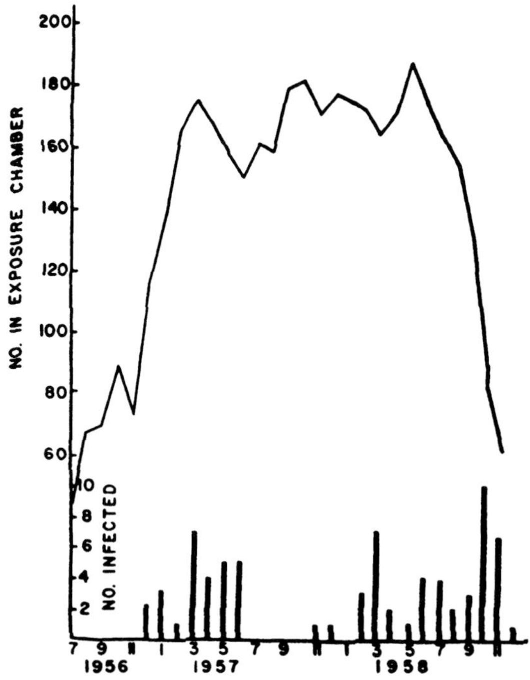

FIGURE 1. Average number of guinea pigs in exposure chamber each month and number identified as having tuberculosis each month. Note that the scale of the ordinate differs in each case. Note also that the date of infection was, on the average, about 6 or 7 weeks earlier than the date of identification of infection. The vertical bars in the figure are placed at the latter date.

on the ward of initial treatment cases with drug-susceptible organisms. These patients were receiving drug therapy. There is a possibility, to be discussed below, that the viability of the tubercle bacilli in the air-borne state was reduced under these circumstances. Infections in guinea pigs again occurred when the patients with chronic disease and drug-resistant organisms were placed on the ward.

The distribution of tuberculous infections throughout the exposure chamber was studied by comparing the number of times that infections occurred in each of the 36 cages. Ward air was distributed radially through the cages before being vented to the outdoors. This

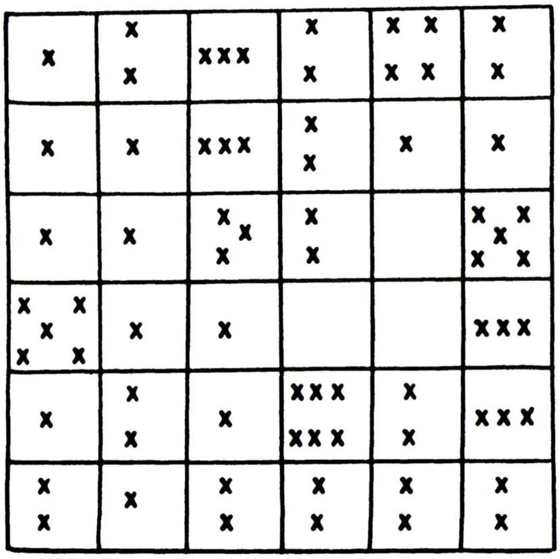

FIGURE 2. Distribution of tuberculosis infections by cages. The frame holding the 36 cages rotates, so each cage maintained a constant position relative to the others but not relative to the room. All 71 infections are included.

design prevented air from passing from one cage to another. Throughout most of the 2-year period there were 4 or 5 guinea pigs in each cage. When for any reason an animal was removed from a cage, he was replaced by an uninfected animal. The total census of animals thus remained fairly constant (see figure 1 and table 1). The distribution of tuberculous infections by cages did not differ significantly from theoretical random when tested by χ^2, the probability that the observed deviation from theoretical random would arise by chance being 0.7 +. This distribution is shown in figure 2.

At autopsy single primary pulmonary tubercles were usually found in the lungs of infected guinea pigs. The anatomical location of all these individual tubercles is shown in table 2. The distribution suggests that air-borne tuberculosis goes wherever the inspired air goes within the lung. It is consistent with the random distribution of infection in different cages, since in this in-

stance also air-borne tuberculosis appears to have gone wherever the air went.

The pathological findings in the guinea pigs which were sacrificed because of conversion of the tuberculin test to positive are shown in table 3. Tuberculous infection was found in 71 of the 77. The results of the tuberculin tests will be reported in detail elsewhere. A primary pulmonary tubercle was found in 51 of the 71 infected animals. Tubercle bacilli were seen on microscopic examination of the hilar lymph nodes in every case except two (no. 13 and 19). The finding of no pulmonary tubercle in 20 out of 71 cases, even though the lung was presumably the portal of entry in all cases, suggests that tuberculosis may sometimes be carried rapidly to the lymph nodes without leaving macroscopic evidence in the lungs. This possibility has been discussed by Lurie (3).

Identification of the patient who infected a particular guinea pig was possible in 21 of the 22 instances in which complete tests of drug susceptibility were carried out. Organisms from the sputum of patients occupying the ward were completely studied in all cases, but up to March 31, 1958, organisms from only about half of the infected guinea pigs (22 out of 46) were cultured and tested for drug susceptibility. The cultural characteristics of the organisms

TABLE 2

Distribution of primary tubercles in guinea-pig lungs

Right upper lobe	12
Left upper lobe	18
Right lower lobe	10
Left lower lobe	10
Left pyramidal lobe*	1
	—
Total	51

* The left pyramidal lobe is very small.

TABLE 3

Pathological findings in guinea pigs sacrificed because of conversion of tuberculin test to positive

No.	Date	Pulm. tubercles No.	Pulm. tubercles Position	Bacilli seen in:* Hilar nodes	Bacilli seen in:* Spleen	Remarks
	1956					
1	12/4	1	R.L.L.			
2	12/4	1	R.L.L.			
	1957					
3	1/2	1	L.U.L.			
4	1/2	0		+		
5	1/2	1	R.U.L.	+		
6	1/28	1	R.L.L.	+		
7	2/25	1	R.U.L.	+		
8	2/25	1	R.U.L.	+		
9	2/25	1	R.U.L.	+		
10	2/25	0		−		No tbc demonstrated
11	2/25	1	L.L.L.	+		
12	2/25	1	L.U.L.	+		
13	2/25	1	L.L.L.	−		
14	2/25	0		+		
15	3/27	1	R.L.L.	+		
16	3/27	1	R.U.L.	+		
17	3/27	1	L.U.L.	+		
18	3/27	0				No tbc demonstrated
19	3/27	0		−	+	
20	4/29	1	L.L.L.	+	−	
21	4/29	1	R.U.L.	+	−	
22	4/29	0		+		
23	4/29	1	L.U.L.	+		
24	4/29	1	L.U.L.	+	+	
25	6/3	1	L.U.L.	+		
26	6/3	0		+	−	
27	6/3	1	L.P.L.	+	−	
28	6/3	0		+		
29	6/3	1	R.U.L.	+		
30	8/5	0		−		No tbc demonstrated
31	8/5	0		−		No tbc demonstrated
32	10/30	1	R.L.L.	+	−	
33	10/30	0		+		No tbc demonstrated
34	11/25	0		+		
	1958					
35	2/3	0		+	+	
36	2/3	1	L.U.L.	+	+	
37	3/3	0		+	+	
38	3/3	0		+	+	
39	3/3	2	R.U.L.	+	+	1 tubercle larger
40	3/3	1+	L.U.L.	+	+	Secondary spread
41	3/3	0		+	+	
42	3/3	1	R.U.L.	+	−	
43	3/3	1	L.L.L.	+	+	
44†	2/19	1	L.U.L.			Natural death
45	3/31	1+	R.U.L.	+	+	Secondary spread
46	3/31	1	L.U.L.	+	+	
47	4/28	2	L.L.L.	+	+	Tubercles contiguous
48	5/27	1‡	R.L.L.	+	+	
49	5/27	1	L.L.L.	+	+	
50	5/27	1	L.U.L.	+	+	
51	6/25	0		+	+	
52	6/25	0		+	+	
53	6/25	1	L.L.L.	+	+	
54	6/25	1+‡	L.U.L.	+	+	Secondary spread
55	7/14	1‡	L.U.L.	+	+	Natural death
56	7/22	1+	L.U.L.	+	+	Secondary spread
57	7/22	0		+	+	
58	8/28	0		+	+	
59	8/28	0				No tbc demonstrated
60	8/28	0		+	+	
61	9/23	0		+	+	
62	9/23	1	R.U.L.	+	+	
63	9/23	1+	L.L.L.	+	+	
64	9/23	1+	R.L.L.	+	+	Secondary spread
65	9/23	1+	R.L.L.	+	+	Secondary spread
66	9/23	1	L.L.L.	+	+	
67	9/23	0		+	+	
68	9/23	0		+	+	
69	9/23	1+	L.U.L.	+	+	Secondary spread
70	9/23	1	L.U.L.	+	+	
71	10/15	1+	L.L.L.	+	+	Secondary spread
72	10/15	1	R.L.L.	+	+	
73	10/15	1‡	L.U.L.	+	+	
74	10/15	1‡	L.U.L.	+	−	
75	10/15	0		+	+	
76	11/10	1	R.U.L.	+	+	
77	12/10	1	R.L.L.	+	+	

* A blank indicates that the tissue was not examined histologically.

† Tuberculin test ± since 3/26/57.

‡ No tuberculosis seen histologically in pulmonary tubercle.

                    R. L. RILEY ET AL.

TABLE 4

*Time relationships, pathological findings, cultural characteristics and identification
of sources of 22 air-borne infections in guinea pigs*

No. infected	Guinea pig			Guinea pig and patient			Patient	
	Autopsy			Strain of tuberculosis				
	Animal number (see table 3)	Date infection discovered	No. with tuberculosis in spleen	Catalase	Resistant to:		On ward	Name
12	1, 3, 5, 6, 7, 8, 11, 12, 17, 23, 25, 27	Between 12/4/56 and 6/3/57	0	Neg.	SM-INH-PAS		10/5/56 to 5/1/57	Art
7	36, 39, 40, 41, 43, 45, 46	2/3/58 and 3/31/58	7	Pos.	SM-PAS		12/21/57 to 2/26/58	Jon
1	29	6/3/57	0	Neg.	SM-INH		4/20/57 to 5/2/57	Smi
1	24	4/29/57	1	Pos.	No resistance		?	?*
1	42	3/3/58	0	Pos.	PAS		11/22/57 to 1/15/58	Dag†

* A number of patients receiving original treatment produced organisms susceptible to all 3 drugs. It was impossible to decide which of these patients infected guinea pig no. 24.

† Occupancy of the ward by Dag terminated 7 weeks earlier than the date on which infection in the guinea pig was discovered. This time lag is possible because conversion of the tuberculin test to positive occurs about 3 weeks after infection in the guinea pig, and the tuberculin testing was performed at intervals of approximately 4.5 weeks. Hence an infection 2.5 weeks old, missed at the time of one test, would be about 7 weeks old when detected at the time of the next testing.

from a guinea pig matched those from the sputum of a patient occupying the ward at the time the guinea pig was infected in 21 out of 22 instances. The correspondence of both cultural characteristics and time of exposure provided presumptive identification of a specific patient as the one whose air-borne organisms infected a specific guinea pig. This work has been reported in detail by Dr. Louise U. Sultan (4). It is summarized in table 4.

There have been interesting changes in the pathological findings in the infected guinea pigs during the course of the study. Since the beginning of 1958 the spleens have almost always been infected, and in recent months there has been an increase in the number of times that secondary hematogenous infection of the lungs has been found (see tables 3 and 4). It seems

probable that these pathological differences can be correlated with differences in the strain of tubercle bacillus with which the guinea pigs were infected. For example, among the 22 cultures shown in table 4 there were 5 different bacterial strains. Twelve of the 22 were of one strain, 7 of another, and 1 each of the remaining 3 strains. In none of the 12 guinea pigs which were infected with the first strain was the spleen involved with tuberculosis (gross examinations only in most of these cases) and in none was there secondary spread to the lungs. The spleen was involved in every one of the 7 guinea pigs which were infected with the second strain, and in 2 animals there was secondary spread to the lungs.

The aerial infectivity of artificially atomized sputum from patients occupying the pilot ward was studied on 35

TABLE 5

Infectivity of aerosolized sputum

Patient	Date	Tubercle count	Patient	Date	Tubercle count
Art	11/7/56	180	Smi	1/10/58	38
	3/7/57	140			17
		93		2/4/58	9†
					24
Lay	10/28/57	49			
		54	Bra	2/4/58	3
					14
Law	11/12/57	5		2/17/58	26
		7			21
Bro	1/10/58	11	Rut	2/4/58	194
					200
McM	12/20/57	0		5/12/58	106
		0			143
Nea	12/20/57	0	Jon	1/10/58	376
				2/17/58	442
Mas	1/22/58	40			282
		30			
			Kel	6/13/58	0
Bes	1/22/58	1			1
		1			
	2/17/58	0	Wil	6/13/58	167
		0		5/12/58	204
					180
Erm	9/30/57	248			
	11/12/57	2	Bai	6/13/58	81
		1			89
Rob	9/30/57	>100*	Hin	6/20/58	315
	11/18/58	29			409
		28			
			Bro	6/20/58	17
Cou	12/20/57	8			22
		4			37
Acr	12/20/57	28	Ruc	5/12/58	1
		32			4
Dag	12/20/57	27	Mar	5/12/58	0
		32			2
Lan	3/7/57	12			
	11/12/57	8†			
		13			

* Animal was autopsied 38 days after infection and possibility of secondary spread exists.
† Tubercles in right lung only.

occasions. The sputum was diluted with water, homogenized in a Waring blendor, and atomized into a small chamber in which, in most instances, two guinea pigs were exposed. The design of this exposure chamber has been described by Wells (5). The dilution of the sputum with water (ordinarily 1 to 5), the rate of atomization (0.2 ml per minute), the dilution of the droplet nuclei with air (approximately 1.5 cubic feet per minute), and the duration of exposure of the guinea pigs (ordinarily 15 minutes) were known. The pulmonary ventilation of all guinea pigs was assumed to be approximately the same. After allowing time for tubercles to develop in the guinea pigs' lungs (ordinarily 25 to 31 days), the animals were sacrificed and the tubercles on the surface of the lungs counted. This provided a quantitative comparison of the infectivity of aerosolized sputum from different patients.

The results are shown in table 5. Four features deserve special mention: (a) The guinea-pig lung provided a selective medium in which the tubercle bacillus flourished and all contaminating organisms were rejected. Only one animal out of 65 died of nontuberculous infection following exposure to atomized sputum. (b) The number of tubercles developing in the lungs of guinea pigs exposed simultaneously was approximately the same, in conformity with the belief that the number of tubercles depends on the number of infectious droplet nuclei inhaled. (c) In general, the distribution of tubercles throughout the lungs was random, indicating that airborne tuberculosis went wherever the air went. (d) There was great variation in the infectivity for guinea pigs of aerosolized sputum from different patients. The two men who infected the largest number of guinea pigs under natural conditions (Art and Jon, table 4) also showed very highly infectious sputum when this was tested by atomization. With the exception of Rut, the other patients who consistently showed highly infectious sputum occupied the ward at a later date and the number of guinea pigs which they infected under natural conditions is not yet known.

Discussion

It is essential to establish beyond reasonable doubt that the guinea-pig infections reported herein were transmitted aerially from the patients on the ward and not acquired in some other way. The experimental design was such that tuberculosis had direct access to the exposure chamber only via air vented from the ward. However, other conceivable sources of infection included animal caretakers, food, water and cross infection between animals. For the following reasons it seems unlikely that infection occurred in these indirect ways: (a) The two people who handled the animals have never had active tuberculosis, although both are tuberculin positive. Furthermore, when the exposure chamber is opened in order to care for the animals, the flow of air is away from the animals and toward the caretaker, and the entire area surrounding the exposure chamber is intensely irradiated with ultraviolet light from unshielded tubes.

(b) Infection of the guinea pigs by contaminated food or water would imply enteric infection, and the disease observed in the lungs and hilar nodes would have to have been caused by secondary spread. The characteristic lesion in the lungs of infected animals was a single tubercle, a lesion unlikely to have been caused by secondary spread.

(c) Strong evidence against cross infection between guinea pigs is supplied by the random distribution of infections in the different cages. Cross infection would cause an increased incidence of disease in cage-mates, and this was not seen. On the other hand, one would expect infection transmitted by air from the ward to be randomly distributed throughout the 36 cages since, if air-borne tuberculosis goes wherever the air goes, the animals in each cage should stand an equal chance of becoming infected. This was the type of distribution which was in fact seen.

Other factors minimized the possibility of cross infection. Tuberculous guinea pigs were removed from the exposure chamber before their disease reached a stage in which tubercle bacilli are discharged. Tuberculous cavities in the lungs, which constitute the chief source of organisms expelled from the respiratory tract, were never seen. Although complete studies of the kidneys and the gut were not made, it is unlikely that the disease ever progressed in our animals to the stage where tubercle bacilli were discharged in the urine or feces. The danger of cross infection from contamination of this sort was further minimized by keeping the guinea pigs on wire, without any bedding, so that no potentially infectious dust was present. Urine and feces fell through the wire mesh on to a pan below where contact between animals and excreta was impossible.

The nature of the pulmonary lesions seen in our guinea pigs suggests that the tuberculous infection, from whatever source, was transmitted aerially. Other mechanisms of pulmonary infection would be unlikely to produce a single pulmonary tubercle, located by chance in any lobe. Hematogenous seeding, for example, while equally random, would not consistently produce single tubercles since organisms are not ordinarily liberated into the blood stream one at a time. Air-borne organisms, on the other hand, are expected to come one at a time, both on theoretical grounds and on the basis of our findings. On an average, only 3 animals out of 156 inhaled an infectious dose of tuberculosis organisms in any one month. Since the vast majority of animals remained uninfected, it seems unlikely that the infected ones received more than a single infectious dose. Although on two occasions (guinea pigs no. 39 and 47, table 3) two tubercles were found in the lungs of a single animal on post-mortem examination, this finding seems to us more logically explained by local spread of infection after initial implantation than by chance deposition of two separate infectious air-borne particles in close proximity. Under the conditions of our experiment, the characteristic lesion resulting from aerial infection was the single pulmonary tubercle, attributable to a single air-borne particle.

The matching of the cultural characteristics of organisms isolated from the single tubercle of a guinea pig with those in the sputum of a patient provides presumptive identification of that particular patient as the source of the infection. Since there was no contact between patient and guinea pig other than through the air, there can be little doubt that the infection was introduced into the exposure chamber by the aerial route. While this evidence does not of itself preclude the possibility of subsequent cross infection between animals, the observed time relationships reduce this possibility to the vanishing point. All of the 21 infections for which patient sources were identified occurred while the patients occupied the ward and usually occurred within too short a time period for cross infection. If present, cross infections would have continued to appear after the patient had left the ward, but this did not occur.

The weight of evidence, in our opinion, justifies the conclusion that all 71 guinea pigs were infected by aerial contamination produced by patients occupying the tuberculosis ward.

Variability in the infectivity of different patients was far greater than we realized at the time of the previous report. It is now apparent that a statistical mean infectivity for far advanced tuberculosis cannot be approximated by taking the average infectivity of any 6 patients in this stage of the disease. Two of our patients produced 19 out of 22 infections in guinea pigs even though 62 patients occupied the ward during the period under consideration. The astounding infectivity of these two patients in comparison with the others was related in part to the infectivity of their sputum. The number of organisms seen on smear was high and the infectivity for guinea pigs exposed to artificially atomized sputum was also high. We have no information regarding the physical characteristics of the sputum from different patients which might affect the ease with which the sputum could be atomized under natural conditions in the upper respiratory tract. Observation of the social behavior of some of the highly infectious patients suggests that they may have been careless in covering their mouths when they coughed. By contrast, one quiet and well mannered patient (Rut) had highly infectious sputum but probably infected only two guinea pigs. Still another factor in infectivity may have been variability in the resistance of different organisms to the rigors of atomization and dehydration in the air.

On physical grounds it seems inescapable that the evaporation of droplets to droplet nuclei must vastly increase the concentration of all constituents of the original droplets. Since detectable amounts of antibiotics are known to exist in the sputum of patients who are being treated, relatively high concentrations must have existed after evaporation in the droplet nuclei produced by coughing. Possibly in this way all organisms except those which were the most drug resistant may ordinarily have been inactivated in the air-borne state. This might account for the preponderance of drug-resistant strains found in the guinea pigs infected by breathing ward air (table 4). We do not know whether untreated tuberculous patients would exhibit as much variability in infectivity as the group under study. This important matter is being investigated.

A statement published in 1910 in Chapin's "Sources and Modes of Infection" (6) is peculiarly pertinent to the studies which we have reported: ". . . a quantitative examination of the floating bacteria is necessary if we wish to determine the real danger from the inhalation of the air. No such enumeration of tubercle bacilli seems to have been made, and the difficulty of finding them suggests that they are not very numerous in the vicinity of patients, and that perhaps the air of a room is not always dangerous to breathe even if tubercle bacilli can be found in the settled dust."

It now appears that viable air-borne tubercle bacilli are, as Chapin suspected, not very numerous and that the quantitative study of the floating bacteria had to await development of a sufficiently sensitive method of detection. In the present study the large number of animals and the controlled ventilation provided a sufficiently sensitive method. Over the 2-year period the guinea pigs breathed, and hence sampled between 1,000,000 and 1,500,000 cubic feet of ward air. The 71 animals which became infected demonstrated that 71 units of air-borne tuberculosis had been extracted from this huge volume of air. On the average, each unit was suspended in 15,000 to 20,000 cubic feet of noninfectious ward air (1,000,000 or 1,500,000 ÷ 71).

The average rate at which air-borne infection was produced by the patients occupying the pilot ward can be estimated. Since the amount of air vented from the closed circuit system supplying the ward was approximately 213 cubic feet per minute (1) or about 300,000 cubic feet per day, the number of units of air-borne tuberculosis introduced into the system per day was, on the average, 300,000 divided by 15,000 to 20,000, or 15 to 20 units. This daily quota was produced by 6 patients, so the daily quota for each patient averaged only 2 to 3 units. Chapin was quite correct when he said, ". . . the difficulty of finding them suggests that they are not very numerous even in the vicinity of patients."

Nearly half a century has elapsed since the publication of Chapin's book, and in the light of present knowledge we would shift the emphasis of that part of his statement which says, ". . . perhaps the air of a room is not always dangerous to breathe even if tubercle bacilli can be found in the settled dust." Since a nurse working on our experimental ward and breathing about a third of a cubic foot a minute would take the better part of a year to breathe the 15,000 to 20,000 cubic feet of air required, on the average, for infection, it is true that the air is not very dangerous to breathe for a short time. How-

ever, numerous epidemiologic studies have shown that it ordinarily takes a tuberculin-negative nurse at least a year to convert to positive, regardless of the manner in which she becomes infected (2). Thus the amount of air-borne tuberculosis in the vicinity of patients, though small, appears to be enough to account for the observed rate of infection, at least in nurses.

This simple arithmetic, which is used to convert infections in guinea pigs to infections which would have occurred in human beings if they had breathed the same air as the guinea pigs, involves the assumption that an infectious dose for a human being is the same as that for a guinea pig. Although this assumption cannot be tested directly, we believe it to be valid, at least to a first approximation. The situation is further complicated, however, by the fact that the studies of tuberculin conversion in nurses were performed before the era of chemotherapy whereas all the patients involved in the present study were receiving drug treatment. It is possible, as suggested above, that the presence of antituberculous drugs in air-borne droplet nuclei may inactivate some of the tubercle bacilli which are not highly drug resistant. If this occurred, the infectious droplet nuclei remaining would be fewer in number than would have been the case in the absence of drug therapy. In other words, more of our guinea pigs might have been infected if all the circumstances had been the same except that the patients had not received drug therapy. If this were the case, it would require revision of our estimate of the time needed for a nurse breathing the air of our ward to convert her tuberculin test to positive, or else modification of our assumption of equal susceptibility on the part of nurses and guinea pigs. Such refinements must await further evidence. It does not seem likely, however, that these refinements will alter the major conclusion that there is enough air-borne tuberculosis in spaces occupied by patients to account for the spread of pulmonary tuberculosis in human beings.

Tuberculosis is characterized by chronicity, and over a period of years a contagious patient may share atmospheres with susceptible people in large numbers and for prolonged periods. It is in terms of these dimensions that the sanitarian must qualify Chapin's statement. The air of a room occupied by a tuberculous patient may not be very dangerous to breathe for a short time, yet the product of aerial infectivity times duration of infectivity appears to be high enough to account for the epidemiologic pattern of pulmonary tuberculosis. This evidence, which we submit in extension of less direct evidence available in the literature (7, 8), suggests that pulmonary tuberculosis is a classic example of air-borne contagion.

Summary

1. An average of 156 guinea pigs was exposed continuously for 2 years to air from a 6-bed ward occupied by patients with active pulmonary tuberculosis.

2. Seventy-one guinea pigs became infected with tuberculosis.

3. The characteristic pulmonary lesion in infected animals was a single tubercle.

4. In 20 out of 71 cases, no pulmonary tubercle was found, but the hilar lymph nodes were infected. This suggests that tuberculosis may sometimes be carried to the lymph nodes without leaving macroscopic evidence in the lungs.

5. In 21 out of 22 instances in which complete tests of drug susceptibility were carried out, identification of the specific patient who infected a particular guinea pig was possible.

6. There was great variation in the infectivity of different patients even though the clinical and laboratory findings were similar.

7. On 35 occasions the aerial infectivity of artificially atomized sputum from patients occupying the pilot ward was studied.

8. Quantitative considerations lead to the conclusion that the amount of air-borne tuberculosis in spaces occupied by patients, though small, is enough to account for the spread of pulmonary tuberculosis in human beings.

References

1. Riley, R. L., Wells, W. F., Mills, C. C., Nyka, W., and McLean, R. L. Air hygiene in tuberculosis: quantitative studies of infectivity and control in a pilot ward. Amer. Rev. Tuberc. and Pulm. Dis., 1957, 75: 420–431.
2. Riley, R. L. The J. Burns Amberson Lecture: Aerial dissemination of pulmonary tuberculosis. Amer. Rev. Tuberc. and Pulm. Dis., 1957, 76: 931–941.
3. Lurie, M. Experimental epidemiology of tuberculosis. The route of infection in naturally acquired tuberculosis of the guinea pig. Jour. Exper. Med., 1930, 51: 769–776.
4. Sultan, L. U. Identification of patient sources of infection. VA–AF Conference on Chemotherapy of Tuberculosis, St. Louis, February, 1959.
5. Wells, W. F. Airborne Contagion and Air Hygiene. Published for the Commonwealth Fund by Harvard University Press, Cambridge, Mass., 1955. P. 112.
6. Chapin, C. V. The Sources and Modes of Infection. John Wiley and Sons, New York, 1910. P. 249.
7. Medlar, E. M. The pathogenesis of minimal pulmonary tuberculosis. A study of 1,225 necropsies in cases of sudden and unexpected death. Amer. Rev. Tuberc., 1948, 58: 583–611.
8. Rich, A. R. The Pathogenesis of Tuberculosis. Second edition. Charles C. Thomas, Springfield, Ill., 1951.

Antibiotic Susceptibility Testing by a Standardized Single Disk Method

A. W. BAUER, W. M. M. KIRBY, J. C. SHERRIS, AND M. TURCK

In the words of the authors, the paper by A. W. Bauer et al., from the University of Washington in Seattle, on a standardized single-disk method for antibiotic susceptibility testing "... consolidate(s) and update(s) previous descriptions of the method and provide(s) a concise outline for its performance and interpretation." Clinical microbiologists were relieved that finally a disk diffusion method had been standardized, could be used with ease, and provided reliable results as compared with minimum inhibitory concentration tests. The pivotal role of Hans Ericsson's theoretical and practical studies (H. Ericsson and G. Svartz-Malmberg, *Antibiot. Chemother.* **6:**41–74, 1959), as well as earlier reports by some of the authors of the publications cited, must be mentioned as a matter of fairness. Most of the recommendations given are still valid today even though some of the antimicrobial agents are obsolete, new ones have been added, some zone sizes had to be modified, and new media were designed for *Haemophilus influenzae* and *Neisseria gonorrhoeae*. Recommendations of the National Committee for Clinical Laboratory Standards continue to be based on this publication; the "Kirby-Bauer" method is, among the many disk methods used in other countries, still the one that has been researched most thoroughly and updated continuously.

ALEXANDER VON GRAEVENITZ

THE AMERICAN JOURNAL OF CLINICAL PATHOLOGY
Copyright © 1966 by The Williams & Wilkins Co.
Vol. 45, No. 4   *Printed in U.S.A.*

Reprinted from TECHNICAL BULLETIN OF THE
REGISTRY OF MEDICAL TECHNOLOGISTS
Vol. 36, No. 3, 1966

ANTIBIOTIC SUSCEPTIBILITY TESTING BY A STANDARDIZED SINGLE DISK METHOD

A. W. BAUER, M.D., W. M. M. KIRBY, M.D., J. C. SHERRIS, M.D., AND
M. TURCK, M.D.

*Departments of Microbiology and Medicine, University of Washington, School of Medicine,
Seattle, Washington 98105*

Most clinical microbiologic laboratories in this country now use the paper disk method for determining susceptibility of bacteria to antibiotics and chemotherapeutic agents. A number of modifications of the test are employed. When this type of test was first developed, only 1 disk was used for each agent to be tested,[7, 10] but subsequently it became common practice to use 2 or more disks of different potency and to judge susceptibility on the basis of the presence or absence of growth around the disks. Our approach has been to continue to develop a single disk method based on measurement of sizes of zones. We believe that this is rational in theory and that it correlates better with the results of dilution technics.

A number of reports on the technical details, experimental basis, and interpretative standards of the single disk method have been published,[1, 4-6, 13] and recently some of the theoretical aspects have been reviewed in more detail.[2, 3, 11] The purpose of the present communication is to consolidate and update previous descriptions of the method and provide a concise outline for its performance and interpretation.

METHOD

Rapidly Growing Pathogens Such as Staphylococci and Enterobacteriaceae

A few colonies (3 to 10) of the organism to be tested are picked with a wire loop from the original culture plate and introduced into a test tube containing 4 ml. of tryptose phosphate or trypticase soy broth. These tubes are then incubated for 2 to 5 hr., to produce a bacterial suspension of moderate cloudiness. The suspension is then diluted, if necessary, with water or saline solution to a density visually equivalent to that of a standard prepared by adding 0.5 ml. of 1 per cent $BaCl_2$ to 99.5 ml. of 1 per cent H_2SO_4 (0.36 N). An alternative procedure is to dilute broth cultures overnight to the density of the opacity standard (10- to 100-fold). For the sensitivity plates, large (15-cm.) Petri dishes are used with Mueller-Hinton agar (5 to 6 mm. in depth). Plates are dried for about 30 min. before inoculation and are used within 4 days of preparation.

The bacterial broth suspension is streaked evenly in 3 planes onto the surface of the medium with a cotton swab (not a wire loop or glass rod). Surplus suspension is removed from the swab by being rotated against the side of the tube before the plates are seeded. After the inoculum has dried (3 to 5 min.), the disks are placed on the agar with flamed forceps or a single disk applicator and gently pressed down to ensure contact. Plates are incubated immediately, or within 30 min. The large Petri dishes are spacious enough to accommodate about 9 disks in an outer ring, and 3 or 4 more in the center. It is advantageous to place antibiotics which diffuse well in the outer circle and disks which produce smaller inhibition zones, such as vancomycin and polymyxin-B, in the central area of the plate.

After overnight incubation, the zone diameters (including the 6-mm. disk) are measured with a ruler on the undersurface of the Petri dish or with calipers near the agar surface. A reading of 6 mm. indicates no zone. The end point is taken as complete inhibition of growth as determined by the naked eye, except in the case of sulfonamides, where organisms grow through several generations before inhibition takes effect.

Received, August 17, 1965.

Diagnostic Microbiology and Epidemiology   41

TABLE 1

ZONE SIZES AND THEIR INTERPRETATION FOR FREQUENTLY USED CHEMOTHERAPEUTICS

Antibiotic or Chemotherapeutic Agent	Disk Potency	Inhibition Zone Diameter to Nearest Millimeter		
		Resistant	Intermediate	Sensitive
Ampicillin				
S. aureus	10 μg.	20 or less	21–28	29 or more
All other organisms	10 μg.	11 or less	12–13	14 or more
Bacitracin	10 units	8 or less	9–12	13 or more
Cephalothin	30 μg.	14 or less	15–17	18 or more
Chloramphenicol	30 μg.	12 or less	13–17	18 or more
Colistin	10 μg.	8 or less	9–10	11 or more
Erythromycin	15 μg.	13 or less	14–17	18 or more
Kanamycin	30 μg.	13 or less	14–17	18 or more
Lincomycin§	2 μg.			17 or more
Methicillin	5 μg.	9 or less	10–13	14 or more
Nalidixic acid*	30 μg.	13 or less	14–18	19 or more
Neomycin	30 μg.	12 or less	13–16	17 or more
Nitrofurantoin*	300 μg.	14 or less	15–16	17 or more
Novobiocin†	30 μg.	17 or less	18–21	22 or more
Oleandomycin	15 μg.	11 or less	12–16	17 or more
Penicillin-G	10 units	20 or less	21–28	29 or more
Polymyxin-B	300 units	8 or less	9–11	12 or more
Streptomycin	10 μg.	11 or less	12–14	15 or more
Sulfonamides‡	300 μg.	12 or less	13–16	17 or more
Tetracycline	30 μg.	14 or less	15–18	19 or more
Vancomycin	30 μg.	9 or less	10–11	12 or more

* Standards apply to urinary tract infections only.

† Zone sizes not applicable when blood is added to medium.

‡ Any of the commercially available 300- or 250-μg. sulfonamide disks may be used with the same standards of zone interpretation.

§ Tentative standard.

Slight growth (80 per cent or more inhibition) with sulfonamides is therefore disregarded, and the margin of heavy growth read to determine the zone size.[6] Swarming of Proteus strains is not inhibited by all antibiotics, and a veil of swarming into an inhibition zone is also ignored. If several colonies are seen within a zone of inhibition, the strain should be checked for purity and retested. If they are still present, the colonies are regarded as significant growth; this is not a common occurrence. The zone diameters are recorded and interpreted according to Table 1, and the results are reported to the clinician. When they are needed, zone diameters may be read after incubation for 6 to 8 hr. Standard control organisms of known susceptibility should be employed at least once a week as a check on the activity of the disks and on the reproducibility of the test.

The technic has proved satisfactory for sulfonamide susceptibility testing of all organisms examined except Group A β-hemolytic streptococci; these do not show clear-cut-zones and the method should not be used for the organism.

Slower Growing and Fastidious Organisms

The standards in Table 1 have been developed for rapidly growing pathogens, for which the test is highly satisfactory. More slowly growing organisms, such as Hemophilus, have been less completely studied, but zone sizes are somewhat larger for an equivalent minimum inhibitory concentration than they are with rapid growers. For testing such organisms, 5 per cent sheep or human blood may be added to the medium, which may also be "chocolatized" when indicated.

Organisms of this type, with zone sizes in

the resistant range, may be safely reported as resistant. Those giving zones in the intermediate range or within 2 mm. of the borderline between sensitive and intermediate should be regarded as of uncertain susceptibility and tested by a dilution method if one is indicated.

The test cannot be regarded as more than a very rough guide for organisms requiring more than 24 hr. to yield macroscopic colonies. It should not be used for sensitivities on *Neisseria gonorrhoeae* or for sulfonamide sensitivities on meningococci until further data become available.

DISCUSSION

The numerical values in the table have been established by comparing zone sizes with a large series of tube or plate dilution tests and by relating these to blood levels found with frequently used dose schedules. In the case of nitrofurantoin and nalidixic acid, levels in the urinary tract have been taken into account in establishing standards. Confirmation of the validity of these standards has also been obtained from curves showing the distribution of susceptibilities of large numbers of individual strains of various species.[3] With many species from which resistant mutants have emerged, such distribution curves show 2 rather clearly separated sensitive and resistant populations, with few strains falling in the intermediate zone. For example, not more than 3 to 5 per cent of all strains of *Staphylococcus aureus* fall into the intermediate range with tetracycline, chloramphenicol, or erythromycin.

The technic should be used exactly as described, because although it has considerable flexibility, changes in conditions may combine to produce inaccuracies; it should be particularly stressed that the zone sizes in the table are *only applicable to the use of Mueller-Hinton medium* and to disk contents listed in the table. Undiluted overnight broth cultures should never be used as an inoculum, but diluted at least 10-fold, or preferably to a density equivalent to the barium sulfate standard. Growth by this method should be confluent with a prop-

erly standardized inoculum, and the test must be repeated if this is not achieved.

The choice of antibiotics to be tested depends on a number of factors, such as the type of practice of the laboratory and the local preference for a particular agent. For most Gram-positive organisms, the disks routinely used in our laboratories are penicillin-G, methicillin, tetracycline, erythromycin, streptomycin, chloramphenicol, kanamycin, and, occasionally, sulfamethizole and ampicillin. Gram-negative rods are tested with chloramphenicol, tetracycline, streptomycin, kanamycin, polymyxin-B, sulfamethizole, ampicillin, and cephalothin, with added nitrofurantoin for urinary tract infections. Only 1 penicillinase-resistant penicillin and 1 of the tetracyclines are tested routinely because of essential cross-resistance within these groups. Some organisms may be tested only against antibiotics to which resistance has developed; thus Group A β-hemolytic streptococci and pneumococci may need to be tested only against tetracycline. Disks from first line supply houses which are tested and certified for potency by the Food and Drug Administration have proven reliable. They should be stored exactly as recommended.

It will be noted that the interpretative zone diameters are different for each agent, not only because the disk potencies vary, but also because the diffusion and solubility properties of the drugs in Mueller-Hinton medium are different and characteristic for each agent. Obviously, the disk producing the largest inhibition zone does not necessarily indicate the antibiotic of choice for a given pathogen. If, for instance, a strain of *Escherichia coli* shows a zone of 22 mm. around a 30-μg. tetracycline disk and a zone of 26 mm. around a 30-μg. chloramphenicol disk, all that can be said is that the strain is susceptible to both compounds. From the agents reported as being effective against a given organism, the clinician may select the one most appropriate for therapy of the particular case. Pharmacologic and toxicologic properties, as well as the bacteriostatic or bactericidal actions of the drugs, will be taken into account in making this selection. A report of "intermediate" or "resistant" does not necessarily imply that treatment may not be successful with unusually high dosage. Useful predictions of the efficacy of such treatment, however, can only be obtained from comparisons of the results of dilution technics with anticipated chemotherapeutic levels at the sites of infection.

It will also be seen that the sensitivity criteria given for *S. aureus* with ampicillin are different from those of other organisms, because some weakly penicillinase-producing strains of staphylococci may give zone sizes of more than 14 mm., although they should be regarded as clinically resistant. All non-penicillinase-producing staphylococci have zones similar to those seen with penicillin.

The importance of standardizing test conditions as far as possible in practice cannot be overstressed. For this reason, we do not recommend routine direct application of disks to plates seeded with clinical material because of problems of inoculum control and mixed cultures. In emergency situations, such as those in which direct smears of some cerebrospinal fluid and urine specimens indicate that a pure culture may be anticipated, we make direct tests and issue a tentative report, which is always then confirmed with the standard method. Similar emphasis on standardization has been made by Ericsson[8, 9] in his description of an excellent single disk technic which is widely used in Scandinavia, and has been stressed in the Second Report of the Expert Committee on Antibiotics of the World Health Organization,[12] which is sponsoring further studies towards the possible establishment of internationally acceptable standards. Such standards are highly desirable and may lead us to some modifications of our method. In the interim, however, we have found the technic to be easily performed, reproducible, and of great value as a guide to therapy.

REFERENCES

1. Anderson, K. N., Kennedy, R. P., Plorde J. J., Shulman, J. A., and Petersdorf, R. G.: Effectiveness of ampicillin against Gram negative bacteria. J. A. M. A., *187:* 555–561, 1964.
2. Bauer, A. W.: The significance of bacterial inhibition zone diameters. Stuttgart: IIIrd International Congress of Chemotherapy, 1964, p. 466.
3. Bauer, A. W.: The two definitions of bacterial resistance. Stuttgart: IIIrd International Congress of Chemotherapy, 1964, p. 484.
4. Bauer, A. W., Perry, D. M., and Kirby, W. M. M.: Single disc antibiotic sensitivity testing of staphylococci. A. M. A. Arch. Int. Med., *104:* 208–216, 1959.
5. Bauer, A. W., Roberts, C. E., and Kirby, W. M. M.: Single disc versus multiple disc and plate dilution techniques for antibiotic sensitivity testing. *In* Antibiotics Annual, No. 7, 1959–1960. New York: Medical Encyclopedias, Inc., 1960, pp. 574–580.
6. Bauer, A. W., and Sherris, J. C.: The determination of sulfonamide susceptibility of bacteria. Chemotherapia, *9:* 1–19, 1964.
7. Bondi, A., Spaulding, E. H., Smith, D. E., and Dietz, C. C.: A routine method for the rapid determination of susceptibility to penicillin and other antibiotics. Am. J. M. Sc., *213:* 221–225, 1947.
8. Ericsson, H.: Rational use of antibiotics in hospitals. Scandinav. J. Clin. & Lab. Invest., *12:* (Suppl. 50): 1–55, 1960.
9. Ericsson, H., and Svartz-Malmberg, G.: Determination of bacterial sensitivity *in vitro* and its clinical evaluation. Antibiotica et Chemotherapia *6:* 41–74 1959.
10. Morley, D. C.: A simple method of testing the sensitivity of wound bacteria to penicillin and sulfathiazole by use of impregnated blotting paper discs. J. Path. & Bact. *57:* 379–382, 1945.
11. Petersdorf, R. G., and Sherris, J. C.: Methods and significance of *in vitro* testing of bacterial sensitivity to drugs. Am. J. Med., *39:* 766–779, 1965.
12. Second Report of the Expert Committee on Antibiotics. Standardization of Methods for Conducting Microbic Sensitivity Tests. Geneva World Health Organization Technical Report Series, No. 210. 1961, pp. 1–24.
13. Turck, M., Lindemeyer, R. I., and Petersdorf, R. G.: Comparison of single disc and tube dilution techniques in determining antibiotic sensitivities of Gram negative pathogens. Ann. Int. Med., *58:* 56–65, 1963.

Polynucleotide Sequence Relationships among Members of Enterobacteriaceae

D. J. BRENNER, G. R. FANNING, K. E. JOHNSON, R. V. CITARELLA, AND S. FALKOW

The history of microbial taxonomy goes back to the 19th century. Until the late 1960s, bacteria used to be classified mainly by their morphological, tinctorial, and biochemical characteristics. The more these traits became known, the more difficult classification became. Numerical taxonomy sought to offer a solution but faced, among others, the problems of bias in strain and test selection, reproducibility, and dependence of test results on growth factors and growth rate, stability of traits, and failure of certain organisms to grow on artificial media. What may now be called an ideal taxonomy, i.e., one based on bacterial phylogeny, became possible only after groundwork in microbial genetics had been laid by E. T. Bolton, P. Doty, S. Falkow, M. Mandel, J. Marmur, B. J. McCarthy, C. L. Schildkraut, and S. Spiegelman (see *J. Mol. Biol.* **3**:585–617, 1961; *J. Mol. Biol.* **5**:109–118, 1962; *J. Mol. Biol.* **12**:829–842, 1965; *J. Bacteriol.* **84**:1313–1312, 1962; *Proc. Natl. Acad. Sci. USA* **48**:1390–1397, 1962; *Proc. Natl. Acad. Sci. USA* **50**:156–164, 1963; *Progr. Nucleic Acid Res.* **1**:231–300, 1963; *Annu. Rev. Microbiol.* **23**:239–274, 1969). Building on their work and on his own earlier studies on nucleic acid reassociation, Don Brenner and his colleagues determined DNA relatedness by reacting denatured, labeled DNA fragments of one organism with similarly prepared unlabeled fragments of another. Later, the definition of a species was made based on ≥70% DNA-DNA relatedness and a change in melting temperature of ≤5°C.

The results of three methods (agar, membrane filter, and hydroxyapatite methods) applied to various genera and species—as defined in 1969 by "traditional" methods—confirmed that the species examined actually showed the known degree of relatedness except for *Escherichia coli* and *Shigella flexneri,* which were closer relatives.

Later studies on many other species have proven the value of DNA-DNA hybridization, which by now is the standard method used to outline a species.

ALEXANDER VON GRAEVENITZ

JOURNAL OF BACTERIOLOGY, May 1969, p. 637–650
Copyright © 1969 American Society for Microbiology

Vol. 98, No. 2
Printed in U.S.A.

Polynucleotide Sequence Relationships among Members of *Enterobacteriaceae*

DON J. BRENNER, GEORGE R. FANNING, KARL E. JOHNSON, R. V. CITARELLA,
AND STANLEY FALKOW

*Division of Biochemistry and Department of Bacterial Immunology, Walter Reed Army Institute of Research,
Washington, D.C. 20012, and Department of Microbiology, Schools of Medicine and Dentistry,
Georgetown University, Washington, D.C. 20007*

Received for publication 6 February 1969

Polynucleotide relationships were examined among many representatives of the *Enterobacteriaceae* by means of agar, membrane filter, and hydroxyapatite procedures. The amount of deoxyribonucleic acid (DNA) that reassociated was dependent, especially in interspecific reactions, on the annealing temperature. In only three cases: *Escherichia coli-Shigella flexneri*, *Salmonella typhimurium-S. typhi*, and *Proteus mirabilis-P. vulgaris*, was relative interspecific duplex formation 80% or higher. In most cases interspecies DNA duplex formation was 40% or less of that obtained from intraspecies DNA reassociation reactions. The stability of *E. coli-S. flexneri* DNA duplexes formed at either 60 or 75 C was virtually identical to that of homologous *E. coli* DNA duplexes, and the degree of interspecies duplex formation was minimally affected by the temperature increase (86% at 60 C; 77% at 75 C). The thermal stability of DNA duplexes formed at 60 C between DNA from *E. coli* and DNA from strains of *Aerobacter aerogenes, S. typhimurium, S. typhi*, and *P. mirabilis* was about 12 to 14 C below that of reassociated *E. coli* DNA. At 75 C, the formation of the interspecific DNA duplexes was markedly decreased, but the stability of the DNA able to reassociate at this temperature approximated that of reassociated *E. coli* DNA. The degree of reassociation and the thermal stability of *E. coli-S. flexneri* DNA duplexes suggests relatively little evolutionary divergence in these organisms. The other enterobacteria tested, however, have diverged to a point where less than one-half of their DNA can reanneal with *E. coli* DNA at 60 C and less than 10% reacts at 75 C. The degree of divergence between various enterobacteria does not appear to be uniform along the DNA molecule. Ribosomal ribonucleic acid (RNA)-specific sequences are conserved among most enterobacteria. An examination of messenger RNA relatively specific for the lactose operon suggests that specific chromosomal genes may diverge more or less than the genome as a whole.

Ideally, microbial taxonomy should be based on phylogenetic relationships. This type of taxonomy has, until recently, been impossible, largely owing to the lack of a fossil record and to the relatively few morphological features available for study. Recent advances in protein and nucleic acid biochemistry, and in microbial genetics, allow a preliminary appraisal of relatedness at the molecular level.

Several years ago, it was noted that there was virtually no information concerning the genetic organization of most enteric organisms (38). This remains essentially the case. It still is not known whether the close identity of gross chromosome organization in strains of *Escherichia coli* (49) to that in strains of *Salmonella* (18, 25, 26, 45) and *Shigella* (20, 48) species extends to strains of *Aerobacter, Serratia,* or *Proteus* species. In addition, there is little information about the pattern of nucleotide sequence divergence among enteric bacteria. Several factors which are barriers to recombination make comparative genetic studies difficult among many enteric bacteria. These include restriction and modification, surface incompatibility, the presence of independent fertility groups, different specificity of recombination enzymes (10, 16, 51), and dissimilarity in deoxyribonucleic acid (DNA) base sequence.

Diagnostic Microbiology and Epidemiology 47

Nucleic acid reassociation experiments, however, allow one to circumvent the problems of mating incompatibility, and to assess both overall interspecific relatedness and relatedness within specific portions of the genome. The nucleic acid reassociation experiments reported here examine quantitative relationships among many species of enteric organisms. In several cases, the thermal stability of reassociated DNA was determined for reactions carried out at different incubation temperatures to allow assessment of different levels of interspecies relationships.

MATERIALS AND METHODS

Organisms and media. The strains used in the present investigation are listed in Table 1. Antibiotic medium no. 3 (Penassay Broth, Difco), meat extract-agar, and Brain Heart Infusion (Difco) were used for routine cultivation of organisms. For the purpose of labeling DNA, log-phase cells were suspended in a tris(hydroxymethyl)aminomethane (Tris)-glucose salt medium lacking phosphate salts and containing 0.5 to 1.0% Brain Heart Infusion. Carrier-free $H_2{}^{32}PO_4$ (New England Nuclear Corp., Boston, Mass.) was added to this medium, and the cultures were incubated at 37 C overnight. Media used in labeling of ribonucleic acid (RNA) are listed below.

Preparation of DNA. Cells were harvested from broth by centrifugation. DNA used in some of the earlier DNA-agar experiments and in the RNA-DNA reactions carried out on membrane filters was extracted from bacterial strains as described by Marmur (36). These DNA preparations were sheared, when desired, by sonic treatment at 4 C in a Branson Sonifier (Heat Systems Co., Melville, N. Y.) to an average molecular weight of 2.5×10^5 to 3.5×10^5 daltons (17).

In many of the DNA-agar experiments and in all of the experiments utilizing hydroxyapatite, DNA was prepared by a modification of the method of Berns and Thomas (4). Bacterial cells were suspended at an approximate 1:50 dilution (w/v) in a solution containing 0.1 M sodium chloride, 0.1 M ethylenediaminetetraacetate (EDTA; 0.05 M EDTA was used in later experiments), 0.05 M Tris hydrochloride buffer, pH 8.2 (Tris buffer), and self-digested (37 C, 2 hr) Pronase (Calbiochem, Los Angeles, Calif.) at 50 μg/ml. Sodium lauryl sulfate (SLS) was added to a final concentration of 0.5%, and the suspension was incubated at 37 C overnight to maximize cell lysis and Pronase action. The SLS concentration was increased to 1%, and an equal volume of phenol was added to the suspension. The phenol and aqueous layers were well mixed and then gently shaken for several minutes. The mixture was centrifuged to separate the phase. The aqueous phase was collected, and sodium perchlorate was added to a 1 M concentration. An equal volume of chloroform was added to the aqueous

TABLE 1. *Bacterial strains employed*

Organism	Source
Escherichia coli K-12-1	University of Washington culture collection
E. coli K-12 200 μ *lac⁻(z⁺y⁻)*	WRAIR[a] (25)
E. coli K12-1485	(17)
E. coli K12 *lac*Δ	A. D. Pardee
Salmonella typhi 643	WRAIR (25)
S. typhi 643 *lac₃*	WRAIR (25)
S. typhi 643 X30T	WRAIR (25)
S. typhi 643 X30W	WRAIR (25)
S. typhimurium LT2	T. Theodore
S. typhimurium 7823	ATCC
Salmonella sp. (ser.) *ballerup*	Brown University culture collection
Salmonella sp. (ser.) *arizonae*	W. H. Ewing
Shigella flexneri 2a 24570	WRAIR (48)
Serratia marcescens	(17)
Aerobacter aerogenes	University of Washington culture collection
Proteus mirabilis-1	(21)
P. mirabilis-1 F-*lac⁺*	(21)
P. morganii	C. A. Stuart (19)
P. rettgeri	C. A. Stuart (19)
P. vulgaris	C. A. Stuart (19)
P. inconstans (Providence 29911)	C. A. Stuart (19)
Bethesda-10	W. H. Ewing
Neisseria perflava	D. Kingsbury
N. gonorrhoeae	D. Kingsbury
N. catarrhalis	D. Kingsbury
Pasteurella pestis	WRAIR
Pseudomonas aeruginosa	University of Washington culture collection

[a] Walter Reed Army Institute of Research.

phase, and the mixture was shaken and centrifuged. The aqueous phase was recovered, and the chloroform treatment was repeated once or twice until the interphase material was largely removed. Two volumes of cold 95% ethyl alcohol were added to the aqueous phase and the resultant DNA precipitate was loosely spooled around a glass rod, washed in ethyl alcohol, and suspended in SSC/100 (SSC = 0.15 M sodium chloride + 0.015 M sodium citrate). The DNA was repeatedly (three or four times) precipitated (after the addition of NaCl to 0.1 M final concentration) with 95% ethyl alcohol and suspended in SSC/100 until the precipitate was translucent. The DNA solution was made 0.1 M to NaCl and 0.05 M to EDTA and Tris buffer, and was incubated with 25 μg of pancreatic ribonuclease per ml (crystallized; Worthington Biochemical Corp., Freehold, N.J.) at 37 C or, in more recent preparations, at 60 C for 1 hr. SLS was added to 0.5%, and the DNA was incubated overnight at 37 C in the presence of 50 μg of Pronase per ml. The SLS concentration was then increased to 1%, and the DNA was again treated once with phenol and twice with chloroform. The DNA was then repeatedly (three or four times) precipitated with 2-ethoxyethanol and suspended in SSC/100. It should be noted that dilute DNA solutions are poorly precipitated by 2-ethoxyethanol, and the majority of the DNA may be lost (29). We try to have DNA solutions at a concentration of 1 mg/ml or higher for precipitation by 2-ethoxyethanol. When desired, the DNA was fragmented by mechanical shear at 50,000 psi to a molecular weight of approximately 2×10^5 (9) and was filtered through Metricel filter discs (0.45 μm pore size, Gelman Instrument Co., Ann Arbor, Mich.). Labeled DNA fragments were denatured by heating; they were further purified (for hydroxyapatite experiments) by passing them through a hydroxyapatite column equilibrated with 0.12 M PB (PB = phosphate buffer, an equimolar mixture of NaH_2PO_4 and Na_2HPO_4, pH 6.8) and were held at 60 C. Material that bound to the column under these conditions was discarded. This procedure decreases the "zero time" binding (label bound to the column immediately after the DNA has been denatured) from about 2 to 2.5% to about 0.5 to 0.7% (Kohne and Britton, *unpublished data*).

Preparation of DNA-agar and DNA-agar reactions. DNA was denatured and embedded in Oxoid Ionagar No. 2 as described by Bolton and McCarthy (5). It was then pressed through a stainless-steel screen (35 mesh), extensively washed with 2× SSC at 66 C, and assayed optically (260 nm) for DNA content by dissolving the agar in 5 M sodium perchlorate. Agar in amounts of 0.2 to 0.5 g, containing between 300 and 1,000 μg of DNA per g, was incubated for 15 to 18 hr at 60 or 66 C with denatured, ^{32}P-labeled DNA fragments contained in a volume of 2× SSC equal to the weight of agar employed. The ratio of DNA in agar to DNA fragments was 200:1,000. The DNA-agar was then transferred to a glass tube fitted with a Saran screen bottom (3) and was washed with ten 15-ml portions of 2× SSC at the incubation temperature to remove unbound DNA. Labeled DNA fragments bound to the agar were recovered by washing the agar with five 15-ml portions of water at 75 C.

All wash fluids were assayed for trichloroacetic acid-precipitable radioactivity. The percentage of DNA fragments bound to the DNA-agar was calculated as bound counts/total counts × 100.

Preparation of labeled 23S RNA. Cells were grown [by the method of McCarthy and Bolton (33)] in glucose-salts medium containing ^{3}H-uridine (2 μc/ml), extracted, and chromatographed on methylated albumin-Kieselguhr columns (35). The peak containing 23S RNA was pooled and rechromatographed before use.

Preparation of pulse-labeled RNA. Pulse-labeled RNA was prepared by growing *E. coli* K-12 W1485 F$^-$ in 500 ml of TCG (50) medium to a cell density of about 2.5×10^8 cells/ml and then adding 1 μc of ^{3}H-uridine per ml (2.5 c/mmole). After 90 sec of incorporation, the culture was poured over an equal volume of crushed, frozen medium. The cells were collected by centrifugation at 4 C, washed with TM-buffer (10^{-2} M Tris, pH 7.4, containing 10^{-3} M Mg^{++}). RNA was extracted by freezing and thawing the cells three times in the presence of 200 μg of lysozyme per ml and 50 μg of deoxyribonuclease per ml (both from Worthington Biochemical Corp.). The lysate was brought to room temperature, and lysis completed with 0.4% SLS for 5 min. The lysate was chilled and shaken three times with water-saturated phenol. The aqueous phase was then extracted three times with cold, anhydrous ether to remove the phenol. Traces of ether remaining in the aqueous phase were removed by bubbling nitrogen through the solution. The RNA was precipitated with three volumes of cold ethyl alcohol. The precipitate was dissolved in TM buffer and again precipitated.

Preparation of lactose messenger RNA (mRNA). The method used was similar to that described by Naono, Rouviere, and Gros (43), which exploits the preferential transcription of the lactose operon during the diauxic growth of *E. coli*. A strain of *E. coli* K-12 200 μ *lac$^-$* (z^+y^-) was infected with the F merogenote F-*lac$^+$* (z^+y^+) and grown overnight in minimal salts medium containing 0.2% glucose. The culture was centrifuged in the cold and suspended to an optical density (OD) at 420 nm of 0.500 in minimal medium containing 0.03% glucose and 0.2% lactose. The culture was incubated at 37 C with shaking for 120 min, at which time growth from glucose stopped (OD increased from 0.50 to 1.05). After a stationary phase of approximately 20 min, the second phase of growth on lactose started, and at this time 1 μc of ^{3}H-uridine per ml (13 c/mmole; Schwarz BioResearch Inc., Orangeburg, N.Y.) was added to the culture for 60 sec. The labeled cells were washed with TM buffer, and the RNA was extracted as described for pulse-labeled RNA. Approximately 2 mg of the RNA (6.5×10^5 counts per min per mg) was chromatographed on a methylated albumin-Kieselguhr column (35) with the use of a linear gradient of 0.3 to 0.8 M NaCl. Pilot experiments were carried out by use of a mixture of ^{14}C-labeled RNA pulsed just before adaptation to lactose under diauxie growth conditions and ^{3}H-RNA prepared as lactose adaption took place. The ^{3}H-RNA contained a significant percentage (10% or more) eluting at about 0.7 M NaCl (about 27S)

that could duplex with DNA derived from *P. mirabilis* F-*lac*$^+$ DNA, whereas only a negligible fraction (about 1%) of the ^{14}C-labeled RNA reacted with this DNA. Approximately 1.3×10^5 counts/min of ^{3}H-RNA in $2\times$ SSC, containing most of the hybridizable F-*lac*$^+$ material, was incubated repeatedly with 100-µg samples of denatured DNA from *E. coli lac* Δ, an Hfr strain carrying a deletion of the lactose operon. The remaining RNA contained about 3×10^4 counts/min, and was filtered through a type A (coarse) filter (Schleicher & Schuell Co., Keene, N.H.). Samples of this RNA were incubated overnight at 67 C with membrane filters containing 100 µg of appropriate DNA preparations. The filters were treated with 30 µg of ribonuclease per ml (60 min in $2\times$ SSC at 26 C), washed, dried, and counted.

Immobilization of DNA on filters. DNA was denatured just before use by heating at 100 C for 10 min in SSC/10 at a concentration not exceeding 50 µg/ml. Alternatively, DNA was denatured with alkali as described by Gillespie and Spiegelman (22). Denatured DNA solutions were diluted with $6\times$ SSC to approximately five times their original volume, and the DNA was immobilized on type B6 nitrocellulose filters (Schleicher & Schuell Co.) according to the method of Gillespie and Spiegelman (22). The amount of DNA retained by the filters was determined by monitoring the OD at 260 nm before and after filtration of the denatured DNA solution through the filters. The amount of DNA fixed to a single 25-mm filter was 100 µg $\pm$ 5 µg.

Binding of mRNA to DNA immobilized on membrane filters. Membrane filters to which 100 µg of denatured DNA were fixed by heat (22) in vacuo were incubated with an appropriate quantity of mRNA (usually 12 to 50 µg) in 5 ml of $2\times$ SSC for 20 hr at 60 or 67 C in a glass vial. The membrane filters were washed with $2\times$ SSC, 50 ml per side, and were incubated for 60 min at 26 C in 5 ml of $2\times$ SSC containing 20 µg of pancreatic ribonuclease per ml (Worthington Biochemical Corp.). The filters were washed with cold $2\times$ SSC, 50 ml per side, dried in an oven, and counted. Some experiments were also performed with 10 µg of T1 ribonuclease per ml (Miles Laboratories, Inc., Elkhart, Ind.) and a combination of T1 and pancreatic ribonuclease. There was, however, no significant difference in the degree of relative binding compared with those obtained by use of pancreatic ribonuclease alone.

Separation of single- and double-stranded DNA on hydroxyapatite. The effectiveness of hydroxyapatite for fractionating native and denatured DNA was first demonstrated by Bernardi (3). Miyazawa and Thomas (41) developed a technique for fractionating double-stranded DNA bound to hydroxyapatite. A modification of this procedure (9) was used to determine the thermal stability of reassociated bacterial DNA. In this procedure, thermally denatured, ^{32}P-labeled *E. coli* DNA fragments were incubated at 60, 66, or 75 C with a 4,000- to 7,500-fold excess of denatured unlabeled DNA fragments in 1 ml of 0.12 M PB. The length of incubation was chosen so that unlabeled DNA fragments would be at least 80% reassociated, whereas the reassociation of labeled

DNA fragments with one another would be only 1 to 2%. After incubation, the samples were quickly cooled in an ice bath and frozen until use. Each sample was subsequently thawed and passed through a 10-ml hydroxyapatite (BioRad Laboratories, Richmond, Calif.) column held at the temperature at which the fragments had been incubated. In some later experiments, a batch procedure (7) which allowed six or more samples to be processed simultaneously was employed. The columns were then washed with 15-ml portions of 0.12 M PB at temperatures increasing in increments of 2.5 C to 100 C. As the column temperature exceeded the dissociation temperature of DNA duplexes bound to the hydroxyapatite, the resultant single-stranded DNA was eluted from the column. The column was finally washed with two or three 15-ml portions of 0.4 M PB to elute any remaining material bound to the hydroxyapatite. It should be noted that the capacity of hydroxyapatite to adsorb double-stranded DNA varies from lot to lot. A typical lot of hydroxyapatite, in our hands, bound approximately 100 to 250 µg of double-stranded DNA/ml. It is desirable to use an amount of hydroxyapatite two to three times higher than that necessary to adsorb the amount of DNA employed, to insure against exceeding the column capacity. It has also been noted that zero-time binding of single-stranded DNA often increases markedly (up to 10 to 15%) in old batches of hydroxyapatite (after 4 months or longer). This difficulty can be overcome by boiling the hydroxyapatite before use.

When only the percentage of reassociated DNA fragments, rather than the thermal stability of the reassociated product, was of interest, the column was washed with 15-ml portions of 0.12 M PB as above, then with one 15-ml 0.12 M PB wash at 95 C and one at 100 C, and finally with three washes with 0.4 M PB.

Radioactivity assay. Samples from agar experiments, as well as many of the samples from hydroxyapatite experiments, were precipitated in 5% trichloroacetic acid in the presence of approximately 100 µg of yeast RNA carrier. The precipitates were collected on membrane filters, dried, and placed in counting vials. A 15-ml amount of scintillation fluid was added, and the samples were counted in either a Packard or a Nuclear-Chicago liquid scintillation spectrometer. Alternatively, eluates from hydroxyapatite experiments were placed directly into counting vials and assayed by Cerenkov counting (11).

RESULTS

Quantitative nucleic acid relationships among the Enterobacteriaceae. An expanding body of data has accumulated which attempts to define the relationships between bacteria by a study of heteroduplex DNA molecules—or mRNA-DNA hybrid molecules. In a reassociation experiment, nucleic acids are incubated under a set of conditions such that single-stranded DNA fragments (or mRNA molecules) may collide with unlabeled DNA in free solution (hydroxyapatite method), immobilized in agar (5), or on filters

(13, 22). Double-stranded duplexes are formed as bases in one strand pair with their complementary bases in the other strand. The reassociation of DNA is affected by several different parameters among which are the following.

(i) The base composition (39): guanine and cytosine (G + C) base pairs exhibit greater thermal stability than adenine + thymine (A + T) base pairs. The greater the fraction of G + C base pairs within a given DNA duplex, the higher is the thermal stability of that duplex.

(ii) The size of nucleic acid fragment (9, 32, 34): with very small DNA fragments, no specificity is to be expected since any particular short base sequence may be found in all DNA molecules. The minimal specific DNA fragment for bacteria is about 15 nucleotides (34). The length of DNA fragments affects DNA reassociation in free solution (9). The fragments used in this study were about 500 to 600 and 1,000 nucleotides in length. The larger fragments reassociate about 25% faster than the smaller ones (9).

(iii) The ionic strength (9): both the rate of DNA reassociation and the thermal stability of reassociated DNA increase as the ionic strength is increased. For any given set of experiments reported below, the ionic strength was held as a constant parameter.

(iv) The temperature of incubation (6, 27, 33): the optimal temperature for reassociation is some 30 C below the temperature at which a DNA becomes denatured (37, 39). Temperatures significantly lower than the optimum permit distantly related (6, 32) and nonspecific sequences to reassociate (27, 32). Incubation temperatures employed in this study were never less than 30 C below the melting temperature of the DNA under study.

(v) DNA concentration and the time of incubation (9, 33): the concentration of labeled DNA (or mRNA) and unlabeled DNA employed, as well as the time of incubation, must be carefully chosen to obtain meaningful reassociation data. Generally, in agar and filter reactions, the ratio of unlabeled to labeled DNA in a reaction was at least 100:1. This large ratio provided an excess of available sites with which any related labeled nucleic acid sequence could reassociate. The incubation time was chosen so that reassociation between labeled DNA (or mRNA) and unlabeled DNA was maximal. DNA concentrations and incubation time are particularly critical in free solution reactions, since labeled fragments reassociated with one another cannot be distinguished from the desired reassociation product of labeled with unlabeled DNA fragments. In hydroxyapatite experiments, therefore, the labeled DNA concentration was small enough to insure little or no label-label reassociation during the course of incubation. Concomitantly, the unlabeled DNA concentration was large enough to permit the reassociation of labeled with unlabeled DNA fragments to proceed to virtual completion. The term "duplex" is used interchangeably with "fragments bound," as we assume that bound fragments are in a duplex arrangement.

Table 2 presents a compilation of reassociation studies performed according to several different criteria, done by agar, filter, and hydroxyapatite techniques, and with measurement of both DNA-DNA and mRNA-DNA duplex formation. The nature of Table 2 precludes detailed comments; however, several general features of these data are evident. Essentially the same results were obtained whether a filter, agar, or hydroxyapatite method was employed (compare columns 1 and 4 and columns 2, 3, and 5). Greater than 80% relative reassociation was obtained in only three interspecific reactions: *E. coli-S. flexneri*, *S. typhi-S. typhimurium*, and *P. mirabilis-P. vulgaris*. Where tested, this high level of relative binding proved to be essentially temperature-insensitive. The vast majority of DNA-DNA and mRNA-DNA reactions showed 40% or less relative binding according the most relaxed criteria (60 C in 2× SSC or 0.12 M PB), less than 30% relative binding with intermediate criteria (66 or 67 C in 2× SSC or 66 C in 0.12 M PB), and 10% or less relative binding with the most stringent criterion employed (75 C in 0.12 M PB). Even DNA from so-called intermediate forms, such as the classical *Salmonella* sp. (ser.) *ballerup*, a strain of the Bethesda group, and *Salmonella* sp. (ser.) *arizonae* showed comparatively little binding with both *S. typhimurium* mRNA and *E. coli* mRNA. DNA from several organisms, presumably unrelated to enterobacteria, was included in these studies as a control for reaction specificity. No DNA from the *Neisseria* species tested showed appreciable reaction with *E. coli* DNA. Marginal reassociation was observed between DNA of *Pasteurella pestis* or *Pseudomonas aeruginosa* and that of *E. coli* (Table 2).

These experiments were designed to examine possible nucleic acid sequence similarities between broad, general groups of organisms. Quantitative reassociation was studied in most detail in the genus *Proteus*. Aside from the high relative relatedness between DNA preparations from strains of *P. mirabilis* and *P. vulgaris*, *P. mirabilis* DNA showed 16% or less reaction with DNA from any other *Proteus* species, or

TABLE 2. *Relative nucleic acid binding relationships among enterobacteria*[a]

Nucleic acid reaction	DNA-agar, 60 C	DNA-agar, 66 C	mRNA*/DNA filters, 67 C	Hydroxyapatite		
				60 C	66 C	75 C
	%	%	%	%	%	%
E. coli*/S. typhimurium LT2	38 ± 2 (8)	26 ± 4 (6)	25 ± 4 (6)	39 ± 6 (10)	30 ± 4 (4)	9 ± 2 (5)
E. coli*/S. typhimurium 7823	—	—	—	38 (1)	—	9 (1)
E. coli*/A. aerogenes	41 ± 3 (3)	—	22 ± 2 (6)	37 ± 4 (3)	23 (1)	8 ± 2 (4)
E. coli*/S. typhi 643	—	—	27 ± 1 (3)	43 ± 4 (5)	—	9 ± 3 (4)
E. coli*/P. mirabilis 1	5 (1)	—	5 ± 1 (5)	6 ± 2 (3)	—	1 ± 1 (6)
E. coli*/S. flexneri 24570	82 (1)	—	82 ± 7 (6)	86 ± 7 (5)	—	80 ± 3 (3)
E. coli*/Bethesda-10	—	—	24 ± 1 (3)	—	—	—
E. coli*/Salmonella sp. (ser.) ballerup	—	—	29 ± 2 (3)	—	—	—
E. coli*/Salmonella sp. (ser.) arizonae	—	—	23 ± 0 (3)	—	—	—
E. coli*/S. marcescens	—	—	9 ± 3 (6)	—	—	—
E. coli*/P. morganii	—	—	6 ± 1 (3)	—	—	—
E. coli*/P. aeruginosa	—	—	—	4 (1)	—	0 (2)
E. coli*/P. pestis	4 (1)	—	—	—	—	—
E. coli*/N. perflava	2 (1)	—	—	—	—	—
E. coli*/N. gonorrhoeae	—	—	—	0 (3)	—	0 (3)
E. coli*/N. catarrhalis	—	—	—	1 (1)	—	0 (1)
S. typhimurium*/S. typhi	—	—	87 ± 3 (3)	—	—	—
S. typhimurium*/E. coli	—	—	30 ± 3 (3)	—	—	—
S. typhimurium*/A. aerogenes	—	—	34 ± 3 (3)	—	—	—
S. typhimurium*/Sh. flexneri	—	—	26 ± 4 (3)	—	—	—
S. typhimurium*/Bethesda-10	—	—	26 ± 2 (3)	—	—	—
S. typhimurium*/Salmonella sp. (ser.) ballerup	—	—	34 ± 1 (3)	—	—	—
S. typhimurium*/Salmonella sp. (ser.) arizonae	—	—	42 ± 2 (3)	—	—	—
S. typhimurium*/S. marcescens	—	—	14 ± 4 (3)	—	—	—
S. typhimurium*/P. mirabilis	—	—	8 ± 1 (3)	—	—	—
S. typhimurium*/P. morganii	—	—	12 ± 2 (3)	—	—	—
P. mirabilis*/P. vulgaris	92 (1)	—	—	—	—	—

[a] DNA-agar reactions and mRNA*/DNA filter reactions were carried out in 2 × SSC. Hydroxyapatite reactions were carried out in 0.12 M PB. In all cases the intraspecies reaction (e.g., *E. coli*/E. coli*) was arbitrarily designated as 100%. All other reactions percentages are relative values. The amount of reassociation (binding) observed for intraspecies reactions was as follows: DNA-agar, 60 C = 30 to 45%; DNA-agar, 66 C = 20 to 30%; RNA*/DNA filters, 67 C = 20 to 30% of total RNA bound to homologous DNA; hydroxyapatite, 60 C = 82 to 97%; hydroxyapatite, 66 C = 80%, hydroxyapatite, 75 C = 75 to 92%. Asterisks designate the source of radioactive fragments. Numbers in parentheses show the number of experiments.

TABLE 2.—*Continued*

Nucleic acid reaction	DNA-agar, 60 C	DNA-agar, 66 C	mRNA*/DNA filters, 67 C	Hydroxyapatite		
				60 C	66 C	75 C
	%	%	%	%	%	%
P. mirabilis/*P. rettgeri*	6 (1)	—	—	—	—	—
P. mirabilis/*P. inconstans*	16 (1)	—	—	—	—	—
P. mirabilis/*E. coli* K-12	5 (1)	—	—	7 ± 1 (2)	—	1 ± 0.5 (2)
P. mirabilis/*S. typhimurium* LT2	—	—	—	5 (1)	—	—
P. mirabilis/*S. typhimurium* 7823	—	—	—	4 (1)	—	0.5 (1)
P. mirabilis/*S. typhi* 643	—	—	—	6 (1)	—	0.7 (1)
P. mirabilis/*A. aerogenes*	—	—	—	5 (1)	—	0.3 (1)
Sh. flexneri/*E. coli* K-12	—	—	77 ± 6 (5)	—	—	—
Sh. flexneri/*S. typhimurium*	—	—	20 ± 0 (2)	—	—	—
Sh. flexneri/*S. typhi* 643	—	—	21 ± 2 (5)	—	—	—
S. typhi/*S. typhimurium*	—	—	89 ± 1 (2)	—	—	—
S. typhi/*Sh. flexneri*	—	—	30 ± 1 (2)	—	—	—
S. typhi/*E. coli* K-12	—	—	21 ± 1 (2)	—	—	—

indeed with any other enteric organism. These observations, coupled with the generally low values obtained for interspecies DNA reassociation with other genera, are evidence for marked nucleotide sequence divergence in the *Enterobacteriaceae*.

Thermal stability of interspecies DNA duplexes formed at increasing incubation temperatures. Variation in the degree of reassociation with increased incubation temperatures is assumed to provide information as to the relative amounts of closely and partially related base sequences between the DNA of one organism and that of another. A corollary to this assumption is that the thermal stability of DNA duplexes formed at various incubation temperatures will be an index of the extent of base pairing. It was recently reported that the thermal stability of *E. coli-S. typhimurium* DNA duplexes increased despite decreased binding when the reactions were incubated at 66 C instead of at 60 C (6).

The thermal stability of interspecies enteric DNA duplexes was studied in 60 and 75 C reactions by measuring the thermal release of reassociated DNA fragments bound to hydroxyapatite. It should be noted that the method of thermal elution is a process of dissociating strand pairs with steps of increasing temperature and assaying (by radioactivity) the eluted fractions. The melting temperature ($T_{m(e)}$) obtained by the release of DNA fragments due to complete strand separation is not theoretically nor experimentally equivalent to the T_m measured optically by following the disruption of the helical content of DNA (9, 37). As a practical matter, however, the optical T_m (89 C ± 1 C, in 0.12 M PB) agrees rather closely with the elution $T_{m(e)}$ (87.5 C ± 1 C) for *E. coli* DNA.

Table 3 shows the stability of some *E. coli*-enterobacterial DNA duplexes. In the intraspecies *E. coli* reassociation experiments, the absolute binding was approximately 90% (as high as 97% in certain experiments) for a 60 C reaction and some 5 to 7% less in 75 C reactions. This binding was arbitrarily designated 100%. The reaction between the DNA of *E. coli* and the DNA of *S. flexneri* exhibited some 85% relative reassociation at the 60 C incubation temperature and about 80% at the 75 C incubation temperature. Similarly, the thermal stability of the *E. coli-S. flexneri* DNA reassociation products was the same at both temperatures and was very similar to the stability of the *E. coli-E. coli* DNA reassociation products. The reaction between *E. coli* DNA and *S. typhimurium* showed about 40% relative reassociation at the 60 C

TABLE 3. *Thermal elution midpoints of enterobacterial DNA duplexes formed at 60 or 75 C[a]*

Nucleic acid reaction	$T_{m(e)}$ from 60 C incubation	Relative binding	$T_{m(e)}$ from 75 C incubation	Relative binding
	C	%	C	%
E. coli*/E. coli	87.5 ± 1.0	100	87.5 ± 1.5	100
E. coli*/S. typhimurium LT2	74.0 ± 1.0	39	84.5 ± 1.5	9
E. coli*/S. typhimurium 7823	74.0^b	38	83.0^b	9
E. coli*/S. typhi 643	75.0 ± 1.0	43	83.0 ± 1.0	9
E. coli*/A. aerogenes	75.0 ± 1.0	37	85.5 ± 1.5	8
E. coli*/P. mirabilis-1	73.5 ± 1.5	6	$—^c$	1
E. coli*/P. mirabilis-1 F-lac⁺	80.0 ± 1.0	8	87.0 ± 0	3
E. coli*/S. flexneri	86.5 ± 0.5	86	86.5 ± 0.5	80
E. coli*/S. typhi 643 lac₃	76.0 ± 1.0	52	84.0 ± 1.0	16
E. coli*/S. typhi 643 X30T	80.0 ± 1.0	59	87.0 ± 1.0	32
E. coli*/S. typhi 643 X30W	83.0 ± 2.0	58	87.5 ± 0.5	38

[a] All reactions were carried out with hydroxyapatite. Asterisks designate the source of radioactive fragments.

[b] Only one experiment. All other reactions were carried out from 3 to 20 times.

[c] Too little reassociation to allow accurate assay.

incubation temperature, but only about 9% at the 75 C temperature. The duplexes formed between DNA from *E. coli* and *S. typhimurium* at 60 C had a $T_{m(e)}$ some 13.5 C below that of *E. coli* DNA duplexes. The product formed at 75 C, however, exhibited a $T_{m(e)}$ only about 3 C lower than that of *E. coli* DNA duplexes. Similar changes in relative binding and $T_{m(e)}$ were also noted in the interspecies DNA duplexes formed between *E. coli* and *S. typhi*, or *E. coli* and *A. aerogenes*. In the case of *E. coli-P. mirabilis* duplexes, at the 60 C incubation temperature there was approximately 6% relative reassociation with a $T_{m(e)}$ some 14 C below that of *E. coli* DNA. At 75 C, the relative reassociation was about 1%, which was too little to permit an accurate assay of thermal stability.

The thermal elution profiles show that interspecies duplexes were formed at the 60 C incubation temperature, as well as at the 75 C temperature (Fig. 1). The difference was that virtually all of the reaction product was quite stable in a 75 C incubation (Fig. 1A), whereas at 60 C duplexes with a markedly decreased thermal stability were also permitted. In fact, reassociated nucleotide sequences with low thermal stability make up the bulk of the reaction product formed at 60 C (Fig. 1B) in the case of duplexes formed between DNA from *E. coli* and DNA from *Salmonella*, *Aerobacter*, and *Proteus* species. In contrast, as shown in Table 3 and Fig. 2, at both the 60 and 75 C incubation temperatures, the *E. coli* and *S. flexneri* reaction product was highly stable and contained only a very small proportion of DNA that eluted at low temperatures.

Studies with synthetic polynucleotides have

demonstrated that the thermal stability of double-stranded nucleic acid complexes is very sensitive to quite small proportions of unpaired bases (2, 9, 30). The presence of approximately 1% unpaired bases within a reassociated nucleotide sequence lowers its thermal stability by 1 C (2). On this basis, under our experimental conditions, the degree of unpaired bases tolerated in *E. coli-S. typhimurium* duplexes would be about 1 in 7 at an incubation temperature of 60 C, about 1 in 9 at 66 C, and less than 1 in 30 at 75 C. Therefore, the higher the temperature of incubation (i.e., the stricter the criterion of reassociation) presumably the greater is the discrimination and complementarity of reassociated nucleotide sequences. The alternative, but less likely, explanation that decreased thermal stability is predominantly due to a preferentially high percentage of A + T base pairs within interspecies duplexes is not considered in the argument. This alternative has been discussed elsewhere (6, 28).

Detection of E. coli-specific DNA in the genomes of heterologous organisms. It was assumed that the presence of increasing amounts of *E. coli*-specific nucleotide sequences in the genome of a heterologous organism would be detectable in reassociation and thermal stability experiments with *E. coli* DNA. Two kinds of genetic hybrids were employed to test this assumption. Three *S. typhi-E. coli* hybrids were used (1, 18): *S. typhi* 643 lac₃, which had substituted 7 to 13% of the *E. coli* chromosome for its own; *S. typhi* 643 X30T, which had substituted between 20 and 28% of the *E. coli* chromosome; and *S. typhi* 643 X30W, which contained between 39 and 44% of the *E. coli* chromosome.

[32]P-labeled *E. coli* DNA fragments (0.1 μg) were incubated with approximately 400 μg of unlabeled DNA fragments from *E. coli, S. typhi, S. typhi* 643 lac₃, *S. typhi* 643 X30T, and *S. typhi* 643 X30W contained in 1 ml of 0.12 M PB. Labeled *E. coli* fragments incubated with *N. gonorrhoeae* DNA fragments and labeled *E. coli* fragments incubated in the absence of unlabeled DNA served as controls. Both the 60 and 75 C reactions were incubated for 24 hr. The reaction mixtures were passed through hydroxyapatite columns, and reassociated DNA

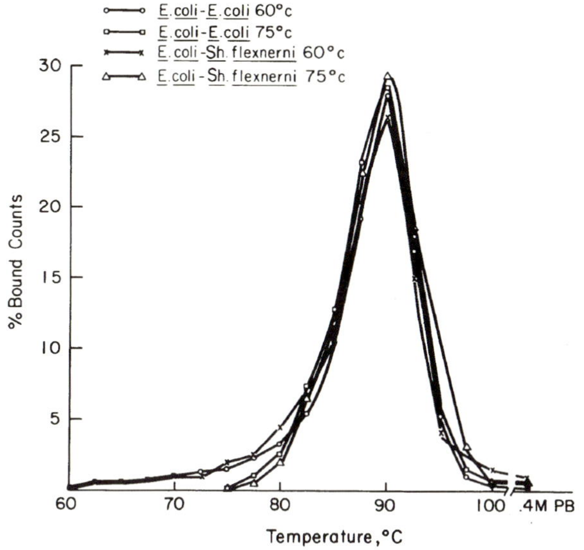

FIG. 2. *Thermal elution profiles of interspecies E. coli-S. flexneri DNA duplexes formed at 60 and 75 C. The reactions were carried out identically to those described in Fig. 1.*

was eluted as described in Materials and Methods.

The ability of the unlabeled DNA fragments to reassociate under our experimental conditions was confirmed by optically assaying the amount of input DNA sticking to the column after incubation. In reactions carried out at 60 C, unlabeled DNA fragments, regardless of origin, were 80 to 95% reassociated. An average 5 to 7% decrease in reassociation of unlabeled DNA fragments was observed in samples incubated at 75 C. Control reactions, in which 0.1 μg of labeled *E. coli* DNA fragments was incubated at 60 and 75 C, showed 2.0 and 1.8% binding to hydroxyapatite. Relative binding percentages and the thermal stability of duplexes formed between *E. coli* DNA and DNA from these *Salmonella* strains are listed in Table 3. Reaction specificity was assured by the inability of *E. coli* fragments to reassociate with *N. gonorrhoeae* DNA, in excess of control values, at either reaction temperature. The base composition of both the *E. coli* and *N. gonorrhoeae* DNA is approximately 50 mole per cent A + T (Kingsbury and Brenner, *unpublished data*).

The thermal elution profile obtained after reassociation of *E. coli* DNA fragments at 60 C (Fig. 1B) tends toward a Gaussian distribution with a $T_{m(e)}$ of about 88 C. In contrast, a profile obtained from an *E. coli-S. typhi* reaction at 66 C was very broad and exhibited a $T_{m(e)}$ some 13 C below that of the intraspecific 60 C *E. coli* reaction. The presence of increasing amounts of

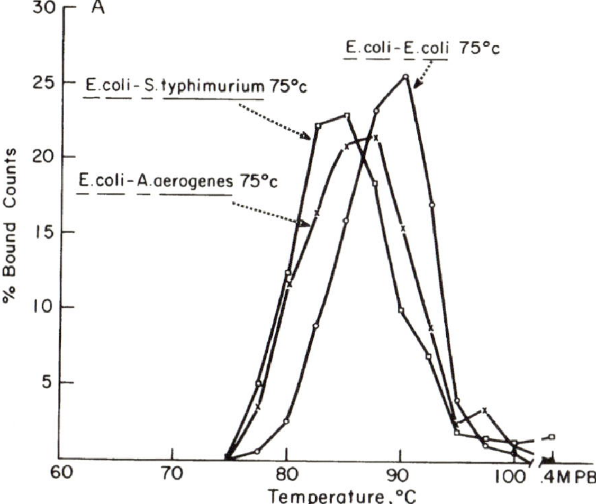

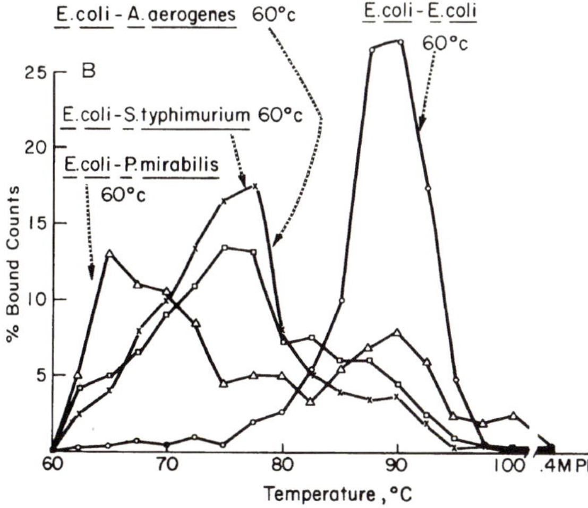

FIG. 1. *Thermal elution profiles of reassociated interspecies enterobacterial DNA duplexes. (A) Profiles obtained from 75 C reactions. (B) Profiles obtained from 60 C reactions. A 0.1-μg amount of* [32]*P-labeled E. coli DNA fragments was incubated with approximately 500 μg of unlabeled DNA fragments from the indicated organisms in 1 ml of 0.12 M PB for 18 hr at 60 or 75 C. The samples were loaded onto hydroxyapatite columns and eluted and assayed as described in Materials and Methods.*

E. coli-specific DNA in *S. typhi* strains lac₃, X30T, and X30W was evident from the increased relative percentage of reassociation and the increased $T_{m(e)}$ (Table 3). Thermal elution profiles from *E. coli*-*S. typhi* hybrid reactions at 60 C were biphasic in nature, with the appearance of a thermally stable *E. coli*-like peak (Fig. 3A).

An increase in thermal stability was apparent in elution profiles obtained from all interspecies reactions at 75 C (Table 3). These increases were due to the loss of reassociated sequences which were stable at 60 C, but could not withstand the more stringent incubation temperature. The loss of these sequences resulted in sharper, more stable elution profiles (Fig. 3B). Table 4 summarizes the results obtained with *S. typhi-E. coli* hybrid strains. The data indicate that the degree of reassociation at 75 C may be used to obtain a reasonable estimate as to the degree of specific *E. coli* genetic substitution in these strains.

Approximately 9% relative reassociation occurred between DNA from *E. coli* and *S. typhi* at 75 C. This "background" problem was largely avoided by using a strain of *P. mirabilis* that carried the extrachromosomal element F-*lac*⁺ (21). The F-*lac*⁺ element is equivalent to about 2.5% of the *E. coli* genome and is present in *Proteus* as an addition to the *Proteus* genetic material (17). These experiments were carried out under conditions identical to those described for the *S. typhi* hybrid strains. Control binding values for *E. coli* labeled DNA fragments incubated in the absence of unlabeled DNA were

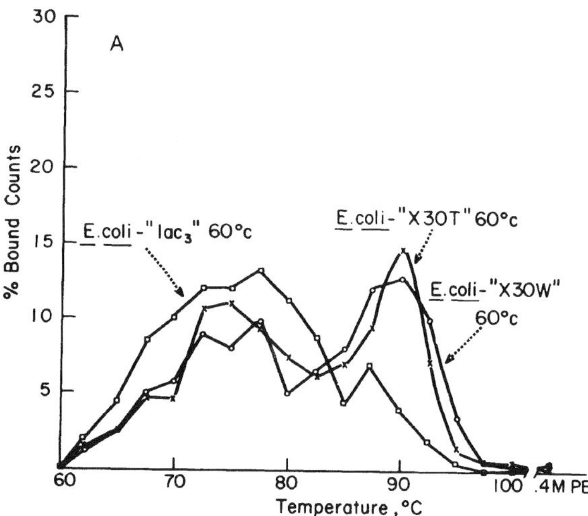

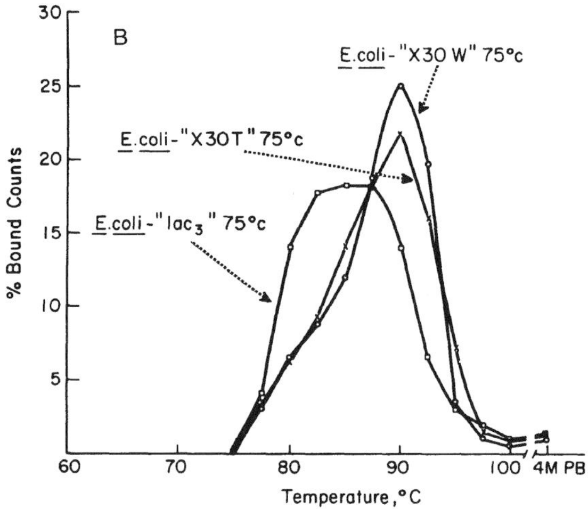

FIG. 3. *Detection of E. coli-specific DNA in the genome of S. typhi genetic hybrids.* ³²P-labeled E. coli DNA fragments were incubated with unlabeled DNA from the indicated organisms as described in the text. The thermal elution and assay procedures are described in Materials and Methods. (A) Profiles obtained from 60 C reactions. (B) Profiles obtained from 75 C reactions.

TABLE 4. *Detection of E. coli-specific DNA in the genomes of S. typhi hybrid strains*

Source of unlabeled DNA	Estimated *E. coli* genetic substitution[a]	Relative binding of *E. coli* DNA		Estimated *E. coli* DNA substitution[b]
		60 C	75 C	
	%	%	%	%
S. typhi 643	0	43 ± 4	9 ± 3	0
S. typhi 643 lac₃	7–13	52 ± 2	16 ± 1	7
S. typhi 643 X30T	20–28	59 ± 1	32 ± 3	25
S. typhi 643 X30W	39–44	58 ± 5	38 ± 2	33

[a] The estimate of the degree of *E. coli* substitution was calculated on the basis of *E. coli* characters stably present in *E. coli* × *S. typhi* hybrid assuming uniform substitution of *E. coli* material into the genome of *S. typhi*. It was assumed that the chromosomes of *E. coli* K-12 and *S. typhi* 643 are both 2.5×10^9 daltons. The genetic analysis of these hybrids was presented in detail by Falkow, Rownd, and Baron (18).

[b] The estimate of the degree of *E. coli*-specific nucleic acid substitution into the *S. typhi* chromosome was calculated by subtracting the "background" reaction between *E. coli* and *S. typhi* at 75 C from the percentage of *E. coli* DNA bound by the *S. typhi* hybrid strains at this temperature. It was also assumed that the "background" binding is distributed at random along the *S. typhi* chromosome and that the genetic substitution replaces a proportional amount of this "background" reaction. All reactions were carried out with hydroxyapatite. *E. coli**/*E. coli* reactions at 60 and 75 C were arbitrarily designated 100%; all binding percentages are relative to these intraspecies reactions.

2.1% at 60 C and 1.5% at 75 C. The presence of F-*lac*+ in *P. mirabilis* did indeed increase the relative reassociation of *E. coli* DNA and *P. mirabilis* DNA by about 2% and, concurrently, significantly increased the thermal stability of this reaction product (Table 3). *P. mirabilis* F-*lac*+ formed duplexes with about 3% of the *E. coli* DNA, and the $T_{m(e)}$ of this reaction product was similar to that of reassociated *E. coli* DNA duplexes (Table 3).

Martin and Hoyer (40) used the ratio of binding in agar at different incubation temperatures to determine relationships between DNA preparations from different animals. The ratio of relative reassociation at 75 and 60 C, which we have called the thermal binding index (TBI), has been useful in gauging the presence or absence of highly related genetic material in interspecies DNA reassociation reactions. The TBI values of *E. coli* DNA with DNA from strains of *Salmonella*, *Aerobacter*, and *Proteus* species were all about 0.2 (Table 5), whereas that with *S. flexneri* DNA was 0.93. The increased amounts of *E. coli* material in *P. mirabilis* F-*lac*+ and in the *E. coli*-*S. typhi* genetic hybrids were readily apparent from the TBI.

Conservation of ribosomal RNA genes. Although the degree of nucleic acid binding is a measure of overall relatedness between organisms, it is clear that different parts of the chromosome may not have evolved at the same rate. In several groups of bacteria, it has already been demonstrated that the DNA sequences that code for ribosomal RNA (rRNA) appear to be conserved in base sequence as compared with the rest of the genome (14, 15, 29, 42). Using rRNA (42) and using isolated and purified DNA sequences that code for rRNA (29), other investi-

gators have shown that the rRNA cistrons in enterobacteria are more highly conserved and form more stable interspecies hybrids than does bulk DNA from these organisms. DNA sequences that code for soluble RNA also appear to be more highly conserved than the bulk of enterobacterial DNA (24). Table 6 summarizes similar data obtained with labeled 23S rRNA from *E. coli* reacted with unlabeled DNA molecules immobilized on nitrocellulose filters. In every case, the relative binding of *E. coli* 23S rRNA to DNA from a heterologous organism was greater than the relative binding of *E. coli* DNA or mRNA to DNA from the heterologous organism. When 23S rRNA was tested on DNA from organisms previously studied (42), the two sets of results were in excellent agreement.

Divergence of the lactose operon among enteric bacteria. In the near future, it should be feasible to study the distribution of other well-defined parts of the chromosome, rather than depending upon bulk mRNA or the entire DNA complement. DNA fractions isolated from intergenetic hybrids (23) or from synchronous populations (12) should be useful in carrying out such experiments. As a first approximation, pulse-labeled RNA was isolated from *E. coli* under conditions where there was a preferential synthesis of β-galactosidase (43). A portion of the bulk pulse-labeled RNA was incubated with filter-immobilized DNA from several enterobacteria for measurement of the overall relative binding of the RNA preparation. Another portion of the RNA preparation was repeatedly reacted with the DNA of an *E. coli* strain with a genetic deletion which included the entire lactose (*lac*+) operon. The RNA that remained after reaction with the deletion strain (about 3% of the original

TABLE 5. *Thermal binding index*[a]

Nucleic acid reaction	Thermal binding index
*E. coli**/*S. typhimurium* LT2	.23
*E. coli**/*S. typhimurium* 7823	.24
*E. coli**/*S. typhi* 643	.21
*E. coli**/*A. aerogenes*	.22
*E. coli**/*P. mirabilis*-1	.17
*E. coli**/*P. mirabilis*-1 F-*lac*+	.38
*E. coli**/*S. flexneri*	.93
*E. coli**/*S. typhi* 643 *lac*₃	.31
*E. coli**/*S. typhi* 643 X30T	.54
*E. coli**/*S. typhi* 643 X30W	.66

[a] The data used in this table were taken from the hydroxyapatite experiments shown in Table 3. Thermal binding index = relative binding at 75 C/relative binding at 60 C. Asterisks designate source of labeled DNA.

TABLE 6. *Relative binding of E. coli 23S RNA with the DNA of other enterobacteria*[a]

Source of unlabeled DNA	Relative binding of *E. coli* 23S RNA
	%
E. coli	100
A. aerogenes	80
S. typhi	79
S. marcescens	74
P. mirabilis	76
P. morganii	52
S. flexneri	92

[a] Membrane filters containing 100 μg of the indicated DNA were incubated with 30 μg of 23S ribosomal RNA (4.3×10^4 counts per min per μg) overnight at 75 C in 2 × SSC. After incubation, the samples were treated with ribonuclease (60 min; in 2 × SSC; at 26 C).

radioactivity) was presumed to be relatively enriched for *lac*[+] mRNA and was tested for its binding to heterologous DNA species. In the main, the per cent relative binding with the *lac*[+] mRNA fraction was lower than the binding of the bulk mRNA (Table 7). In the case of *S. flexneri* DNA, the relative binding decreased by 13%. A decrease was expected, since it has been reported that the *y* region of the *lac*[+] operon is deleted in strains of *Shigella* species (31). The *lac*[+] fraction did not react measurably with DNA from *P. mirabilis*, *P. morganii*, or *S. marcescens*. An unexpected finding was that the reaction with the *lac*[+] RNA fraction was about 70% higher with *Salmonella* sp. (ser.) *ballerup* and about 10% higher with *S. typhimurium* than was the bulk mRNA reaction with DNA from these organisms. Most genetic data (45) suggest that the *lac*[+] region is deleted from *S. typhimurium*. Several years ago, however, it was reported that *lac*[+] variants of *S. typhimurium* could be selected (46). Similar mutants have been isolated in this laboratory and are currently being characterized genetically. Some preliminary evidence suggests that there may be at least part of the structural gene for β-galactosidase present in *S. typhimurium*. There is no evidence, however (on the basis of complementation with *lac*[+] episomes) for functional *i* or *y* loci. A study of the *lac*[+] region of *Salmonella* sp. (ser.) *ballerup* has not been attempted. No data have been obtained to support the experimental inference that the *lac*[+] operon of *Salmonella* sp. (ser.) *ballerup* is essentially identical to that of *E. coli*, although the strain employed here fermented lactose slowly.

It should also be emphasized that the *lac*[+] operon was only part of the genetic deletion in the strain employed in these experiments. Portions of the genome adjacent to the *lac*[+] operon may account, therefore, for part (or all) of the observed reaction. The deletion, however, was not more than 2%, as determined by interrupted mating experiments. At any rate, it does appear that, assuming a common ancestor, particular regions of the genome of enteric organisms have diverged at quite different rates.

DISCUSSION

We presume that the interactions we measure at the level of DNA reasonably reflect both genetic and evolutionary relationships among enterobacteria. If DNA from two organisms is unable to reassociate under ideal conditions, it is clear that these organisms are no longer related. The ability to form specific DNA duplexes between two organisms is evidence for relatedness.

The degree of interspecies DNA reassociation and the stability of these duplexes varies greatly as the experimental conditions are altered. The case of *E. coli* and *S. typhimurium* can be used to illustrate this point. Initially, the degree of DNA reactions between these organisms was determined by allowing a mixture of the two types of DNA, one of which was uniformly substituted with a heavy isotope, to reassociate (47). The presence of reassociated duplexes composed of one "light" strand and one "heavy" strand in CsCl density gradients was the criterion employed to determine DNA similarity. Under these experimental conditions duplex formation was not detected between DNA from *E. coli* and *S. typhimurium*. A more quantitative estimate of nucleic acid similarity between these organisms was obtained by measuring the binding of DNA fragments (as well as mRNA) to DNA immobilized in agar (33). Under these conditions, *E. coli* and *S. typhimurium* DNA showed approximately 70% relative relatedness. A more recent study (6), and our results obtained by use of hydroxyapatite, filter, and DNA-agar techniques, show that the relative relatedness of *E. coli* to *S. typhimurium* varied between 40 and 9% as the temperature was varied.

One should question (indeed, one must consider) just what a DNA reassociation experiment

TABLE 7. *Binding of pulse-labeled E. coli "lac*[+]*" RNA to enterobacterial DNA*[a]

Source of unlabeled DNA on filter	Bulk mRNA		Adsorbed mRNA	
	Ribonuclease-resistant counts/min bound	Relative binding	Ribonuclease-resistant counts/min bound	Relative binding
		%		%
E. coli K-12 *lac*[+]..	39,267	100	1,320	100
E. coli K-12 *lac*Δ.	39,156	99	27	2
P. mirabilis.	1,149	3	28	2
A. aerogenes.	7,945	20	190	14
S. typhi.	8,599	22	220	17
S. typhimurium.	8,864	22	460	34
S. flexneri.	32,149	82	910	69
S. marcescens.	3,215	8	29	2
Bethesda-10.	8,864	22	340	26
Salmonella sp. (ser.) *ballerup*.	8,334	21	1,230	93
Salmonella sp. (ser.) *arizonae*..	9,868	25	270	20
P. morganii.	2,464	6	26	2
None.	130	1	28	2

[a] See preparation of lactose mRNA in Materials and Methods.

implies and which is the "correct" degree of binding between two organisms? In fact, each of the results (as long as the reaction conditions are stringent to rule out nonspecific reassociation) is the "correct" answer for the criteria employed. It is simply that the percentage of DNA fragments bound in any experiment does not by itself yield enough information concerning the amount of paired bases present in the reassociated DNA duplexes.

The optimal temperature for DNA reassociation is approximately 30 C below the T_m of that DNA (37, 39). If we apply this rule to the enteric bacteria and to our data, then the optimal temperature for these reactions would be about 60 C. Yet, at this temperature, the reassociated duplexes formed between the DNA of *E. coli* and the DNA of several other enteric species had $T_{m(e)}$ values 13 to 14 C below that of the homologous reaction. It was not until these reactions were carried out at 75 C that the DNA heteroduplexes showed thermal stabilities close to that of reassociated *E. coli* DNA duplexes. In DNA reactions between *E. coli* and *S. flexneri*, the situation seems reasonably straightforward. The majority of their DNA was quite similar and formed thermally stable duplexes according to any criteria employed. Less than 15% of the DNA sequences of these organisms have diverged to a point where they cannot reassociate. Approximately an additional 5% of their DNA has diverged sufficiently to prevent interspecies duplex formation at the stringent 75 C criterion. The data clearly indicate that the vast majority of the nucleotide sequence of strains of *E. coli* and *P. mirabilis* or *S. marcescens* were not related by any criteria employed. Furthermore, with few exceptions, most nucleotide sequences held in common among enteric organisms show significant divergence, as judged by the low degree of interspecies duplex formation and by the decreased stability of interspecies duplexes.

There are no formal guidelines available with which to correlate genetic and molecular information to taxonomic groupings. Therefore, any attempt to equate the extent and stability of interspecies duplexes with lines of speciation would be somewhat premature and certainly arbitrary. A reasonable rule of thumb is to use less stringent, but specific, reaction conditions to detect overall relatedness between species. Highly related or specific interspecies DNA duplexes should be studied under stringent, in this case 75 C, reaction conditions. The data obtained in this study correlate well with existing taxonomic groupings. The exceptions which occur will, it is hoped, present the taxonomist with a new view.

The use of recombinant strains of *S. typhi* and *P. mirabilis* merodiploids allows an assessment of similarities or differences within specific portions of the genome. *Proteus* strains carrying specific regions of the *E. coli* chromosome (23) or specific episomes as in *P. mirabilis* F-*lac*⁺ appear to be ideal for these experiments because of the very slight relatedness between parental *P. mirabilis* and *E. coli* in 75 C reactions.

The experiments carried out with isolated RNA from the lactose operon of *E. coli* provide a preliminary assessment of its distribution among several enterobacteria. It is not surprising that the distribution of this operon was found to be quite variable. The stability of these interspecies lactose operon duplexes remains to be tested. We feel that, even with the relatively crude techniques now at our disposal, nucleic acid reassociation experiments may be used to great advantage in determining relationships between specific genetic regions.

ACKNOWLEDGMENTS

We are grateful to L. S. Baron, R. J. Britten, D. B. Cowie, P. Gemski, Jr., D. H. Haapala, E. M. Johnson, D. E. Kohne, and J. R. Wohlhieter for their continued interest in this work. We thank R. J. Britten for the use of his pressure cell to shear DNA samples and D. E. Kohne and J. A. Chiscon for providing some of the ³²P-labeled DNA preparations. We are indebted to V. B. D. Skerman for reviewing this manuscript and for taking the time to educate us in some of the complexities and subtleties of bacterial taxonomy.

This project was supported by National Science Foundation grant GB-6137, as well as under the sponsorship of the Commission on Enteric Infections of the Armed Forces Epidemiological Board in conjunction with research contract C-7061 of the U.S. Army Medical Research and Development Command, Department of the Army.

LITERATURE CITED

1. Baron, L. S., P. Gemski, Jr., E. M. Johnson, and J. A. Wohlhieter. 1968. Intergeneric bacterial matings. Bacteriol. Rev. 32:362–369.
2. Bautz, E. K. F. 1965. Evolutionary aspects of the distribution of nucleotides in DNA and RNA, p. 419–433. *In* V. Bryson and H. J. Vogel (ed.), Evolving genes and proteins. Academic Press Inc., New York.
3. Bernardi, G. 1965. Chromatography of nucleic acids on hydroxyapatite. Nature 206:779–783.
4. Berns, K. I., and C. A. Thomas. 1965. Isolation of high molecular weight DNA from *Hemophilus influenzae*. J. Mol. Biol. 11:476–490.
5. Bolton, E. T., and B. J. McCarthy. 1962. A general method for the isolation of RNA complementary to DNA. Proc. Natl. Acad. Sci. U.S. 48:1390–1397.
6. Brenner, D. J., and D. B. Cowie. 1968. Thermal stability of *Escherichia coli-Salmonella typhimurium* deoxyribonucleic acid duplexes. J. Bacteriol. 95:2258–2262.
7. Brenner, D. J., G. R. Fanning, A. Rake, and K. E. Johnson. 1969. A batch procedure for thermal elution of DNA from hydroxyapatite. Anal. Biochem., *in press*.
8. Brenner, D. J., M. A. Martin, and B. H. Hoyer. 1967. Deoxyribonucleic acid homologies among some bacteria. J. Bacteriol. 94:486–487.
9. Britten, R. J., and D. E. Kohne. 1966. Nucleotide sequence

repetition in DNA. Carnegie Inst. Wash. Yearbook 65:78–106.

10. Clark, A. J. 1967. The beginning of a genetic analysis of recombination proficiency. J. Cell. Physiol. 70(Suppl. 1):165–180.

11. Clausen, T. 1968. Measurement of ^{32}P activity in a liquid scintillation counter without the use of scintillator. Anal. Biochem. 22:70–73.

12. Cutler, R. G., and J. E. Evans. 1967. Isolation of selected segments from the genome of Hfr *Escherichia coli*. J. Mol. Biol. 26:81–90.

13. Denhardt, D. T. 1966. A membrane filter technique for the detection of complementary DNA. Biochem. Biophys. Res. Commun. 23:641–646.

14. Doi, R. H., and R. T. Igarashi. 1965. Conservation of ribosomal and messenger ribonucleic acid cistrons in *Bacillus* species. J. Bacteriol. 90:384–390.

15. Dubnau, D., I. Smith, P. Morell, and J. Marmur. 1965. Gene conservation in *Bacillus* species. I. Conserved genetic and nucleic acid base sequence homologies. Proc. Natl. Acad. Sci. U.S. 54:491–498.

16. Echols, H., and R. Gingery. 1968. Mutants of bacteriophage λ defective in vegetative genetic recombination. J. Mol. Biol. 34:239–249.

17. Falkow, S., and R. V. Citarella. 1965. Molecular homology of F-merogenote DNA. J. Mol. Biol. 12:138–151.

18. Falkow, S., R. Rownd, and L. S. Baron. 1962. Genetic homology between *Escherichia coli* K-12 and *Salmonella*. J. Bacteriol. 84:1303–1312.

19. Falkow, S., I. R. Ryman, and O. Washington. 1962. Deoxyribonucleic acid base composition of *Proteus* and Providence organisms. J. Bacteriol. 83:1318–1321.

20. Falkow, S., H. Schneider, L. S. Baron, and S. B. Formal. 1963. Virulence of *Escherichia-Shigella* genetic hybrids for the guinea pig. J. Bacteriol. 86:1251–1258.

21. Falkow, S., J. A. Wohlhieter, R. V. Citarella, and L. S. Baron. 1964. Transfer of episomic elements to *Proteus*. I, Transfer of F-linked chromosomal determinants. J. Bacteriol. 87:209–219.

22. Gillespie, D., and S. Spiegelman. 1965. A quantitative assay for DNA/RNA hybrids with DNA immobilized on a membrane. J. Mol. Biol. 12:829–842.

23. Gemski, P., Jr., J. A. Wohlhieter, and L. S. Baron. 1967. Chromosome transfer between *E. coli* Hfr strains and *Proteus mirabilis*. Proc. Natl. Acad. Sci. U.S. 58:1461–1467.

24. Goodman, H. M., and A. Rich. 1962. Formation of a DNA-soluble RNA hybrid and its relation to the origin, evolution and degeneracy of soluble RNA. Proc. Natl. Acad. Sci. U.S. 48:2101–2109.

25. Johnson, E. M., S. Falkow, and L. S. Baron. 1964. Recipient ability of *Salmonella typhosa* in genetic crosses with *Escherichia coli*. J. Bacteriol. 87:54–60.

26. Johnson, E. M., S. Falkow, and L. S. Baron. 1964. Chromosome transfer kinetics of *Salmonella* Hfr strains. J. Bacteriol. 88:395–400.

27. Johnson, J. L., and E. J. Ordal. 1968. Deoxyribonucleic acid homology in bacterial taxonomy: effect of incubation temperature on reaction specificity. J. Bacteriol. 95:893–900.

28. Kingsbury, D. T., G. R. Fanning, K. E. Johnson, and D. J. Brenner. 1969. Thermal stability of interspecies *Neisseria* DNA duplexes. J. Gen. Microbiol., *in press*.

29. Kohne, D. E. 1968. Isolation and characterization of bacterial ribosomal RNA cistrons. Biophys. J. 8:1104–1118.

30. Kotaka, T., and R. L. Baldwin. 1964. Effects of nitrous acid on the dAT copolymer as a template for DNA polymerase. J. Mol. Biol. 9:323–339.

31. Luria, S. E., M. J. Adams, and R. C. Ting. 1960. Transduction of lactose-utilizing ability among strains of *E. coli* and *S. dysenteriae* and the properties of the transducing phage particles. Virology 12:348–390.

32. McCarthy, B. J. 1967. Arrangement of base sequences in deoxyribonucleic acid. Bacteriol. Rev. 31:215–229.

33. McCarthy, B. J., and E. T. Bolton. 1963. An approach to the measurement of genetic relatedness among organisms. Proc. Nat. Acad. Sci., U. S. 50:156–164.

34. McConaughy, B. L., and B. J. McCarthy. 1967. The interaction of oligodeoxyribonucleotides with denatured DNA. Biochem. Biophys. Acta. 149:180–189.

35. Mandell, J. D., and A. D. Hershey. 1960. A fractionating column for analysis of nucleic acids. Anal. Biochem. 1:66–77.

36. Marmur, J. 1961. A procedure for the isolation of deoxyribonucleic acid from micro-organisms. J. Mol. Biol. 5:109–118.

37. Marmur, J., and P. Doty. 1961. Thermal renaturation of DNA. J. Mol. Biol. 3:585–594.

38. Marmur, J., S. Falkow, and M. Mandel. 1963. New approaches to bacterial taxonomy. Ann. Rev. Microbiol. 17:329–372.

39. Marmur, J., R. Rownd, and C. R. Schildkraut. 1963. Denaturation and renaturation of DNA. Progr. Nucleic Acid Res. 1:231–300.

40. Martin, M. A., and B. H. Hoyer. 1966. Thermal stabilities and species specificities of reannealed animal deoxyribonucleic acids. Biochemistry 5:2706–2713.

41. Miyazawa, Y., and C. A. Thomas. 1965. Composition of short segments of DNA molecules. J. Mol. Biol. 11:223–237.

42. Moore, R. L., and B. J. McCarthy. 1967. Comparative study of ribosomal ribonucleic acid cistrons in enterobacteria and myxobacteria. J. Bacteriol. 94:1066–1074.

43. Naono, S., J. Rouviere, and F. Gros. 1965. Preferential transcription of the lactose operon during the diauxic growth of *Escherichia coli*. Biochem. Biophys. Res. Commun. 18:664–674.

44. Pittard, J. 1964. Effect of phage-controlled restriction on genetic linkage in bacterial crosses. J. Bacteriol. 87:1256–1257.

45. Sanderson, K. E. 1967. Revised linkage map of *Salmonella typhimurium*. Bacteriol. Rev. 31:354–372.

46. Schäfler, S., L. Mintzer, and C. Schäfler. 1960. Acquisition of lactose fermenting properties by salmonellae. II. Role of the medium. J. Bacteriol. 79:203–212.

47. Schildkraut, C. L., J. Marmur, and P. Doty. 1961. The formation of hybrid DNA molecules and their use in studies of DNA homologies. J. Mol. Biol. 3:595–617.

48. Schneider, H., and S. Falkow. 1964. Characterization of an Hfr strain of *Shigella flexneri*. J. Bacteriol. 88:682–689.

49. Taylor, A. L., and C. D. Trotter. 1967. Revised linkage map of *Escherichia coli*. Bacteriol. Rev. 31:332–353.

50. Thomas, C. A., Jr., and J. Abelson. 1966. The isolation and characterization of DNA from bacteriophage, p. 553–561. *In* G. L. Cantoni and D. R. Davies (ed.), Procedures in nucleic acid research. Harper and Row, New York.

51. Weil, J., and E. R. Signer. 1968. Recombination in bacteriophage λ. II. Site-specific recombination promoted by the integration system. J. Mol. Biol. 34:273–279.

Legionnaires' Disease. Isolation of a Bacterium and Demonstration of Its Role in Other Respiratory Disease

J. E. McDade, C. C. Shepard, D. W. Fraser, T. R. Tsai, M. A. Redus, W. R. Dowdle, and the Laboratory Investigation Team

This paper by McDade and his numerous collaborators from the Centers for Disease Control appeared $1\,{}^{1}/_{2}$ years after a cluster of severe respiratory illnesses had been observed in elderly men who had attended the 1976 American Legion convention in Philadelphia. Public worries about the mysterious disease had kept the spotlight on the investigative team, which spared no effort to identify the causative agent, later called *Legionella pneumophila,* in record time. Practically all routine methods used for isolation in clinical bacteriology, mycology, and virology, as well as toxicological ones (in a search for metallic elements), were used. In the end, animal experiments gave the clue, but the authors eventually succeeded in culturing the organism as well. By using an indirect fluorescent antibody technique, they also found that isolated outbreaks in the District of Columbia in 1965, and in Pontiac, Mich., in 1968, were caused by the same organism, although these outbreaks were characterized only by fever, chills, and minor respiratory symptoms.

This paper and its companion in the same journal (which reported clinical and epidemiological details) laid the groundwork for the large body of knowledge on legionellae amassed in subsequent years. We now know that there are over 40 species of *Legionella,* that the organisms occur widely in water sources, that human infections often occur endemically and in nosocomial settings, and that these infections can be effectively treated with antimicrobial agents. Although the mode of transmission in the original epidemic was clearly airborne, it still needs to be elucidated for many endemic cases.

Alexander von Graevenitz

LEGIONNAIRES' DISEASE

Isolation of a Bacterium and Demonstration of Its Role in Other Respiratory Disease

JOSEPH E. MCDADE, PH.D., CHARLES C. SHEPARD, M.D., DAVID W. FRASER, M.D., THEODORE R. TSAI, M.D., MARTHA A. REDUS, WALTER R. DOWDLE, PH.D., AND THE LABORATORY INVESTIGATION TEAM[*]

Abstract   To identify the etiologic agent of Legionnaires' disease, we examined patients' serum and tissue specimens in a search for toxins, bacteria, fungi, chlamydiae, rickettsiae and viruses. From the lungs of four of six patients we isolated a gram-negative, non-acid-fast bacillus in guinea pigs. The bacillus could be transferred to yolk sacs of embryonated eggs. Classification of this organism is incomplete. We used yolk-sac cultures of the bacillus as antigen to survey suspected serum specimens, employing antihuman-globulin fluorescent antibody. When compared to controls, specimens from 101 of 111 patients meeting clinical criteria of Legionnaires' disease showed diagnostic increases in antibody titers. Diagnostic increases were also found in 54 recent sporadic cases of severe pneumonia and, retrospectively, in stored serum from most patients in two other previously unsolved outbreaks of respiratory disease. We conclude that Legionnaires' disease is caused by a gram-negative bacterium that may be responsible for widespread infection. (N Engl J Med 297:1197-1203, 1977)

AN outbreak of severe respiratory illness occurred in the summer of 1976 in Pennsylvania, chiefly among persons who attended a state American Legion Convention. An estimated 182 cases of pneumonia occurred, and 29 people died. A detailed description of the outbreak, including its clinical presentation, is reported in the companion paper.[1] An extensive laboratory investigation was undertaken; technics for detection of a variety of toxins and for identification of infections caused by bacteria, chlamydiae, fungi, mycoplasmas, parasites, rickettsiae and viruses were used. A bacterium was isolated in guinea pigs from the lung tissues in four fatal cases. The etiologic role of this organism was demonstrated by indirect fluorescent-antibody tests of survivors' serum specimens.

MATERIALS AND METHODS

Tissue samples were analyzed for abnormal concentration of more than 30 metallic elements by one or more of the following technics: atomic absorption spectrophotometry; neutron activation; electron-induced x-ray fluorescence; and proton-induced x-ray fluorescence. Appropriate tissue and urine samples were also examined for a broad spectrum of organic toxic substances. Technics employed were high-pressure liquid chromatography, gas chromatography and mass spectroscopy, with various combinations of sample extraction and preparation. In addition, specific assays were performed for organic compounds suggested by the Legionnaires' symptoms — for example, 1,1'-dimethyl-4-4'-dipyridilium (Paraquat).[*]

The microbiologic and serologic methods used and the specimens examined in the search for an etiologic agent are summarized in tables on deposit.[†] These tests involved attempts by fluorescent-antibody and eight other methods to visualize the agent, isolation attempts on 14 bacteriologic and mycologic mediums and 13 virologic host systems, and tests of patients' serums against 77 infectious agents.

The rickettsiologic technics employed were those in regular use in the Leprosy and Rickettsia Branch. Adult male guinea pigs (weighing approximately 600 g) were inoculated intraperitoneally with 1 ml of 10 per cent suspensions of patients' tissues prepared in sucrose-phosphate-glutamate buffer, pH 7.2.[2] Fresh autopsy materials from patients with Legionnaires' disease were used in three isolation attempts, and tissues that had been stored at −70°C for four

From the Leprosy and Rickettsia Branch, Virology Division, Bureau of Laboratories, Center for Disease Control, Public Health Service, U.S. Department of Health, Education, and Welfare, Atlanta, GA 30333, where reprint requests should be addressed (attention of Dr. Shepard).

[*]*Bacteriology:* Albert Balows, Ph.D.; William B. Cherry, Ph.D.; James C. Feeley; Charles Hatheway, Ph.D.; George L. Lombard, Dr.P.H.; Donald C. Mackel; George K. Morris, Ph.D.; Dwane L. Rhoden; Catherine Sulzer; Peter B. Smith, Ph.D.; and Robert Weaver, M.D., Ph.D. *Mycology:* Libero Ajello, Ph.D.; William Kaplan, D.V.M.; and Leo Kaufman, Ph.D. *Parasitology:* Henry M. Mathews, Ph.D.; and Alexander Sulzer, Ph.D. *Pathology:* John A. Blackmon, M.D.; Francis W. Chandler, Ph.D.; and Martin D. Hicklin, M.D. *Toxicology:* David D. Bayse, Ph.D.; Renate Kimbrough, M.D.; John A. Liddle, Ph.D.; and Larry L. Needham, Ph.D. *Virology:* Jared J. Gardner, M.D.; Milford H. Hatch, Sc.D.; John Hierholzer, Ph.D.; Karl M. Johnson, M.D.; Harold S. Kaye, M.S.; Alan P. Kendal, Ph.D.; Vester J. Lewis, Ph.D.; Frederick A. Murphy, D.V.M., Ph.D.; Verne F. Newhouse, Ph.D.; Gary R. Noble, M.D.; Erskine L. Palmer, Ph.D.; and Herta T. Wulff, Ph.D.

[*]Use of trade names is for identification only and does not constitute endorsement by the Public Health Service or by the U.S. Department of Health, Education, and Welfare.

[†]For six pages of supplementary material, order NAPS Document 03184 from ASIS/NAPS, c/o Microfiche Publications, P.O. Box 3513, Grand Central Station, New York, NY 10017. Remit, in advance, $3 for each microfiche-copy reproduction or $5 for each photocopy. Outside the United States and Canada, postage is $3 for a photocopy or $1 for a microfiche. Foreign orders add $3 for postage and handling. Make checks payable to Microfiche Publications.

months in three other attempts. Animals were monitored by daily rectal temperatures and clinical observation, and those that became ill were killed on the second or third day after the onset of fever. Pieces of guinea-pig spleen, liver and lung were ground separately with alundum in a mortar and pestle, and 10 per cent suspensions of each were prepared in sucrose-phosphate-glutamate buffer. Aliquots of these suspensions were tested for the presence of bacteria by inoculation onto trypticase soy agar and sheep-blood agar and into thioglycollate broth. Half-milliliter aliquots of the tissue suspensions were inoculated into the yolk sacs of embryonated hens' eggs six to seven days old from antibiotic-free flocks. The eggs were incubated at 35°C and candled daily. Those that died before the third day were considered to be contaminated with extraneous bacteria and were discarded. Yolk sacs from those that died after the third day were smeared, stained by the Giménez method,[3] and examined microscopically.

Indirect fluorescent-antibody tests were carried out on smears of 5 per cent suspensions of infected yolk sacs prepared in phosphate-buffered saline, pH 7.2. Microhematocrit tubes were used to place microdrops of the suspensions (one drop of each of two isolates, side by side) in each 5-mm well of a 12-well microscope slide. Slides were prepared fresh daily from small aliquots stored at −70°C, and, after the smears had dried, they were fixed in acetone at room temperature for 15 minutes. Twofold dilutions of each patient's serum, made in 10 per cent normal yolk sac in phosphate-buffered saline, were placed on the smears and incubated at 37°C for 30 minutes in a humid atmosphere. The slides were washed with phosphate-buffered saline and then incubated for an additional 30 minutes with the optimal dilution of rabbit antihuman conjugate capable of staining human IgG and IgM. The titer of the serum was taken as the reciprocal of the highest serum dilution that gave distinct staining of the organism. Nearly all serums were tested as coded unknowns, and a positive control serum was included in each test. The microscope used epi-illumination, which greatly facilitates accurate estimation of the degree of fluorescence. Serum end points tend not to be sharp with indirect fluorescent antibody, so that whenever feasible, serial dilutions of the serum specimens from the same patient were run in the same test, to ensure accurate comparison of the brightness of staining at particular dilutions.

RESULTS

Toxicologic Studies

Because the clinical symptoms exhibited by the patients with Legionnaires' disease were nonspecific and could have been caused by a variety of agents, toxicologic studies were initiated along with microbiologic studies on the first specimens having sufficient quantities. Comparison of case and control data failed to show the consistent presence of any unusual components or elevated levels of toxins that might be related to the epidemic.

Microbiologic Studies

Inoculation of specimens on bacteriologic and mycologic mediums resulted in the isolation of a number of organisms commonly found in normal flora or in the flora of patients under treatment with broad-spectrum antibiotics, and otherwise formed no particular pattern of isolation. Cultures for mycoplasma and spiroplasma were negative.

With the exception of one herpesvirus, no agent was isolated in the cell cultures, eggs or mice. Primary monkey-kidney, human embryonic lung fibroblasts, HEp-2, human embryonic kidney and Vero-cell cultures inoculated with specimens from the respiratory tract were challenged in the terminal (fourth or fifth) blind passage with Coxsackie A-9 virus in an effort to detect an agent that might be demonstrated by viral interference. None was found.

Thin-section electron-microscope studies were carried out on lung tissues in 10 fatal cases. Thin-walled bacteria were found in specimens from seven of 10 patients. No other microbial agents were observed. Direct fluorescent-antibody staining of lung tissue with antiserums to 13 known agents failed to show the presence of any microbial agents, with the exception of specimens from one patient with a secondary candida infection whose lung stained positively for a candida species by the fluorescent-antibody technic.*

Serum specimens from serial bleedings were examined by complement fixation, indirect fluorescent-antibody, immunoprecipitin, and indirect hemagglutination tests for antibodies to a large variety of antigens.* Substantial (fourfold or greater) increases in antibody titers were detected for *Mycoplasma pneumoniae* (in one patient), herpesvirus (in two patients), and *Blastomyces dermatitides* — in one patient, who had evidence of disseminated candidiasis at death.

Isolation and Preliminary Description of the Agent

An agent was isolated in guinea pigs from four of six lung specimens collected on autopsy. Three isolates were from patients classified epidemiologically as having Legionnaires' disease. The other isolate was from a patient who had Broad Street pneumonia[1] — that is, the patient had not attended the convention and had not entered the hotel about which the epidemic appeared to center, but had been within one block of the hotel during the epidemic period. In the four successful isolation attempts, a febrile illness characterized by watery eyes and eventual prostration developed in guinea pigs. With two isolates, fever (temperatures ranging from 39.5° to 41.0°C) developed in guinea pigs as early as 18 hours after inoculation, and with the two other isolates similar fever developed after an incubation period of one or two days. Impression smears of guinea-pig liver and spleen obtained on the second day of fever, stained by the Giménez method, contained scattered bacilli. When the disease was allowed to progress in guinea pigs, the animals became moribund three to six days after the onset of fever. An exudate containing numerous bacilli was observed in the peritoneum, especially on the liver and spleen of animals killed when moribund.

Spleen, liver and lung-tissue suspensions from the affected guinea pigs, when inoculated into embryonated eggs, caused death in four to seven days. Eggs inoculated with spleen tissue died earliest, and eggs

*Further information is on file with the National Auxiliary Publications Service. See footnote on page 1197.

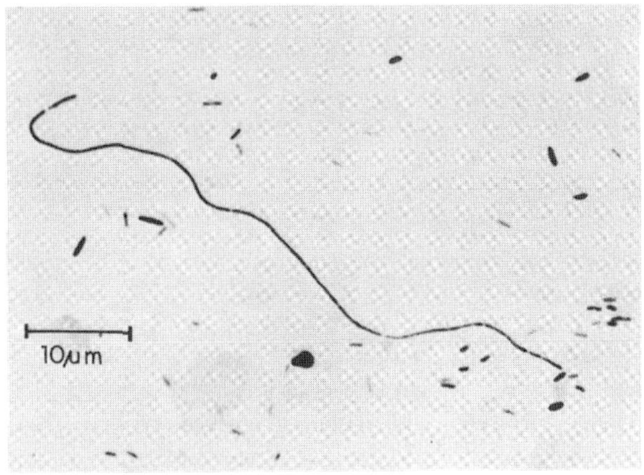

Figure 1. Photomicrograph of the Bacteria of Legionnnaires' Disease in Yolk Sac Stained by the Giménez Method.

inoculated with lung tissue died last, indicating, as did smears and histologic sections, that the spleen contained a greater number of organisms than the lung tissue. Smears of yolk sacs from embryos that died, when stained by the Giménez method (Fig. 1), showed many bacilli. The bacilli were 0.3 to 0.4 μm in width and usually 2 to 3 μm long, but bacilli 8 to 20 μm long were often observed, and forms longer than 50 μm were occasionally seen (Fig. 1). The sides of the bacteria were sometimes not parallel, and the ends often were somewhat pointed. We found the organism to be gram-negative and not acid fast. By thin-section electron microscopy of infected yolk sacs we found the bacteria to have structure typical of gram-negative rods. The organism grew in primary chick-embryo cell cultures and was nonpathogenic for mice.

The organism did not grow on trypticase soy or blood agar or in thioglycollate broth, but as described in another report,[4] it has been successfully cultured on Mueller–Hinton agar containing 1 per cent hemoglobin and 1 per cent Isovitalex (BBL) in 5 per cent carbon dioxide. (Supplemented Mueller–Hinton agar was not used in the initial isolation attempts.) We subsequently isolated the organism on supplemented Mueller–Hinton agar directly from human-

lung tissue in two of six attempts. The latter isolates reacted with convalescent-phase serums from patients with Legionnaires' disease in indirect fluorescent-antibody tests. Also, they produced the characteristic pattern of disease when inoculated into guinea pigs and were reisolated from infected guinea-pig tissues both by cultivation in embryonated eggs and by direct inoculation onto supplemented Mueller–Hinton agar. Serums from convalescent guinea pigs specifically stained the organism in indirect fluorescent-antibody tests; preinfection serums from the same guinea pigs did not. Guinea-pig convalescent-phase serum did not react in complement-fixation tests with standard rickettsial antigens prepared from *Rickettsia rickettsii, R. prowazekii, R. mooseri (typhi)* or *Coxiella burnetii.*

Etiologic Role of the Yolk-Sac Isolates in the Pennsylvania Outbreak

Serologic evidence for the etiologic role of the bacterium was obtained by indirect fluorescent-antibody staining. Results with some of the first serum specimens tested are shown in Table 1. With certain patients, antibody rises were observed; in others, the first specimen appeared to have been taken too late for an antibody rise to be demonstrated, but high titers were obtained. The brightness of staining and the changes in titers were very similar to those observed in other infectious diseases — for example, Rocky Mountain spotted fever, with which we have had extensive experience.

Most patients from whom suitable serum specimens were available could be classified as having "seroconversion" (with titer increases ranging from fourfold to 128-fold and reaching a minimum titer of 64) or as having "positive" results (minimum titer of 128). The titers observed in these two groups are shown in Figure 2. Titers for a few patients were elevated six or seven days after onset. Their titers rose rapidly in the next two to three weeks to a maximum at about the fifth week. The proportion of serums with titers greater than 64 was largest in the interval from 22 to 60 days, and titers decreased somewhat during the next few months. The trends for individual patients were similar, but the time at which elevated titers ap-

Table 1. Representative Results of Indirect Fluorescent-Antibody Tests with Bacterial Agent on Serum Specimens from Patients with Legionnaires' Disease.

Case No.	Days after Onset	Reciprocal Dilution of Serum*							Titer†	Interpretation
		16	32	64	128	256	512	1024		
1	31		1/1	±/±	0/0	0/0	0/0	0/0	32/32	Negative
	163		1/1	±/±	0/0	0/0	0/0	0/0	32/32	
2	1	1/1	±/±	0/0	0/0	0/0	0/0		16/16	Conversion
	23	3/3	3/3	2/3	1/2	±/1	0/0		128/256	
3	13	3/3	3/3	3/3	2/3	1/2	1/1	0/±	512/512	Positive
	30	3/3	3/3	3/3	2/3	1/2	1/1	0/±	512/512	

*Numbers indicate brightness of fluorescence from 0 (no staining) through ± (questionable staining), 1+ (minimal but definite staining) to 4+ (maximally bright). Staining with isolate 1/staining with isolate 2.

†Highest dilution of patient's serum giving definite staining. Titer with isolate 1/titer with isolate 2.

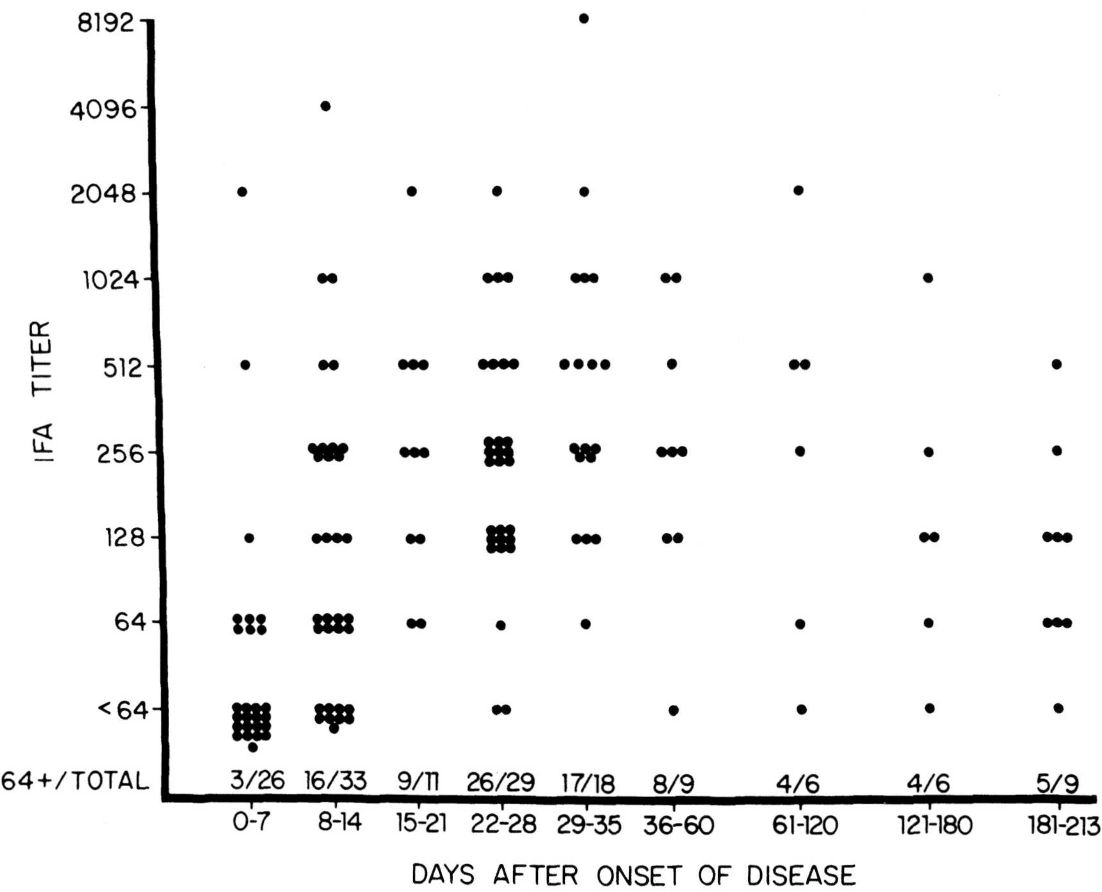

Figure 2. Results of Indirect Fluorescent-Antibody (IFA) Tests with Bacterial Agent on Serum Specimens from Patients with Legionnaires' Disease Classified as Having Seroconversion or a Positive Result.

peared varied. Table 2 shows the results of the serologic tests of case serums. Of the 111 patients with suitably timed specimens, 62 could be placed in the group with seroconversion and 39 in the group with positive results. The maximum indirect fluorescent-antibody titers obtained with the various patient categories are also shown in Table 2.

Table 2. Results of Indirect Fluorescent-Antibody Tests with Bacterial Agent of Serum Specimens from Patients with Legionnaires' Disease.

SEROLOGICAL CATEGORY	NO. OF PATIENTS	HIGHEST TITER OBSERVED FOR EACH PATIENT								
		<32	32	64	128	256	512	1024	2048	4096	8192
Seroconversions*	62			4	11	14	16	9	5	2	1
Positive only†	39				10	14	13	1	1		
Negative‡	10	4	2	4							
Questionable§	25	9	7	9							
Total patients tested	136										

*Increase in titer of at least 2 dilutions with highest titer 64 or greater.

†No seroconversion but highest titer ≥128.

‡Did not meet qualification as seroconversion or positive only even though specimen was available from interval 22–60 days after onset.

§Did not meet qualification as seroconversion or positive only but specimen was not available from interval 22–60 days after onset.

The basis for the selection of the minimum titer requirement for serologic diagnosis is given in Table 3, in which results with various control groups are shown. In Group 3, a titer of 256 was found in convalescent-phase serum from a patient with pneumonia who came from Michigan and in another from Vermont; acute-phase serums showed titers of <16 and <32, respectively. In Group 4, one patient from California had a titer of 2048; the acute-phase serum titer was 16. The convalescent-phase serums from these three patients are considered below, along with other isolated cases of pneumonia presumed to be due to the agent of Legionnaires' disease. The Pennsylvania health-department personnel (Groups 5 and 6) were tested shortly after they had begun handling specimens associated with the epidemic of Legionnaires' disease and again several months later. No clinical illness had been observed in this group,[1] and the results provided no evidence of inapparent infections. Serum from a psittacosis outbreak (Groups 7 and 8) provided a stringent test for observation of antigenic relation of the agent of Legionnaires' disease to *Chlamydia psittaci*. Even though all the specimen pairs test-

Table 3. Results of Indirect Fluorescent-Antibody Tests with Bacterial Agent on Various Control Serums.

LINE	GROUP	TITER				
		<64	64	128	256	2048
1	Rickettsia, serologic test negative*	38	2			
2	Pneumonia,† serologic test negative	81	10	2		
3	Pneumonia,† serologic test positive	68	5	2	2‡	
4	Q fever negative§	41	6	1		1‡
5	Health Dept personnel,¶ Aug. 1976	30				
6	Health Dept personnel,¶ Feb. 1977	31				
7	Psittacosis outbreak,‖ S 1	20		1		
8	Psittacosis outbreak,‖ S 2	19	2			
9	Psittacosis, serologic test positive**	7	8			
10	Rheumatoid arthritis & SLE††	10				
11	Legionnaires, not attending‡‡	9	2			
	Totals§§	304	35	5	2	1

*Convalescent-phase serum with request for rickettsial tests. Rickettsial serologic reaction was negative.

†Convalescent-phase serum specimens from patients reported to have pneumonia. Specimens were tested against viral & mycoplasmal antigens, & they are separated here according to the serologic results with these antigens.

‡Convalescent-phase serum specimens from patients with pneumonia, very probably caused by the agent of Legionnaires' disease (see Table 5).

§Convalescent-phase serum submitted for Q-fever serologic test; all were negative with this antigen.

¶Health Department personnel handling specimens from Philadelphia outbreak.

‖Acute-phase & convalescent-phase serum specimens from patients with diagnostic rises in complement-fixing antibody to chlamydia group antigen during outbreak of psittacosis in turkey-processing plant.

**Convalescent-phase serum specimens from other patients with diagnostic rises in complement-fixing antibody to chlamydia-group antigen.

††Group of serum specimens with rheumatoid factor & antinuclear antibody.

‡‡Pennsylvania Legionnaires who did not attend convention in Philadelphia.

§§Not including lines 5 & 7.

ed had distinct antibody rises to psittacosis antigen, none had any noteworthy change in titer to the agent of Legionnaires' disease. One patient had a titer of 128 in the first specimen and 64 in the second. Another had a titer of 32 in the first specimen and 64 in the second.

On the basis of the results presented in Table 3, a minimum titer requirement for serologic diagnosis by indirect fluorescent antibody was set at 64 for patients who had a fourfold rise in antibody titer and 128 for patients who did not have such a rise. The data also indicate that with a titer of 128, the expected number of false-positive results would be about five in 344. With a titer of 256, the expected number of false-positive results would be less than one in 344. If, however, a titer of 256 was set as the minimum requirement, in the absence of a fourfold or greater rise in titer, 10 of 39 patients with Legionnaires' disease in the group with positive results only (Table 2) would have been called seronegative.

Causes of Two Previous Outbreaks of Respiratory Disease

Epidemiologically, the outbreak of Legionnaires' disease was similar in many respects to two large outbreaks of febrile disease, one in 1965 (District of Columbia)[5] and the other in 1968 (Pontiac, Michigan).[6] Despite intensive investigation, the cause of these two outbreaks had not been determined before the present investigation. Serum specimens stored in the serum bank at the Center for Disease Control from these earlier investigations were examined for evidence of antibody to the bacterium from Legionnaires' disease.

The District of Columbia outbreak involved patients in a large psychiatric hospital in July, 1965, in which there were 81 cases and 12 deaths.[5] The clinical picture was similar to that observed in Pennsylvania Legionnaires in 1976. Appropriately timed acute-phase and convalescent-phase serum specimens from 23 patients were tested against the agent of Legionnaires' disease (Table 4). Of the 23 patients, 21 had serologic results characteristic of Legionnaires' disease.

The Pontiac outbreak of acute febrile illness involved personnel and visitors in an office of the county health department in July, 1968.[6] There were 144 cases and no deaths. Typically, there was an acute onset of fever, with chills, myalgia and minor respiratory symptoms. Pneumonia was not seen. Paired serum specimens from patients and controls were tested for antibodies to the Legionnaire isolate (Table 4). Of 37 cases, 31 had seroconversion and one was seropositive only. All 10 control serums collected from workers in an unaffected county health-department building were negative. The titers observed and the brightness of staining at low dilutions of positive serums were similar to those observed with patients in the pneumonia outbreaks in Pennsylvania and the District of Columbia.

Table 4. Results of Indirect Fluorescent-Antibody Tests with Bacterial Agent on Paired Serum Specimens from Patients Involved in Other Outbreaks.

INTERPRETATION OF TITERS	OUTBREAK	
	WASHINGTON, DC	PONTIAC, MI
Seroconversions*	17	31
Positive only†	4	1
Negative	2	5‡
Total patients tested	23	37

*Increase in titer of at least 2 dilutions with highest titer ≥64.

†No seroconversion, but highest titer ≥128.

‡Control serums collected at the same time as case serums, but from 10 workers in an unaffected county health-department building, were negative.

Isolated Cases of Pneumonia Presumed to be Due to the Agent of Legionnaires' Disease

Serologic results with individual cases found thus far have been arranged chronologically in Table 5 according to date of onset of illness. Cases 5 and 8 had apparently not been out of their home state during the estimated incubation period of two to 10 days. Case 6 was a merchant seaman who became ill aboard ship en route to Alaska; the ship had left southern California nine days before the onset of his illness. Case 7 had been on a Caribbean cruise and had returned to Tennessee eight days before onset. Case 9 was a truck driver who became sick and collasped at the wheel while driving from California to Ohio. The results with serum specimens from Cases 1, 2 and 6 were obtained in the process of screening serum from 170 patients with a clinical diagnosis of pneumonia and 49

Table 5. Sporadic Cases of Pneumonia with Seroconversions to the Bacterial Agent of Legionnaires' Disease.

Case No.	Age (Yr)	Sex	State of Residence	Date of Onset	Serum 1		Serum 2		Clinical Course
					titer*	day	titer*	day	
1	31	F	Vermont	8/10/76	<32/<32	8	256/256	31	Respiratory failure; death on 35th day.
2	62	M	Massachusetts	8/18/76	32/32	7	≥1,024/≥1,024	23	Fever (of 40.6°C) for 11 days
3	34	M	Michigan	8/19/76	<16/<16	3	≥512/≥512	21	Fever (to 41.7°C); respiratory-distress syndrome; recovery.
4	55	F	Massachusetts	9/07/76	<32/<32	3	512/512	50	Fever (of 40.6°C) for 5 days; respiratory & cardiac arrest.
5	32	M	Indiana	10/05/76	16/16	10	64/128	14	Pneumonia; respiratory failure; death on 14th day.
6	60	M	California†	1/03/77	16/16	10	2,048/2,048	20	Diffuse interstitial pneumonia, with renal failure; recovery.
7	60	M	Tennessee†	2/11/77					Respiratory & renal failure; death on 10th day.‡
8	31	M	Vermont	3/06/77	64/256	4	≥1,024/≥1,024	17	Fulminant broncho-pneumonia; temperature of 40.6°C; recovery with ventilatory support.
9	43	M	California†	4/01/77	64/64	5	1,024/1,024	15	Gastroenteritis; trilobar pneumonia; mechanical ventilation for 3 days; recovery.

*Titer against isolate 1/titer against isolate 2.

†Because these patients traveled during their incubation periods, they could have acquired the infection outside state of residence.

‡Bacterium isolated from lung specimen.

patients with a clinical diagnosis of Q fever (presumably, many of them also had pneumonia); thus, the proportion of the pneumonias caused by the bacterium in such patients was only 1 to 2 per cent. With the other serums, the bacterium of Legionnaires' disease was suspected as the cause because of the severity of the illness.

DISCUSSION

The etiologic role of the organism in Legionnaires' disease seems proved by its isolation from the tissues of four of six patients who died and by the very high proportion of seroconversion to the agent among cases. The importance of these findings is greatly increased, however, by the negative results obtained in very extensive investigations in this institution and elsewhere of the possible role of other infectious agents and toxic materials. The bacterium isolated from Legionnaires' disease was indistinguishable on electron microscopy from that initially observed in the lungs in seven fatal cases. Although initial conventional bacterial and fungal stains did not consistently demonstrate an agent in paraffin-embedded lung sections, more recent studies with a silver-impregnation stain[7] have shown the bacterium to be easily detected and present in large numbers in the alveoli, both within inflammatory cells and extracellularly.[8] These histologic findings and the early observations by electron microscopy confirm the presence of extensive numbers of bacteria in the lungs of patients who died of Legionnaires' disease. The organisms were morphologically identical to those seen in the peritoneal exudate of inoculated guinea pigs.

At this initial stage of investigation, the agent of Legionnaires' disease has been distinguished by the following properties: the characteristic disease that it produces in guinea pigs; the characteristic death pattern of chick embryos after yolk-sac inoculation; morphology of the organism in smears of infected yolk sacs and guinea-pig peritoneal exudates; gram-negative characteristics; failure to grow on ordinary bacteriologic mediums such as trypticase soy agar, blood agar, and thioglycollate broth; and specific immunofluorescent staining with convalescent-phase serums of patients with Legionnaires' disease. As mentioned above, the organism has since been cultivated on supplemented Mueller–Hinton medium, and studies of its classification are in progress.[4] We cannot say as yet whether the organism has previously been isolated from other sources, but it seems clear that this is the first demonstration of its role in human disease. The properties listed above differentiate it from other known microbes causing human disease.

Several clinical and epidemiologic manifestations of infection appear to be involved. The most dramatic one is severe pneumonia with 15 to 20 per cent mortality, as exemplified by the epidemic associated with the Legionnaires' convention in Philadelphia in 1976 and the epidemic in the psychiatric hospital in the District of Columbia in 1965.[5] The clinical picture suggested severe viral pneumonia.

Another entity is sporadic severe pneumonia, as represented by the nine isolated cases in Table 5. In addition, there was an isolation directly on chocolate agar from the lung tissue in a fatal case of pneumonia in Michigan in December, 1976.[9] This organism appears to be the same as the isolates in Legionnaires' disease, since it is stained specifically in the indirect fluorescent-antibody test with serum from patients with Legionnaires' disease. As with the outbreaks in Philadelphia and the District of Columbia, in none of these isolated cases was there evidence of secondary spread. Moreover, the clinical picture appeared to be the same as that in those two outbreaks. Three of the sporadic cases were found in a survey of 170 paired serum specimens, submitted for virologic and mycoplasmal diagnosis, from patients with clinical pneumonia (Table 3); thus, the bacterium appears to have been the cause of only 1 to 2 per cent of cases of such pneumonia.

Finally, there is the entity of a sharp outbreak of acute febrile illness without pneumonia, as represented by Pontiac fever.[6] Convalescent-phase serum specimens from the Pontiac outbreak reacted with the Philadelphia organism to the same degree as serum from patients with Legionnaires' disease did, and this finding suggests that the etiologic agent of Pontiac fever is antigenically similar, if not identical, to the Legionnaire bacterium. The differences in the clinical picture do not necessarily suggest that the agent of Pontiac fever has a different pathogenicity for man, but rather that it could be related to dose response or to host factors. The organism that apparently caused Pontiac fever has recently been isolated from lung tissues preserved at −60°C from guinea pigs exposed to the air in the health department. Those isolates are now being compared to the isolates obtained in the fatal cases of Legionnaires' disease.

Since the manuscript was first submitted for publication, the indirect fluorescent-antibody test has been used for the laboratory diagnosis of a total of 54 isolated cases of pneumonia in the United States, two cases of pneumonia in Scots who had vacationed in Benidorm, Spain,[10] and in outbreaks of pneumonia in Ohio, Vermont and Tennessee that are still being investigated.

We are indebted to numerous staff members of the following institutions for making this study possible: Philadelphia Health Department, Pennsylvania Department of Health, St. Elizabeth's Hospital, District of Columbia Department of Human Resources, Armed Forces Institute of Pathology, Oakland County Health Department, Michigan Department of Public Health and the Center for Disease Control.

REFERENCES

1. Fraser DW, Tsai T, Orenstein W, et al: Legionnaires' disease: description of an epidemic of pneumonia. N Engl J Med 297:1189-1197, 1977
2. Bovarnick MR, Miller JC, Snyder JC: The influence of certain salts, amino acids, sugars, and proteins on the stability of rickettsiae. J Bacteriol 59:509-522, 1950
3. Giménez DF: Staining rickettsiae in yolk-sac cultures. Stain Technol 39:135-140, 1964
4. Follow-up on respiratory disease — Pennsylvania. Morbid Mortal Weekly Rep 26:93, 1977
5. Institutional outbreak of pneumonia. Morbid Mortal Weekly Rep 14:265-286, 1965
6. Epidemic of obscure illness — Pontiac, Michigan. Morbid Mortal Weekly Rep 17:315-320, 1968
7. Dieterle RR: Method for demonstration of *Spirochaeta pallida* in single microscopic sections. Arch Neurol Psychiatry 18:73-80, 1927
8. Chandler FW, Hicklin MD, Blackmon JA: Demonstration of the agent of Legionnaires' disease in tissue. N Engl J Med 297:1218-1220, 1977
9. Follow-up on Legionnaires' disease. Morbid Mortal Weekly Rep 26:111-112, 1977
10. Respiratory illness (Benidorm episode/Legionnaires' disease). Weekly Rep Commun Dis Scotland 77/33:i, 1977

SHELDON WOLF, M.D.

Lyme Disease—A Tick-Borne Spirochetosis?

W. Burgdorfer, A. G. Barbour, S. F. Hayes, J. L. Benach,
E. Grunwaldt, and J. P. Davis

The 1982 article from *Science* by Willy Burgdorfer, from the Rocky Mountain Laboratories, and his associates suggested, for the first time, a connection between Lyme disease and a "treponema-like spirochete" described by the authors. They were able to isolate the organism and show antibody formation in patients with clinically diagnosed Lyme disease, by means of indirect immunofluorescence.

Case descriptions of what we strongly suspect was Lyme disease had been published in Europe since 1883, by A. Buchwald, K. Herxheimer (who had coined the term acrodermatitis chronica atrophicans), A. Afzelius (describing erythema chronicum migrans), B. Lipschütz, C. Garin and Bujadoux, C. Hövelborn, A. Bannwarth, and B. Bäverstedt (see G. Stanek et al., *Wien. Klin. Wochenschr.* **108**:741–747, 1996). Later, a combination of meningitis, polyneuritis, and radiculitis was found to be associated with tick bites and erythema chronicum migrans and was called Bannwarth syndrome. But it was only after 1975, when A. C. Steere and his collaborators at Yale (A. C. Steere, *Arth. Rheum.* **20**:7–17, 1977) observed a cluster of cases of inflammatory arthritis in the town of Old Lyme, Conn., and were able to show epidemiological evidence for *Ixodes* ticks as vectors, that Lyme disease came into focus as an entity. The causative agent was described in 1984 (A. C. Steere et al., *N. Engl. J. Med.* **308**:733–740, 1983) as *Borrelia burgdorferi* in honor of Willy Burgdorfer (later, more species were added to the genus). In view of the clinical and epidemiological importance of Lyme borreliosis, with its various stages and protean symptoms, the paper presented here was a true trailblazer in clinical microbiology and infectious diseases.

Alexander von Graevenitz

Reprinted with permission from *Science* 216:1317–1319. Copyright © 1982. American Association for the Advancement of Science.

Lyme Disease—A Tick-Borne Spirochetosis?

Abstract. *A treponema-like spirochete was detected in and isolated from adult* Ixodes dammini, *the incriminated tick vector of Lyme disease. Causally related to the spirochetes may be long-lasting cutaneous lesions that appeared on New Zealand White rabbits 10 to 12 weeks after infected ticks fed on them. Samples of serum from patients with Lyme disease were shown by indirect immunofluorescence to contain antibodies to this agent. It is suggested that the newly discovered spirochete is involved in the etiology of Lyme disease.*

Lyme disease is an epidemic inflammatory disorder that usually begins with a skin lesion called erythema chronicum migrans (ECM). Weeks to months later the lesion may be followed by neurologic or cardiac abnormalities, migratory polyarthritis, intermittent attacks of oligoarticular arthritis, or chronic arthritis in the knees (*1*).

Although in the United States cases of ECM were first reported from Wisconsin (*2*) and southeastern Connecticut (*3*), Lyme disease as a new form of inflammatory arthritis was first recognized in 1975 in Lyme, Connecticut (*4*). It has since been reported from other northeastern, midwestern, and western states (*5*).

Epidemiologic evidence suggests that Lyme disease is caused by an infectious agent transmitted by ticks of the genus *Ixodes*. In the Northeast and Midwest *Ixodes dammini* and, in the West, *I. pacificus* have been incriminated as potential vectors (*6, 7*). Until recently, all attempts to isolate the causative agent either from ticks or from patients were unsuccessful.

Recently we isolated from *I. dammini* a spirochete that binds immunoglobulins of patients convalescing from Lyme disease. We also recorded the development of lesions resembling ECM in New Zealand White rabbits on which ticks harboring this spirochete had fed.

Adult *I. dammini* were collected in late September and early October 1981 by flagging lower vegetation on Shelter Island, New York—a known endemic focus of Lyme disease (*8*). Of 126 such ticks that were dissected, 77 (61 percent; 65 males and 12 females) contained spirochetes. The spirochetes were distributed mainly in the midgut but were occasionally also seen in the hindgut and rectal ampule. No other tissues, including the salivary glands, contained spirochetes. The organisms stained moderately well with Giemsa (Fig. 1); in wet preparations examined by dark-field mi-

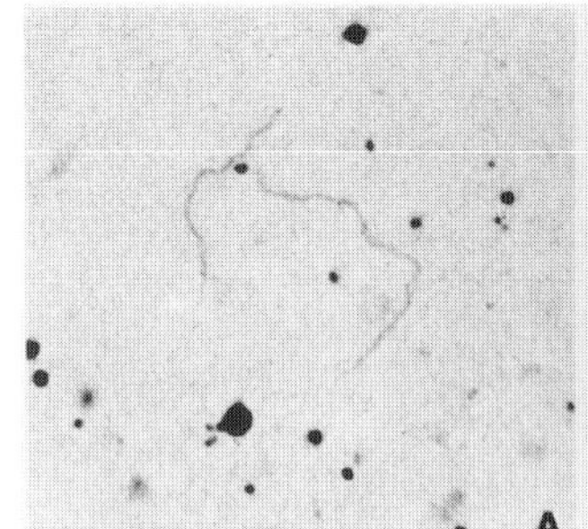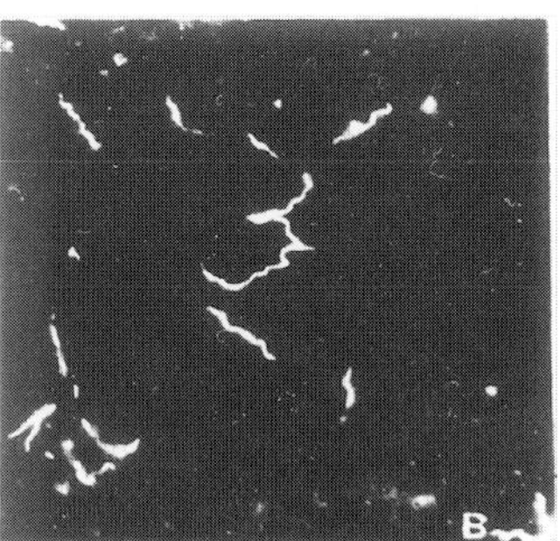

Fig. 1. *Ixodes dammini* spirochetes in midgut tissues of its tick vector. (A) Giemsa staining (×1200). (B) Serum of patient J.G. examined by indirect immunofluorescence (×570).

Table 1. Serologic evaluation (indirect immunofluorescence) of serum from persons with Lyme disease.

Pa-tient*	Disease contracted	Serum collected	Serum dilution end point
B.B.	May 1978	September 1978	1:1280
B.Br.	July 1980	July 1980	1:240
E.D.	July 1980	July 1980	1:80
C.G.	June 1979	March 1980	1:640
J.G.	June 1979	March 1980	1:1280
L.H.	June 1980	September 1980	1:640
J.S.	July 1979	January 1982	1:640
A.S.	July 1977	March 1980	1:80
C.T.	June 1979	March 1980	1:320
Controls: Four samples from New York and ten from Montana			≦1:20

*Diagnosed by E.G. except for J.S., whose serum was submitted to the New York State Health Department. All patients contracted the disease while visiting Shelter Island, New York.

croscopy they moved sluggishly and rotated slowly. The degree of infection varied; some ticks contained only a few spirochetes, others contained large numbers often to the extent that clumps of spirochetes were present throughout the midgut.

Electron microscopy (*9*) of midgut diverticula revealed spirochetes closely associated with the microvillar brush border of the gut epithelium (Fig. 2). Fine structural features of the organism were similar to those reported for *Treponema* species (*10*). Irregularly coiled, the spirochetes range from 10 to 30 μm in length and from 0.18 to 0.25 μm in diameter. The ends appear tapered with four to eight filaments inserted subterminally at each end. Insertion points of the filaments are in a row paralleling the cell's long axis. Cross sections of the cells show six to eight filaments interspersed between the outer membrane and the cytoplasmic membrane in the asymmetric region of the section profile (Fig. 2).

The *I. dammini* spirochete was isolated by inoculating 0.1 ml of a suspension prepared from midgut tissues of four infected ticks into 8.5 ml of modified Kelly's medium (*11*). After 5 days of incubation at 35°C, all the culture tubes contained spirochetes that could be regu-

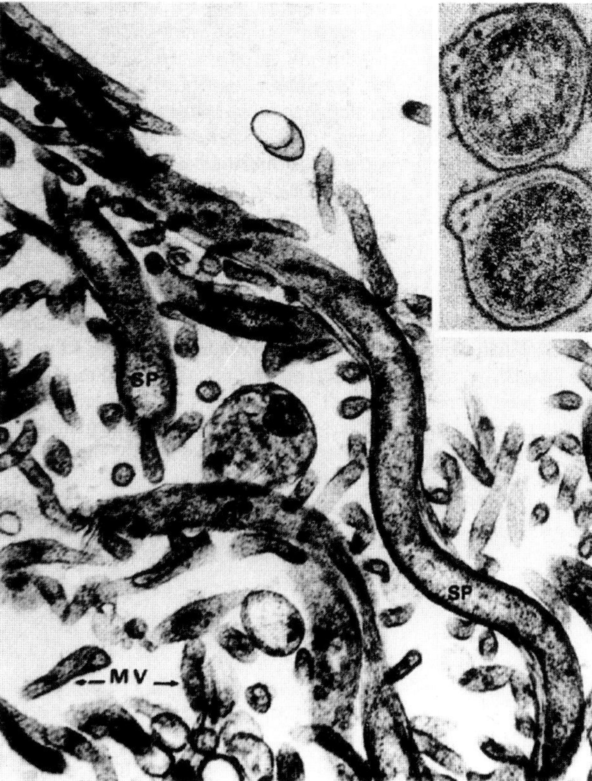

Fig. 2. Electron micrograph of *I. dammini* spirochetes (*SP*) associated with microvillar brush border (*MV*) of the tick's midgut ($\times$55,440). Inset shows cross section of spirochetes ($\times$122,100).

larly subcultured and maintained at 35°C.

When about 300 *I. dammini* were allowed to feed on eight New Zealand White rabbits (*12*), they appeared to have no immediately adverse effects. Blood smears examined daily for 14 days after placement of the ticks were negative for spirochetes. However, 10 to 12 weeks after the ticks had engorged, up to 15 small (2 to 3 mm in diameter) macules and papules appeared in the skin of the back and lateral trunk of each rabbit. Within 3 to 5 days, these lesions had enlarged (up to 5 cm in diameter) to slightly elevated annular or oval lesions with bright red to reddish-violet margins. Similar lesions on the abdomen, the site of tick attachment, were recorded on only one of the eight rabbits. All lesions persisted for at least 8 weeks.

Sections of biopsy specimens were stained with hematoxylin and eosin. These sections showed that the skin lesions consisted of a thickened, slightly hyperkeratotic epidermis with the dermis showing dense mononuclear cell infiltration and edema of the superficial layer. Limited attempts to isolate spirochetes from suspensions of biopsied skin lesions in Kelly's medium were negative.

Even though microscopic examination of repleted *I. dammini* showed that at least two ticks harboring spirochetes had fed on each rabbit, we are not certain whether the described skin reaction on the rabbits is causally related to the spirochetes or is due to other factors associated with the ticks' feeding process.

When tested by an indirect immunofluorescence method (*13*), antibodies to the spirochetes in titers of ≧ 1:1280 were present in the serum of all rabbits on which ticks had fed 30 and 60 days earlier. The serum of rabbits that had not been exposed to ticks did not react at dilutions of > 1:20.

That the *I. dammini* spirochete is antigenically related to the etiologic agent of Lyme disease was suggested by the positive reactions we obtained when we examined serum samples from nine patients with clinically diagnosed Lyme disease by means of indirect immunofluorescence (Fig. 1) (*14*). Antibody titers ranging from 1:80 to 1:1280 were recorded for persons who had Lyme disease currently or as many as 32 months previously (Table 1). In contrast, serum samples from four people from New York and ten from Montana with no history of the disease did not react with the spirochete in titers higher than 1:20.

Our observations suggest that the

treponema-like organism isolated from *I. dammini* may be involved in the etiology of Lyme disease. It is interesting that organisms presenting the morphological characteristics of spirochetes were said to be associated with ECM in Europe as early as 1948 (*15*). Although this was never confirmed, a recent study (*16*) showing that resolution of lesions and concurrent symptoms occurs faster in patients treated with penicillin suggests a penicillin-susceptible bacterium as an etiologic agent of Lyme disease.

Our results establish the susceptibility of the domestic rabbit to the *I. dammini* spirochete and demonstrate the possible value of the indirect immunofluorescence test as a diagnostic tool for Lyme disease. They also suggest the need for additional investigations not only into the epidemiology and ecology of Lyme disease and related disorders, such as ECM of Europe (*17*), but also into the relations between the spirochete and its vector *I. dammini*.

WILLY BURGDORFER
Epidemiology Branch, Rocky Mountain Laboratories, National Institute of Allergy and Infectious Diseases, Hamilton, Montana 59840
ALAN G. BARBOUR
Laboratory of Microbial Structure and Function, Rocky Mountain Laboratories
STANLEY F. HAYES
Rocky Mountain Operations Branch, Rocky Mountain Laboratories
JORGE L. BENACH
State of New York Department of Health and *Department of Pathology, State University of New York, Stony Brook 11794*
EDGAR GRUNWALDT
44 South Ferry Road Shelter Island, New York 11964
JEFFREY P. DAVIS
Department of Health and Social Services, Madison, Wisconsin 53701

References and Notes

1. A. C. Steere *et al.*, *Ann. Intern. Med.* **93**, 8 (1980).
2. R. J. Scrimenti, *Arch. Dermatol.* **102**, 104 (1970).
3. W. E. Mast and W. M. Burrows, Jr., *J. Am. Med. Assoc.* **236**, 859 (1976).
4. A. C. Steere, S. E. Malawista, J. A. Hardin, S. Ruddy, P. W. Askenase, W. A. Andiman, *Ann. Intern. Med.* **86**, 685 (1977).
5. A. C. Steere and S. E. Malawista, *ibid.* **91**, 730 (1979).
6. A. C. Steere, T. F. Broderick, S. E. Malawista, *Am. J. Epidemiol.* **108**, 312 (1978).
7. R. C. Wallis, S. E. Brown, K. O. Kloter, A. J. Main, Jr., *ibid.* **108**, 322 (1978).
8. The ticks were first examined by the hemolymph test [W. Burgdorfer, *Am. J. Trop. Med. Hyg.* **19**, 1010 (1970)]. Subsequently they were dissected for the preparation of multiple smears from gut, malpighian tubules, salivary glands, central ganglion, and testes or ovary. Smears were stained according to the Giménez method [D. F. Giménez, *Stain Technol.* **39**, 135 (1964)] or with Giemsa. Once spirochetes were detected, wet preparations of tissues were examined also under dark field.
9. For electron microscopy, diverticula of midgut were removed by dissection and were processed according to S. F. Hayes and W. Burgdorfer [*J. Bacteriol.* **137**, 605 (1979)].
10. K. Hovind-Hougen, *Acta Pathol. Microbiol. Scand. Sect. B Suppl. No. 225* (1976).
11. Kelly's medium [R. Kelly, *Science* **173**, 443 (1971)] modified by addition of CMRL medium 1066 (Gibco No. 330-1540) and Yeastolate (Difco) for final concentrations of 5 and 0.2 percent, respectively (H. G. Stoenner, in preparation).
12. Fifteen to twenty *I. dammini* females and equal numbers of males for mating (males may ingest small amounts of blood) were placed on each of eight rabbits. The ticks were contained in metal capsules attached by adhesive tape to the shaved abdomen of each rabbit.
13. In accordance with the data of R. N. Philip, E. A. Casper, R. A. Ormsbee, M. G. Peacock, and W. Burgdorfer [*J. Clin. Microbiol.* **3**, 51 (1976)] midgut smears of infected ticks or cultured spirochetes were used as antigen. Fluorescein isothiocyanate–conjugated goat antibody to rabbit immunoglobulin (Chappel Laboratories) was used at a 1:50 dilution in phosphate-buffered saline with 1 percent bovine serum albumin.
14. Fluorescein isothiocyanate–conjugated goat antibody to human immunoglobulin (BBL, Cockeysville, Md.) was used at 1:100 dilution in phosphate-buffered saline with 1 percent bovine serum albumin.
15. C. Lennhoff, *Acta Derm. Venereol.* **28**, 295 (1948).
16. A. C. Steere, S. E. Malawista, J. H. Newman, P. N. Spieler, N. H. Bartenhagen, *Ann. Intern. Med.* **93**, 1 (1980).
17. Since submission of this manuscript, microscopic examination by one of us (W.B.) of midgut smears from *Ixodes pacificus* from Oregon and of *I. ricinus* from Switzerland also revealed, in some instances, the presence of spirochetes.
18. We thank the Nature Conservancy Incorporation for permission to collect ticks in their Shelter Island Preserve. We also thank E. Bosler, S. Guirgis, D. Massey, and J. Coleman for their assistance in collecting ticks. Special thanks also to W. H. Hadlow, Epidemiology Branch, Rocky Mountain Laboratories, for the histologic characterization of the rabbit lesions.

26 February 1982; revised 20 April 1982

Hemorrhagic Colitis Associated with a Rare Escherichia coli Serotype

L. W. Riley, R. S. Remis, S. D. Helgerson, H. B. McGee,
J. G. Wells, B. R. Davis, R. J. Hebert, E. S. Olcott,
L. M. Johnson, N. T. Hargrett, P. A. Blake, and M. L. Cohen

The paper by Riley et al., from the Centers for Disease Control and state and local health departments in Oregon and Michigan, described for the first time what was later called EHEC (enterohemorrhagic *Escherichia coli*) diarrhea. The authors could exclude all known infectious diarrheagenic agents except *E. coli* serotype O157:H7 as the possible cause of this infection. It was characterized by abdominal cramps, initially watery and later bloody diarrhea, occasional fever, and blood (but not intestinal) leukocytosis. The symptoms lasted from 3 to >7 days and were not influenced by antimicrobial agents. The same serotype was detected in beef patties consumed by the patients but not in healthy individuals or in patients with diarrhea caused by other agents. Infant rabbits infected with *E. coli* O157:H7 exhibited diarrhea, albeit nonbloody diarrhea.

In the years after the report, the unique character of this gastrointestinal disorder became clear. By now, ca. 160 serovars of *E. coli* are known to produce so-called verotoxins (VT1 and VT2), which produce hemorrhagic lesions in the distal ileum and colon. Their detection is best done today by PCR or colony hybridization methods. The source is the gastrointestinal tract of cattle, of other farm animals, and occasionally of humans. The clinical and epidemiological importance of EHEC has become clear in the past decade. Undercooked beef and raw milk are the main sources of human infections, and 5 to 10% of the latter may give rise to serious complications in children, i.e., hemolytic uremic syndrome and thrombotic thrombocytopenic purpura.

Alexander von Graevenitz

Reprinted with permission from *New England Journal of Medicine* 308:681–685. Copyright © 1983. Massachusetts Medical Society. All rights reserved.

HEMORRHAGIC COLITIS ASSOCIATED WITH A RARE *ESCHERICHIA COLI* SEROTYPE

Lee W. Riley, M.D., Robert S. Remis, M.D., M.P.H., Steven D. Helgerson, M.D., M.P.H., Harry B. McGee, M.P.H., Joy G. Wells, M.S., Betty R. Davis, M.S., Richard J. Hebert, M.D., Ellen S. Olcott, R.N., Linda M. Johnson, R.N., M.S., Nancy T. Hargrett, Ph.D., Paul A. Blake, M.D., M.P.H., and Mitchell L. Cohen, M.D.

Abstract   We investigated two outbreaks of an unusual gastrointestinal illness that affected at least 47 people in Oregon and Michigan in February through March and May through June 1982. The illness was characterized by severe crampy abdominal pain, initially watery diarrhea followed by grossly bloody diarrhea, and little or no fever.

It was associated with eating at restaurants belonging to the same fast-food restaurant chain in Oregon (P<0.005) and Michigan (P = 0.0005) and with eating any of three sandwiches containing three ingredients in common (beef patty, rehydrated onions, and pickles).

Stool cultures did not yield previously recognized pathogens. However, a rare *Escherichia coli* serotype, 0157:H7, that was not invasive or toxigenic by standard tests was isolated from 9 of 12 stools collected within four days of onset of illness in both outbreaks combined, and from a beef patty from a suspected lot of meat in Michigan. The only known previous isolation of this serotype was from a sporadic case of hemorrhagic colitis in 1975. This report describes a clinically distinctive gastrointestinal illness associated with *E. coli* 0157:H7, apparently transmitted by undercooked meat. (N Engl J Med. 1983; 308:681-5.)

IN the first half of 1982 two outbreaks of an unusual gastrointestinal illness characterized by sudden onset of severe abdominal cramps and grossly bloody diarrhea, with no fever or low-grade fever, occurred in Oregon and Michigan. Isolated cases of a similar illness had recently been reported from Japan[1] and the United States,[2-5] but the etiologic agent had not been identified. The outbreaks in Oregon and Michigan led to intensive epidemiologic and laboratory investigations. In this report we describe the illness and the evidence that it is associated with a rare serotype of *Escherichia coli* that is neither invasive nor enterotoxigenic according to standard tests and is not a recognized enteropathogenic *E. coli*.

METHODS

Epidemiologic Investigation

We defined as a case an illness characterized by severe abdominal cramps, grossly bloody diarrhea, and stool examinations that did not yield shigella, salmonella, campylobacter, ova, or parasites.

To find cases, we contacted local physicians, reviewed records of chief complaints in emergency rooms and discharge records from December 1981 to February 1982 in parts of Oregon, and from May to June 1982 in parts of Michigan, and began active surveillance in all hospitals in the affected areas and nearby counties. In both states

From the Enteric Diseases Branch and Statistical Services Activity, Division of Bacterial Diseases, Center for Infectious Diseases, Atlanta; Field Services Division, Epidemiology Program Office, Centers for Disease Control, Atlanta; the Department of Human Resources, Oregon State Health Division, Portland, Ore.; Laboratory and Epidemiological Administration, Michigan Department of Public Health, Lansing, Mich.; Jackson County Health Department, Medford, Ore.; and Tri-County Health Department, Traverse City, Mich. Address reprint requests to Dr. Riley at CID:DBD:EDB 1-5428, Centers for Disease Control, Atlanta, GA 30333.

we conducted case–control studies with either one or two age-matched and neighborhood-matched controls for each case, using a questionnaire developed after intensive interviews of reported cases. We examined specific food exposures at restaurants implicated by the Michigan neighborhood case–control study, by comparing foods eaten by cases and by controls selected from persons who had visited the restaurants with the cases and had remained well. Food-handling procedures, food delivery and turnover, and employee records were reviewed at the implicated restaurants. In Michigan, grill temperature was measured in one implicated restaurant by means of a rapid readout surface pyrometer (Pyrcon, type 4000A, Alnor Instruments). Logistic regression analysis, the binomial test, and the Pike–Morrow extension of the Mantel–Haenszel test were used for statistical analyses.[6]

Laboratory Investigation

Stool specimens from cases were examined at the local hospital and state health-department laboratories for salmonella, shigella, campylobacter, ova, and parasites in both states and for *Yersinia enterocolitica* in Oregon. Stool specimens from some cases and controls were frozen ($-70°C$) until examination at the Centers for Disease Control (CDC). In Oregon, stool specimens were also collected from 45 persons who visited emergency rooms because of nonbloody diarrhea. Environmental samples, including food, were collected at implicated food establishments.

At the CDC, stool specimens from the outbreaks were examined for salmonella, shigella, pathogenic vibrios, *Y. enterocolitica*, campylobacter species, bacillus species, *Staphylococcus aureus*, enterotoxigenic and enteroinvasive *E. coli*, and anaerobes (including *Clostridium difficile* and toxin).[7] Five *E. coli* isolates from each stool were serotyped. The stool specimens were also examined for viruses by electron microscopy, by immunoelectron microscopy with acute-phase and convalescent-phase serum, and by culture in rhesus-monkey and human-fibroblast tissue cells.[8,9] The 45 diarrheal stool specimens obtained at emergency rooms were examined for *E. coli* 0157 and klebsiella. The foods were examined for *E. coli* 0157 and *Bacillus pumilus*.

E. coli 0157:H7, *Klebsiella oxytoca*, and *B. pumilus* isolates were tested for invasiveness by the Sereny test, for heat-stable toxin production by the suckling-mouse assay, and for heat-labile toxin production by the Y-1 adrenal-cell test.[10] *E. coli* 0157:H7 was tested in an infant-rabbit assay (Potter ME: personal communication).

The *E. coli* serotyping records of the U.S. Department of Agriculture Animal Laboratories at Ames, Iowa, the Pennsylvania State University Veterinary Science Laboratory, and the CDC Enteric Reference Laboratory were reviewed for previous identifications of *E. coli* 0157:H7.

CASE REPORT

The following case report is typical of the cases seen in both outbreaks. A 56-year-old man was awakened by severe abdominal cramps in the right lower quadrant. Later the same morning, watery diarrhea developed, with bowel movements every 15 to 30 minutes. The patient initially noted small amounts of blood, but later the same day the diarrhea became grossly bloody, with bright-red blood, described as "all blood and no stool." He had slight nausea but no vomiting. He was hospitalized on the following day with continuous crampy abdominal pain and frequent bloody diarrhea. He was afebrile and on abdominal examination had no guarding, rebound tenderness, or distension. The white-cell count was 17,900 with a slight shift to the left. A barium enema revealed edema of the ascending and transverse colon, with areas of spasm. Examinations of three stool specimens collected within three days after the onset of illness did not detect salmonella, shigella, campylobacter, yersinia,

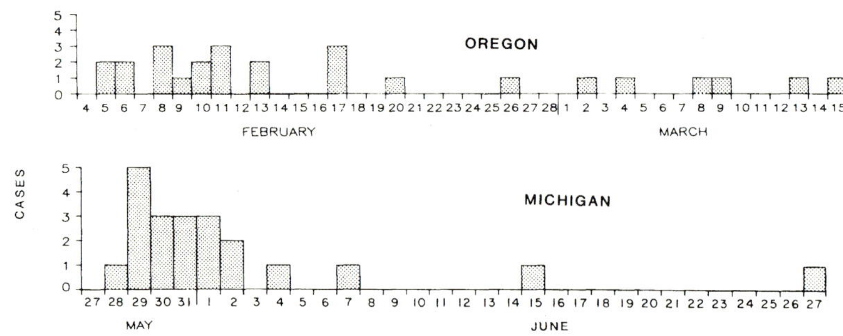

Figure 1. Cases of Hemorrhagic Colitis, According to Date of Onset (1982).

ova, or parasites. The patient was treated with intravenous fluids and doxycycline, and the bloody diarrhea subsided by the fifth hospital day. He was discharged the next morning.

RESULTS

Clinical and Epidemiologic Investigations

In Oregon, 25 persons from seven municipalities in one county and one person from an adjacent county became ill between February 5 and March 15 (Fig. 1). The median age was 28 years, with a range of 8 to 76 years; there were 16 males and 10 females. All consulted a physician, and 19 (73 per cent) were hospitalized. The duration of illness ranged from two to nine days, with a median of four days. In Michigan, 21 persons had onset of illness between May 28 and June 27 (Fig. 1). The median age was 17 years, with a range of 4 to 58 years. All consulted a physician, and 14 (67 per cent) were hospitalized. The illness lasted from three to more than seven days.

The symptoms in both outbreaks are shown in Table 1. Three patients in both outbreaks had temperatures of 38.5°C or more. The white-cell count ranged from 7600 to 19,600 (mean, 14,000) in Oregon and from 7600 to 17,400 (mean, 13,000) in Michigan, with a slight to moderate shift to the left. The erythrocyte sedimentation rates, serum electrolyte concentrations, liver-function tests, prothrombin times, and urinalyses were normal in all patients in whom these tests were done. Sigmoidoscopy performed in 10 patients revealed moderately hyperemic mucosa in 3. In six of seven patients barium enemas demonstrated marked submucosal edema with spasm and a "thumb-printing" pattern in the ascending and transverse colon (Fig. 2). In Oregon, 11 of 23 patients whose treatment histories were available received tetracycline compounds (eight patients) or erythromycin (three patients). The mean duration of illness of the group treated with antimicrobials was not significantly different from that of the untreated group. No transfusions were administered. There were no deaths, complications, or sequelae in any of the cases.

In Oregon, 25 of 26 cases and 47 neighborhood controls were interviewed. During the two weeks before onset of illness, 21 of 25 cases (84 per cent) but only 13 of 47 controls (28 per cent) had eaten at Restaurant 1, one of a chain of fast-food restaurants (Chain A) (P<0.005 by logistic regression analysis). Three of the four who did not recall having eaten at Restaurant 1 had eaten at another Chain A restaurant within a week before the onset of illness.

The patients who ate at one of the three restaurants of Chain A in the county were more likely (21 of 24) to have eaten one of the chain's specialty hamburgers than were neighborhood controls (11 of 20) who had eaten at the same restaurants (P<0.05, logistic regression analysis). The three patients who did not eat a specialty hamburger ate a regular hamburger (one patient) or a cheeseburger (two patients). The three types of sandwiches shared three ingredients, which were always served together — i.e., reconstituted dehydrated onions, standard-size hamburger meat patties, and pickles. Each of these three ingredients was eaten by a higher proportion of cases than of neighborhood controls (P<0.05, Pike–Morrow extension of the Mantel–Haenszel test), but no single ingredient could be independently associated with disease, because they had always been served together.

In Michigan, 18 of 21 cases and their age-matched neighborhood controls were interviewed. (Matched controls could not be found for two cases.) Seventeen of 18 cases and 4 of 16 controls had eaten at either Restaurant 2 or Restaurant 3 of Chain A within 10 days before the onset of illness (P = 0.0005, binomial test). No other exposures were significantly associated with illness. Again, cases were more likely than their restaurant controls to have eaten the same three food items implicated in Oregon (17 of 17 vs. 12 of 19; P<0.05, Pike–Morrow extension of the Mantel–Haenszel test). The mean period between single exposures to the implicated foods and onset of crampy abdominal pain was 3.9 days in Oregon and 3.8 days in Michigan. The attack rate for persons eating sandwiches that included the three ingredients was estimated to be about 1 case per 1000 sandwiches in Oregon, and 1.8 cases per 1000 for specialty hamburgers in Michigan, and 0.6 case per 1000 regular hamburgers and cheeseburgers in Michigan. The specialty hamburger had twice the quantity of meat (two patties vs. one) and onions as the regular hamburger or cheeseburger.

Only one case of bloody diarrhea occurred among employees at the three restaurants. None of the family members of the cases in Oregon and Michigan had bloody diarrhea. In Michigan, 4 of 13 persons who had accompanied the cases to the implicated restaurants and had eaten one of the implicated foods had cramps and diarrhea without blood in one to seven days, whereas none of 12 who had not eaten these foods were ill (P = 0.06, Fisher's exact test, one-tailed).

There were no obvious defects in equipment or food-handling practices in the Oregon or Michigan restaurants. In Michigan, inadequate stock rotation of some foods was observed in Restaurant 2, and during busy periods certain parts of the grill were cooler than the temperature standard established by Chain A.

Laboratory Investigation

E. coli 0157:H7, *B. pumilus*, and *K. oxytoca* were the only bacteria isolated from three or more cases in either outbreak. *E. coli* 0157:H7 was recovered from stool of three of six Oregon cases and none of 10 neighborhood controls (P = 0.03, Fisher's exact test, one-

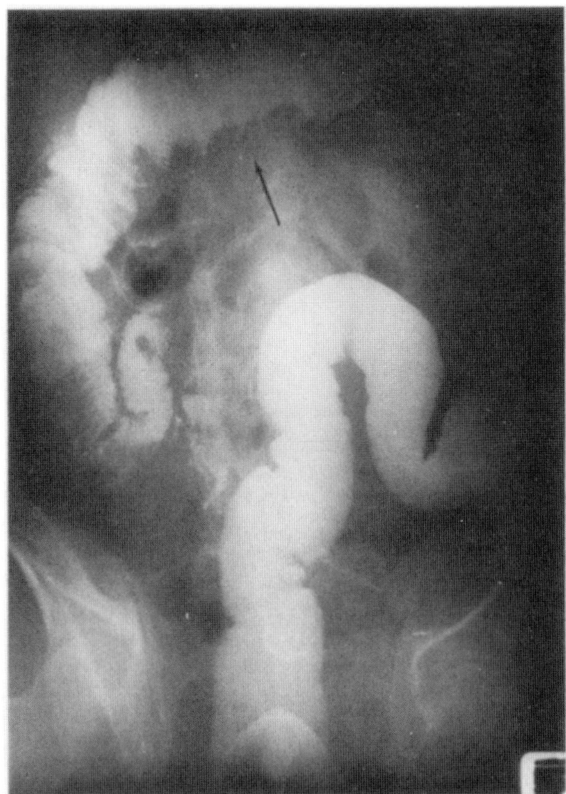

Figure 2. Barium-Enema Radiograph in a Case of Hemorrhagic Colitis.

The area of "thumbprinting" (arrow) in the transverse colon suggests submucosal edema, a characteristic finding in this disease.

Table 1. Distribution of Symptoms in 43 Cases of Hemorrhagic Colitis.

Symptom	Oregon	Michigan	Combined
	(N = 25)	(N = 18)	(N = 43)
	per cent with symptoms *		
Bloody diarrhea †	100	100	100
Abdominal cramps †	100	100	100
Nausea	60	67	63
Vomiting	44	61	49
Chills	28	33	30
URI symptoms ‡	12	28	19
Fever (>38°C)	8	6	7

*Based on all cases with complete clinical history.

†Part of case definition.

‡Upper-respiratory-tract infection.

tailed). An additional patient with *E. coli* 0157:H7 was identified by screening *E. coli* isolates sent to the CDC from Oregon. *K. oxytoca* was isolated from two cases and no controls (it was isolated in Oregon from one of the controls); *B. pumilus* was isolated from two cases and one control. *Y. enterocolitica* was recovered from one case only on cold enrichment, suggesting small numbers of organisms. Examination of the 45 control emergency-room diarrheal stool specimens yielded *K. oxytoca* from two samples and *E. coli* 0157:H7 from none. In the virologic studies we did not detect viral particles on electron microscopy or immunoelectron microscopy or in tissue cultures.

E. coli 0157:H7 was recovered from 6 of 14 specimens from cases and none of 4 specimens from controls in Michigan. *K. oxytoca* and *B. pumilus* were each isolated from three cases. In both states combined, *E. coli* 0157:H7 was isolated from 9 of 12 stools collected within four days after the onset of illness, but from none of 7 stools collected seven or more days after onset (P = 0.002, Fisher's exact test, one-tailed). In both outbreaks 0157:H7 was the predominant *E. coli* isolated (median of four of five isolates serotyped). The serotypes of *K. oxytoca* from patients' stools were different. The *E. coli* and *K. oxytoca* were sensitive to all antimicrobial agents tested, and the *B. pumilus* had two different antimicrobial-susceptibility patterns.

E. coli 0157:H7 was not isolated from food samples collected in Oregon but was isolated from a frozen, raw, standard-size hamburger patty from a suspected lot used at the Michigan restaurants during the outbreak period. This patty had been stored at a processing plant in another state, as part of a quality-control program, and had never been in either restaurant.

E. coli 0157:H7 did not produce either heat-labile or heat-stable enterotoxin, nor were isolates invasive on Sereny testing or tissue-culture assays. The organism, however, did produce a nonbloody diarrhea in infant rabbits. The laboratories of the U.S. Department of Agriculture and Pennsylvania State University reported no *E. coli* 0157:H7 from animals in the United States. The CDC laboratory detected only one strain of *E. coli* 0157:H7 among over 3000 *E. coli* organisms serotyped since 1973; it was isolated from a 50-year-old California woman in 1975 during an acute, self-limited afebrile illness with severe abdominal cramps followed by grossly bloody diarrhea.

Discussion

Two outbreaks of a clinically distinctive diarrheal illness occurred three months apart in two widely separated areas of the country among persons who had eaten at restaurants of a single fast-food chain. A rare *E. coli* serotype, 0157:H7, was isolated from ill patients in both outbreaks and from a retained specimen of hamburger patty from a suspected lot in the Michigan outbreak. We hypothesize that on two occasions, *E. coli* 0157:H7 contaminated the meat before it was made into hamburger patties, survived the cooking procedures at the restaurants, and caused illness

among some people who ate the meat. Apparently, failure to use meat by lots permitted cases to occur during a prolonged period in both outbreaks. The low attack rate in both outbreaks suggested a low level of contamination of a large volume of raw meat, reduction of the inoculum by cooking, or unknown host susceptibility factors. The pathogenesis of the illness and the source of contamination of the raw meat by this *E. coli* serotype are not known.

The evidence that *E. coli* 0157:H7 was the etiologic agent in these outbreaks may be summarized as follows. In two outbreaks this serotype was isolated from ill persons but not from healthy persons, ill persons with other forms of diarrhea, or patients who had completely recovered from bloody diarrhea. The only other isolate of *E. coli* 0157:H7 identified at the CDC was from a patient who had an identical disease in 1975. Preliminary studies indicate that this *E. coli* produces nonbloody diarrhea in infant rabbits, whereas *E. coli* isolates from two human control stool specimens do not (Potter ME: personal communication).

The clinical presentation of this illness may be distinguished from that of the bloody diarrhea or dysentery described in shigellosis, amebiasis, campylobacteriosis, or invasive *E. coli* gastroenteritis by the lack of fever and the bloody discharge resembling lower gastrointestinal bleeding. However, like other causes of bloody diarrhea, *E. coli* 0157:H7 may produce a spectrum of illness. In Michigan there was evidence of nonbloody diarrhea among persons who ate the same foods implicated at restaurants of Chain A. Our case definition, which required the presence of bloody diarrhea, would have excluded milder cases of this illness.

The case report of bloody diarrhea associated with *E. coli* 0157:H7 in 1975 suggests that this disease has occurred sporadically in the past. Descriptions of sporadic cases of a similar illness, frequently associated with antimicrobial use and distinct from pseudomembranous colitis, have been reported from several countries, including Japan and the United States.[1-5] Sakurai et al. reported that colonoscopy revealed areas of diffuse mucosal hemorrhage or erosion, mostly in the right colon; biopsy specimens of the mucosa showed little or no inflammatory change.[1] In the Oregon and Michigan outbreaks, only one patient in Oregon had been receiving an antibiotic (penicillin) before the onset of illness. Other reports have described cases of hemorrhagic colitis not associated with antimicrobial agents[1] or cases of so-called ischemic colitis in young adults.[11] Some of these may represent sporadic cases of the same illness that we have described in these outbreaks.

Acute hemorrhagic enterocolitis has been reported in patients with *K. oxytoca* isolated from their stools,[12] but no control stool cultures were obtained. In the two outbreaks that we investigated, strains of *K. oxytoca* were of several serotypes and were isolated from cases and controls. Although *B. pumilus* can induce enterocolitis in guinea pigs given clindamycin,[13] the *B. pumilus* isolates obtained in the two outbreaks were isolated from both cases and controls and appeared to include more than one strain.

E. coli can cause diarrhea by direct invasion of the intestinal mucosa and by elaboration of heat-stable enterotoxins or of heat-labile resembling cholera toxin.[14-16] Our laboratory studies have shown that strain 0157:H7 does not cause disease by these mechanisms. Strains of *E. coli* that cause diarrhea by poorly defined mechanisms have been studied, but they do not typically cause bloody diarrhea.[17-19] *E. coli* 0157:H7 may cause diarrhea by an as yet unknown mechanism, perhaps by the production of previously unrecognized enterotoxins.

Isolation of *E. coli* 0157:H7, a rare serotype, from cases in two outbreaks of bloody diarrhea, from the suspected vehicle for the outbreaks, and from a sporadic case of bloody diarrhea in 1975 strongly suggests, but does not prove, that it caused the illness; proof may require studies using animal models and perhaps human volunteers. Similarly, the epidemiology, clinical spectrum, and pathogenesis of this unusual illness and the reservoir of the putative etiologic agent are still poorly understood or unknown and require continued clinical, epidemiologic, and laboratory studies.

We are indebted to Dr. Laurence Foster and Mr. Robert Sokolow of the Oregon State Division of Health; to Dr. William Hall of the Michigan Department of Public Health and to Mrs. Connie Courtade, Mr. Thomas Roberts, Mr. Gary Stevens, and Dr. Taira Fukushima of the county health departments; to Dr. John Walker, Dr. David Martin, and other members of the hospital staff who cared for the patients, for their assistance in the epidemiologic investigation; to Dr. George Morris, Ms. Cheryl Bopp, Dr. J. J. Farmer III, Ms. Nancy Puhr, Ms. Janice Haney, Dr. George Lombard, Dr. Otto Nunez, Mr. George Marchetti, Dr. Milford Hatch, Dr. William Gary, and Dr. Morris Potter at the CDC, for laboratory assistance; to Ms. Barbara Strickland for data programming; to Mrs. Dot Anderson and Ms. Jesse Furman for the preparation of the manuscript; and to Dr. Roger A. Feldman for advice and criticism.

REFERENCES

1. Sakurai Y, Tsuchiya H, Ikegami F, Funatomi T, Takasu S, Uchikoshi T. Acute right-sided hemorrhagic colitis associated with oral administration of ampicillin. Dig Dis Sci. 1979; 24:910-5.
2. Toffler RB, Pingoud EG, Burrell MI. Acute colitis related to penicillin and penicillin derivatives. Lancet. 1978; 2:707-9.
3. Barrison IG, Kane SP. Penicillin-associated colitis. Lancet. 1978; 2:843.
4. Dickinson RJ, Meyer P, Warren RE. Hemorrhagic colitis. Dig Dis Sci. 1982; 27:187.
5. Pittman FE, Pittman JC, Humphrey CD. Colitis following oral lincomycin therapy. Arch Intern Med. 1974; 134:368-72.
6. Breslow NE, Day NE. Statistical methods in cancer research. Lyons: International Agency for Research on Cancer, 1980:84-119. (IARC scientifiic publication no. 32).
7. Lennette EH, Balows A, Hausler WJ Jr, Truant JP, eds. Manual of clinical microbiology. 3d ed. Washington, D.C.: American Society for Microbiology, 1980.
8. Melnick JL, Wenner HA, Phillips CA. Enteroviruses. In: Lennette EH, Schmidt NJ, eds. Diagnostic procedures for viral, rickettsial, and chlamydial infections. 5th ed. Washington, D.C.: American Public Health Association, 1979:471-534.
9. Kapikian AZ, Yolken RH, Greenberg HB, et al. Gastroenteritis viruses. In: Lennette EH, Schmidt NJ, eds. Diagnostic procedures for viral, rickettsial, and chlamydial infections. 5th ed. Washington, D.C.: American Public Health Association, 1979:927-95.
10. Morris GK, Merson MH, Sack DA, et al. Laboratory investigation of diarrhea in travelers to Mexico: evaluation of methods for detecting enterotoxigenic *Escherichia coli*. J Clin Microbiol. 1976; 3:486-95.
11. Clark AW, Lloyd-Mostyn RH, Sadler MR de C. "Ischaemic" colitis in young adults. Br Med J. 1972; 4:70-2.
12. Totani T. [Acute hemorrhagic enteritis by *Klebsiella oxytoca*.] Nippon Rinsho. 1978; 36:1308-9. (Japanese).
13. Brophy PF, Knoop FC. *Bacillus pumilus* in the induction of clindamycin-associated enterocolitis in guinea pigs. Infect Immun. 1982; 35:289-95.
14. DuPont HL, Formal SB, Hornick RB, et al. Pathogenesis of *Escherichia coli* diarrhea. N Engl J Med. 1971; 285:1-9.
15. Guerrant RL, Moore RA, Kirschenfeld PM, Sande MA. Role of toxigenic and invasive bacteria in acute diarrhea of childhood. N Engl J Med. 1975; 293:567-73.
16. Rudoy RC, Nelson JD. Enteroinvasive and enterotoxigenic *Escherichia coli*: occurrence in acute diarrhea of infants and children. Am J Dis Child. 1975; 129:668-72.
17. Ulshen MH, Rollo JL. Pathogenesis of *Escherichia coli* gastroenteritis in man — another mechanism. N Engl J Med. 1980; 302:99-101.
18. Cantey RJ, Blake RK. Diarrhea due to *Escherichia coli* in the rabbit: a novel mechanism. J Infect Dis. 1977; 135:454-62.
19. Levine MM, Bergquist EJ, Nalin DR, et al. *Escherichia coli* strains that cause diarrhoea but do not produce heat-labile or heat-stable enterotoxins and are non-invasive. Lancet. 1978; 1:1119-22.

Unidentified Curved Bacilli in the Stomach of Patients with Gastritis and Peptic Ulceration

B. J. MARSHALL AND J. R. WARREN

In this paper, Marshall and Warren confirmed an infectious etiology for diseases long thought to be expressions primarily of psychosomatic disorders: antral gastritis and gastric and duodenal ulcers. The authors were careful enough in their report to postulate an important role for "a new species related to the genus *Campylobacter*" in the etiology of these diseases. Actually, Robin Warren had noticed curved bacteria in gastric biopsy specimens as early as 1979, and similar organisms had occasionally been observed by European pathologists. Warren and a resident in internal medicine, Barry Marshall, undertook the first systematic study of the phenomenon and noticed a significant association between duodenal and gastric ulcers, as well as chronic active gastritis and "unidentified curved bacilli" in the mucus layer overlying the gastric mucosa (*Lancet* **i:**1273–1275, 1983). The paper cited here confirmed this association and the lack of bacteria in histologically normal biopsies, and it reported the successful cultivation of a *Campylobacter*-like organism under microaerophilic conditions after 3 days of incubation—something no one had done earlier.

Later, experimental infections, positive serologies, and elimination of the organism through antimicrobial therapy proved the etiological role of what is now called *Helicobacter pylori*. The discovery of this bacterium not only led to the cure of thousands of ulcer patients, but also spawned a search for other *Helicobacter* species and has recently led to the recognition of *H. pylori* as a carcinogen associated with adenocarcinoma, non-Hodgkin's lymphoma, and mucosa-associated lymphoid tissue lymphoma of the stomach.

ALEXANDER VON GRAEVENITZ

The Lancet · Saturday 16 June 1984

UNIDENTIFIED CURVED BACILLI IN THE STOMACH OF PATIENTS WITH GASTRITIS AND PEPTIC ULCERATION*

BARRY J. MARSHALL   J. ROBIN WARREN

*Departments of Gastroenterology and Pathology,
Royal Perth Hospital, Perth, Western Australia*

Summary   Biopsy specimens were taken from intact areas of antral mucosa in 100 consecutive consenting patients presenting for gastroscopy. Spiral or curved bacilli were demonstrated in specimens from 58 patients. Bacilli cultured from 11 of these biopsies were gram-negative, flagellate, and microaerophilic and appeared to be a new species related to the genus *Campylobacter*. The bacteria were present in almost all patients with active chronic gastritis, duodenal ulcer, or gastric ulcer and thus may be an important factor in the aetiology of these diseases.

Introduction

GASTRIC spiral bacteria have been repeatedly observed, reported, and then forgotten for at least 45 years.[1-3] In 1940 Freedburg and Barron stated that "spirochaetes" could be found in up to 37% of gastrectomy specimens,[4] but examination of gastric suction biopsy material failed to confirm these findings.[5] The advent of fibreoptic biopsy techniques permitted biopsy of the antrum, and in 1975 Steer and Colin-Jones observed gram-negative bacilli in 80% of patients with gastric ulcer.[6] The curved bacilli they illustrated were said to be *Pseudomonas,* possibly a contaminant, and the bacteria were once more forgotten. The repeated demonstration of these bacteria in inflamed gastric antral mucosa[7] prompted us to do a pilot study in twenty patients. Typical curved bacilli were present in over half the biopsy specimens and the number of bacteria was closely related to the severity of the gastritis. The present study was designed to confirm the association between antral gastritis and the bacteria, to discover associated gastrointestinal diseases, to culture and identify the bacteria, and to find factors predisposing to infection.

---

*Based on paper read at Second International Workshop on Campylobacter Infections (Brussels, 1983).

Patients and Methods

Patients

All patients referred for gastroscopy on clinical grounds were eligible for the study which continued until there were 100 participants who gave informed consent and in whom biopsy was considered to be safe. The study was approved by our hospital's human rights committee.

Questionnaire

Where possible patients completed a clinical questionnaire designed to detect a source of infection or show any relationship with "known" causes of gastritis or *Campylobacter* infection, rather than give a detailed account of each patient's history. The emphasis was on animal contact, travel, diet, dental hygiene, and drugs, rather than symptoms.

Endoscopy

The gastroscopies were done by colleagues at the Royal Perth Hospital. Participants fasted for at least 4 h before endoscopy. An Olympus GIF-K fibreoptic gastroduodenoscope was used. Routine biopsies were done when indicated. For the study two extra specimens were taken from an area of intact antral mucosa, at a distance from any focal lesion such as an antral ulcer. When the mucosa appeared inflamed the specimens were taken from a red area, otherwise any part of the antrum was used. One biopsy was immediately fixed in phosphate-buffered formalin for histological examination, the other was placed in chilled anaerobic transport medium and taken to the microbiology laboratory within 1 h. In a few cases an extra specimen was taken for ultrastructural examination.

The gastroenterologist dictated his report soon after the endoscopy. We had not planned to analyse these reports so a standard terminology was not used and no special attention was paid to minor endoscopic lesions. Findings of doubtful clinical significance, such as mild endoscopic gastritis or duodenogastric bile reflux, may thus have been under-reported. (Hereafter the term "gastritis" refers to a histological grade of chonic gastritis unless stated otherwise.) Before we analysed the data, the endoscopy reports were coded for the major diagnoses.

Histopathology

Sections were stained with haematoxylin and eosin (H & E) and graded for gastritis (by J. R. W.) as 0 (normal), inflammatory cells rarely seen; 1 (normal), lymphoid cells present but within normal limits and with no other evidence of inflammation (see below); 2 (chronic), chronic gastritis; or 3 (active), active chronic gastritis.

8390 © The Lancet Ltd, 1984

Gradings were based solely on the type of inflammatory cells. Other types of mucosal change, such as gland atrophy or intestinal metaplasia, were noted separately, but were not used as evidence of inflammation. "Chronic gastritis" indicated inflammation with no increase in polymorphonuclear leucocytes (PMNs). There were either increased numbers of lymphoid cells or normal cell numbers with other evidence of inflammation such as oedema, congestion, or cell damage. The term "active" was used to indicate an increase in PMNs.[8] The gastritis was considered active if a few PMNs infiltrated one gland neck or pit, if occasional PMNs were scattered throughout the superficial epithelium, or if there was an obvious increase in PMNs in the lamina propria.

Later, sections stained with Warthin-Starry silver stain were examined for small curved bacilli on the surface epithelium. Numbers of bacteria were graded as 0, no characteristic bacteria; 1, occasional spiral bacteria found after searching; 2, scattered bacteria in most high-power fields or occasional groups of numerous bacteria; or 3, numerous bacteria in most high-power fields.

Microbiology

Tissue smears were Gram stained and examined for curved bacilli resembling *Campylobacter*. The remaining tissue was minced, plated on non-selective blood and chocolate agar, and cultured at 37°C under microaerophilic conditions as used for *Campylobacter* isolation.[9] At first plates were discarded after 2 days but when the first positive plate was noted after it had been left in the incubator for 6 days during the Easter holiday, cultures were done for 4 days.

Analysis of Results

Questionnaires, gastroscopy reports, and histopathology and microbiology results were coded independently in separate departments. Complete results for individual patients were not known until the statistician had received all the data. The findings were tested for positive correlation with the presence of either bacteria or gastritis, by the chi-squared method. Fisher's exact test of significance was used for all the 2 × 2 tables in this paper.

Results

In 12 weeks 184 patients were examined by the gastroenterology unit. Of the 84 patients excluded, 5 refused consent, 4 had contraindications to biopsy, and 75 patients, mostly unbooked cases, could not be invited to participate. These patients closely matched the study group for age, sex, and incidence of peptic ulcers (table I).

Questionnaires

99 patients completed the questionnaires. The only symptom which correlated with gastritis or bacteria was "burping" which was more common in patients with bacteria ($p = 0.03$) or gastritis ($p = 0.007$). This association remained when patients with peptic ulcer were excluded. None of the other questionnaire responses showed any relationship to the presence of gastric bacteria or gastritis.

Endoscopy

There was a very close correlation between both gastric ulcer and duodenal ulcer and the presence of the bacteria (table II). Most patients with peptic ulcer also had gastritis ($29/31$; $p = 0.0002$).

TABLE I—COMPARISON OF PARTICIPANTS WITH EXCLUDED PATIENTS

—	Study group (n = 100)	Exclusions (n = 84)
Mean age (range)	55 (20–88) yr	57 (18–88) yr
Males	63 (63%)	55 (65%)
Females	37 (37%)	29 (35%)
Gastric ulcer	22 (22%)	19 (23%)
Duodenal ulcer	13 (13%)	8 (10%)

TABLE II—ASSOCIATION OF BACTERIA WITH ENDOSCOPIC DIAGNOSES

Endoscopic appearance*	Total	With bacteria	p
Gastric ulcer	22	18 (77%)	0·0086
Duodenal ulcer	13	13 (100%)	0·00044
All ulcers	31	27 (87%)	0·00005
Oesophagus abnormal	34	14 (41%)	0·996
Gastritis†	42	23 (55%)	0·78
Duodenitis†	17	9 (53%)	0·77
Bile in stomach	12	7 (58%)	0·62
Normal	16	8 (50%)	0·84
Total	100	58 (58%)	

*More than one description applies to several patients (eg, 4 patients had both gastric and duodenal ulcers).
†Refers to endoscopic appearance, not histological inflammation.

TABLE III—HISTOLOGICAL GRADING OF GASTRITIS AND BACTERIA

Gastritis	Bacterial grade				
	Nil	1+	2+	3+	Total
Normal*	29	2	0	0	31
Chronic	12†	9	7	1	29
Active	2	5	15	18	40
Total	43	16	22	19	100

*Gastritis grades 0 and 1 normal.
†1 case showed bacteria on gram stained smear.

TABLE IV—RELATION BETWEEN GASTRITIS AND BACTERIA IN PATIENTS WITHOUT PEPTIC ULCER

Gastritis	Bacteria		
	No	Yes	Total
Normal	28	1	29
Chronic	8	12	20
Active	2	18	20
Total	38	31	69

Histopathology

Gastritis could usually be graded with confidence at low magnification. There was some difficulty with about 25 cases where the changes were mild or the specimens were small, superficial, or distorted. To ensure that gradings were reliable, single H & E sections from the last 40 cases were examined "blind" by another pathologist who agreed with the presence or absence of gastritis in 36 cases (90%), and gave an identical grading in 32.

Gradings for bacteria by silver staining were more straightforward. The bacteria stained well and were easily differentiated from contaminant bacteria or debris. Silver staining was the most sensitive method of detecting the spiral bacteria. Silver stained sections and Gram stained smears were both done in 96 cases and spiral bacteria were seen in 56 of them; 32 with both stains, 23 with silver alone, and 1 case with the Gram stain alone.

The correlation between gastritis and bacteria, defined by Gram and/or by silver staining, was remarkable (table III). Gastritis was present in 55/57 biopsy specimens with bacteria ($p = 2 \times 10^{-12}$). When the 31 patients with peptic ulcer were excluded, the correlation persisted, implying that the presence of bacteria was not secondary to an ulcer crater (table IV).

Microbiology

Specimens for culture were received from 96 patients and 11 were culture positive, all being seen with Gram and silver staining also. No spiral bacteria were grown from the first 34 cases, probably because the cultures were discarded too soon.

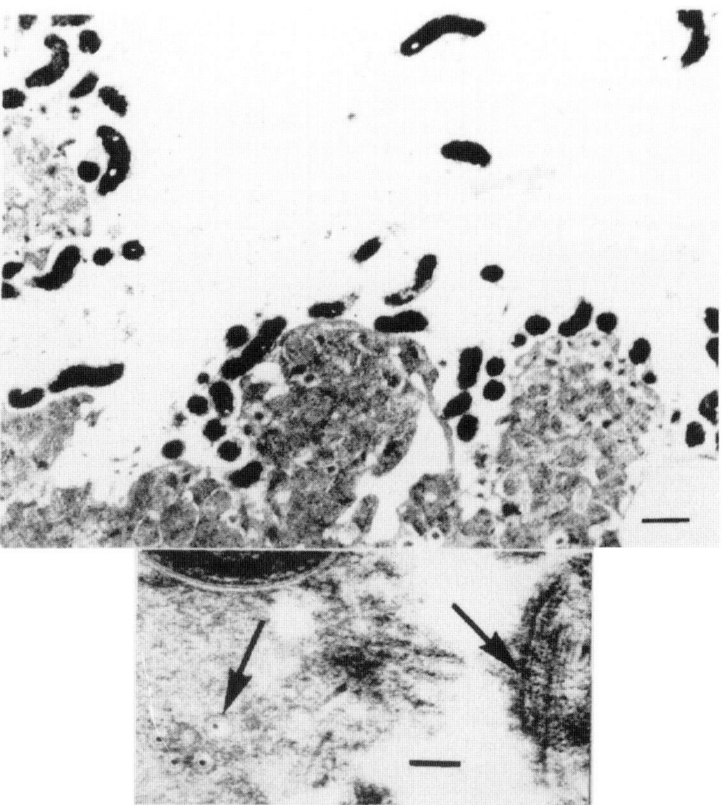

Electron micrograph from a mucosal biopsy with active chronic gastritis.

Upper: many profiles of sectioned pyloric campylobacter are located on the luminal aspect of mucus-secreting epithelial cells; plasma membranes are intact, but indented and almost devoid of microvilli (bar = 1 μm).

Lower: at higher magnification groups of transversely and longitudinally cut sheathed flagella are visible (arrows; bar = 100 nm).

The bacteria were S-shaped or curved gram-negative rods, 3 μm × 0·5 μm, with up to 1½ wavelengths. In electron micrographs they had smooth coats and there were usually four sheathed flagella arising from one end of the cell. They grew best in a microaerophilic atmosphere at 37°C; a campylobacter gas generating kit was sufficient (Oxoid BR56). Moist chocolate or blood agar was the preferred medium. Growth was evident in 3 days as 1 mm diameter non-pigmented colonies. In artificial media the bacteria were usually larger and less curved than those seen on Gram stains of fresh tissue. They formed coccoid bodies in old cultures. The bacteria were oxidase +, catalase +, H_2S +, indole −, urease −, nitrate −, and did not ferment glucose. They were sensitive to tetracycline, erythromycin, kanamycin, gentamicin and penicillin, and resistant to nalidixic acid. DNA base analysis gave a guanine + cytosine content of 36 mol%, a value in the range for campylobacters.

Sources of Bias

The patient sample was from a defined population with gastric symptoms expected to have some gastroenterological abnormality. The biopsy tissue studied was from apparently intact mucosa—ie, not the sort of specimen a pathologist usually sees. We attempted to limit bias by making the study consecutive and blind, and were partly successful. The study was not strictly consecutive since 84 patients had to be excluded. However, gastroscopy reports and laboratory investigations were completed serially and usually independently ("blind") except that clinically relevant material was sent (to J. R. W.) with study biopsies, mainly from cases of gastric ulcer. However, an independent blind assessment of gastritis in 40 cases matched the study results well.

Discussion

The spiral bacteria of the human gastric antrum have never been cultured before, and their association with active chronic gastritis has not been described. They are a new species closely resembling campylobacters morphologically and in respect of atmospheric requirements and DNA base composition, but their flagellar morphology is not that of the genus *Campylobacter*.[9] Campylobacters have a single unsheathed flagellum at one or both ends of the cell whereas the new organism has four sheathed flagella at one end.[7,10] If it is premature to talk of "*Campylobacter pyloridis*"[11] perhaps the name "pyloric campylobacter" will do to define the site where these organisms are commonly found and to indicate the similarity to known *Campylobacter* spp.

There was no well-defined clinical syndrome associated with pyloric campylobacter. Only "burping" was significantly associated. Others have described this symptom in patients with non-ulcer dyspepsia and PMN infiltration of the antrum is also common in such patients.[12,13] We expected abdominal pain to correlate with pyloric campylobacter or gastritis, but it did not. Perhaps, since most patients undergoing gastroscopy have pain (75% in our study) the question "Do you have abdominal pain—yes or no?" was too general.

Diagnostic Microbiology and Epidemiology 83

Much of the questionnaire was designed to select likely sources or causes of pyloric campylobacter infection. For example, bacteria might have colonised patients who already had gastritis and were taking antacids, milk, or cimetidine, thus impairing their "gastric acid barrier" and predisposing them to infection.[14] Animal contact and carious teeth were also considered as sources of infection. Campylobacters are commensals of domestic and farm animals (*C coli*, *C jejuni*), and they also inhabit the human mouth (*C sputorum* ss *sputorum*).[15] We found no evidence that any of these factors predisposed to the infection.

The absence of a relation between "known causes" of gastritis and the presence of histological gastritis has been noted by others. For example, analgesic abusers often have no gastritis, even when a gastric ulcer is present;[16] alcohol consumption is not clearly related to gastritis;[17] the quantity of bile in the stomach (duodenogastric reflux) is not obviously related to the state of gastric mucosa;[18] autoimmune disease is an unlikely cause, since gastric autoantibodies are uncommon except in pernicious anaemia, where the main histological changes are in the body of the stomach, not the antrum.[19] Gastric ulcer seems an unlikely primary cause of antral gastritis because the gastritis remains after successful treatment of the ulcer with cimetidine or carbenoxolone, and gastritis is just as common in patients with duodenal ulcer as with gastric ulcer.[6,20-23] Thus, the aetiology of chronic gastritis remains uncertain.

We have found a close association between pyloric campylobacter and antral gastritis. When PMN infiltrated the mucosa the bacteria were almost always present (38/40). In the absence of inflammation they were rare (2/31), suggesting that they are not commensals. The bacteria were not cultured unless the patient had histological evidence of both gastritis and pyloric campylobacter. We know of no other disease state where, in the absence of complicating factors such as ulceration (table IV), bacteria and PMNs are so intimately related without the bacteria being pathogenic.

How does pyloric campylobacter survive? The bacteria were usually in close contact with the mucosa, often in grooves between cells, within acinus-like infoldings of the epithelium or within the mucosal pits (figure). The surface mucus coating was superficial to the bacteria and any foreign material or organisms from the oral flora were present above the mucus, rarely mixed with it, and not beneath it: the mucus appeared to form a stable layer over the spiral bacteria. The antrum secretes mainly mucus, and the deeper levels of the surface mucus coating are slightly alkaline.[24] Thus pyloric campylobacter grows in a near-neutral environment, in close contact with the mucosa and protected from the bactericidal gastric juice. The absence of these bacteria from past reports of gastric microbiology may be because only gastric juice was cultured.[25,26] Even salmonellae cannot survive the low intragastric pH for more than a few minutes.[14] Where gastric biopsy material has been cultured,[6,27,28] microaerophilic techniques were not used and pyloric campylobacter did not grow.

Peptic ulcer was the only endoscopic finding associated with histological gastritis and pyloric campylobacter. This was surprising since the bacteria were not prominent on gastric ulcer borders and in duodenal ulcer no correlation would be expected. Perhaps the mucus coating is deficient or unstable near ulcer borders, thus allowing damage to the bacteria as well as the mucosa. Within a few millimetres of an ulcer, both pyloric campylobacter and gastritis were usually present. Other studies have shown continuing gastritis after ulcer healing with cimetidine and we have observed the persistence of pyloric campylobacter colonisation in such patients. The failure of the H_2 receptor antagonists to prevent ulcer relapse is attributed to an underlying ulcer diathesis which is unaffected by therapy. A bacterial aetiology, with continuing gastritis, could be the explanation. The diathesis may be a myth. Of ulcer-healing agents the only one thought to improve relapse rates is tripotassium dicitrato-bismuthate.[29] This compound is bactericidal to pyloric campylobacter and in patients treated with it the gastritis improved and the bacteria disappeared.[30]

The aetiology of peptic ulceration is unknown but until now a bacterial cause has not really been considered. We have found colonisation of the gastric antrum with pyloric campylobacter in over half of a series of cases at routine endoscopy. The bacteria were present almost exclusively in patients with chronic antral gastritis and were also common in those with peptic ulceration of the stomach or duodenum. Although cause-and-effect cannot be proved in a study of this kind, we believe that pyloric campylobacter is aetiologically related to chronic antral gastritis and, probably, to peptic ulceration also.

We thank Dr T. E. Waters, Dr C. R. Sanderson, and the gastroenterology unit staff for the biopsies, Miss Helen Royce and Dr D. I. Annear for the microbiological studies, Mr Peter Rogers and Dr L. Sly for supplying the G & C data, Dr J. A. Armstrong for the electron microscopy, Dr R. Glancy for reviewing slides, Miss Joan Bot for the silver stains, Mrs Rose Rendell of Raine Medical Statistics Unit UWA, and Ms Maureen Humphries, secretary, and, for travel support, Fremantle Hospital.

Correspondence should be addressed to: B. M., Department of Microbiology, Fremantle Hospital, PO Box 480, Fremantle 6160, Western Australia.

REFERENCES

1. Doenges JL. Spirochaetes in gastric glands of macacus rhesus and humans without definite history of related disease. *Proc Soc Exp Biol Med* 1938; **38**: 536–38.
2. Ito S. Anatomic structure of the gastric mucosa. In: Heidel US, Cody CF, eds. Handbook of physiology, section 6: Alimentary canal, vol II: secretion. Washington, DC: American Physiological Society, 1967: 705–41.
3. Fung WP, Papadimitriou JM, Matz LR. Endoscopic, histological and ultrustructural correlations in chronic gastritis. *Am J Gastroenterol* 1979; **71**: 269–79.
4. Freedburg AS, Barron LE. The presence of spirochaetes in human gastric mucosa. *Am J Dig Dis* 1940; **7**: 443–45.
5. Palmer ED. Investigation of the gastric spirochaetes of the human. *Gastroenterology* 1954; **27**: 218–20.
6. Steer HW, Colin-Jones DG. Mucosal changes in gastric ulceration and their response to carbenoxolone sodium. *Gut* 1975; **16**: 590–97.
7. Warren JR, Marshall B. Unidentified curved bacilli on gastric epithelium in active chronic gastritis. *Lancet* 1983; i: 1273–75.
8. Whitehead R, Truelove SC, Gear MWL. The histological diagnosis of chronic gastritis in fibreoptic gastroscope biopsy specimens. *J Clin Pathol* 1972; **25**: 1–11.
9. Kaplan RL. Campylobacter. In: Lenette E, Balows A, Hausler WJ, Truant JP, eds. Manual of clinical microbiology, 3rd ed. Washington, DC: American Society for Microbiology, 1980: 235–41.
10. Pead PJ. Electron microscopy of *Campylobacter jejuni*. *J Med Microbiol* 1979; **12**: 383–85.
11. Skirrow MB. Taxonomy and biotyping: Morphological aspects. In: Pearson AD, Skirrow MB, Rowe B, Davies JR, Jones DM, eds. Campylobacter II: Proceedings of the Second International Workshop on Campylobacter Infections. London: Public Health Laboratory Service, 1983: 36.
12. Crean GP, Card WI, Beattie AD, Holden RJ, James WB, Knill-Jones RP, Lucas RW, Spiegelhalter D. Ulcer-like dyspepsia. *Scand J Gastroenterol* 1982; **17** (suppl 79): 9–15.
13. Greenlaw R, Sheahan DG, Deluca V, Miller D, Myerson D, Myerson P. Gastroduodenitis: a broader concept of peptic ulcer disease. *Dig Dis Sci* 1980; **25**: 660–72.
14. Giannela RA, Broitman SA, Zamcheck N. Gastric acid barrier to ingested microorganisms in man: Studies in vivo and in vitro. *Gut* 1972; **13**: 251–56.
15. Blaser MJ, Reller LB. Campylobacter enteritis. *N Engl J Med* 1981; **305**: 1444–52.
16. MacDonald WC. Correlation of mucosal histology and aspirin intake in chronic gastric ulcer. *Gastroenterology* 1973; **65**: 381–89.
17. Wolff G. Does alcohol cause chronic gastritis? *Scand J Gastroenterol* 1970; **5**: 289–91.
18. Goldner FH, Boyce HW. Relationship of bile in the stomach to gastritis. *Gastrointest Endosc* 1976; **22**: 197–99.
19. Whitehead R. Mucosal biopsy of the gastrointestinal tract. In: Bennington JL, ed. Major problems in pathology: Vol III, 2nd ed. Philadelphia: WB Saunders, 1979: 15.
20. Gilmore HM, Forrest JAH, Pettes MR, Logan RFA, Heading RC. Effect of short and long term cimetidine on histological duodenitis and gastritis. *Gut* 1978; **19**: 981.
21. McIntrye RLE, Piris J. Truelove SC. Effect of cimetidine on chronic gastritis in gastric ulcer patients. *Aust NZ J Med* 1982; **12**: 106.
22. Schrager J, Spink R, Mitra S. The antrum in patients with duodenal and gastric ulcers. *Gut* 1967; **8**: 497–508.

References continued at foot of next column

B. J. MARSHALL AND J. R. WARREN: REFERENCES—*continued*

23. Magnus HA. Gastritis. In: Jones FA, ed. Modern trends in gastroenterology. London: Butterworth, 1952: 323–51.
24. Allen A, Garner G. Mucus and bicarbonate secretion in the stomach and their possible role in mucosal protection. *Gut* 1980; 21: 249–62.
25. Draser BS, Shiner M, McLeod GM. Studies on the intestinal flora I: The bacterial flora of the gastrointestinal tract in healthy and achlorhydric persons. *Gastroenterology* 1969; 56: 71–79.
26. Enander LK, Nilsson F, Ryden AC, Schwan A. The aerobic and anaerobic flora of the gastric remnant more than 15 years after Billroth II resection. *Scand J Gastroenterol* 1982; 17: 715–20.
27. Mackay IR, Hislop IG. Chronic gastritis and gastric ulcer. *Gut* 1966; 7: 228–33.
28. Rollason TP, Stone J, Rhodes JM. Spiral organisms in endoscopic biopsies of the human stomach. *J Clin Pathol* 1984; 37: 23–26.
29. Martin DF, May SJ, Tweedle DE, Hollanders D, Ravenscroft MM, Miller JP. Difference in relapse rates of duodenal ulcer after healing with cimetidine or tripotassium di-citrato bismuthate. *Lancet* 1980, i: 7–10.
30. Marshall B, Hislop I, Glancy R, Armstrong J. Histological improvement of active chronic gastritis in patients treated with De-Nol. *Aust NZ J Med* (in press) (abstr)

The Agent of Bacillary Angiomatosis. An Approach to the Identification of Uncultured Pathogens

D. A. RELMAN, J. S. LOUTIT, T. M. SCHMIDT, S. FALKOW, AND L. S. TOMPKINS

Although this paper appeared only 9 years ago, it seemed to us that it is a seminal one in that the authors used two molecular methods, 16S rRNA analysis and PCR, to "identify an uncultured pathogen." This development was fruitful not only for the elucidation of the causative agent of bacillary angiomatosis and cat-scratch disease (so far, *Bartonella henselae* and *Bartonella quintana* have been implicated) but also in the search for other microorganisms possibly involved in the etiology of other disorders (e.g., Whipple's disease).

The study proceeded from the assumption that a bacterial etiology of bacillary angiomatosis was very likely because the lesions found in this disease contained clusters of bacilli that stained with the Warthin-Starry silver stain and thus resembled organisms found in lesions of cat-scratch disease. The latter, a benign illness in immunologically normal hosts, may disseminate in immunocompromised ones, e.g., those with human immunodeficiency virus (HIV) infection. Tissues from three patients with bacillary angiomatosis yielded a 16S rRNA gene sequence that was most closely related to that of *Rochalimaea quintana*.

A companion paper in the same issue (p. 1587–1593) by Leonard Slater and collaborators reported the isolation of a motile, curved, gram-negative rod as the cause of fever and bacteremia in two patients with HIV infection. After lysis-centrifugation of the blood specimen, colonies were seen in 5 to 15 days on chocolate agar. They were biochemically inert and resembled *R. quintana* in their fatty acid composition.

The agents found in the two studies were later recognized as belonging to one genus, i.e., *Bartonella*. The fact that they can now be cultivated does not take away the importance of the paper by Relman et al., particularly because culturing is very time-consuming and has not always been successful.

ALEXANDER VON GRAEVENITZ

Reprinted with permission from *New England Journal of Medicine* 323:1573–1580. Copyright © 1990. Massachusetts Medical Society. All rights reserved.

The New England
Journal of Medicine

©Copyright, 1990, by the Massachusetts Medical Society

| Volume 323 | DECEMBER 6, 1990 | Number 23 |

THE AGENT OF BACILLARY ANGIOMATOSIS

An Approach to the Identification of Uncultured Pathogens

David A. Relman, M.D., Jeffery S. Loutit, M.B., Ch.B., Thomas M. Schmidt, Ph.D.,
Stanley Falkow, Ph.D., and Lucy S. Tompkins, M.D., Ph.D.

Abstract *Background.* Bacillary angiomatosis is an infectious disease causing proliferation of small blood vessels in the skin and visceral organs of patients with human immunodeficiency virus infection and other immunocompromised hosts. The agent is often visualized in tissue sections of lesions with Warthin–Starry staining, but the bacillus has not been successfully cultured or identified. This bacillus may also cause cat scratch disease.

Methods. In attempting to identify this organism, we used the polymerase chain reaction. We used oligonucleotide primers complementary to the 16S ribosomal RNA genes of eubacteria to amplify 16S ribosomal gene fragments directly from tissue samples of bacillary angiomatosis. The DNA sequence of these fragments was determined and analyzed for phylogenetic relatedness to other known organisms. Normal tissues were studied in parallel.

Results. Tissue from three unrelated patients with bacillary angiomatosis yielded a unique 16S gene sequence. A sequence obtained from a fourth patient with bacillary angiomatosis differed from the sequence found in the other three patients at only 4 of 241 base positions. No related 16S gene fragment was detected in the normal tissues. These 16S sequences associated with bacillary angiomatosis belong to a previously uncharacterized microorganism, most closely related to *Rochalimaea quintana.*

Conclusions. The cause of bacillary angiomatosis is a previously uncharacterized rickettsia-like organism, closely related to *R. quintana.* This method for the identification of an uncultured pathogen may be applicable to other infectious diseases of unknown cause. (N Engl J Med 1990; 323:1573-80.)

BACILLARY angiomatosis (also called epithelioid angiomatosis) and cat scratch disease are important clinical disorders that are presumed to be infectious on the basis of direct visualization of microorganisms in diseased tissues. Because these organisms cannot be grown reproducibly, the identity of the presumed pathogen or pathogens remains unknown.

Bacillary angiomatosis is a distinct vascular proliferative disease initially described in the skin and lymph nodes of patients seropositive for the human immunodeficiency virus (HIV).[1,2] Lesions from this disorder have been found to contain clusters of bacilli with positive results on Warthin–Starry staining that resemble those found in cat scratch disease,[3,4] although the histologic appearance of the disease differs from that of classic cat scratch disease.[5] Bacillary angiomatosis is known to cause disseminated visceral disease in HIV-seropositive patients and other immunocompromised hosts,[6-8] and it has recently been found to do so in an immunocompetent host.[9] Cat scratch disease is usually a benign granulomatous disease involving skin and regional lymph nodes[10,11]; however, disseminated disease has been described that involves visceral organs, particularly in immunocompromised hosts.[12-14] The causative agent of cat scratch disease was first visualized in 1983 with the Warthin–Starry silver stain[15] but has resisted identification despite reports of isolation in culture.[16] Distinguishing between disseminated bacillary angiomatosis and cat scratch disease in the immunocompromised host is difficult[17]; the ability to identify the causative bacteria would help resolve this difficulty as well as improve diagnosis and treatment and perhaps lead to a better understanding of pathogenesis.

Two recent experimental developments — analysis of 16S ribosomal RNA (rRNA)[18,19] and the polymerase chain reaction (PCR)[20-22] — have allowed a novel approach to the identification of disease-associated, uncultured microorganisms. All cells contain multiple copies of a gene encoding a 16S or 16S-like small subunit of rRNA, in which hypervariable sequences are interspersed with regions of highly conserved

From the Department of Microbiology and Immunology (D.A.R., S.F., L.S.T.) and the Division of Infectious Diseases, Department of Medicine (D.A.R., J.S.L., S.F., L.S.T.), Stanford University, Stanford, Calif., and the Department of Biology, Indiana University, Bloomington (T.M.S.). Address reprint requests to Dr. Relman at the Department of Microbiology and Immunology, Fairchild Building D309, Stanford University, Stanford, CA 94305-5402.

Supported by Public Health Service grants (AI 26195 [Dr. Falkow] and AI 23796 [Dr. Tompkins]) from the National Institutes of Health, a contract (N14-87-K-0813 [Dr. Schmidt]) with the Office of Naval Research, and an unrestricted gift from Praxis Biologics, Rochester, N.Y. Dr. Relman is a recipient of a Lucille P. Markey Scholar Award in Biomedical Science.

sequence.[19] The analysis of the variable portions permits the determination of phylogenetic and evolutionary relations among organisms and now forms the basis of a revised system of natural classification.[19] Sequences conserved throughout each or all three kingdoms (eukaryotes, archaeobacteria, and eubacteria) can serve as targets for primer-directed DNA amplification or hybridization.[23]

The PCR can be used to amplify a 16S-like eukaryotic rRNA sequence from purified genomic DNA.[24] It permits rapid amplification of any DNA region of interest if the sequences flanking the segment of interest are known, so that two amplification primers can be designed, and if there is an appropriate pool of DNA containing the target region.[21,25] Therefore, one might be able to identify an unknown eubacterium with primer-directed, amplified variable regions of its 16S rRNA sequence by using primers that anneal to flanking sequences conserved within the kingdom of eubacteria.[26] Although pure cultured organisms have usually been the source of the 16S ribosomal DNA (rDNA) target in PCR, infected human tissue could also be used.

A recent case of disseminated bacillary angiomatosis at our institution[8] provided us with the appropriate target DNA needed for the execution of this strategy. Using broad-range eubacterial 16S rDNA primers, we amplified, cloned, and sequenced a portion of eubacterial 16S rRNA gene directly from tissue infected with the presumed etiologic agent. Subsequent analysis of this sequence and study of infected tissues from other patients suggest that bacillary angiomatosis is caused by a rickettsia-like organism that is most closely related to *Rochalimaea quintana*.

METHODS

Patients

Patient 1 was a 38-year-old woman in whom fever, chills, headache, myalgias, nausea, and vomiting developed five years after cardiac transplantation while she was following a maintenance regimen of immunosuppressive drugs.[8] Three months earlier she had required surgical repair of a cat scratch. Abdominal CT scanning revealed hepatosplenomegaly with multiple areas of low attenuation and surrounding enhancement. Exploratory laparotomy and splenectomy were performed. Histologic examination of liver, spleen, and splenic hilar lymph node demonstrated a proliferation of histiocytoid endothelial cells forming vascular channels. Numerous bacillary organisms were seen on Warthin–Starry staining and electron microscopy (Fig. 1). Routine bacterial cultures were negative. A diagnosis of bacillary angiomatosis was made. The patient was treated with an eight-week course of erythromycin, with complete resolution of her symptoms.

Patient 2 was a 24-year-old HIV-seropositive man who had a two-month history of fever, chills, sweating, abdominal pain, anorexia, and weight loss.[8] He had had several cat scratches in the previous three months. Physical examination revealed a 1.0-cm purplish nodule on his left elbow, left-axillary lymphadenopathy, and hepatosplenomegaly. A skin biopsy, bone marrow aspiration and biopsy, and core liver biopsy were performed. Pathological examination of the skin was suggestive of pyogenic granuloma; however, Warthin–Starry staining revealed bacillary organisms. Sections of bone marrow showed focal areas of epithelioid angiomatosis, but there was insufficient material for Warthin–Starry staining. Histologic findings in the liver were consistent with epithelioid angioma-

tosis, with Warthin–Starry staining positive for bacillary organisms. All cultures were negative. Treatment with doxycycline led to prompt clinical improvement.

Patient 3 was a 54-year-old woman infected with HIV. A biopsy of a cutaneous heel lesion was performed, and Warthin–Starry staining demonstrated clumps of bacilli. Histologic findings suggested a diagnosis of bacillary angiomatosis. (This case has been described elsewhere by LeBoit et al. [Case 8].[5])

Patient 4 was a 29-year-old woman with the acquired immunodeficiency syndrome in whom diffuse lymphadenopathy developed. She had no skin lesions. Biopsy of an inguinal lymph node showed histologic features consistent with epithelioid angiomatosis. Clusters of bacilli were seen on Warthin–Starry staining.

DNA Extraction

DNA was extracted from a frozen splenic nodule and a frozen piece of splenic hilar lymph node from Patient 1, a frozen sample of bone marrow from Patient 2, paraffin-embedded, formalin-fixed tissue from a skin lesion from Patient 3, and a paraffin-embedded, formalin-fixed lymph node from Patient 4. Each of these patients had the biopsy performed at a different hospital. As a control for DNA carryover contamination, for every biopsy sample from the patients we prepared DNA extracts from at least one of the following tissues: frozen spleen from a patient with idiopathic thrombocytopenic purpura, with no clinical or cultural evidence of infection; a paraffin-embedded, formalin-fixed spleen sample obtained by splenectomy in a trauma victim; and frozen tonsil from two patients with recurrent tonsillitis.

Frozen Tissue

The methods used were modified from those described previously.[27,28] Approximately 100 mg of each tissue was rinsed in sterile water, minced, and placed in 5 ml of digestion buffer (500 mM TRIS, pH 9; 20 mM EDTA; 10 mM sodium chloride; and 1 percent sodium dodecyl sulfate) with fresh proteinase K (Sigma, St. Louis) at a final concentration of 1 mg per milliliter. The samples were incubated at 60°C for 48 to 72 hours. Ribonuclease A (Sigma) was added to a final concentration of 4 μg per milliliter; the samples were incubated at room temperature for two hours and then extracted three times with equal volumes of TRIS-buffered phenol and chloroform–isoamyl alcohol (24:1). The DNA was precipitated with ethanol according to standard techniques[29] and resuspended in 50 μl of sterile water.

Paraffin-Embedded Tissue

The method used was a modification of that published by Shibata et al.[30] and described by Wright and Manos.[31] Ten-micrometer sections from the various paraffin-embedded blocks of tissue were trimmed, treated with octane and ethanol, and then placed in 200 μl of digestion buffer as described.[31] After digestion and heat inactivation of proteinase K, the reaction mixture was centrifuged and the supernatant was used directly in amplification reactions.

Oligonucleotide Primers

The sequences of primers p11E and p13B were adapted with modifications from Chen et al.[26] Primers p93E, p24E, and p12B were designed on the basis of the 16S rRNA sequence determined in this study (Table 1). Each included either an *Eco*RI or *Bam*HI restriction-endonuclease site at the 5′ end to facilitate the cloning of the resultant PCR product. All primers were synthesized by Operon Technologies (Alameda, Calif.).

PCR Amplification

DNA extracts from frozen tissue were adjusted to a concentration of approximately 400 to 500 μg of total DNA per milliliter; 1 μl was used in a 100-μl reaction volume. Ten microliters (5 percent) of the paraffin-embedded DNA extract was sufficient for the amplification of human β-globin genes with primers PC04 and GH20 (Perkin-Elmer–Cetus, Norwalk, Conn.); the same amount was used for the amplification of eubacterial 16S rRNA. Each reaction contained 20

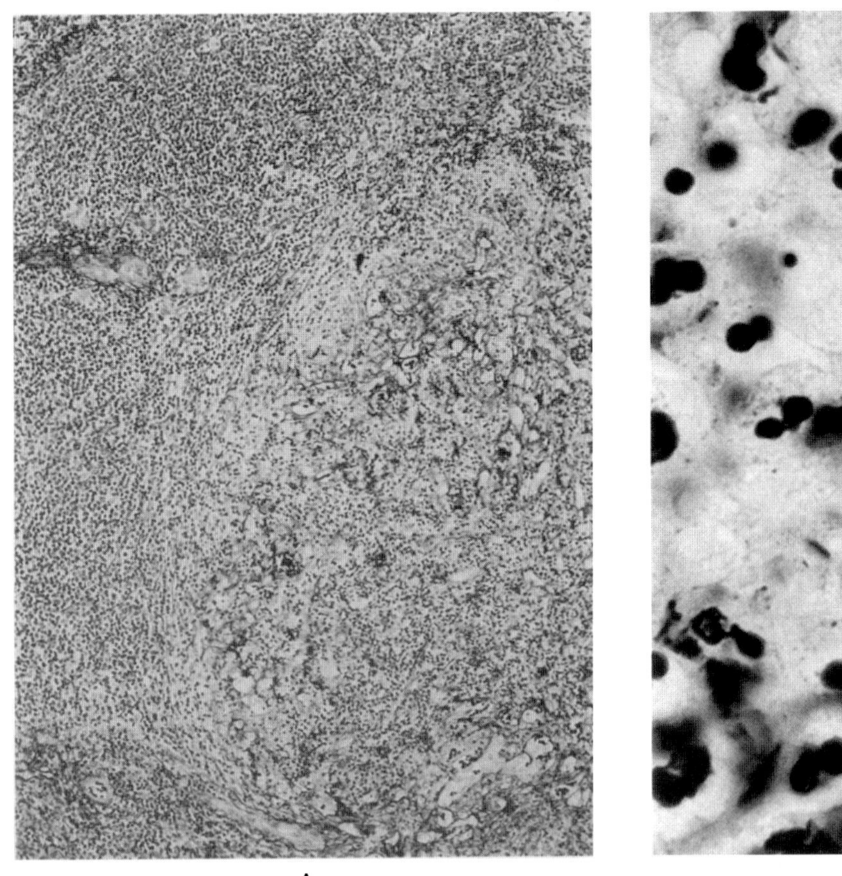
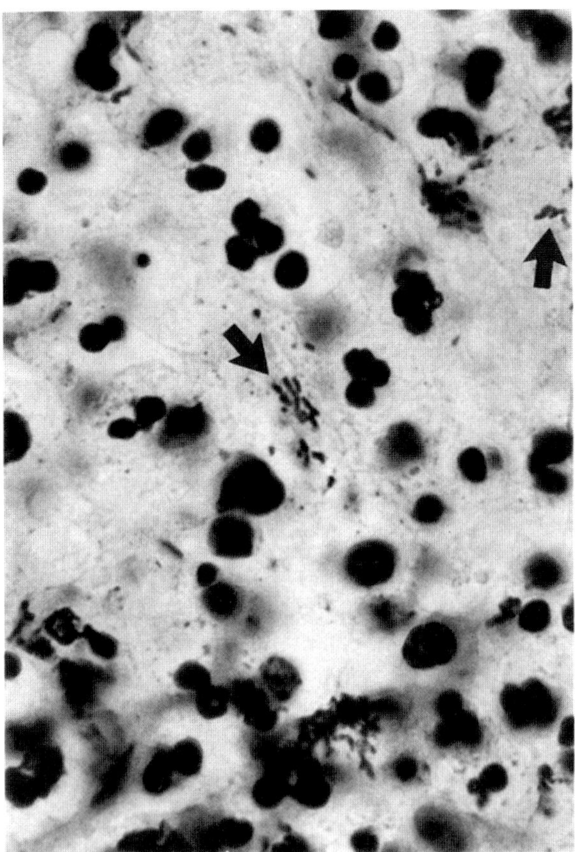

A   B

Figure 1. Splenic Nodule from Patient 1 after Warthin–Starry Staining.

In Panel A, inflammatory cells, focal areas of necrosis, and numerous epithelioid endothelial cells forming vascular channels can be seen (×670). In Panel B, at a higher magnification (×3300), pleomorphic bacillary organisms are apparent and often are in clumps (arrows). Similar organisms were seen in samples of splenic hilar lymph node and liver from the same patient.

pmol of each primer and standard amounts of Gene-Amp reagents (Perkin-Elmer–Cetus), except that 2 mM magnesium chloride was used for amplification with extracts of paraffin-embedded tissue. This mixture was irradiated with ultraviolet light to reduce contamination of reagents. An overlay of sterile mineral oil was added to the tubes, followed by the DNA template. PCR cycles consisted of 1 minute of denaturation at 94°C, 1 minute of annealing at 55°C, and 1.5 minutes of extension at 72°C in a programmable thermal controller (MJ Research, Watertown, Mass.). For amplification with extracts of paraffin-embedded tissue, we used a step file in an automated DNA thermal cycler (Perkin-Elmer–Cetus). Negative controls were included during every amplification and consisted of reaction mixtures without the DNA template, as well as reaction mixtures with DNA extracts from nondiseased tissues. The product of PCR was detected by electrophoresing 10 μl of reaction solution in a 1.2 percent agarose gel containing ethidium bromide; a 1-kilobase (kb) DNA ladder (GIBCO BRL, Gaithersburg, Md.) was used as the DNA size standard on each gel.

Purification, Cloning, and Sequencing of the PCR Product

Half the PCR product was extracted with phenol and chloroform, precipitated with ethanol, and pelleted. The 3' ends of DNA strands were filled in with DNA polymerase I large (Klenow) fragment.[29] The product was purified by low-melting-point agarose-gel electrophoresis, cut with a combination of *Eco*RI and *Bam*HI (New England Biolabs, Beverly, Mass.), and then ligated with pBluescript II KS phagemid vector (Stratagene, La Jolla, Calif.), which had been

cut with *Eco*RI and *Bam*HI.[29] *Escherichia coli* DH5α (Bethesda Research Laboratories, Bethesda, Md.) was electro-transformed[32] with the recombinant plasmids with a Bio-Rad gene pulser (Richmond, Calif.). After screening of the ampicillin-resistant transformants for the appropriate recombinant plasmid, both strands of the plasmid inserts were sequenced with double-stranded DNA templates (Sequenase; U.S. Biochemical, Cleveland) and the dideoxy chain-termination method.[33]

Data Analysis

Sequences were aligned manually on the basis of conserved regions and secondary structural elements. Regions of ambiguous alignment were omitted from the phylogenetic analysis. The evolutionary distance separating two sequences was computed from the percent similarities[34]; a least-squares method was used to infer the phylogenetic tree most consistent with the calculated distance.[35] All reference 16S rRNA sequences were obtained from the GenBank (Los Alamos, N.M.)/European Molecular Biology Laboratory (Heidelberg, Germany) data bases.

RESULTS

Amplification of 16S rRNA Sequence from Tissue Extract Using Broad-Range Eubacterial Primers

Frozen splenic tissue from Patient 1 was chosen as the initial target for 16S rRNA amplification be-

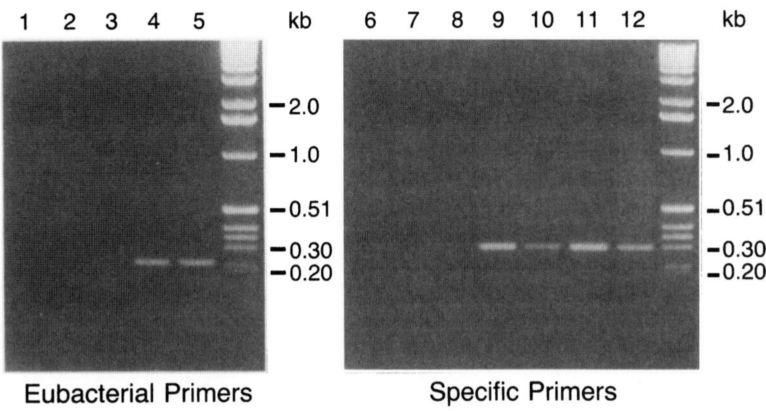

Figure 2. Agarose-Gel Electrophoresis of PCR-Amplified 16S rDNA Fragments from Tissue DNA Extracts.

Broad-range eubacterial 16S rDNA primers — p11E and p13B — were used for the reactions in lanes 1 through 5. They prime the amplification of a 232-bp fragment. Specific primers — p24E and p12B — were used for the reactions corresponding to lanes 6 through 12; they generate a 296-bp fragment. Lanes 1 and 6 contain no added DNA extract; lanes 2 and 7, a spleen sample from a patient with idiopathic thrombocytopenic purpura; lanes 3 and 8, tonsillar tissue from a patient with recurrent tonsillitis; lanes 4 and 9, a spleen sample from Patient 1; lane 5, a sample of splenic hilar lymph node from Patient 1; lane 10, bone marrow from Patient 2; lane 11, tissue from a skin-lesion biopsy from Patient 3; and lane 12, lymph-node tissue from Patient 4. The size of some of the DNA standards is indicated in kilobases.

cause of the relatively large number of bacteria seen with Warthin–Starry staining (Fig. 1B). Eubacterial broad-range primers p11E and p13B were designed from 16S rDNA gene sequences known to be conserved in all eubacteria[26] but not found in eukaryotic 16S-like genes. Previous studies have demonstrated that picogram amounts of *E. coli* 16S rDNA could be detected with the use of these primers as hybridization probes in the presence of a large excess of eukaryotic genomic DNA.[26]

After 25 cycles of PCR amplification using p11E and p13B and splenic or splenic hilar lymph-node DNA from Patient 1 as the target, a product of approximately 200 base pairs (bp) was detected reproducibly with agarose-gel electrophoresis (Fig. 2, lanes 4 and 5). No DNA band could be amplified from the spleen sample from a patient with idiopathic thrombocytopenic purpura or in the absence of added tissue extract (Fig. 2, lanes 2 and 1, respectively). With DNA from "normal" human tonsillar tissue, a DNA species of similar size was sometimes barely seen although not easily reproduced (Fig. 2, lane 3). This might be expected since such tissue was obtained from a site contaminated with normal mucosal flora. After 35 or more cycles of PCR amplification with p11E and p13B, faint bands could sometimes be seen in all reaction mixtures.

Sequence Determination and Development of Specific Primers

The DNA fragment amplified from the splenic extract from Patient 1 was purified, cloned, and then sequenced. Comparison of this sequence with all known DNA sequences in the GenBank/European Molecular Biology Laboratory data libraries showed that this 232-bp fragment could readily be aligned with eubacterial 16S rRNA genes but was not identical to any known sequence. The new organism represented by this sequence was designated strain BA-TF. A new primer was designed, p93E, that corresponds to a highly conserved 16S rRNA sequence at an upstream site. With the use of this primer and p13B, a 480-bp DNA product was amplified from the splenic extract from Patient 1, but not from that from the patient with idiopathic thrombocytopenic purpura or from other nondiseased splenic DNA extracts (data not shown). On the basis of the sequence of this 480-bp fragment, two BA-TF–specific primers — p24E and p12B — were designed (Table 1). These two primer sites are separated by 241 bp and are maximally divergent from known 16S rRNA sequences of eubacteria in those regions.

The two specific primers reproducibly amplified a DNA fragment of the expected size from a spleen sample (Fig. 2, lane 9) and a sample of splenic hilar lymph node from Patient 1 after 35 cycles of PCR. Reaction mixtures without the DNA template, as well as with DNA from tonsillar tissues from two patients and splenic tissue from patients without bacillary angiomatosis, were consistently negative. The absence of detectable product with bacterially contaminated tonsillar tissue demonstrated the specificity of these primers (Fig. 2, lane 8).

Analysis of a Novel 16S rRNA Sequence Found in Unrelated Patients with Bacillary Angiomatosis

To determine whether strain BA-TF was associated with other cases of bacillary angiomatosis, we obtained diseased tissue from three other patients, each seen at other institutions — frozen bone marrow from

Table 1. Oligonucleotide Primers Used for Amplification of 16S rDNA.*

PRIMER	SEQUENCE (5'→3')	*E. coli* 16S rRNA POSITIONS†
p11E	GGAATTCCGAGGAAGGTGGGGATGACG	1175–1193
p13B	CGGGATCCCAGGCCCGGGAACGTATTCAC	1390–1371
p93E	GGAATTCCCGCACAAGCGGTGGAGCA	930–950
p24E	GGAATTCCCTCCTTCAGTTAGGCTGG	~1017–1036
p12B	CGGGATCCCGAGATGGCTTTTGGAGATTA	~1298–1278

*Primers p11E and p13B were used as eubacterial broad-range primers (see Methods); p24E and p12B were designed as primers specific for strain BA-TF. The underlined nucleotides are tails at the 5' end of the primers that contain *Eco*RI or *Bam*HI restriction-endonuclease recognition sequences.

†Positions within the *E. coli* 16S rRNA sequence that correspond to the 5' and 3' ends of each primer. The sequences of p24E and p12B differ by 16 and 10 nucleotides, respectively, from the corresponding *E. coli* sequence.

Patient 2, paraffin-embedded, formalin-fixed tissue from Patient 3, and paraffin-embedded, formalin-fixed lymph-node tissue from Patient 4. All three samples yielded a PCR product of the same expected size when the specific primers were used (Fig. 2, lanes 10 through 12). Forty cycles of amplification were performed with DNA extracted from formalin-fixed, paraffin-embedded tissues. These DNA products were cloned and sequenced. To minimize any chance of carryover contamination from our laboratory, tissue from Patients 3 and 4 was studied at a site distant from the Stanford campus.

The DNA sequence determined from the frozen tissue from Patient 2 was identical to that of BA-TF throughout the 241 base positions in the amplified fragment. One of two recombinant clones from the formalin-fixed tissue from Patient 4 had the same sequence as BA-TF; the other clone had a sequence that differed from BA-TF at 2 of 241 positions. Two recombinant clones from Patient 3 were sequenced; these sequences were found to be identical and differed from the BA-TF sequence at only 4 positions. To determine whether this sequence heterogeneity might be found with frozen tissue samples as well, the product amplified from the frozen splenic hilar lymph-node tissue from Patient 1 with the specific primers was cloned. In comparison with the original splenic BA-TF sequence, there was 1 nucleotide substitution of a total of 1350 positions from eight clones. This result is consistent with previously published error frequencies for thermostable *Taq* polymerase during amplification of undamaged human genomic DNA.[22]

Optimized sequence alignments were prepared with use of the 480-bp 16S rRNA sequence of BA-TF (excluding the primer sequences) and those of other organisms. These alignments were used to construct a phylogenetic tree according to standard techniques.[36,37] As shown in Figure 3, strain BA-TF falls within the α subdivision of purple eubacteria and specifically within the *R. quintana–Agrobacterium tumefaciens* complex. This tree is consistent with the evolutionary relation of the purple eubacteria and the tribe Rickettsieae, on the basis of comparisons of full-length 16S rRNA gene sequences.[38] Table 2 shows the percent sequence similarities between BA-TF and other eubacteria. Even though the data in the table are based on selected 16S rRNA sequence positions, calculations of sequence similarities using the entire available BA-TF sequence gave equivalent results (data not shown). In a similar fashion we analyzed the sequence from Patient 3 and the deviant sequence from Patient 4. The results for each showed no significant difference from the analysis for BA-TF: the same relations to other organisms were confirmed. The sequence from Patient 3 was even more closely related to *R. quintana;* this limited portion of the 16S rRNA sequence was identical to the corresponding region of the *R. quintana* 16S rRNA sequence. Phylogenetic trees, constructed from each of these two deviant sequences, were nearly

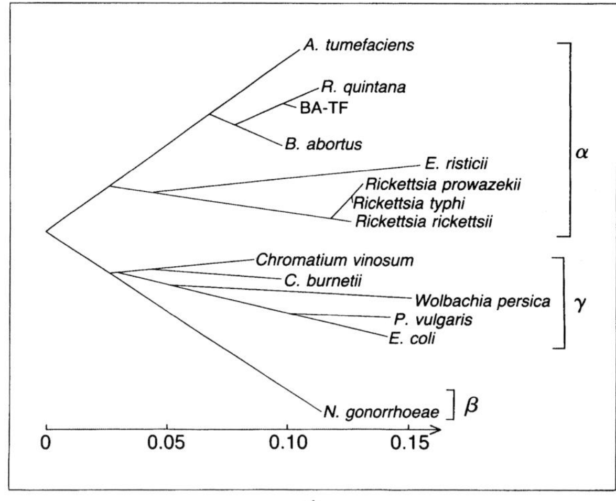

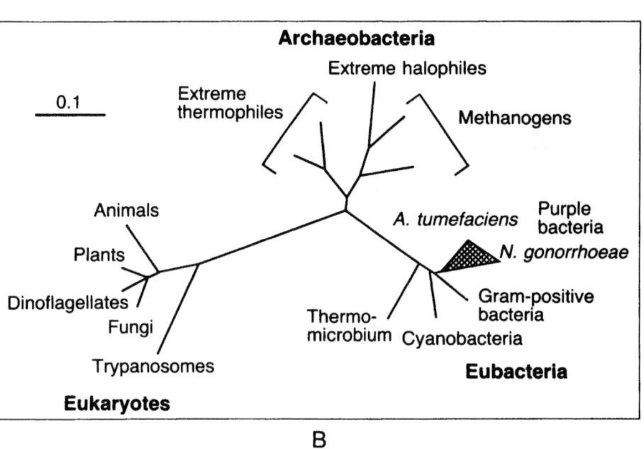

Figure 3. Phylogenetic Trees.

The tree in Panel A, based on 16S rDNA sequences, indicates evolutionary relations between the new organism, BA-TF, and representative purple eubacteria (see Methods for details).

The evolutionary distance is proportional only to the horizontal scale, expressed as the number of point mutations per sequence position. The tree was constructed with the use of sequences from *Pseudomonas testosteroni, Rhodocyclus gelatinosa,* and *Rhodopseudomonas marina,* which are not shown. Panel B shows an "unrooted" phylogenetic tree with the three kingdoms — eukaryotes, archaeobacteria, and eubacteria. In this tree the length of the line segments reflects the evolutionary distance. Crosshatching indicates the location of the purple eubacterial phylum. The scale bar represents 0.1 point mutation per sequence position. (Adapted from Turner et al.[37])

identical to the original (data not shown). The 480-bp 16S rRNA sequence from BA-TF has been deposited in the GenBank/European Molecular Biology Laboratory data libraries.

Using the broad-range primers p93E and p13B, no irradiation of reagents, and 40 cycles of PCR, we occasionally obtained a visible DNA product when no DNA template or extract was added to the reaction mixture. The DNA product from two such reactions was cloned and sequenced. In these cases the sequences were identical to each other and were most closely related to *Pseudomonas cepacia,* a β purple eu-

Table 2. 16S rRNA Sequence Similarities between the New Organism, BA-TF, and Selected Rickettsia and Other Eubacteria.

SPECIES	SEQUENCE SIMILARITY TO BA-TF*
	%
Rochalimaea quintana	98.3
Brucella abortus	95.7
Agrobacterium tumefaciens	93.5
Rickettsia rickettsii	84.5
Rickettsia prowazekii	83.6
Neisseria gonorrhoeae	81.2
Ehrlichia risticii	80.5
Coxiella burnetii	80.2
Proteus vulgaris	79.0
Escherichia coli	78.6

*Based on calculations in which only selected positions were used (see Methods).

bacterium commonly found in aquatic environments (data not shown). These findings strengthen the specificity of the association between the BA-TF sequence and bacillary angiomatosis.

Many of the human pathogens closely related to strain BA-TF are in the tribe Rickettsieae, although this tribe is phylogenetically dispersed throughout the purple eubacterial phylum.[38] *Brucella abortus* is the second most closely related organism to BA-TF among those whose 16S rRNA sequences have been characterized. According to our analysis, BA-TF is a unique rickettsia-like organism that is most closely related to *R. quintana*, the causative agent of trench fever. *A. tumefaciens* and other bacteria in this complex are plant pathogens and soil organisms, suggesting that BA-TF, as well as some other rickettsia, may have arisen as a plant-associated organism.[39]

DISCUSSION

Our understanding of microbial pathogens and the diseases they cause is limited by our ability to cultivate these organisms. A number of human diseases are linked to microbes that can be consistently visualized but not grown. Their identity has been a mystery. We have applied a technique that circumvents the absolute need to isolate or grow a disease-associated organism but that nonetheless may reveal its phylogenetic relation. Such relations may predict important and useful features of an organism, since evolutionarily related organisms are expected to share some characteristics.

Bacillary angiomatosis is a well-defined, increasingly common infection of the skin and visceral organs, most often seen in immunocompromised hosts, especially patients with HIV infection.[8,40] It is potentially fatal if not treated. The numerous bacilli found in the lesions of this disease have been compared to those seen in lesions from cat scratch disease, although the vascular proliferative response in bacillary angiomatosis is not a common feature of cat scratch disease. The

bacilli in each disease are pleomorphic, measure 0.2 to 0.5 μm by 1 to 3 μm, grow in clumps, and have trilaminar walls.[4] These bacilli have resisted identification. We have used eubacterial broad-range primers to amplify a novel 16S rDNA sequence from diseased tissue from four unrelated patients with bacillary angiomatosis. The sequence obtained from three of these patients was identical; we designated it the reference sequence, BA-TF. The sequences found in the third patient and in one clone from the fourth differed from BA-TF only minimally. None of these sequences were detected in tissues from patients without bacillary angiomatosis. On the basis of our phylogenetic analysis of these sequences, we conclude that bacillary angiomatosis is caused by a rickettsia-like organism, most closely related to *R. quintana*. Even though the sequence in the third patient was identical to the corresponding region of the *R. quintana* 16S rRNA gene, the limited amount of available sequences from this patient and the predominance of the BA-TF sequence in the other patients with bacillary angiomatosis suggest that BA-TF more accurately reflects the true 16S rRNA sequence of the pathogen. Given the paucity of α purple bacterial 16S rRNA sequences, it is not yet possible to assign BA-TF to a genus or species; however, the sequence similarity between BA-TF and *R. quintana* is the same as that between *Rickettsia typhi* and *Rickettsia rickettsii* (98.3 percent).

PCR is an extremely sensitive method of detecting target DNA sequences. Sequences of 16S rRNA can be amplified and compared with a reference-sequence data base in order to infer phylogenetic relations.[24,35] The use of broad-range or "universal" 16S rRNA primers,[23] however, presents special problems. Contamination of PCR reagents and target with a few bacteria, bacterial genomes, or very small amounts of eukaryotic DNA is virtually impossible to prevent and can lead to false positive results. Exquisite care must be taken to keep contamination to a minimum.[41,42] We found that the use of no more than 25 PCR cycles with broad-range primers prevented the detection of product with uninfected tissues (unpublished data). Ultraviolet irradiation has been shown to reduce carryover contamination of PCR reagents and was useful in our 40-cycle amplifications.[43,44] The use of specific primers reduces the need for such stringency. Our association of the BA-TF sequence with tissue from patients with bacillary angiomatosis was strengthened by the independent processing and amplification of the various samples in different laboratories. In addition, samples of internal tissue from patients without bacillary angiomatosis never produced a visible amplification product.

Despite the small number of nucleotide differences between BA-TF and the 16S rRNA sequence of Patient 3 and of a single clone from Patient 4, these sequences are all related to 16S rRNA sequences of the same organisms. On the basis of the 16S rRNA nucleotide positions included in our phylogenetic analysis, the sequences from these two patients differed by only 0.7 percent from the BA-TF sequence. However, these differences do raise some important issues. Could sequence heterogeneity indicate the presence of multiple but related strains within the diseased tissue — i.e., a mixed infection? This is unlikely for two reasons. First, at rRNA positions known to participate in the formation of base pairs in secondary structures, we have seen nucleotide substitutions without compensatory substitutions at the corresponding paired position; such changes would be disruptive to the rRNA secondary structure. Second, sequence heterogeneity was apparent only in Patients 3 and 4, for whom formalin-fixed tissue was studied. Another possibility is that the heterogeneity is due in part to the amplification of different copies of the 16S rRNA gene from the same organism. A small number of nucleotide differences are known to exist within the seven copies of the *E. coli* rRNA genes.[45,46] This possibility cannot be ruled out for BA-TF; however, the explanation we most favor for the observed sequence differences is *Taq* polymerase error. *Taq* polymerase can mistakenly incorporate nucleotides, particularly during the amplification of damaged DNA.[47,48] Formalin fixation can damage nucleic acids at particular positions and therefore contributes to higher polymerase-error rates. Under these conditions, when there are few original templates in a tissue sample, it is imperative to sequence multiple clones of the amplified product.

Complete 16S rRNA sequences are not necessary for the construction of meaningful phylogenetic trees[23]; however, the amount of partial sequence available and the composition of the reference 16S rRNA sequence collection are important variables. Our evolutionary-distance tree duplicated the relations among the Rickettsieae and other purple bacteria established by Weisburg et al. using full 16S rRNA sequences.[38] Thus, we are confident about our phylogenetic analysis of strain BA-TF. In addition, a rickettsial origin is consistent with previous morphologic descriptions of the bacillus causing bacillary angiomatosis.

What are the implications of the relation between BA-TF, *R. quintana*, and other α purple bacteria? *R. quintana* is transmitted by the body louse and is the cause of trench fever, a systemic illness characterized by persistent or recurrent fever, bone and muscle pain, rash, and splenomegaly.[49] *R. quintana* is sensitive in vitro to erythromycin and the tetracyclines,[50] agents that have been used successfully to treat bacillary angiomatosis.[40] Although trench fever and disseminated bacillary angiomatosis share some clinical features, the vascular lesions of bacillary angiomatosis distinguish it from trench fever. A separate member of the order Rickettsiales, *Bartonella bacilliformis*, can produce lesions (verruga peruana) that are indistinguishable from those caused by bacillary angiomatosis, and this has led to suggestions that it might be the cause of this disease.[4,51] There are no 16S rRNA sequence data

available for *Bart. bacilliformis;* therefore, its relation to BA-TF remains unclear. *B. abortus* has features in common with the agent of bacillary angiomatosis: both are transmitted to humans from domestic animals by percutaneous inoculation, frequently involve the reticuloendothelial system, and can cause chronic febrile illnesses. On the other hand, infrequent skin involvement, a granulomatous rather than a vascular proliferative host-tissue response, a gram-negative staining pattern, and 16S rRNA sequence distinguish brucellosis from bacillary angiomatosis.[52]

The grouping of *R. quintana* and BA-TF with the agrobacteria and rhizobacteria suggests a plant or soil origin.[39] Humans are the only known reservoir for *R. quintana*, but the other characterized member of this genus, *R. vinsonii*, is carried by voles.[53] The grouping of BA-TF with these organisms would be consistent with its transmission from the environment to humans by means of a break in the skin. What role, if any, cats have, and the exact relation of BA-TF to the bacillus of cat scratch disease must await further analysis of the latter organism. We are currently using the techniques described in this report to identify the cat-scratch-disease bacillus.

R. quintana is the only pathogenic rickettsia to grow on cell-free medium.[54] The relation of BA-TF to this organism may help us develop appropriate culture conditions for the growth of this new pathogen. Growth of BA-TF would facilitate the investigation of the angioproliferative response it elicits in the host.[40] Our results also have implications for the diagnosis of bacillary angiomatosis: divergent portions of the BA-TF 16S rRNA gene or adjacent regions of the genome may serve as probes for in situ hybridization. If such probes were labeled with fluorescent dyes, hybridization to the agent of bacillary angiomatosis could be visualized with epifluorescence.[55] Alternatively, an in situ PCR-based technique might be feasible.

The discovery of this novel rickettsia-like organism in patients with bacillary angiomatosis is one example of the application of a potentially powerful technique. We expect that this approach will be applicable to other infectious diseases with unclear causes and will expand our understanding of interactions between humans and microbes.

We are indebted to Drs. Carol Kemper and Stanley Deresinski and our other clinical colleagues for their support and enthusiasm; to Dr. Norman Pace, Indiana University, for advice, constructive suggestions, and access to data bases; to Dr. Donald Regula, Department of Pathology, Stanford University, for providing invaluable assistance with tissue samples from Patient 1 and photographs; to Dr. Kemper, Santa Clara Valley Medical Center, San Jose, for providing tissue from Patient 2; to Dr. Philip LeBoit, Department of Pathology, University of California, San Francisco, for providing tissue from Patient 3; to Dr. Ronald F. Dorfman, Department of Pathology, Stanford University, and Dr. Klaus J. Lewin, Department of Pathology, University of California, Los Angeles, for providing tissue from Patient 4; and to Dr. Michele Manos and Ms. Catherine Greer, Cetus Corporation, Emeryville, California, for giving us valuable advice about and assistance with the extraction of DNA from formalin-fixed, paraffin-embedded tissue.

REFERENCES

1. Stoler MH, Bonfiglio TA, Steigbigel RT, Pereira M. An atypical subcutaneous infection associated with acquired immune deficiency syndrome. Am J Clin Pathol 1983; 80:714-8.
2. Cockerell CJ, Whitlow MA, Webster GF, Friedman-Kien AE. Epithelioid angiomatosis: a distinct vascular disorder in patients with the acquired immunodeficiency syndrome or AIDS-related complex. Lancet 1987; 2:654-6.
3. Angritt P, Tuur SM, Macher AM, et al. Epithelioid angiomatosis in HIV infection: neoplasm or cat-scratch disease? Lancet 1988; 1:996.
4. LeBoit PE, Berger TG, Egbert BM, et al. Epithelioid haemangioma-like vascular proliferation in AIDS: manifestation of cat scratch disease bacillus infection? Lancet 1988; 1:960-3.
5. LeBoit PE, Berger TG, Egbert BM, Beckstead JH, Yen TS, Stoler MH. Bacillary angiomatosis: the histopathology and differential diagnosis of a pseudoneoplastic infection in patients with human immunodeficiency virus disease. Am J Surg Pathol 1989; 13:909-20.
6. Koehler JE, LeBoit PE, Egbert BM, Berger TG. Cutaneous vascular lesions and disseminated cat-scratch disease in patients with the acquired immunodeficiency syndrome (AIDS) and AIDS-related complex. Ann Intern Med 1988; 109:449-55.
7. Milam MW, Balerdi MJ, Toney JF, Foulis PR, Milam CP, Behnke RH. Epithelioid angiomatosis secondary to disseminated cat scratch disease involving the bone marrow and skin in a patient with acquired immune deficiency syndrome: a case report. Am J Med 1990; 88:180-3.
8. Kemper CA, Lombard CM, Deresinski SC, Tompkins LS. Visceral bacillary epithelioid angiomatosis: possible manifestations of disseminated cat scratch disease in the immunocompromised host: a report of two cases. Am J Med 1990; 89:216-22.
9. Cockerell CJ, Bergstresser PR, Myrie-Williams C, Tierno PM. Bacillary epithelioid angiomatosis occurring in an immunocompetent individual. Arch Dermatol 1990; 126:787-90.
10. Carithers HA, Carithers CM, Edwards RO Jr. Cat-scratch disease: its natural history. JAMA 1969; 207:312-6.
11. Carithers HA. Cat-scratch disease: an overview based on a study of 1,200 patients. Am J Dis Child 1985; 139:1124-33.
12. Margileth AM, Wear DJ, English CK. Systemic cat scratch disease: report of 23 patients with prolonged or recurrent severe bacterial infection. J Infect Dis 1987; 155:390-402.
13. Lenoir AA, Storch GA, DeSchryver-Kecskemeti K, et al. Granulomatous hepatitis associated with cat scratch disease. Lancet 1988; 1:1132-6.
14. Delahoussaye PM, Osborne BM. Cat-scratch disease presenting as abdominal visceral granulomas. J Infect Dis 1990; 161:71-8.
15. Wear DJ, Margileth AM, Hadfield TL, Fischer GW, Schlagel CJ, King FM. Cat scratch disease: a bacterial infection. Science 1983; 221:1403-5.
16. English CK, Wear DJ, Margileth AM, et al. Cat-scratch disease: isolation and culture of the bacterial agent. JAMA 1988; 259:1347-52.
17. Schlossberg D, Morad Y, Krouse TB, Wear DJ, English CK. Culture-proved disseminated cat-scratch disease in acquired immunodeficiency syndrome. Arch Intern Med 1989; 149:1437-9.
18. Fox GE, Stackebrandt E, Hespell RB, et al. The phylogeny of prokaryotes. Science 1980; 209:457-63.
19. Woese CR. Bacterial evolution. Microbiol Rev 1987; 51:221-71.
20. Saiki RK, Scharf S, Faloona F, et al. Enzymatic amplification of β-globin genomic sequences and restriction site analysis for diagnosis of sickle cell anemia. Science 1985; 230:1350-4.
21. Mullis KB, Faloona FA. Specific synthesis of DNA in vitro via a polymerase-catalyzed chain reaction. In: Methods in enzymology. Vol. 155. Wu R, ed., Recombinant DNA, Part F. San Diego: Academic Press, 1987:335-50.
22. Saiki RK, Gelfand DH, Stoffel S, et al. Primer-directed enzymatic amplification of DNA with a thermostable DNA polymerase. Science 1988; 239:487-91.
23. Lane DJ, Pace B, Olsen GJ, Stahl DA, Sogin ML, Pace NR. Rapid determination of 16S ribosomal RNA sequences for phylogenetic analyses. Proc Natl Acad Sci U S A 1985; 82:6955-9.
24. Medlin L, Elwood HJ, Stickel S, Sogin ML. The characterization of enzymatically amplified eukaryotic 16S-like rRNA-coding regions. Gene 1988; 71:491-9.
25. Eisenstein BI. The polymerase chain reaction: a new method of using molecular genetics for medical diagnosis. N Engl J Med 1990; 322:178-83.
26. Chen K, Neimark H, Rumore P, Steinman CR. Broad range DNA probes for detecting and amplifying eubacterial nucleic acids. FEMS Microbiol Lett 1989; 48:19-24.
27. Goelz SE, Hamilton SR, Vogelstein B. Purification of DNA from formaldehyde fixed and paraffin embedded human tissue. Biochem Biophys Res Commun 1985; 130:118-26.

28. Dubeau L, Chandler LA, Gralow JR, Nichols PW, Jones PA. Southern blot analysis of DNA extracted from formalin-fixed pathology specimens. Cancer Res 1986; 46:2964-9.
29. Sambrook J, Fritsch EF, Maniatis T. Molecular cloning: a laboratory manual. 2nd ed. Cold Spring Harbor, N.Y.: Cold Spring Harbor Laboratory, 1989.
30. Shibata DK, Arnheim N, Martin WJ. Detection of human papilloma virus in paraffin-embedded tissue using the polymerase chain reaction. J Exp Med 1988; 167:225-30.
31. Wright DK, Manos MM. Sample preparation from paraffin-embedded tissues. In: Innis MA, Gelfand DH, Sninsky JJ, White TJ, eds. PCR protocols: a guide to methods and applications. San Diego, Calif.: Academic Press, 1990:153-8.
32. Dower WJ, Miller JF, Ragsdale CW. High efficiency transformation of E. coli by high voltage electroporation. Nucleic Acids Res 1988; 16:6127-45.
33. Sanger F, Nicklen S, Coulson AR. DNA sequencing with chain-terminating inhibitors. Proc Natl Acad Sci U S A 1977; 74:5463-7.
34. Jukes TH, Cantor CR. Evolution of protein molecules. In: Munro HN, ed. Mammalian protein metabolism. Vol. 3. New York: Academic Press, 1969:21-132.
35. Olsen GJ. Earliest phylogenetic branchings: comparing rRNA-based evolutionary trees inferred with various techniques. Cold Spring Harb Symp Quant Biol 1987; 52:825-37.
36. Fitch WM, Smith TF. Optimal sequence alignments. Proc Natl Acad Sci U S A 1983; 80:1382-6.
37. Turner S, Burger-Wiersma T, Giovannoni SJ, Mur LR, Pace NR. The relationship of a prochlorophyte Prochlorothrix hollandica to green chloroplasts. Nature 1989; 337:380-2.
38. Weisburg WG, Dobson ME, Samuel JE, et al. Phylogenetic diversity of the Rickettsiae. J Bacteriol 1989; 171:4202-6.
39. Weisburg WG, Woese CR, Dobson ME, Weiss E. A common origin of rickettsiae and certain plant pathogens. Science 1985; 230:556-8.
40. Cockerell CJ, LeBoit PE. Bacillary angiomatosis: a newly characterized, pseudoneoplastic, infectious, cutaneous vascular disorder. J Am Acad Dermatol 1990; 22:501-12.
41. Kwok S, Higuchi R. Avoiding false positives with PCR. Nature 1989; 339:237-8. [Erratum, Nature 1989; 339:490.]
42. Kitchin PA, Szotyori Z, Fromholc C, Almond N. Avoidance of false positives. Nature 1990; 344:201.
43. Sarkar G, Sommer SS. Shedding light on PCR contamination. Nature 1990; 343:27.
44. Cimino GD, Metchette K, Isaacs ST, Zhu YS. More false-positive problems. Nature 1990; 345:773-4.
45. Brownlee GG, Sanger F, Barrell BG. The sequence of 5s ribosomal ribonucleic acid. J Mol Biol 1968; 34:379-412.
46. Jinks-Robertson S, Nomura M. Ribosomes and tRNA. In: Neidhardt FC, ed. Escherichia coli and Salmonella typhimurium: cellular and molecular biology. 2. Washington, D.C.: American Society for Microbiology, 1987:1358-85.
47. Pääbo S, Irwin DM, Wilson AC. DNA damage promotes jumping between templates during enzymatic amplification. J Biol Chem 1990; 265:4718-21.
48. Ennis PD, Zemmour J, Salter RD, Parham P. Rapid cloning of HLA-A,B cDNA by using the polymerase chain reaction: frequency and nature of errors produced in amplification. Proc Natl Acad Sci U S A 1990; 87:2833-7.
49. Vinson JW, Varela G, Molina-Pasquel C. Trench fever. 3. Induction of clinical disease in volunteers inoculated with Rickettsia quintana propagated on blood agar. Am J Trop Med Hyg 1969; 18:713-22.
50. Myers WF, Grossman DM, Wisseman CL Jr. Antibiotic susceptibility patterns in Rochalimaea quintana, the agent of trench fever. Antimicrob Agents Chemother 1984; 25:690-3.
51. Kreier JP, Ristic M. The biology of hemotrophic bacteria. Annu Rev Microbiol 1981; 35:325-38.
52. Spink WW. The nature of brucellosis. Minneapolis: University of Minnesota Press, 1956.
53. Weiss E, Dasch GA. Differential characteristics of strains of Rochalimaea: Rochalimaea vinsonii sp. nov., the Canadian vole agent. Int J Syst Bacteriol 1982; 32:305-14.
54. Varela G, Vinson JW, Molina-Pasquel C. Trench fever. II. Propagation of Rickettsia quintana on cell-free medium from the blood of two patients. Am J Trop Med Hyg 1969; 18:708-12.
55. DeLong EF, Wickham GS, Pace NR. Phylogenetic stains: ribosomal RNA-based probes for the identification of single cells. Science 1989; 243:1360-3. [Erratum, Science 1989; 245:1312.]

Pathogenesis and Host Response Mechanisms

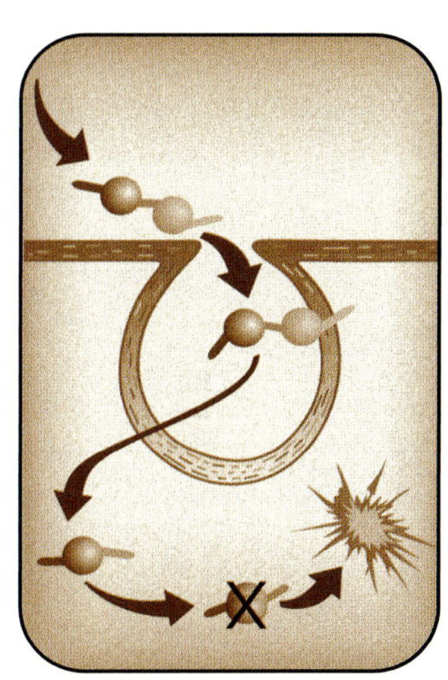

Pathogenesis and Host Response Mechanisms

The Division II committee members who selected the papers for this section included Gordon Archer, M.D., Emil Gotschlich, M.D., Charles Janeway, M.D., John Mekalanos, Ph.D., and Alison O'Brien, Ph.D. (Committee Chair); this preface and some of the introductions are by Dr. O'Brien. The mandate for our committee was to pick landmark papers published during the past 100 years (with emphasis on the last 60 years) that reflect milestones in the general area of pathogenesis and host response to infection. Specifically, we were asked to choose articles that represent areas of expertise within Division II, i.e., antimicrobial agents and chemotherapy, immunology, microbial pathogenesis, general medical microbiology, mycoplasmology, and clinical and diagnostic immunology. The major difficulty we had in fulfilling our directive was to narrow our final choices to a strictly limited number of pages of text. Initially, the five of us independently nominated a total of 22 papers (271 pages of text), of which only 1 was picked twice. Of these 21 top papers, 8 were then chosen by consensus or by "I can't decide; you pick." It should be noted that two of us felt strongly that Koch's article (R. Koch, Die Aetiologie der Tuberkulose, *Mittheilung. Kais. Gesundheitsamt.* **2**:1–88, 1884) ranked as the seminal paper in the area of microbial pathogenesis, but we eliminated it from our short list because the date of its publication fell outside the suggested time line for this project.

1929

On the Antibacterial Action of Cultures of a Penicillium, with Special Reference to Their Use in the Isolation of B. influenzae   A. Fleming

Reprinted with permission from *British Journal of Experimental Pathology* 10:226–236. (Now *International Journal of Experimental Pathology.*) Copyright © 1929. Blackwell Science Ltd.

1940

Penicillin as a Chemotherapeutic Agent
E. Chain, H. W. Florey, A. D. Gardner, N. G. Heatley, M. A. Jennings, J. Orr-Ewing, and A. G. Sanders

Reprinted with permission from *Lancet* ii:226–228. Copyright © 1940. The Lancet Ltd.

An Enzyme from Bacteria Able To Destroy Penicillin   E. P. Abraham and E. Chain

Reprinted by permission from *Nature* 146:837. Copyright © 1940. Macmillan Magazines Ltd.

The first, oft-cited publication by Fleming is believed to be the first report of the antibacterial action of a fungal product. This paper is particularly pleasing to the bench microbiologists among us, who teach our students to be alert for the unexpected. Fleming made the observation that when bacterial culture plates streaked with *Staphylococcus* were left on a lab bench for an extended period, mold, identified subsequently as *Penicillium*, developed on the plates and caused lysis of the staphylococcal colonies. He showed that broth in which the *Penicillium* had been grown at room temperature for 2 weeks was bacteriocidal and bacteriolytic against certain pyogenic cocci (e.g., *Streptococcus pyogenes*, *Streptococcus pneumoniae*, and *Staphylococcus aureus*) as well as the diphtheria bacillus (now called *Corynebacterium diphtheriae*) but was inactive against many gram-negative organisms. Fleming named the antibacterial substance penicillin, and he showed that it was not toxic to animals when given at very high doses. In his discussion, Fleming proposed that penicillin might be applied to the dressing of wounds. However, he foresaw the major immediate use of penicillin as a means of reducing the growth of penicillin-susceptible flora from sputum to facilitate isolation of such penicillin-resistant agents as Pfiffers bacillus of influenza (now called *Haemophilus influenzae*).

In the second publication, Chain and coworkers followed up on Fleming's observations with the clear intent of investigating the use of penicillin as a therapeutic agent to treat systemic bacterial infections. They resuscitated his mold cultures, purified penicillin from broth supernatants, and showed that the substance could be used to treat infections in animals. Systemic infections in mice caused by staphylococci, streptococci, and clostridia were dramatically cured by penicillin.

Finally, a footnote to these dramatic discoveries is the letter to the editor of *Nature* from Abraham and Chain describing a substance in *B. coli* (now *Escherichia coli*) that could inactivate penicillin. Penicillinase was the first bacterial product isolated that mediated resistance to an antibacterial agent, and this paper foreshadowed the era, 15 years later, of widespread resistance of staphylococci to penicillin. For their seminal discoveries, Fleming, Chain, and Florey were corecipients of the Nobel Prize in Physiology and Medicine in 1945.

Gordon Archer and Alison O'Brien

ON THE ANTIBACTERIAL ACTION OF CULTURES OF A PENICILLIUM, WITH SPECIAL REFERENCE TO THEIR USE IN THE ISOLATION OF *B. INFLUENZÆ*.

ALEXANDER FLEMING, F.R.C.S.

From the Laboratories of the Inoculation Department, St. Mary's Hospital, London.

Received for publication May 10th, 1929.

WHILE working with staphylococcus variants a number of culture-plates were set aside on the laboratory bench and examined from time to time. In the examinations these plates were necessarily exposed to the air and they became contaminated with various micro-organisms. It was noticed that around a large colony of a contaminating mould the staphylococcus colonies became transparent and were obviously undergoing lysis (see Fig. 1).

Subcultures of this mould were made and experiments conducted with a view to ascertaining something of the properties of the bacteriolytic substance which had evidently been formed in the mould culture and which had diffused into the surrounding medium. It was found that broth in which the mould had been grown at room temperature for one or two weeks had acquired marked inhibitory, bactericidal and bacteriolytic properties to many of the more common pathogenic bacteria.

CHARACTERS OF THE MOULD.

The colony appears as a white fluffy mass which rapidly increases in size and after a few days sporulates, the centre becoming dark green and later in old cultures darkens to almost black. In four or five days a bright yellow colour is produced which diffuses into the medium. In certain conditions a reddish colour can be observed in the growth.

In broth the mould grows on the surface as a white fluffy growth changing in a few days to a dark green felted mass. The broth becomes bright yellow and this yellow pigment is not extracted by $CHCl_3$. The reaction of the broth becomes markedly alkaline, the pH varying from 8·5 to 9. Acid is produced in three or four days in glucose and saccharose broth. There is no acid production in 7 days in lactose, mannite or dulcite broth.

Growth is slow at 37°C. and is most rapid about 20°C. No growth is observed under anaerobic conditions.

In its morphology this organism is a penicillium and in all its characters it most closely resembles *P. rubrum*. Biourge (1923) states that he has never found *P. rubrum* in nature and that it is an "animal de laboratoire." This penicillium is not uncommon in the air of the laboratory.

PENICILLIN.

IS THE ANTIBACTERIAL BODY ELABORATED IN CULTURE BY ALL MOULDS ?

A number of other moulds were grown in broth at room temperature and the culture fluids were tested for antibacterial substances at various intervals up to one month. The species examined were: *Eidamia viridiscens, Botrytis cineria, Aspergillus fumigatus, Sporotrichum, Cladosporium, Penicillium,* 8 strains. Of these it was found that only one strain of penicillium produced any inhibitory substance, and that one had exactly the same cultural characters as the original one from the contaminated plate.

It is clear, therefore, that the production of this antibacterial substance is not common to all moulds or to all types of penicillium.

In the rest of this article. allusion will constantly be made to experiments with filtrates of a broth culture of this mould, so for convenience and to avoid the repetition of the rather cumbersome phrase " Mould broth filtrate," the name " penicillin " will be used. This will denote the filtrate of a broth culture of the particular penicillium with which we are concerned.

METHODS OF EXAMINING CULTURES FOR ANTIBACTERIAL SUBSTANCE.

The simplest method of examining for inhibitory power is to cut a furrow in an agar plate (or a plate of other suitable culture material), and fill this in with a mixture of equal parts of agar and the broth in which the mould has grown. When this has solidified, cultures of various microbes can be streaked at right angles from the furrow to the edge of the plate. The inhibitory substance diffuses very rapidly in the agar, so that in the few hours before the microbes show visible growth it has spread out for a centimetre or more in sufficient concentration to inhibit growth of a sensitive microbe. On further incubation it will be seen that the proximal portion of the culture for perhaps one centimetre becomes transparent, and on examination of this portion of the culture it is found that practically all the microbes are dissolved, indicating that the anti-bacterial substance has continued to diffuse into the agar in sufficient concentration to induce dissolution of the bacteria. This simple method therefore suffices to demonstrate the bacterio-inhibitory and bacterio-lytic properties of the mould culture, and also by the extent of the area of inhibition gives some measure of the sensitiveness of the particular microbe tested. Fig. 2 shows the degree of inhibition obtained with various microbes tested in this way.

The inhibitory power can be accurately titrated by making serial dilutions of penicillin in fresh nutrient broth, and then implanting all the tubes with the same volume of a bacterial suspension and incubating them. The inhibition can then readily be seen by noting the opacity of the broth.

For the estimation of the antibacterial power of a mould culture it is unnecessary to filter as the mould grows only slowly at 37°C., and in 24 hours, when the results are read, no growth of mould is perceptible. Staphylococcus is a very suitable microbe on which to test the broth as it is hardy, lives well in culture, grows rapidly, and is very sensitive to penicillin.

The bactericidal power can be tested in the same way except that at intervals measured quantities are explanted so that the number of surviving microbes can be estimated.

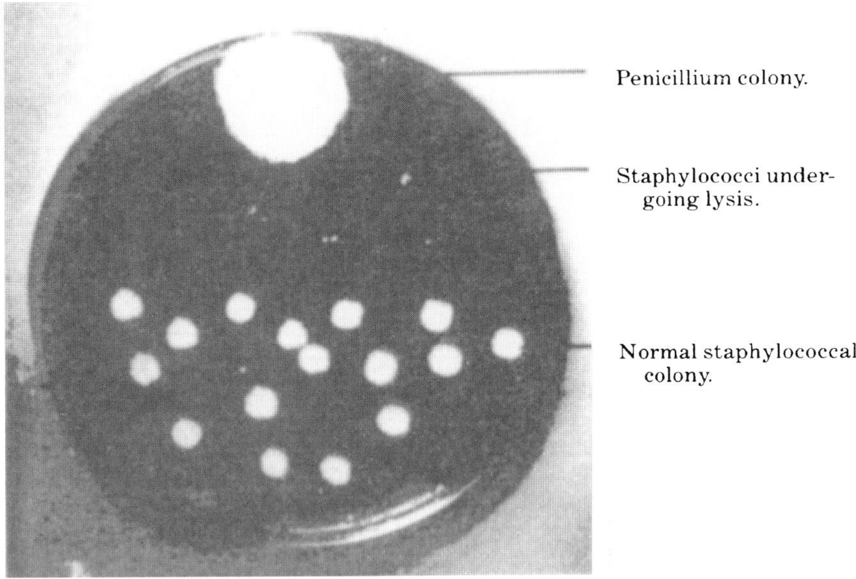

FIG. 1.—Photograph of a culture-plate showing the dissolution of staphylococcal colonies in the neighbourhood of a penicillium colony.

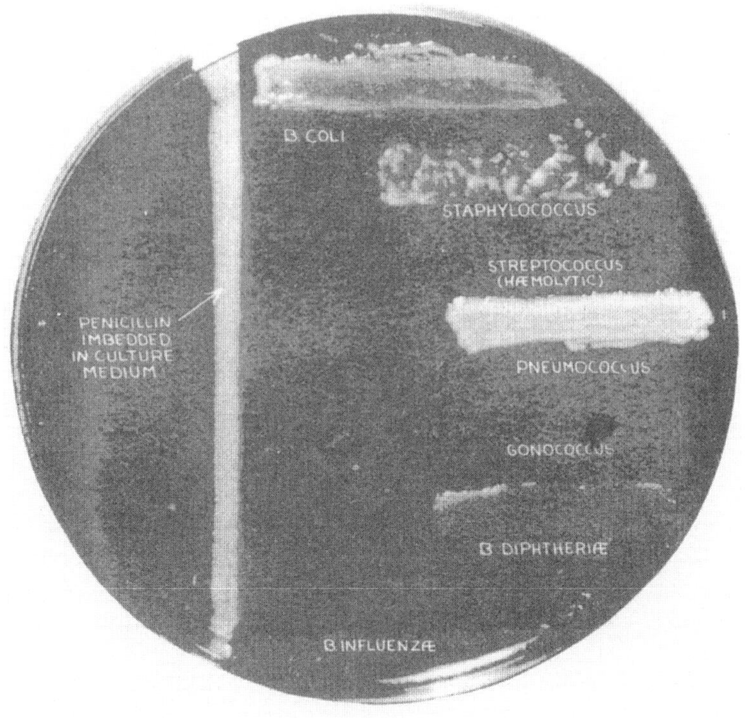

FIG. 2.

Fleming.

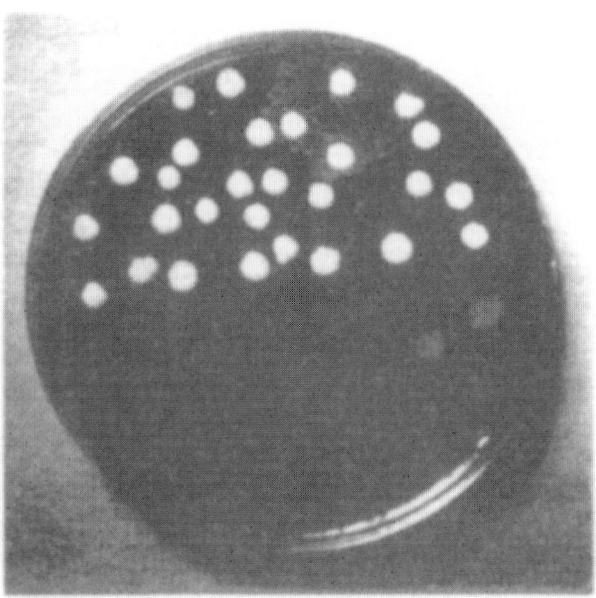

FIG. 3.—Photograph of a culture-plate (Fildes medium) which had been evenly planted with a mixture of staphylococci and *B. influenzæ*. Six drops of penicillin were then spread over the lower half of the plate. Note complete inhibition of staphylococci in the penicillin treated area with resultant pure culture of *B. influenzæ*.

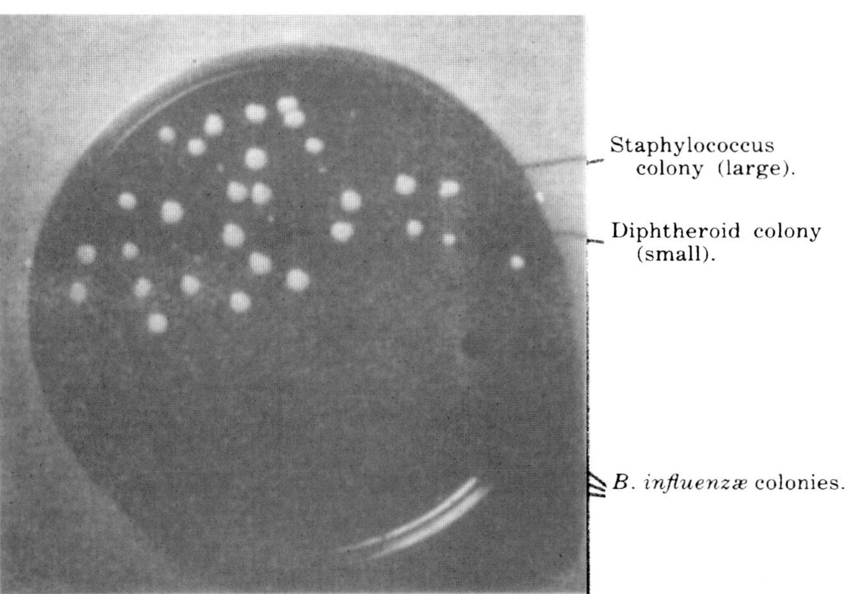

FIG. 4.—Photograph of a culture-plate (Fildes medium) which had been evenly planted with nasal mucus from an individual suffering from a "cold." Six drops of penicillin were spread over the lower half of the plate before incubation. Note profuse growth of staphylococci and diphtheroid bacilli in untreated half, whereas in treated half only some three colonies of *B. influenzæ* are seen.

Fleming.

A. FLEMING.

PROPERTIES OF THE ANTIBACTERIAL SUBSTANCE.

Effect of heat.—Heating for 1 hour at 56° or 80° C. has no effect on the antibacterial power of penicillin. Boiling for a few minutes hardly affects it (see Table II). Boiling for 1 hour reduces it to less than one quarter its previous strength if the fluid is alkaline, but if it is neutral or very slightly acid then the reduction is much less. Autoclaving for 20 minutes at 115° C. practically destroys it.

Effect of filtration.—Passage through a Seitz filter does not diminish the antibacterial power. This is the best method of obtaining sterile active mould broth.

Solubility.—It is freely soluble in water and weak saline solutions. My colleague, Mr. Ridley, has found that if penicillin is evaporated at a low temperature to a sticky mass the active principle can be completely extracted by absolute alcohol. It is insoluble in ether or chloroform.

Rate of development of inhibitory substance in culture.—A 500 c.c. Erlenmeyer flask containing 200 c.c. of broth was planted with mould spores and incubated at room temperature (10° to 20° C.). The inhibitory power of the broth to staphylococcus was tested at intervals.

After 5 days complete inhibition in 1 in 20 dilution.

,, 6 ,,   ,,   ,,   ,, 1 in 40   ,,
,, 7 ,,   ,,   ,,   ,, 1 in 200   ,,
,, 8 ,,   ,,   ,,   ,, 1 in 500   ,,

Grown at 20° C. the development of the active principle is more rapid and a good sample will completely inhibit staphylococci in a 1 in 500 or 1 in 800 dilution in 6 or 7 days. As the culture ages the antibacterial power falls and may in 14 days at 20° C. have almost disappeared.

The antibacterial power of penicillin falls when it is kept at room temperature. The rate of this fall can be seen from Table I.

TABLE I.—*Effect of Keeping at Room Temperature on the Anti-Staphylococcal Power of Penicillin.*

Growth of staphylococcus in dilutions of penicillin as under.

	1/20.	1/40.	1/60.	1/80.	1/100.	1/200.	1/300.	1/400.	1/600.	1/800.	1/1000.	Control.
At time of filtration	−	−	−	−	−	−	−	−	−	±	+ +	+ +
After 4 days	−	−	−	−	−	−	−	−	−	±	+ +	+ +
,, 7 ,,	−	−	−	−	−	−	−	±	+	+	+ +	+ +
,, 9 ,,	−	−	−	−	−	−	−	±	+	+	+ +	+ +
,, 13 ,,	−	−	−	−	−	+	+	+	+	+	+ +	+ +
,, 15 ,,	−	:	+	+	+	+	+	+	+	+	+ +	+ +

If the reaction of penicillin is altered from its original pH of 9 to a pH of 6·8 it is much more stable.

The small drops of bright yellow fluid which collect on the surface of the mould may have a high antibacterial titre. One specimen of such fluid completely inhibited the growth of staphylococci in a dilution of 1 in 20,000 while the broth in which the mould was growing, tested at the same time, inhibited staphylococcal growth in 1 in 800.

If the mould is grown on solid medium and the felted mass picked off and

PENICILLIN.

TABLE II.— *Inhibitory Power of Penicillin (Heated and Unheated) on Various Microbes (Agar Plate Method).*

Type of microbe.	Extent of inhibition in mm. from penicillin embedded in agar, serum agar, or blood agar plates.	
	Unheated.	Boiled for 1 minute.
Experiment 1 :		
Staphylococcus pyogenes	23	21
Streptococcus ,,	17	17
,, *viridans* (mouth)	17	15
Diphtheroid bacillus	27	22
Sarcina	10	10
Micrococcus lysodeikticus	6	7
,, from air (1)	20	16
,, ,, (2)	4	9
B. anthracis	0	0
B. typhosus	0	0
Enterococcus	0	0
Experiment 2 :		
Staphylococcus pyogenes	24	…
Streptococcus ,,	30	…
,, *viridans* (mouth)	25	…
Pneumococcus	30	…
Diphtheroid bacillus	35	…
B. pyocyaneus	0	…
B. pneumoniæ (Friedlander)	0	…
B. coli	0	…
B. paratyphosus A	0	…
Experiment 3 :		
Staphylococcus pyogenes	16	…
Gonococcus	16	…
Meningococcus	17	…
Experiment 4 :		
Staphylococcus pyogenes	17	…
,, *epidermidis*	18	…
Streptococcus pyogenes	15	…
,, *viridans* (fæces)	5	…
B. diphtheriæ (2 strains)	14	…
Diphtheroid bacillus	10	…
Gram-negative coccus from the mouth (1)	12	…
,, ,, ,, (2)	0	…
B. coli	0	…
B. influenzæ (Pfeiffer) 6 strains	0	…

TABLE III.—*Inhibitory Power of Penicillin on Different Bacteria.*

Dilution of penicillin in broth.

	1/5.	1/10.	1/20.	1/40.	1/80.	1/100.	1/200.	1/400.	1/800.	1/1600.	1/3200.	Control.
Staphylococcus aureus	0	0	0	0	0	0	0	0	±	+ +	+ +	+ +
,, *epidermidis*	0	0	0	0	0	0	0	0	±	+ +	+ +	+ +
Pneumococcus	0	0	0	0	0	0	0	0	0	+ +	+ +	+ +
Streptococcus (hæmolytic)	0	0	0	0	0	0	0	0	0	±	+ +	+ +
,, *viridans* (mouth)	0	0	0	0	0	0	±	+ +	+ +	+ +	+ +	+ +
,, *fæcalis*	+ +	+ +	+ +	+ +	+ +	+ +	+ +	+ +	+ +	+ +	+ +	+ +
B. anthracis	0	0	+	+	+ +	+ +	+ +	+ +	+ +	+ +	+ +	+ +
B. pseudo-tuberculosis rodentium	+	+	+ +	+ +	+ +	+ +	+ +	+ +	+ +	+ +	+ +	+ +
B. pullorum	+	+	+ +	+ +	+ +	+ +	+ +	+ +	+ +	+ +	+ +	+ +
B. dysenteriæ	+	+ +	+ +	+ +	+ +	+ +	+ +	+ +	+ +	+ +	+ +	+ +
B. coli	+ +	+ +	+ +	...	...	...	...	...	...	...	...	+ +
B. typhosus	+ +	+ +	+ +	...	...	...	...	...	...	...	...	+ +
B. pyocyaneus	+ +	+ +	+ +	...	...	...	...	...	...	...	...	+ +
B. proteus	+ +	+ +	+ +	...	...	...	...	...	...	...	...	+ +
V. choleræ	+ +	+ +	+ +	...	...	...	...	...	...	...	...	+ +

	1/60.	1/120.	1/300.	1/600.	Control.
B. diphtheriæ (3 strains)	0	±	+ +	+ +	+ +
Streptococcus pyogenes (13 strains)	0	0	0	+ +	+ +
,, ,, (1 ,,)	0	0	±	+ +	+ +
,, *fæcalis* (11 ,,)	+ +	+ +	+ +	+ +	+ +
,, *viridans* at random from fæces (1 strain)	0	0	0	+ +	+ +
,, ,, ,, ,, (2 strains)	0	0	±	+ +	+ +
,, ,, ,, ,, (1 strain)	0	±	+ +	+ +	+ +
,, ,, ,, ,, (1 ,,)	+	+ +	+ +	+ +	+ +
,, ,, ,, ,, (1 ,,)	+ +	+ +	+ +	+ +	+ +
,, ,, at random from mouth (1 ,,)	0	±	+ +	+ +	+ +
,, ,, ,, ,, (2 strains)	0	0	+ +	+ +	+ +
,, ,, ,, ,, (1 strain)	0	0	0	+ +	+ +

0 = no growth; ± = trace of growth; + = poor growth; + + = normal growth.

extracted in normal salt solution for 24 hours it is found that the extract has bacteriolytic properties.

If this extract is mixed with a thick suspension of staphylococcus suspension and incubated for 2 hours at 45° C. it will be found that the opacity of the suspension has markedly diminished and after 24 hours the previously opaque suspension will have become almost clear.

Influence of the medium on the antibacterial titre of the mould culture. — So far as has been ascertained nutrient broth is the most suitable medium for the production of penicillin. The addition of glucose or saccharose, which are fermented by the mould with the production of acid, delays or prevents the appearance of the antibacterial substance. Dilution of the broth with water delays the formation of the antibacterial substance and diminishes the concentration which is ultimately reached.

INHIBITORY POWER OF PENICILLIN ON THE GROWTH OF BACTERIA.

Tables II and III show the extent to which various microbes, pathogenic and non-pathogenic, are inhibited by penicillin. The first table shows the inhibition by the agar plate method and the second shows the inhibitory power when diluted in nutrient broth.

Certain interesting facts emerge from these Tables. It is clear that penicillin contains bacterio-inhibitory substance which is very active towards some microbes while not affecting others. The members of the coli-typhoid group are unaffected as are other intestinal bacilli such as *B. pyocyaneus*, *B. proteus* and *V. cholerae*. Other bacteria which are insensitive to penicillin are the enterococcus, some of the Gram-negative cocci of the mouth, Friedländer's pneumobacillus, and *B. influenzae* (Pfeiffer), while the action on *B. dysenteriae* (Flexner), and *B. pseudo-tuberculosis rodentium* is almost negligible. The anthrax bacillus is completely inhibited in a 1 in 10 dilution but in this case the inhibitory influence is trifling when compared with the effect on the pyogenic cocci.

It is on the pyogenic cocci and on bacilli of the diphtheria group that the action is most manifest.

Staphylococci are very sensitive, and the inhibitory effect is practically the same on all strains, whatever the colour or type of the staphylococcus.

Streptococcus pyogenes is also very sensitive. There were small differences in the titre with different strains, but it may be said generally that it is slightly more sensitive than staphylococcus.

Pneumococci are equally sensitive with *Streptococcus pyogenes*.

The green streptococci vary very considerably, a few strains being almost unaffected while others are as sensitive as *S. pyogenes*. Gonococci, meningococci, and some of the Gram-negative cocci found in nasal catarrhal conditions are about as sensitive as are staphylococci. Many of the Gram-negative cocci found in the mouth and throat are, however, quite insensitive.

B. diphtheriæ is less affected than staphylococcus but is yet completely inhibited by a 1% dilution of a fair sample of penicillin.

It may be noted here that penicillin, which is strongly inhibitory to many bacteria, does not inhibit the growth of the original penicillium which was used in its preparation.

A. FLEMING.

The Rate of Killing of Staphylococci by Penicillin.

Some bactericidal agents like the hypochlorites are extremely rapid in their action, others like flavine or novarsenobillon are slow. Experiments were made to find into which category penicillin fell.

To 1 c.c. volumes of dilutions in broth of penicillin were added 10 c.mm. volumes of a 1 in 1000 dilution of a staphylococcus broth culture. The tubes were then incubated at 37°C. and at intervals 10 c.mm. volumes were removed and plated with the following result :

	Number of colonies developing after sojourn in penicillin in concentrations as under:				
	Control.	1/80.	1/40.	1/20.	1/10.
Before . . .	27	27	27	27	27
After 2 hours . .	116	73	51	48	23
,, 4½ ,, . .	∝	13	1	2	5
,, 8 ,, . .	∝	0	0	0	0
,, 12 ,, . .	∝	0	0	0	0

It appears, therefore, that penicillin belongs to the group of slow acting antiseptics, and the staphylococci are only completely killed after an interval of over 4½ hours even in a concentration 30 or 40 times stronger than is necessary to inhibit completely the culture in broth. In the weaker concentrations it will be seen that at first there is growth of the staphylococci and only after some hours are the cocci killed off. The same thing can be seen if a series of dilutions of penicillin in broth are heavily infected with staphylococcus and incubated. If the cultures are examined after four hours it may be seen that growth has taken place apparently equally in all the tubes but when examined after being incubated overnight, the tubes containing penicillin in concentrations greater than 1 in 300 or 1 in 400 are perfectly clear while the control tube shows a heavy growth. This is a clear illustration of the bacteriolytic action of penicillin.

TOXICITY OF PENICILLIN.

The toxicity to animals of powerfully antibacterial mould broth filtrates appears to be very low. Twenty c.c. injected intravenously into a rabbit were not more toxic than the same quantity of broth. Half a c.c. injected intraperitoneally into a mouse weighing about 20 gm. induced no toxic symptoms. Constant irrigation of large infected surfaces in man was not accompanied by any toxic symptoms, while irrigation of the human conjunctiva every hour for a day had no irritant effect.

In vitro penicillin which completely inhibits the growth of staphylococci in a dilution of 1 in 600 does not interfere with leucocytic function to a greater extent than does ordinary broth.

USE OF PENICILLIN TO DEMONSTRATE OTHER BACTERIAL INHIBITIONS.

When materials like saliva or sputum are plated it is not uncommon to see, where the implant is thick, an almost pure culture of streptococci and

PENICILLIN.

pneumococci, and where the implant is thinner and the streptococcal colonies are more widely separated, other colonies appear, especially those of Gram-negative cocci. These Gram-negative cocci are inhibited by the streptococci (probably by the peroxide they produce in their growth) and it is only when the mass effect of the streptococci is reduced that they appear in the culture.

Penicillin may be used to give a striking demonstration of this inhibition of bacteria by streptococci and pneumococci. Sputum is spread thickly on a culture plate, and then 5 or 6 drops of penicillin is spread over one half of it. After incubation it may be seen that on the half untreated with penicillin there is a confluent growth of streptococci and pneumococci and nothing else, while on the penicillin-treated half many Gram-negative cocci appear which were inhibited by the streptococci and pneumococci, and can only flourish when these are themselves inhibited by the penicillin.

If some active penicillin is embedded in a streak across an agar plate planted with saliva an interesting growth sometimes results. On the portion most distal from the penicillin there are many streptococci, but these are obscured by coarsely growing cocci, so that the resultant growth is a copious confluent rough mass. These coarse growing cocci are extremely penicillin sensitive and stop growing about 25 mm. from the embedded penicillin. Then there is a zone of about 1 cm. wide of pure streptococci, then they are inhibited by the penicillin, and as soon as that happens Gram-negative cocci appear and grow right up to the embedded penicillin. The three zones of growth produced in this way are very striking.

USE OF PENICILLIN IN THE ISOLATION OF *B. INFLUENZÆ* (PFEIFFER) AND OTHER ORGANISMS.

It sometimes happens that in the human body a pathogenic microbe may be difficult to isolate because it occurs in association with others which grow more profusely and which mask it. If in such a case the first microbe is insensitive to penicillin and the obscuring microbes are sensitive, then by the use of this substance these latter can be inhibited while the former are allowed to develop normally. Such an example occurs in the body, certainly with *B. influenzæ* (Pfeiffer) and probably with Bordet's whooping-cough bacillus and other organisms. Pfeiffer's bacillus, occurring as it does in the respiratory tract, is usually associated with streptococci, pneumococci, staphylococci and Gram-negative cocci. All of these, with the exception of some of the Gram-negative cocci, are highly sensitive to penicillin and by the addition of some of this to the medium they can be completely inhibited while *B. influenzæ* is unaffected. A definite quantity of the penicillin may be incorporated with the molten culture medium before the plates are made, but an easier and very satisfactory method is to spread the infected material, sputum, nasal mucus, etc., on the plate in the usual way and then over one half of the plate spread 2 to 6 drops (according to potency) of the penicillin. This small amount of fluid soaks into the agar and after cultivation for 24 hours it will be found that the half of the plate without the penicillin will show the normal growth while on the penicillin treated half there will be nothing but *B. influenzæ* with Gram-negative cocci and occasionally some other microbe. This makes it infinitely easier to isolate these penicillin-insensitive organisms, and repeatedly

B. influenzæ has been isolated in this way when they have not been seen in films of sputum and when it has not been possible to detect them in plates not treated with penicillin. Of course if this method is adopted then a medium favourable for the growth of *B. influenzæ* must be used, e.g. boiled blood agar, as by the repression of the pneumococci and the staphylococci the symbiotic effect of these, so familiar in cultures of sputum on blood agar, is lost and if blood agar alone is used the colonies of *B. influenzæ* may be so minute as to be easily missed.

Figs. 3 and 4 are photographs of culture-plates made after the method described above. On the plate shown in Fig. 3 a mixture of staphylococci and *B. influenzæ* was spread over the whole plate of Fildes medium (Fildes, 1921), then 6 drops of penicillin were spread over the lower half of the plate. The upper half shows the mixed culture while the lower half gives a pure culture of *B. influenzæ*. Fig. 4 represents a culture of nasal mucus from a "cold" made on the same medium. Here, on the upper half (untreated with penicillin) staphylococci and diphtheroid bacilli grow abundantly, while on the treated (lower) half only some three or four colonies of *B. influenzæ* appear.

In conjunction with my colleague, Dr. McLean, a series of cultures were made from the throats of 25 nurses warded for "influenza." The swabs were planted on boiled blood agar and over half of each plate was spread 3 or 4 drops of penicillin. The results are set forth in Table IV.

TABLE IV.—*Summary of Results obtained from Post-Nasal Swabs in 25 Consecutive Cases of "Influenza."*

No.	Without penicillin			With penicillin		
	Pneumococcus or Streptococcus	B. influenzæ	Gram-negative cocci	Pneumococcus or Streptococcus	B. influenzæ	Gram-negative cocci
1.	+ +	+	+	−	+ +	+
2.	+ +	+ +	+ +	−	+ +	+
3.	+ +	+ +	+	−	+	−
4.	+	−	−	−	+	+
5.	+ +	−	−	−	+ +	−
6.	+ +	−	−	−	+ +	+ +
7.	+ +	−	+ +	−	+	+
8.	+	+	−	−	+	−
9.	+ +	−	−	−	+	+
10.	+ +	−	−	−	+	−
11.	+ +	−	−	−	+	−
12.	+ +	+	+ +	−	+	+
13.	+	+ +	+ +	−	+	+ +
14.	+ +	−	−	−	+	−
15.	+ +	−	−	−	+	+
16.	+ +	−	−	−	+	+
17.	+ +	−	−	−	+	+
18.	+ +	−	−	−	+	+
19.	+ +	+	−	−	+	+ +
20.	+ +	−	−	−	−	+
21.	+ +	−	−	−	+	+
22.	+ +	−	−	−	+	−
23.	+	−	+	−	+	+
24.	+ +	+ +	−	−	+ +	−
25.	+ +	−	+ +	−	−	−

In the above Table account has only been taken of the common microbes found in these cultures. In some there were a few diphtheroid bacilli which were always penicillin sensitive, and in others there were Gram-negative bacilli which were penicillin insensitive, although they were inhibited by streptococci or pneumococci. Pneumococci and streptococci were classed together, as complete tests were not made to differentiate one from the other.

(From the appearance of the colonies and the morphological characters pneumococci were evidently present in most cases in much larger numbers than were streptococci.)

The swabs were generally planted thickly and in some cases where the growth on the portion of the plate without penicillin was almost confluent, the cultures were sampled by taking smears from thick portions of the growth. In these cases it is possible that the results given do not give a quite complete picture of the cultures. This, however, does not affect the present argument that by the addition of penicillin to the culture medium, and the consequent inhibition of the pyogenic cocci, the isolation of *B. influenzæ* is very much easier. And in a number of cases it was isolated when it was completely missed in the cultures without penicillin.

It is quite immaterial how many pneumococci and streptococci are present in a specimen—they are completely inhibited—and even a few *B. influenzæ* can be isolated from a mixture with an enormous number of these cocci.

From a number of observations which have been made on sputum, postnasal and throat swabs it seems likely that by the use of penicillin, organisms of the *B. influenzæ* group will be isolated from a great variety of pathological conditions as well as from individuals who are apparently healthy.

DISCUSSION.

It has been demonstrated that a species of penicillium produces in culture a very powerful antibacterial substance which affects different bacteria in different degrees. Speaking generally it may be said that the least sensitive bacteria are the Gram-negative bacilli, and the most susceptible are the pyogenic cocci. Inhibitory substances have been described in old cultures of many organisms; generally the inhibition is more or less specific to the microbe which has been used for the culture, and the inhibitory substances are seldom strong enough to withstand even slight dilution with fresh nutrient material. Penicillin is not inhibitory to the original penicillium used in its preparation.

Emmerich and other workers have shown that old cultures of *B. pyocyaneus* acquire a marked bacteriolytic power. The bacteriolytic agent, pyocyanase, possesses properties similar to penicillin in that its heat resistance is the same and it exists in the filtrate of a fluid culture. It resembles penicillin also in that it acts only on certain microbes. It differs however in being relatively extremely weak in its action and in acting on quite different types of bacteria. The bacilli of anthrax, diphtheria, cholera and typhoid are those most sensitive to pyocyanase, while the pyogenic cocci are unaffected, but the percentages of pyocyaneus filtrate necessary for the inhibition of these organisms was 40, 33, 40 and 60 respectively (Bocchia, 1909). This degree of inhibition is hardly comparable with 0·2% or less of penicillin which is necessary to completely inhibit the pyogenic cocci or the 1% necessary for *B. diphtheriæ.*

Penicillin, in regard to infections with sensitive microbes, appears to have some advantages over the well-known chemical antiseptics. A good sample will completely inhibit staphylococci, *Streptococcus pyogenes* and pneumococcus in a dilution of 1 in 800. It is therefore a more powerful inhibitory agent than

18

is carbolic acid and it can be applied to an infected surface undiluted as it is non-irritant and non-toxic. If applied, therefore, on a dressing, it will still be effective even when diluted 800 times which is more than can be said of the chemical antiseptics in use. Experiments in connection with its value in the treatment of pyogenic infections are in progress.

In addition to its possible use in the treatment of bacterial infections penicillin is certainly useful to the bacteriologist for its power of inhibiting unwanted microbes in bacterial cultures so that penicillin insensitive bacteria can readily be isolated. A notable instance of this is the very easy isolation of Pfeiffers bacillus of influenza when penicillin is used.

In conclusion my thanks are due to my colleagues, Mr. Ridley and Mr. Craddock, for their help in carrying out some of the experiments described in this paper, and to our mycologist, Mr. la Touche, for his suggestions as to the identity of the penicillium.

SUMMARY.

1. A certain type of penicillium produces in culture a powerful antibacterial substance. The antibacterial power of the culture reaches its maximum in about 7 days at 20° C. and after 10 days diminishes until it has almost disappeared in 4 weeks.

2. The best medium found for the production of the antibacterial substance has been ordinary nutrient broth.

3. The active agent is readily filterable and the name " penicillin " has been given to filtrates of broth cultures of the mould.

4. Penicillin loses most of its power after 10 to 14 days at room temperature but can be preserved longer by neutralization.

5. The active agent is not destroyed by boiling for a few minutes but in alkaline solution boiling for 1 hour markedly reduces the power. Autoclaving for 20 minutes at 115° C. practically destroys it. It is soluble in alcohol but insoluble in ether or chloroform.

6. The action is very marked on the pyogenic cocci and the diphtheria group of bacilli. Many bacteria are quite insensitive, *e.g.* the coli-typhoid group, the influenza-bacillus group, and the enterococcus.

7. Penicillin is non-toxic to animals in enormous doses and is non-irritant. It doses not interfere with leucocytic function to a greater degree than does ordinary broth.

8. It is suggested that it may be an efficient antiseptic for application to, or injection into, areas infected with penicillin-sensitive microbes.

9. The use of penicillin on culture plates renders obvious many bacterial inhibitions which are not very evident in ordinary cultures.

10. Its value as an aid to the isolation of *B. influenzæ* has been demonstrated.

REFERENCES.

BIOURGE.—(1923) ' Des moissures du group *Penicillium* Link.' Louvain, p. '72.
EMMERICH, LOEUW AND KORSCHUN.—(1902) *Zbl. Bakt.*, 30, 1.
BOCCHIA.—(1909) *Ibid.*, 50, 220.
FILDES, P.—(1920) *Brit. J. Exp. Path.*, 1, 129.

PENICILLIN AS A CHEMOTHERAPEUTIC AGENT

BY

E. CHAIN, PH.D. CAMB.

H. W. FLOREY,
M.B. ADELAIDE,

A. D. GARDNER,
D.M. OXFD, F.R.C.S.

N. G. HEATLEY, PH.D. CAMB.

M. A. JENNINGS,
B.M. OXFD,

J. ORR-EWING,
B.M. OXFD,

A. G. SANDERS,
M.B. LOND.

(*From the Sir William Dunn School of Pathology, Oxford*)

IN recent years interest in chemotherapeutic effects has been almost exclusively focused on the sulphonamides and their derivatives. There are, however, other possibilities, notably those connected with naturally occurring substances. It has been known for a long time that a number of bacteria and moulds inhibit the growth of pathogenic micro-organisms. Little, however, has been done to purify or to determine the properties of any of these substances. The antibacterial substances produced by *Pseudomonas pyocyanea* have been investigated in some detail, but without the isolation of any purified product of therapeutic value.

Recently, Dubos and collaborators (1939, 1940) have published interesting studies on the acquired bacterial antagonism of a soil bacterium which have led to the isolation from its culture medium of bactericidal substances active against a number of gram-positive micro-organisms.[1] Pneumococcal infections in mice were successfully treated with one of these substances, which, however, proved to be highly toxic to mice (Hotchkiss and Dubos 1940) and dogs (McLeod et al. 1940).

Following the work on lysozyme in this laboratory it occurred to two of us (E. C. and H. W. F.) that it would be profitable to conduct a systematic investigation of the chemical and biological properties of the antibacterial

---

1. See *Lancet*, 1940, 1, 1172.

RESULTS OF THERAPEUTIC TESTS ON MICE INFECTED WITH *Strep. pyogenes*, *Staph. aureus* AND *Cl. septique*

In the survivors columns, *6, 12, 24* are **hours** and *2, 3, 4, 5, 6, 7, 8, 9, 10* are **days** under the heading "Survivors at end of—".

Expt.	—	Dose of infecting culture (c.cm.)	Interval before starting treatment (hrs.)	Duration of treatment	Single dose (mg.)	Total dose (mg.)	No. of mice	6	12	24	2	3	4	5	6	7	8	9	10
Strep. pyogenes—Lancefield, Gp. A.																			
1	Controls	0·5	..	..	..	..	25	..	15	9	8	6	..	5	..	4	..	..	4[1]
	Treated	0·5	1	12 hrs.	2	10·0	50	..	..	..	49	42	..	34	30	28	..	26	25
2	Controls	0·5[2]	..	..	..	..	25	24	3	0[3]	..	..	..	..	..	..	..	..	0
	Treated	0·5	2	45 hrs.	0·5	7·5	25	24	..	..	..	..	..	..	..	..	..	..	24
Staph. aureus[4]																			
1	Controls	1·0	..	..	..	..	24	21	1	0[3]	..	..	..	..	..	..	..	..	0
	Treated	1·0	1	55 hrs.	0·5	9·0	25	25	12	..	11	..	10	..	..	..	..	..	8
2	Controls	0·2[2]	..	..	..	..	24	23	15	5	0	..	..	..	..	..	..	..	0
	Treated	0·2	1	4 days	0·5	11·5	24	..	23	22	..	..	21	..	..	..	..	..	21
Cl. septique																			
1	Controls	see text	..	..	..	..	25	..	21	0[5]	..	..	..	..	..	..	..	..	0
	Treated	..	1	10 days	0·5	19	25	..	..	24	21	..	18	..	..	..	..	..	18
		..	1	10 ,,	1·0	38	25	..	..	..	..	..	24	..	..	..	..	..	24

1. A control mouse which was killed by mistake at 24 hrs. is counted as a survivor. Heart-blood culture strongly positive.
2. Between experiments 1 and 2 the virulence of the organism was raised by passage.
3. Controls all dead within 16 hrs.
4. A bovine strain exceptionally virulent to mice kindly supplied by Dr. H. J. Parish of the Wellcome Laboratories.
5. Controls all dead within 17 hrs.

substances produced by bacteria and moulds. This investigation was begun with a study of a substance with promising antibacterial properties, produced by a mould and described by Fleming (1929). The present preliminary report is the result of a coöperative investigation on the chemical, pharmacological and chemotherapeutic properties of this substance.

Fleming noted that a mould produced a substance which inhibited the growth, in particular, of staphylococci, streptococci, gonococci, meningococci and *Corynebacterium diphtheriæ*, but not of *Bacillus coli*, *Hæmophilus influenzæ*, *Salmonella typhi*, *P. pyocyanea*, *Bacillus proteus* or *Vibrio choleræ*. He suggested its use as an inhibitor in the isolation of certain types of bacteria, especially *H. influenzæ*. He also noted that the injection into animals of broth containing the substance, which he called " penicillin," was no more toxic than plain broth, and he suggested that the substance might be a useful antiseptic for application to infected wounds. The mould is believed to be closely related to *Penicillium notatum*. Clutterbuck, Lovell and Raistrick (1932) grew the mould in a medium containing inorganic salts only and isolated a pigment—chrysogenin—which had no antibacterial action. Their culture media contained penicillin but this was not isolated. Reid (1935) reported work on the inhibitory substance produced by Fleming's mould. He did not isolate it but noted some of its properties.

During the last year methods have been devised here for obtaining a considerable yield of penicillin, and for rapid assay of its inhibitory power. From the culture medium a brown powder has been obtained which is freely soluble in water. It and its solution are stable for a considerable time and though it is not a pure substance, its anti-bacterial activity is very great. Full details will, it is hoped, be published later.

EFFECTS ON NORMAL ANIMALS

Various tests were done on mice, rats and cats. There is some œdema at the site of subcutaneous injection of strong solutions (e.g. 10 mg. in 0·3 c.cm.). This may well be due to the hypertonicity of the solution. No sloughing of skin or suggestion of serious damage has ever been encountered even with the strongest solutions or after repeated injections into the same area.

Intravenous injections showed that the penicillin preparation was only slightly, if at all, toxic for mice. An intravenous injection of as much as 10 mg. (dissolved in 0·3 c.cm. distilled water) of the preparation we have used for the curative experiments did not produce any observable toxic reactions in a 23 g. mouse. It was subsequently found that 10 mg. of a preparation having twice the penicillin content of the above was apparently innocuous to a 20 g. mouse.

Subcutaneous injections of 10 mg. into two rats at 3-hourly intervals for 56 hours did not cause any obvious change in their behaviour. They were perhaps slightly less lively than normal rats but they continued to eat their food. Their blood showed a fall of total leucocytes after 24 hours, but after 48 hours the count had risen again to about the original total. There was, however, a relative decrease in the number of polymorphs, but the normal number was restored 24 hours after stopping the administration of the substance. One of these two rats was killed for histological examination ; there was some evidence that the tubule cells of the kidney were damaged. The other has remained perfectly well, and its weight increased from 76 to 110 g. in 23 days. It is to be noted that these rats received, weight for weight, about five times the dose of penicillin used in the curative experiments in mice. No evidence of toxic effects was obtained from the treated mice, which received penicillin for many days.

Other pharmacological effects.—On the blood-pressure, heart-beat and respiration of cats no effects have been observed after intravenous injection of 40 mg.—enough to bring up the concentration in the blood just after injection to 1/5000. *Perfusion* of the isolated cat's heart, with Ringer-Locke solution containing 1/5000 penicillin produced progressive slowing during 15 minutes and at the end of that time the heart looked as though it would stop beating ; however, it was quickly revived by perfusing with Ringer-Locke solution alone. The same depressant action was seen at 1/10,000 dilution but the effect was less than at 1/5000. Solutions are absorbed from the intestine in the rat without causing any observable damage to the mucosa. They are also readily absorbed after subcutaneous injection and the substance can be detected in the blood. It is excreted by the kidneys, the urine becoming bright yellow. At least 40-50% appears in the urine in a still active form. *Human leucocytes* remain active in a 1/1000 solution for at least 3 hours.

It must be emphasised that the results of these preliminary tests have been obtained with an impure substance and such slight toxic effects as have been noted may possibly be due, in part at least, to these impurities.

EFFECTS ON BACTERIA IN VITRO

In view of this slight evidence of tissue toxicity it is all the more striking that the substance in a dilution of one in several hundred thousand inhibits in vitro the growth of many micro-organisms, including anaerobes. Of those so far tested in this laboratory the following are sensitive to the inhibiting action of our preparation : *Clostridium welchii* (2 strains) ; *Cl. septique* (1 strain, Nat. coll. type-cultures No. 458) ; *Cl. œdematiens* (1 strain, N.C.T.C. No. 277) ; *C. diphtheriæ* (1 strain, mitis type) ; *Streptococcus pyogenes* (Lancefield group A) ; *Str. viridans* (1 strain from tooth) ; *Str. pneumoniæ* (type 8) ; staphylococci (3 strains). Penicillin is not immediately bactericidal but seems to interfere with multiplication.

THERAPEUTIC EFFECTS

From all the above tests it was clear that this substance possessed qualities which made it suitable for trial as a chemotherapeutic agent. Therapeutic tests were therefore done on mice infected with streptococci, staphylococci and *Cl. septique* ; the results are summarised in the accompanying table, the preliminary trials on small numbers of mice being omitted.

Bacteriological methods : streptococcus and staphylococcus.—The staphylococcus was kept in culture in meat extract broth and the streptococcus in the same with serum. Both organisms were repeatedly passed through mice and were used for experiment 2–4 days after the last passage. The experimental infection was induced with 20–24 hour broth cultures injected intraperitoneally. According to opacity measurements with Brown's tubes, doses of 450 and 350 million cocci, living and dead, were given respectively in the two streptococcal experiments, and doses of 760 and 200 million in the staphylococcal experiments.

Pathogenic anaerobes.—The therapeutic effects were tried in mice infected with spores of *Cl. septique* in the manner described by Henderson and Gorer (1940). Attempts to carry out similar tests with *Cl. welchii* were temporarily abandoned owing to the difficulty of establishing in mice a type of infection which is both certainly fatal and allows adequate time before death for treatment to take effect. A spore suspension of *Cl. septique* was made by anaerobic growth at 37° C. for 48 hours on Dorset's egg slopes, suspension of the growth of each slope (surface about 15 sq. cm.) in about 1 c.cm. of sterile distilled water, centrifugalisation, once washing with distilled water, recentrifugalisation and resuspension in the original volume of distilled water. The suspension was then heated to 75° C. for one hour. The virulence of the suspension, when injected with an equal volume of 5% calcium chloride was roughly titrated by injection into the thigh muscle of mice, and a dose was chosen for the therapeutic experiment such as would kill for certain without being unnecessarily severe, viz., 0·05 c.cm. of the suspension diluted one in three, mixed with 0·05 c.cm. of the calcium chloride solution. This was injected into the thigh muscles of the 75 mice whose subsequent fate is recorded in the table.

Treatment.—The general principle has been to keep up an inhibitory concentration of the substance in the tissues of the body throughout the period of treatment by repeated subcutaneous injections. No extended tests have been made to determine the minimum effective quantities or the longest intervals possible between injections. The doses employed have been effective and not toxic ; they may have been excessive. The solution contained 10 mg. per c.cm. of substance.

In the first streptococcal experiment the treatment was continued for 12 hours only. That this was inadequate was shown by deaths occurring during the arbitrarily chosen 10-day period. In the second experiment the time of treatment was lengthened, with improved results. In this and the two staphylococcal experiments injections were given 3-hourly for the first 32–37 hours, then at longer intervals.

In preliminary experiments with *Cl. septique* it was found that the infection could be satisfactorily held in check so long as penicillin was being given (i.e., for 2 days) but when the administration was stopped the infection developed. In the experiment quoted, therefore, the injections were given for 10 days, 3-hourly for 41 hours, then at longer intervals, and twice daily for the last 2 days of the period. No deaths have subsequently occurred (22 days after beginning of experiment).

The behaviour of the mice infected with streptococci and staphylococci was interesting. For some hours after the start of treatment they looked sick—some even appeared to be dying—but as the experiment went on they progressively improved till at the end of 24 hours in the case of the streptococci and about 36 to 48 hours with the staphylococci it was difficult or impossible to distinguish them from normal mice. The survivors of the *Cl. septique* infection on the other hand remained well throughout, except for a few in which leg lesions appeared near the site of injection and cleared up in a few hours.

Summarising the data given in the table we see that in the final streptococcus experiment (no. 2) whereas 25/25 controls died, 24/25 treated animals survived. With *Staphylococcus aureus* the final experiment (no. 2) shows 24/24 deaths of the controls and 21/24 survivals among the treated. Lastly, with *Cl. septique* when the larger doses of penicillin were given (bottom line) the figures are 25/25 control deaths and 24/25 treatment survivals.

CONCLUSIONS

The results are clear cut, and show that penicillin is active in vivo against at least three of the organisms inhibited in vitro. It would seem a reasonable hope that all organisms inhibited in high dilution in vitro will be found to be dealt with in vivo. Penicillin does not appear to be related to any chemotherapeutic substance at present in use and is particularly remarkable for its activity against the anaerobic organisms associated with gas gangrene.

In addition to the facilities provided by the university we have had financial assistance from the Rockefeller Foundation, the Medical Research Council and the Nuffield Trust. N. G. Heatley has held a Rockefeller fellowship. To all of these we wish to express our thanks. We also wish to acknowledge the technical assistance of D. S. Callow, G. Glister, S. A. Creswell, J. A. Kent, and E. Vincent.

REFERENCES

Clutterbuck, P. W., Lovell, R. and Raistrick, H. (1932) *Biochem. J.* 26, 1907.
Dubos, R. J. and Cattaneo, C. J. (1939) *J. exp. Med.* 70, 249.
Fleming, A. (1929) *Brit. J. exp. Path.* 10, 226.
Henderson, D. W. and Gorer, P. A. (1940) *J. Hyg., Camb.* 40, 345.
Hotchkiss, R. D. and Dubos, R. J. (1940) *J. Biol. Chem.* 132, 791, 793.
McLeod, C. M., Mirick, G. S. and Curnen, E. I. (1940) *Proc. Soc. exp. Biol.* 43, 461.
Reid, R. D. (1935) *J. Bact.* 29, 215.

LETTERS TO THE EDITORS

An Enzyme from Bacteria able to Destroy Penicillin

FLEMING[1] noted that the growth of *B. coli* and a number of other bacteria belonging to the coli-typhoid group was not inhibited by penicillin. This observation has been confirmed. Further work has been done to find the cause of the resistance of these organisms to the action of penicillin.

An extract of *B. coli* was made by crushing a suspension of the organisms in the bacterial crushing mill of Booth and Green[2]. This extract was found to contain a substance destroying the growth-inhibiting property of penicillin. The destruction took place on incubating the penicillin preparation with the bacterial extract at 37°, or at room temperature for a longer time. The following is a typical experiment showing the penicillin-destroying effect of *B. coli* extracts. A solution of 1 mgm. penicillin in 0·8 c.c. of water was incubated with 0·2 c.c. of centrifuged and dialysed bacterial extract at 37° for 3 hours, in the presence of ether, and a control solution of penicillin of equal concentration was incubated without enzyme for the same time. (The penicillin used was extracted from cultures of *Penicillium notatum* by a method to be described in detail later. It possessed a degree of purity similar to that of the samples used in the chemotherapeutic experiments recorded in a preliminary report[3].) The growth-inhibiting activity of the solutions was then tested quantitatively on agar plates against *Staphylococcus aureus*. The penicillin solution incubated with the enzyme had entirely lost its growth-inhibiting activity, whereas the control solution had retained its full strength.

The conclusion that the active substance is an enzyme is drawn from the fact that it is destroyed by heating at 90° for 5 minutes and by incubation with papain activated with potassium cyanide at *p*H 6, and that it is non-dialysable through 'Cellophane' membranes. It can be precipitated by 2 volumes of alcohol, but much of its activity is lost during this operation. The activity of the enzyme, which we term penicillinase, is slight at *p*H 5, but increases considerably towards the alkaline range of *p*H. It is very active at *p*H 8 and 9. Higher *p*H's could not be tested as penicillin is unstable above *p*H 9.

The mechanism of the enzymatic inactivation of penicillin is being studied. No oxygen uptake occurs during the reaction, and the inactivation proceeds with equal facility under aerobic and anaerobic conditions. No appearance of acid groups could be detected by *p*H measurement with the hydrogen electrode. Extracts of a number of other microorganisms, made by crushing the bacteria in the bacterial grinding mill, were tested for penicillinase. The enzyme was absent from extracts of the penicillin-sensitive *Staphylococcus aureus*, of yeast and of *Penicillium notatum*. It was present in a Gram-negative rod, insensitive to penicillin, found as a contaminant of some Penicillium cultures. Unlike *B. coli*, it was not necessary to crush the organism in the bacterial mill in order to obtain the enzyme from it ; the latter appeared in the culture fluid. The enzyme was also found in *M. lysodeikticus*, an organism sensitive to the action of penicillin, though less so than *Staphylococcus aureus*. Thus, the presence or absence of the enzyme in a bacterium may not be the sole factor determining its insensitivity or sensitivity to penicillin.

The tissue extracts and tissue autolysates that have been tested were found to be without action on the growth-inhibiting power of penicillin. Prof. A. D. Gardner has found staphylococcal pus to be devoid of inhibiting action, but has demonstrated a slight inhibition by the pus from a case of *B. coli* cystitis. The bacteriostatic action of the sulphonamide drugs is known to be inhibited in the presence of tissue constituents and pus.[4]   That the anti-bacterial activity of penicillin is not affected under these conditions gives this substance a definite advantage over the sulphonamide drugs from the chemotherapeutic point of view. The fact that a number of bacteria contain an enzyme acting on penicillin points to the possibility that this substance may have a function in their metabolism.

E. P. ABRAHAM.
E. CHAIN.

Sir William Dunn School of Pathology,
  Oxford.
  Dec. 5.

[1] Fleming, A., *Brit. J. Exp. Path.*, **10**, 226 (1929).
[2] Booth, V. H., and Green, D. E., *Biochem. J.*, **32**, 855 (1938).
[3] Chain, E., Florey, H. W., Gardner, A. D., Heatley, N. G., Jennings, M. A., Orr-Ewing, J., and Sanders, A. G., *Lancet*, 226 (1940).
[4] MacLeod, C., *J. Exp. Med.*, **72**, 217 (1940).

Studies on the Chemical Nature of the Substance Inducing Transformation of Pneumococcal Types. Induction of Transformation by a Desoxyribonucleic Acid Fraction Isolated from Pneumococcus Type III

O. T. AVERY, C. M. MACLEOD, AND M. MCCARTY

The authors of this paper were the first to report that the biochemical material responsible for the transfer of genetic information was deoxyribonucleic acid (DNA). The observations presented in this manuscript laid the foundation for the evolution of our understanding of the molecular function of DNA and for the development of Watson and Crick's Nobel Prize-winning model of that structure. In this study, Avery and colleagues pursued an observation reported by F. Griffith (*J. Hyg.* **27**:113–159, 1928) that, until that time, had not been fully appreciated. Griffith found that when a culture of avirulent, live, rough *Streptococcus pneumoniae* type II was mixed with virulent, heat-killed smooth-type III pneumococci and injected subcutaneously into mice, the animals frequently died. Smooth-type III *S. pneumoniae* were isolated from the heart blood of dying animals. Avery et al. made the critical connection that transformation of *S. pneumoniae* from an avirulent phenotype to a virulent one was a consequence of transfer of DNA from dead smooth organisms to live rough ones. The authors correctly concluded that the "... chemically induced alterations in cellular structure and function are predictable, type-specific, and transmissible."

ALISON O'BRIEN

STUDIES ON THE CHEMICAL NATURE OF THE SUBSTANCE INDUCING TRANSFORMATION OF PNEUMOCOCCAL TYPES

INDUCTION OF TRANSFORMATION BY A DESOXYRIBONUCLEIC ACID FRACTION ISOLATED FROM PNEUMOCOCCUS TYPE III

By OSWALD T. AVERY, M.D., COLIN M. MacLEOD, M.D., AND MACLYN McCARTY,* M.D.

(From the Hospital of The Rockefeller Institute for Medical Research)

PLATE 1

(Received for publication, November 1, 1943)

Biologists have long attempted by chemical means to induce in higher organisms predictable and specific changes which thereafter could be transmitted in series as hereditary characters. Among microörganisms the most striking example of inheritable and specific alterations in cell structure and function that can be experimentally induced and are reproducible under well defined and adequately controlled conditions is the transformation of specific types of Pneumococcus. This phenomenon was first described by Griffith (1) who succeeded in transforming an attenuated and non-encapsulated (R) variant derived from one specific type into fully encapsulated and virulent (S) cells of a heterologous specific type. A typical instance will suffice to illustrate the techniques originally used and serve to indicate the wide variety of transformations that are possible within the limits of this bacterial species.

Griffith found that mice injected subcutaneously with a small amount of a living R culture derived from Pneumococcus Type II together with a large inoculum of heat-killed Type III (S) cells frequently succumbed to infection, and that the heart's blood of these animals yielded Type III pneumococci in pure culture. The fact that the R strain was avirulent and incapable by itself of causing fatal bacteremia and the additional fact that the heated suspension of Type III cells contained no viable organisms brought convincing evidence that the R forms growing under these conditions had newly acquired the capsular structure and biological specificity of Type III pneumococci.

The original observations of Griffith were later confirmed by Neufeld and Levinthal (2), and by Baurhenn (3) abroad, and by Dawson (4) in this laboratory. Subsequently Dawson and Sia (5) succeeded in inducing transformation *in vitro*. This they accomplished by growing R cells in a fluid medium containing anti-R serum and heat-killed encapsulated S cells. They showed that in the test tube as in the animal body transformation can be selectively induced, depending on the type specificity of the S cells used in the reaction system. Later, Alloway (6) was able to cause

---

* Work done in part as Fellow in the Medical Sciences of the National Research Council.

137

specific transformation *in vitro* using sterile extracts of S cells from which all formed elements and cellular debris had been removed by Berkefeld filtration. He thus showed that crude extracts containing active transforming material in soluble form are as effective in inducing specific transformation as are the intact cells from which the extracts were prepared.

Another example of transformation which is analogous to the interconvertibility of pneumococcal types lies in the field of viruses. Berry and Dedrick (7) succeeded in changing the virus of rabbit fibroma (Shope) into that of infectious myxoma (Sanarelli). These investigators inoculated rabbits with a mixture of active fibroma virus together with a suspension of heat-inactivated myxoma virus and produced in the animals the symptoms and pathological lesions characteristic of infectious myxomatosis. On subsequent animal passage the transformed virus was transmissible and induced myxomatous infection typical of the naturally occurring disease. Later Berry (8) was successful in inducing the same transformation using a heat-inactivated suspension of washed elementary bodies of myxoma virus. In the case of these viruses the methods employed were similar in principle to those used by Griffith in the transformation of pneumococcal types. These observations have subsequently been confirmed by other investigators (9).

The present paper is concerned with a more detailed analysis of the phenomenon of transformation of specific types of Pneumococcus. The major interest has centered in attempts to isolate the active principle from crude bacterial extracts and to identify if possible its chemical nature or at least to characterize it sufficiently to place it in a general group of known chemical substances. For purposes of study, the typical example of transformation chosen as a working model was the one with which we have had most experience and which consequently seemed best suited for analysis. This particular example represents the transformation of a non-encapsulated R variant of Pneumococcus Type II to Pneumococcus Type III.

EXPERIMENTAL

Transformation of pneumococcal types *in vitro* requires that certain cultural conditions be fulfilled before it is possible to demonstrate the reaction even in the presence of a potent extract. Not only must the broth medium be optimal for growth but it must be supplemented by the addition of serum or serous fluid known to possess certain special properties. Moreover, the R variant, as will be shown later, must be in the reactive phase in which it has the capacity to respond to the transforming stimulus. For purposes of convenience these several components as combined in the transforming test will be referred to as the *reaction system*. Each constituent of this system presented problems which required clarification before it was possible to obtain consistent and reproducible results. The various components of the system will be described in the following order: (1) nutrient broth, (2) serum or serous fluid, (3) strain of R Pneumococcus, and (4) extraction, purification, and chemical nature of the transforming principle.

1. Nutrient Broth.—Beef heart infusion broth containing 1 per cent neopeptone with no added dextrose and adjusted to an initial pH of 7.6–7.8 is used as the basic medium.   Individual lots of broth show marked and unpredictable variations in the property of supporting transformation.   It has been found, however, that charcoal adsorption, according to the method described by MacLeod and Mirick (10) for removal of sulfonamide inhibitors, eliminates to a large extent these variations; consequently this procedure is used as routine in the preparation of consistently effective broth for titrating the transforming activity of extracts.

2. Serum or Serous Fluid.—In the first successful experiments on the induction of transformation *in vitro*, Dawson and Sia (5) found that it was essential to add serum to the medium.   Anti-R pneumococcal rabbit serum was used because of the observation that reversion of an R pneumococcus to the homologous S form can be induced by growth in a medium containing anti-R serum.   Alloway (6) later found that ascitic or chest fluid and normal swine serum, all of which contain R antibodies, are capable of replacing antipneumococcal rabbit serum in the reaction system.   Some form of serum is essential, and to our knowledge transformation *in vitro* has never been effected in the absence of serum or serous fluid.

In the present study human pleural or ascitic fluid has been used almost exclusively. It became apparent, however, that the effectiveness of different lots of serum varied and that the differences observed were not necessarily dependent upon the content of R antibodies, since many sera of high titer were found to be incapable of supporting transformation.   This fact suggested that factors other than R antibodies are involved.

It has been found that sera from various animal species, irrespective of their immune properties, contain an enzyme capable of destroying the transforming principle in potent extracts.   The nature of this enzyme and the specific substrate on which it acts will be referred to later in this paper.   This enzyme is inactivated by heating the serum at 60°–65°C., and sera heated at temperatures known to destroy the enzyme are often rendered effective in the transforming system.   Further analysis has shown that certain sera in which R antibodies are present and in which the enzyme has been inactivated may nevertheless fail to support transformation.   This fact suggests that still another factor in the serum is essential.   The content of this factor varies in different sera, and at present its identity is unknown.

There are at present no criteria which can be used as a guide in the selection of suitable sera or serous fluids except that of actually testing their capacity to support transformation.   Fortunately, the requisite properties are stable and remain unimpaired over long periods of time; and sera that have been stored in the refrigerator for many months have been found on retesting to have lost little or none of their original effectiveness in supporting transformation.

The recognition of these various factors in serum and their rôle in the reaction system has greatly facilitated the standardization of the cultural conditions required for obtaining consistent and reproducible results.

3. The R Strain (R36A).—The unencapsulated R strain used in the present study was derived from a virulent "S" culture of Pneumococcus Type II. It will be recalled that irrespective of type derivation all "R" variants of Pneumococcus are characterized by the lack of capsule formation and the consequent loss of both type specificity and the capacity to produce infection in the animal body. The designation of these variants as R forms has been used to refer merely to the fact that on artificial media the colony surface is "rough" in contrast to the smooth, glistening surface of colonies of encapsulated S cells.

The R strain referred to above as R36A was derived by growing the parent S culture of Pneumococcus Type II in broth containing Type II antipneumococcus rabbit serum for 36 serial passages and isolating the variant thus induced. The strain R36A has lost all the specific and distinguishing characteristics of the parent S organisms and consists only of attenuated and non-encapsulated R variants. The change S → R is often a reversible one provided the R cells are not too far "degraded." The reversion of the R form to its original specific type can frequently be accomplished by successive animal passages or by repeated serial subculture in anti-R serum. When reversion occurs under these conditions, however, the R culture invariably reverts to the encapsulated form of the same specific type as that from which it was derived (11). Strain R36A has become relatively fixed in the R phase and has never spontaneously reverted to the Type II S form. Moreover, repeated attempts to cause it to revert under the conditions just mentioned have in all instances been unsuccessful.

The reversible conversion of S⇌R within the limits of a single type is quite different from the transformation of one specific type of Pneumococcus into another specific type through the R form. Transformation of types has never been observed to occur spontaneously and has been induced experimentally only by the special techniques outlined earlier in this paper. Under these conditions, the enzymatic synthesis of a chemically and immunologically different capsular polysaccharide is specifically oriented and selectively determined by the specific type of S cells used as source of the transforming agent.

In the course of the present study it was noted that the stock culture of R36 on serial transfers in blood broth undergoes spontaneous dissociation giving rise to a number of other R variants which can be distinguished one from another by colony form. The significance of this in the present instance lies in the fact that of four different variants isolated from the parent R culture only one (R36A) is susceptible to the transforming action of potent extracts, while the others fail to respond and are wholly inactive in this regard. The fact that differences exist in the responsiveness of different R variants to the same specific stimulus enphasizes the care that must be exercised in the selection of a suitable R variant for use in experiments on transformation. The capacity of this R strain (R36A) to respond to a variety of different transforming agents is shown by the readiness with which it can be transformed to Types I, III, VI, or XIV, as well as to its original type (Type II), to which, as pointed out, it has never spontaneously reverted.

Although the significance of the following fact will become apparent later on, it must be mentioned here that pneumococcal cells possess an enzyme capable of destroying the activity of the transforming principle. Indeed, this enzyme has been found to be present and highly active in the autolysates of a number of different strains. The fact that this intracellular enzyme is released during autolysis may explain, in part at least, the observation of Dawson and Sia (5) that it is essential in bringing about transformation in the test tube to use a small inoculum of young and actively growing R cells. The irregularity of the results and often the failure to induce transformation when large inocula are used may be attributable to the release from autolyzing cells of an amount of this enzyme sufficient to destroy the transforming principle in the reaction system.

In order to obtain consistent and reproducible results, two facts must be borne in mind: first, that an R culture can undergo spontaneous dissociation and give rise to other variants which have lost the capacity to respond to the transforming stimulus; and secondly, that pneumococcal cells contain an intracellular enzyme which when released destroys the activity of the transforming principle. Consequently, it is important to select a responsive strain and to prevent as far as possible the destructive changes associated with autolysis.

Method of Titration of Transforming Activity.—In the isolation and purification of the active principle from crude extracts of pneumococcal cells it is desirable to have a method for determining quantitatively the transforming activity of various fractions.

The experimental procedure used is as follows: Sterilization of the material to be tested for activity is accomplished by the use of alcohol since it has been found that this reagent has no effect on activity. A measured volume of extract is precipitated in a sterile centrifuge tube by the addition of 4 to 5 volumes of absolute ethyl alcohol, and the mixture is allowed to stand 8 or more hours in the refrigerator in order to effect sterilization. The alcohol precipitated material is centrifuged, the supernatant discarded, and the tube containing the precipitate is allowed to drain for a few minutes in the inverted position to remove excess alcohol. The mouth of the tube is then carefully flamed and a dry, sterile cotton plug is inserted. The precipitate is redissolved in the original volume of saline. Sterilization of active material by this technique has invariably proved effective. This procedure avoids the loss of active substance which may occur when the solution is passed through a Berkefeld filter or is heated at the high temperatures required for sterilization.

To the charcoal-adsorbed broth described above is added 10 per cent of the sterile ascitic or pleural fluid which has previously been heated at 60°C. for 30 minutes, in order to destroy the enzyme known to inactivate the transforming principle. The enriched medium is distributed under aseptic conditions in 2.0 cc. amounts in sterile tubes measuring 15 $\times$ 100 mm. The sterilized extract is diluted serially in saline neutralized to pH 7.2–7.6 by addition of 0.1 N NaOH, or it may be similarly diluted in M/40 phosphate buffer, pH 7.4. 0.2 cc. of each dilution is added to at least 3 or 4 tubes of the serum medium. The tubes are then seeded with a 5 to 8 hour blood broth culture of R36A. 0.05 cc. of a 10^{-4} dilution of this culture is added to each tube, and the cultures are incubated at 37°C. for 18 to 24 hours.

The anti-R properties of the serum in the medium cause the R cells to agglutinate during growth, and clumps of the agglutinated cells settle to the bottom of the tube leaving a clear supernatant. When transformation occurs, the encapsulated S cells, not being affected by these antibodies, grow diffusely throughout the medium. On the other hand, in the absence of transformation the supernatant remains clear, and only sedimented growth of R organisms occurs. This difference in the character of growth makes it possible by inspection alone to distinguish tentatively between positive and negative results. As routine all the cultures are plated on blood agar for confirmation and further bacteriological identification. Since the extracts used in the present study were derived from Pneumococcus Type III, the differentiation between the colonies of the original R organism and those of the transformed S cells is especially striking, the latter being large, glistening, mucoid colonies typical of Pneumococcus Type III. Figs. 1 and 2 illustrate these differences in colony form.

A typical protocol of a titration of the transforming activity of a highly purified preparation is given in Table IV.

Preparative Methods

Source Material.—In the present investigation a stock laboratory strain of Pneumococcus Type III (A66) has been used as source material for obtaining the active principle. Mass cultures of these organisms are grown in 50 to 75 liter lots of plain beef heart infusion broth. After 16 to 18 hours' incubation at 37°C. the bacterial cells are collected in a steam-driven sterilizable Sharples centrifuge. The centrifuge is equipped with cooling coils immersed in ice water so that the culture fluid is thoroughly chilled before flowing into the machine. This procedure retards autolysis during the course of centrifugation. The sedimented bacteria are removed from the collecting cylinder and resuspended in approximately 150 cc. of chilled saline (0.85 per cent NaCl), and care is taken that all clumps are thoroughly emulsified. The glass vessel containing the thick, creamy suspension of cells is immersed in a water bath, and the temperature of the suspension rapidly raised to 65°C. During the heating process the material is constantly stirred, and the temperature maintained at 65°C. for 30 minutes. Heating at this temperature inactivates the intracellular enzyme known to destroy the transforming principle.

Extraction of Heat-Killed Cells.—Although various procedures have been used, only that which has been found most satisfactory will be described here. The heat-killed cells are washed with saline 3 times. The chief value of the washing process is to remove a large excess of capsular polysaccharide together with much of the protein, ribonucleic acid, and somatic "C" polysaccharide. Quantitative titrations of transforming activity have shown that not more than 10 to 15 per cent of the active material is lost in the washing, a loss which is small in comparison to the amount of inert substances which are removed by this procedure.

After the final washing, the cells are extracted in 150 cc. of saline containing sodium desoxycholate in final concentration of 0.5 per cent by shaking the mixture mechanically 30 to 60 minutes. The cells are separated by centrifugation, and the extraction process is repeated 2 or 3 times. The desoxycholate extracts prepared in this manner are clear and colorless. These extracts are combined and precipitated by the addition of 3 to 4 volumes of absolute ethyl alcohol. The sodium desoxycholate being soluble in alcohol remains in the supernatant and is thus removed at this step. The precipitate forms a fibrous mass which floats to the surface of the alcohol and can be removed directly by lifting it out with a spatula. The excess alcohol is drained from the precipitate which is then redissolved in about 50 cc. of saline. The solution obtained is usually viscous, opalescent, and somewhat cloudy.

Deproteinization and Removal of Capsular Polysaccharide.—The solution is then deproteinized by the chloroform method described by Sevag (12). The procedure is repeated 2 or 3 times until the solution becomes clear. After this preliminary treatment the material is reprecipitated in 3 to 4 volumes of alcohol. The precipitate obtained is dissolved in a larger volume of saline (150 cc.) to which is added 3 to 5 mg. of a purified preparation of the bacterial enzyme capable of hydrolyzing the Type III capsular polysaccharide (13). The mixture is incubated at 37°C., and the destruction of the capsular polysaccharide is determined by serological tests with Type III antibody solution prepared by dissociation of immune precipitate according to the method described by Liu and Wu (14). The advantages of using the antibody solution for this purpose are that it does not react with other serologically active substances in the extract and that it selectively detects the presence of the capsular polysaccharide in dilutions as high as 1:6,000,000. The enzymatic breakdown of the polysaccharide is usually complete within 4 to 6 hours, as evidenced by the loss of serological reactivity. The digest is then precipitated in 3 to 4 volumes of ethyl alcohol, and the precipitate is redissolved in 50 cc. of saline. Deproteinization by the chloroform process is again used to remove the added enzyme protein and remaining traces of pneumococcal protein. The procedure is repeated until no further film of protein-chloroform gel is visible at the interface.

Alcohol Fractionation.—Following deproteinization and enzymatic digestion of the capsular polysaccharide, the material is repeatedly fractionated in ethyl alcohol as follows. Absolute ethyl alcohol is added dropwise to the solution with constant stirring. At a critical concentration varying from 0.8 to 1.0 volume of alcohol the active material separates out in the form of fibrous strands that wind themselves around the stirring rod. This precipitate is removed on the rod and washed in a 50 per cent mixture of alcohol and saline. Although the bulk of active material is removed by fractionation at the critical concentration, a small but appreciable amount remains in solution. However, upon increasing the concentration of alcohol to 3 volumes, the residual fraction is thrown down together with inert material in the form of a flocculent precipitate. This flocculent precipitate is taken up in a small volume of saline (5 to 10 cc.) and the solution again fractionated by the addition of 0.8 to 1.0 volume of alcohol. Additional fibrous material is obtained which is combined with that recovered from the original solution. Alcoholic fractionation is repeated 4 to 5 times. The yield of fibrous material obtained by this method varies from 10 to 25 mg. per 75 liters of culture and represents the major portion of active material present in the original crude extract.

Effect of Temperature.—As a routine procedure all steps in purification were carried out at room temperature unless specifically stated otherwise. Because of the theoretical advantage of working at low temperature in the preparation of biologically active material, the purification of one lot (preparation 44) was carried out in the cold. In this instance all the above procedures, with the exception of desoxycholate extraction and enzyme treatment were conducted in a cold room maintained at 0–4°C. This preparation proved to have significantly higher activity than did material similarly prepared at room temperature.

Desoxycholate extraction of the heat-killed cells at low temperature is less efficient and yields smaller amounts of the active fraction. It has been demonstrated that higher temperatures facilitate extraction of the active principle, although activity is best preserved at low temperatures.

Analysis of Purified Transforming Material

General Properties.—Saline solutions containing 0.5 to 1.0 mg. per cc. of the purified substance are colorless and clear in diffuse light. However, in strong transmitted light the solution is not entirely clear and when stirred exhibits a silky sheen. Solutions at these concentrations are highly viscous.

Purified material dissolved in physiological salt solution and stored at 2–4°C. retains its activity in undiminished titer for at least 3 months. However, when dissolved in distilled water, it rapidly decreases in activity and becomes completely inert within a few days. Saline solutions stored in the frozen state in a CO_2 ice box (−70°C.) retain full potency for several months. Similarly, material precipitated from saline solution by alcohol and stored under the supernatant remains active over a long period of time. Partially purified material can be preserved by drying from the frozen state in the lyophile apparatus. However, when the same procedure is used for the preservation of the highly purified substance, it is found that the material undergoes changes resulting in decrease in solubility and loss of activity.

The activity of the transforming principle in crude extracts withstands heating for 30 to 60 minutes at 65°C. Highly purified preparations of active material are less stable, and some loss of activity occurs at this temperature. A quantitative study of the effect of heating purified material at higher temperatures has not as yet been made. Alloway (6), using crude extracts prepared from Type III pneumococcal cells, found that occasionally activity could still be demonstrated after 10 minutes' exposure in the water bath to temperatures as high as 90°C.

The procedures mentioned above were carried out with solutions adjusted to neutral reaction, since it has been shown that hydrogen ion concentrations in the acid range result in progressive loss of activity. Inactivation occurs rapidly at pH 5 and below.

Qualitative Chemical Tests.—The purified material in concentrated solution gives negative biuret and Millon tests. These tests have been done directly on dry material with negative results. The Dische diphenylamine reaction for desoxyribonucleic acid is strongly positive. The orcinol test (Bial) for ribonucleic acid is weakly positive. However, it has been found that in similar concentrations pure preparations of desoxyribonucleic acid of animal origin prepared by different methods give a Bial reaction of corresponding intensity.

Although no specific tests for the presence of lipid in the purified material have been made, it has been found that crude material can be repeatedly extracted with alcohol and ether at $-12°C$. without loss of activity. In addition, as will be noted in the preparative procedures, repeated alcohol precipitation and treatment with chloroform result in no decrease in biological activity.

Elementary Chemical Analysis.[1]—Four purified preparations were analyzed for content of nitrogen, phosphorus, carbon, and hydrogen. The results are presented in Table I. The nitrogen-phosphorus ratios vary from 1.58 to 1.75 with an average value of 1.67 which is in close agreement with that calculated

TABLE I

Elementary Chemical Analysis of Purified Preparations of the Transforming Substance

Preparation No.	Carbon	Hydrogen	Nitrogen	Phosphorus	N/P ratio
	per cent	*per cent*	*per cent*	*per cent*	
37	34.27	3.89	14.21	8.57	1.66
38B	—	—	15.93	9.09	1.75
42	35.50	3.76	15.36	9.04	1.69
44	—	—	13.40	8.45	1.58
Theory for sodium desoxyribonucleate.....	34.20	3.21	15.32	9.05	1.69

on the basis of the theoretical structure of sodium desoxyribonucleate (tetranucleotide). The analytical figures by themselves do not establish that the substance isolated is a pure chemical entity. However, on the basis of the nitrogen-phosphorus ratio, it would appear that little protein or other substances containing nitrogen or phosphorus are present as impurities since if they were this ratio would be considerably altered.

Enzymatic Analysis.—Various crude and crystalline enzymes[2] have been tested for their capacity to destroy the biological activity of potent bacterial extracts. Extracts buffered at the optimal pH, to which were added crystalline trypsin and chymotrypsin or combinations of both, suffered no loss in activity following treatment with these enzymes. Pepsin could not be tested because

[1] The elementary chemical analyses were made by Dr. A. Elek of The Rockefeller Institute.

[2] The authors are indebted to Dr. John H. Northrop and Dr. M. Kunitz of The Rockefeller Institute for Medical Research, Princeton, N. J., for the samples of crystalline trypsin, chymotrypsin, and ribonuclease used in this work.

extracts are rapidly inactivated at the low pH required for its use. Prolonged treatment with crystalline ribonuclease under optimal conditions caused no demonstrable decrease in transforming activity. The fact that trypsin, chymotrypsin, and ribonuclease had no effect on the transforming principle is further evidence that this substance is not ribonucleic acid or a protein susceptible to the action of tryptic enzymes.

In addition to the crystalline enzymes, sera and preparations of enzymes obtained from the organs of various animals were tested to determine their effect on transforming activity. Certain of these were found to be capable of completely destroying biological activity. The various enzyme preparations tested included highly active phosphatases obtained from rabbit bone by the method of Martland and Robison (15) and from swine kidney as described by

TABLE II

The Inactivation of Transforming Principle by Crude Enzyme Preparations

Crude enzyme preparations	Enzymatic activity			
	Phosphatase	Tributyrin esterase	Depolymerase for desoxyribonucleate	Inactivation of transforming principle
Dog intestinal mucosa.....................	+	+	+	+
Rabbit bone phosphatase.................	+	+	−	−
Swine kidney "	+	−	−	−
Pneumococcus autolysates................	−	+	+	+
Normal dog and rabbit serum.............	+	+	+	+

H. and E. Albers (16). In addition, a preparation made from the intestinal mucosa of dogs by Levene and Dillon (17) and containing a polynucleotidase for thymus nucleic acid was used. Pneumococcal autolysates and a commercial preparation of pancreatin were also tested. The alkaline phosphatase activity of these preparations was determined by their action on β-glycerophosphate and phenyl phosphate, and the esterase activity by their capacity to split tributyrin. Since the highly purified transforming material isolated from pneumococcal extracts was found to contain desoxyribonucleic acid, these same enzymes were tested for depolymerase activity on known samples of desoxyribonucleic acid isolated by Mirsky[3] from fish sperm and mammalian tissues. The results are summarized in Table II in which the phosphatase, esterase, and nucleodepolymerase activity of these enzymes is compared with their capacity to destroy the transforming principle. Analysis of these results shows that irrespective of the presence of phosphatase or esterase only those

[3] The authors express their thanks to Dr. A. E. Mirsky of the Hospital of The Rockefeller Institute for these preparations of desoxyribonucleic acid.

preparations shown to contain an enzyme capable of depolymerizing authentic samples of desoxyribonucleic acid were found to inactivate the transforming principle.

Greenstein and Jenrette (18) have shown that tissue extracts, as well as the milk and serum of several mammalian species, contain an enzyme system which causes depolymerization of desoxyribonucleic acid. To this enzyme system Greenstein has later given the name desoxyribonucleodepolymerase (19). These investigators determined depolymerase activity by following the reduction in viscosity of solutions of sodium desoxyribonucleate. The nucleate and enzyme were mixed in the viscosimeter and viscosity measurements made at intervals during incubation at 30°C. In the present study this method was used in the measurement of depolymerase activity except that incubation was carried out at 37°C. and, in addition to the reduction of viscosity, the action of the enzyme was further tested by the progressive decrease in acid precipitability of the nucleate during enzymatic breakdown.

The effect of fresh normal dog and rabbit serum on the activity of the transforming substance is shown in the following experiment.

Sera obtained from a normal dog and normal rabbit were diluted with an equal volume of physiological saline. The diluted serum was divided into three equal portions. One part was heated at 65°C. for 30 minutes, another at 60°C. for 30 minutes, and the third was used unheated as control. A partially purified preparation of transforming material which had previously been dried in the lyophile apparatus was dissolved in saline in a concentration of 3.7 mg. per cc. 1.0 cc. of this solution was mixed with 0.5 cc. of the various samples of heated and unheated diluted sera, and the mixtures at pH 7.4 were incubated at 37°C. for 2 hours. After the serum had been allowed to act on the transforming material for this period, all tubes were heated at 65°C. for 30 minutes to stop enzymatic action. Serial dilutions were then made in saline and tested in triplicate for transforming activity according to the procedure described under Method of titration. The results given in Table III illustrate the differential heat inactivation of the enzymes in dog and rabbit serum which destroy the transforming principle.

From the data presented in Table III it is evident that both dog and rabbit serum in the unheated state are capable of completely destroying transforming activity. On the other hand, when samples of dog serum which have been heated either at 60°C. or at 65°C. for 30 minutes are used, there is no loss of transforming activity. Thus, in this species the serum enzyme responsible for destruction of the transforming principle is completely inactivated at 60°C. In contrast to these results, exposure to 65°C. for 30 minutes was required for complete destruction of the corresponding enzyme in rabbit serum.

The same samples of dog and rabbit serum used in the preceding experiment were also tested for their depolymerase activity on a preparation of sodium desoxyribonucleate isolated by Mirsky from shad sperm.

A highly viscous solution of the nucleate in distilled water in a concentration of 1 mg. per cc. was used. 1.0 cc. amounts of heated and unheated sera diluted in saline as shown in the preceding protocol were mixed in Ostwald viscosimeters with 4.0 cc.

TABLE III

Differential Heat Inactivation of Enzymes in Dog and Rabbit Serum Which Destroy the Transforming Substance

	Heat treatment of serum	Dilution*	Triplicate tests					
			1		2		3	
			Diffuse growth	Colony form	Diffuse growth	Colony form	Diffuse growth	Colony form
Dog serum	Unheated	Undiluted	−	R only	−	R only	−	R only
		1:5	−	R "	−	R "	−	R "
		1:25	−	R "	−	R "	−	R "
	60°C. for 30 min.	Undiluted	+	SIII	+	SIII	+	SIII
		1:5	+	SIII	+	SIII	+	SIII
		1:25	+	SIII	+	SIII	+	SIII
	65°C. for 30 min.	Undiluted	+	SIII	+	SIII	+	SIII
		1:5	+	SIII	+	SIII	+	SIII
		1:25	+	SIII	+	SIII	+	SIII
Rabbit serum	Unheated	Undiluted	−	R only	−	R only	−	R only
		1:5	−	R "	−	R "	−	R "
		1:25	−	R "	−	R "	−	R "
	60°C. for 30 min.	Undiluted	−	R only	−	R only	−	R only
		1:5	−	R "	−	R "	−	R "
		1:25	−	R "	−	R "	−	R "
	65°C. for 30 min.	Undiluted	+	SIII	+	SIII	+	SIII
		1:5	+	SIII	+	SIII	+	SIII
		1:25	+	SIII	+	SIII	+	SIII
Control (no serum)	None	Undiluted	+	SIII	+	SIII	+	SIII
		1:5	+	SIII	+	SIII	+	SIII
		1:25	+	SIII	+	SIII	+	SIII

* Dilution of the digest mixture of serum and transforming substance.

of the aqueous solution of the nucleate. Determinations of viscosity were made immediately and at intervals over a period of 24 hours during incubation at 37°C.

The results of this experiment are graphically presented in Chart 1. In the case of unheated serum of both dog and rabbit, the viscosity fell to that of water in 5 to 7 hours. Dog serum heated at 60°C. for 30 minutes brought about no significant reduction in viscosity after 22 hours.   On the other hand, heating rabbit serum at 60°C. merely reduced the rate of depolymerase action, and after 24 hours the viscosity was brought to the same level as with the unheated serum.   Heating at 65°C., however, completely destoyed the rabbit serum depolymerase.

Thus, in the case of dog and rabbit sera there is a striking parallelism between the temperature of inactivation of the depolymerase and that of the enzyme which destroys the activity of the transforming principle.   The fact that this difference in temperature of inactivation is not merely a general property of all enzymes in the sera is evident from experiments on the heat inactivation of

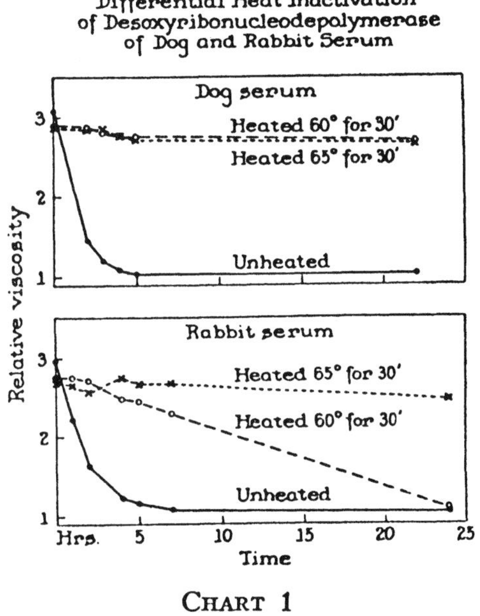

CHART 1

tributyrin esterase in the same samples of serum.   In the latter instance, the results are the reverse of those observed with depolymerase since the esterase of rabbit serum is almost completely inactivated at 60°C. while that in dog serum is only slightly affected by exposure to this temperature.

Of a number of substances tested for their capacity to inhibit the action of the enzyme known to destroy the transforming principle, only sodium fluoride has been found to have a significant inhibitory effect.   Regardless of whether this enzyme is derived from pneumococcal cells, dog intestinal mucosa, pancreatin, or normal sera its activity is inhibited by fluoride.   Similarly it has been found that fluoride in the same concentration also inhibits the enzymatic depolymerization of desoxyribonucleic acid.

The fact that transforming activity is destroyed only by those preparations containing depolymerase for desoxyribonucleic acid and the further fact that in both instances the enzymes concerned are inactivated at the same temperature and inhibited by fluoride provide additional evidence for the belief that the active principle is a nucleic acid of the desoxyribose type.

Serological Analysis.—In the course of chemical isolation of the active material it was found that as crude extracts were purified, their serological activity in Type III antiserum progressively decreased without corresponding loss in biological activity. Solutions of the highly purified substance itself gave only faint trace reactions in precipitin tests with high titer Type III antipneumococcus rabbit serum.[4] It is well known that pneumococcal protein can be detected by serological methods in dilutions as high as 1:50,000 and the capsular as well as the somatic polysaccharide in dilutions of at least 1:5,000,000. In view of these facts, the loss of serological reactivity indicates that these cell constituents have been almost completely removed from the final preparations. The fact that the transforming substance in purified state exhibits little or no serological reactivity is in striking contrast to its biological specificity in inducing pneumococcal transformation.

Physicochemical Studies.[5]—A purified and active preparation of the transforming substance (preparation 44) was examined in the analytical ultracentrigue. The material gave a single and unusually sharp boundary indicating that the substance was homogeneous and that the molecules were uniform in size and very asymmetric. Biological activity was found to be sedimented at the same rate as the optically observed boundary, showing that activity could not be due to the presence of an entity much different in size. The molecular weight cannot be accurately determined until measurements of the diffusion constant and partial specific volume have been made. However, Tennent and Vilbrandt (20) have determined the diffusion constant of several preparations of thymus nucleic acid the sedimentation rate of which is in close agreement with the values observed in the present study. Assuming that the asymmetry of the molecules is the same in both instances, it is estimated that the molecular weight of the pneumococcal preparation is of the order of 500,000.

Examination of the same active preparation was carried out by electrophoresis in the Tiselius apparatus and revealed only a single electrophoretic component of relatively high mobility comparable to that of a nucleic acid. Transforming activity was associated with the fast moving component giving the

[4] The Type III antipneumococcus rabbit serum employed in this study was furnished through the courtesy of Dr. Jules T. Freund, Bureau of Laboratories, Department of Health, City of New York.

[5] Studies on sedimentation in the ultracentrifuge were carried out by Dr. A. Rothen; the electrophoretic analyses were made by Dr. T. Shedlovsky, and the ultraviolet absorption curves by Dr. G. I. Lavin. The authors gratefully acknowledge their indebtedness to these members of the staff of The Rockefeller Institute.

optically visible boundary.   Thus in both the electrical and centrifugal fields, the behavior of the purified substance is consistent with the concept that biological activity is a property of the highly polymerized nucleic acid.

Ultraviolet absorption curves showed maxima in the region of 2600 Å and minima in the region of 2350 Å.   These findings are characteristic of nucleic acids.

Quantitative Determination of Biological Activity.—In its highly purified state the material as isolated has been found to be capable of inducing transformation in amounts ranging from 0.02 to 0.003 μg.   Preparation 44, the purification of which was carried out at low temperature and which had a nitrogen-phosphorus

TABLE IV

Titration of Transforming Activity of Preparation 44

| Transforming principle Preparation 44* | | Quadruplicate tests | | | | | | | |
| | | 1 | | 2 | | 3 | | 4 | |
Dilution	Amount added	Diffuse growth	Colony form	Diffuse growth	Colony form	Diffuse growth	Colony form	Diffuse growth	Colony form
	$\mu g.$								
10^{-2}	1.0	+	SIII	+	SIII	+	SIII	+	SIII
$10^{-2.5}$	0.3	+	SIII	+	SIII	+	SIII	+	SIII
10^{-3}	0.1	+	SIII	+	SIII	+	SIII	+	SIII
$10^{-3.5}$	0.03	+	SIII	+	SIII	+	SIII	+	SIII
10^{-4}	0.01	+	SIII	+	SIII	+	SIII	+	SIII
$10^{-4.5}$	0.003	−	R only	+	SIII	−	R only	+	SIII
10^{-5}	0.001	−	R "	−	R only	−	R "	−	R only
Control	None	−	R "	−	R "	−	R "	−	R "

*Solution from which dilutions were made contained 0.5 mg. per cc. of purified material. 0.2 cc. of each dilution added to quadruplicate tubes containing 2.0 cc. of standard serum broth.   0.05 cc. of a 10^{-4} dilution of a blood broth culture of R36A is added to each tube.

ratio of 1.58, exhibited high transforming activity.   Titration of the activity of this preparation is given in Table IV.

A solution containing 0.5 mg. per cc. was serially diluted as shown in the protocol. 0.2 cc. of each of these dilutions was added to quadruplicate tubes containing 2.0 cc. of standard serum broth.   All tubes were then inoculated with 0.05 cc. of a 10^{-4} dilution of a 5 to 8 hour blood broth culture of R36A.   Transforming activity was determined by the procedure described under Method of titration.

The data presented in Table IV show that on the basis of dry weight 0.003 μg. of the active material brought about transformation.   Since the reaction system containing the 0.003 μg. has a volume of 2.25 cc., this represents a final concentration of the purified substance of 1 part in 600,000,000.

DISCUSSION

The present study deals with the results of an attempt to determine the chemical nature of the substance inducing specific transformation of pneumococcal types. A desoxyribonucleic acid fraction has been isolated from Type III pneumococci which is capable of transforming unencapsulated R variants derived from Pneumococcus Type II into fully encapsulated Type III cells. Thompson and Dubos (21) have isolated from pneumococci a nucleic acid of the ribose type. So far as the writers are aware, however, a nucleic acid of the desoxyribose type has not heretofore been recovered from pneumococci nor has specific transformation been experimentally induced *in vitro* by a chemically defined substance.

Although the observations are limited to a single example, they acquire broader significance from the work of earlier investigators who demonstrated the interconvertibility of various pneumococcal types and showed that the specificity of the changes induced is in each instance determined by the particular type of encapsulated cells used to evoke the reaction. From the point of view of the phenomenon in general, therefore, it is of special interest that in the example studied, highly purified and protein-free material consisting largely, if not exclusively, of desoxyribonucleic acid is capable of stimulating unencapsulated R variants of Pneumococcus Type II to produce a capsular polysaccharide identical in type specificity with that of the cells from which the inducing substance was isolated. Equally striking is the fact that the substance evoking the reaction and the capsular substance produced in response to it are chemically distinct, each belonging to a wholly different class of chemical compounds.

The inducing substance, on the basis of its chemical and physical properties, appears to be a highly polymerized and viscous form of sodium desoxyribonucleate. On the other hand, the Type III capsular substance, the synthesis of which is evoked by this transforming agent, consists chiefly of a non-nitrogenous polysaccharide constituted of glucose-glucuronic acid units linked in glycosidic union (22). The presence of the newly formed capsule containing this type-specific polysaccharide confers on the transformed cells all the distinguishing characteristics of Pneumococcus Type III. Thus, it is evident that the inducing substance and the substance produced in turn are chemically distinct and biologically specific in their action and that both are requisite in determining the type specificity of the cell of which they form a part.

The experimental data presented in this paper strongly suggest that nucleic acids, at least those of the desoxyribose type, possess different specificities as evidenced by the selective action of the transforming principle. Indeed, the possibility of the existence of specific differences in biological behavior of nucleic acids has previously been suggested (23, 24) but has never been experimentally demonstrated owing in part at least to the lack of suitable biological methods.

The techniques used in the study of transformation appear to afford a sensitive means of testing the validity of this hypothesis, and the results thus far obtained add supporting evidence in favor of this point of view.

If it is ultimately proved beyond reasonable doubt that the transforming activity of the material described is actually an inherent property of the nucleic acid, one must still account on a chemical basis for the biological specificity of its action. At first glance, immunological methods would appear to offer the ideal means of determining the differential specificity of this group of biologically important substances. Although the constituent units and general pattern of the nucleic acid molecule have been defined, there is as yet relatively little known of the possible effect that subtle differences in molecular configuration may exert on the biological specificity of these substances. However, since nucleic acids free or combined with histones or protamines are not known to function antigenically, one would not anticipate that such differences would be revealed by immunological techniques. Consequently, it is perhaps not surprising that highly purified and protein-free preparations of desoxyribonucleic acid, although extremely active in inducing transformation, showed only faint trace reactions in precipitin tests with potent Type III antipneumococcus rabbit sera.

From these limited observations it would be unwise to draw any conclusion concerning the immunological significance of the nucleic acids until further knowledge on this phase of the problem is available. Recent observations by Lackman and his collaborators (25) have shown that nucleic acids of both the yeast and thymus type derived from hemolytic streptococci and from animal and plant sources precipitate with certain antipneumococcal sera. The reactions varied with different lots of immune serum and occurred more frequently in antipneumococcal horse serum than in corresponding sera of immune rabbits. The irregularity and broad cross reactions encountered led these investigators to express some doubt as to the immunological significance of the results. Unless special immunochemical methods can be devised similar to those so successfully used in demonstrating the serological specificity of simple non-antigenic substances, it appears that the techniques employed in the study of transformation are the only ones available at present for testing possible differences in the biological behavior of nucleic acids.

Admittedly there are many phases of the problem of transformation that require further study and many questions that remain unanswered largely because of technical difficulties. For example, it would be of interest to know the relation between rate of reaction and concentration of the transforming substance; the proportion of cells transformed to those that remain unaffected in the reaction system. However, from a bacteriological point of view, numerical estimations based on colony counts might prove more misleading than enlightening because of the aggregation and sedimentation of the R cells agglutinated by the antiserum in the medium. Attempts to induce transformation in suspensions of resting cells held under conditions inhibiting growth and multiplication have thus far proved unsuccessful, and it seems probable that transformation occurs only during active reproduction of the cells. Important in this connection is the fact that the R cells, as well as those that have undergone transformation, presumably also all other variants and types of pneumococci, contain an intracellular enzyme which is released during autolysis and in the free state is capable of rapidly and completely destroying the activity of the transforming agent. It would appear, therefore, that during the logarithmic phase of growth when cell division is most active and autolysis least apparent, the cultural conditions are optimal for the maintenance of the balance between maximal reactivity of the R cell and minimal destruction of the transforming agent through the release of autolytic ferments.

In the present state of knowledge any interpretation of the mechanism involved in transformation must of necessity be purely theoretical. The biochemical events underlying the phenomenon suggest that the transforming principle interacts with the R cell giving rise to a coordinated series of enzymatic reactions that culminate in the synthesis of the Type III capsular antigen. The experimental findings have clearly demonstrated that the induced alterations are not random changes but are predictable, always corresponding in type specificity to that of the encapsulated cells from which the transforming substance was isolated. Once transformation has occurred, the newly acquired characteristics are thereafter transmitted in series through innumerable transfers in artificial media without any further addition of the transforming agent. Moreover, from the transformed cells themselves, a substance of identical activity can again be recovered in amounts far in excess of that originally added to induce the change. It is evident, therefore, that not only is the capsular material reproduced in successive generations but that the primary factor, which controls the occurrence and specificity of capsular development, is also reduplicated in the daughter cells. The induced changes are not temporary modifications but are permanent alterations which persist provided the cultural conditions are favorable for the maintenance of capsule formation. The transformed cells can be readily distinguished from the parent R forms not alone by serological reactions but by the presence of a newly formed and visible capsule which is the immunological unit of type specificity and the accessory structure essential in determining the infective capacity of the microorganism in the animal body.

It is particularly significant in the case of pneumococci that the experimentally induced alterations are definitely correlated with the development of a new morphological structure and the consequent acquisition of new antigenic and invasive properties. Equally if not more significant is the fact that these changes are predictable, type-specific, and heritable.

Various hypotheses have been advanced in explanation of the nature of the changes induced. In his original description of the phenomenon Griffith (1) suggested that the dead bacteria in the inoculum might furnish some specific protein that serves as a "pabulum" and enables the R form to manufacture a capsular carbohydrate.

More recently the phenomenon has been interpreted from a genetic point of view (26, 27). The inducing substance has been likened to a gene, and the capsular antigen which is produced in response to it has been regarded as a gene product. In discussing the phenomenon of transformation Dobzhansky (27) has stated that "If this transformation is described as a genetic mutation—and it is difficult to avoid so describing it—we are dealing with authentic cases of induction of specific mutations by specific treatments. . . ."

Another interpretation of the phenomenon has been suggested by Stanley (28) who has drawn the analogy between the activity of the transforming agent and that of a virus. On the other hand, Murphy (29) has compared the causative agents of fowl tumors with the transforming principle of Pneumococcus. He has suggested that both these groups of agents be termed "transmissible mutagens" in order to differentiate them from the virus group. Whatever may prove to be the correct interpretation, these differences in viewpoint indicate the implications of the phenomenon of transformation in relation to similar problems in the fields of genetics, virology, and cancer research.

It is, of course, possible that the biological activity of the substance described is not an inherent property of the nucleic acid but is due to minute amounts of some other substance adsorbed to it or so intimately associated with it as to escape detection. If, however, the biologically active substance isolated in highly purified form as the sodium salt of desoxyribonucleic acid actually proves to be the transforming principle, as the available evidence strongly suggests, then nucleic acids of this type must be regarded not merely as structurally important but as functionally active in determining the biochemical activities and specific characteristics of pneumococcal cells. Assuming that the sodium desoxyribonucleate and the active principle are one and the same substance, then the transformation described represents a change that is chemically induced and specifically directed by a known chemical compound. If the results of the present study on the chemical nature of the transforming principle are confirmed, then nucleic acids must be regarded as possessing biological specificity the chemical basis of which is as yet undetermined.

SUMMARY

1. From Type III pneumococci a biologically active fraction has been isolated in highly purified form which in exceedingly minute amounts is capable under appropriate cultural conditions of inducing the transformation of unencapsulated R variants of Pneumococcus Type II into fully encapsulated cells of the

same specific type as that of the heat-killed microorganisms from which the inducing material was recovered.

2. Methods for the isolation and purification of the active transforming material are described.

3. The data obtained by chemical, enzymatic, and serological analyses together with the results of preliminary studies by electrophoresis, ultracentrifugation, and ultraviolet spectroscopy indicate that, within the limits of the methods, the active fraction contains no demonstrable protein, unbound lipid, or serologically reactive polysaccharide and consists principally, if not solely, of a highly polymerized, viscous form of desoxyribonucleic acid.

4. Evidence is presented that the chemically induced alterations in cellular structure and function are predictable, type-specific, and transmissible in series. The various hypotheses that have been advanced concerning the nature of these changes are reviewed.

CONCLUSION

The evidence presented supports the belief that a nucleic acid of the desoxyribose type is the fundamental unit of the transforming principle of Pneumococcus Type III.

BIBLIOGRAPHY

1. Griffith, F., *J. Hyg.*, Cambridge, Eng., 1928, **27**, 113.
2. Neufeld, F., and Levinthal, W., *Z. Immunitätsforsch.*, 1928, **55**, 324.
3. Baurhenn, W., *Centr. Bakt., 1. Abt., Orig.*, 1932, **126**, 68.
4. Dawson, M. H., *J. Exp. Med.*, 1930, **51**, 123.
5. Dawson, M. H., and Sia, R. H. P., *J. Exp. Med.*, 1931, **54**, 681.
6. Alloway, J. L., *J. Exp. Med.*, 1932, **55**, 91; 1933, **57**, 265.
7. Berry, G. P., and Dedrick, H. M., *J. Bact.*, 1936, **31**, 50.
8. Berry, G. P., *Arch. Path.*, 1937, **24**, 533.
9. Hurst, E. W., *Brit. J. Exp. Path.*, 1937, **18**, 23. Hoffstadt, R. E., and Pilcher, K. S., *J. Infect. Dis.*, 1941, **68**, 67. Gardner, R. E., and Hyde, R. R., *J. Infect. Dis.*, 1942, **71**, 47. Houlihan, R. B., *Proc. Soc. Exp. Biol. and Med.*, 1942, **51**, 259.
10. MacLeod, C. M., and Mirick, G. S., *J. Bact.*, 1942, **44**, 277.
11. Dawson, M. H., *J. Exp. Med.*, 1928, **47**, 577; 1930, **51**, 99.
12. Sevag, M. G., *Biochem. Z.*, 1934, **273**, 419. Sevag, M. G., Lackman, D. B., and Smolens, J., *J. Biol. Chem.*, 1938, **124**, 425.
13. Dubos, R. J., and Avery, O. T., *J. Exp. Med.*, 1931, **54**, 51. Dubos, R. J., and Bauer, J. H., *J. Exp. Med.*, 1935, **62**, 271.
14. Liu, S., and Wu, H., *Chinese J. Physiol.*, 1938, **13**, 449.
15. Martland, M., and Robison, R., *Biochem. J.*, 1929, **23**, 237.
16. Albers, H., and Albers, E., *Z. physiol. Chem.*, 1935, **232**, 189.
17. Levene, P. A., and Dillon, R. T., *J. Biol. Chem.*, 1933, **96**, 461.
18. Greenstein, J. P., and Jenrette, W. Y., *J. Nat. Cancer Inst.*, 1940, **1**, 845.

19. Greenstein, J. P., *J. Nat. Cancer Inst.*, 1943, **4,** 55.
20. Tennent, H. G., and Vilbrandt, C. F., *J. Am. Chem. Soc.*, 1943, **65,** 424.
21. Thompson, R. H. S., and Dubos, R. J., *J. Biol. Chem.*, 1938, **125,** 65.
22. Reeves, R. E., and Goebel, W. F., *J. Biol. Chem.*, 1941, **139,** 511.
23. Schultz, J., in Genes and chromosomes. Structure and organization, Cold Spring Harbor symposia on quantitative biology, Cold Spring Harbor, Long Island Biological Association, 1941, **9,** 55.
24. Mirsky, A. E., in Advances in enzymology and related subjects of biochemistry, (F. F. Nord and C. H. Werkman, editors), New York, Interscience Publishers, Inc., 1943, **3,** 1.
25. Lackman, D., Mudd, S., Sevag, M. G., Smolens, J., and Wiener, M., *J. Immunol.*, 1941, **40,** 1.
26. Gortner, R. A., Outlines of biochemistry, New York, Wiley, 2nd edition, 1938, 547.
27. Dobzhansky, T., Genetics and the origin of the species, New York, Columbia University Press, 1941, 47.
28. Stanley, W. M., in Doerr, R., and Hallauer, C., Handbuch der Virusforschung, Vienna, Julius Springer, 1938, **1,** 491.
29. Murphy, J. B., *Tr. Assn. Am. Physn.*, 1931, **46,** 182; *Bull. Johns Hopkins Hosp.*, 1935, **56,** 1.

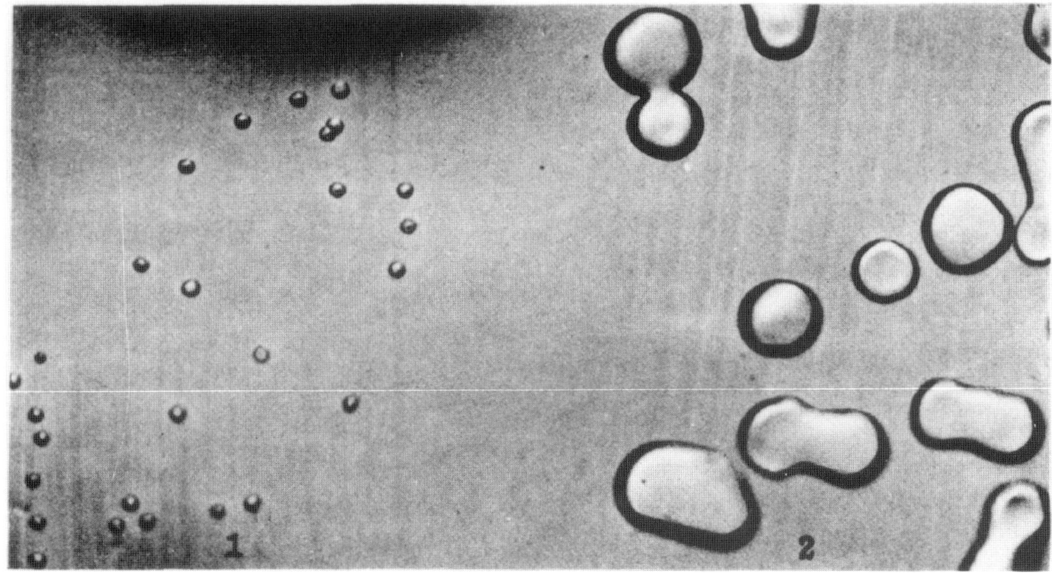

(Avery *et al.*: Transformation of pneumococcal types)

EXPLANATION OF PLATE 1

The photograph was made by Mr. Joseph B. Haulenbeek.

FIG. 1. Colonies of the R variant (R36A) derived from Pneumococcus Type II. Plated on blood agar from a culture grown in serum broth in the absence of the transforming substance.   ×3.5.

FIG. 2. Colonies on blood agar of the same cells after induction of transformation during growth in the same medium with the addition of active transforming principle isolated from Type III pneumococci.   The smooth, glistening, mucoid colonies shown are characteristic of Pneumococcus Type III and readily distinguishable from the small, rough colonies of the parent R strain illustrated in Fig. 1.   ×3.5.

Initiation of Immune Responses by Small Lymphocytes

J. L. GOWANS, D. D. MCGREGOR, D. M. COWEN, AND C. E. FORD

Before the discoveries of Sir James L. Gowans, the small lymphocyte was a cell without a function. Small lymphocytes could be enumerated: Winston Churchill, on seeing a lymphocyte count in his hospital chart, is reported to have asked, "What are these small lymphocytes?"

"We don't know, Prime Minister," his physician replied.

"Then why do you count them?"

That was the state of knowledge until the studies of Gowans. Gowans became interested in the behavior of lymphocytes because of their dynamic characteristics; the thoracic duct, which he could cannulate, produced enough small lymphocytes in a day to create a hefty dose of cells, but their function was unknown. It was Gowans who figured out that these small, featureless cells could mount both cellular and humoral immune responses to specific antigens. That is, they were the units of selection in Burnet's theory of clonal selection. Since Gowans' paper was published, the number of papers containing the word "lymphocyte" has doubled annually for many years, leading to the present-day understanding of adaptive immunity based on the clonal selection of small, resting lymphocytes. It is difficult to imagine that any other paper could have such enormous impact.

CHARLES JANEWAY

INITIATION OF IMMUNE RESPONSES BY SMALL LYMPHOCYTES

By Prof. J. L. GOWANS, Dr. D. D. McGREGOR and DIANA M. COWEN
Sir William Dunn School of Pathology, Oxford

AND

Dr. C. E. FORD
Radiobiological Research Unit, Harwell

ANIMALS may develop a fatal wasting disease if they are injected with adult lymphoid cells which lack antigens present in the tissues of the host. The ability of a particular cell-type to induce such a disease is an index of its 'immunological competence', in the sense that it must be able to react with antigens in the tissues of the host and initiate an immune response against them[1,2]. Work on the graft-versus-host reactions which occur when thoracic duct cells from genetically appropriate donor rats are injected into new-born rats[3], adult F_1 hybrid rats[4,5] and

lethally X-irradiated mice[5] has shown that the small lymphocyte is an immunologically competent cell. These investigations have also shown unequivocally that small lymphocytes can change rapidly into dividing cells with new morphological characteristics[4,5]. The change occurs in a proportion of the donor's small lymphocytes during their first 24-h residence in the lymphoid tissue of the host: the small lymphocytes enlarge, develop prominent nucleoli and a cytoplasm which stains strongly with pyronin; they then begin to divide. Such 'large pyroninophilic 'cells' do not appear when syngeneic (isologous) small lymphocytes are injected into normal hosts.

The large pyroninophilic cells which arise from small lymphocytes in graft-versus-host reactions resemble morphologically the cells which appear in regional lymphoid tissue during the reaction against homografts of skin[6] and during antibody formation[7,8,21]. This resemblance, together with the demonstration that the small lymphocyte is immunologically competent, suggest that primary immune responses in general may be initiated by the interaction of antigens with small lymphocytes. The experiments recorded here provide some support for this view: they show that, in the rat, small lymphocytes probably initiate the reaction against first-set homografts of skin and the primary antibody response to a particulate and a soluble antigen.

Two groups of experiments were performed. First, the identity of the cell-type formed by the dividing large pyroninophilic cells was determined by studying the fate of rat small lymphocytes in the spleen of lethally irradiated mice. The significance of these experiments for the homograft reaction was then examined. Secondly, rats were rendered immunologically unresponsive to skin homografts, sheep erythrocytes and tetanus toxoid. Suspensions of small lymphocytes from other rats were then injected intravenously to test their ability to abolish the unresponsive states.

Small lymphocytes. Almost pure suspensions of small lymphocytes were prepared by incubating rat thoracic duct cells *in vitro* for 24 h at 37° C in a conventional tissue culture medium[5]. This procedure kills most of the large lymphocytes in the culture.

In each experiment the proportion of small lymphocytes in the inoculum was estimated in smears from a count on 2,000 cells; the proportion ranged from 99·4 to more than 99·95 per cent.

Three methods were used for inducing immunological unresponsiveness in rats belonging to highly inbred strains.

(1) Immunological tolerance to skin homografts was induced in new-born albino rats by an intravenous injection of 10–40 million spleen cells from adult F_1 (albino × hooded) hybrid rats. A full thickness graft of F_1 skin was applied to the lateral thoracic wall when the injected rats were 6–8 weeks of age and they were regarded as highly tolerant if the grafts survived in good condition for more than 50 days. Four out of 40 grafted rats were highly tolerant, and these were used in the experiments described here.

Immunological tolerance to sheep erythrocytes was induced by an injection of $2·5 \times 10^9$ erythrocytes on the day of birth; half the dose was given intravenously and half intraperitoneally. Twice-weekly intraperitoneal injections of $2·5 \times 10^9$ erythrocytes were then given for 12 weeks, and the continued absence of hæmolysin from the serum was taken as evidence of tolerance. Cells were collected from the thoracic duct of these rats 4 days after the last injection of erythrocytes.

(2) Adult rats were made unresponsive to sheep erythrocytes and to tetanus toxoid by chronic drainage of cells from a thoracic duct fistula[9]. Lymph and cells were allowed to drain away for 5 days from the thoracic duct of restrained, unanæsthetized rats. They were then released from restraint and their fistulæ closed. 'Closure of the fistula' was performed by gently pulling the cannula out through the skin; a transient chylous ascites developed, but within a week normal lymph drainage from the intestine had been re-established. In all experiments antigen was given immediately after the fistula was closed.

(3) Exposure to 500 rads of whole-body X-irradiation rendered adult rats unresponsive to sheep erythrocytes.

Production of New Cells from Small Lymphocytes

Chromosome analyses have shown that when rat small lymphocytes are injected into a lethally irradiated mouse some of them colonize the white pulp of the spleen, develop into large pyroninophilic cells and start to divide[5]. The fate of these dividing cells was determined autoradiographically by labelling them in the mouse with a single pulse of tritiated thymidine. Individual mice were then killed at intervals after the single injection of thymidine and each spleen was divided into two parts: one part was sectioned and examined by stripping film autoradiography to identify the progeny of the originally labelled cells; the other part was used for a chromosome analysis to ensure that the labelled cells were 'rat' in origin.

Male *CBA* mice weighing 20–30 g were given an intravenous injection of 85 million rat small lymphocytes 24 h after 1,000 rads of whole-body X-irradiation. The inocula were contaminated with less than 0·05 per cent of any other cell-type. A single injection of tritiated thymidine (1 μc./g body-weight; specific activity, 4·4 c./m.mole) was given 48 h after the injection of cells. In the spleen, which was removed 12 h after the injection of thymidine, 28 mitotic cells were scored; all were 'rat'. Autoradiographs of this spleen showed heavy labelling over mitotic figures and over the nuclei of many large pyroninophilic cells in the white pulp. Other labelled cells in the white pulp possessed a less intensely pyroninophilic cytoplasm, no obvious nucleolus and a more pachychromatic nucleus; they were identified as 'large lymphocytes'. Labelled small lymphocytes were very rare at this time.

Cell division in the mouse spleens which were removed 37 and 61 h after the injection of thymidine was again confined almost exclusively to rat cells: 70 out of 71 and 72 out of 74 mitotic cells, respectively, were scored as 'rat'. At 61 h the white pulp of the spleen contained many labelled cells of which the great majority were lightly labelled small lymphocytes. No labelled mature plasma cells were found. It was concluded that small lymphocytes can form more small lymphocytes through the intermediary of the large pyroninophilic cell. The grain counts over the 'large lymphocytes' at 37 and 61 h suggest that they are intermediates between the large pyroninophilic cell and the small lymphocyte.

It could be objected that the labelled small lymphocytes were not derived from the originally injected rat cells but from mouse cells which had divided in the lymph nodes and had migrated by way of the blood into the spleen. This is very unlikely since, in another experiment, no dividing mouse cells were found in the mesenteric and brachial lymph nodes of *CBA* mice on the third, fifth and sixth days after 1,000 rads of X-irradiation. Rat small lymphocytes had again been injected 24 h after the irradiation.

Skin Homograft Reaction

During the response to a first-set homograft of skin in the rat the regional (axillary) lymph node becomes populated by large pyroninophilic cells which are identical in appearance to those described by Scothorne[6] in the regional node draining a homograft in the rabbit and to those which have been shown to develop from rat small lymphocytes[5]. The demonstration that the large pyroninophilic cell can itself generate more small lymphocytes suggests a possible origin for the small lymphocytes which heavily infiltrate first-set homografts in the rat: some small lymphocytes in the regional node react to antigens from the graft by differentiating into large pyroninophilic cells; these produce a specific crop of new small lymphocytes which leave the node, circulate in the blood and invade the graft. Two further experiments support this possibility. First, it has been shown that virtually all the small lymphocytes which infiltrate a first-set homograft are formed after the application of the graft. Secondly, injections of thoracic duct lymphocytes from normal rats will cause the breakdown of a long-standing homograft of skin in an immunologically tolerant rat. The small lymphocytes which invade such grafts are not those which were injected; they are also newly formed cells.

Full thickness grafts of skin were exchanged between rats belonging to a randomly bred albino colony. The first signs of necrosis in such grafts occur at about 14 days. At the time of grafting and every 12 h thereafter tritiated thymidine was injected intraperitoneally (0·5 μc./g body-weight; specific activity, 4·4 c./m.mole). Biopsies of the graft were taken at 7 and 10 days and examined by stripping film autoradiography. At both times the dermis was heavily infiltrated with small lymphocytes and in autoradiographs exposed for 30 days more than 90 per cent of them were labelled. At 7 days about 7 per cent of the small lymphocytes in the blood were labelled, while at 10 days the medullary sinuses of the regional (axillary) lymph node contained about 16 per cent of labelled small lymphocytes. Thus, the small lymphocytes which invade a homograft are not a random sample from the total lymphocyte pool. It is likely that they are a specific group of new cells formed in response to the graft.

Destruction of Grafts in Immunologically Tolerant Rats

The immunological inadequacy of a tolerant animal can be corrected by an injection of lymph node cells from a normal (non-sensitized) animal[10]. It can also be corrected by an injection of cells from the thoracic duct[4,11]. Thus, the injection of three fully tolerant albino rats with 700–1,000 million thoracic duct cells (99·6 per cent small lymphocytes) from normal rats of the same inbred strain destroyed healthy F_1 skin homografts in 8–10 days; the tempo of destruction was as rapid as that of first-set grafts on normal recipients. In a fourth tolerant rat twice-daily injections of tritiated thymidine were started immediately after the injection of lymphocytes and were continued for 7 days. Again, autoradiographs showed that more than 90 per cent of the small lymphocytes which were present in the graft at 7 days were labelled.

The small lymphocytes which were injected in these experiments did not pass into the graft and generate new cells within it. By labelling the small lymphocytes *in vitro* with tritiated adenosine[5] before their injection it was shown autoradiographically that they 'homed' rapidly into all the lymph nodes of the animal. The basis of this homing is the special affinity of small lymphocytes for the endothelium of the post-capillary venules in the cortex of the nodes[12]. An electron microscope study has shown that small lymphocytes pass from the blood into the nodes by

actually entering the endothelial cells and traversing their cytoplasm[13].

The ability of albino-strain small lymphocytes to initiate a graft-versus-host reaction in F_1 (albino × hooded) hybrid rats[5] makes it very probable that they also initiate the destruction of F_1 skin grafts in tolerant rats. It seems likely that small lymphocytes interact with antigen derived from the F_1 graft and generate within lymphoid tissue a specific group of immunologically activated small lymphocytes through the intermediary of the large pyroninophilic cell. Soon after their formation the new cells are released into the blood. Consistent with this view is the demonstration that immunologically activated cells can be detected in the blood of homografted rats at approximately the same time as they appear in the regional lymph nodes[11].

Antibody Formation

Agents such as X-irradiation and cortisone which damage lymphoid tissue *in vivo*, and neonatal thymectomy which prevents its normal development[14], have provided circumstantial evidence that small lymphocytes are involved in primary immune responses. Animals subjected to these measures become severely depleted of small lymphocytes and this can be associated with a failure to produce circulating antibody[15]. However, the idea that the immunological deficiency is due solely to a lack of small lymphocytes would be strengthened if the unresponsiveness could be corrected by injections of small lymphocytes from normal animals. In the experiments reported here, adult rats have been depleted of lymphocytes by chronic drainage of cells from a thoracic duct fistula[9]. The primary antibody response of such animals is severely depressed but it can be restored by injections of small lymphocytes from other rats of the same highly inbred strain.

The loss of about 2×10^9 small lymphocytes over a 5-day period of drainage from a thoracic duct fistula is reflected in a reduction in the weight of all the lymph nodes, a gross depletion of small lymphocytes in the cortex of the nodes and in the white pulp of the spleen, and in a lymphopænia. In the rat, small lymphocytes normally recirculate from blood to lymph through all the lymph nodes[12,16]; thus, drainage from the thoracic duct withdraws lymphocytes from all the nodes and not merely from those in the splanchnic bed.

Fig. 1 shows that lymphocyte-depleted rats cannot be sensitized by a primary injection of tetanus toxoid. In a normal rat the second of two doses of toxoid given three weeks apart results in a brisk antibody response; but if the first dose is given immediately after lymph drainage then a second dose three weeks later elicits no response. On the other hand, the secondary response is not affected by lymphocyte-depletion: when the second dose is given after drainage from the thoracic duct a normal response is obtained. Thus, the cells which mediate the secondary response are 'fixed' within lymphoid tissue and cannot be withdrawn by drainage from the thoracic duct.

Fig. 2 shows that the primary immune response to sheep erythrocytes becomes progressively less as the period of lymph drainage is increased. The impairment of response is not due to stress or operative trauma[9] but to the progressive loss of small lymphocytes from the fistula: the unresponsiveness can be corrected by an injection of small lymphocytes from normal (non-immunized) rats of the same inbred

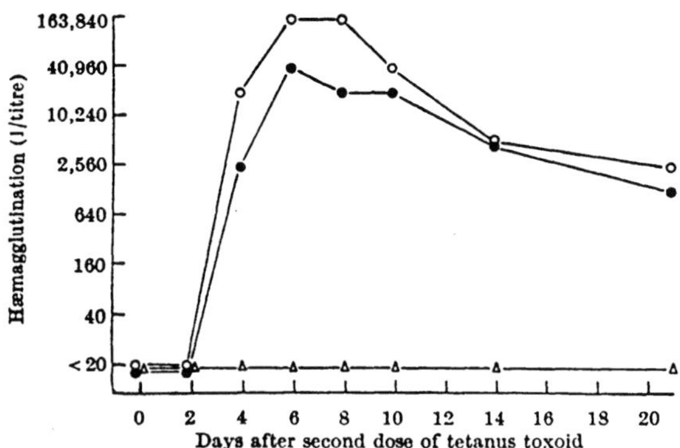

Fig. 1. Response to tetanus toxoid in a normal rat and rats depleted of lymphocytes by 5 days' drainage of cells from the thoracic duct. Each animal received two injections of 10 Lf of soluble toxoid 3 weeks apart, the first intraperitoneally, the second intravenously
●, Response of a normal rat; △, response of a rat given the first dose of toxoid immediately after the closure of a thoracic duct fistula; ○, response of a rat given the second dose of toxoid immediately after the fistula was closed. Antibody assayed by tanned red cell hæmagglutination (ref. 22)

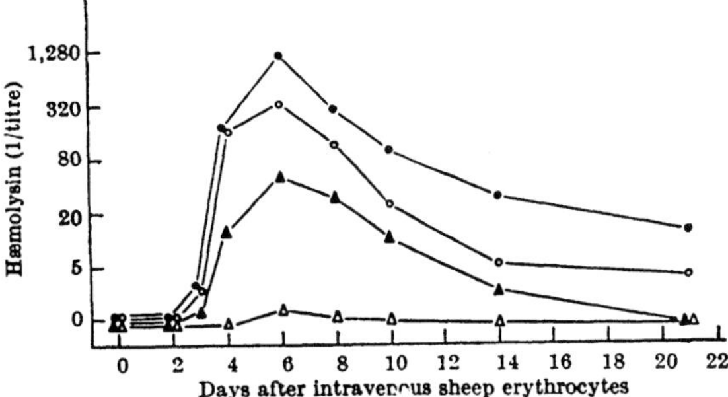

Fig. 2. Hæmolysin response to a single intravenous injection of 10⁸ sheep erythrocytes in normal rats and rats depleted of lymphocytes by drainage of cells from a thoracic duct fistula.
●, Mean response of 14 normal rats; ○, mean response of 4 rats after 2 days' drainage; ▲, mean response of 3 rats after 3 days' drainage; △, mean response of 10 rats after 5 days' drainage

strain. This restorative effect is related to the number of small lymphocytes injected. Thus, the injection of 10⁹ thoracic duct cells (99·75 per cent small lymphocytes) restored the response of 5-day drained rats to approximately the level which was observed after 2 days drainage (Fig. 3). About 10⁹ small lymphocytes emerge from a thoracic duct fistula between the second and fifth days after cannulation.

'Immunologically Tolerant' Small Lymphocytes

The hæmolysin response of rats to a single dose of 10⁸ sheep erythrocytes is abolished by prior administration of 500 rads of whole-body X-irradiation. This unresponsiveness could be corrected with smaller doses of lymphocytes than were necessary in rats which had been depleted by drainage from the thoracic duct. Fig. 4 shows that an intravenous injection of 2·5 × 10⁸ thoracic duct cells from a normal rat (99·4 per cent small lymphocytes) restored the hæmolysin response of an X-irradiated rat but that an equal number of fresh thoracic duct cells (not incubated *in vitro*; 95 per cent small lymphocytes) from a rat immunologically tolerant to sheep erythrocytes was without effect. This finding is important for two reasons. First, it shows that the restorative power of small lymphocytes is not merely a non-specific effect due to the injection of any cellular material. Secondly, it illus-

trates that immunological tolerance is a property of a population of small lymphocytes. Tolerant populations of cells have been described by others[17],[18]. The present experiments identify a tolerant cell-type; they do not, of course, distinguish a population which is tolerant because reactive cells have been eliminated from one in which tolerant cells have been created.

Discussion

There is strong evidence that small lymphocytes can initiate graft-versus-host reactions[5]. Our experiments show that they may also initiate the response to homografts of skin and, by first differentiating into large pyroninophilic cells, give rise to the immunologically activated small lymphocytes which invade the graft. The sequence, small lymphocyte → large pyroninophilic cell → small lymphocyte has been demonstrated in a graft-versus-host reaction. This scheme may imply that there are two classes of small lymphocytes: those which are immunologically 'committed', in that they, or their precursors, have already reacted to antigen; and those which are 'uncommitted'. The latter class can be proposed without prejudice to the idea that the range of response in individual cells may be restricted in the sense suggested by Burnet[19]. The recirculation of small lymphocytes[12],[16] could enable specifically reactive cells of either class to be selected from the total recirculating pool into any regionally stimulated node.

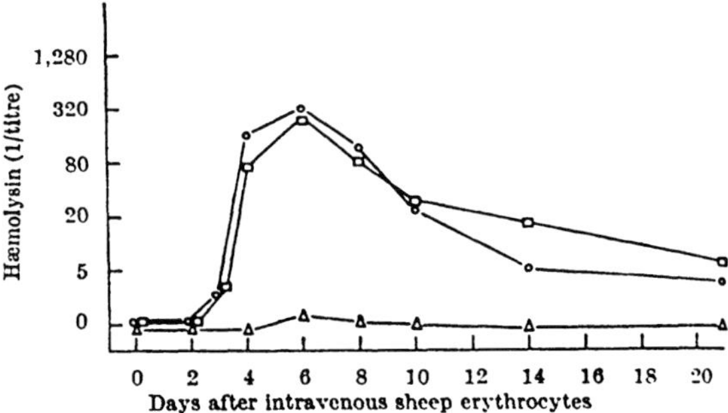

Fig. 3. Hæmolysin response to a single intravenous injection of 10⁸ sheep erythrocytes in lymphocyte-depleted rats
○, Mean response of 4 rats after 2 days' drainage from a thoracic duct fistula; △, mean response of 10 rats after 5 days' drainage; □, mean response of 4 rats after 5 days' drainage; 10⁹ small lymphocytes were injected intravenously when the fistula was closed and sheep erythrocytes were injected 48 h later

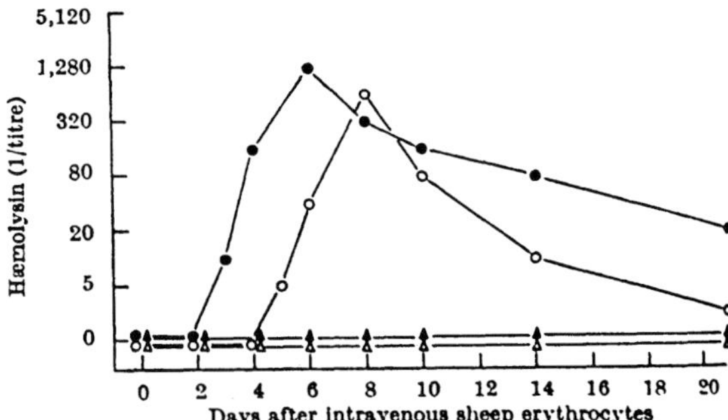

Fig. 4. Hæmolysin response to a single intravenous injection of 10⁸ sheep erythrocytes in a normal rat and rats exposed to 500 rads of total body X-irradiation; sheep erythrocytes were injected 24 h after irradiation
●, Response of a normal rat; △, response of an irradiated rat; ○, response of an irradiated rat injected intravenously 2 h after irradiation with 2·5 × 10⁸ small lymphocytes from a normal rat; ▲, response of an irradiated rat injected intravenously 2 h after irradiation with 2·5 × 10⁸ lymphocytes from a rat immunologically tolerant to sheep erythrocytes

The demonstration that a population of small lymphocytes can carry the property of immunological tolerance is strong evidence that they are involved in the inductive phase of antibody formation[18]. The simplest view would be that small lymphocytes interact with antigen (or with antigen which has been 'processed' by reticuloendothelial phagocytes), become fixed in lymphoid tissue and give rise to a dividing cell-line of the kind identified by Nossal and Mäkelä[20]. This cell-line generates the cells which eventually synthesize antibody. Although there is strong morphological evidence that small lymphocytes are the ultimate precursors of antibody-forming cells[5,21], it must be emphasized that this has not yet been unequivocally demonstrated.

We are very grateful to Dr. A. J. Fulthorpe for help with the estimation of tetanus antitoxin. One of us (D. D. McG.) was supported by the Medical Research Council of Canada.

[1] Simonsen, M., *Progress in Allergy*, edit. by Kallós, P., and Waksman, B. H., **6**, 349 (1962).
[2] Billingham, R. E., and Brent, L., *Phil. Trans. Roy. Soc.*, B, **242**, 439 (1959).
[3] Billingham, R. E., Defendi, V., Silvers, W. K., and Steinmuller, D., *J. Nat. Cancer Inst.*, **28**, 365 (1962).
[4] Gowans, J. L., Gesner, B. M., and McGregor, D. D., in *Biological Activity of the Leucocyte* (Ciba Foundation Study Group No. 10), 32 (London, 1961).
[5] Gowans, J. L., *Ann. N.Y. Acad. Sci.*, **99** (in the press).
[6] Scothorne, R. J., *Ann. N.Y. Acad. Sci.*, **64**, 1028 (1957).
[7] Fagraeus, A., *Acta Med. Scand.*, Supp. 204 (1948).
[8] Coons, A. H., Leduc, E. H., and Connolly, J. M., *J. Exp. Med.*, **102**, 49 (1955).
[9] McGregor, D. D., and Gowans, J. L. (in the press).
[10] Billingham, R. E., Brent, L., and Medawar, P. B., *Phil. Trans. Roy. Soc.*, B, **239**, 357 (1956).
[11] Billingham, R. E., Silvers, W. K., and Wilson, D. B., *Lancet*, i, 512 (1962).
[12] Gowans, J. L., and Knight, E. J. (in the press).
[13] Marchesi, V. T., and Gowans, J. L. (in the press).
[14] Miller, J. F. A. P., *Proc. Roy. Soc.*, B, **156**, 415 (1962).
[15] Janković, B. D., Waksman, B. H., and Arnason, B. G., *J. Exp. Med.*, **116**, 159 (1962).
[16] Gowans, J. L., *J. Physiol.*, **146**, 54 (1959).
[17] Michie, D., Woodruff, M. F. A., and Zeiss, I. M., *Immunology*, **4**, 413 (1961).
[18] Smith, R. T., *Advances in Immunology*, edit. by Taliaferro, W. H., and Humphrey, J. H., 1, 67 (1961).
[19] Burnet, F. M., *The Clonal Selection Theory of Acquired Immunity* (Cambridge, 1959).
[20] Nossal, J. G. V., and Mäkelä, O., *J. Exp. Med.*, **115**, 209 (1962).
[21] Langevoort, H. L., Keuning, F. J., Meer, J. v. d., Nieuwenhuis, P., and Oudendijk, P., *Proc. Acad. Sci. Amst.*, C, **64**, S97 (1961).
[22] Fulthorpe, A. J., *J. Hyg.* (Camb.), **55**, 382 (1957).

Effect of Diphtheria Toxin on Protein Synthesis: Inactivation of One of the Transfer Factors

R. J. COLLIER

The committee hailed this publication as the first report of the precise molecular function of a bacterial protein virulence factor. This study most certainly served as an inspiration to a number of scientists working on the modes of action of other bacterial toxins such as cholera toxin and *Pseudomonas* exotoxin A, and it heralded the beginning of the field of molecular and biochemical analysis of bacterial pathogenic components. In this publication, Collier clearly demonstrated that diphtheria toxin, which was already known to inhibit protein synthesis in cell-free systems, could, in the presence of the cofactor nicotinamide adenine dinucleotide (NAD), inactivate one of the supernatant factors required for the transfer of amino acids from aminoacyl-sRNA (now called tRNA) to ribosomes. He accomplished this remarkable biochemical feat by hypothesizing that diphtheria toxin must irreversibly inactivate one of the macromolecular components involved in protein synthesis and then systematically testing the effect of toxin on ribosomes, aminoacyl-sRNA, and the transfer factors. He designated as "factor II" the component that was inactivated; it is now called elongation factor 2 (EF-2). The understanding of how bacterial toxins act on eucaryotic cells has not only furthered the study of bacterial pathogenesis but has also led to the use of these substances as tools to dissect the biology of eucaryotic cells at the molecular level.

ALISON O'BRIEN

J. Mol. Biol. (1967) **25**, 83–98

Effect of Diphtheria Toxin on Protein Synthesis: Inactivation of One of the Transfer Factors

R. J. COLLIER†

Institut de Biologie Moléculaire
Université de Genève
Genève, Switzerland

(*Received 24 August 1966*)

The mechanism by which diphtheria toxin inhibits amino acid transfer from aminoacyl-sRNA to ribosomes has been studied in the reticulocyte cell-free system. In the presence of the required cofactor NAD, toxin has been found to inactivate one of the supernatant transfer factors by 65 to 85%; the complementary factor is not affected. Removal of NAD by gel filtration does not reverse the inactivation. Aminoacyl-sRNA and ribosomes treated with toxin plus NAD are fully active after removal of NAD. Toxin does not break down polysomes to single ribosomes and does not release nascent polypeptide chains from ribosomes.

1. Introduction

Diphtheria toxin, a protein of molecular weight about 70,000, has been found to inhibit protein synthesis in cell-free systems from mammalian cells (Kato, 1962; Collier & Pappenheimer, 1964b; Pappenheimer, Collier & Miller, 1963). NAD is specifically required for the inhibition by toxin, and in its presence, cell-free amino-acid incorporation is inhibited up to 95% by toxin concentrations on the order of 1 μg/ml.

Toxin has been found to exert its inhibitory effect in cell-free systems at the level of the transfer of amino acids from aminoacyl-sRNA to ribosomes (Collier & Pappenheimer, 1964b). In the present work, experiments have been carried out to determine the precise point of action of toxin on the transfer reaction. We have assumed that toxin irreversibly inactivates one of the macromolecular components of this reaction, and have examined the effect of toxin in turn upon ribosomes, aminoacyl-sRNA, and the transfer factors. The results show that neither ribosomes nor aminoacyl-sRNA are affected, and that toxin specifically inactivates one of the supernatant protein factors required for the transfer reaction.

2. Materials and Methods

(a) *Diphtheria toxin*

This was prepared according to a modification (Collier & Pappenheimer, 1964a) of the method of Yoneda (1957). The final product contained about 22 guinea-pig minimum lethal doses per μg protein.

† Present address: Department of Bacteriology, University of California, Los Angeles, Calif. 90024, U.S.A.

(b) *Reagents and radioactive compounds*

NAD and ATP were obtained from P-L Biochemicals, Inc., Milwaukee, Wisconsin. Boehringer & Soehne, Mannheim, Germany, supplied GTP and GSH, and poly U was purchased from Miles Chemical Company, Elkhart, Indiana. Puromycin was obtained from Nutritional Biochemicals Corp., Cleveland, Ohio, [³H]leucine, [¹⁴C]leucine and [¹⁴C]algal protein hydrolysate from New England Nuclear Corp., Boston, Mass., and [¹⁴C]phenylalanine from Radiochemical Centre, Amersham, England.

(c) *Reticulocyte fractions*

These were prepared by a modification of the method of Allen & Schweet (1962). Reticulocytes from phenylhydrazine-poisoned rabbits were washed twice with 0·14 M-NaCl containing 5 mM-KCl and 1·5 mM-MgCl₂, and lysed by adding 2 vol. of 2 mM-MgCl₂. 0·6 vol. of 1·5 M-sucrose containing 0·15 M-KCl was added after 2 min to stop lysis. After a centrifugation at 15,000 *g* for 15 min to remove unlysed cells and cell-membrane fragments, the ribosomes were sedimented by centrifugation at 78,000 *g* for 2·5 hr. The ribosomes were washed as described by Allen & Schweet (1962) and stored frozen. Ribosomes to be used in experiments involving poly U were further treated with 0·1 M-KCl and deoxycholate according to the method of Arlinghaus, Shaeffer & Schweet (1964).

The ribosome-free supernatant fraction was treated with protamine sulfate to remove nucleic acid; this step was often omitted, however. The supernatant solution was fractionated by precipitation with ammonium sulfate; the fraction precipitating between 35 and 55% saturation (SF-55) or between 40 and 70% saturation (SF-70) at 4°C was collected. The precipitated fraction from 100 ml. cells was routinely dissolved in 100 ml. of 0·01 M-Tris–HCl buffer (pH 7·2) containing 1 mM-GSH and 0·1 mM-EDTA, and reprecipitated by addition of an appropriate amount of saturated ammonium sulfate. The final precipitate was dissolved, dialyzed and stored according to the method of Allen & Schweet (1962). For convenience, SF-70 and SF-55 are referred to as *supernatant* in the text; the fraction used is specified in each experiment. A suffix of P indicates that the fraction had been treated with protamine sulfate.

(d) *[¹⁴C]Phenylalanyl-sRNA*

This was prepared from rabbit liver according to the method of Moldave (1963) and from *Escherichia coli* by the method of Nathans & Lipmann (1961). The final product from liver contained 17,500 cts/min/mg RNA and that from *E. coli* 140,000 cts/min/mg RNA.

(e) *Assay of amino acid incorporation in cell-free systems*

Two procedures were used for this purpose.

System A. Incorporation of free radioactive amino acids was carried out in the reaction mixture as described by Allen & Schweet (1962). SF-70 contained sufficient sRNA to permit incorporation of amino acids, and supplementary sRNA was not added. Incorporation of radioactive amino acids was assayed according to the method of Mans & Novelli (1960).

System B. Transfer of [¹⁴C]phenylalanine from [¹⁴C]phenylalanyl-sRNA to ribosomes stimulated by poly U was assayed in a reaction mixture of 0·25 ml. containing the following components: 12·5 μmoles Tris–HCl (pH 7·4, 30°C); 25 μmoles KCl; 2·5 μmoles MgCl₂; 0·1 μmole GTP; 4 μmoles GSH; 0·1 μmole phenylalanine; 40 μg poly U; 20 μg ribosomes; 1000 cts/min (60 μg) [¹⁴C]phenylalanyl-sRNA; enzyme, as indicated. After 30 min at 37°C, 2 ml. of 10% trichloroacetic acid and 50 μl. of 1% bovine serum albumin were added. The precipitate was heated for 20 min at 90°C and filtered on a cellulose nitrate membrane filter, which was then glued to a planchet with whole milk and counted on a low-background gas-flow counter. The amount of radioactive material incorporated was found to be proportional to the amount of supernatant enzyme added up to about 400 cts/min; a maximum of about 600 cts/min was incorporated in the presence of excess enzyme, representing about 60% of the added radioactive material.

(f) *Sucrose gradients*

Sucrose gradients were prepared in Spinco SW25 tubes from either 15 and 30% (w/w) sucrose or 10 and 25% sucrose in the standard buffer of Warner, Knopf & Rich (1963). After centrifugation, a small hypodermic needle was inserted in the bottom of the tube, and fractions were collected. The optical density of each fraction was measured at 260 mμ, and the radioactivity was determined by precipitating each sample with 5% trichloroacetic acid in the presence of 100 μg bovine serum albumin carrier protein. After heating at 90°C for 20 min, the fractions were filtered on membrane filters, which were then glued to planchets and counted. Tritium was measured by spotting samples of each fraction on filter disks (see system A above), which were washed and counted according to the method of Mans & Novelli (1960).

(g) *Labeling of nascent polypeptide chains*

Labeling of nascent polypeptide chains with [^{14}C]algal protein hydrolysate for assay of release of nascent chains by toxin was carried out in a complete incubation mixture (system A) containing SF-70 and reticulocyte ribosomes. After incubation at 37°C, the mixture was layered on a 10 to 25% sucrose gradient and centrifuged for 8 hr at 64,000 g, which was sufficient to sediment the ribosomes to the bottom of the tube. After removal of the supernatant solution, the ribosomal pellet was rinsed and resuspended in 0·25 M-sucrose.

(h) *Removal of dialyzable products*

Removal of NAD and other dialyzable products from supernatant after treatment with toxin was accomplished by filtration of each sample through a column of Sephadex G25 or G50. The volume of each column was 5 to 10 times the volume of the sample, and all columns were washed and equilibrated beforehand with 0·05 M-Tris–HCl buffer (pH 7·5) containing 2 mM-GSH and 1 mM-EDTA. After filtration through Sephadex, the protein concentrations of the samples in a series were measured and equalized by appropriate dilution.

(i) *Chromatography*

Chromatography on Sephadex G100 was carried out using 50- to 90-cm columns equilibrated with 0·05 M-Tris–HCl buffer (pH 7·4) containing 0·1 M-KCl, 2 mM-GSH and 1 mM-EDTA. The sample, the volume of which was usually 1% of the bed volume of the column, was brought to a concentration of 10% sucrose by addition of solid sucrose and layered under the buffer at the top of the column. The column was allowed to flow at a rate determined by the hydrostatic pressure head, which was kept at 12 cm.

(j) *Binding assays*

Binding assays of [^{14}C]phenylalanyl-sRNA to ribosomes were carried out essentially according to the method of Nirenberg & Leder (1964). An incubation mixture identical to system B was used, except that the ribosome concentration was 10 times higher. After 20 min at 37°C, each sample was diluted with 10 vol. of cold 0·05 M-Tris–HCl buffer (pH 7·4) containing 0·1 M-KCl and 0·01 M-MgCl$_2$ and filtered through a type HA Millipore filter. After washing with three 4-ml. portions of the same buffer, each filter was glued to a planchet and counted.

(k) *Protein estimations*

Protein estimations were carried out by the method of Lowry, Rosebrough, Farr & Randall (1951).

3. Results

(a) *Studies on ribosomes treated with toxin*

(i) *Sucrose gradient patterns of polysomes and incorporation of [^{14}C]leucine into nascent polypeptide chains*

Two cell-free amino acid-incorporation mixtures were prepared containing reticulocyte ribosomes and supernatant (SF-70), and [^{14}C]leucine. Saturating levels of

R. J. COLLIER

toxin and NAD were added to one of the mixtures prior to addition of ribosomes. After incubation at 37°C for 30 minutes, each mixture was layered on a sucrose gradient and centrifuged 2·5 hours at 64,000 g. The results are shown in Fig. 1.

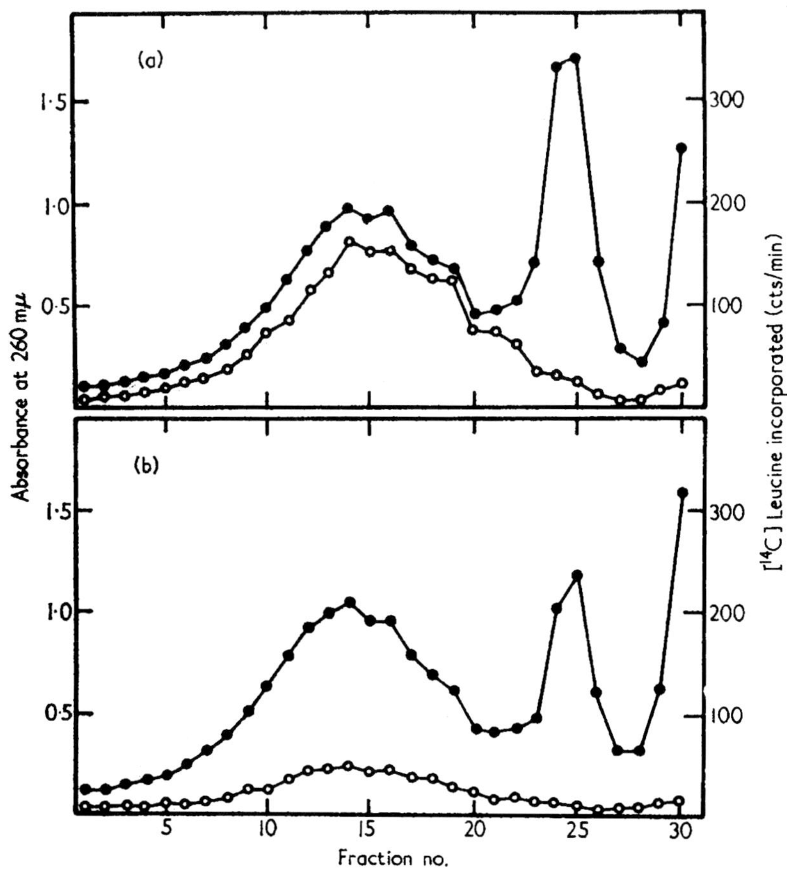

FIG. 1. Effect of toxin on patterns of ribosomes and [¹⁴C]leucine incorporation into nascent polypeptide chains.

Each of two system A amino acid-incorporation mixtures (0·7 ml.) contained 1 mg ribosomes, 1·5 mg SF-70 and 0·6 μc [¹⁴C]leucine (110 μc/μmole).

(a) Control, no toxin or NAD. (b) 30 μg toxin plus 100 μg NAD added before initiation of incorporation reaction. After 30 min at 37°C, each mixture was layered on a 25-ml. 15 to 30% sucrose gradient and centrifuged. The gradients were analyzed as described under Materials and Methods. —●—●—, Optical density at 260 mμ; —○—○—, cts/min ¹⁴C incorporated into polypeptides. The bottom of the gradients is to the left.

Toxin inhibited incorporation of [¹⁴C]leucine into nascent polypeptide chains by about 80%. The distribution of the incorporated [¹⁴C]leucine not inhibited by toxin was identical to that in the control, indicating that toxin did not preferentially inhibit incorporation into certain classes of polysomes. In addition, it is clear from the optical density patterns at 260 mμ in the two gradients that toxin did not accelerate the breakdown of polysomes to single ribosomes. Indeed the proportion of polysomes was consistently found to be higher (about 5%), and that of single ribosomes lower, in mixtures incubated with toxin. Ribosome and nascent chain patterns in mixtures containing either toxin or NAD alone were found to be identical with the control.

Since polysome breakdown was not observed under conditions where protein synthesis was strongly inhibited by toxin, it is unlikely that the inhibition was due to the action of ribonuclease in the toxin. This confirmed our earlier results, which indicated that toxin contained little or no ribonuclease (Collier & Pappenheimer, 1964*b*).

(ii) *Toxin does not release nascent polypeptide chains from ribosomes*

Puromycin, an antibiotic which strongly inhibits protein synthesis, releases nascent polypeptide chains from ribosomes (Morris & Schweet, 1961). Although toxin, being a protein, would not be expected to act by the same chemical mechanism as puromycin, it is conceivable that it might bring about release of nascent chains by a different mechanism.

Washed polysomes, the nascent polypetide chains of which had been labeled by incubation with [^{14}C]amino acids in a cell-free system, were incubated in a complete amino acid-incorporation mixture (containing only [^{12}C]amino acids) together with either puromycin or toxin plus NAD. Each mixture was layered on a sucrose gradient and centrifuged three hours at 64,000 *g*, and the distribution of ribosomes and labeled polypeptides in each gradient was determined. In contrast to puromycin, toxin caused no increase in the release of nascent polypeptide chains from the polysomes and no conversion of polysomes to single ribosomes.

(iii) *Toxin does not inactivate ribosomes*

To test the possibility that toxin might inactivate ribosomes by some mechanism not reflected in polysome breakdown or release of nascent polypeptides, an experiment was designed in which the activity of polysomes was assayed after a short incubation in the presence of toxin and NAD. Two complete amino acid-incorporation mixtures were prepared containing reticulocyte supernatant (SF-70) and polysomes, and [^{3}H]leucine; one of the mixtures contained excess toxin and NAD. After ten minutes at 37°C, each mixture was layered on a sucrose gradient and centrifuged for 2·5 hours at 64,000 *g*. It is clear that the ribosomes, after centrifugation through the sucrose gradient, were free from soluble factors present in the initial incubation mixture, including toxin, NAD and [^{3}H]leucine. The remaining amino acid-incorporation activity of the polysomes in each gradient fraction was assayed by measuring the incorporation of [^{14}C]leucine into protein in the presence of fresh SF-70, ATP and the other components of a complete incorporation mixture. Figure 2(b) shows that the ribosomes previously exposed to toxin and NAD actually showed a higher incorporation of [^{14}C]leucine than the control ribosomes. The inhibition of [^{3}H]leucine incorporation in the pre-incubation step (Fig. 2(a)) confirms that the toxin was active and that the ribosomes had been exposed to intoxicating conditions. Thus incubation under conditions where protein synthesis was strongly inhibited by toxin did not irreversibly damage the ribosomes.

(b) *Studies on supernatant treated with toxin*

(i) *Toxin inactivates supernatant*

To test for an effect of toxin on the supernatant, an ammonium sulfate-precipitated fraction (SF-70) of ribosome-free supernatant, was incubated with excess toxin and

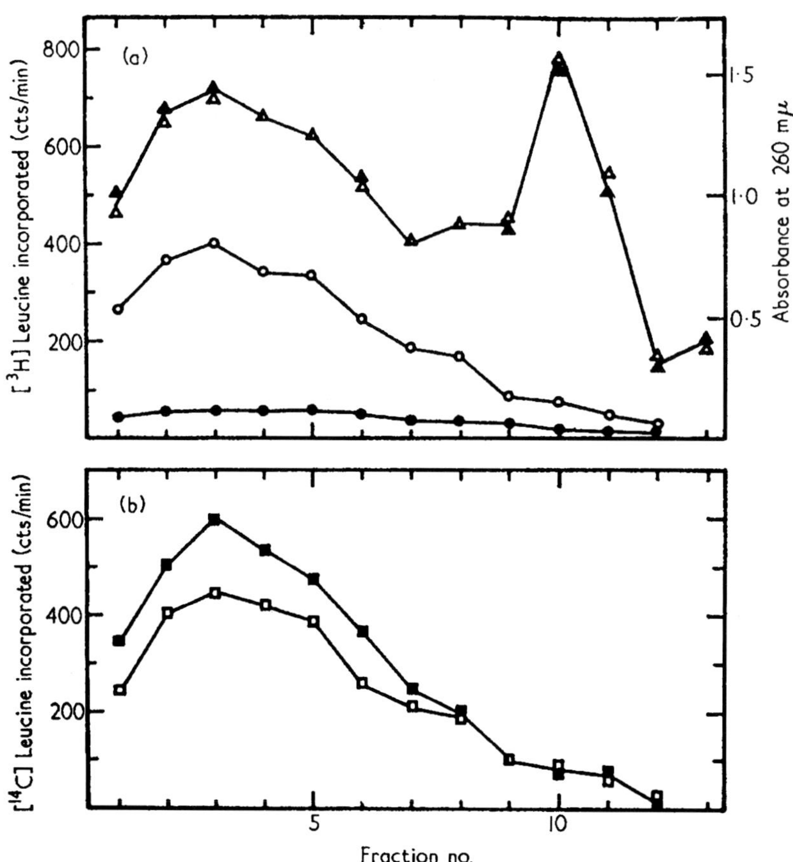

Fig. 2. Incorporation of [¹⁴C]leucine by ribosomes pre-incubated in the presence of toxin.

Pre-incubation step: Each of two system A amino acid-incorporation mixtures (0·55 ml.) contained 1·5 mg SF-70, 3 mg ribosomes and 7 µc (22 mµmoles) [³H]leucine. One mixture was a control; the other contained 25 µg toxin plus 70 µg NAD. After 10 min at 37°C, each was layered on a 25-ml. 10 to 25% sucrose gradient and centrifuged 3 hr at 64,000 *g*.

(a) Optical density at 260 mµ: —△—△—, control; —▲—▲—, toxin. ³H incorporated into polypeptides: —○—○—, control; —●—●—, toxin.

Assay of ribosome activity: 250 µl. from each sucrose gradient fraction was incubated in a system A amino acid-incorporation mixture (0·385 ml.) containing 0·75 mg SF-70 and 0·2 µc (4 mµmoles) [¹⁴C]leucine. After 60 min at 37°C, the [¹⁴C]leucine incorporated into protein was assayed as described under Materials and Methods.

(b) —□—□—, Cts/minute [¹⁴C]leucine incorporated, control gradient; —■—■—, cts/minute [¹⁴C]leucine incorporated, toxin gradient.

NAD, plus all the components of a normal amino acid-incorporation mixture except ribosomes and the [¹⁴C]amino acid. After 40 minutes at 37°C, the mixture was passed through a small column of Sephadex G25 to remove NAD and all other low molecular weight components of the mixture. The toxin, which remained with the macromolecular components of the supernatant, was without inhibitory activity in the subsequent assay of the supernatant due to the absence of NAD. The activity of the toxin-treated supernatant was measured in the usual system with [¹⁴C]leucine. Table 1 shows that the activity of supernatant treated with toxin plus NAD (referred to hereafter as *toxin-treated* supernatant) was inhibited by about 75% compared with controls treated with toxin or NAD alone. When excess toxin was added to the

TABLE 1

Amino acid-incorporation activity of supernatant pre-treated with toxin

Step 1	Step 2				
Pre-incubation	Assay of pre-treated supernatant				
Supernatant pre-treated with:	Pre-treated supernatant	Additions to assay mixtures:			
		None	NAD	Toxin	NAD + toxin
I. NAD alone	I	3134	3416	3293	286
II. Toxin + NAD	II	833	262	725	201
III. Toxin alone	III	3460	597	3377	236
	Mixed I + II	4023	831	—	—

Step 1: 7·5 mg SF-70 was incubated in a system A reaction mixture (0·5 ml.) without ribosomes or [^{14}C]amino acids and containing either 50 μg NAD (I), 38 μg toxin (III), or both NAD and toxin (II). After 20 min at 37°C, NAD was removed by passage of each sample through a separate column of Sephadex G25.

Step 2: The activity of each supernatant was measured in a system A incorporation mixture (0·25 ml.) containing 400 μg polysomes, 0·12 μc (2·5 mμmoles) [^{14}C]leucine, and 200 μg pre-treated SF-70. NAD (25 μg) and/or toxin (7·5 μg) were added where indicated. Incubation was for 40 min at 37°C. Numbers are expressed as cts/min [^{14}C]leucine incorporated into hot trichloroacetic acid-precipitable material.

various supernatants before assay, no change occurred in incorporation, demonstrating that the NAD had been completely removed by gel filtration. However, when excess NAD was added, the toxin remaining from the pre-incubation step became active, and inhibition of [^{14}C]leucine incorporation was observed in those samples which had been pre-incubated with toxin. The activity of mixed control and toxin-treated supernatants was approximately additive, showing that (1) excess toxin remaining in the toxin-treated supernatant is not active after removal of NAD, and (2) toxin-treated supernatant does not inhibit the activity of normal supernatant. These results indicate that the effect of toxin is to inactivate some component of supernatant, rather than to produce an inhibitory product. It is clear that the removal of NAD does not reactivate the supernatant.

Subsequent experiments showed that the inactivation of supernatant by toxin does not require an energy-generating system and occurs in a medium containing only Tris buffer and NAD. Furthermore, addition of EDTA to a final concentration of 0·01 M does not inhibit the inactivation reaction. Thus neither divalent cations nor cofactors other than NAD are required for the inactivation of supernatant. The inactivation reaction is complete within five minutes at 37°C in the presence of 100 μg NAD per ml. and 10 μg toxin per ml.

(ii) *Toxin does not inactivate aminoacyl-sRNA*

An inhibition of amino acid-incorporation with toxin-treated supernatant was also observed when aminoacyl-sRNA was used as substrate instead of the free amino acid. [^{14}C]Phenylalanine transfer from [^{14}C]phenylalanyl-sRNA to ribosomes was measured in a system completely dependent upon added supernatant, poly U and GTP (Table 2, exp. 1). As in the preceding experiment, toxin-treated supernatant was only about 25% as active as control supernatant, and the activities of control and toxin-treated supernatant were roughly additive. The fact that substrate [^{14}C]phenylalanyl-sRNA was not present during treatment of the supernatant with toxin strongly suggested that the inhibitory action of toxin was not directed against aminoacyl-sRNA. The finding that [^{14}C]phenylalanine transfer from *E. coli* sRNA was also inhibited, was further evidence for this, since cell-free systems from this organism are insensitive to toxin (Collier & Pappenheimer, 1964*b*).

TABLE 2

Transfer activity of toxin-treated supernatant and aminoacyl-sRNA

Source of sRNA	Assay conditions	Incorporated radioactivity
Experiment 1:		
Rabbit liver	Complete system	346
	Minus poly U	7
	Minus GTP	2
	Minus supernatant	2
	Minus ribosomes	3
	Control supernatant (a)	335
	Toxin-treated supernatant (b)	100
	(a) + (b)	401
E. coli	Control supernatant	217
	Toxin-treated supernatant	73
Experiment 2:		
Rabbit liver	Control [^{14}C]phenylalanyl-sRNA	332
	Toxin-treated [^{14}C]phenylalanyl-sRNA	342

Experiment 1: Transfer of [^{14}C]phenylalanine from [^{14}C]phenylalanyl-sRNA to polyphenyl-alanine was measured in system B incorporation mixtures. 50 μg of either control or toxin-treated supernatant was added, followed by [^{14}C]phenylalanyl-sRNA (1000 cts/min) to initiate the transfer reaction. Samples were incubated for 30 min at 37°C, precipitated with trichloroacetic acid, and processed for counting as described under Materials and Methods. Numbers are expressed as cts/min incorporated into hot trichloroacetic acid-precipitable material.

Experiment 2: [^{14}C]Phenylalanyl-sRNA from rabbit liver was incubated in the presence of 0·05 M-Tris–HCl (pH 7·4), 0·1 M-KCl, 10 mM-MgCl$_2$ and 0·1 mM-NAD. Toxin (30 μg) was added to one such mixture containing 0·7 mg of [^{14}C]phenylalanyl-sRNA, and after 15 min at 37°C this and the control mixture were filtered through separate 4-ml. Sephadex G50 columns equilibrated with 1 mM-potassium acetate buffer (pH 5·2). The transfer activity of samples (500 cts/min) of the [^{14}C]phenylalanyl-sRNA from each column was measured in a system B mixture containing excess supernatant. About 65% of the added radioactive material was transferred in both control and toxin-treated samples.

The lack of a direct effect of toxin on aminoacyl-sRNA was demonstrated by treating [^{14}C]phenylalanyl-sRNA with toxin plus NAD, and testing for transfer of the [^{14}C]phenylalanine after removal of NAD. Experiment 2 in Table 2 shows that [^{14}C]phenylalanine transfer from toxin-treated [^{14}C]phenylalanyl-sRNA to ribosomes was identical to the control.

All subsequent assays of amino acid-incorporation were carried out using [^{14}C]phenylalanyl-sRNA as substrate and poly U as messenger RNA.

(c) *Inactivation of a transfer factor by toxin*

(i) *Separation and assay of transfer factors*

Gasior & Moldave (1965) have reported that two complementary transfer factors from rat liver can be separated in a single operation by passage of supernatant through a column of Sephadex G100 or G200. We have applied this method to reticulocyte supernatant and have found that the transfer factors from this system also can be separated by Sephadex chromatography. Figure 3(a) shows the results of chromatography of an ammonium sulfate-precipitated fraction of reticulocyte supernatant (SF-55P) on Sephadex G100. When each fraction from the column was tested for ability to catalyze the transfer of phenylalanine from [^{14}C]phenylalanyl-

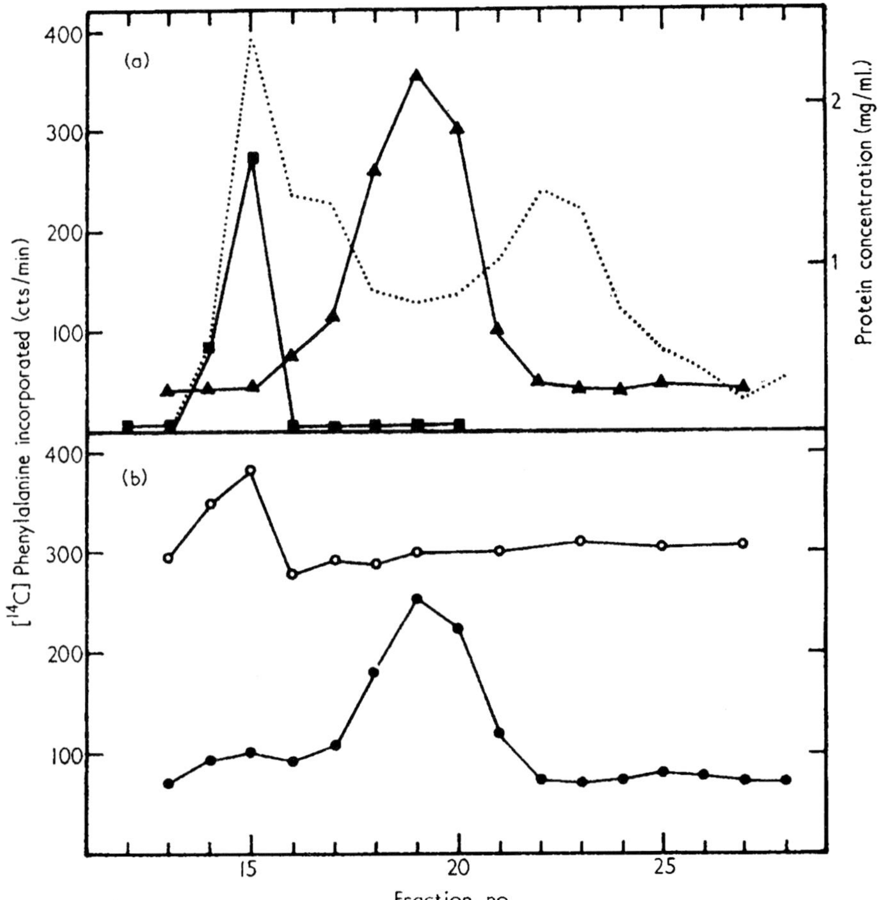

FIG. 3. Fractionation of reticulocyte supernatant by column chromatography on Sephadex G100.

3 ml. of SF-55P (75 mg protein) was applied to a 90 cm × 2·7 cm column and fractionated as described under Materials and Methods.

(a) Distribution of transfer factors in the column fractions. Transfer activity was measured in system B mixtures as described. Assays for factor I, —■—■—, contained 5 μl. of the fraction assayed plus 10 μl. fraction 19; those for factor II, —▲—▲—, contained 5 μl. of the fraction assayed plus 10 μl. fraction 15. Protein content of the fractions is indicated by the dotted line.

(b) Stimulation of the transfer activity of control and intoxicated supernatant by addition of column fractions. Transfer activity was measured in system B mixtures, containing 5 μl. of the fraction assayed plus 5 μl. (50 μg) of either control supernatant, —●—●—, or toxin-treated supernatant, —○—○—.

sRNA to polyphenylalanine, only fraction 15 had any measurable activity, and the activity of this fraction was very low (22 cts/min). However, a strong synergistic effect was observed in the presence of fraction 15 together with fractions in the region of fraction 19, indicating the presence of complementary factors in these regions. The precise distribution of each of the two factors was determined by assaying a constant volume from each fraction in the presence of a constant amount of protein from either fraction 15 or 19 (Fig. 3(a)). These factors have been designated transfer factors I and II, according to the order of their appearance from a column of Sephadex G100. The measurable activity of fraction 15 alone may be due to slight overlap of factors I and II.

It has been generally assumed that both of the complementary transfer factors are proteins, although purification has not proceeded to the point where this has definitely been proved. The following results are in agreement with this hypothesis. (1) Both factors I and II are inactivated by greater than 97% after heating for five minutes at 60°C and are completely destroyed by heating for one minute at 100°C. (2) Factor I is eluted from DEAE-cellulose at a chloride ion concentration of 0·16 M, and factor II at 0·1 M-chloride. RNA would not be expected to be eluted below 0·5 M-chloride. (3) Neither factor is removed from supernatant by treatment with protamine sulfate.

(ii) *Restoration of activity of intoxicated supernatant by addition of factor II*

If toxin inactivates a single component of the supernatant, it would be expected that after removal of NAD the inhibited transfer activity of toxin-treated supernatant could be restored by addition of that specific component from untreated supernatant. Figure 3(b) shows the results of assays of the transfer activity of each fraction from the Sephadex column of Fig. 4(a) in the presence of a constant amount

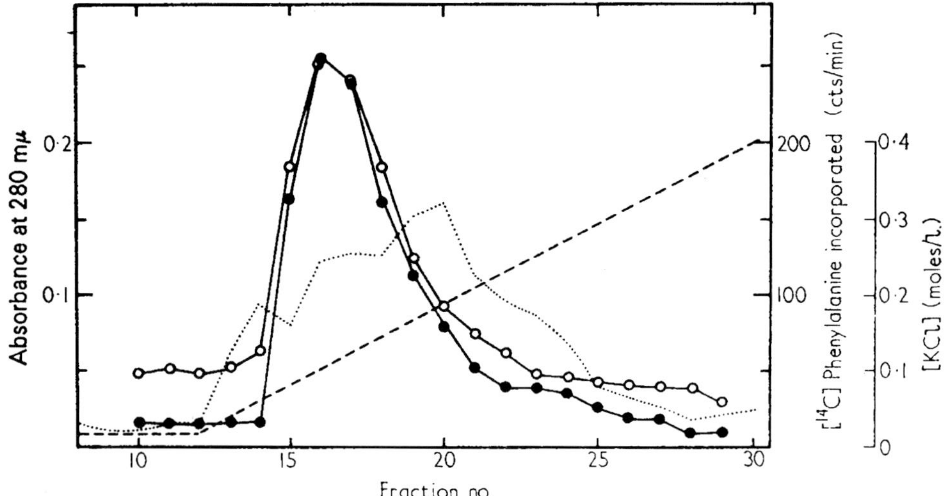

FIG. 4. Elution of factor II from DEAE-cellulose.

Factor II from fractionation of untreated supernatant on Sephadex G100 was dialyzed and applied to a 2-ml. column of DEAE-cellulose equilibrated with 0·02 M-Tris–HCl buffer (pH 7·5), 1 mM-GSH, 0·5 mM-EDTA. After washing with the same buffer, elution was performed with a gradient of KCl in buffer. Transfer was measured in system B mixtures containing 20 μl. of column fraction plus either intoxicated supernatant (50 μg), —○—○—, or factor I (40 μg), —●—●—. Protein concentration is indicated by the dotted line and KCl concentration by the dashed line.

of either toxin-treated supernatant or control supernatant. The transfer activity of toxin-treated supernatant was found to be greatly enhanced by addition of a factor the distribution of which coincided with transfer factor II. The transfer activity of control supernatant, on the other hand, was enhanced by the addition of fractions containing factor I, but not by those containing factor II. These results suggest that transfer factor I was limiting in control supernatant and transfer factor II became limiting after treatment with toxin.

The activity of toxin-treated supernatant could be restored completely to the level of control supernatant by addition of larger amounts (15 to 20 μg protein) of the peak fraction containing factor II. The activity of control supernatant was enhanced slightly by increasing amounts of this fraction.

(iii) *Elution of factor II from DEAE-cellulose*

The coincidence of the fractions on the Sephadex column, shown in Fig. 3, of transfer factor II and the factor which restored transfer activity to toxin-treated supernatant, suggested that these factors were identical. Evidence to support this suggestion was obtained by the following experiment. The peak fractions containing factor II from chromatography of supernatant on Sephadex G100 were combined, dialyzed against 0·02 M-Tris buffer (pH 7·5) containing 1 mM-GSH and 0·5 mM-EDTA, and applied to a 2-ml. column of DEAE-cellulose. The column was washed with the same buffer medium, and elution was carried out with a gradient of KCl in buffer. Each fraction was assayed for (1) protein content, (2) activity of transfer factor II, and (3) ability to restore transfer activity to toxin-treated supernatant. Figure 4 shows that activities (2) and (3) coincided exactly. These results support the hypothesis that a single component is responsible for both activities.

(iv) *Fractionation of toxin-treated supernatant on Sephadex G100*

If toxin specifically inactivates transfer factor II, it would be expected that fractionation of toxin-treated supernatant on Sephadex G100 would reveal a large decrease in the activity of this factor. Two samples of supernatant, one a control containing only NAD and the other containing toxin and NAD, were incubated for 20 minutes at 30°C and then applied to two identical Sephadex G100 columns and fractionated. Assays of transfer factors I and II were carried out on the fractions from both columns in the presence of the respective untreated complementary factors. Free NAD is strongly retarded on Sephadex G100 and was not present in any of the fractions assayed. Figure 5 shows that the activity of factor II from the toxin-treated supernatant was inhibited by about 80%, while no decrease occurred in the activity of factor I.

(v) *Inactivation of factor II by toxin in the absence of factor I*

Peak fractions of transfer factors I and II from a Sephadex G100 column were treated with either toxin, NAD, or toxin plus NAD. After removal of NAD by passage of each mixture through a small column of Sephadex G25, the activity of each factor was assayed in the presence of the untreated, complementary transfer factor. The results, shown in Table 3, indicate that factor II was strongly inhibited by treatment with toxin plus NAD, whereas factor I was unaffected. Neither toxin nor NAD alone inhibited factor II. These results indicate that the presence of factor I is not required for the action of toxin on factor II.

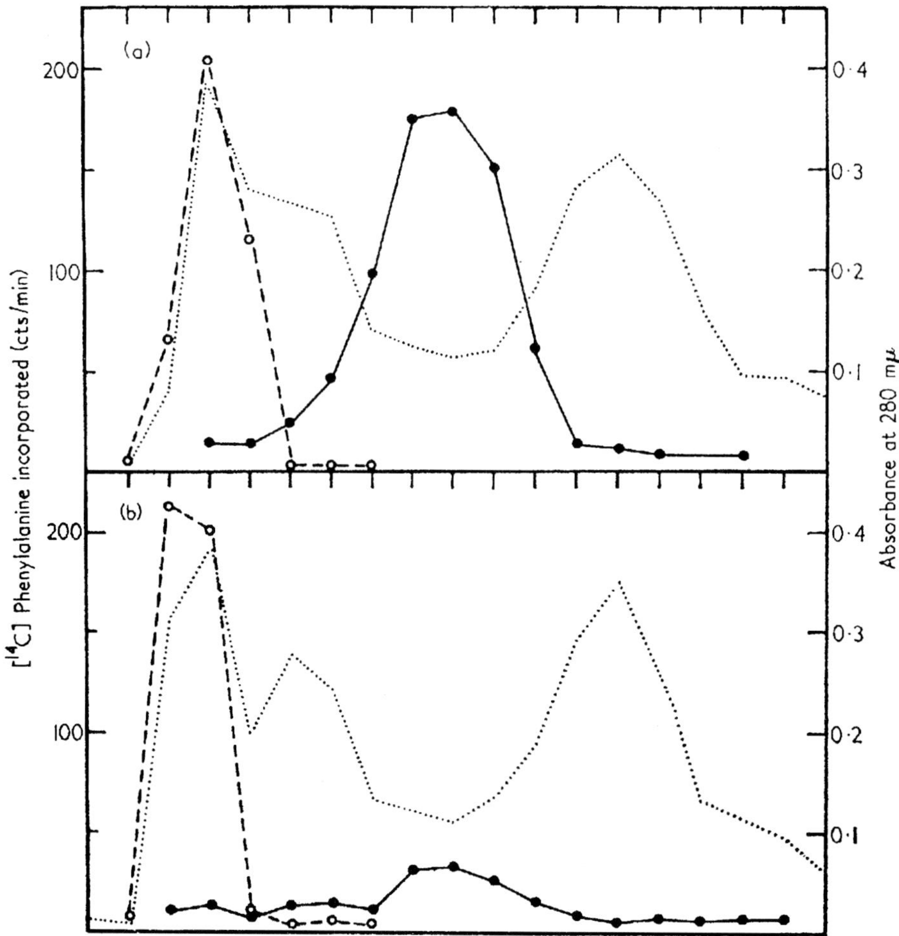

FIG. 5. Fractionation of normal and toxin-treated supernatant on Sephadex G100.

Each of two reaction mixtures (0·7 ml.) contained 0·05 M-Tris–HCl (pH 7·5), 0·1 M-KCl, 20 mM-GHS, 10 mM-EDTA, 70 μg NAD and 15 mg SF-55P. Toxin (30 μg) was added to one, and after 20 min at 30°C each was layered on a 50 cm × 1·7 cm column of Sephadex G100 and fractionated.

(a) Column fractions from control supernatant, and (b) from intoxicated supernatant. Protein concentration (dotted lines) and activity of each transfer factor were measured for each fraction. Factor I, --○--○--, was assayed in system B mixtures containing 10 μl. from the fraction assayed plus a constant amount of factor II from a separate column. Factor II, —●—●—, was assayed in the presence of 10 μl. from the column fraction assayed plus factor I from a separate column.

(vi) *Lack of effect of toxin on the binding of aminoacyl-sRNA to ribosomes*

Arlinghaus *et al.* (1964) have found that one of the two transfer factors which they have separated from reticulocyte supernatant, TF-1, catalyzes binding of aminoacyl-sRNA to ribosomes in the presence of GTP. The other factor, TF-2, is thought to catalyze the formation of the peptide bond, although there is no direct evidence for this. From comparison of our data with those given by Arlinghaus *et al.* (1964) on the salt concentrations required to elute TF-1 and TF-2 from DEAE-cellulose, it has been tentatively concluded that our factor I is identical to TF-1 and factor II identical to TF-2. If this assignment is correct, it would be expected that our factor I would catalyze binding of aminoacyl-sRNA to ribosomes, and the binding reaction would not be inhibited by toxin. The results shown in Table 4 are in agreement with these expectations. The binding of [14C]phenylalanyl-sRNA to ribosomes in the presence of poly U was stimulated about seven fold by addition of transfer factor I, whereas factor II caused no increase in binding. Addition of toxin plus NAD did not inhibit the binding catalyzed by factor I. In addition the binding reaction was studied kinetically, and no effect of toxin was observed on the rate of the reaction. No polyphenylalanine formation was observed unless both transfer factors were present.

TABLE 3

Effect of toxin pre-treatment on separated transfer factors

| Transfer factor | Pre-treatment with: | | | |
	—	NAD	Toxin	Toxin & NAD
I	133	—	—	131
II	143	145	131	29

Factor I (1 mg protein) and factor II (700 μg protein) were treated for 30 min at 37°C in mixtures (0·1 ml.) containing 0·05 M-Tris–HCl buffer (pH 7·5) and 20 mM-GSH. NAD (10 μg) and toxin (15 μg) were added as indicated. NAD was removed by filtration of each sample through a separate column of Sephadex G25, and the protein concentrations of the samples in each series were equalized by appropriate dilution. The activity of each sample was then measured in a system B incorporation mixture containing an excess of untreated complementary transfer factor. Numbers are expressed as cts/minute [14C]phenylalanine transferred to hot trichloroacetic acid-precipitable material.

TABLE 4

Effect of toxin on binding of [14C]phenylalanyl-sRNA to ribosomes

| Transfer factor added | Binding of [14C]phenylalanyl-sRNA to ribosomes | | Synthesis of polyphenylalanine | |
	Control	Toxin + NAD	Control	Toxin + NAD
None	22	23	7	7
Factor I	142	140	15	14
Factor II	19	17	6	6
Factors I + II	—	—	438	135

Binding of [14C]phenylalanyl-sRNA to ribosomes was measured as described under Materials and Methods. Factor I (45 μg) and factor II (75 μg) were added as indicated. Numbers are expressed as cts/min [14C]phenylalanyl-sRNA bound to ribosomes.

Incorporation of [14C]phenylalanine into polyphenylalanine was measured by precipitating an identical series of samples with 10% trichloroacetic acid. After heating to 90°C for 20 min, each sample was filtered, washed and counted as described. Numbers are expressed as cts/min [14C]phenylalanine incorporated into hot trichloroacetic acid-precipitable material.

4. Discussion

The data presented indicate that diphtheria toxin blocks protein synthesis in cell-free systems by specifically inactivating one of the supernatant transfer factors. Separation of two complementary factors has been achieved by chromatography of reticulocyte supernatant on Sephadex G100, and one of these, designated factor II, has been shown to be inhibited by about 80% after treatment with toxin in the presence of NAD. The inactive factor is not reactivated upon removal of NAD. Neither ribosomes nor aminoacyl-sRNA are affected by toxin.

Although it has not yet been clearly demonstrated that factor II contains only a single component required for the transfer reaction, the evidence presented suggests that this is the case. The peak of factor II activity might be a region of overlap of two (or more) factors needed for transfer, one of which is presumed to be inactivated by toxin. However, the peaks for factor II activity and ability to restore toxin-treated supernatant coincide exactly both on Sephadex G100 and DEAE-cellulose, suggesting that a single factor is responsible for both activities.

The conclusion that diphtheria toxin specifically inactivates one of the supernatant transfer factors is consistent with all of the observations made so far in cell-free systems. In earlier work it was found that toxin did not affect the formation of aminoacyl-sRNA, but rather, inhibited transfer of amino acids from aminoacyl-sRNA to ribosomes (Collier & Pappenheimer, 1964*b*). Furthermore, the synthesis of polyphenylalanine directed by poly U was inhibited by toxin, suggesting that the action of toxin probably was not specific for any single amino acid (Pappenheimer *et al.*, 1963). In the present work, it has been shown that toxin does not inactivate reticulocyte ribosomes; indeed, the presence of toxin plus NAD in cell-free systems actually retarded both the inactivation and the breakdown of polysomes. These results can be explained if by blocking the action of a transfer factor, toxin merely retards the simultaneous growth of polypeptide chains and movement of ribosomes along the messenger RNA. Release of single ribosomes from messenger RNA would be expected to be inhibited by toxin; and the level of polysomes, which is related to incorporation activity, should remain higher than in control mixtures where amino acid incorporation occurs normally.

The explanation of the residual 20 to 25% activity of transfer factor II remaining after treatment with excess toxin is not clear. Factor II might exist in two states, only one of which is sensitive to the toxin. It is also conceivable that a presumed toxin-inactivated form of the factor may retain partial activity.

Since the function of factor II in the transfer reaction is not yet known, it is not possible to state exactly which reaction in protein synthesis is blocked by toxin. However, evidence has been presented to show that factor I catalyzes binding of aminoacyl-sRNA to ribosomes, and it is reasonable to think that factor II is involved in, and thus toxin blocks, a step subsequent to this reaction. The TF-2 transfer factor prepared by Arlinghaus *et al.* (1964), which may be identical to our factor II, has been postulated to be an enzyme which catalyzes the actual formation of the peptide bond. It is anticipated that toxin may be useful in studying the roles in protein synthesis of the transfer factors.

Neither the mechanism by which toxin inactivates transfer factor II nor the role of NAD in the inactivation reaction has yet been clearly demonstrated. It has been found that NADP will not replace NAD (Collier & Pappenheimer, 1964*b*), and Goor,

Pappenheimer & Ames (1966) have recently reported that NADH is also inactive. It is assumed, therefore, that the added NAD is utilized in the intoxication reaction in unaltered form. No evidence is yet available concerning the possibility that NAD may be altered during the course of the reaction. Data reported here show that removal of free NAD does not reverse the inactivation of factor II, and in addition that the inactivation reaction does not require divalent cations or any cofactor other than NAD. Assuming a direct interaction between toxin and factor II, these results are compatible with either of two mechanisms of inactivation. (1) It is conceivable that toxin inactivates transfer factor II by, for example, chemically altering an amino acid side-chain on the enzyme; the requirement for NAD would suggest that such a reaction might involve oxidation. (2) It would also seem possible that toxin might inactivate factor II by binding to it to form an inactive complex. In that case, NAD might bind to toxin to bring about a conformational change in its structure. Goor *et al.* (1966) have recently found that the intoxication reaction can be reversed by addition of nicotinamide. This supports the latter of these two mechanisms, since it is difficult to envision a mechanism by which nicotinamide could reverse a chemical change in the structure of the transfer enzyme. It is believed that nicotinamide may displace NAD bound as an essential component of a NAD–toxin–transfer factor complex (Pappenheimer, personal communication).

The inactivation of the transfer factor shown above in cell-free experiments is a likely explanation of the cessation of protein synthesis in HeLa cells after addition of toxin (Strauss & Hendee, 1959; Strauss, 1960; Kato & Pappenheimer, 1960).

The experiment concerning the effect of toxin and puromycin on release of nascent polypeptide chains, as well as that showing that toxin does not break down polysomes, was carried out in the laboratory of Dr A. M. Pappenheimer, Jr., at Harvard University. This portion of the work was supported jointly by funds from the National Science Foundation, the Commission on Immunization of the Armed Forces Epidemiological Board, and the Office of the Surgeon General, Department of the Army.

The rest of the work was carried out during tenure of a National Science Foundation Postdoctoral Fellowship at the University of Geneva, with the aid of funds from the Swiss National Fund for Scientific Research.

I wish to thank Drs Tissières and Pappenheimer, and also Drs R. Traut, P.-F. Spahr and R. Gesteland for helpful advice.

REFERENCES

Allen, E. S. & Schweet, R. S. (1962). *J. Biol. Chem.* **237**, 760.

Arlinghaus, R., Shaeffer, J. & Schweet, R. (1964). *Proc. Nat. Acad. Sci., Wash.* **51**, 1291.

Collier, R. J. & Pappenheimer, A. M., Jr. (1964a). *J. Exp. Med.* **120**, 1007.

Collier, R. J. & Pappenheimer, A. M., Jr. (1964b). *J. Exp. Med.* **120**, 1019.

Gasior, E. & Moldave, K. (1965). *J. Biol. Chem.* **240**, 3346.

Goor, R., Pappenheimer, A. M., Jr. & Ames, E. (1966). *Fed. Proc.* **25**, 775.

Kato, I. (1962). *Japan J. Exp. Med.* **32**, 355.

Kato, I. & Pappenheimer, A. M., Jr. (1960). *J. Exp. Med.*, **112**, 329.

Lowry, O. H., Rosebrough, N. J., Farr, A. L. & Randall, R. J. (1951). *J. Biol. Chem.* **193**, 265.

Mans, R. J. & Novelli, G. D. (1960). *Biochem. Biophys. Res. Comm.* **3**, 540.

Moldave, K. (1963). In *Methods of Enzymology*, ed. by S. P. Colowick & N. O. Kaplan, vol. 6, p. 757. New York: Academic Press.

Morris, A. & Schweet, R. (1961). *Biochim. biophys. Acta*, **47**, 415.

Nathans, D. & Lipmann, F. (1961). *Proc. Nat. Acad. Sci., Wash.* **47**, 497.

Nirenberg, M. & Leder, P. (1964). *Science*, **145**, 1399.

Pappenheimer, A. M., Jr., Collier, R. J. & Miller, P. A. (1963). In *Cell Culture in the Study of Bacterial Disease*, ed. by M. Solotorovsky, p. 21. New Brunswick: Rutgers University Press.

Strauss, N. (1960). *J. Exp. Med.* **112**, 351.

Strauss, N & Hendee, E. D. (1959). *J. Exp. Med.* **109**, 145.

Warner, J. P., Knopf, P. M. & Rich, A. (1963). *Proc. Nat. Acad. Sci., Wash.* **49**, 122.

Yoneda, M. (1957). *Brit. J. Exp. Path.* **38**, 190.

A Restriction Enzyme from Hemophilus influenzae. I. Purification and General Properties

H. O. SMITH AND K. W. WILCOX

The major breakthrough in molecular biology that permitted the "cutting and pasting" of DNA fragments was the discovery by Smith and Wilcox of a restriction enzyme (they called it endonuclease R) from *Hemophilus influenzae* (now called *Haemophilus influenzae*). These investigators reported the isolation of an enzyme that was capable of cleaving double-stranded native T7 phage DNA and other foreign double-stranded DNA but not native DNA from *H. influenzae* or denatured foreign DNA. In the discussion section of their paper, the authors speculated that (i) the native host DNA was probably protected by modifications to the DNA and (ii) enzyme R recognized specific DNA sequences and that these sequences had to be 5 to 6 bases in length. These hypotheses were later proven to be correct. The impact of this discovery was recognized quickly, and Smith was awarded the Nobel Prize in Physiology and Medicine in 1978.

ALISON O'BRIEN

Reprinted from *Journal of Molecular Biology* 51:379–391. Copyright © 1970, by permission of the publisher, Academic Press.

J. Mol. Biol. (1970) **51**, 379–391

A Restriction Enzyme from *Hemophilus influenzae*

I. Purification and General Properties

HAMILTON O. SMITH AND K. W. WILCOX

Department of Microbiology
Johns Hopkins University, School of Medicine
Baltimore, Md. 21205, U.S.A.

(Received 15 September 1969)

Extracts of *Hemophilus influenzae* strain Rd contain an endonuclease activity which produces a rapid decrease in the specific viscosity of a variety of foreign native DNA's; the specific viscosity of *H. influenzae* DNA is not altered under the same conditions. This "restriction" endonuclease activity has been purified approximately 200-fold. The purified enzyme contains no detectable exo- or endonucleolytic activity against *H. influenzae* DNA. However, with native phage T7 DNA as substrate, it produces about 40 double-strand 5′-phosphoryl, 3′-hydroxyl cleavages. The limit product has an average length of about 1000 nucleotide pairs and contains no single-strand breaks. The enzyme is inactive on denatured DNA and it requires no special co-factors other than magnesium ions.

1. Introduction

A number of bacteria are capable of recognizing and degrading ("restricting") foreign DNA, such as the DNA of a virus grown on another bacterial strain. The DNA of the host is protected by a "host-controlled modification" (Arber, 1965). Recently, Meselson & Yuan (1968) have purified a restriction endonuclease from *Escherichia coli* K12. The enzyme has the interesting properties: (1) that it is site-specific in action, producing only a limited number of double-strand breaks in unmodified DNA, and (2) that it requires adenosine triphosphate and *S*-adenosyl methionine in addition to magnesium ions.

We have made the chance discovery of what appears to be a similar type of enzyme in *Hemophilus influenzae*, strain Rd. In the course of some experiments in which competent *H. influenzae* cells were incubated with radioactively labeled DNA from the *Salmonella* phage P22, we found that this DNA was apparently degraded since it could not be recovered in cesium chloride density gradients. It seemed likely that the effect was one of restriction. We were able to show the presence in crude extracts of an endonuclease activity which produced a rapid decrease in viscosity of foreign DNA preparations and which was without effect on the *H. influenzae* DNA. We describe in this report the purification and properties of the endonuclease. As with the *E. coli* restriction enzyme, our enzyme produces double-strand breaks in a limited number of specific sites. The enzyme requires only magnesium ions as a co-factor, unlike the *E. coli* enzyme. A preliminary report has been published (Smith & Wilcox, 1969).

2. Materials and Methods

(a) *Bacterial and phage strains*

H. influenzae strain Rd was obtained from Dr Roger Herriot as a frozen culture. The phage P22 c_2 clear plaque mutant (Levine, 1957) grown on *S. typhimurium* LT2 was used as a source of phage P22 DNA. Phage T7 and its host *E. coli* B were obtained from Dr Bernard Weiss.

(b) *Enzymes and standards*

Bacterial alkaline phosphatase was obtained from Worthington Biochemical Corp. Polynucleotide kinase, 5000 units/ml. (Richardson, 1965) and rechromatographed bacterial alkaline phosphatase, 20 units/ml. (Weiss, Live & Richardson, 1968) for use in the ^{32}P terminal labeling procedure were kindly supplied by Dr Bernard Weiss.

Bovine serum albumin and sperm whale myoglobin with molecular weights of 67,000 and 17,800, respectively, were obtained in a molecular weight marker kit from Mann Research Laboratories.

(c) *Nucleic acids and nucleotides*

Phage P22 was purified from L broth lysates (Levine, 1957) by differential centrifugation and banding in a CsCl step gradient (Thomas & Abelson, 1966). The DNA was extracted with 1 vol. of cold, redistilled phenol and then precipitated with 2 vol. of cold ethanol and redissolved in 1 vol. of NaCl–Tris buffer (0·05 M-NaCl, 0·01 M-Tris–HCl, pH 7·4, formerly ST buffer). The precipitation was repeated and the DNA was finally redissolved in the above buffer at 1·30 mg/ml. This DNA stock was used for the viscometry assays (see below).

Unlabeled phage T7 DNA was similarly prepared from phage grown on *E. coli* B. Phage T7 labeled with ^{32}P was prepared from phage grown in synthetic medium containing 2 μg phosphorus/ml. (Smith, 1968) and 5 μc of carrier-free [^{32}P]orthophosphate/ml. (New England Nuclear Corp.). The labeled phage were purified as described above and then extracted once with cold phenol. The DNA-containing aqueous phase was removed and dialysed for 20 hr against three 500-ml. changes of NaCl–Tris buffer.

Unlabeled *H. influenzae* DNA was extracted by the procedure of Marmur (1961) from a culture grown to saturation in Difco brain–heart infusion supplemented with NAD, 2 μg/ml., and hemin (Eastman), 10 μg/ml. *H. influenzae* DNA labeled with [^{3}H]thymidine was prepared from cells grown in a synthetic medium as described by Carmody & Herriot (1970).

[γ-^{32}P]ATP was obtained from Dr Bernard Weiss. The method of preparation has been described (Weiss *et al.*, 1968).

(d) *Zone sedimentation of DNA in sucrose density-gradients*

Native DNA was sedimented in linear gradients of 5 to 20% (w/v) sucrose in NaCl–Tris buffer. Denatured DNA was centrifuged in similar gradients containing 0·1 M-NaOH. The DNA was denatured by addition of 0·1 vol. of 1 N-NaOH for 5 min at room temperature followed by neutralization with 0·1 vol. of 1·1 N-HCl, 0·2 M-Tris as described by Studier (1965). Centrifugation was carried out in an SW50 rotor at 4°C in a Spinco model L centrifuge. Approximately 30 ten-drop fractions were collected from the bottom of the tube directly into scintillation vials containing 15 ml. of scintillation medium (2-methoxy-ethanol, 373 ml.; PPO, 9·3 g; dimethyl-POPOP, 30 mg; toluene to 1 liter) and counted in a Packard scintillation spectrometer.

(e) *Assay of the enzymic activity by DNA viscometry*

DNA viscosity measurements were performed at 30°C in an Ostwald viscometer having a flow-time for water of approximately 60 sec. The viscometer was filled with 3·5 ml. of phage P22 DNA solution, 40 μg/ml. in Tris–Mg–mercaptoethanol buffer (6·6 mM each of Tris buffer, pH 7·4, MgCl$_2$, and mercaptoethanol). Several flow-time measurements were taken after the DNA solution had reached thermal equilibrium and these were generally reproducible to within 0·1 sec. Five to 50 μl. of extract or purified enzyme was then introduced into the reservoir bulb of the viscometer and rapidly mixed by blowing air retrograde into the bulb. Flow-time measurements were taken as rapidly as was practical. The DNA viscosity was expressed as specific viscosity, $\eta_{sp} = (t/t_D) - 1$, where t_D represents the flow time at the end of the experiment, 5 min after addition of 50 μg of pancreatic DNase (this corresponded to the flow-time for pure solvent within the accuracy of the measurements). Specific viscosity measurements were plotted against time on semi-logarithm paper as fractional values of the zero-time value. One unit of enzyme activity is defined as that amount which produces a decrease in the DNA specific viscosity of 25% in 1 min under the conditions described above. It should be pointed out that the viscometric assay is valid even on crude extracts since, as will be shown in the Results section, no activity is found in crude extracts against homologous DNA.

(f) *Purification of endonuclease R*

H. influenzae cells were grown in 12 l. of brain–heart infusion, supplemented with 10 μg hemin/ml. and 2 μg NAD/ml. to O.D.$_{650}$ = 0·7, harvested by centrifugation, washed once, and resuspended in 20 ml. of 0·05 M-Tris (pH 7·4), 0·001 M-glutathione. The cells were disrupted by sonication for 4 min at 8 A output on a Bronson sonicator while being cooled in an ice–salt water bath. All subsequent operations were carried out at 0 to 4°C. Cellular debris was removed by centrifugation for 30 min at 100,000 g. The supernatant (27 ml.) was brought to 1 M-NaCl by addition of 1·58 g of NaCl and layered onto a 2·5 cm × 49 cm Bio-Gel A 0·5 M (200 to 400 mesh) column prewashed with 10 vol. of 1 M-NaCl, 0·02 M-Tris–HCl, pH 7·4, 0·01 M-mercaptoethanol. Elution was carried out at 1·2 ml./min using the same buffer solution and 6-ml. fractions were collected. Fractions 18 to 28, containing nearly all the activity and only 10% of the O.D.$_{260}$ absorbing material, were pooled. The pooled fractions (65 ml.) were diluted with 140 ml. of 0·02 M-Tris, pH 7·4, and stirred in an ice bath. Ammonium sulfate, 64·3 g, was added slowly over a 30-min period. The precipitate (0 to 50%) was removed by centrifugation. The supernatant solution was precipitated with ammonium sulfate, 15·8 g, to obtain a 50 to 60% precipitate. The supernatant was again reprecipitated with 16·8 g of ammonium sulfate and this precipitate (60 to 70%) was combined with the 50 to 60% precipitate and dissolved in 20 ml. of 0·05 M-NaCl, 0·02 M-Tris (pH 7·4), 0·001 M-mercaptoethanol.

Part of the 50 to 70% ammonium sulfate fraction was further purified by column chromatography. A phosphocellulose (Whatman, P11) column, 0·5 cm × 9·5 cm, was equilibrated with 100 ml. of 0·01 M-potassium phosphate buffer, pH 7·4. 6 ml. of the ammonium sulfate fraction, containing 60 mg. of protein, was diluted with 54 ml. of 0·01 M-phosphate buffer, pH 7·4, and loaded onto the column at a flow rate of 5 ml./hr at 4°C. Protein was then eluted stepwise with 5-ml. portions of 0·01 M-potassium phosphate buffer, pH 7·4, containing increasing molarities of KCl as follows: 0·0, 0·1, 0·2, 0·3 and 0·4 M. Fractions of 1 ml. were collected. The bulk of the activity was eluted at 0·2 M-KCl. The first two fractions at 0·3 M-KCl contained a small amount of activity and were combined with the 0·2 M-KCl fractions.

The enzyme was finally concentrated by precipitation of the combined phosphocellulose fractions (7 ml.) with 3·3 g of ammonium sulfate. The precipitate was redissolved in 1·5 ml. of 0·20 M-NaCl, 0·02 M-Tris (pH 7·4) containing bovine serum albumin, 0·3%, at a final activity of 16 units/ml. The activity is stable at 4°C in this solution for at least 8 months. Table 1 summarizes the purification procedure. The significant increase in total enzyme units observed following ammonium sulfate precipitation is not explained, but could be due to removal of interfering activities.

(g) ^{32}P-*labeling of the 5′-end of the DNA*

The 5′-phosphoryl ends of DNA were first dephosphorylated by incubation with alkaline phosphatase and then rephosphorylated with [γ-^{32}P]ATP in the polynucleotide kinase reaction as described by Weiss *et al.* (1968). The reactions were carried out as follows: DNA, 20 mμmoles, in 0·26 ml. 0·1 M-Tris–HCl, pH 8·0, was incubated with 5 μl. of bacterial alkaline phosphatase (20 units/ml.) at 37°C for 30 min to remove the terminal 5′-phosphoryl groups. A 0·035-ml. volume of a kinase reaction mixture (containing 0·3 M-MgCl$_2$; 0·03 M-potassium phosphate buffer, pH 7·4; [γ-^{32}P]ATP, 1 μmole/ml., 2·2 × 10^8 cts/min/ μmole; and 1·0 M-dithiothrietol, in a ratio of 2 : 2 : 2 : 1 by vol.), and 5 μl. of polynucleo- tide kinase, 5000 units/ml., were then added and the reaction mixture was incubated for 30 min at 37°C. The terminally labeled DNA was precipitated at 0°C by adding 0·3 ml. of 0·1 M-sodium pyrophosphate followed by 2·5 ml. of cold 6% trichloroacetic acid con- taining 0·01 M-sodium pyrophosphate. The precipitate was collected on a glass filter, washed with nine 2-ml. portions of 6% trichloroacetic acid containing pyrophosphate, and two 2-ml. portions of ethanol. After drying, the filters were counted in a scintillation counter. Under these conditions labeling takes place only at the ends of native DNA. Internal single-strand breaks (nicks) are not labeled because at 37°C the phosphatase is inactive at the nicks (Weiss *et al.*, 1968). If the DNA is denatured with alkali before labeling, then the nicks are exposed and can be labeled by the above procedure.

(h) *Molecular weight estimation by gel filtration*

The molecular weight of purified enzyme was estimated by filtration through a 0·8 cm × 17·2 cm column of superfine Sephadex G200 (Pharmacia Fine Chemicals). The column was washed with several volumes of NaCl–Tris buffer at a flow rate of 0·5 ml./hr under 10 cm of hydrostatic pressure at 4°C. A mixture containing 250 μg of bovine serum albumin, 250 μg of myoglobin, 0·5 optical density unit (at a wavelength of 650 nm) of dextran blue 2000 (Pharmacia Fine Chemicals), and 25 μl. of purified *H. influenzae* enzyme (16 units/ml.) in a volume of 0·1 ml. was layered onto the column and eluted with NaCl–Tris buffer under the above conditions. Forty 10-drop fractions (0·23 ml.) were collected. Dextran blue was measured by absorbance at 650 nm and the bovine serum albumin and myoglobin by absorbance at 230 nm. The enzyme activity was measured in the following way. Ten μl. of each fraction was incubated for 30 min at 37°C with 25 μl, of ^{32}P-labeled T7 DNA (28 μg/ml., 6·5 × 10^6 cts/min/ml.), 1 μl. of bacterial alkaline phosphatase, 57 units/ml., and 0·2 ml. of Tris–Mg–mercaptoethanol buffer containing 0·05 M-NaCl. The reaction tubes were chilled and 0·1 ml. vol. of salmon sperm DNA, 500 μg/ml. was added as carrier. The DNA was precipitated with 0·3 ml. of chilled 10% trichloroacetic acid. After 5 min the tubes were centrifuged at 4000 *g* for 10 min and 0·5 ml. of each supernatant was removed and counted in scintillation medium. This assay procedure gives results similar to the more cumbersome viscometry method but can only be used with the purified enzyme. The phosphatase serves to cleave off ^{32}P groups exposed by the enzyme digestion and these then appear as trichloroacetic acid-soluble radioactivity.

3. Results

(a) *Detection of an* H. influenzae *nuclease specific for foreign DNA*

H. influenzae extracts appear to contain no detectable endonuclease activity against native *H. influenzae* DNA by the viscometric assay (see Materials and Methods). Addition of 20 μl. of an extract from sonicated cells (containing 16 mg protein/ml.) to a viscometer containing *H. influenzae* DNA caused no decrease in η_{sp} during 60 minutes at 30°C (Fig. 1). However, the η_{sp} of phage P22 DNA was significantly decreased by as little as 10 μl. of extract under the same conditions. The extract thus apparently contains a nuclease with specificity toward the foreign DNA. Addition of 50 μl. of extract produced approximately a fivefold greater rate of fall of the η_{sp} of the phage DNA. Fractional decrease in η_{sp} proceeded logarithmically with time until the value was below 0·7, after which the decrease became less rapid. The proportionality between initial rate of decrease in η_{sp} and the amount of extract added provides a quantitative assay. A unit of the enzyme activity can be defined as that amount which produces a 25% decrease in η_{sp} in one minute.

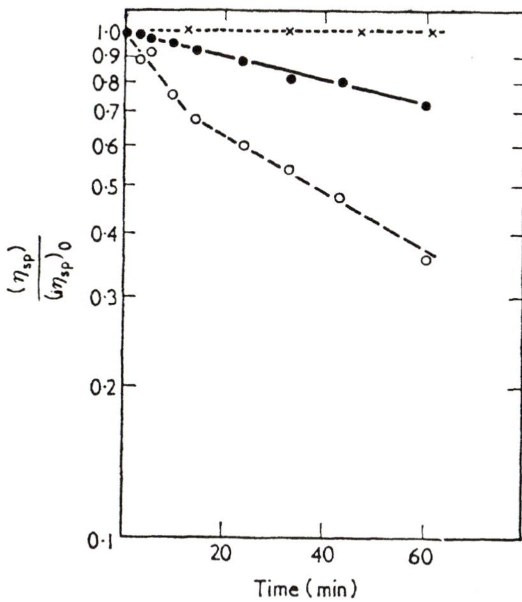

FIG. 1. Effect of a sonicated cell extract from *H. influenzae* on the specific viscosity of phage P22 and *H. influenzae* DNA.

Two viscometers containing 3·5 ml. of phage P22 DNA at a concentration of 40 μg/ml. in Tris–Mg–mercaptoethanol buffer were equilibrated at 30°C. At zero time, 10 μl. (—●—●—) of sonicated cell extract from *H. influenzae*, containing 16 mg protein/ml. was added to the first viscometer and 50 μl. (--○--○--) was added to the second. Measurements of the specific viscosity (η_{sp}) were then taken at intervals and were plotted as fractional values of the zero time specific viscosity ($\eta_{sp})_0$. Control measurements were obtained from a viscometer containing 3·5 ml. of *H. influenzae* DNA at a concentration of 40 μg/ml. in the same buffer to which 20 μl. (--×--×--) of cell extract was added at zero time. (The initial specific viscosity of the *H. influenzae* DNA solution was comparable to that of the phage P22 DNA solution.)

TABLE 1

Purification procedure

	Total (units)	Total (mg)	Spec. act. (units/mg)
100,000 *g* supernatant	135	1350	0.10
Bio-Gel column	185	650	0.28
Ammonium sulfate (50 to 70% ppt)	455	200	2.2
Phosphocellulose column†	144	5.5	26.0

† Entries are calculated assuming that all of the ammonium sulfate fraction was purified through the phosphocellulose step.

(b) *Properties of the purified nuclease*

A preparation of the *H. influenzae* nuclease approximately 200-fold purified from the sonicated cell extract was obtained as described in Materials and Methods.

(i) *Magnesium ion requirements*

Enzymic activity is dependent on the presence of Mg^{2+} ions (Table 2). The purified enzyme was optimally active when assayed at $6·6 \times 10^{-3}$ M-Mg^{2+} in 0·06 M-NaCl. The activity was decreased by a factor of about 50 when assayed at 10^{-4} M-Mg^{2+} and was unmeasureable at 10^{-5} M-Mg^{2+}. About 40% of the maximum activity was obtained at 10^{-3} M-Mg^{2+}.

TABLE 2

Effects of sodium chloride and magnesium ion concentration on the nuclease activity

Mg^{2+} (M)	NaCl (M)	Relative activity
$6{\cdot}6 \times 10^{-3}$	0.00	1.0
$6{\cdot}6 \times 10^{-3}$	0.02	2.3
$6{\cdot}6 \times 10^{-3}$	0.04	2.9
$6{\cdot}6 \times 10^{-3}$	0.06	3.4
$6{\cdot}6 \times 10^{-3}$	0.08	3.2
$6{\cdot}6 \times 10^{-3}$	0.10	2.2
10^{-5}	0.06	< 0.01
10^{-4}	0.06	0.06
10^{-3}	0.06	1.3

DNA viscosity measurements were carried out with phage P22 DNA, 40 μg/ml. in 3·5 ml. of solvent containing 6·6 mM-Tris–HCl, pH 7·4, and 6·6 mM-mercaptoethanol at the various listed NaCl and Mg^{2+} concentrations. Activities are expressed relative to that obtained under standard assay conditions in Tris–Mg–mercaptoethanol solvent.

(ii) *Salt requirements*

Optimum activity was obtained in 0·06 M-NaCl (Table 2). At this molarity an approximately threefold increase was obtained over that found under the standard assay conditions in Tris–Mg–mercaptoethanol buffer.

(iii) *Molecular weight estimate*

In the agarose column purification step the nuclease activity was recovered in fractions corresponding to an approximate molecular weight of 80,000. Gel filtration

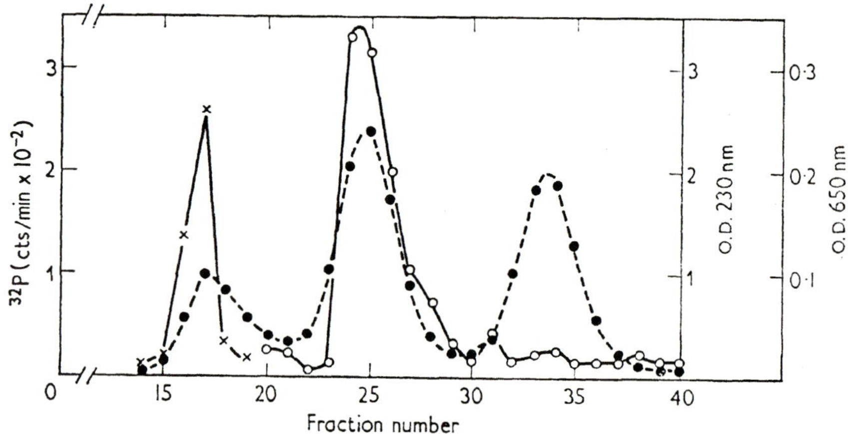

FIG. 2. Gel filtration of the purified nuclease on a Sephadex G200 column. Dextran blue was measured by absorbancy at 650 nm (— ×— ×—). Bovine serum albumin and myoglobin were measured by absorbancy at 230 nm (--●--●--). The nuclease activity was measured as released ^{32}P activity as described in Materials and Methods (—○—○—).

of the purified enzyme on Sephadex G200 was carried out with bovine serum albumin and myoglobin as known molecular weight markers. Dextran blue 2000 was used to measure the exclusion volume. The nuclease activity was found in a peak closely associated with the albumin (67,000 molecular weight). The enzyme is thus approximately the same molecular weight (Fig. 2).

(iv) *Foreign DNA* versus H. influenzae *DNA as substrate*

A mixture of native [³H]thymidine-labeled *H. influenzae* DNA and native ³²P-labeled phage T7 DNA was incubated at 37°C for 30 minutes with an excess of enzyme. Samples were removed from the reaction mixture at zero time (before enzyme addition), five minutes and 30 minutes for assay of trichloroacetic acid-soluble radioactivity and additional fractions were removed for sucrose gradient analysis. One-half of each of the latter fractions was zone sedimented on a neutral sucrose density-gradient and the other half was first denatured in alkali and then sedimented on an alkaline sucrose gradient. At zero time both phage and bacterial DNA sedimented approximately to mid-position in the neutral gradient tube (Fig. 3). The bacterial DNA was more heterogeneous in size and produced a broad band in comparison to the phage DNA. In the alkaline gradient the phage DNA showed a trailing shoulder of smaller pieces but appeared greater than 50% intact. The bacterial peak was slightly broader than in the neutral gradient. After five minutes of treatment with the *H. influenzae* nuclease, the bacterial DNA peaks in both neutral and alkaline gradients were unaltered but the phage DNA was degraded to an average molecular weight of $1{\cdot}45 \times 10^6$ in the neutral gradient and a molecular weight of $0{\cdot}77 \times 10^6$ in the alkaline gradient (calculated according to Studier, 1965) as compared to a molecular weight of $26{\cdot}4 \times 10^6$ for the intact molecule (Studier, 1965). No apparent decrease in size over that found at five minutes was obtained after 30 minutes. The ³²P-labeled trichloroacetic acid-soluble radioactivity was $< 0{\cdot}1\%$, $<0{\cdot}1\%$ and $0{\cdot}26\%$ at 0, 5 and 30 minutes, respectively. No trichloroacetic acid-soluble ³H radioactivity was detectable. The nuclease is thus inactive on homologous native DNA, producing neither double- nor single-strand breaks. On heterologous native DNA it appears to act by making a limited number of double-strand breaks. Since the denatured product is approximately one-half the molecular size of the native product, no nicks appear to be produced over and above the double-strand cleavages. The enzyme will be referred to in the remainder of this report as endonuclease R.

(v) *Absence of exonucleolytic activity*

In the above experiment, little, if any, ³²P radioactivity was released as trichloroacetic acid-soluble material during the digestion with endonuclease R. Therefore it appears unlikely that the enzyme itself has an exonuclease activity or that it is contaminated with a significant exonuclease activity. In order to substantiate this further the digest was examined for released nucleotides. A solution of ³²P-labeled phage T7 DNA was extensively digested with an excess of endonuclease R. A sample was then mixed with unlabeled marker nucleotides and chromatographed in two dimensions. As seen in Table 3, essentially no activity was found associated with the nucleotide spots.

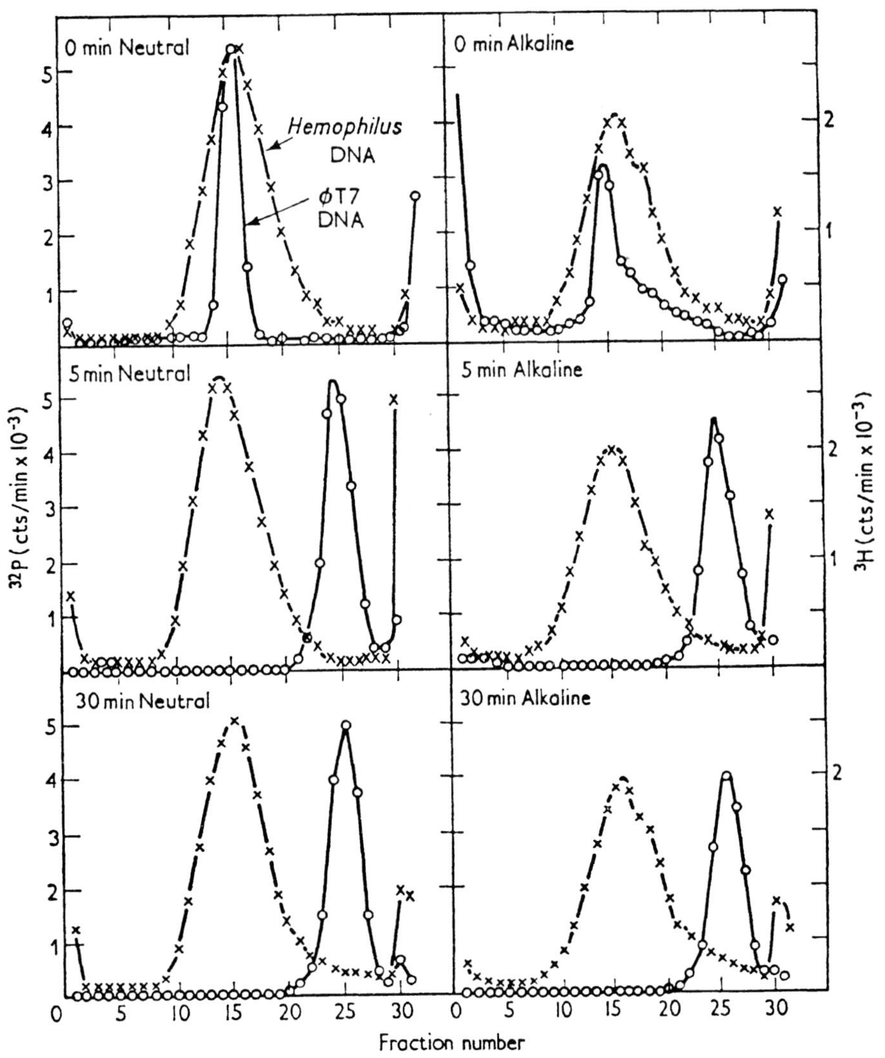

Fig. 3. Zone sedimentation in neutral and alkaline sucrose gradients of the products of the *H. influenzae* nuclease digestion of phage T7 DNA.

The reaction mixture (0·975 ml.) contained ^{3}H-labeled *H. influenzae* DNA, 17·1 μg, $8·7 \times 10^4$ cts/min; ^{32}P-labeled phage T7 DNA, 0·94 μg, $6·4 \times 10^4$ cts/min; 50 mM-NaCl; 6·6 mM-Tris–HCl, pH 7·4; 6·6 mM-MgCl$_2$; 6·6 mM-mercaptoethanol. At zero time 0·1 ml. was removed for trichloroacetic acid precipitation and 0·15 ml. was pipetted into 0·15 ml. of 0·1 M-EDTA, pH 8·0, on ice for sucrose gradient analysis. Purified *H. influenzae* nuclease, 5 μl., was then added and the reaction mixture was incubated at 37°C. Additional samples for trichloroacetic acid precipitation (0·1 ml.) and sucrose analysis (0·15 ml.) were removed at 5 and 30 min. A 0·1-ml. portion of each of the samples that had been taken for sucrose gradient analysis was layered on a neutral gradient and centrifuged at 50,000 rev./min for 2 hr at 4°C. The remaining 0·2 ml., was alkali-denatured by addition of 10 μl. of 4 M-NaOH and 0·1 ml. was then centrifuged at 50,000 rev./min for 2·5 hr. Liquid fractions collected drop wise from the bottom of the centrifuge tubes were counted directly in scintillation medium. (— × — × —), ^{3}H radioactivity; (—○—○—), ^{32}P radioactivity. Trichloroacetic acid precipitation was carried out by addition of 0·2 ml. of salmon sperm DNA (500 μg/ml.) and 0·3 ml. of 10% trichloroacetic acid on ice. The samples were centrifuged at 4500 g for 10 min and 0·3 ml. of the supernatant was counted in scintillation medium. The results are reported in the text.

TABLE 3

Release of nucleotides by endonuclease R digestion

Nucleotide species	Counts/50 min	%
dAMP	38	0·0004
dTMP	121	0·001
dGMP	0	0
dCMP	0	0
origin	$1·04 \times 10^7$	100

A reaction mixture (0·1 ml.) containing ^{32}P-labeled phage T7 DNA 78 mμmoles, $5·5 \times 10^4$ cts/min/mμmole in Tris–Mg–mercaptoethanol buffer was incubated with 5 μl. of endonuclease R for 120 min at 37°C and for 42 min with an additional 5 μl. of enzyme. Five μl. were then removed, mixed with the 4 standard 5′-nucleotides, spotted on the corner of a thin layer PEI cellulose square and chromatographed in the first dimension with 1 M-formic acid followed by a second dimension with 1 M-LiCl using methods described by Kelly & Smith (1970). The marker spots were located with a shortwave ultraviolet mineralight, cut out and counted in toluene scintillation medium. The origin including the surrounding 1 cm of thin layer material was also counted. The results are presented as 50-min counts corrected for background.

(vi) *Native* versus *denatured DNA as substrate*

Endonuclease R is active only on native DNA. No decrease in the size of denatured T7 DNA was demonstrable by zone sedimentation on alkaline sucrose gradients after incubation with enzyme (Fig. 4(a) and (b)), whereas native DNA was degraded (Fig. 4(c)).

(vii) *3′-Hydroxyl, 5′-phosphoryl cleavage*

Polynucleotide kinase catalyses the transfer of a ^{32}P-labeled phosphoryl group from [γ-^{32}P]ATP to the 5′ terminus of the polynucleotide chain. This reaction is known to depend upon the presence of a free hydroxyl group at the 5′ terminus of the substrate and will not occur if a 5′-phosphoryl group is present (Weiss *et al.*, 1968). It follows that if pretreatment of the substrate with bacterial alkaline phosphatase is required in order to obtain transfer then the 5′ termini of the substrate must be phosphorylated. As seen in Table 4, phage P22 DNA which has been digested to completion with endonuclease R, incubated with phosphatase, and then with kinase and [γ-^{32}P]ATP, incorporated 3740 cts/min/20 mμmole of phage DNA whereas omission of the phosphatase treatment results in only 136 cts/min incorporation. Thus, endonuclease R produces a 3′-hydroxyl, 5′-phosphoryl cleavage. Without endonuclease R treatment only the ends of the intact DNA molecules are labeled. Subtracting the blank of 66 cts/min obtained when kinase is omitted, 114 cts/min are incorporated per 20 mμmole of phage DNA. Using this and the known specific activity of the [γ-^{32}P]ATP, the molecular weight of the phage DNA is calculated as 26×10^6 (see legend of Table 3 for method of calculation). This figure agrees well with the measured molecular weight of $26·3 \times 10^6$ obtained by Rhoades, MacHattie & Thomas (1968). The average calculated molecular weight of the limit product DNA fragments is $0·81 \times 10^6$. Approximately 32 breaks are produced per phage P22 DNA molecule by the enzyme.

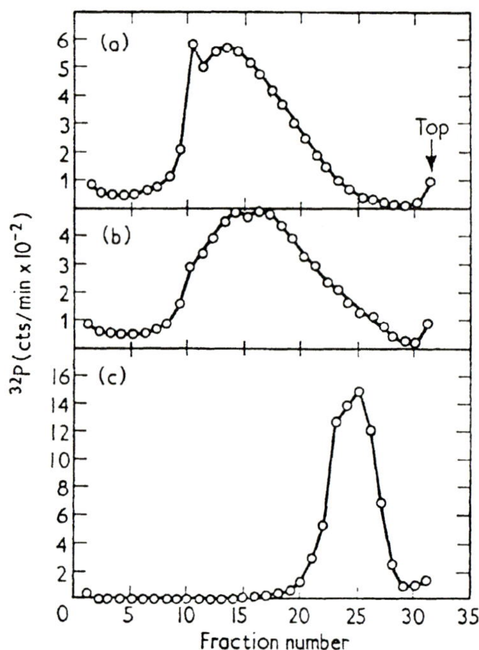

Fig. 4. Nuclease activity on denatured and native DNA.

Three reaction mixtures (a), (b) and (c) were set up in the following way. A control reaction mixture (a) contained 10 μl. of alkali-denatured ^{32}P-labeled phage T7 DNA, $1 \cdot 15 \times 10^4$ cts/min/mμmole, 100 mμmole/ml. and 50 μl. of Tris–Mg–mercaptoethanol buffer containing $0 \cdot 04$ M-NaCl. Reaction mixture (b) contained 5 μl. of purified endonuclease R in addition to the above ingredients. Reaction mixture (c) was the same as in (b) except that 10 μl. of native ^{32}P-labeled T7 DNA was used. The three reaction mixtures were incubated for 15 min at 37°C after which 5 μl. of $0 \cdot 5$ M-EDTA was added and then each was alkali-denatured. Each mixture was then layered onto an alkaline sucrose gradient and centrifuged for $2 \cdot 5$ hr at 50,000 rev./min. Fractions were collected and counted as in Fig. 2. (The ^{32}P-labeled T7 DNA used in this experiment had suffered considerable radiation damage due to storage, and this accounts for the broad peaks obtained.)

TABLE 4

^{32}P-labeling of the 5'-end of endonuclease R-treated phage P22 DNA

DNA treatment	Cts/min	Calc. mol. wt
+ Bacterial alkaline phosphatase + kinase	180	26×10^6
Endonuclease + Bacterial alkaline phosphatase	66	
Endonuclease + kinase	136	
Endonuclease + Bacterial alkaline phosphatase + kinase	3740	$0 \cdot 81 \times 10^6$

The reaction mixture for endonuclease R digestion ($0 \cdot 45$ ml.) contained 180 mμmoles of phage P22 DNA, 50 mM-NaCl, and $6 \cdot 6$ mM each of Tris–HCl (pH $7 \cdot 4$), MgCl$_2$, and mercaptoethanol. At zero time, endonuclease R (5 μl.) was added and the mixture was incubated at 37°C for 30 min. Samples of $0 \cdot 05$ ml. were removed at zero time (before addition of enzyme) and after the completion of digestion and treated with the various listed combinations of phosphatase or polynucleotide kinase. The [γ-^{32}P]ATP used in the kinase labeling reaction had a specific activity of $2 \cdot 25 \times 10^5$ cts/min/mμmole. The amount of ^{32}P incorporated terminally was measured as trichloroacetic acid-precipitable counts.

Molecular weight = $660\, ma/(c - b)$, where m is the mμmoles of DNA, c is the incorporated terminal ^{32}P radioactivity, b is the blank in which kinase was absent, a is the specific activity of [γ-^{32}P]ATP, and the molecular weight of the base pair is 660.

(viii) *Mechanism of double-strand cleavage*

The sucrose gradient experiments indicate that native foreign DNA possesses a limited number of substrate sites which are subject to double-strand cleavage by endonuclease R. The question arises as to whether the introduction of the second single-strand break within a site is independent of the first. If so, then a limited digestion with endonuclease R should result in a large excess of nicks over duplex breaks. On the other hand, if the introduction of the second nick is coupled to that of the first, then duplex breaks should appear in significant amount early in the reaction.

To determine by which mechanism endonuclease R acts, ^{32}P–5′ end-labeling techniques were again used. It is possible, as described in Materials and Methods, to discriminate nicks from ends by this procedure taking advantage of the fact that nicks will be labeled only if the DNA is denatured before phosphatase treatment. Separate reaction mixtures were set up containing phage T7 DNA and various concentrations of endonuclease R ranging from an amount capable of producing relatively very few breaks during the incubation to an amount which was saturating. After incubation with the enzyme the DNA was labeled at the 5′-termini with ^{32}P either before or after denaturation (Fig. 5). The data show that with incomplete digestion, more label

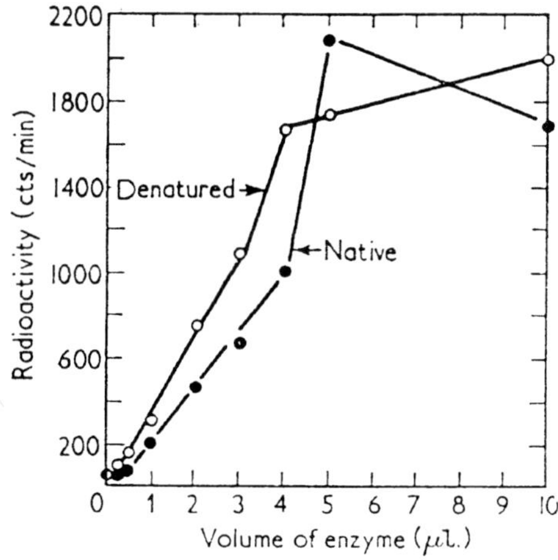

FIG. 5. Production of nicks and double-strand breaks in phage T7 DNA.
The reaction mixtures for endonuclease R digestion each contained unlabeled phage T7 DNA, 66 mμmoles and ^{3}H-labeled *H. influenzae* DNA, 0·23 mμmole, 1·5 × 10⁴ cts/min/mμmole in a 0·15-ml. vol of Tris–Mg–mercaptoethanol buffer containing 0·06 M-NaCl. To each reaction mixture the amount of enzyme shown in the Figure was added and incubation was carried out for 20 min at 37°C. Endonuclease R was then inactivated by treatment at 65°C for 10 min. The DNA in each tube was precipitated by 2 ml. of 70% ethanol and washed with 2 ml. of acetone. The DNA was redissolved in 0·15 ml. of NaCl–Tris buffer. An 0·5 ml. sample from each digest was then terminally labeled by the kinase reaction with or without alkali denaturation preceding the phosphatase step. The ^{3}H-labeled DNA was used to monitor recovery from the precipitation step. Recovery averaged about 60% and corrections were made to 100% recovery.

was incorporated into the DNA after denaturation than before, indicating that some nicks are produced first. However, the early appearance of a significant proportion of breaks clearly suggests an association between the two nicking events which result in a duplex break. This association might be explained by either of two mechanisms. Either the enzyme binds to a site and breaks first one strand and then the other without becoming detached, or else it introduces nicks in separate binding events with the probability for rebinding being much greater following the first nick. In the latter case we might imagine that the presence of the first nick makes the other chain more accessible.

An additional result of the experiments is that the DNA fragments produced as a limit product with high levels of enzyme contained no nicks, within the limits of error of the terminal labeling procedure, since both curves reached a plateau together at an average of about 1875 cts/min incorporated terminally. Since about 45 cts/min were incorporated into the undigested intact DNA, about 40 to 45 breaks were produced per phage T7 molecule.

4. Discussion

Endonuclease R produces double-strand, 5′-phosphoryl, 3′-hydroxyl cleavages at a limited number of sites on foreign native DNA. In addition to phage T7 and phage P22 DNA, we have tested *S. typhimurium* DNA, salmon sperm DNA and *Bacillus subtilis* DNA. All are degraded to about the same extent. The fragments produced in the limit digest of phage T7 DNA contain no excess of nicks, and essentially no trichloroacetic acid-soluble products or nucleotides are released by the digestion. The enzyme has no demonstrable activity on native *H. influenzae* DNA or on denatured foreign DNA. There are no special co-factor requirements for activity other than magnesium ions.

The similarity of endonuclease R to the restriction enzyme purified from *E. coli* by Meselson & Yuan (1968) is considerable. Both enzymes produce a small number of specific double-strand breaks in foreign DNA and are inactive on DNA of the bacterial strain from which they have been purified. An interesting feature of the *E. coli* enzyme is that it requires *S*-adenosyl methionine and adenosine triphosphate as co-factors in addition to magnesium ions. This suggests a connection between the restriction activity and the DNA modification activity which is known to protect the host DNA from cleavage. Arber (1968) has identified the DNA modification as methylation of adenine to form 6-methyl amino purine and recently Kühnlein, Linn & Arber (1969) have demonstrated modification *in vitro*. We have not as yet investigated DNA modification in *H. influenzae*.

We have been particularly interested in the ability of endonuclease R to "recognize" only a few specific sites on rather large foreign DNA molecules. It appears likely that this recognition specificity resides in the base sequence of the sites. An estimate of the size of the base sequence can be made. For phage T7 DNA we observed 40 to 45 breaks per molecule. Since T7 DNA is about 40,000 base pairs in length, the average fragment is about 1000 base pairs in length. For phage P22 DNA, the average fragment is approximately 1300 base pairs in length. To attain this degree of specificity a site would have to be five to six bases in length providing that the enzyme recognizes a completely unique sequence. In the accompanying paper (Kelly & Smith, 1970) the base sequence recognized by endonuclease R is completely identified and provides confirmation of this estimate.

We wish to acknowledge especially the considerable help received from Dr Bernard Weiss. He was very generous in supplying us with labeled ATP, several enzymes and much good advice. We thank Dr Paul Englund who donated enzymes and good advice; Dr Nagaraja Rao who helped during the initial stages of the work and Dr Thomas J. Kelly, Jr. who gave helpful suggestions during preparation of the manuscript.

This work was supported by U.S. Public Health Service grant no. AI-07875.

One of the authors (H.O.S.) holds a U.S. Public Health Service Career Development Award no. AI-17902.

REFERENCES

Arber, W. (1965). *Ann. Rev. Microbiol.* **19**, 365.
Arber, W. (1968). *18th Symp. Soc. Gen. Microbiol.: Molecular Biology of Viruses*, p. 295. London: Cambridge University Press.
Carmody, J. & Herriot, R. M. (1970). *J. Bact.* **101**, 525.
Kelly, T. J. & Smith, H. O. (1970). *J. Mol. Biol.* **51**, 393.
Kühnlein, U., Linn, S. & Arber, W. (1969). *Proc. Nat. Acad. Sci., Wash.* **63**, 556.
Levine, M. (1957). *Virology*, **3**, 203.
Marmur, J. (1961). *J. Mol. Biol.* **3**, 208.
Meselson, M. & Yuan, R. (1968). *Nature*, **217**, 1110.
Rhoades, M., MacHattie, L. A. & Thomas, C. A., Jr. (1968). *J. Mol. Biol.* **37**, 21.
Richardson, C. C. (1965). *Proc. Nat. Acad. Sci., Wash.* **54**, 158.
Smith, H. O. (1968). *Virology*, **34**, 203.
Smith, H. O. & Wilcox, K. W. (1969). *Fed. Proc.* **28**, 465.
Studier, F. W. (1965). *J. Mol. Biol.* **11**, 373.
Thomas, C. A., Jr. & Abelson, J. (1966). *Procedures in Nucleic Acid Research*, ed. by G. Cantoni & D. Davies, p. 553. New York: Harper & Row.
Weiss, B., Live, T. R. & Richardson, C. C. (1968). *J. Biol. Chem.* **243**, 4530.

Construction of Biologically Functional Bacterial Plasmids In Vitro

S. N. COHEN, A. C. Y. CHANG, H. W. BOYER, AND R. B. HELLING

The innovative methodology described in this paper, along with the discovery of the *Haemophilus influenzae* restriction enzyme by Smith and Wilcox (see above), set the stage for recombinant DNA technology as we know it today. Cohen and colleagues were the first to construct a recombinant plasmid. They showed that ligation of two *Eco*RI-digested DNA fragments from separate antibiotic resistance plasmids resulted in a functional hybrid replicon when transformed into *Escherichia coli* C600. Furthermore, Cohen et al. showed that the recombinant plasmids retained the genetic nucleotides from both of the parental DNA sources. It should be pointed out that the demonstration of nucleotide sequence homology between parent and recombinant plasmid DNA was a much harder undertaking in 1973 than it is today. Indeed, to provide such proof, Cohen and colleagues had to show by electron microscopic analyses that heteroduplex formation occurred between pSC109 and each of its component plasmids. The Southern hybridization method and techniques to sequence segments of DNA were not published until 1975 and 1977, respectively (E. Southern, *J. Mol. Biol.* **98**:503–517, 1975; F. Sanger, S. Nicklen, and A.R. Coulson, *Proc. Natl. Acad. Sci. USA* **74**:5463–5467, 1975; and A. M. Maxam and W. Gilbert, *Proc. Natl. Acad. Sci. USA* **74**:560–564, 1977). The process of cleaving DNA with an enzyme, joining the restricted fragments to form a plasmid, transforming the recombinant plasmid into a bacterial host, and permitting growth of the transformants with subsequent amplification of the recombinant plasmid, constitutes the bread and butter of molecular-cloning technology. As one of our committee members asked about the methods described in this paper, "What technique could have possibly had a bigger single impact on how we do molecular analyses of microbes today?"

ALISON O'BRIEN

Reprinted from *Proceedings of the National Academy of Sciences USA* 70:3240–3244, Copyright © 1973, by permission of the authors.

Reprinted from

Proc. Nat. Acad. Sci. USA
Vol. 70, No. 11, pp. 3240–3244, November 1973

Construction of Biologically Functional Bacterial Plasmids *In Vitro*

(R factor/restriction enzyme/transformation/endonuclease/antibiotic resistance)

STANLEY N. COHEN*, ANNIE C. Y. CHANG*, HERBERT W. BOYER†, AND ROBERT B. HELLING†

* Department of Medicine, Stanford University School of Medicine, Stanford, California 94305; and † Department of Microbiology, University of California at San Francisco, San Francisco, Calif. 94122

Communicated by Norman Davidson, July 18, 1973

ABSTRACT   **The construction of new plasmid DNA species by *in vitro* joining of restriction endonuclease-generated fragments of separate plasmids is described. Newly constructed plasmids that are inserted into *Escherichia coli* by transformation are shown to be biologically functional replicons that possess genetic properties and nucleotide base sequences from both of the parent DNA molecules. Functional plasmids can be obtained by reassociation of endonuclease-generated fragments of larger replicons, as well as by joining of plasmid DNA molecules of entirely different origins.**

Controlled shearing of antibiotic resistance (R) factor DNA leads to formation of plasmid DNA segments that can be taken up by appropriately treated *Escherichia coli* cells and that recircularize to form new, autonomously replicating plasmids (1). One such plasmid that is formed after transformation of *E. coli* by a fragment of sheared R6-5 DNA, pSC101 (previously referred to as Tc6-5), has a molecular weight of 5.8×10^6, which represents about 10% of the genome of the parent R factor. This plasmid carries genetic information necessary for its own replication and for expression of resistance to tetracycline, but lacks the other drug resistance determinants and the fertility functions carried by R6-5 (1).

Two recently described restriction endonucleases, *Eco*RI and *Eco*RII, cleave double-stranded DNA so as to produce short overlapping single-stranded ends. The nucleotide sequences cleaved are unique and self-complementary (2–6) so that DNA fragments produced by one of these enzymes can associate by hydrogen-bonding with other fragments produced by the same enzyme. After hydrogen-bonding, the 3′-hydroxyl and 5′-phosphate ends can be joined by DNA ligase (6). Thus, these restriction endonucleases appeared to have great potential value for the construction of new plasmid species by joining DNA molecules from different sources. The *Eco*RI endonuclease seemed especially useful for this purpose, because on a random basis the sequence cleaved is expected to occur only about once for every 4,000 to 16,000 nucleotide pairs (2); thus, most *Eco*RI-generated DNA fragments should contain one or more intact genes.

We describe here the construction of new plasmid DNA species by *in vitro* association of the *Eco*RI-derived DNA fragments from separate plasmids. In one instance a new plasmid has been constructed from two DNA species of entirely different origin, while in another, a plasmid which has itself been derived from *Eco*RI-generated DNA fragments of a larger parent plasmid genome has been joined to another replicon derived independently from the same parent plasmid. Plasmids that have been constructed by the *in vitro* joining of

*Eco*RI-generated fragments have been inserted into appropriately-treated *E. coli* by transformation (7) and have been shown to form biologically functional replicons that possess genetic properties and nucleotide base sequences of both parent DNA species.

MATERIALS AND METHODS

E. coli strain W1485 containing the RSF1010 plasmid, which carries resistance to streptomycin and sulfonamide, was obtained from S. Falkow. Other bacterial strains and R factors and procedures for DNA isolation, electron microscopy, and transformation of *E. coli* by plasmid DNA have been described (1, 7, 8). Purification and use of the *Eco*RI restriction endonuclease have been described (5). Plasmid heteroduplex studies were performed as previously described (9, 10). *E. coli* DNA ligase was a gift from P. Modrich and R. L. Lehman and was used as described (11). The detailed procedures for gel electrophoresis of DNA will be described elsewhere (Helling, Goodman, and Boyer, in preparation); in brief, duplex DNA was subjected to electrophoresis in a tube-type apparatus (Hoefer Scientific Instrument) (0.6×15-cm gel) at about 20° in 0.7% agarose at 22.5 V with 40 mM Tris–acetate buffer (pH 8.05) containing 20 mM sodium acetate, 2 mM EDTA, and 18 mM sodium chloride. The gels were then soaked in ethidium bromide (5 µg/ml) and the DNA was visualized by fluorescence under long wavelength ultraviolet light ("black light"). The molecular weight of each fragment in the range of 1 to 200×10^5 was determined from its mobility relative to the mobilities of DNA standards of known molecular weight included in the same gel (Helling, Goodman, and Boyer, in preparation).

RESULTS

R6-5 and pSC101 plasmid DNA preparations were treated with the *Eco*RI restriction endonuclease, and the resulting DNA products were analyzed by electrophoresis in agarose gels. Photographs of the fluorescing DNA bands derived from these plasmids are presented in Fig. 1*b* and *c*. Only one band is observed after *Eco*RI endonucleolytic digestion of pSC101 DNA (Fig. 1*c*), suggesting that this plasmid has a single site susceptible to cleavage by the enzyme. In addition, endonuclease-treated pSC101 DNA is located at the position in the gel that would be expected if the covalently closed circular plasmid is cleaved once to form noncircular DNA of the same molecular weight. The molecular weight of the linear fragment estimated from its mobility in the gel is 5.8×10^6, in agreement with independent measurements of the size of the intact molecule (1). Because pSC101 has a single *Eco*RI cleavage site and is derived from R6-5, the equivalent DNA sequences of

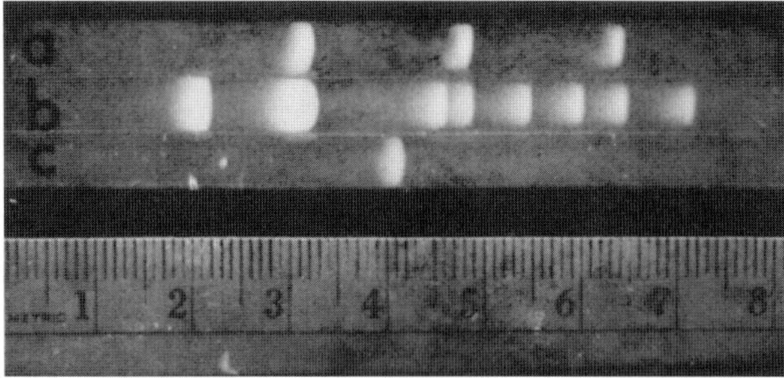

FIG. 1. Agarose-gel electrophoresis of *Eco*RI digests. (*a*) pSC102. The three fragments derived from the plasmid correspond to fragments III, V, and VIII of R6-5 (Fig. 1*b* below) as shown here and as confirmed by electrophoresis in other gels (see *text*). (*b*) R6-5. The molecular weights calculated for the fragments, as indicated in *Methods*, are (from *left* to *right*) I, 17.0; II & III (double band), 9.6 and 9.1; IV, 5.2; V, 4.9; VI, 4.3; VII, 3.8; VIII, 3.4; IX, 2.9. All molecular weight values have been multiplied by 10^{-6}. (*c*) pSC101. The calculated molecular weight of the single fragment is 5.8×10^6. Migration in all gels was from *left* (cathode) to *right*; samples were subjected to electrophoresis for 19 hr and 50 min. ·

the parent plasmid must be distributed in two separate *Eco*RI fragments.

The *Eco*RI endonuclease products of R6-5 plasmid DNA were separated into 12 distinct bands, eight of which are seen in the gel shown in Fig. 1*b*; the largest fragment has a molecular weight of 17×10^6, while three fragments (not shown in Fig. 1*b*) have molecular weights of less than 1×10^6, as determined by their relative mobilities in agarose gels. As seen in the figure, an increased intensity of fluorescence, of the second band suggests that this band contains two or more DNA fragments of almost equal size; when smaller amounts of *Eco*RI-treated R6-5 DNA are subjected to electrophoresis for a longer period of time, resolution of the two fragments (i.e., II and III) is narrowly attainable. Because 12 different *Eco*RI-generated DNA fragments can be identified after endonuclease treatment of covalently closed circular R6-5, there must be at least 12 substrate sites for *Eco*RI endonuclease present on this plasmid, or an average of one site for every 8000 nucleotide pairs. The molecular weight for each fragment shown is given in the caption to Fig. 1. The sum of the molecular weights of the *Eco*RI fragments of R6-5 DNA is 61.5×10^6, which is in close agreement with independent estimates for the molecular weight of the intact plasmid (7, 10).

The results of separate transformations of *E. coli* C600 by endonuclease-treated pSC101 or R6-5 DNA are shown in Table 1. As seen in the table, cleaved pSC101 DNA transforms *E. coli* C600 with a frequency about 10-fold lower than was observed with covalently closed or nicked circular (1) molecules of the same plasmid. The ability of cleaved pSC-101 DNA to function in transformation suggests that plasmid DNA fragments with short cohesive endonuclease-generated termini can recircularize in *E. coli* and be ligated *in vivo*; since the denaturing temperature (T_m) for the termini generated by the *Eco*RI endonuclease is 5–6° (6) and the transformation procedure includes a 42° incubation step (7), it is unlikely that the plasmid DNA molecules enter bacterial cells with their termini already hydrogen-bonded. A corresponding observation has been made with *Eco*RI endonuclease-cleaved

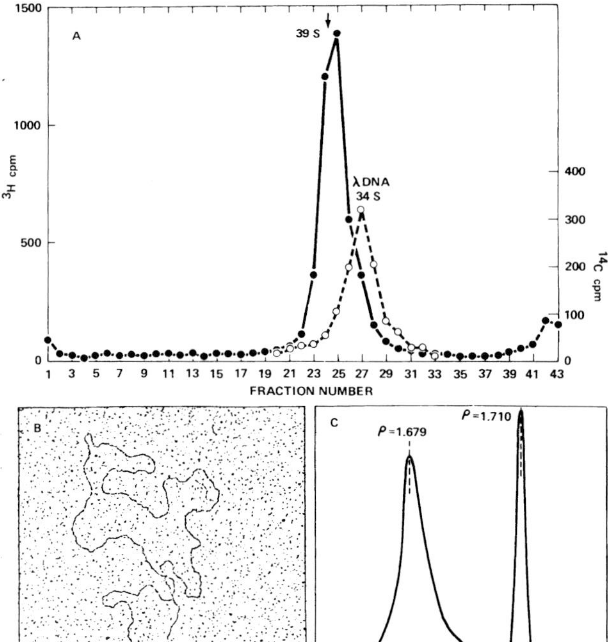

FIG. 2. Physical properties of the pSC102 plasmid derived from *Eco*RI fragments of R6-5. (*A*) Sucrose gradient centrifugation analysis (1, 8) of covalently closed circular plasmid DNA (●———●) isolated from an *E. coli* transformant clone as described in text. 34 S linear [^{14}C]DNA from λ was used as a standard (○– – –○). (*B*) Electron photomicrograph of nicked (7) pSC102 DNA. The length of this molecule is approximately 8.7 μm. (*C*) Densitometer tracing of analytical ultracentrifugation (8) photograph of pSC102 plasmid DNA. Centrifugation in CsCl ($\rho = 1.710$ g/cm^3) was carried out in the presence of d(A-T)$_n$·d(A-T)$_n$ density marker ($\rho = 1.679$ g/cm^3).

SV40 DNA, which forms covalently closed circular DNA molecules in mammalian cells *in vivo* (6).

Transformation for each of the antibiotic resistance markers present on the R6-5 plasmid was also reduced after treatment of this DNA with *Eco*RI endonuclease (Table 1). Since the pSC101 (tetracycline-resistance) plasmid was derived from R6-5 by controlled shearing of R6-5 DNA (1), and no tetracycline-resistant clone was recovered after transformation by the *Eco*RI endonuclease products of R6-5, [whereas tetracycline-resistant clones are recovered after transformation with intact R6-5 DNA (1)], an *Eco*RI restriction site may separate the tetracycline resistance gene of R6-5 from its replicator locus. Our finding that the linear fragment produced by treatment of pSC101 DNA with *Eco*RI endonuclease does not correspond to any of the *Eco*RI-generated fragments of R6-5 (Fig. 1) is consistent with this interpretation.

A single clone that had been selected for resistance to kanamycin and which was found also to carry resistance to neomycin and sulfonamide, but not to tetracycline, chloramphenicol, or streptomycin after transformation of *E. coli* by *Eco*RI-generated DNA fragments of R6-5, was examined further. Closed circular DNA obtained from this isolate (plasmid designation pSC102) by CsCl–ethidium bromide gradient

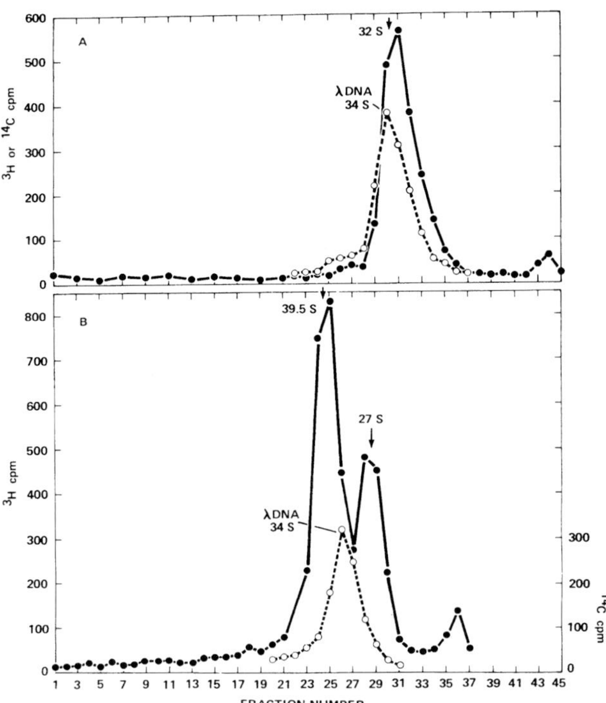

FIG. 3.   Sucrose gradient centrifugation of DNA isolated from *E. coli* clones transformed for both tetracycline and kanamycin resistance by a mixture of pSC101 and pSC102 DNA. (*A*) The DNA mixture was treated with *Eco*RI endonuclease and was ligated prior to use in the transformation procedure. Covalently closed circular DNA isolated (7, 8) from a transformant clone carrying resistance to both tetracycline and kanamycin was examined by sedimentation in a neutral 5–20% sucrose gradient (8). (*B*) Sucrose sedimentation pattern of covalently closed circular DNA isolated from a tetracycline and kanamycin resistant clone transformed with an *untreated* mixture of pSC101 and pSC102 plasmid DNA.

centrifugation has an S value of 39.5 in neutral sucrose gradients (Fig. 2*A*) and a contour length of 8.7 μm when nicked (Fig. 2*B*). These data indicate a molecular weight

TABLE 1.   *Transformation by covalently closed circular and Eco RI-treated plasmid DNA*

Plasmid DNA species	Transformants per μg DNA		
	Tetracycline	Kanamycin (neomycin)	Chloramphenicol
pSC101 covalently closed circle	3×10^5	—	—
*Eco*RI-treated	2.8×10^4	—	—
R6-5 covalently closed circle	—	1.3×10^4	1.3×10^4
*Eco*RI-treated	<5	1×10^2	4×10^1

Transformation of *E. coli* strain C600 by plasmid DNA was carried out as indicated in *Methods*. The kanamycin resistance determinant of R6-5 codes also for resistance to neomycin (15). Antibiotics used for selection were tetracycline (10 μg/ml), kanamycin (25 μg/ml) or chloramphenicol (25 μg/ml).

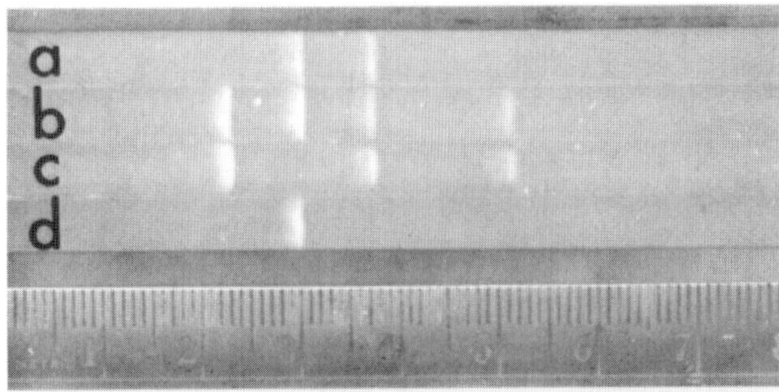

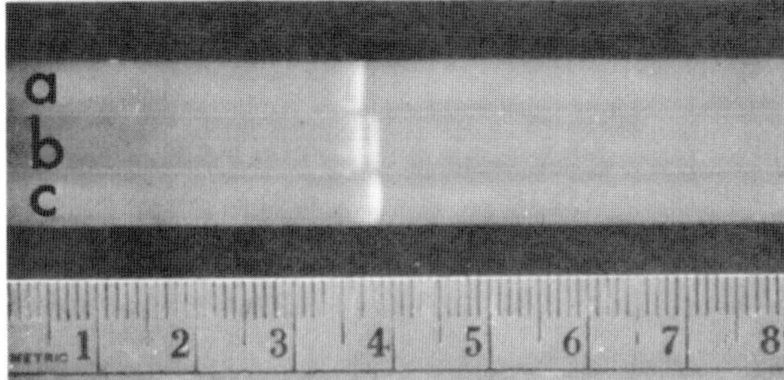

FIGS. 4 and 5.   Agarose-gel electrophoresis of *Eco*RI digests of newly constructed plasmid species.   Conditions were as described in *Methods*.

FIG. 4.   (*top*) Gels were subjected to electrophoresis for 19 hr and 10 min. (*a*) pSC105 DNA. (*b*) Mixture of pSC101 and pSC102 DNA. (*c*) pSC102 DNA. (*d*) pSC101 DNA.

FIG. 5.   (*bottom*) Gels were subjected to electrophoresis for 18 hr and 30 min. (*a*) pSC101 DNA. (*b*) pSC109 DNA. (*c*) RSF1010 DNA. Evidence that the single band observed in this gel represents a linear fragment of cleaved RSF1010 DNA was obtained by comparing the relative mobilities of *Eco*RI-treated DNA and untreated (covalently closed circular and nicked circular) RSF1010 DNA in gels. The molecular weight of RSF-1010 calculated from its mobility in gels is 5.5×10^6.

about 17×10^6. Isopycnic centrifugation in cesium chloride of this non-self-transmissible plasmid indicated it has a buoyant density of 1.710 g/cm³ (Fig. 2*C*). Since the nucleotide base composition of the antibiotic resistance determinant (R-determinant) segment of the parent R factor is 1.718 g/cm³ (8), the various component regions of the resistance unit must have widely different base compositions, and the pSC102 plasmid must lack a part of this unit that is rich in high buoyant density G+C nucleotide pairs. The existence of such a high buoyant density *Eco*RI fragment of R6-5 DNA was confirmed by centrifugation of *Eco*RI-treated R6-5 DNA in neutral cesium chloride gradients (Cohen and Chang, unpublished data).

Treatment of pSC102 plasmid DNA with *Eco*RI restriction endonuclease results in formation of three fragments that are separable by electrophoresis is agarose gels (Fig. 1*a*); the estimated molecular weights of these fragments determined by gel mobility total 17.4×10^6, which is in close agreement with the molecular weight of the intact pSC102 plasmid determined by sucrose gradient centrifugation and electron microscopy (Fig. 2). Comparison with the *Eco*RI-generated fragments of R6-5 indicates that the pSC102 fragments correspond to fragments III (as determined by long-term electrophoresis in gels containing smaller amounts of DNA), V, and VIII of the parent plasmid (Fig. 1*b*). These results suggest that *E. coli* cells transformed with *Eco*RI-generated DNA fragments of R6-5

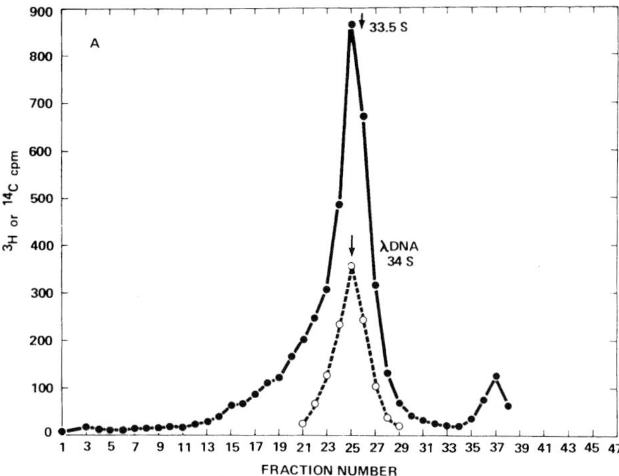

FIG. 6. Sucrose gradient sedimentation of covalently closed circular DNA representing the pSC109 plasmid derived from RSF1010 and pSC101.

TABLE 2. *Transformation of E. coli C600 by a mixture of pSC101 and pSC102 DNA*

Treatment of DNA	Transformation frequency for antibiotic resistance markers		
	Tetracycline	Kanamycin	Tetracycline + kanamycin
None	2×10^5	1×10^5	2×10^2
*Eco*RI	1×10^4	1.1×10^3	7×10^1
*Eco*RI + DNA ligase	1.2×10^4	1.3×10^3	5.7×10^2

Transformation frequency is shown in transformants per μg of DNA of each plasmid species in the mixture. Antibiotic concentrations are indicated in legend of Table 1.

can ligate reassociated DNA fragments *in vivo*, and that reassociated molecules carrying antibiotic resistance genes and capable of replication can circularize and can be recovered as functional plasmids by appropriate selection.

A mixture of pSC101 and pSC102 plasmid DNA species, which had been separately purified by dye–buoyant density centrifugation, was treated with the *Eco*RI endonuclease, and then was either used directly to transform *E. coli* or was ligated prior to use in the transformation procedure (Table 2). In a control experiment, a plasmid DNA mixture that had not been subjected to endonuclease digestion was employed for transformation. As seen in this table, transformants carrying resistance to both tetracycline and kanamycin were isolated in all three instances. Cotransformation of tetracycline and kanamycin resistance by the untreated DNA mixture occurred at a 500- to 1000-fold lower frequency than transformation for the individual markers. Examination of three different transformant clones derived from this DNA mixture indicated that each contained two separate covalently closed circular DNA species having the sedimentation characteristics of the pSC101 and pSC102 plasmids (Fig. 3*B*). The ability of two plasmids derived from the same parental plasmid (i.e., R6-5) to exist stably as separate replicons (12) in a single

bacterial host cell suggests that the parent plasmid may contain at least two distinct replicator sites. This interpretation is consistent with earlier observations which indicate that the R6 plasmid dissociates into two separate compatible replicons in *Proteus mirabilis* (8). Cotransformation of tetracycline and kanamycin resistance by the *Eco*RI treated DNA mixture was 10- to 100-fold lower than transformation of either tetracycline or kanamycin resistance alone, and was increased about 8-fold by treatment of the endonuclease digest with DNA ligase (Table 2). Each of four studied clones derived by transformation with the endonuclease-treated and/or ligated DNA mixture contained only a single 32S covalently closed circular DNA species (Fig. 3*A*) that carries resistance to both tetracycline and kanamycin, and which can transform *E. coli* for resistance to both antibiotics. One of the clones derived from the ligase-treated mixture was selected for further study, and this plasmid was designated pSC105.

When the plasmid DNA of pSC105 was digested by the *Eco*RI endonuclease and analyzed by electrophoresis in agarose gels, two component fragments were identified (Fig. 4); the larger fragment was indistinguishable from endonuclease-treated pSC101 DNA (Fig. 4*d*) while the smaller fragment corresponded to the 4.9×10^6 dalton fragment of pSC102 plasmid DNA (Fig. 4*c*). Two endonuclease fragments of pSC102 were lacking in the pSC105 plasmid; presumably the sulfonamide resistance determinant of pSC102 is located on one of these fragments, since pSC105 does not specify re-

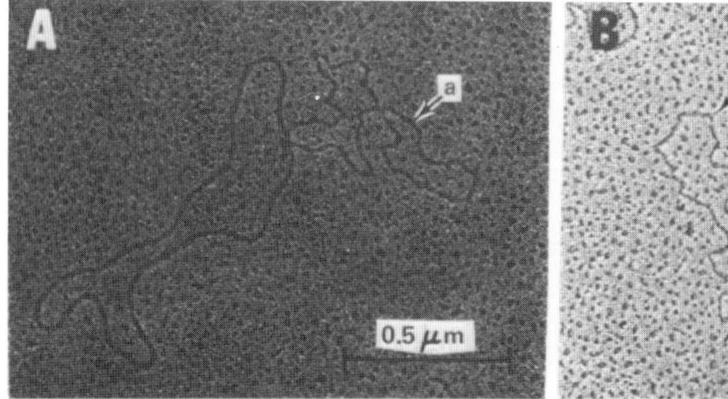

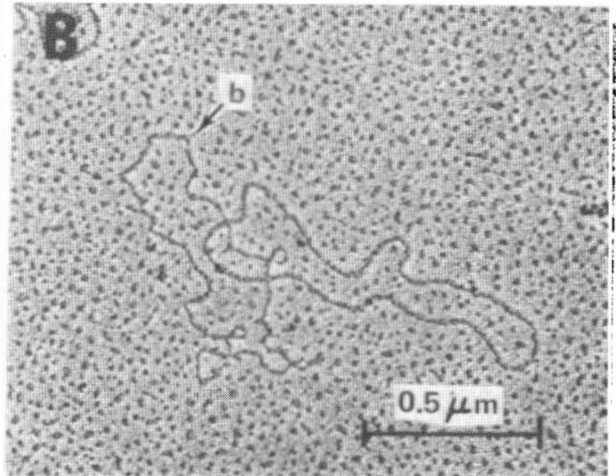

FIG. 7. (*A*) Heteroduplex of pSC101/pSC109. The single-stranded DNA loop marked by *a* represents the contribution of RSF1010 to the pSC109 plasmid. (*B*) Heteroduplex of RSF1010/pSC109. The single-stranded DNA loop marked by *b* represents the contribution of pSC101 to the pSC109 plasmid. pSC101 and RSF1010 homoduplexes served as internal standards for DNA length measurements. The scale is indicated by the bar on each electron photomicrograph.

sistance to this antibiotic. Since kanamycin resistance *is* expressed by pSC105, we conclude that this resistance gene resides on the 4.9 × 10^6 dalton fragment of pSC102 (fragment V of its parent, R6-5). The molecular weight of the pSC105 plasmid is estimated to be 10.5 × 10^6 by addition of the molecular weights of its two component fragments; this value is consistent with the molecular weight determined for this recombinant plasmid by sucrose gradient centrifugation (Fig. 3A) and electron microscopy. The recovery of a biologically functional plasmid (i.e., pSC105) that was formed by insertion of a fragment of another plasmid fragment into pSC101 indicates that the *Eco*RI restriction site on pSC101 does not interrupt the genetic continuity of either the tetracycline resistance gene or the replicating element of this plasmid.

We also constructed new biologically functional plasmids *in vitro* by joining cohesive-ended plasmid DNA molecules of entirely different origin. RSF1010 is a streptomycin and sulfonamide resistance plasmid which has a 55% G+C nucleotide base composition (13) and which was isolated originally from *Salmonella typhimurium* (14). Like pSC101, this non-self-transmissible plasmid is cleaved at a single site by the *Eco*RI endonuclease (Fig. 5c). A mixture of covalently closed circular DNA containing the RSF1010 and pSC101 plasmids was treated with the *Eco*RI endonuclease, ligated, and used for transformation. A transformant clone resistant to both tetracycline and streptomycin was selected, and covalently closed circular DNA (plasmid designation pSC109) isolated from this clone by dye–buoyant density centrifugation was shown to contain a single molecular species sedimenting at 33.5 S, corresponding to an approximate molecular weight of 11.5 × 10^6 (Fig. 6). Analysis of this DNA by agarose gel electrophoresis after *Eco*RI digestion (Fig. 5b) indicates that it consists of two separate DNA fragments that are indistinguishable from the *Eco*RI-treated RSF1010 and pSC101 plasmids (Fig. 5a and c).

Heteroduplexes shown in Fig. 7A and B demonstrate the existence of DNA nucleotide sequence homology between pSC109 and each of its component plasmids. As seen in this figure, the heteroduplex pSC101/pSC109 shows a double-stranded region about 3 µm in length and a slightly shorter single-stranded loop, which represents the contribution of RSF1010 to the recombinant plasmid. The heteroduplex formed between RSF1010 and pSC109 shows both a duplex region and a region of nonhomology, which contains the DNA contribution of pSC101 to pSC109.

SUMMARY AND DISCUSSION

These experiments indicate that bacterial antibiotic resistance plasmids that are constructed *in vitro* by the joining of *Eco*RI-treated plasmids or plasmid DNA fragments are bio-

logically functional when inserted into *E. coli* by transformation. The recombinant plasmids possess genetic properties and DNA nucleotide base sequences of both parent molecular species. Although ligation of reassociated *Eco*RI-treated fragments increases the efficiency of new plasmid formation, recombinant plasmids are also formed after transformation by *unligated Eco*RI-treated fragments.

The general procedure described here is potentially useful for insertion of specific sequences from prokaryotic or eukaryotic chromosomes or extrachromosomal DNA into independently replicating bacterial plasmids. The antibiotic resistance plasmid pSC101 constitutes a replicon of considerable potential usefulness for the selection of such constructed molecules, since its replication machinery and its tetracycline resistance gene are left intact after cleavage by the *Eco*RI endonuclease.

We thank P. A. Sharp and J. Sambrooke for suggesting use of ethidium bromide for staining DNA fragments in agarose gels. These studies were supported by Grants AI08619 and GM14378 from the National Institutes cf Health and by Grant GB-30581 from the National Science Foundation. S.N.C. is the recipient of a USPHS Career Development Award. R.B.H. is a USPHS Special Fellow of the Institute of General Medical Sciences on leave from the Department of Botany, University of Michigan.

1. Cohen, S. N. & Chang, A. C. Y. (1973) *Proc. Nat. Acad. Sci. USA* **70**, 1293–1297.
2. Hedgepeth, J., Goodman, H. M. & Boyer, H. W. (1972) *Proc. Nat. Acad. Sci. USA* **69**, 3448–3452.
3. Bigger, C. H., Murray, K. & Murray, N. E. (1973) *Nature New Biol.*, **224**, 7–10.
4. Boyer, H. W., Chow, L. T., Dugaiczyk, A., Hedgepeth, J. & Goodman, H. M. (1973) *Nature New Biol.*, **224**, 40–43.
5. Greene, P. J., Betlach, M. C., Goodman, H. M. & Boyer, H. W. (1973) "DNA replication and biosynthesis," in *Methods in Molecular Biology*, ed. Wickner, R. B. Marcel Dekker, Inc. New York), Vol. 9, in press.
6. Mertz, J. E. & Davis, R. W. (1972) *Proc. Nat. Acad. Sci. USA* **69**, 3370–3374.
7. Cohen, S. N., Chang, A. C. Y. & Hsu, L. (1972) *Proc. Nat. Acad. Sci. USA* **69**, 2110–2114.
8. Cohen, S. N. & Miller, C. A. (1970) *J. Mol. Biol.* **50**, 671–687.
9. Sharp, P. A., Hsu, M., Ohtsubo, E. & Davidson, N. (1972) *J. Mol. Biol.* **71**, 471–497.
10. Sharp, P. A., Cohen, S. N. & Davidson, N. (1973) *J. Mol. Biol.* **75**, 235–255.
11. Modrich, P. & Lehman, R. L. (1973) *J. Biol. Chem.*, in press.
12. Jacob, F., Brenner, S. & Cuzin, F. (1963) *Cold Spring Harbor Symp. Quant. Biol.* **23**, 329–484.
13. Guerry, P., van Embden, J., & Falkow, S. (1973) *J. Bacteriol.*, in press.
14. Anderson, E. S. & Lewis, M. J. (1965) *Nature* **208**, 843–849.
15. Davies, J., Benveniste, M. S. & Brzezinka, M. (1971) *Ann. N.Y. Acad. Sci.* **182**, 226–233.

General and Applied Microbiology

From more than 100 papers suggested by about 30 microbiologists, the 10 that follow were selected by a panel consisting of the following people: Harold Drake, Ph.D., Robert P. Gunsalus, Ph.D., John R. Sokatch, Ph.D., Judy D. Wall, Ph.D., William B. Whitman, Ph.D., and Lars G. Ljungdahl, Ph.D. (Committee Chair). The latter three were also responsible for the introductions.

Eenheid en Verscheidenheid in de Stofwisseling der Microben

A. J. KLUYVER

This paper by Kluyver, "Uniformity and diversity in metabolism of microorganisms," discussed the state of microbiology at the beginning of the 20th century. First, Kluyver dealt with the general public conception that microorganisms were irreconcilable enemies of humans, animals, and plants and that they must be controlled or killed. On the contrary, Kluyver pointed out that life on Earth without microorganisms would not be possible; they are useful, and they could be—and already were at that time—used in industrial settings. He mentioned, as examples of environmentally important processes, nitrogen fixation and the circulation of elements on Earth. He distinguished between assimilatory and dissimilatory processes, fermentation and respiration, and anaerobes and aerobes, and discussed the relationships between these concepts. Kluyver also covered the question of energy generation and pointed out that autotrophic organisms need to have assimilatory as well as dissimilatory metabolism. He noted that many metabolic processes of microorganisms and higher forms of life are similar, but that there are also many differences. He postulated that different chemical pathways for substrates, intermediates, and products exist in microorganisms and that this is reflected by or correlated with energy production and growth yields. One may consider Kluyver's paper as postulating the future of microbial research. He foresaw that microbial research would evolve from a descriptive state to yield results at both very high and fundamental levels.

LARS G. LJUNGDAHL

Reprinted from *Chemisch Weekblad* 21:266–277, copyright © 1924, by permission of the publisher.

576.8 : 612.015

EENHEID EN VERSCHEIDENHEID IN DE STOFWISSELING DER MICROBEN

door

A. J. KLUYVER. [1]

> My daughter Alice is a student in High School. One of the prescribed courses is General Science. The section on Bacteria left her with a vague impression of a world teeming with deadly germs awaiting an opportunity to infect mankind. It seems probable that this malignant conception of bacteria is very generally held.
> In reality civilization owes much to the microbe.
>
> Uit het voorbericht van A. I. Kendall.
> Civilization and the Microbe. Boston 1923.

Toen ik eenigen tijd geleden van het Bestuur der Nederlandsche Chemische Vereeniging de vereerende uitnoodiging mocht ontvangen een voordracht te houden over een onderwerp van biochemischen aard, heb ik aan die uitnoodiging gaarne gevolg gegeven. Immers een rechtgeaard microbioloog mag nimmer de gelegenheid laten voorbijgaan om mede te werken aan het eerherstel der kleinste levende wezens. Maar al te zeer verbreid is toch nog steeds de opvatting, volgens welke de microben, als onverzoenlijke vijanden van mensch, dier en plant, slechts belangstelling verdienen om ze des te beter te kunnen bestrijden.

Hoe verklaarbaar een dergelijk inzicht ook is, wanneer men denkt aan de zegenrijke gevolgen, welke uit de schitterende ontdekkingen over de rol der microben bij tallooze ziekteprocessen zijn voortgekomen, dit neemt niet weg, dat het toch een uiterst verwrongen beeld van de werkelijkheid geeft.

Intusschen wil ik den mij gegunden tijd niet besteden om U te laten zien, hoe zonder de aanwezigheid van microben de voorwaarden voor het menschelijk leven op aarde al spoedig niet meer zouden zijn gerealiseerd en de mensch in het microbenrijk dus minstens evenveel weldoeners als vijanden heeft. Ik wil mij er toe beperken — gebruik makende van het feit, dat ik voor een chemisch geschoold auditorium spreek — te trachten een gevoelige snaar in Uw gemoed te treffen door U het een en ander mede te deelen over de chemische verrichtingen der microben. Ik heb eenig vertrouwen, dat een sober relaas van deze prestaties voldoende zal zijn om te maken, dat U in de toekomst de kleinste levende wezens met een ander, welwillender, oog zult aanzien.

Men kan zeggen, dat als meest specifieke eigenschap der levende stof geldt de stofwisseling, d. w. z. het ervaringsfeit, dat voor de instandhouding van het leven noodig is geregelde toevoer van bepaalde chemische stoffen, welke in de levende cel omzettingen ondergaan en in veranderden vorm de cel althans ten deele weer verlaten.

Voor zoover het de hoogere organismen betreft, zijt U allen hiermede afdoende vertrouwd; dat dit voor de microben evenzeer geldt, blijkt welhaast zonder meer daaruit, dat in zoovele gevallen de aanwezigheid van microben zich aan U opdringt juist door het feit, dat U chemische veranderingen ziet optreden. Wanneer melk begint zuur te worden, wanneer in suiker-, of eiwithoudende vloeistoffen gistingen intreden, dan is het de stofwisseling, welke Uw aan-

---

[1] Voordracht, gehouden op de Algemeene Vergadering der Nederl. Chem. Vereeniging van 24 April 1924.

dacht trekt en U — na ruim een halve eeuw van microbiologisch onderzoek — secundair tot de aanwezigheid van microben zal doen besluiten.

Zooals U bekend is, is het mijn bedoeling U heden te spreken over de eenheid en de verscheidenheid in de stofwisseling der microben, maar laat ik hieraan dadelijk toevoegen, dat ik in afwijking van dezen titel in de eerste plaats zal spreken over de verscheidenheid, om eerst daarna te trachten U iets te doen zien van de eenheid, die deze verscheidenheid bindt. Zoodoende volg ik tevens in hoofdzaak de historische ontwikkeling der microbiologie.

Toen dank zij de baanbrekende onderzoekingen van Pasteur het inzicht ingang had gevonden, dat gistings-, rottings- en mineralisatieprocessen niets anders waren dan stofwisselingsprocessen van microscopisch kleine wezens, lag het voor de hand, dat het streven der beoefenaren der algemeene microbiologie in de eerste plaats er op gericht was een overzicht te verkrijgen van de werkzame agentia in hun volle verscheidenheid.

Van de chemische zijde bezien, openbaarde deze verscheidenheid zich in twee opzichten, eenerzijds in een verscheidenheid van de in een bepaald voedingsmedium gevormde stofwisselingsproducten, anderzijds in de verscheidenheid van de eischen, welke door de uiteenloopende microben aan de chemische samenstelling van hun voedsel worden gesteld.

Laat ik U dit met eenige voorbeelden mogen toelichten, in de eerste plaats wat betreft de uiteenloopende stofwisselingsproducten.

Wanneer ik een aantal op de gebruikelijke wijze met watten kiemdicht afgesloten kolfjes neem, waarin ik breng een bijvoorbeeld vijfprocentige oplossing van glucose in een waterig extract van gist — om te zorgen voor de behalve de C, H en O benoodigde elementen — en ik ent deze tevoren kiemvrij gemaakte oplossingen respectievelijk met reincultures van een kaamgist (Mycoderma cerevisiae), en van veel voorkomende schimmelsoorten als Aspergillus niger en Citromyces glaber, dan zal het chemisch onderzoek van de voedingvloeistof na ingetreden ontwikkeling der organismen verschillende uitkomsten opleveren. Wanneer het onderzoek zich in de eerste plaats beperkt tot het verdwijnen van de suiker en de daaruit klaarblijkelijk gevormde producten, dan zal blijken, dat de kaamgist de suiker heeft geoxydeerd tot koolzuur en water en de totale stofwisseling van dit organisme zal dus aan die der dierlijke organismen herinneren.

Daarentegen zal men kunnen vaststellen, dat in het met Aspergillus niger geënte kolfje uit de suiker onder meer ook een niet onbelangrijke hoeveelheid oxaalzuur is gevormd, terwijl Citromyces glaber naast het koolzuur ook citroenzuur heeft doen ontstaan.

Hier openbaart zich dus reeds, een weliswaar nog slechts bescheiden, verscheidenheid in stofwisseling. Duidelijker al komt deze tot uiting, wanneer men de proefneming nog uitbreidt, door een nieuwe reeks van kolfjes met denzelfden inhoud te nemen en deze — ditmaal onder meer of minder volledige uitsluiting van vrije zuurstof — ent met reincultures van de volgende organismen: Saccharomyces cerevisiæ (de persgist), Lactobacillus Delbrücki (de zoogenaamde cultuur-melkzuurbacterie), Lactobacillus fermentum (een „wilde" melkzuurbacterie), Bacterium coli, Bacterium aërogenes, Bacterium typhosum, Granulobacter sac-

charobutyricum en Granulobacter butylicum. Wanneer men nu door een hier niet ter zake doende passende regeling der temperaturen de ontwikkeling van al deze organismen mogelijk maakt, dan zal men na eenigen tijd kunnen vaststellen, dat S. cerevisiae de suiker voor een belangrijk deel heeft omgezet in alcohol en koolzuur, L. Delbrücki in melkzuur, L. fermentum in melkzuur, azijnzuur, alcohol en koolzuur, B. coli in melkzuur, azijnzuur, barnsteenzuur, koolzuur en waterstof, B. aërogenes in dezelfde producten, maar bovendien in 2—3 butyleenglycol, B. typhosum in mierenzuur, azijnzuur, melkzuur en alcohol, Gr. saccharobutyricum in boterzuur, azijnzuur, koolzuur en waterstof, Gr. butylicum in butylalcohol, aceton, koolzuur en waterstof.

Hiermede openbaart zich dus reeds een verrassende verscheidenheid in stofwisseling der microben. Maar hoe nietig is nog deze verscheidenheid berustende op de verschillen in stofwisselingsproducten uit éénzelfde voedingsstof, wanneer we nu aandacht gaan schenken aan de verscheidenheid in de door de verschillende microben gestelde eischen, wat betreft de samenstelling der als voedsel deugdelijke stoffen.

De eerste microbiologen waren, wat aangaat de door hen te bestudeeren organismen, min of meer afhankelijk van het toeval, dat hun nu eens het eene, dan weer het andere organisme in handen speelde. Langzamerhand ontging het hun evenwel niet, dat er een zeker verband bestond tusschen de aanvankelijke chemische samenstelling van het milieu, waarin zich de omzettingen afspeelden en de daarbij optredende micro-organismen. Om slechts één voorbeeld te noemen: het trad al spoedig duidelijk aan het licht, dat ware gistsoorten slechts in suikerhoudende vloeistoffen optraden. Nadat de invoering der reincultuurmethoden een nadere studie van de stofwisselingsprocessen had mogelijk gemaakt, bleek steeds duidelijker, hoe stoffen, die voor één bepaald micro-organisme een uitnemend voedsel vormden, voor andere dit in veel mindere mate of in het geheel niet waren. Toen dit gezichtspunt eenmaal verkregen was, lag het voor de hand, dat het besef wakker werd, dat de onderzoeker hierin een machtig hulpmiddel had om de ontwikkeling van bepaalde microben te bevorderen ten koste van andere, mede in het uitgangsmateriaal aanwezige, soorten.

Hoewel deze methode der electieve cultuur in laatste instantie is terug te voeren op Pasteur en Raulin, lijdt het geen twijfel of de schoonste vruchten heeft zij afgeworpen in de handen van twee onderzoekers: Winogradsky en Beijerinck. Haast duizelingwekkend is de verscheidenheid in stofwisseling, die door de consequente toepassing van de door hen aangegeven „ophoopingsmethoden" is geopenbaard.

Ter toelichting enkele voorbeelden.

Voortbouwende op de waarnemingen van Berthelot, welke aantoonde, dat de toeneming van het gehalte aan gebonden stikstof van braakliggende gronden buiten twijfel een biologisch proces is, slaagde Winogradsky in 1893 er in één der daarbij werkzame bacteriënsoorten door hem Clostridium pasteurianum genoemd, door toepassing der electieve cultuurmethode te isoleeren. [2]) Het bleek een organisme te zijn,

---

[2]) Voor nadere bijzonderheden zij verwezen naar een meer uitvoerige publicatie van Winogradsky in Centr. Bakt. Parasitenk. 2. Abt. 9, 43 (1902).

dat slechts bij afwezigheid van vrije zuurstof zich ontwikkelen kan. Op de merkwaardige stofwisseling van dit organisme werpt de volgende verrassende proefneming een helder licht. [3]) Men brengt sporen van het organisme in een vloeistof, die naast glucose alle voor den groei benoodigde elementen, met uitzondering van de stikstof, bevat. Nu plaatst men de cultuurkolf in een afgesloten ruimte, welke geëvacueerd wordt om den schadelijken invloed van de luchtzuurstof te elimineeren. Ook na geruimen tijd ziet men nauwelijks eenige verandering optreden. Thans vult men het vacuum aan met zorgvuldig van de laatste sporen zuurstof en van stikstofverbindingen bevrijd stikstofgas. Na 24 uur ziet men nu een krachtige gisting intreden, de aanwezige suiker verdwijnt binnen enkele dagen tot op de laatste resten, er blijkt krachtige bacteriën-ontwikkeling plaats te hebben gevonden en stikstof te zijn gebonden. Men heeft hier dus het merkwaardige verschijnsel van een levend organisme in rusttoestand, dat door het zoo inerte stikstofgas tot nieuw leven werd gewekt!

Aan dit merkwaardige organisme sluit zich nauw aan het enkele jaren later door Beijerinck en van Delden geïsoleerde organisme, *Azotobacter chroococcum*, dat eveneens en wel in nog sterker mate het vermogen bezit om de vrije stikstof te binden, maar dat in tegenstelling met *Cl. pasteurianum* juist alleen hiertoe in staat is, indien ruime toetreding van vrije zuurstof een krachtige oxydatie van organische stof mogelijk maakt. [4])

Kan het verwondering wekken, dat de verrichtingen van deze organismen de afgunst van een Haber hebben opgewekt?

Maar andere specialiteiten onder de microben vragen al weer onze aandacht. Welke merkwaardige groep van organismen vormen niet de ureumbacteriën, welke, in tegenstelling met zoovele andere bacteriënsoorten, met meerdere of mindere intensiteit ureum in ammoniumcarbonaat omzetten. Een speciale vermelding verdient dan in dit verband zeker nog een bacterie als *Urobacillus pasteurii*, welke onder daartoe gunstige voorwaarden nog in 10 % ureum-oplossingen alle ureum in ammoniumcarbonaat omzet en welke dus de alkalische reactie van meer dan 10-procentige ammoniumcarbonaat-oplossingen straffeloos verdraagt.

Niet voorbijgaan mag ik een bacterie als *Bacillus cligocarbophilus*, waarvoor Beijerinck en van Delden konden aantoonen, dat zij zich onder meer met de sporen organische stof, welke in de laboratoriumlucht nimmer schijnen te ontbreken, voedt. [5]) Evenmin

mogen onvermeld blijven de reeks bacteriën, waarvoor Söhngen kon aantoonen, dat voor haar benzine- en petroleumdampen een voedsel vormden. [6])

Maar ook al deze verscheidenheid is nog in één opzicht beperkt. De voor de ontwikkeling der tot nu toe genoemde organismen onmisbare koolstof blijkt slechts in organische binding — zij het dan ook onderling zeer uiteenloopende — als voedsel te kunnen dienst doen. Dit wil dus zeggen, dat deze organismen in laatste instantie hun voedsel aan andere organismen ontleenen, weshalve men ze dan ook als heterotrophe organismen pleegt aan te duiden.

Bijna veertig jaar geleden heeft intusschen Winogradsky al de meening uitgesproken, dat ook in dit opzicht de verscheidenheid in stofwisseling der microben geen beperking ondervindt. Voor de chlorophylhoudende micro-organismen stond het uiteraard reeds geruimen tijd vast, dat zij, dank zij de door hen vastgelegde zonne-energie, uit koolzuur en verdere anorganische stoffen hun celbestanddeelen konden opbouwen. Dat intusschen ook kleurlooze microben hiertoe in staat zouden zijn en wel geheel afgescheiden van alle stralende energie, is een gedachte zoo stoutmoedig, dat men zich nog heden ten dage moet verwonderen, dat een mensch haar heeft durven uiten.

Maar een genie als Winogradsky aarzelde niet om reeds in 1887 op grond van zijn even eenvoudige als vernuftige proefnemingen te besluiten, dat het in zwavelwaterstofhoudende bronwateren veelvuldig voorkomende kleurlooze organisme, *Beggiatoa alba*, zich in het donker kon vermeerderen in een zuiver anorganisch milieu. [7])

Uit koolzure zouten zouden de organische celbestanddeelen worden opgebouwd en dit zou mogelijk worden gemaakt, doordat de daarvoor benoodigde energie werd ontleend aan de oxydatie van zwavelwaterstof in de eerste plaats tot zwavel, in de tweede plaats tot sulfaten. Behalve voor *Beggiatoa* zou dit ook gelden voor een geheele groep van ten deele kleurlooze, ten deele purper gekleurde organismen, welke als gemeenschappelijk kenmerk hadden de aanwezigheid van vrije zwavel in hun cellen.

De latere proefnemingen hebben de juistheid van Winogradsky's theorie afdoende bewezen. Doelbewust opgezette ophoopingsproeven hebben aan het licht gebracht, dat er naast de den natuuronderzoekers welbekende relatief groote organismen, welke de zwavel in kolloïdalen toestand als druppels in hun cellen als reservestof afscheiden, nog een groote groep van kleine bacteriën bestaat, die bij cultuur in zwavelwaterstofhoudend milieu de zwavel uitsluitend extracellulair afscheiden, maar die toch op grond van haar. met die van *Beggiatoa* overeenstemmende, stofwisseling, terecht ook tot de physiologische groep der zwavelbacteriën worden gerekend. Het zijn deze door Nathansohn ontdekte, doch vooral door Beijerinck. Lieske [8]) en Jacobsen [9]) bestudeerde bacteriën geweest, welke juist in de laatste jaren in Amerika veel van zich hebben doen spreken en die mij in staat stellen U nog eens een sprekend voorbeeld te geven van de chemische vaardigheden der microben. In een reeks verhandelingen is door den Amerikaanschen onder-

---

[3]) Hoewel het feit van de assimilatie der vrije stikstof door *Cl. pasteurianum* reeds geruimen tijd buiten twijfel was gesteld, is de elegante inrichting der demonstratie, waarbij een scherpe scheiding wordt verkregen tusschen de schadelijke werking der vrije zuurstof en de zegenrijke werking van het vrije stikstofgas, voor zoover bekend, eerst onlangs door Ir. H. J. L. Donker, assistent voor de microbiologie aan de Technische Hoogeschool, aangegeven. Voor degenen, die deze proefneming zouden willen herhalen, moge worden vermeld, dat toevoeging van een kleine hoeveelheid humaat aan de aanvankelijke cultuurvloeistof onmisbaar is voor het welslagen. De strekking van de proef wordt hierdoor niet veranderd.

[4]) Voor nadere bijzonderheden aangaande de door Beijerinck bestudeerde organismen moge met één enkele verwijzing naar zijn in 1922 verschenen „Verzamelde Geschriften" worden volstaan.

[5]) Latere onderzoekers zijn geneigd hierbij in de eerste plaats aan sporen koolmonoxyde te denken, welk voor de hoogere dieren zoo vergiftig gas in ieder geval door *B. oligocarbophilus* als eenig organisch voedsel kan worden gebruikt. Centr. Bakt. Parasitenk. 2. Abt. **57**, 309 (1922).

[6]) Centr. Bakt. Parasitenk. 2e Abt. **37**, 595 (1903).

[7]) De lezing van de betreffende klassieke verhandeling in Botan. Ztg. **45**, (1887) zij een ieder ten zeerste aanbevolen.

[8]) Sitzb. Heidelb. Akad. Wiss. B. 1912

[9]) Folia Microbiolog. **1**, 487 (1912) en ibid. **3**, 155 (1914).

zoeker Waksman[*]) verslag uitgebracht over de wonderbaarlijke eigenschappen van een door hem *Thiobacillus thiooxydans* genoemd organisme, dat eveneens zwavelpoeder krachtig oxydeert tot zwavelzuur en met koolzuur als eenig koolstofvoedsel toekomt. Van de eerder door Beijerinck en door Jacobsen beschreven *Thiobacillus thioparis* onderscheidt de eerstgenoemde bacterie zich vooral door haar groote ongevoeligheid voor hooge zuurgraden, welke zoo ver gaat, dat in bepaalde cultuurvloeistoffen een waterstofionen concentratie overeenkomende met een p_H van 0.6 wordt bereikt. Ja, zelfs deelt J. G. Lipman, de bekende hoofdredacteur van het tijdschrift „Soil Science", in een vrij recente verhandeling mee, dat nog in oplossingen, welke 5 % zwavelzuur bevatten, een zeer merkbare zwaveloxydatie door *Thiobacillus thiooxydans* wordt bewerkstelligd.

Het zal U niet verwonderen, dat van deze eigenschap om op zoo krachtige wijze zwavelzuur te produceeren in Amerika in verschillende richtingen profijt is getrokken. Slechts een enkel voorbeeld daarvan wil ik hier nog vermelden, omdat het in zekeren zin een aanslag beteekent op een der takken van chemische industrie. Lipman en Waksman hebben reeds op vrij groote schaal proeven genomen teneinde vast te stellen in hoeverre het mogelijk was de fabriekmatige superphosphaatbereiding over te slaan, door een mengsel van het natuurlijke phosphaat en vrije zwavel in den grond te brengen en nu onder den invloed van *Th. thiooxydans* ter plaatse zwavelzuur en dus ook superphosphaat te laten ontstaan. Op bepaalde gronden schijnt men hiermede alleszins succes te hebben gehad.

In indrukwekkendheid van chemische prestatie wordt laatstgenoemde bacterie intusschen nog geslagen door Beijerinck's *Thiobacillus denitrificans*, die in een milieu van zwavel, salpeter en krijt en kleine hoeveelheden kaliumphosphaat bij volkomen uitsluiting van vrije zuurstof en gasvormig koolzuur zich ontwikkelt en die dus zelfs in diepgelegen aardlagen uit de genoemde algemeen voorkomende anorganische stoffen organische stof zou kunnen opbouwen.

Nog tal van andere specialiteiten zou ik voor U kunnen laten optreden, die met de zwavelbacteriën overeenkomen in dit opzicht, dat zij zich eveneens met enkel anorganische stoffen kunnen voeden. De tijd ontbreekt mij lang bij haar stil te staan, laat ik nog slechts eenige voorbeelden aanstippen, zooals het nitraatferment dat nitrieten tot nitraten oxydeert, het nitrietferment, dat ammoniumzouten tot nitrieten oxydeert en waarvoor een stof als glucose bijna even vergiftig is als voor andere organismen sublimaat. Laat ik dan tenslotte nog wijzen op de verschillende soorten organismen, voor wie knalgas of methaan en zuurstof of stikstofoxydule en waterstof een voedsel is.

Wanneer wij dit alles eens overwegen, dan kan het wel haast niet anders, of wij komen onder den indruk van de geweldige verscheidenheid in stofwisseling der microben.

Overzien wij dan verder den ontwikkelingsgang der microbiologie in de laatste decennia, dan kan men zich niet los maken van de gedachte, dat het streven tot demonstratie dezer verscheidenheid het onderzoek in hooge mate heeft beheerscht.

[*]) Men zie o.m.: J. Bact. **7**, 231 (1922); ibid. **7**, 239 (1922); ibid. **7**, 605 (1922); ibid. **7**, 609 (1922); Soil Science **13**, 329 (1922) en J. Biol. Chem. **50**, 35 (1922).

Niet de micro-organismen zelf vormden het uitgangspunt voor het onderzoek, neen, het enkel vermoeden, dat bepaalde chemische omzettingen zich in de natuur zouden afspelen, was voldoende om de hypothese op te stellen, dat er microben zouden zijn, welke deze omzettingen zouden versnellen. En in de handen van hen, die het wapen der electieve cultuur hadden geleerd te hanteeren, bevestigden de proefnemingen nagenoeg steeds de juistheid der hypothese. De simpele overweging, dat de kringloop der elementen op aarde gesloten is, en dat dus ook stoffen als benzine en petroleum in de natuur op den duur in koolzuur en water worden overgevoerd, bracht Söhngen tot de ontdekking en isoleering van een belangrijke reeks tot het geslacht *Mycobacterium* behoorende organismen, welke in staat zijn genoemde stoffen als koolstofvoedsel te aanvaarden.

Of om een nog sprekender voorbeeld te geven. De studie der zooeven besproken autotrophe bacteriën, waarvan de zwavelbacteriën, de nitriet- en nitraatfermenten wel de best bekende voorbeelden zijn, deed Nathansohn onbewust den koenen sprong wagen om aan te nemen, dat ook nog andere bij gewone temperatuur zeer langzaam verloopende reacties als basis van een microbenstofwisseling zouden kunnen dienst doen en dit zelfs wanneer de reageerende stof, voor zoover althans bekend, niet in de natuur wordt aangetroffen. Dit leidde hem tot het vaststellen van het feit, dat voor bepaalde soorten van zwavelbacteriën het alleen door menschenhand bereide natriumthiosulfaat als hoofdvoedsel naast enkele andere anorganische stoffen kon dienst doen.

Deze voorbeelden zullen voldoende zijn om te doen inzien, hoe het onderzoek steeds meer in dienst kwam van, wat men zou mogen noemen, een microbiologisch imperialisme. Het werd een streven om steeds maar weer nieuwe, oogenschijnlijk ontoegankelijke, gebieden voor de microben te ontsluiten. De eene specialiteit onder de microben was nog niet ontdekt, of een andere, die nog verrassender kunststukken verrichtte, werd reeds aangekondigd: het microbiologisch schouwtooneel geleek één grootsch natuurwetenschappelijk variété!

Met hoeveel eerbied wij thans ook opzien tegen hen, die deze vlucht der algemeene microbiologie door hun geniale intuïtie en scherpzinnige experimenteerkunst hebben geleid en hoe onmisbaar deze grondige exploratie van het microbenrijk voor de verdere ontwikkeling der microbiologie ook is geweest, op den duur kon deze gang van zaken geen bevrediging geven.

Al mocht dan de groote stroom van het microbiologisch onderzoek op de verscheidenheid zijn gericht, daarnaast ontwikkelde zich geleidelijk een studierichting, die in de verscheidenheid der verschijnselen naar de eenheid zocht.

Het zou onbillijk zijn het voor te stellen, alsof deze laatste richting geheel een product van den allerlaatsten tijd was. Neen, vele klassieke figuren uit de microbiologie — en zeker haar grondlegger Pasteur niet in de laatste plaats — mogen ook in dit verband met groote eere worden genoemd.

Maar toch kan men zeggen, dat het streven om de stofwisselingsprocessen der microben te bezien in het licht van de uitkomsten van het algemeen physiologisch onderzoek eerst in den aanvang van de 20ste eeuw door den bekenden Berlijnschen physioloog en hygiënist Rubner consequent is aangepakt. En hieraan mag wel dadelijk worden toegevoegd, dat dit streven tot dusver nog maar zeer weinig weerklank heeft gevonden.

Dit betreurenswaardige feit vindt een gereede verklaring in de omstandigheid, dat de microbiologie tot dusver in hoofdzaak zich als toegepaste wetenschap heeft ontwikkeld. De overgroote meerderheid der medische, technische en landbouwkundige microbiologen bestudeerden de rol der microben in de groote practische problemen, welke hun aandacht vroegen. Van het betrekkelijk zeer geringe aantal onderzoekers, die gelegenheid en liefde hadden om de microben om haarzelfswille te bestudeeren, waren dan — en hoe begrijpelijk, gezien de verrassende bekoring, welke van dit onderzoek uitging — de meesten nog in den ban der verscheidenheid.

Wanneer ik dan nu er toe overga te trachten U iets te doen zien van de eenheid, die in de stofwisseling der microben is te vinden, dan ben ik mij er ten volle van bewust, dat het een sober relaas is, wat ik U bied. Dat ik het toch gewaagd heb Uwe aandacht hiervoor een oogenblik te vragen, komt, omdat ik zoodoende gelegenheid zal krijgen U te doen gevoelen, dat er hier nog een onafzienbaar arbeidsveld braak ligt. De oplossing van de daarin aanwezige problemen zal intusschen, naar het mij voorkomt, onmisbaar zijn om de microbiologie uit haar staat van in hoofdzaak beschrijvenden tak van wetenschap tot hooger plan op te heffen. Maar tevens onmisbaar in het belang van de toepassingen der microbiologie op zoo uiteenloopend gebied.

En naar mijn vaste overtuiging zal slechts nauwe samenwerking tusschen microbiologen en pur-sang chemici in deze vooruitgang kunnen brengen.

Laat ik dan in de eerste plaats U mogen toelichten, hoe de eenheid in de zoo uiteenloopende stofwisselingsprocessen der microben zich hierin uit, dat men er dezelfde groote trekken in terugvindt, welke bij het onderzoek van de stofwisseling der hoogere organismen zoo duidelijk aan het licht zijn getreden.

De gelegenheid ontbreekt hier om op een en ander uitvoerig in te gaan, de volgende beschouwingen mogen het bovengenoemde gezichtspunt intusschen toelichten. Het onderzoek van de stofwisseling der hoogere organismen heeft onmiskenbaar geleerd, dat zich daarbij ten allen tijde twee processen laten onderscheiden. Een gedeelte van het voedsel toch wordt in celbestanddeelen in het organisme omgezet, hetzij om te dienen voor groei, hetzij voor de remplaceering van tegronde gegaan celmateriaal. Een ander deel van het voedsel ontleent blijkbaar zijn beteekenis vóór alles aan de daarin voorhanden chemische energie, welke door de levende stof ontketend wordt, haar in staat stelt energievragende functies te verrichten en welke het organisme tenslotte in gedegradeerden toestand, in den regel als warmte, verlaat. Men onderscheidt deze beide processen als assimilatie en dissimilatie. [11])

[11]) De in de Duitsche literatuur gebruikelijke aanduidingen „Baustoffwechsel" en „Betriebsstoffwechsel" geven het onderscheid wellicht juister weer.
Voor een nadere beschouwing van deze problemen zij o.m. verwezen naar M. Rubner, Kraft und Stoff im Haushalte der Natur (1909) en ook naar het door C. Oppenheimer bewerkte hoofdstuk : „Energetik der lebenden Substanz" in C. Oppenheimer. Handbuch der Biochemie der Menschen und der Tiere, 2. Aufl. Bd. II p. 223 (1923).

Het lijdt geen twijfel, dat het dissimilatieproces het meest essentieele kenmerk is van alle leven. Andere typische levenskenmerken als groei, voortplanting, inwendige en uitwendige beweging zullen niet zelden ontbreken, de dissimilatie zal alleen niet aan te toonen zijn in sommige stadia van latent leven, bij indroging en dergelijke.

Voor de hoogere organismen was dus komen vast te staan, dat het voortbestaan van het leven gebonden was aan een voortdurende omzetting van chemische energie in andere energievormen. Daarbij kon door Rubner de geldigheid van de 1ste hoofdwet der mechanische warmtetheorie voor de energie-omzettingen in het dierlijk organisme, door Atwater ook voor die in den mensch, worden vastgesteld.

Wat nu den aard van de energieleverende omzettingen aangaat, was reeds door Lavoisier gewezen op de beteekenis van het ademhalingsproces, dat wil dus zeggen van het zich in de levende organismen afspelende langzame verbrandingsproces, dat onder meer in de behoefte dier organismen aan vrije zuurstof tot uiting kwam. Scherper nog trad de gegrondheid van deze zienswijze aan het licht, toen Rubner de juistheid van het door hem opgestelde principe der „isodynamische vervangbaarheid" proefondervindelijk kon aantoonen. Voor het volwassen dierlijk organisme bewees Rubner namelijk, dat bij gelijk blijvende levensuitingen, tot op zekere hoogte, eiwitten, vetten en koolhydraten elkaar wederzijds kunnen vervangen in een gewichtsverhouding, welke omgekeerd evenredig is met de verbrandingswaarden dezer stoffen. Met andere woorden hoe gelijke chemische energie in het voedsel — binnen zekere grenzen — ook gelijke voedingswaarde beteekent.

Reeds in 1885 wees Rubner er op, hoe naar alle waarschijnlijkheid ook in de stofwisseling der microben zou zijn aan te wijzen een omzetting, welke haar beteekenis voor het organisme geheel aan de daarbij vrijkomende energie zou ontleenen.

Nu had Pasteur reeds omstreeks 1860, toen hij het eerste geval ontdekte van organismen, welke zich bij volledige uitsluiting van vrije zuurstof konden ontwikkelen, met ware intuitie doorvoeld het verband, dat er klaarblijkelijk bestond tusschen het wegvallen van het ademhalingsproces bij deze organismen en het voor deze leefwijze typeerende gistingsproces.

Intusschen werd deze wederzijdsche vervangbaarheid van ademhaling en gisting aanvankelijk meer van de stoffelijke zijde beschouwd, met dien verstande, dat men van de gisting sprak als een ademhaling met gebonden zuurstof.

De noodzakelijkheid om deze vervangbaarheid vóór alles van energetische zijde te leeren bezien is eerst door Rubner met voldoende scherpte geformuleerd, als een logisch uitvloeisel van zijn streven om de stofwisseling van alles wat leeft, van uit één standpunt te bezien. Doch eerst in 1902 werd een aanvang gemaakt met de reeks van microcalorimetrische proefnemingen over de stofwisseling van verschillende microben waarbij de juistheid van deze hypothesen bevestigd werd.

Sinds dezen tijd is nu met afdoende zekerheid komen vast te staan, dat ook in de stofwisseling van iederen microbe naast het proces van den opbouw van nieuw celmateriaal verloopt een dissimilatieproces, dat daardoor gekenmerkt is, dat chemische energie door de levende microbencel wordt verbruikt, daarin energievragende functies verricht en tenslotte als warmte vrijkomt.

Een aantal voorbeelden van de belangrijkste dissimilatieprocessen, welke bij uiteenloopende microbengroepen worden aangetroffen, zijn in Tabel I weergegeven.

Tabel I.

DISSIMILATIEPROCESSEN DER MICROBEN. [12])

A. *Oxydatieve.*

Voorbeelden van organismen :

I. $C_6H_{12}O_6 + 6 O_2 = 6 CO_2 + 6 H_2O + 676$ cal. — Schimmelsoorten.

II. $C_2H_5OH + 3 O_2 = 2 CO_2 + 3 H_2O + 325^{1}/_2$ cal. — Kaamgist.

III. $CH_2NH_2COOH + 1^{1}/_2 O_2 = 2 CO_2 + H_2O + NH_3 + 142$ cal. — Vele aërobe bacteriën.

IV. $C_2H_5OH + O_2 = CH_3COOH + H_2O + 116^{1}/_2$ cal. — Azijnbacteriën.

V. $CH_4 + 2 O_2 = CO_2 + 2 H_2O + 210$ cal. — Methaanoxyd. bacteriën.

VI. $H_2 + {}^{1}/_2 O_2 = H_2O + 68^{1}/_2$ cal. — Waterstofoxyd. bact.

VII. $NH_3 + 1^{1}/_2 O_2 = HNO_2 + H_2O + 79$ cal. — Nitrietferment.

VIII. $KNO_2 + {}^{1}/_2 O_2 = KNO_3 + 22$ cal. — Nitraatferment.

IX. $H_2S + {}^{1}/_2 O_2 = H_2O + S + 67$ cal. $\Big\}$

X. $S + 1^{1}/_2 O_2 + H_2O = H_2SO_4 + 141$ cal. $\Big\}$ Zwavelbacteriën.

B. *Fermentatieve.*

XI. $C_6H_{12}O_6 = 2 C_2H_5OH + 2 CO_2 + 25$ cal. — Gisten.

XII. $C_6H_{12}O_6 = 2 C_3H_6O_3 + 28$ cal. — Melkzuurbacteriën.

XIII. $C_6H_{12}O_6 = C_4H_8O_2 + 2 CO_2 + 2 H_2 + 15$ cal. — Boterzuurbacteriën.

XIV. $C_6H_{12}O_6 + {}^{1}/_2 H_2O = CH_3CHOHCOOH + CO_2 + H_2 + {}^{1}/_2 CH_3COOH + {}^{1}/_2 C_2H_5OH + 8$ cal. — Bact. der coligroep.

XV. $(CH_3COO)_2 Ca + H_2O = CaCO_3 + CO_2 + 2 CH_4 + 19$ cal. — Methaanfermenten.

XVI. $O=C(NH_2)_2 + 2 H_2O = (NH_4)_2 CO_3 + 6$ cal. — Ureumbacteriën.

XVII. $C_2H_5OH + 2,4 KNO_3 = 1,2 N_2 + 1,2 K_2CO_3 + 0,8 CO_2 + 3 H_2O + 293$ cal. — Denitrific. bact.

XVIII. $C_2H_5OH + 1^{1}/_2 CaSO_4 = 1^{1}/_2 H_2S + 1^{1}/_2 CaCO_3 + {}^{1}/_2 CO_2 + 1^{1}/_2 H_2O + 14$ cal. — Sulfaatreduc. bact.

gegeven. [13]) In de eerste plaats een tiental oxydatieve dissimilatieprocessen, waarvan de eerste drie ook bij de stofwisseling der hoogere dierlijke organismen zijn aangetroffen, terwijl de overige de typische dissimilatieprocessen zijn van reeds eerder genoemde specialiteiten onder de microben. De onder het opschrift „fermentatieve" genoemde dissimilatieprocessen, zijn voorbeelden van omzettingen, welke voorzien in de energetische behoefte van organismen, die tijdelijk of blijvend onder uitsluiting van vrije zuurstof leven.

[12]) Bij het samenstellen van deze tabel is geenszins naar volledigheid gestreefd. In het bijzonder is de lijst van oxydatieve dissimilatieprocessen met organische substraten voor onbeperkte uitbreiding vatbaar. Wat de fermentatieve dissimilatieprocessen aangaat, zijn de belangrijkste der tot nu toe bekende typen vertegenwoordigd met uitzondering van de anaerobe ontleding van eiwitsplitsingsproducten. Deze laatste processen laten zich namelijk zeer bezwaarlijk in eenvoudige symbolen weer geven. Men zie evenwel bijv.: Arch. Hyg. **66**, 209 (1908).

[13]) Bij de vaststelling van het meerendeel der in deze Tabel opgenomen calorische effecten, had ik het voorrecht te mogen profiteeren van de zoo deskundige voorlichting van collega Verkade, wien ik ook hier ter plaatse mijn oprechten dank betuig. Er moge worden opgemerkt, dat er naar gestreefd is zoo goed mogelijk te becijferen het verschil in verbrandingswaarden van het in water opgeloste voedsel en van de eveneens in water opgeloste stofwisselingsproducten.

Deze fermentatieve dissimilatieprocessen zijn, zooals begrijpelijk is, gekenmerkt door een belangrijk lager calorisch effect en de juistheid van de energetische beschouwingswijze dezer omzettingen weerspiegelt zich in het feit, dat per gewichtseenheid van werkzaam organisme, bij de fermentatieve leefwijze zulke naar verhouding zeer groote hoeveelheden voedsel worden verbruikt en stofwisselingsproducten worden gevormd.

In het bijzonder treedt dit verschijnsel duidelijk aan het licht bij organismen, welke al naar gelang der omstandigheden hun energie ontleenen aan een oxydatieve, dan wel aan een fermentatieve dissimilatie. Zoo bijv. bij de gist, waarvoor Pasteur reeds vaststelde, dat het suikerverbruik per gewichtseenheid bij anaërobe leefwijze aanmerkelijk grooter was dan bij cultuur aan de lucht, in welk geval naast de gisting steeds ook een verbranding van een deel van de suiker intreedt.

Het is nu aan geen twijfel onderhevig, dat de energetische beschouwingswijze der stofwisseling ook in andere richting verhelderend en ordenend kan werken.

In de eerste plaats dwingt het den onderzoeker zich een scherpere voorstelling van de stofwisseling van het te onderzoeken organisme te maken dan gemeenlijk gebruikelijk is. Om een voorbeeld te geven. Voor een zoo veelvuldig onderzochte bacterie als *B. coli* vindt men in den regel zonder meer aangegeven, dat zij zich zoowel op koolhydraatvrije als koolhydraathoudende vleeschextractvoedingsbodems ontwikkelt en dat zij facultatief anaëroob is, waarmede men dan wil zeggen, dat zij zoowel bij aan- als afwezigheid van lucht kan groeien.

Menig microbioloog zal intusschen niet beseffen, dat de dissimilatieprocessen van deze bacterie eenerzijds bestaan uit een oxydatieve verwerking van in hoofdzaak eiwitsplitsingsproducten, anderzijds uit een fermentatieve ontleding van koolhydraten (of suikeralcoholen) met dien verstande dat de aanwezigheid van deze stoffen dan ook conditio sine qua non is voor het anaerobe leven. Dit in tegenstelling voor de bacteriën der *Proteus*-groep, waarvoor men in de literatuur meest dezelfde gegevens vindt, maar die bovendien het vermogen blijken te bezitten de benoodigde energie aan een vergisting van eiwitsplitsingsproducten te ontleenen en die dus ook bij afwezigheid van koolhydraten een anaëroob leven kunnen leiden.

Blijkt uit dit voorbeeld al hoe de vage aanduiding van een organisme als facultatief anaëroob in het belang van een juist inzicht der ontwikkelingsvoorwaarden behoort te worden vervangen door een rekenschap afleggen van de mogelijke dissimilatieprocessen, ook het volgende voorbeeld kan dit nog verduidelijken.

De organismen, welke zich laten samenbrengen in de groep der ware melkzuurbacteriën, vindt men niet zelden aangeduid als facultatief anaerobe organismen. In zooverre is dit ook wel toelaatbaar, dat het meerendeel dier organismen inderdaad zoowel bij aan- als afwezigheid van luchtzuurstof kan groeien. Intusschen zou men zich deerlijk vergissen, indien men meende, dat deze organismen evenals de beide hierboven genoemde facultatief anaërobe bacteriënsoorten zoowel over een oxydatief als over een fermentatief dissimilatieproces beschikten. Integendeel staat vast, dat deze ware melkzuurbacteriën ook bij aanwezigheid van lucht nimmer eenige vrije zuurstof in hun stofwisseling betrekken. Het zijn dus zuiver fermentatieve organismen, voor wier ontwikkeling koolhydraten (of suikeralcoholen) levensvoorwaarde zijn en die van verschillende andere zuiver fermentatieve microben slechts verschillen, doordat voor de melkzuurbacteriën de vrije zuurstof de ontwikkeling niet geheel remt.

In dit verband is het niet zonder belang er op te wijzen, dat het ontbreken van een oxydatief dissimilatieproces bij de melkzuurbacteriën ongetwijfeld samenhangt met een eigenschap, welke Beijerinck reeds in 1893 voor alle melkzuurbacteriën vaststelde, namelijk het ontbreken van het in alle cellen van hoogere planten en dieren en op dat tijdstip in alle daarop onderzochte microben aangetroffen enzym katalase, dat waterstofsuperoxyde in water en zuurstof ontleedt. Later is door Orla Jensen er op gewezen, dat ook de uitsluitend anaëroob levende boterzuurbacteriën de katalase misten, terwijl wij dit ook voor de anaërobe eiwitrottende bacteriën konden vaststellen. De veelal aangenomen doch aan den anderen kant nog steeds heftig bestreden hypothese [14]), dat de katalase bij alle aëroob levende organismen, zoowel hoogere als lagere een rol speelt bij de overbrenging der vrije zuurstof op het ademhalingssubstraat, vindt in deze feiten ongetwijfeld een naderen steun.

Omgekeerd lijkt het aannemelijk, dat het onderzoek naar de aanwezigheid van katalase in een nieuw geïsoleerde microbe met vrucht zal kunnen worden gebruikt om een oxydatieve dissimilatie aan te toonen.

Het groote belang van een vaststellen van den aard van het dissimilatieproces (of processen) van een microbe, als zijnde het in de eerste plaats karakteristieke deel van de stofwisseling, is verder vooral ook gelegen in het feit, dat de natuurlijke verwantschap der microben ongetwijfeld ook in de stofwisseling tot uiting komt. [15]) Dit gezichtspunt is reeds in 1909 door Orla Jensen op voortreffelijke wijze toegepast in zijn ontwerp eener algemeene bacteriënsystematiek. [16]) Door nu ieder te onderzoeken organisme in een der geleidelijk zich steeds duidelijker afteekenende natuurlijke groepen onder te brengen, zal men niet zelden er in kunnen slagen ook eigenschappen van dit organisme op grond van zijn verwantschap te voorspellen, en voorts zal het mogelijk worden, meer rationeele cultuurvoorwaarden voor deze organismen aan te geven. Wanneer men bijv. zoo dikwijls klachten leest over de moeilijkheid om tal van streptococcenstammen te cultiveeren, dan ligt dit zeker niet in de laatste plaats aan het feit, dat de betrokken onderzoekers geen oogenblik realiseeren, dat zij te doen hebben met organismen uit de natuurlijke groep der ware melkzuurbacteriën.

Aangaande de verspreiding der dissimilatieprocessen over enkele belangrijke microbengroepen kan Tabel II een denkbeeld geven. Daaruit kan blijken, dat de oxydatieve processen in den regel weinig specifiek zijn wat betreft den aard van het voor oxydatie in aanmerking komende materiaal. Niet zelden zal het oxydeerend vermogen zich zoowel tot N-houdende verbindingen als tot koolhydraten (en suikeralcoho-

len) en tot organische zouten uitstrekken, maar in de reeks azijnbacteriën, schimmels, aërobe sporenvormers en de *Pseudomonas*-groep [17]) laat zich een afnemende tendentie voor koolhydraatverbranding en toenemende tendentie tot oxydatie van stikstofhoudende verbindingen signaleeren. Daarentegen zijn in den regel de fermentatieve processen veel meer aan een bepaald substraat gebonden, hoewel bij sommige groepen meerdere mogelijkheden zijn gerealiseerd.

Maar nog andere beschouwingen dringen zich bij een nadere overweging van de beteekenis der energetische zijde van het stofwisselingsproces aan ons op.

Dat een mensch of een hooger dier energie behoeft is iets, wat als het ware direct tot ons gemoed spreekt. Het handhaven van de lichaamstemperatuur, het verrichten van arbeid, zoowel inwendige als uitwendige, is zonder energietoevoer niet te denken. Wanneer men zich nu evenwel afvraagt, waarvoor dient bij een op constante temperatuur gehouden onbewegelijken microbe, waarvan ook het inwendige geenerlei beweging verraadt, de energiestroom, die toch blijkens de ervaring voor de instandhouding van het leven onmisbaar is, dan zijn er eigenlijk maar twee antwoorden denkbaar. Eenerzijds kan men aannemen, dat de voortdurende energieomzetting nu eenmaal zonder meer een bestaansvoorwaarde voor de levende substantie als zoodanig is. Anderzijds is het verleidelijk om een eng verband aan te nemen tusschen dit voortdurende verbruik van chemische energie en het andere deel van het stofwisselingsproces, namelijk de vorming van nieuwe celbestanddeelen, de assimilatie dus.

Dit overwegende dringt zich in de eerste plaats de vraag op, of dit niet zou kunnen blijken, indien men er in zou slagen den omvang van de assimilatie der micro-organismen tot een minimum terug te brengen, zoodanig namelijk, dat zichtbare nieuwe vorming van celmateriaal, d. w. z. celvermeerdering is uitgesloten. Wij stuiten dan dus op het probleem of het mogelijk is microbencellen in leven te houden door het toedienen van een zoodanig voedselrantsoen, dat celvermeerdering juist is uitgesloten. Rubner heeft een poging gedaan, deze vraag voor de persgist experimenteel te onderzoeken en komt daarbij tot de slotsom, dat deze mogelijkheid niet bestaat. [18]) Tegen de door hem gevolgde techniek zijn intusschen ernstige bezwaren aan te brengen, zoodat wij moeten constateeren, dat een dergelijke voor de microbiologie fundamenteele kwestie nog steeds onopgelost is.

Maar ook in die gevallen, waarin zichtbare groei achterwege zou blijven, heeft ongetwijfeld toch altijd nog een remplaceering van celbestanddeelen plaats, zoodat ook dan nog de dissimilatie in dienst van assimilatie zou kunnen worden gedacht.

Aan den anderen kant staat het wel vast, dat althans een deel van de energie der dissimilatie onmisbaar is om assimilatie mogelijk te maken voor zoover wij namelijk kunnen vaststellen, dat deze laatste met een duidelijke energiebinding gepaard gaat. Een goed voorbeeld in dit opzicht vormen de autotrophe bacteriën, waarbij uit koolzuur en verdere anorganische stoffen organische stof als bijv. bacteriëneiwit wordt

---

[14]) Zie bijv. H. D. Dakin, Oxidations and Reductions in the Animal Body, London 1922, pag. 19.

[15]) Hieruit volgt natuurlijk niet, dat gelijke stofwisselingsprocessen ook niet in onafhankelijke phylogenetische ontwikkelingsreeksen kunnen optreden, zooals ook Orla Jensen terecht aanneemt.

[16]) Centr. Bakt. Parasitenk. 2e Abt. **22**, 305 (1909).

[17]) De in deze groep samengevatte organismen vertoonen wel physiologische overeenkomst, doch vormen zeker *geen* natuurlijke groep. Zie noot 15 vorige pagina.

[18]) Men zie hiervoor: M. Rubner, Die Ernährungsphysiologie der Hefezelle bei alkoholischer Gärung; Leipzig, 1913.

TABEL II.

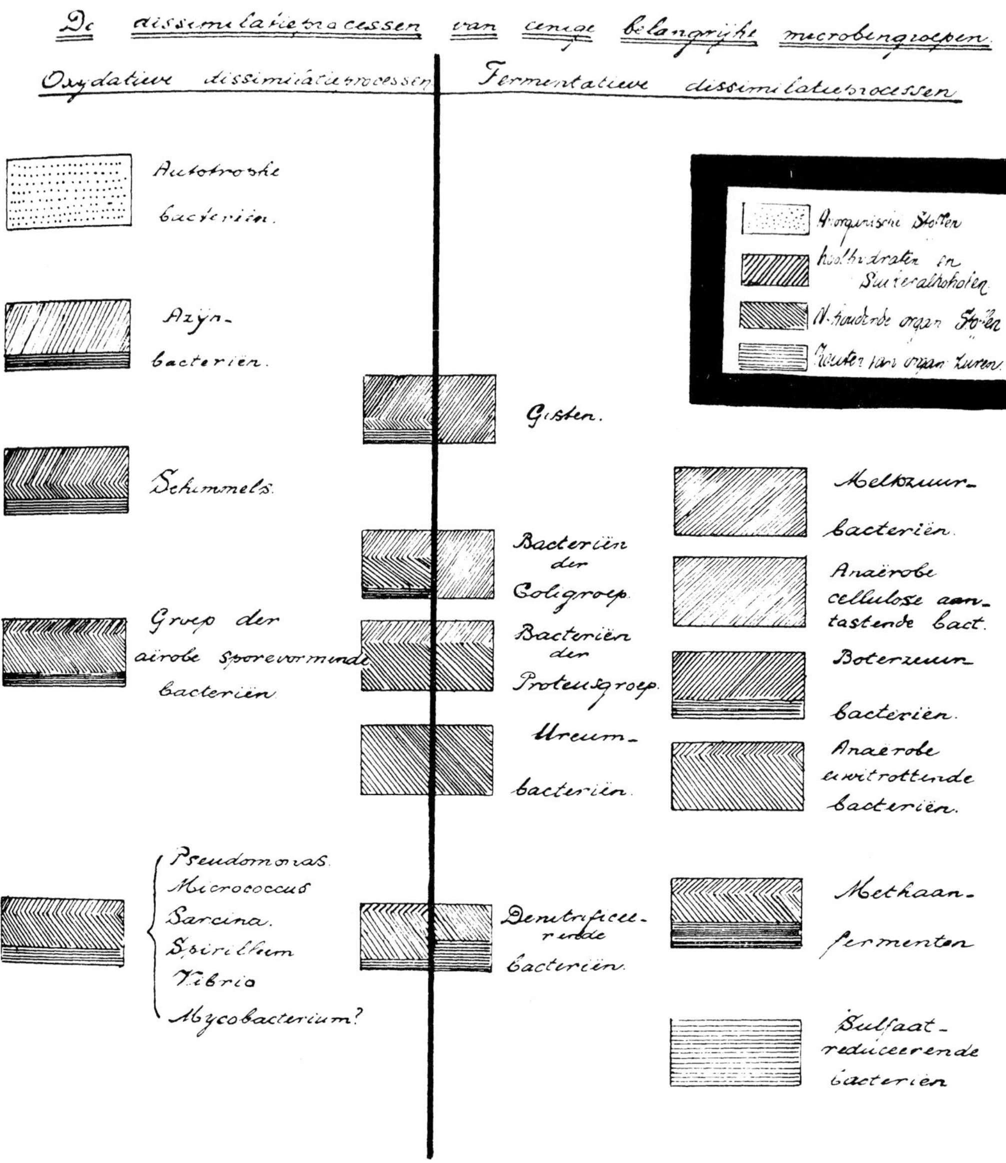

gevormd. Voor dergelijke — ten onrechte ook wel als chemosynthesische onderscheiden — bacteriën is dan ook ten allen tijde de energieleverende reactie door de microbiologen welbewust naar voren gebracht.

Wanneer men intusschen bedenkt, dat bijv. bij de zwavelbacteriën de ervaring leert, dat om 1 deel koolstof in bacteriënmateriaal gebonden te krijgen, 32 deelen zwavel tot zwavelzuur moeten worden geoxydeerd (dissimilatieproces X van Tabel I) dan blijkt daaruit wel zonder meer, dat ook bij de autotrophe bacteriën slechts een fractie van de aan de dissimilatie ontleende energie in de geassimileerde stof wordt vastgelegd.

Dit gezichtspunt komt bij de op organische voeding aangewezen microben nog duidelijker tot uiting. Uit Rubner's calorimetrische waarnemingen is overtuigend gebleken, dat de autolyse van gistcellen — evenals trouwens de meeste enzymatische hydrolysen — slechts met zeer geringe warmte-ontwikkeling verloopt. Omgekeerd staat dus vast, dat de opbouw van gistcellen uit de autolyseproducten met zeer geringe energie-vastlegging gepaard gaat. Toch is het energieverbruik der dissimilatie, wanneer men vrij ver geautolyseerde gist door plotselinge suikertoevoeging zich laat restaureeren, vrij aanzienlijk.

Ook op grond hiervan zou men geneigd zijn te besluiten, dat de vorming van celbestanddeelen een proces is, dat in energetisch opzicht met een zeer gering nuttig effect verloopt, m. a. w. dat de energie, welke aan een of andere chemische omzetting wordt ontleend, slechts op zoodanige wijze kan ingrijpen in een tweede reactie, dat gelijktijdig een belangrijk deel van de energie tot warmte wordt gedegradeerd.

Wanneer we dan er toe overgaan dit vraagstuk scherper te formuleeren, dan moet worden vastgesteld, dat de — naar wij mogen aannemen isotherm verloopende — stofwisseling der levende cel ons te zien geeft een reactie, welke onder toeneming van vrije energie verloopt, de assimilatie dus, wat uiteraard alleen denkbaar is doordat gelijktijdig een tweede reactie zich afspeelt, welke een grooter verlies aan vrije energie met zich voert, met dien verstande dat de vrije energie van het totale systeem belangrijk afneemt. Daarbij is dan te bedenken, dat als regel moet worden aangenomen, dat de beide reacties geen stoffelijken component gemeen hebben.

Ik zou nu den physico-chemici onder mijn toehoorders gaarne de vraag willen voorleggen, in hoeverre hun uit de doode natuur voorbeelden eener dergelijke wijze van reactiekoppeling bekend zijn en in hoeverre er bij hen principieele bezwaren zijn om het bestaan van een dergelijke energetische koppeling bij de niet-levende stof te accepteeren.

Door Euler en Af Ugglas is deze vraag, wellicht in wat minder scherpen vorm, reeds in 1911 opgeworpen.[19] Zij meenen, dat een dergelijke energetische beïnvloeding van twee stoffelijk onafhankelijke reacties denkbaar is, wanneer beide reacties door éénzelfden katalysator worden versneld, wat zij nader verduidelijken door te wijzen op de evenwichtsverschuivingen, welke noodzakelijk moeten intreden, wanneer de katalysator zoowel met de beide substraten als met de reactieproducten verbindingen aangaat.

Zonder hier in een waardeering van deze bijzonderheden te treden, volgt uit de noodzakelijkheid om een koppeling van dissimilatie en assimilatie aan te nemen toch wel, dat het volslagen ondenkbaar is, dat het energieleverende dissimilatieproces zou verloopen onder den invloed van een van het verdere deel der levende cel los te denken katalysator.

Een dergelijke opvatting, die sinds Buchner's zymasevondst door chemici maar al te lichtzinnig wordt gepropageerd, kan alleen verkondigd worden door hen, die blind zijn voor de biologische beteekenis van het gistingsproces, als zijnde *het* dissimilatieproces van de gist en dat mitsdien in samenhang met de assimilatie moet worden beschouwd.

Duidelijk weerspiegelt de bijzondere plaats, die de dissimilatiereacties in de cel innemen in vergelijking met de andere het voedsel voorbereidende omzettingen, zich in het feit, dat de laatste verloopen onder den invloed van in vele gevallen bijna quantitatief van de cel af te scheiden enzymen, terwijl de eerste daarentegen met groote hardnekkigheid weerstand geboden hebben aan de pogingen om het daarbij werkzame enzym te isoleeren. Zeker, eindelijk is Buchner er in geslaagd uit enkele bijzondere gistsoorten met een zeer gering rendement iets van de zymatische kracht los van de levende cel te verkrijgen, maar er zijn verschillende argumenten, welke er voor pleiten, dat in die gevallen in de levende cel naast een onder den invloed van het levende protoplasma verloopende op den voorgrond tredende dissimilatie een gering deel

van de suiker een fermentatieve omzetting ondergaat onder den invloed van een in kleine hoeveelheid afgesplitste zymase-functie. In deze opvatting wordt men nog versterkt, wanneer men de min of meer lachwekkend lage rendementen ziet, welke Buchner en zijn medewerkers Gaunt en Meisenheimer verkregen bij hun pogingen om uit melkzuurbacteriën en azijnzuurbacteriën respectievelijk een lactozymase en een alcoholoxydase te isoleeren, welke uitkomsten zij ondanks alles toch nog als een bevestiging van hun enzymatische stofwisselingstheorie durven interpreteeren.[20] Ook de oude strijd of bij de ureumbacteriën de ureumomzetting al of niet onder den invloed van een oplosbaar enzym, de urease, verloopt, verschijnt in dit verband in een nieuw licht. Bij de ware ureumbacteriën, waarvoor de ureumomzetting ongetwijfeld ook dissimilatieproces is, is afscheiding der urease, zooals ook Beijerinck bewees, niet of practisch niet te bereiken. Voor andere organismen, schijnt de ureumomzetting tengevolge van de aanwezigheid van andere dissimilatieprocessen haar energetische beteekenis vrijwel te hebben verloren en zoo wordt het verklaarbaar, dat Jacoby[21] er in slaagt uit *Proteus*-bacteriën oplosbare urease te isoleeren, iets wat hem uit *Urobacillus pasteurii* niet zou gelukken.[22]

De gelegenheid ontbreekt mij verder nog lang stil te staan bij de gezichtspunten, welke uit de energetische beschouwing der stofwisseling voortvloeien. Slechts moge worden aangestipt, dat het ongetwijfeld in hooge mate loonend moet zijn, het gansche vraagstuk der stofwisseling der microben ook van het standpunt der 2de hoofdwet der mechanische warmtetheorie aan een nieuwe bewerking te onderwerpen. Van uit algemeen physiologisch standpunt is op het groote belang van deze kwestie in 1906 door onzen landgenoot. Zwaardemaker reeds uitvoerig gewezen.[23] Met de toepassing van deze beginselen op de relatief zoo eenvoudige verhoudingen in de microbiologie is nog nauwelijks een begin gemaakt en toch moet dit haast onvermijdelijk tot belangrijke uitkomsten leiden.

Het is bijv. in hooge mate verleidelijk om te trachten een verband te leggen tusschen de quantitatieve waarden der dissimilatie-reacties der verschillende microben en de voor iedere microbe typeerende assimilatieprocessen. Zou het niet mogelijk zijn, dat men langs dezen weg zou kunnen leeren begrijpen, waarom de boterzuurgisting als dissimilatiereactie de boterzuurbacteriën wel in staat stelt hun lichaamseiwit uit ammoniak op te bouwen, terwijl de melkzuurgisting de melkzuurbacteriën hiertoe niet in staat stelt, maar wel uit in het voedsel aangeboden peptonen? Het is duidelijk, dat het calorisch effect der dissimilatieprocessen in dit opzicht geen juisten maatstaf biedt, maar dat het daarentegen er op aan komt de afneming van de vrije energie bij deze dissimilatieprocessen te bepalen.

Terloops moge er op worden gewezen, dat het om deze reden volmaakt denkbaar is, dat er organismen

[19] Z. allgem. Physiol. **12**, 374 (1911).

[20] Ann. **349**, 125 (1906); ibid. **349**, 140 (1906).
[21] Bioch. Z. **84**, 355 (1917). Zie ook: ibid. **140**, 168 (1923).
[22] De gemakkelijke oplosbaarheid van de urease der sojaboonen hangt ongetwijfeld ten nauwste samen met het feit, dat de ureumomzetting voor de cellen dezer hoogere planten in verband met haar geprononceerde oxydatieve leefwijze energetisch van geen belang is.
[23] Ergebnisse Physiol. **5**, II, 108 (1906).

zouden zijn. wier dissimilatie-reactie een vanzelf-verloopend endotherm proces is en die dus in tegenstelling met alle bekende organismen geen warmte zouden produceeren, doch deze uit de omgeving zouden opnemen!

Door Baron en Polanyi is er op gewezen, dat voor de met gering calorisch effect verloopende omzettingen de afneming der vrije energie geenszins met het calorisch effect overeenkomt, doch dat bij de oxydatieve processen de afwijking tusschen beide grootheden niet zeer belangrijk is. [24]

Indien dit laatste zoo is en wij tevens moeten besluiten, dat de energieomzettingen der dissimilatie dienstbaar zijn aan het verloop der assimilatie, zouden wij mogen verwachten, dat voor éénzelfde of voor nauwverwante organismen, een gelijke dissimilatorische energieomzetting zich ook in een gelijk assimilatorisch effect zou uiten.

De gelegenheid om dit te controleeren biedt zich aan bij het nitriet- en nitraatferment, organismen met een zeer verwante stofwisseling, die intusschen hun energie ontleenen aan dissimilatieprocessen met uiteenloopend calorisch effect. Het is nu ongetwijfeld belangwekkend te constateeren, dat, terwijl de calorische effecten van deze processen respectievelijk zijn 79 en 22, de verhouding geoxydeerde stikstof tot in de bacteriën vastgelegde koolstof voor beide organismen respectievelijk zijn: 35 : 1 en 135 : 1, dus inderdaad overeenkomstig de verwachting practisch omgekeerd evenredig aan de calorische effecten.

Doch te lang reeds stond ik stil bij de energetische eenheid in de stofwisseling. Het zou mij verheugen, indien U den indruk mocht hebben gekregen, dat de energetische beschouwingswijze weliswaar hier en daar reeds aardige kijkjes opent, doch dat aan den anderen kant er nog nauwelijks een begin is gemaakt met het nadere onderzoek.

Maar niet alleen uit energetisch oogpunt is uit de stofwisseling der microben eenheid te vinden. Ook wat de stoffelijke zijde aangaat is de eenheid in verschillende stofwisselingsprocessen, naar uit latere onderzoekingen is gebleken, veel grooter dan nog niet zoo lang geleden werd aangenomen. De gelegenheid ontbreekt al weer om een en ander uitvoerig te documenteeren.

Maar er moge op worden gewezen, dat de onderzoekingen van Neuberg en zijn medewerkers toch wel aannemelijk hebben gemaakt, dat acetaldehyde zoowel bij de gewone alcoholgisting [25], als bij de gisting door colibacteriën [26], boterzuurbacteriën [27] en cellulose-aantastende bacteriën [28] als tusschenproduct der stofwisseling optreedt: En ook dat deze tusschenproductvorming bijvoorbeeld bij de boterzuurbacteriën zoowel uit substraten van de C_6-reeks als van de C_3-reeks plaats heeft, iets dat het merkwaardige feit van de vorming van het boterzuur met zijn vier C-atomen uit verbindingen van het C_3-type althans meer begrijpelijk maakt.

Ook het optreden van pyrodruivenzuur als tusschenproduct bij verschillende gistingsprocessen is geschikt om de eenheid in stoffelijk opzicht van op het eerste gezicht geheel uiteenloopende processen te demonstreeren.

Sterker evenwel nog dringt de stoffelijke eenheid in de stofwisselingsprocessen zich aan ons op, wanneer wij nu het oog richten op die microben, welke gekenmerkt zijn door het bezit van typisch oxydatieve dissimilatieprocessen.

Zooals ook uit Tabel II blijkt, wordt deze voor alle hoogere planten en dieren kenmerkende wijze om in de energetische behoefte der cellen te voorzien, ook in het microbenrijk bij relatief zeer uiteenloopende groepen van microben aangetroffen. In tegenstelling met wat voor de gistingsstofwisselingsprocessen geldt, heeft het chemisch onderzoek aan deze aërobe processen intusschen steeds relatief weinig aandacht geschonken. De verklaring hiervan ligt voor de hand: in den regel toch zullen deze aërobe organismen van het hun aangeboden energetisch voedsel maximaal profijt trekken, de oxydatie zal dus zoover mogelijk worden doorgezet, d. w. z. tot het voedsel is omgezet in koolzuur, water en eventueel ammoniak.

Toch waren op dezen regel reeds van oudsher uitzonderingen bekend.

Zoo was het onmiddellijk na de ontdekking der azijnbacteriën duidelijk, dat men hier een voorbeeld had van organismen, die gekenmerkt waren door een onvolledig oxydatieve stofwisseling. Dit uitte zich niet alleen in de onvolledige verbranding van alcohol tot azijnzuur, maar ook in die van glucose tot gluconzuur.

Maar ook onder de andere groepen van aërobe organismen begonnen zich geleidelijk eenige specialiteiten te vertoonen. Zoo verkreeg een schimmel als *Aspergillus niger* een zekere reputatie als verwekker van wat men met een zeer ongelukkige benaming noemde: een oxaalzuurgisting, d. w. z. men stelde vast, dat bij cultuur van dit organisme op koolhydraatrijken voedingsbodem niet onbelangrijke hoeveelheden oxaalzuur werden gevormd. Een aantal andere, nauw aan de bekende *Penicillium glaucum* verwante, schimmels bleek onder soortgelijke voorwaarden citroenzuur te produceeren, een omstandigheid, welke men voldoende achtte om deze specialiteiten in een afzonderlijk geslacht *Citromyces* te vereenigen.

Voorts had reeds in 1886 de Fransche chemicus Boutroux een onder den invloed van een, door hem niet nader beschreven, aërobe bacterie verloopende omzetting gevonden, waarbij glucose bij aanwezigheid van krijt werd overgevoerd in calcium-oxygluconaat. [29]

Omstreeks 1900 verschenen dan de terecht ook bij de organici veel opzien barende publicaties van Bertrand over de biochemie der zoogenaamde sorbosebacterie (*Acetobacter xylinum*). [30] Hoewel het van den aanvang af duidelijk was, dat laatstgenoemde bacterie ook in de groep der azijnbacteriën moest worden ondergebracht, vertoonde haar stofwisseling aan den anderen kant toch zeer belangrijke afwijkingen van die der bacteriën. welke bij de bereiding van snelazijn en bierazijn werkzaam zijn. Dit bleek onder meer uit het feit. dat cultuur der sor-

---

[24] Biochem. Z. **53**, 1 (1913).
[25] Men zie bijv. de samenvatting van W. Fuchs, Der gegenwärtige Stand der Gärungsproblems; Stuttgart, 1922.
[26] Biochem. Z. **96**, 133 (1919).
[27] Biochem. Z. **117**, 269 (1921).
[28] Biochem. Z. **139**, 527 (1923).

[29] Compt. rend. **102**, 924 (1886).
[30] Men vergelijkende de samenvattende verhandeling in: Ann. Chim. Phys. 8e Série, **3**, 18 (1904).

bosebacterie op sorbiet aanleiding gaf tot de vorming van belangrijke hoeveelheden sorbose, cultuur op manniet leverde laevulose, cultuur op glycerine gaf dioxyaceton, alle omzettingen waartoe de gewone azijnbacteriën niet in staat bleken te zijn.

Zoo ziet men, dat geleidelijk ook onder de op het eerste gezicht uit stofwisselingsoogpunt zoo uniforme categorie der microben met aërobe leefwijze zich specialiteiten begonnen af te teekenen, specialiteiten, waarvan de respectievelijke ontdekkers niet moe werden den lof te verkondigen.

Het is nu zeker alleszins belangwekkend, dat onderzoekingen uit den laatsten tijd leeren, dat ook de verrichtingen van al deze oogenschijnlijk zoo uiteenloopende specialiteiten steeds meer van één gezichtspunt zijn samen te vatten.

In de eerste plaats bleek uit de proefnemingen van Currie [31]), en vooral ook uit die van Molliard [32]), dat men door wijziging van de cultuurvoorwaarden er in kon slagen een organisme als *Aspergillus niger* een belangrijk deel van de glucose te doen omzetten, al naar wensch in gluconzuur, citroenzuur of oxaalzuur. Zoodoende kon men dus dit organisme, dat altijd als oxaalzuurspecialiteit was bekend geweest, stoffen laten voortbrengen, welke tot dusver alleen als speciale producten respectievelijk van azijnbacteriën en *Citromyces*-soorten bekend waren. Dit feit maakte het nu in hooge mate waarschijnlijk, dat men in genoemde stoffen niets anders had te zien dan tusschenproducten van de oxydatieve verwerking van glucose tot koolzuur en water, waarbij het verschil van de specialiteiten alleen daarin zou bestaan, dat het eene organisme bij een eerderen tusschentrap der oxydatie halt hield dan het andere.

Mogelijk zal het U toeschijnen, dat de opvatting van stofwisselingsproducten als citroenzuur en oxaalzuur (en vanzelfsprekend van gluconzuur) als zijnde normale tusschenproducten der biologische glucoseoxydatie wel altijd voor de hand heeft gelegen. Toch had dit inzicht geenszins algemeen ingang gevonden, wat blijkt uit het feit, dat een onderzoeker als Butkewitsch, welke in de laatste jaren uitgebreide experimenteele onderzoekingen over de vorming van citroenzuur en oxaalzuur door schimmels heeft verricht, slechts zeer onlangs onafhankelijk tot dezelfde conclusie kwam. [33])

Tot dezelfde opvatting van het trapsgewijze verloop der oxydatieve stofwisselingsprocessen leiden nu ook de in het laatste jaar in het Delftsche laboratorium verrichte onderzoekingen over de stofwisseling van de verschillende azijnzuurbacteriën. In de eerste plaats werd daar door de Leeuw naar alle waarschijnlijkheid teruggevonden de verloren gegane bacterie van Boutroux, welke een door ons *Acetobacter suboxydans* genoemde azijnbacterie-soort bleek te zijn, die tevens groote verwantschap met Bertrand's sorbosebacterie vertoonde, doch daarvan toch in enkele opzichten duidelijk verschilde. Het onderzoek leerde namelijk, dat *A. suboxydans* dezelfde voorzichtige oxydaties van uiteenloopende suikeralcoholen tot stand bracht als hierboven voor de sorbosebacterie werden vermeld. Maar ook bleek,

dat *A. suboxydans* een nog zwakkere oxydatie-intensiteit heeft dan *A. xylinum*, wat o. m. hieruit bleek, dat eerstgenoemde bacterie calciumgluconaat slechts in calciumoxygluconaat omzet, de tweede deze stof ook door verbrandt tot calciumcarbonaat, terwijl het bijv. niet gelukte om *A. suboxydans* stoffen als dioxyaceton en kalium-oxygluconaat verder te laten oxydeeren, iets waartoe *A. xylinum* wel in staat is. Laatstgenoemde bacterie blijkt dus zijn karakteristieke stofwisselingsproducten onder daartoe gunstige voorwaarden weer verder te verbranden, iets waaruit dus ook weer het tusschenproductkarakter van deze onvolledige oxydaties blijkt.

Maar duidelijker nog treedt de juistheid van deze opvattingen naar voren uit het feit, dat het ons gelukte door toepassing van soortgelijke maatregelen als Molliard bij *Aspergillus niger* trof, ook de azijnbacteriën van de snelazijngroep er toe te brengen, uit stoffen als manniet en glycerine de bovenvermelde tusschenproducten te vormen, ondanks het feit, dat zij onder meer normale cultuurvoorwaarden de genoemde stoffen volledig verbranden. [34])

Een enkel woord moge dan tenslotte nog worden gewijd aan de middelen, welke Molliard en ook wij toepasten om de oxydatie bij de verschillende tusschentrappen te doen halt houden.

Door Molliard werd daarbij in de eerste plaats aandacht geschonken aan de quantitatieve regulatie van de hoeveelheden stikstof, phosphor en kalium, in de cultuurvloeistof. Door ons werd, in aansluiting met wat reeds door Beijerinck en Hoyer was opgemerkt, gebruik gemaakt van het feit, dat de aard van het stikstofvoedsel in vele gevallen beslissend is, of een bepaalde stof, dus ook een bepaald tusschenproduct, al of niet verder zal worden geoxydeerd.

Ook hierin openbaart zich een zeer sprekende samenhang van dissimilatie en assimilatie en misschien is ook hier juist een weg te vinden om dieper in deze vraagstukken door te dringen.

Het is toch wel zeer merkwaardig, dat *A. suboxydans* zich uitstekend vermeerdert in een minerale voedingsvloeistof met eenige procenten manniet, indien voor stikstof in den vorm van een ammoniumzout wordt zorg gedragen. Deze bacterie kan dus de ammoniakstikstof zeker voor den opbouw van haar eiwit bezigen. Nu blijkt evenwel, dat dezelfde bacterie in dezelfde oplossing, waarin alleen de manniet door glucose is vervangen, zich niet vermeerdert, ofschoon toch bij aanwezigheid van peptonstikstof de glucose uiterst krachtig wordt geoxydeerd, als energetisch voedsel dus alleszins bruikbaar is en er dan ook een krachtige vermeerdering intreedt. We zien dus hoe twee stoffen, die ieder in hun eigen functie bruikbaar zijn, toch een als voedsel ondeugdelijke combinatie opleveren. Zouden wij ook dit zoo onverwachte verschijnsel misschien terzijnertijd kunnen terugvoeren op een mogelijk verschil in afneming in vrije energie respectievelijk bij den eersten stap der manniet- en bij die der glucoseoxydatie, een verschil, dat oorzaak zou kunnen zijn. dat in het eerste geval de ammoniakstikstof wel tot

<hr>

[31]) J. Biol. Chem. **31**, 15 (1917).

[32]) Compt. rend. biol. **72**, 479 (1920); ibid. **87**, 967 (1922); Compt. rend. **174**, 881 (1922).

[33]) Men vergelijke voor een en ander: Biochem. Z. **131**, 327 (1922); ibid. **131**, 338 (1922) en ibid. **142**, 195 (1923).

[34]) Voor nadere bijzonderheden aangaande deze kwestie moge worden verwezen naar een verhandeling van F. J. G. de Leeuw en schrijver dezes, welke in het Tijdschrift voor Vergelijkende Geneeskunde (orgaan der Nederl. Vereeniging voor Microbiologie) **10** (1924) is verschenen.

bacteriëneiwit zou kunnen worden opgeheven, in het tweede geval niet?

Wij weten het niet, maar wel leeren wij er uit, hoe het intreden van een biologisch oxydatieproces niet alleen afhangt van den aard en den toestand der levende cellen en de aanwezigheid van een oxydeerbare stof, maar hoe dit intreden afhankelijk is van zeer subtiele wijzigingen in den aard van het medium.

Onvermijdelijk worden wij bij een nadere overweging van deze problemen herinnerd aan die andere verschijnselen van onvolledige en onvoldoende biologische oxydatie, welke kenmerkend zijn voor de als diabetes bekende pathologische afwijkingen in de stofwisseling van mensch en dier.

Is het een gezochte vergelijking, wanneer wij zeggen, dat wij *A. suboxydans* van diabetes genezen, wanneer wij de cellen van dit organisme gesuspendeerd in een glucose en ammoniakstikstofhoudende — blijkens de mannietproef toch op zichzelf wel volledige — voedingsvloeistof tot nieuw leven wekken door toevoeging van kleine hoeveelheden peptonstikstof? En is er geen aanleiding om bij de studie van de zoo zegenrijke doch betrekkelijk nog zoo geheimzinnige insuline-werking, ook aandacht te schenken aan de mogelijkheid, dat onder den invloed van dit hormoon chemisch misschien goed aantoonbare stoffen in de bloedbaan worden gebracht, wier aanwezigheid voldoende is om de oxydatieve verwerking der suiker door de lichaamscellen te doen intreden? [35]) De beantwoording van deze vraag laat ik gaarne aan meer bevoegden over; ik heb evenwel gemeend niet onberechtigd te zijn haar hier op te werpen.

Dit temeer, omdat er steeds meer aanwijzingen komen, dat aan het chemisme der biologische oxydaties een eenheid ten grondslag ligt, die zich uitstrekt tot over alles wat aeroob leeft. Een enkel voorbeeld nog. Het was reeds lang bekend, dat in ranzig cocosvet onder meer ketonen worden aangetroffen, zooals methylnonylketon, methylheptylketon e. a. — Door Dr. Derx werd nu onlangs mijn aandacht er op gevestigd, hoe onderzoekingen van Stokoe [36]) en van hemzelf [37]) bewezen hebben, dat deze verbindingen ontstaan onder den invloed van schimmels, die deze producten blijkbaar weer vormen door onvolledige oxydatie van de uit het vet gevormde vetzuren.

Het is nu toch wel zeer treffend, hoe bij deze door microben bewerkte onvolledige oxydatie uit laurinezuur o. m. het methylnonylketon ontstaat, geheel overeenkomstig de door Knoop voor de stofwisseling der hoogere organismen aangegeven regel, volgens welke de biologische oxydatie der vetzuren aangrijpt in het β-koolstofatoom, waarbij eerst β-ketozuren en hieruit door koolzuurafsplitsing ketonen worden gevormd. Een omzetting, welke bij de onvolledige verbranding in het lichaam van den diabeteslijder aanleiding geeft tot de vorming van verbindingen als acetylazijnzuur en aceton.

Zoo zouden er nog tal van argumenten zijn aan te voeren ten gunste van de opvatting, dat aan het chemisme der oxydatieve dissimilatieprocessen een groote mate van eenheid ten grondslag ligt.

Voor tal van microben met geringe oxydatie-intensiteit lijkt het een uitvoerbare taak voor elk der physiologisch belangrijke stoffen zooals glucose, glycerine e. d. de opeenvolgende tusschentrappen der oxydatie vast te stellen en eveneens zal er naar kunnen worden gestreefd om de gemakkelijkheid, waarmede de verschillende trappen worden afgelegd, ook in quantitatief opzicht vast te leggen.

Er valt nauwelijks aan te twijfelen of een dergelijk onderzoek zal ons inzicht in het wezen der stofwisseling der microben verdiepen, maar tevens dank zij de hierboven gesignaleerde eenheid ook voor de kennis van de stofwisseling van de, voor het experiment veel minder toegankelijke, hoogere organismen ruimschoots nut afwerpen.

Hoe dit ook zijn moge, ik hoop, dat mijn uiteenzettingen U de overtuiging zullen hebben geschonken, dat het onderzoek der stofwisseling van de microben juist èn door de daarin aanwezige verscheidenheid èn door de daarin na te speuren eenheid ook den chemicus menig aantrekkelijk probleem biedt.

---

[35]) Ook al zou men moeten aannemen, dat alle glucose bij de oxydatie, zooals dit voor de omzettingen bij de spiercontractie door de onderzoekingen van Embden, Laquer, Meyerhof e.a. bewezen is, den omweg over glycogeen, lactacidogeen en melkzuur aflegt, dan nog gelden soortgelijke overwegingen voor de laatste, oxydatieve, phase van het proces. Men vergelijke de samenvattende overzichten van F. Laquer, Insulin, Die Naturwissenschaften **12**, 89 (1924) en O. Meyerhof, Die Energie-umwandlungen im Muskel, ibid **12**, 181 (1924).

[36]) J. Soc. Chem. Ind. **40**, 76 (1922).

[37]) Nog niet gepubliceerd.

The Utilisation of CO_2 in the Dissimilation of Glycerol by the Propionic Acid Bacteria

H. G. WOOD AND C. H. WERKMAN

This paper by Wood and Werkman is truly outstanding. First, it is a classical example of determination of the path of a fermentation by careful quantitative analyses of substrates consumed and products formed. Second and most importantly, the results led to the discovery of CO_2 fixation by heterotrophic bacteria. The dogma at the time the work was done and as expressed in textbooks held that only specially adapted chemosynthetic and photosynthetic autotrophic microorganisms, and of course plants, could fix CO_2. The analyses by Wood and Werkman indicated that succinate was the product of the fixation. This was confirmed by Wood et al. in 1941 with the use of $^{13}CO_2$ and mass spectroscopy. The discovery by Wood was extended by him and by others to involve livers from higher animals, and it is now clear that several types of CO_2 fixation reactions occur in most if not all cells. The discovery of heterotrophic CO_2 fixation is important, because this process is needed for converting pyruvate to oxaloacetate (a reaction named the Wood-Werkman reaction long before the actual enzyme catalyzing the reaction was characterized) to connect glycolysis with the citric acid cycle, for synthesis of malonyl-CoA needed for synthesis of lipids, for the function of the urea cycle, and for numerous other metabolic processes including the synthesis of nucleotides, DNA, and RNA. How the discoverers of heterotrophic CO_2 fixation avoided getting a Nobel Prize is difficult to understand.

LARS G. LJUNGDAHL

X. THE UTILISATION OF CO$_2$ IN THE DISSIMILATION OF GLYCEROL BY THE PROPIONIC ACID BACTERIA.

By HARLAND GOFF WOOD AND CHESTER HAMLIN WERKMAN.[1]

From the Department of Bacteriology, Iowa State College, Ames, Iowa, U.S.A.

(*Received December 2nd, 1935.*)

REDTENBACHER [1846] was probably the first to study the fermentation of glycerol by the propionic acid bacteria. Largely by accident he obtained a good growth of these organisms in probably impure culture and found propionic acid as substantially the only product. It remained for Van Niel [1928] to confirm the results of Redtenbacher with pure cultures. He found that glycerol was converted quantitatively into propionic acid under anaerobic conditions with no formation of gas. Van Niel suggested that glyceraldehyde is an intermediate product which is reduced to propionic acid. Pett and Wynne [1933, 1] obtained preliminary evidence for the formation of methylglyoxal and of glyceraldehyde (or dihydroxyacetone) from sodium β-glycerophosphate by fermentation with dried propionic acid organisms. These authors indicate that their identifications are not conclusive. Wood and Werkman [1934] using both CaSO$_3$ and dimedon (dimethyldihydroresorcinol) as fixatives isolated and identified propaldehyde from fermentations of glycerol. They suggested that this compound occurs as an intermediary and is subsequently converted into propionic acid. Pyruvic acid has also been identified (unpublished results).

This brief review summarises our knowledge of the dissimilation of glycerol by the propionic acid bacteria. In the present communication it will be shown that propionic acid is not the only product which may be formed and that the dissimilation is more complex than found by previous investigators.

METHODS.

The volatile acids were determined by the partition method of Osburn *et al.* [1933]. Non-volatile acids were extracted with ethyl ether continuously for 12 hours from the residue of steam-distillations which was taken up in anhydrous Na$_2$SO$_4$. The lactic acid was determined by the method of Friedemann and Kendall [1929]. The succinic acid was obtained by neutralising the hot solution with CaCO$_3$, filtering and precipitating the calcium succinate in 85 % ethyl alcohol. The calcium salt was determined by weight. This acid was also determined by the silver salt method [Moyle, 1924]. The original glycerol was determined by weight. The residual glycerol determination was made on an acidified aliquot part of the medium which was evaporated on a steam-bath to 50–75 ml., neutralised, taken up in plaster of Paris and extracted continuously with acetone for 8 hours. The acetone extract was reduced to 50–75 ml., distilled water added and the acetone removed by further distillation. The glycerol was determined in this preparation by the method of Wagenaar [1911]. The original

[1] Supported in part by a grant from the Industrial Science Research funds of the Iowa State College for the study of the Utilisation of Agricultural By-products.

(48)

CO_2 of the medium was calculated equivalent to the weighed quantity of $CaCO_3$ used in the medium. The liberated CO_2 was determined in two ways. It was either absorbed in soda-lime and weighed or it was absorbed in standard alkali and the excess alkali titrated with standard HCl after the addition of an excess of $BaCl_2$ to precipitate the CO_2. The soda-lime was used in a train of 6-inch U-tubes with $CaCl_2$ as the drying agent. When alkali was used it was contained in two bottles connected by a syphon. The first bottle was attached to the fermentation flask; the second was supplied with a soda-lime tube to prevent entrance of CO_2 from the air. The syphon permitted displacement of the alkali as gas was formed during fermentation. The alkali was transferred to a bottle fitted with a bubble spiral-absorber for the collection of residual CO_2. The residual CO_2 was obtained as follows. The fermentation flask was attached to a reflux condenser which led to the soda-lime train or bubble spiral-absorber and the CO_2 was liberated by adding a known volume of dilute H_2SO_4 slightly in excess of that necessary to react with the $CaCO_3$, a slow current of CO_2-free air was passed through and the contents finally brought to the boil. The CO_2 collected *minus* the CO_2 added as $CaCO_3$ equals the CO_2 produced by the fermentation. The method proved accurate with known quantities of $CaCO_3$.

The media used were as follows. The experiments of Table II were carried out with 700 ml. of medium consisting of glycerol 3 %, $CaCO_3$ 2 % and yeast extract (Difco) 0·4 % in 1 litre Erlenmeyer flasks. These fermentations were maintained anaerobic by continuously bubbling oxygen-free nitrogen through the fermenting medium. The CO_2 was absorbed in soda-lime. The results shown in Table III were obtained with 800 ml. of medium containing glycerol 2 %, $CaCO_3$ 1·14 % and yeast extract 0·4 %. Conditions were made anaerobic by displacing the air in and above the medium with nitrogen immediately after inoculation. The CO_2 was collected in alkali. Constituents of the media were sterilised separately at 20 lb. pressure for half an hour and mixed at the time of inoculation. 3–5-day cultures grown in a yeast extract medium with 0·5 % glucose and equivalent to 5 % by volume were used as inoculum. The flasks were shaken twice daily to aid the buffer action of the carbonate. Purity of cultures before and after fermentation was established by the Kopeloff-Gram stain, absence of growth on aerobic slants and the zone of growth and colony formation in agar shakes. Incubation was at 37°, for the fermentations of Table II and at 30° for those of Table III. The incubation period was 35 days.

The cultures are identified in Table I. Complete descriptions of the species are given by Werkman and Brown [1933].

EXPERIMENTAL.

Results are expressed as millimoles per litre of medium in Table II and per 100 millimoles of fermented glycerol in Table III. The carbon recovery (Table III) is calculated first on the basis of the glycerol fermented and CO_2 utilised and

Table I. Identification of cultures.

Culture no.	Species	Culture no. used by other investigators*
11 W	*P. petersonii*	*Sherman*—22, Van Niel—20
15 W	*P. technicum*	*Sherman*—10
34 W	*P. arabinosum*	*Hitchner*—61
49 W	*P. pentosaceum*	Foote, *Fred* and Peterson—11 Van Niel—4
52 W	*P. shermanii*	Foote, *Fred* and Peterson—19
4875	*P. pentosaceum*	American Type Culture Collection

* Culture received from investigator whose name is italicised.

Table II. *Dissimilation of glycerol by propionic acid bacteria.*

Culture no.	Products per litre			
	Propionic acid mM.	Acetic acid mM.	Succinic acid mM.	CO_2 utilised mM.
4875	101·7	6·5	32·2	46·6
34 W	137·6	2·8	28·8	43·0
49 W	128·0	7·7	19·1	25·6
49 W	168·9	9·3	130·0	—

Table III. *Dissimilation of glycerol by propionic acid bacteria.*

Culture no.	Glycerol fermented per litre mM.	CO_2 utilised per 100 mM. of fermented glycerol mM.	Products per 100 mM. of fermented glycerol			Carbon recovery		Oxidation-reduction index	
			Propionic acid mM.	Acetic acid mM.	Succinic acid‡ mM.	Basis-glycerol *plus* CO_2 %	Basis-glycerol only %	Basis-glycerol *plus* CO_2	Basis-glycerol only
49 W	212·6	37·7	55·8	2·9	42·1	101·2	114·0	1·081	2·550
34 W	209·0	43·2	59·3	2·0	34·5	93·1	106·6	0·925	2·270
52 W*	112·0	20·0	78·4	5·9	8·7	94·6	101·0	0·918	1·386
11 W†	218·4	1·1	89·3	2·6	3·9	96·5	96·8	1·135	1·162
15 W	176·4	12·3	78·4	5·8	7·8	89·1	92·6	1·047	1·376

* 7·0 mM. of lactic acid produced per 100 mM. of fermented glycerol.
† 0·5 mM. of lactic acid produced per 100 mM. of fermented glycerol.
‡ Succinic acid identified by melting-point and mixed melting-point.

secondly on the basis of glycerol alone. The redox indices (Table III) are calculated on the same bases. The redox (oxidation-reduction) index of a fermentation is obtained by (1) multiplying the mM. of each product by its respective oxidation or reduction value based on water as zero (H_2 is equivalent to one atom of oxygen) [*cf.* Johnson *et al.*, 1931]; (2) dividing the sum of oxidation values by reduction values. A perfect balance is represented by an index of 1·0. Propionic acid has a reduction value of 1, succinic an oxidation value 1 and acetic acid is neutral. Glycerol has a reduction value of 1, therefore its utilisation is represented as an oxidation and the glycerol fermented is given an oxidation value 1. Likewise, the utilisation of CO_2, an oxidised compound, represents a reduction and its reduction value is 2.

The glycerol fermented was not determined in Table II. The quantity of non-reducing compounds was not sufficient to influence the fermentation balances.

Discussion.

The most significant fact shown by the data is the apparent utilisation of CO_2 by the propionic acid bacteria. This was evident, since the CO_2 at the conclusion of the fermentation was not equivalent to that of the original medium in the form of $CaCO_3$. This observation has been substantiated by two types of calculation of especial value, *i.e.* carbon recovery and redox index. If CO_2 is utilised, and is in turn (after synthesis) dissimilated, then calculations based on the assumption that glycerol is the sole source of carbon, should show an excess of products, *i.e.* the calculated recovery of carbon will exceed 100 %. Table III shows that this occurred and that calculations based on glycerol *plus* CO_2 are acceptable. The calculation of carbon recovery is not in all cases entirely satisfactory proof of CO_2 utilisation, but the oxidation-reduction balance is convincing. CO_2 contains but one carbon and requires a large utilisation to show a detectable change in the carbon balance; in the oxidation-reduction balance the CO_2 is highly oxidised and therefore has a marked effect. The data show that results calculated on the basis of glycerol *plus* CO_2 are reasonable and acceptable. The fact that the chemical analysis shows a decrease of CO_2 is perhaps proof enough of CO_2 utilisation. However, the carbon and oxidation-reduction balances furnish additional evidence.

The CO_2 analysis has been checked by two different methods, and four species of *Propionibacterium* of known purity, each received from a different source, have shown a distinct utilisation of CO_2 under the conditions of our experiments. Every culture examined with the possible exception of 11W showed ability to utilise CO_2. This behaviour is probably characteristic of the genus *Propionibacterium*.

This observation requires a reinterpretation of previous results. Investigators have not considered the possibility of CO_2 utilisation in constructing schemes of dissimilation. If one considers the limited number of bacteria which have been shown to utilise CO_2 and also that such forms differ markedly from the propionic acid bacteria, failure to consider the possibility of CO_2 utilisation may be understood. It is of interest that a number of investigators have found that growth of bacteria in general is inhibited in the absence of CO_2. Rockwell and Highberger [1927] suggested that CO_2 is utilised by bacteria and expressed the opinion that it is the only direct source of carbon. It is possible that the utilisation of CO_2 as a source of carbon may not be limited to the small group of bacteria now recognised. Since most bacteria produce CO_2 during their growth it is difficult to determine whether they utilise CO_2. In the present case, the propionic acid bacteria utilised more CO_2 than was produced and thus offered direct evidence of CO_2 consumption.

One important problem, which requires consideration in relation to the data presented, is the mechanism of succinic acid formation. A number of investigators working particularly with yeast and fungi [Butkewitsch and Federoff, 1930; Wieland and Sonderhoff, 1933] have reported that succinic acid is formed by a condensation of two molecules of acetic acid. Virtanen [1925; 1934] and Virtanen and Karstrom [1931], however, have suggested that the propionic acid bacteria produce succinic acid from glucose by a 4- and 2-carbon cleavage of the hexose molecule. Virtanen's proposal was prompted by the observation that the propionic acid bacteria in the presence of toluene form succinic and acetic acids from glucose with no gas. This observation appeared incompatible with schemes involving a 3-carbon cleavage and the formation of 2-carbon compounds by a 2- and 1-carbon cleavage. The absence of CO_2 or other 1-carbon compounds appeared conclusive proof against such a scheme. However, the present evidence of the utilisation of CO_2 by the propionic acid bacteria leaves no reason to assume that the 1-carbon compounds should equal the sum of the 2-carbon compounds. It is necessary in the light of our present knowledge to leave open the possibility of a 4- and 2-carbon cleavage.

The mechanism of fermentation (Tables II and III) is of interest but at present must remain largely speculative. The scheme of Van Niel is not complete since it does not include succinic acid. His suggestion that the succinic acid is formed from the yeast extract is excluded in these fermentations as in some a quantity of succinic acid was obtained which exceeded the total quantity of yeast extract used in the medium. The formation of the 4-carbon compound (succinic acid) from the 3-carbon compound (glycerol) is proof of synthesis. It is possible that the synthesis yields a 6-carbon compound which is dissimilated to succinic acid by a 4- and 2-carbon cleavage. However, such a scheme for the

4—2

glycerol fermentation would require a synthesis from a 2-carbon compound since the 2-carbon and 4-carbon compounds do not occur in equivalent quantities. It seems more probable that part of the glycerol is dissimilated to 2- and 1-carbon compounds and that the 1-carbon compound is completely utilised in the synthesis of a compound which is subsequently fermented whilst the 2-carbon compound, probably acetic acid, condenses to yield succinic acid. The evidence presented justifies the assumptions relative to the utilisation of the 1-carbon compound. With regard to the formation of succinic acid by a condensation of 2-carbon compounds more complete evidence will be presented elsewhere. The results obtained by different methods indicate that the propionic acid bacteria can produce succinic acid by a condensation of 2-carbon compounds and it is probable that the condensation involves acetic acid. The formation of succinic acid from glycerol is indirect support of such a mechanism although the possibility cannot be disregarded that other reactions are occurring. It is possible that the succinic acid is formed in more than one way.

Explanation of the formation of propionic acid also offers difficulties. Wood and Werkman [1934] furnished some information by their isolation of propaldehyde but the question arises as to the manner of formation of the propaldehyde. The investigations of Embden *et al.* [1933] and Meyerhof and Kiessling [1933] have emphasised the importance of glycerophosphoric acid and phosphoglyceric acid in the carbohydrate metabolism of yeast and muscle tissue. This suggests the possible rôle of these compounds in the dissimilation of glycerol by the propionic acid bacteria. In the schemes proposed by Embden and Meyerhof the triosephosphate is converted into phosphoglyceric acid. To obtain propaldehyde following such a conversion the carboxyl group would have to be reduced to a carbonyl group. It is doubtful whether the propionic acid bacteria can bring about such a conversion. A more probable series of reactions leading to the formation of propaldehyde and propionic acid is that shown below.

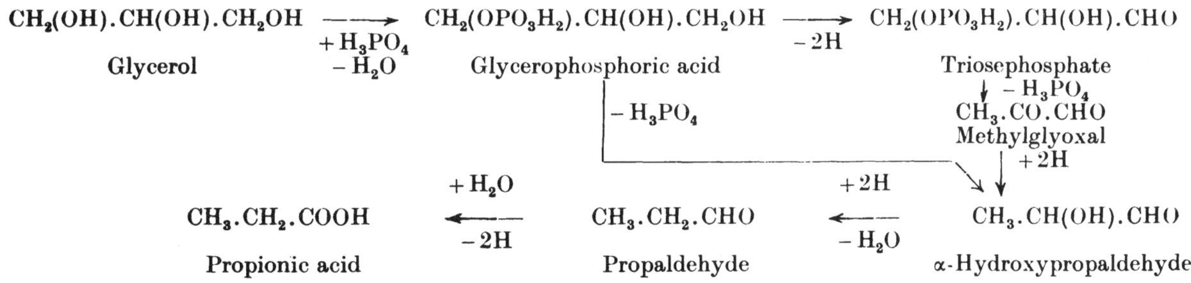

Mechanism of the dissimilation of glycerol to propionaldehyde and propionic acid.

Pett and Wynne [1933, 2] have found that the propionic acid bacteria can dephosphorylate glycerophosphoric acid and also [1933, 1] obtained preliminary evidence of the occurrence of methylglyoxal and glyceraldehyde (or dihydroxyacetone). Reduction of methylglyoxal to propaldehyde might reasonably be expected, for the reduction of pyruvic acid to propionic acid is readily accomplished by these organisms. The proposed scheme involving methylglyoxal appears from our present knowledge to be a logical means of obtaining propaldehyde although other reactions cannot be excluded. For example, the glycerophosphoric acid may go directly to α-hydroxypropaldehyde.

Phosphoglyceric acid may not play a rôle in the formation of propaldehyde although it may be an important intermediary in the fermentation yielding pyruvic acid and subsequently acetic acid, CO_2 and propionic acid. There is no reason apparent why methylglyoxal and phosphoglyceric acid cannot occur in the same fermentation or why a compound cannot be formed by more than one series of reactions.

It is not clear why our fermentations gave results so different from those of Van Niel [1928]. His data in which CaCO$_3$ was used as a buffer give no information relative to the CO$_2$ but show that the dissimilation was not the same, with the possible exception of that obtained with culture 11W. Van Niel accounted quantitatively for his fermented glycerol as propionic acid. The action of CO$_2$ as a hydrogen acceptor is made evident in our fermentations by the production of the oxidised product succinic acid. It is apparent that a large CO$_2$ utilisation is accompanied by a greater production of succinic acid or some other oxidised compound.

SUMMARY.

An investigation has been made of the dissimilation of glycerol by bacteria of the genus *Propionibacterium* and the products have been determined quantitatively. The results obtained support the following conclusions.

1. CO$_2$ obtained from CaCO$_3$ is utilised by the propionic acid bacteria in their fermentation of glycerol.

2. Propionic, acetic, succinic and occasionally lactic acids are produced from the glycerol. CO$_2$ acts as a hydrogen acceptor permitting the formation of oxidised products.

3. The utilisation of CO$_2$ by the propionic acid bacteria necessitates further interpretation of data and places in question the evidence for the 4- and 2-carbon cleavage.

4. The formation of succinic acid from glycerol offers indirect evidence of its formation by a condensation of 2-carbon compounds.

5. The mechanism of glycerol dissimilation is shown to be complex, involving synthetic reactions with subsequent fermentation of a synthesised product.

REFERENCES.

Butkewitsch and Federoff (1930). *Biochem. Z.* **219**, 87.

Embden, Deuticke and Kraft (1933). *Klin. Woch.* **12**, 213.

Friedemann and Kendall (1929). *J. Biol. Chem.* **82**, 23.

Johnson, Peterson and Fred (1931). *J. Biol. Chem.* **91**, 569.

Meyerhof and Kiessling (1933). *Biochem. Z.* **267**, 313.

Moyle (1924). *Biochem. J.* **18**, 351.

Osburn, Wood and Werkman (1933). *Ind. Eng. Chem.* Anal. Ed. **5**, 247.

Pett and Wynne (1933, 1). *Trans. Roy. Soc. Canada*, **27**, 119.

—— —— (1933, 2). *Biochem. J.* **27**, 1660.

Redtenbacher (1846). *Liebig's Ann.* **57**, 174.

Rockwell and Highberger (1927). *J. Inf. Diseases*, **40**, 438.

Van Niel (1928). The propionic acid bacteria. (Haarlem.)

Virtanen (1925). *Soc. Sci. Fennica, Comment. Physic-math.* **2**, 20.

—— (1934). *J. Bact.* **28**, 447.

—— and Karstrom (1931). *Acta Chem. Fennica.* Series B **7**, 17.

Wagenaar (1911). *Pharm. Weekblad.* **48**, 497.

Werkman and Brown (1933). *J. Bact.* **26**, 393.

Wieland and Sonderhoff (1933). *Liebig's Ann.* **503**, 61.

Wood and Werkman (1934). *Proc. Soc. Exp. Biol. Med.* **31**, 938.

Methyl-Vitamin B_{12} as a Source of Methyl Groups for the Synthesis of Acetate by Cell-Free Extracts of Clostridium thermoaceticum

J. M. POSTON, K. KURATOMI, AND E. R. STADTMAN

Acetogenic bacteria are now defined as those that catalyze the reduction of CO_2, CO, and other C-1 compounds by the formation of acetate via the acetyl-coenzyme A (acetyl-CoA) pathway. The Dutch microbiologist K. T. Wieringa, in 1936, discovered *Clostridium aceticum*, the first acetogenic bacterium found to grow with H_2 as an energy source and CO_2 as a carbon source. This bacterium was considered lost until it was rediscovered in 1981. In 1942, F. E. Fontaine et al. discovered *Clostridium thermoaceticum*, recently renamed *Moorella thermoacetica*, which fermented 1 mol of glucose to 3 mol of acetate. Fontaine et al. suggested that 2 mol of the acetate was formed in the fermentation of the glucose and that the third mole was formed by refixation and reduction of CO_2. Further evidence for this was obtained in 1945 by H. A. Barker and M. O. Kamen, who, using $^{14}CO_2$, showed the incorporation of radioactive carbon into both the methyl and the carboxyl groups of acetate. H. G. Wood in 1952, in a classic mass-balance experiment with $^{13}CO_2$, definitely proved the total synthesis of acetate from CO_2. These experiments conducted with isotopes were among the first performed with labeled carbon and were truly pioneering in that they showed the way to use isotopes in metabolic studies. Although the isotope experiments demonstrated the total synthesis of acetate from CO_2, they did not show the pathway for the acetate synthesis. Not much progress was made on the pathway of acetate synthesis from CO_2 until 1963, when it was demonstrated that CO_2 was rapidly converted to formate by extracts of *C. thermoaceticum*. Formate was shown to be a better precursor of the methyl group of acetate than CO_2. J. R. Guest et al., in 1962, showed that methyl-B_{12} served as a precursor of the methyl group of methionine, and this observation prompted E. R. Stadtman and his colleagues to investigate whether methyl-B_{12} could be involved in acetate synthesis. This they showed. It was the first real clue to the acetyl-CoA pathway.

Stadtman and his coworkers did not do much further work on the pathway of acetate synthesis; this was done by Harland G. Wood and his team, who isolated 28 different corrinoids, including some Co-methyl corrinoids from *C. thermoaceticum*, and showed the involvement of tetrahydrofolate in the acetyl-CoA pathway. Other discoveries connected with the acetyl-CoA pathway include the first demonstration of tungsten as a bioactive metal which functions in the formate dehydrogenase from *C. thermoaceticum*. This enzyme also contains selenium and iron. The bifunctional enzyme carbon monoxide dehydrogenase/acetyl-CoA synthase, which catalyzes a reduction of CO_2 to CO and the formation of acetyl-CoA, has gained much interest because of its content of nickel and iron, which form nickel-iron clusters constituting parts of the active sites of the reactions catalyzed by the enzyme. The acetyl-CoA pathway and modifications of it are important not only in acetogenic bacteria but also in methanogenic and sulfate- and sulfur-reducing bacteria. The importance of the acetyl-CoA pathway in the environment is demonstrated by the fact that ca. 100 acetogens have now been isolated from most diverse sources, and it has been estimated that ca. 10 to 20% of Earth's carbon is recycled by the autotrophic acetyl-CoA pathway.

LARS G. LJUNGDAHL

Reproduced from *Annals of the New York Academy of Sciences* 112:804–806, copyright © 1964, by permission of the New York Academy of Sciences.

METHYL-VITAMIN B$_{12}$ AS A SOURCE OF METHYL GROUPS FOR THE SYNTHESIS OF ACETATE BY CELL-FREE EXTRACTS OF *CLOSTRIDIUM THERMOACETICUM*

J. Michael Poston, K. Kuratomi, E. R. Stadtman

Laboratory of Biochemistry, National Heart Institute,
National Institutes of Health, Bethesda, Md.

Glucose fermentation by *Clostridium thermoaceticum* leads to the accumulation of nearly three moles of acetate per mole of substrate decomposed.[1] The possibility that one of the three moles of acetate is derived indirectly from carbon dioxide was suggested by the discovery of Barker and Kamen[2] that acetate, labeled almost equally in both carbon atoms, is produced when glucose is fermented in the presence of added $C^{14}O_2$. This possibility was later confirmed by Wood's demonstration[3] that nearly one-third of the acetate molecules produced during the dissimilation of glucose in the presence of $C^{13}O_2$ is formed by a mechanism in which adjacent carbon atoms are derived from $C^{13}O_2$. Thus, glucose fermentation by *Cl. thermoaceticum* is represented by the following overall reactions:

$$C_6H_{12}O_6 + 2H_2O \rightarrow 2CH_3COOH + 2CO_2 + 8H \qquad (1)$$

$$8H + 2CO_2 \rightarrow CH_3COOH + 2H_2O \qquad (2)$$

$$\text{Sum: } C_6H_{12}O_6 \rightarrow 3CH_3COOH. \qquad (3)$$

As judged by the fact that $C^{14}O_2$ is incorporated into both carbon atoms of acetate during the fermentation of various substrates by other heterotropic anaerobes[4,5,6] it is believed that the ability to catalyze reaction **2** is a fairly common property of anaerobic bacteria. In any event this unusual process is of considerable interest from the biochemical point of view since it involves both reduction and condensation of carbon dioxide. In an effort to establish the mechanism by which CO_2 is converted to acetate, we have recently undertaken a study of this process in cell-free extracts of *Cl. thermoaceticum*. As is shown in TABLE 1, when appropriately supplemented with various cofactors and an electron donor system (isopropanol, DPN, and alcohol dehydrogenase),* cell-free extracts catalyze the incorporation of $C^{14}O_2$ into the methyl and carboxyl groups of acetate. A possible role of vitamin B$_{12}$ co-enzyme in the overall reaction is suggested by the observation that addition of intrinsic factor causes a 16 per cent inhibition* of $C^{14}O_2$ fixation. The further observation that intrinsic factor causes a greater inhibition of $C^{14}O_2$ incorporation into the methyl carbon atom of acetate as compared to the carboxyl carbon, and the recent discovery of Guest *et al.*[9] that methyl-B$_{12}$ serves as a methyl

---

* Preliminary studies in this laboratory by H. Eggerer established the co-factor requirements of this enzyme system.
* In some experiments inhibition as high as 50 per cent has been observed.

804

TABLE 1

INFLUENCE OF INTRINSIC FACTOR ON THE INCORPORATION OF $C^{14}O_2$
INTO ACETATE

	Acetate formed (μMoles)	$C^{14}O_2$ incorporation into acetate			
Conditions		Total Cpm	$-CH_3$ Cpm	$-COOH$ Cpm	$-CH_3/-COOH$ Cpm
− Intrinsic factor	10.9	47,000	15,300	31,700	0.485
+ Intrinsic factor	12.5	39,420	11,100	29,400	0.391

The reaction mixtures contained Tris·HCl buffer, pH 7.4, 100 μMoles; $MgCl_2$, 10 μMoles; 2-mercaptoethanol, 10 μMoles; ATP, 10 μMoles; CoA 1.0 μMole; isopropanol, 10 μMoles; DPN, 1.0 μMole; sodium pyruvate, 30 μMoles; horse liver alcohol deyhdrogenase, 4 μg; $NaHC^{14}O_3$, 56 μMoles, 25 μc; intrinsic factor, when added, 300 μg; French press extract of *C. thermoaceticum*, 5 mg protein. Total volume, 1.25 ml. Gas phase, helium. Incubation,, 40° C., 40 min. The acetate formed was isolated by Celite chromatography[7] and was degraded by Schmidt procedure as described by Phares.[8]

donor in the biosynthesis of methionine, prompted us to investigate the possibility that methyl-B_{12} is involved in acetate formation.

The data of TABLE 2 summarizes the results of some preliminary experi-

TABLE 2

THE INCORPORATION OF ISOTOPE FROM $C^{14}O_2$ AND $C^{14}H_3$-B_{12} INTO ACETATE

	Acetate produced (μMoles)	C^{14} incorporated into acetate			
Conditions		Total cpm	$-CH_3$ cpm	$-COOH$ cpm	$-CH_3/-COOH$ cpm
$C^{14}O_2$, no CH_3-B_{12}	25.5	33,000	6,865	26,135	0.26
CO_2, + $C^{14}H_3$-B_{12}	14.2	2,884	2,869	15	191
CO_2, + $C^{14}H_3$-B_{12}	17.2	2,655	2,599	56	46

The conditions were as described in TABLE 1 except that phosphate buffer, pH 7.4, 100 μMoles replaced Tris·HCl buffer, and where indicated, 1.0 μMole $C^{14}H_3$-B_{12}, (22,000 cpm) and 20 μMoles of $NaHCO_3$ or of $NaHC^{14}O_3$ (10 μc.) were added. Incubation was for 60 minutes at 40° C.

ments showing that the methyl group of synthetic C^{14}-methyl-B_{12}[10,11] is selectively incorporated into the methyl carbon atom of acetate. This incorporation is dependent upon the presence of the bacterial extract and requires in addition, CO_2 and pyruvate. Under these conditions there is essentially no incorporation of labeled carbon into pyruvate.

Although these observations must be interpreted with caution, they suggest the possibility that methyl-B_{12} is an intermediary in the conversion of CO_2 to the methyl group of acetate. Not yet excluded is the possibility that methyl-B_{12} may under the experimental conditions, transfer its methyl group nonenzymatically to a methyl group carrier (free or enzyme bound) that is normally involved in acetate synthesis.

References

1. FONTAINE, F. E., W. H. PETERSON, E. McCOY & M. J. JOHNSON. 1942. J. Bacteriol. **43**: 701.
2. BARKER, H. A. & M. D. KAMEN. 1945. Proc. Nat. Acad. Sci. **31**: 219.
3. WOOD, H. G. 1959. J. Biol. Chem. **194**: 905.
4. BARKER, H. A., M. D. KAMEN & V. HAAS. 1945. Proc. Nat. Acad. Sci. **31**: 355.
5. STADTMAN, T. C. & F. H. WHITE, JR. 1953. J. Bacteriol. **67**: 651.
6. KARLSSON, J. L. & H. A. BARKER, 1949. J. Biol. Chem. **178**: 891.
7. VERNER, J. E. 1957. Methods in Enzymology. Vol. III. P. 397. Academic Press. New York, N. Y.
8. PHARES, E. F. 1951. Arch. Biochem. Biophys. **33**: 173.
9. GUEST, J. R., S. FRIEDMAN, D. D. WOODS & E. L. SMITH. 1962. Nature. **195**: 340.
10. BERNHAUER, K., O. MÜLLER & G. MÜLLER. Biochem. Zeit. **336**: 102.
11. SMITH, E. L., L. MERVYN, A. W. JOHNSON & N. SHAW. 1962. Nature. **194**: 1175.

Life at High Temperatures

T. D. BROCK

Microbial life has now been found in many extreme environments, such as those with high and low pH values, temperatures, salt concentrations, substrate concentrations, and pressure. Although reports about microbial growth at extreme environments have appeared from time to time in the literature, very little research had been directed toward an understanding of how microorganisms could live, grow, and reproduce in extreme environments. Brock's paper reports on studies mostly done of hot springs in Yellowstone National Park, where he observed growth of fungi at temperatures as high as 60°C, of blue-green algae at 75°C, and of bacteria at >90°C. With H. Freeze in 1969, Brock published a description of *Thermus aquaticus* (*J. Bacteriol.* **98**:289–297), which grows at an optimum temperature of 75°C. The findings reported by Brock triggered a renewed interest in thermophilic microorganisms and also in a search for mechanisms explaining thermophily and how biochemicals, especially enzymes, can be stable at temperatures above the boiling point of water. The search for microorganisms in extreme environments, which now has blossomed into a large scientific subject, has yielded information on the diversity of microbial life, has triggered speculation regarding the origin of life, and has led to investigations of extended use of microorganisms and their enzymes for industrial applications.

LARS G. LJUNGDAHL

Reprinted with permission from *Science* 158:1012–1019. Copyright © 1967. American Association for the Advancement of Science.

Life at High Temperatures

Evolutionary, ecological, and biochemical significance
of organisms living in hot springs is discussed.

Thomas D. Brock

Temperature is one of the most important environmental factors controlling the activities and evolution of organisms, and is one of the easiest variables to measure. Not all temperatures are equally suitable for the growth and reproduction of living organisms, and it is, therefore, apt to consider which thermal environments are most fit (1) for living organisms. For such a study, high-temperature environments are of especial interest, in that they reveal the extremes to which evolution has been pushed. The high-temperature environments most useful for study are those associated with volcanic activity, such as hot springs, since these natural habitats have probably existed throughout most of the time in which organisms have been evolving on earth.

A recent survey of the thermal waters of the world (2) provides a broad view of their distribution. The antiquity of many individual springs is noteworthy. Historical records of springs in the Mediterranean go back to the ancient Greeks and Romans (3, 4). Chemical studies on mineral springs were initiated by Robert Boyle in the 17th century, and were greatly systematized in the 19th century by Bunsen (5), who was apparently the first to see the value of chemical studies in interpreting the origin of hot springs. The early geochemical work is reviewed by Allen and Day (6). Chemical analyses have also been made by balneologists seeking an explanation of the reputed curative properties of certain springs (7). Unfortunately for the biologist, many chemical elements of biological significance are not assayed by either the geochemist or balneolo-

gist, but the results of the analyses do show that there are many chemical types of hot springs. The *p*H of hot springs varies and values as low as 1.0 and greater than 9.0 have been recorded (8). Many, but by no means all, hot springs have significant amounts of hydrogen sulfide. The concentration of such interesting elements as fluoride, arsenic, rare earths, and gold varies very much from spring to spring. Many springs are highly radioactive (9), whereas others have no more radioactivity than normal ground waters. Some springs precipitate silica, others deposit travertine ($CaCO_3$), and still others form elemental sulfur. When one considers chemical, hydrologic, thermal, and geographical variation (10), it is clear that every hot spring can be considered as an individual, differing in minor or major ways from other springs. However, many springs are more similar than different. For instance, in the geyser basins of Yellowstone National Park, virtually all springs contain mildly alkaline waters which deposit silica (6), although even these springs show differences in that they contain varying and significant amounts of dissolved gases, including the biologically important gas methane. Some springs have been remarkably constant in thermal, chemical, and hydrologic properties for many years, although minor variations do occur (11). This constancy or steady-state condition (12) attracts the ecologist, because it eliminates many of the complications which interfere with a sophisticated analysis of ecological relationships. Further, many springs form relatively well-defined outflow channels, and, as water cools along these channels, relatively stable thermal gradients are created. In the thermal gra-

dient of a single spring, under appropriate conditions, the only significant variable may be temperature (13). We thus have excellent naturally occurring experimental conditions, and we can ask questions about the ecology and evolutionary relation of organisms at different temperatures along these thermal gradients (14).

Most of the surface of the earth has a moderate temperature, with an average of 12°C (15). Biologists have dealt, for the most part, with organisms living at these temperatures, and most of our ideas of the temperature relations of organisms have been influenced by this fact (16). Especially ideas concerning the thermal instability of enzymes and other proteins probably arise from the fact that those proteins studied have come from organisms that live always at moderately low temperatures.

Upper Temperature of Life

In antiquity, the presence of organisms in hot springs was noted. Pliny the Elder, in his Natural History, noted: "Green plants grow in the hot springs of Padua, frogs in those of Pisa, fishes at Vetulonia in Etruria near the sea" (17). The hot springs at Padua (Abano) still flow, and they are colonized by blue-green algae, perhaps the green plants of Pliny. The algae of hot springs were described by many early workers (18), but it was Ferdinand Cohn (19) who first realized the general biological significance of organisms living in hot springs: "Even simple visual observations of the different colors show that different species exist at different temperatures in the water. Such observations are not merely of passing interest, since even if most aquatic plants and animals cannot live at temperatures above 37°C . . . it is important to know the highest temperature at which organic life, *no matter how organized,* can exist." He did not find any algae growing in Karlsbad waters at temperatures above 55°C, a condition which generally agrees with Agardh's earlier observations and Löwenstein's later ones (20).

Hoppe-Seyler (21) extended Cohn's work and pointed out two basic technical problems: (i) The necessity of showing that the organisms are indeed growing at the temperature in question; and (ii) temperature readings must be taken precisely where the or-

The author is professor of microbiology at Indiana University, Bloomington, 47401.

General and Applied Microbiology   215

ganism is found, since the temperature even a few centimeters away from the organism may be quite different from that of the organism. From his observations on the hot springs of Padua (Abano), Sicily, and Ischia (Bay of Naples), he concluded that the upper temperature for algal growth was just above 60°C. He saw that his observations might have significance for an understanding of the origin of life and speculated that when the earth was cooling, chlorophyll-containing and hence CO_2-fixing and O_2-producing organisms could already live when the temperature was still over 60°C.

Before further discussion, it is necessary to examine what is meant by the question, what is the upper temperature of life? Temperature is only one of many variables influencing the growth of living organisms. It seems reasonable that environmental factors such as pH, nutrient quality or quantity, hydrostatic pressure, salinity, and light intensity (for photosynthetic organisms) will probably influence temperature responses, including the upper temperature at which an organism can grow. Heterotrophic bacteria cannot grow in hot-spring water in the absence of organic matter. We can thus imagine that the highest temperature at which some organism is now living somewhere on earth is not the upper temperature for life, but only the upper temperature at which all conditions for life are possible.

Probably the most careful early observations were those made by Setchell (22); these were apparently never published in detail. He made extensive observations in Yellowstone Park and reported that the upper temperature where algae were visible was 75° to 77°C, and that at which bacteria were found was 89°C. These upper limits occurred in siliceous waters having a slightly alkaline pH. In travertine-depositing springs, the upper temperature for algae was lower by about 10°C. In 1963, Kempner (23) re-examined this question, and cast doubt on Setchell's observations, suggesting that 73°C was the highest temperature at which organisms were found growing, but my own observations in Yellowstone confirm Setchell's conclusions. In the summer of 1966, Castenholz and I, together and separately, made a series of observations in Yellowstone. I have carried these considerably further in the summer of 1967. In the effluent channels of hot springs where the flow and temperature are usually constant, it is quite easy to determine the upper temperature for algal growth, the algal mats forming characteristic V-shaped patterns, which are due to the lower temperatures at the edge than those in the center of the channel (see Fig. 1). We found visible algal growth (of the unicellular blue-green *Synechococcus*) at temperatures up to 73° to 75°C, but not at higher temperatures (24) (Fig. 2). That these algae are growing and not merely existing can be easily shown by darkening the channel; within 5 to 7 days the algal cells have completely disappeared (25). This is due to the fact that a steady-state exists between growth and wash-out of the algal cells. In the absence of light, growth no longer can occur (thus showing that in nature these algae are obligately phototrophic), and the existing algae are quickly washed away (26).

Bacteria are present in some, but by no means all, of the hot springs at

Fig. 1. Lower Geyser Basin, Yellowstone National Park. In the foreground is a small spring showing in its effluent the characteristic V-shaped pattern of the algal mat. The center of the V is devoid of algae. The temperature at which algae are first seen is about 73°C, either at the point or the arms of the V. In the background, Great Fountain Geyser is erupting. [Photo by M. L. Brock]

Organism	Upper limit (°C)
Animals, including protozoa	45–51
Eucaryotic microorganisms (certain fungi and the alga *Cyanidium caldarium*)	56–60
Photosynthetic procaryotes (blue-green algae)	73–75
Nonphotosynthetic procaryotes (bacteria)	>90

temperatures much above 75°C. In the effluents of certain springs in the White Creek area I have found pinkish, yellowish, or whitish masses of filamentous bacteria at temperatures up to 88°C. These bacteria are present in such large amounts that they can be readily seen with the naked eye, and under the microscope are revealed as dense tangles of long filaments (Fig. 3). Spectrophotometry of acetone extracts of these organisms revealed no chlorophyll. In other springs, with temperatures of over 90°C, filamentous and rod-shaped bacteria are visible only microscopically. These organisms are probably similar to those described by Setchell as "filamentous Schizomycetes" although they are not sulfur bacteria, as he erroneously concluded, and are probably members of the Flexibacteria group (*27, 28*). Bacteria at temperatures "near 90°C" were also seen by Van Niel (*6*). Water boils at about 92°C at Yellowstone. That these bacteria are growing at these temperatures is shown by the fact that if an artificial substrate (a glass slide or piece of cotton string) is placed in the pool or effluent channel, it quickly becomes covered with bacterial cells (see Figs. 4–6). We have made a brief survey for bacterial growth in superheated pools of Yellowstone (at temperatures of 93.5° to 95.5°C) by immersing glass slides several feet down in these pools and retrieving them after a week to 10 days. In *every* pool, bacteria were present on the slides, and in over half the pools the bacteria so densely covered the slides as to form a film visible to the naked eye. In one pool, where the pink bacteria are present, string immersed in the pool (at temperatures which never go below 91°C) became covered with macroscopically visible pinkish masses.

In one spring (*Perpetual Spouter,* Norris Geyser Basin) in which bacterial development is very rapid on slides, we have examined the water itself for the presence of bacteria, by staining with acridine orange, filtering through membrane filters, and exam-

ining the filters with incident-light fluorescence microscopy. No bacteria were seen in a 100-milliliter sample. It is thus unlikely that the bacteria have merely become attached passively to the slides from the water. It is also unlikely that these organisms enter the pools from the air, as the organisms in different pools differ morphologically.

We thus see that bacteria are able to grow in Yellowstone at any temperature at which there is liquid water, even in pools which are above the boiling point. It is thus impossible to conclude that there is any "upper temperature of life."

Thermal Limits for Different Taxonomic Groups

We have already noted that nonphotosynthetic organisms grow at temperatures above those of photosynthetic ones. There are two ways in which we can obtain information on the upper temperature limits of different taxonomic groups. One is by looking at various thermal environments throughout the world, and the other is by looking at the species distribution along the thermal gradient in a single spring. In the latter case, we are dealing with organisms all living in water with the same chemical properties.

If we analyze the observations which have been made in all kinds of thermal environments (see *14, 18, 29* for references to the early literature) we can tentatively construct the scheme shown in Table 1.

In evaluating these results, the problem of habitat suitability (other than temperature) and competition with other organisms merits attention. Thus, it is surprising that eucaryotic algae are not common at temperatures above 40°C, whereas eucaryotic fungi are found up to 60°C (*14*). This difference may merely be due to the fact that the niche into which the eucaryotic algae could fit is already occupied by blue-green algae. This idea is strength-

ened by the observation that the eucaryotic alga *Cyanidium caldarium* does live at temperatures near 60°C (*30*), but is restricted to very acid hot springs (*p*H less than 4) where blue-green algae do not grow. Even among the blue-green algae, competition can be seen. For instance, the cosmopolitan thermophile *Mastigocladus laminosus* lives at higher temperatures in Iceland than in many Yellowstone hot springs probably because it does not meet competition in Iceland from *Synechococcus.* In Yellowstone, many of the high-temperature environments in which *Mastigocladus* could grow are already well colonized with *Synechococcus,* and only where water flow is very rapid, and the unicellular *Synechococcus* cannot physically maintain itself, is *Mastigocladus* present in abundance at the higher temperatures (*31*).

Temperature Optima and Environmental Optima

It does not necessarily follow that these organisms actually prefer high temperatures as the name "thermophile" implies, since they may merely be lower temperature forms which have extended their ranges into thermal habitats where competition with less-resistant species is lacking (*32*). However, there has been little work attempting to relate temperature optima of thermophilic forms to environmental temperature. Most physiological work on thermophilic microorganisms has been done with bacteria isolated from soil, compost, or other habitats whose temperatures are variable or ill-defined. Hot springs provide ideal habitats for detailed studies because their temperatures are relatively constant and because the high heat capacity of water insures that the temperature of submerged organisms is the same as that of the water in contact with them.

Blue-green algae provide the best material, as their development in Nature is not as subject to the vagaries of water composition as are the bacteria, and because one of their most fundamental physiological processes, photosynthesis, can be measured easily with isotopes. Quantitative studies of the algal mats along thermal gradients in hot springs have shown a definite correlation between the temperature and the algal biomass (*33*). In the Yellowstone hot springs, maximum algal biomass was found at about 55°C, and

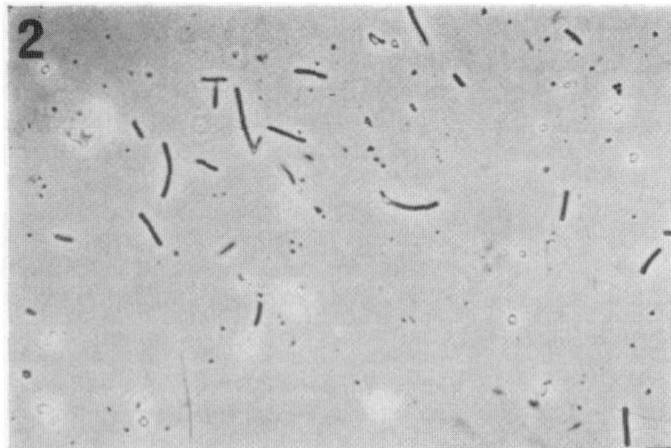

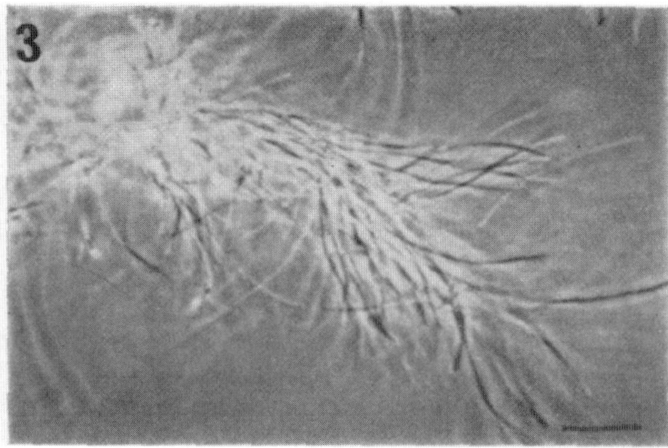

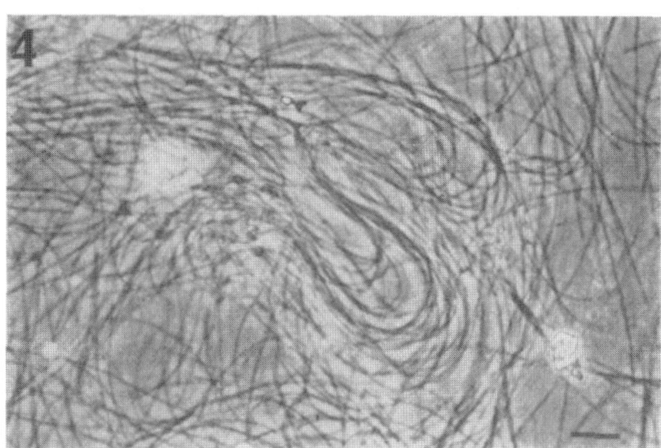

Fig. 2. Phase photomicrograph of microorganisms growing at 70° to 71°C in the outflow channel of Mushroom Spring, Yellowstone Park. The large cells are blue-green algae of the genus *Synechococcus*. There are also several filamentous bacteria visible, plus inorganic precipitate. The marker in the lower right-hand corner is 10 μ.

Fig. 3. Phase photomicrograph of filamentous bacteria taken from a large tuft in rapidly flowing water at 85° to 88°C in pool A. a large spring in the White Creek basin of Yellowstone Park. The marker is 10 μ.

Fig. 4. Phase photomicrograph of filamentous bacteria which grew on a glass slide immersed in Perpetual Spouter at about 90°C for 2 weeks. This field is representative of most areas of the slide. That these bacteria have grown on the slide is indicated by the manner in which the filaments are disposed on the slide. Detectable bacterial growth was seen on slides immersed only for a few hours. The marker is 10 μ.

it falls off sharply as the temperature increases above 55°C (*34*). At temperatures close to the upper temperature for algal growth, the biomass is exceedingly sparse. It was thus of interest to know if the algae living at these high temperatures were optimally adapted to those temperatures. The effect of temperature on the rate of photosynthesis was measured with $C^{14}O_2$ for algae living at different temperatures along the thermal gradient of a single spring (*35*). Typical temperature-rate curves, such as seen with individual species, were obtained, with well-defined optima. The optima were determined for a series of stations along the thermal gradient and then compared with the average environmental temperatures of these stations. In each case, the optimum temperature was quite close to the environmental temperature. This was true even for algae from 72°C, which are living at a temperature close to the upper limit for photosynthetic organisms. Thus, even though this temperature is

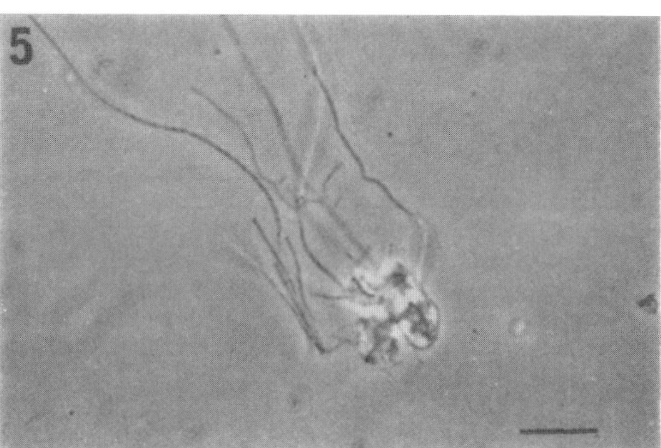

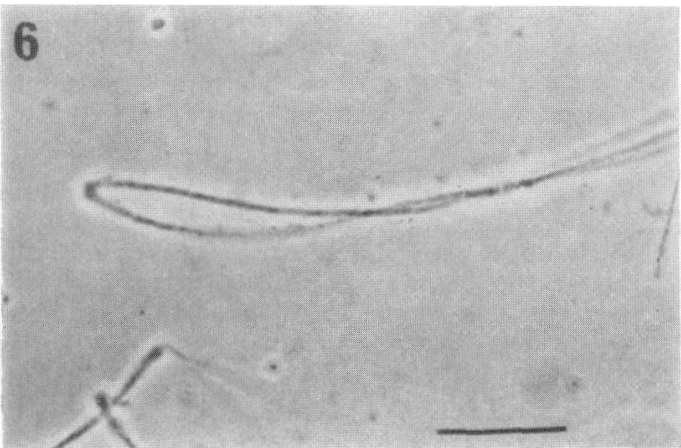

Fig. 5. Phase photomicrograph of bacteria which grew on a glass slide immersed in pool A (see Fig. 3) for 48 hours at 91.5°C. The marker is 10 μ.   Fig. 6. Phase photomicrograph of a bacterial filament which grew on a glass slide immersed in pool A (see Fig. 3) for 48 hours at 91.5°C. The twisted filament is reminiscent of forms seen in some low-temperature Flexibacteria (*27*). The marker is 10 μ.

Table 2. Generation times of bacteria at their optimal growth temperatures.

Organism	Optimum temperature (°C)	Generation time (min)	Ref.
Vibrio marinus	15	81	(54)
Pseudomonas fluorescens	25	52	(55)
Bacillus megaterium	40	22	(56)
B. subtilis	40	26	(56)
Escherichia coli	40	21	(44)
Bacillus sp.	55	16	(57)
B. thermophilus	55	16	(58)
B. stearothermophilus	60 to 65	11	(59)
B. megaterium	70	13	(56)
B. coagulans	70	14	(56)
B. circulans	70	14	(56)

so extreme that algal development is quite sparse, the algae photosynthesize better there than they do at lower temperatures. This remarkable and fundamental observation suggests that the reason photosynthetic algae are not found on earth at temperatures higher than 73° to 75°C is because of some inherent limitation in the organization of their cellular material, a limitation impossible for them to overcome by further evolutionary changes.

Similar temperature-transfer experiments have been done with the bacteria living in association with the algae in a thermal gradient from 30° to 75°C, with analogous results. Studies have not yet been done with the bacteria living at temperatures over 90°C, so that it cannot be stated that these extreme thermophiles are optimally adapted to the temperatures at which they are growing.

Molecular Mechanism of Thermophily

Thus there are microorganisms that not only survive, but actually prefer, high temperatures. How can they live at these temperatures when proteins denature at much lower temperatures? However, concepts of protein denaturation have been developed through studies on proteins derived from mesophiles. Thermophiles have enzymes which, in general, are much more resistant to conditions which cause denaturation (36–39), just as the enzymes of mesophiles are much more stable than those of psychrophiles (40). Organisms seem to have enzymes which are stable throughout the temperature range in which they grow, but not at temperatures too much higher. Indeed, we might well turn the original question around and ask why psychrophiles and mesophiles have heat-labile proteins. There is in fact no reason to expect that a protein with any particular func-

tion could not exist at any high temperature at which its covalent bonds are stable. It is possible that induced fit and allosteric interactions between proteins and small molecules require a certain flexibility of structure which is incompatible with a highly cross-linked, rigid, and hence heat-stable protein. Thermophiles may thus have sacrificed efficiency and control of enzyme function in order to grow at high temperatures. There is some evidence for this idea. Thompson *et al.* (41) have observed that when the heat-stable aldolase of *Bacillus stearothermophilus* was treated with sulfhydryl compounds, the enzyme became considerably more heat sensitive, but at the same time the enzyme became more active at 30°C, although there was little change in the activity at 65°C. A most interesting area for investigation would be the study of the kinetics of a single enzyme common to a variety of organisms with different thermal optima. Finally, there is no reason that completely unfolded (and hence denatured) proteins cannot function as enzymes, and the alpha-amylase of *B. stearothermophilus* seems to be such an enzyme (39).

The stability of the protein-synthesizing machinery is likely to be of more importance than the stability of individual proteins, since there is nothing to replace the protein-synthesizing machinery if it is destroyed. A cell is not a phoenix, able to arise out of its own ashes. There is good evidence (37, 42) that the ribosomes and the protein-synthesizing machinery (as measured in vitro with systems of artificial messengers) of a thermophile are more stable than those of a mesophile. The heat stability of soluble RNA (sRNA) and DNA would never seem to be a problem, since even sRNA and DNA of mesophiles are quite stable in the kind of ionic environment found in the cell. Finally, there is no evidence

that organisms are killed by heat because of the inactivation of proteins or other macromolecules. An analysis of thermal-death curves of various organisms shows that this is a first-order process. Thermal killing due to the inactivation of heat-sensitive enzymes or heat-sensitive ribosomes, of which there are many copies in the cell, should not result in simple first-order kinetics (38). However, first-order kinetics are compatible with an effect of heat on some large structure, such as the cell membrane, since a single hole in the membrane could result in leakage of cell constituents and subsequent death. In psychrophilic bacteria, thermal death (at temperatures around 25° to 30°C) seems to be due to damage to the cell membrane and subsequent lysis (40).

Hence the molecular mechanism of thermophily is possibly related to the stability of the larger membrane structures that are held together by the weak bonds that are likely to be broken by high temperature. The inability of eucaryotes to grow at temperatures as high as procaryotes may be due to the more complicated membrane systems of eucaryotes in their membrane-bound organelles: nuclei, mitochondria, chloroplasts. Extending this further, the reason why photosynthetic procaryotes cannot grow at temperatures as high as nonphotosynthetic ones may be because of the greater complexity of the photosynthetic membrane system. Even among the blue-green algae, the morphologically and biochemically more complicated nitrogen-fixing forms are not found at temperatures as high as non-nitrogen-fixers (31).

Are Life Processes Faster at High Temperatures?

One of the key arguments against the concept of vitalism was that living organisms obeyed the Arrhenius equation relating reaction rate to temperature. However, a point which seems to have been overlooked throughout the history of kinetic biology (43) is that studies with the Arrhenius equation have been performed on single species at temperatures usually below the optimum for growth. If organisms do evolve so that their optima parallel their environmental temperature, as our data above suggest, then the relevant question is whether thermophiles function any faster at their optimum temperatures than mesophiles do at their optima. The answer, from the limited available data, seems to be that thermophiles do not grow as rapidly as one would predict from the Arrhenius equation.

I have summarized the data on the growth rates of various bacteria at their *optimum temperatures* (Table 2). These data were obtained on organisms grown in rich media which probably permitted the maximum growth rate, and in each case the investigator attempted to provide sufficient aeration so that oxygen did not become a limiting factor. The data are plotted in the Arrhenius manner in Fig. 7; also included is the Arrhenius curve for the growth of *Escherichia coli* at various temperatures at and below its optimum (*44*). All of the points for the composite curve fall on a line which generally trends upward, but is much shallower than the curve for *E. coli*. Since Arrhenius curves for both mesophiles and thermophiles alone show not dissimilar slopes (*45*), we can assume that each individual species would give a slope similar to that of *E. coli*. However, since the slope of the collected data is much shallower, we see that thermophiles do not grow as fast at their optima as one would predict (*46*).

We can explain this result in several ways. First, growth may be less rapid than predicted because of continuous thermal destruction of sensitive molecules, so that the organism is always expending considerable energy in re-synthesis. This seems unlikely on the basis of the earlier discussion. Second, the enzymes of thermophiles, because of their increased thermal stability, may be inherently less efficient, and hence less able to take advantage of the increased reaction rates of high-temperature environments (see the earlier discussion of aldolase). Third, there may be some inherent chemical (as opposed to biochemical) process limiting rapid growth. The rate of DNA unwinding provides an upper limit for growth rate (*47*), but from the equation of Freese and Freese (*47*) relating unwinding rate to temperature and molecular weight, and from the recent information on the molecular weight of bacterial DNA (*48*), we can conclude that there is more than enough time for DNA unwinding at any temperature where organisms grow. Thus, the foregoing discussion should dispel the notion that processes are inherently fast at high temperatures. Indeed, at very high temperatures, growth rate for algae even at the optimum is apparently low

(*24*). Since the growth rate and thermal response are genetically fixed (*24*), this suggests that the organisms growing at the thermal limits have had to discard growth efficiency to survive at all.

Thermal Biology and the Origin and Evolution of Life

From the foregoing, we can see that there are definite thermal limits for the evolution of certain kinds of organisms. We also see that certain groups of organisms have evolved members that grow best at high temperature. Further, the temperature optima of organisms are quite stable genetic properties that are not subject to change by selection (*24, 36, 37, 45, 49*). Although temperature-sensitive mutants can be isolated for a variety of microorganisms, the thermal responses of these mutants differ from those of the parent by only a few degrees. To convert an organism with an optimum of 70°C to one with an optimum of 30°C, or the reverse, would probably require a large number of mutations. However, thermophilic species of blue-green algae and bacteria are related taxonomically to the mesophilic species of the same genera, and thus probably evolved from common ancestors.

It has been hypothesized (*21, 29, 50*) that the microorganisms of hot springs are relics of primordial forms of life. Such a speculation does not seem unreasonable when we consider that evidence of hot-spring activity dates back to the Precambrian, and that certain rock formations (for example, the Gunflint chert, 2×10^9 years old, *51*), which probably have been formed in hot-spring deposits (*52*) teem with fossil microorganisms which resemble the Flexibacteria so common in thermal waters today. If organic matter, macromolecules, and primordial organisms arose at high temperatures, low-temperature forms might be derived from them by mutation and selection. In this process, changes in the kinetic properties of enzymes and of the protein-synthesizing machinery could be envisioned, so that a mesophile might sacrifice thermal stability in order to acquire an enzyme of more flexible conformation which would have a higher substrate affinity. In this way, mesophiles would be able to grow and function nearly as fast as thermophiles, even though they live at lower temperatures.

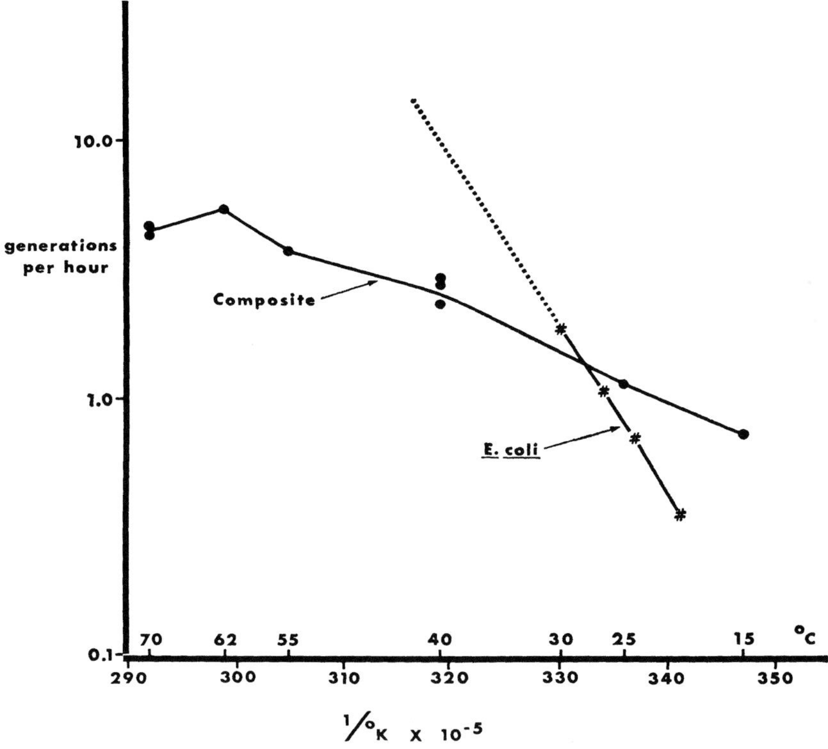

Fig. 7. The data from Table 2 graphed in the Arrhenius manner. The composite represents the growth rates of each organism at its optimum temperature. The data for *E. coli* are from reference (*44*).

Summary

The time is now ripe for a concerted attack on the evolutionary, ecological, and molecular aspects of life at high temperatures. Hot springs provide nearly ideal ecosystems for such study, since they are natural environments of great antiquity and relative constancy, where organisms have evolved to meet the environmental challenges of high temperatures. Even from our present limited knowledge, we can draw a number of conclusions.

1) The upper temperature for life as we know it has not yet been defined. At Yellowstone, some bacteria live and grow essentially at the boiling point. Since increased hydrostatic pressure may permit growth at even higher temperatures (53) there seems to be no reason why bacteria could not live in nature at any temperature where there is liquid water.

2) There is definitely an upper temperature for photosynthetic life, which seems to be at temperatures around 75°C for procaryotic algae. Eucaryotic microorganisms are restricted to lower temperatures (less than 60°C), and unicellular and multicellular animals to still lower temperatures (less than 50°C).

3) Our work in Yellowstone shows that the blue-green algae living at a given temperature are optimally adapted to that temperature, even if it is near the upper limit for algal growth. Thus, these algae are not merely lower temperature forms which have extended their range but are forms which have actually evolved so that their temperature optima resemble their environmental temperatures.

4) The molecular mechanism of thermophily is more likely to be related to the function and stability of cellular membranes than to the properties of specific macromolecules. In bacteria, where species have been studied with optima from 15° to 70°C, we may conclude from the limited data that each organism has many enzymes which are stable at the optimum temperature of the organism, but not at temperatures too much higher. This suggests that organisms do not evolve proteins which are much more stable than they need to be. Thus, there may be some advantage to the organism in synthesizing denaturable proteins. There may be a relation between protein denaturability and current ideas concerning induced fit and allosteric interactions of enzymes. The study of the kinetic properties of a single enzyme which is present in a wide variety of bacteria with differing temperature optima would provide an excellent contribution to our understanding of molecular evolution.

5) Although thermophiles grow somewhat faster at their optima than do mesophiles and psychrophiles at their optima, the increases are considerably less than would be predicted from the Arrhenius equation. This suggests that thermophiles, even though optimally adapted to their environments, are not able to make full use of their thermal environment.

6) In general, the temperature optima of organisms are not easily changed by mutation. On the other hand, thermophilic bacteria and algae are clearly related taxonomically to mesophilic species, and it seems reasonable that one group is derived from the other. One hypothesis has been that thermophilic organisms are relics, and were more widely distributed on the earth in times when it was hotter than it is today. In this respect, it is interesting that fossil microorganisms in ancient rocks, such as the Gunflint Chert (2 × 10⁹ years old) greatly resemble some of the Flexibacterales seen in hot springs today. It is not inconceivable that thermophilic microorganisms are related to primoridal forms which gave rise, through many mutations followed by selection, to mesophilic and psychrophilic forms.

References and Notes

1. L. J. Henderson, *The Fitness of the Environment* (Macmillan, New York, 1913).
2. G. A. Waring, *U.S. Geol. Survey Prof. Papers* **492** (1965).
3. C. M. Kauffmann, *The Baths of Pozzuoli* (Bruno Cassirer, Oxford, 1959); H. E. Sigerist, in *Ciba Symposia* **8**, 302 (1946); S. N. Davis and R. J. M. De Wiest, *Hydrogeology* (Wiley, New York, 1966); J. G. Frazer, *Adonis Attis Osiris* (Macmillan, London, 1914), vol. 1; Aristotle, *On Marvelous Things Heard* (Harvard Univ. Press, Cambridge, 1936).
4. Plinius Secundus, Caius, *Natural History* (Harvard Univ. Press, Cambridge, 1944).
5. R. H. Bunsen, *Liebigs Ann.* **62**, 1 (1847).
6. E. T. Allen and A. L. Day, *Carnegie Inst. Wash. Publ. No.* **466** (1935).
7. G. S. Vinaj and R. Pinali, *Le Acque Minerali* (Umberto Grioni, Milan, 1916), vols. 1 and 2; A. Perone, *Dizionario Universale Topografico Storico Fisico-Chimico Therapeutico* (Angelo Trani, Naples, 1870); H. Vogt and W. Amelung, *Einfuhrung in die Balneologie und Medizinische Klimatologie* (Springer, Berlin, ed. 2, 1952).
8. Y. Yoneda, *Mem. College Agr. Kyoto Univ. Fisheries Ser. No.* **62** (1952).
9. L. Moret, *Les Sources Thermominérales* (Masson, Paris, 1946); A. W. Klement, *Natural Environmental Radioactivity* (Atomic Energy Commission, Washington, D.C., 1965); H. Schlundt and R. B. Moore, *U.S. Geol. Survey, Bull. No.* **395** (1909).
10. T. F. W. Barth, *Carnegie Inst. Wash. Publ. No.* **587** (1950); Y. Uzumasa, *Chemical Investigation of Hot Springs in Japan* (Tsukiji Shokan, Tokyo, 1965); B. N. Thompson, L. U. Kermode, A. Ewart, *New Zealand Dept. Sci. Ind. Res. Inform. Ser.* 50 (1965); A. De Grys, *Water Resources Res.* **1**, 415 (1965); N. D. Stearns, H. T. Stearns, G. A. Waring, *U.S. Geol. Survey Water-Supply Papers* **679-B** (1935); Touring Club Italiano, *Stazioni Idrominerali* (Milan, 1955).
11. V. Munzel, *Die Thermen von Baden* (privately published, Baden, Switzerland, 1947).
12. T. D. Brock, *Bioscience* **17**, 166 (1967); ———, *Principles of Microbial Ecology* (Prentice-Hall, Englewood Cliffs, N.J., 1966).
13. Temperature is not the only variable, since many properties of water itself are affected by temperature, [E. N. Dorsey, *Properties of Ordinary Water Substance* (Reinhold, New York, 1940); M. Falk and G. S. Kell, *Science* **154**, 1013 (1966)], and the solubility of biologically important gases such as oxygen is lowered at high temperature.
14. The only other natural high-temperature environments that have been studied in detail are those associated with decomposing plant materials, such as compost [D. G. Cooney and R. Emerson, *Thermophilic Fungi* (Freeman, San Francisco, 1964)].
15. H. F. Blum, *Time's Arrow and Evolution* (Princeton Univ. Press, Princeton, N.J., 1955).
16. Air temperature ranges from —68° to +58°C [G. Gunter, *Geol. Soc. Amer. Mem.* **67**, part 1, 159 (1957)], but these extremes are rare. The temperature range of the sea is even less, from —1.5°C to 27.9°C [H. V. Sverdrup, M. W. Johnson, R. H. Fleming, *The Oceans* (Prentice-Hall, New York, 1946)] except for local heating in shallow marine bays. Dark soils in full sunlight can also be heated up to about 60°C [J. R. Schramm, *Trans. Amer. Phil. Soc.* **56**, part 1, (1966)], but these are also transitory effects.
17. Plinius Secundus, Caius, ref. *4*, p. 354. "Patavinorum aquis caldis herbae virentes innascuntur, Pisanorum ranae, ad Vetulonius in Etruria non procul a mari pisces."
18. Y. Emoto, *Bot. Mag. (Tokyo)* **47**, 268 (1933); C. T. Brues, *Quart. Rev. Biol.* **2**, 181 (1927); ———, *Proc. Amer. Acad. Arts Sci.* **63**, 139 (1928); ———, *ibid.* **67**, 185 (1932).
19. F. Cohn, *Abhandlungen Schles. Ges. Vaterl. Cultur Breslau, Abt. f. Naturwiss. Med.* **2**, 35 (1862). "Schon der Augsenschein lehrt durch die verschiedene Färbungen, dass in verschieden heissen Theilen des Wassers verschieden Arten sich vorfinden. Solche Beobachtungen haben nicht bloss ein allgemeines Interesse; denn wenn die meisten Wasserpflanzen und Wasserthiere eine Temperature von circa 30°R nicht mehr vertragen und dieselbegewissen Arten ausschliesslich überlassen, so ist es wichtig, zu wissen, bis zu welcher Temperatur überhaupt organisches Leben, wenn auch ausschliesslich dazu organisirt, existiren kann."
20. A. Löwenstein, *Ber. Deutsch. Bot. Gesell.* **21**, 317 (1903).
21. F. Hoppe-Seyler, *Arch. Ges. Physiol.* **11**, 113 (1875).
22. W. A. Setchell, *Science* **17**, 934 (1903).
23. E. S. Kempner, *ibid.* **142**, 1318 (1963).
24. R. W. Castenholz, *Symposium on the Environmental Requirements of Blue-Green Algae* (University of Washington, Seattle, in press).
25. T. D. Brock and M. L. Brock, unpublished observations.
26. I have quantitated this simple procedure in order to measure algal growth rate directly in nature. In one spring, I found a generation or doubling time of 24 hours, and in another of 12 hours, at temperatures of 70° to 72°C.
27. R. A. Lewin, private communication; S. Soriano and R. A. Lewin, *Ant. v. Leeuwenhoek* **31**, 66 (1965).
28. There is confusion concerning the taxonomic status of certain filamentous organisms living at high temperatures. Under the light microscope it is not possible to determine whether organisms around 1 μ in diameter have or do not have chlorophyll. Thus many species described from Yellowstone by Copeland [J. J. Copeland, *Ann. N.Y. Acad. Sci.* **36**, 1 (1936)], to be *Phormidium* and *Oscillatoria* are not algae. We have shown this by use of fluorescence microscopy, since chlorophyll can be detected in single filaments with much more sensitivity than by light microscopy. Further, by autoradiography we can show that these filamentous organisms do not fix CO_2 in the light (T. D. Brock, *Phycologia*, in press). This taxonomic confusion is still occurring in recent literature on Yellowstone material (J. E. Mann and H. F. Schlichting,

Trans. Amer. Microscop. Soc. **86**, 2 (1967).

29. V. Vouk, *Grundriss zu einer Balneobiologie der Thermen* (Birkhäuser, Basel, 1950).
30. W. Doemel and T. D. Brock, unpublished observations.
31. W. D. P. Stewart, *Phycologia*, in press.
32. R. Biebl, *Handb. Protoplasmaforschung* **12**, 161 (1962).
33. T. D. Brock and M. L. Brock, *Nature* **209**, 733 (1966).
34. There is also a decreased biomass at temperatures below 55°C, possibly due to the activities of animal grazers [T. D. Brock, *Ecology* **48**, 566 (1967)].
35. T. D. Brock, *Nature* **214**, 822 (1967).
36. E. Marre, in *Physiology and Biochemistry of the Algae*, R. A. Lewin, Ed. (Academic Press, New York, 1962), pp. 541–550; H. Koffler, *Bacteriol. Rev.* **21**, 227 (1957). Even though some heat labile proteins are known in thermophiles, it is not the stability in cell extracts which is relevant, but the stability within the cell. Since ions, other proteins, and particulate structures will stabilize proteins, it is not unreasonable to think that enzymes will be more stable in vivo than in vitro.
37. G. F. Saunders and L. L. Campbell. *J. Bacteriol.* **91**, 330 (1966).
38. J. L. Ingraham, in *The Bacteria*, I. C. Gunsalus and R. Y. Stanier, Eds. (Academic Press, New York, 1962), vol. 4, pp. 265-296.
39. G. B. Manning, L. L. Campbell, R. J. Foster, *J. Biol. Chem.* **236**, 2958 (1961).
40. R. Y. Morita, *Lectures on Theoretical and Applied Aspects of Modern Microbiology*, University of Maryland, 1965/66.
41. T. L. Thompson, W. E. Militzer, C. E. Georgi, *J. Bacteriol.* **76**, 337 (1958); P. J. Thompson and T. L. Thompson, *ibid.* **84**, 694 (1962). The *B. stearothermophilus* enzyme differs from other microbial aldolases

[W. J. Rutter, *Fed. Proc.* **23**, 1248 (1964)] in its molecular size and lack of metal-ion activation. Recently, H. Freeze in my laboratory has discovered a thermostable aldolase in a non-spore-forming extreme thermophile which is metal-ion activated.
42. S. M. Friedman and I. B. Weinstein, *Biochim. Biophys. Acta* **114**, 593 (1966); M. T. Mangiatini, G. Tecce, G. Toschi, A. Trentalance, *ibid.* **103**, 252 (1965); M. Arca, L. Frontali, G. Tecce, *ibid.* **108**, 326 (1965); B. Pace and L. L. Campbell, *Proc. Nat. Acad. Sci. U.S.* **57**, 1110 (1967).
43. H. Precht, J. Christophersen, H. Hensel, *Temperatur und Leben* (Springer, Berlin, 1955); J. Belehradek, *Temperature and Living Matter* (Gebrüder Bornträger, Berlin, 1935); F. H. Johnson, H. Eyring, M. J. Polissar, *The Kinetic Basis of Molecular Biology* (Wiley, New York, 1954).
44. J. L. Ingraham, *J. Bacteriol.* **76**, 75 (1958).
45. M. B. Allen, *Bacteriol. Rev.* **17**, 125 (1953).
46. I have not included in Table 1 the generation time of 9.8 minutes at 37°C of *Pseudomonas natriegens* [R. G. Eagon, *J. Bacteriol.* **83**, 736 (1962)]; if this datum were included, the lack of correlation between temperature and growth rate would be even more marked.
47. E. B. Freese and E. Freese, *Biochemistry* **2**, 707 (1963).
48. H. R. Massie and B. Zimm, *Proc. Nat. Acad. Sci. U.S.* **54**, 1636 (1965).
49. There is one exception to this statement. Mesophilic aerobic sporeforming bacilli can be adapted to grow at higher temperatures [M. B. Allen (*45*); R. M. Dowben and R. Weidenmüller, *Fed. Proc.* **26**, abstr. No. 1920, (1967)], and conversely, thermophilic bacilli can be adapted to lower temperatures. However, since these organisms are able to form a heat-resistant structure, the bacterial

spore, it is possible that this property is in some way related to their plasticity to temperature adaptation.
50. B. M. Davis, *Science* **6**, 145 (1897).
51. E. S. Barghoorn and S. A. Tyler, *ibid.* **147** 563 (1965).
52. E. S. Barghoorn, Harvard University, personal communication.
53. C. E. ZoBell, *Producers Monthly* **22**, 12 (1958).
54. R. Y. Morita and L. J. Albright, *Can. J. Microbiol.* **11**, 221 (1965).
55. R. H. Olsen and J. J. Jezeski, *J. Bacteriol.* **86**, 429 (1963).
56. L. A. Egorova, *Microbiology* **34**, 865 (1966).
57. P. A. Hansen, *Arch. Mikrobiol.* **4**, 23 (1933).
58. M. M. Mason, *J. Bacteriol.* **29**, 103 (1935).
59. N. E. Neilson, M. F. MacQuillan, J. J. R. Campbell, *Can. J. Microbiol.* **5**, 293 (1959).
60. I thank the NSF (GB-5258) and the PHS under the Research Career Development Award Program (AI-K3-18, 403). In the work in Yellowstone National Park, I received the assistance of M. L. Brock, R. Lenn, J. and S. Murphy, H. Freeze, W. and N. Doemel, and the cooperation and assistance of the National Park Service and John Good, Chief Naturalist. Winter observations in Yellowstone were made possible by the Yellowstone Field Research Expedition, Vincent Schaeffer, leader, supported by the NSF. Study in Iceland was supported by the Atomic Energy Commission through the Surtsey Research Society. Observations in Italy were made while I was a Visiting Investigator, Stazione Zoologica, Naples, through the American Tables Committee of the American Institute of Biological Sciences (NSF). Studies of the Furnas Valley, Sao Miguel, Azores, were made possible through the cooperation of the Commissão Regional de Turismo, Ponta Delgado.

222 Microbiology: A Centenary Perspective

Methanobacillus omelianskii, a Symbiotic Association of Two Species of Bacteria

M. P. BRYANT, E. A. WOLIN, M. J. WOLIN, AND R. S. WOLFE

Although the immediate effect of this landmark work was a reevaluation of the substrate range of the methane-producing Archaea, it has had a much more important impact on our understanding of the role of H_2 in anaerobic metabolism and development of the concepts of interspecies H_2 transfers and H_2 thresholds. These concepts provide the major explanatory paradigms for the interactions of anaerobes during the fermentation of complex organic compounds.

WILLIAM B. WHITMAN

Archiv für Mikrobiologie 59, 20—31 (1967)

Methanobacillus omelianskii,
a Symbiotic Association of Two Species of Bacteria*

M. P. BRYANT, E. A. WOLIN, M. J. WOLIN, and R. S. WOLFE

Departments of Dairy Science and Microbiology, University of Illinois, Urbana,
Illinois

Received April 20, 1967

* Prof. C. B. VAN NIEL has been interested in the biological formation of methane for many years. According to BARKER (1936) he was the first to suggest the "carbon dioxide reduction theory" of methane formation which has subsequently been shown to be generally applicable to biological methane formation from organic compounds other than acetate and methanol (BARKER, 1956). Much of the knowledge presently available on methanogenic bacteria has been accumulated by his students or "Scientific grandchildren". It is a great pleasure for the authors to contribute in his honor an account of the association of two bacterial species in the formation of methane from ethanol and CO_2.

Summary. Two bacterial species were isolated from cultures of *Methanobacillus omelianskii* grown on media containing ethanol as oxidizable substrate. One of these, the *S* organism, is a gram negative, motile, anaerobic rod which ferments ethanol with production of H_2 and acetate but is inhibited by inclusion of 0.5 atm of H_2 in the gas phase of the medium. The other organism is a gram variable, nonmotile, anaerobic rod which utilizes H_2 but not ethanol for growth and methane formation. The results indicate that *M. omelianskii* maintained in ethanol media is actually a symbiotic association of the two species.

Methanobacillus omelianskii Barker (1956) has been described as a strictly anaerobic bacterium which obtains its energy for growth by oxidizing ethanol to acetate. The electrons generated are utilized to reduce carbon dioxide to methane. The fermentation of ethanol is represented by equation (1).

$$2\ CH_3CH_2OH + CO_2 \rightarrow 2\ CH_3COOH + CH_4 \tag{1}$$

The organism also produces methane by the oxidation of H_2 according to equation (2).

$$4\ H_2 + CO_2 \rightarrow CH_4 + 2\ H_2O \tag{2}$$

M. omelianskii is probably one of the more abundant methanogenic bacteria in sewage sludge (HEUKELEKIAN and HEINEMANN, 1939) and has been isolated from fresh water and marine muds. Its metabolism has been studied to a greater extent than that of any other methanogenic bacterium (WOLFE *et al.*, 1966).

The only strain of *M. omelianskii* now in existence is that isolated by BARKER (1940). He isolated the culture from a primary enrichment in an anaerobic mineral-ethanol-carbonate medium. Dilutions were made in shake tubes of a similar anaerobic agar medium. Colonies were picked from these tubes to further dilution series until a pure culture was believed to have been obtained. Although three different colony types were obtained, BARKER found that two of the colony types gave rise to all three types; and the morphology of cells in all types was similar. These results suggested that the different colony types were formed by variants of a single pure strain. Other indications of purity were that growth could not be obtained in a yeast extract-glucose agar under either aerobic or anaerobic conditions or in an ethanol-ammonium nitrogen agar under aerobic conditions.

Some of us have been conducting studies since 1961 on a strain of *M. omelianskii* kindly supplied by BARKER. Microscopic observation suggested that the culture might contain two organism of slightly different shape, and in recent work we found that the organism failed to ferment ethanol after it was grown in media with H_2-CO_2 rather than ethanol-CO_2 as the energy source.

We now present evidence that *M. omelianskii* is actually a mixture of 2 distinct species. Both species are necessary to carry out the methane-forming reaction attributed to the original isolate, i. e., the conversion of ethanol and carbon dioxide to acetic acid and methane.

Materials and Methods

M. omelianskii, obtained through the courtesy of Dr. H. A. BARKER, University of California, Berkeley, was maintained by weekly transfer in tubes of anaerobic ethanol-carbonate-mineral-B vitamin medium of WOLIN *et al.* (1964). This medium will be referred to as the ethanol-carbonate medium. Periodic checks for contamination were made using Oxoid thioglycollate fluid medium, which does not support growth of *M. omelianskii*.

The anaerobic techniques for all other culture work were those of HUNGATE (1950) as modified by BRYANT and ROBINSON (1961).

Rumen fluid media were similiar to medium 98—5 of BRYANT and ROBINSON (1961) except that carbohydrates were deleted and $30^0/_0$ of rumen fluid and $0.2^0/_0$ of trypticase (Baltimore Biological Laboratory) were added. The media contained $0.4^0/_0$ Na_2CO_3 if the gas phase was $100^0/_0$ CO_2 and $0.2^0/_0$ if the gas phase was $50^0/_0$ CO_2. When ethanol was added, a $95^0/_0$ solution was equilibrated well with CO_2 to displace O_2 and $1^0/_0$ (v/v) was added to the autoclaved medium before tubing. When used as roll tubes, the medium contained $2^0/_0$ of agar (Bacto) and was dispensed in 9 ml volumes in sterile 18×150 mm rubber-stoppered culture tubes under the appropriate gas. When used to maintain cultures of methanogenic bacteria, the medium contained $1^0/_0$ agar, $0.2^0/_0$ sodium formate, and no ethanol, and was dispensed in 3.5 ml amounts in 13×100 mm rubber-stoppered culture tubes with a 1:1 CO_2-H_2 gas phase. In this case, the cysteine-sulfide reducing solution was added before the medium was tubed and the medium was solidified as slants. When used as a liquid medium, agar was eliminated and the medium was tubed in 5 ml volumes in 18×150 mm rubber-stoppered culture tubes under the appropriate gas.

Ethanol-trypticase-yeast (ETY) media were used in some experiments on colony counts and for maintenance of the nonmethanogenic bacterium (see below). These media were identical with the rumen fluid media except that rumen fluid was replaced with 0.025 M potassium phosphate, pH 7.0, 0.1% each of trypticase and yeast extract (Difco) were included, and formate was omitted. The ETY agar slant medium for maintenance of the nonmethanogenic bacterium was furnished with a CO_2 gas phase.

When liquid cultures were grown with $1:1$ CO_2-H_2 as the gas phase, the tubes were placed on a reciprocating shaker with a 3.8 cm stroke and about 160 cycles per min. Tubes were inclined at about 30° from horizontal to facilitate mixing. If a large amount of growth was desired, the tubes were flushed with the gas mixture twice each day to replenish H_2 and remove methane.

An estimate of growth in liquid cultures was obtained by determining optical density (O. D.) at 600 mμ with a Bausch and Lomb Spectronic 20 colorimeter.

The anaerobic dilution solution of Bryant and Robinson (1961) was used to dilute cultures for inoculation of media prior to preparing roll tubes. Inoculation of roll tubes was done by pipette rather than by syringe technique. All cultures were incubated at 37—40° C.

Methane and H_2 were determined by gas chromatography as described by Wolin et al. (1964). Total volume of gas in culture tubes was estimated by immersing the tubes in a vessel containing acidified 20% NaCl solution and "pouring" the gas up into a burette filled with the solution.

Results

After receipt of the strain of *M. omelianskii* from Barker in 1961, some of us (E. A. Wolin, R. S. Wolfe, and M. J. Wolin) repeated its purification with essentially the same methods as used by Barker (1940), and obtained similar results. However, in many observations by phase contrast microscopy of cultures grown in the ethanol-carbonate medium we observed cells with two slightly different shapes. The predominating shape was an irregularly curved, slender rod with bluntly rounded ends which occurred singly, in short chains and sometimes as longer filaments. The other shape was a curved rod, wider and shorter and with less bluntly rounded ends than the predominating type.

Effect of H_2 on Growth of M. omelianskii

Experiments indicated that the ability of the strain to utilize ethanol for growth or methane formation was rapidly and irreversibly lost on exposure to 0.5 atm of H_2 in the culture gas phase. The culture from ethanol-carbonate medium grew well (O. D. 0.4 in 3—4 days) and produced a large amount of methane (367 μmoles per 5 ml medium in six days) when transferred to rumen fluid-ethanol medium with a CO_2 gas phase. However, when grown under the same conditions except that two rumen fluid media, with and without ethanol, were modified to contain $1:1$ H_2-CO_2 gas phases, growth was much less (O. D. 0.1—0.12) and methane formation of 116 to 120 μmoles was about equivalent to one-fourth of the H_2 utilized as expected from equation 2 above. This was true whether or not ethanol was present and with incubation times of either 6 or 16 days. About 440 µmoles of H_2 was present in the gas phase of these cultures when inoculated and all was utilized. When cultures grown in the rumen fluid-ethanol medium with H_2-CO_2 gas for only one transfer were inoculated back into the rumen fluid-ethanol medium with CO_2 gas phase, no growth or methane formation occurred even after six days of incubation, although the same source of inoculum grew in the medium with H_2-CO_2 gas phase.

When the culture in the ethanol-carbonate medium was transferred to rumen fluid medium with or without ethanol and with a 1:1 H_2-CO_2 gas phase and the gas phase was replenished twice a day, growth was very good, the O. D. of cultures usually reaching 0.55—0.60 in three days, although at high population, growth was limited by the amount of H_2 made available. After several transfers under these conditions, microscopic observation of wet mounts revealed the presence of only a single type of cell, the slender rod. The wider and shorter rod was not present.

The effect of H_2 on cell shape and on the ability of the organism to grow on ethanol suggested that *M. omelianskii* was actually composed of two kinds of bacteria. One of these would utilize H_2 but not ethanol for methane formation and the other would oxidize ethanol, presumably with H_2 production, but would be inhibited by exposure to 0.5 atm of H_2.

Isolation of Two Species

M. omelianskii, growing in the ethanol-carbonate medium, was serially diluted in anaerobic dilution solution and inoculated into tubes of two rumen fluid-ethanol agar media. One medium contained 1:1 CO_2-H_2 gas phase and the other, 1:1 CO_2-N_2 gas phase. They were prepared as roll tubes and incubated for 23 days before colony counts and gas analyses were made (Table 1). Colony counts in the medium with CO_2-N_2 gas phase (7.4×10^5 per ml) were only 0.6% of those in the medium with CO_2-H_2 gas phase (1.2×10^8 per ml) and the colony types in the two media were quite different.

The methanogenic bacterium. There was no indication of more than one type of colony in the tubes containing the medium with CO_2-H_2 gas. Deep colonies in this medium were greyish-white, roughly round, diffuse and somewhat filamentous. They were similar to the diffuse colony photographed by BARKER (1940) but probably somewhat more dense. Surface colonies were relatively flat with diffuse to filamentous edges, the center often showing a smoother appearance. Colonies in tubes representing 10^{-6} ml of culture were 0.2—0.8 mm in diameter, and about 1.5 mm in diameter in tubes representing 10^{-7} ml of culture. Gas analysis of the

H_2-CO_2 cultures (Table 1) indicate that the colonies form methane from H_2 and not from ethanol as the methane produced accounted for about one fourth of the H_2 which disappeared even in tubes containing many colonies.

Table 1. *Comparison of colony counts and gas analyses of M. omelianskii from ethanol-carbonate medium grown in rumen fluid-ethanol agar tubes containing 1:1 CO_2-N_2 or 1:1 CO_2-H_2 as the gas phase* [1]

Inoculum (ml)	CO_2-H_2 Medium			CO_2-N_2 Medium		
	Colonies	H_2 (μmoles)	CH_4 (μmoles)	Colonies	H_2 (μmoles)	CH_4 (μmoles)
10^{-3}	Many	0	75	Many	0	319
10^{-4}	Many	0	75	48	26	0
10^{-5}	Many	0	76	10	5	0
10^{-6}	117	0	74	0	0	0
10^{-7}	12	248	12	0	0	0

[1] Colony counts are average values from three tubes and gas analyses are average values for two of these tubes. Incubation was for 23 days but colony counts were essentially the same after 14 days of incubation.

Each of four representative colonies from these tubes with H_2-CO_2 gas phase and representing 10^{-7} ml of inoculum was picked and stabbed into the base of a slant of rumen fluid agar with 1:1 H_2-CO_2 gas phase. All four cultures showed growth within eight days. Examination of wet mounts with the phase contrast microscope and of gram stains showed all to be slender, cylindrical, more or less curved rods (Fig. 1 A) which are nonmotile and gram variable. No flagella are observable in chromium shadowed electron micrographs. They resemble in morphology and gram reaction the description given by Barker (1940) for *M. omelianskii* except that neither motility nor spore formation has been observed. They are very strict anaerobes. They grow well and produce methane with H_2 as the electron donor. Neither ethanol nor formate will support growth when substituted for H_2. Three morphologically identical bacteria which utilized H_2 in CH_4 formation also were isolated in each of two similar experiments. In one of these, ETY media rather than rumen fluid media were used. In none of these experiments was there evidence of growth of colonies of any other bacterium in ethanol agar roll tubes containing H_2-CO_2 gas phase.

The S organisms. In the medium with CO_2-N_2 gas (Table 1), only one type of colony was present in the tubes inoculated with 10^{-4} or 10^{-5} ml of culture. These colonies in deep agar were buff to brownish in color, about 0.4 mm in diameter, dense and lens-shaped with somewhat irregular edges. Surface colonies were smooth, entire, slightly convex, trans-

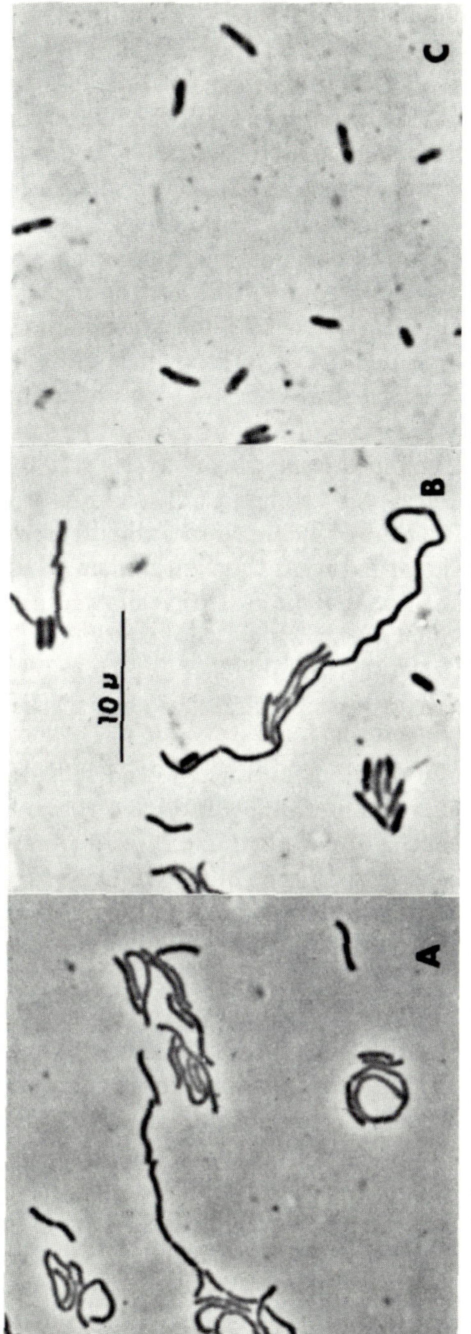

Fig. 1A—C. Photomicrographs of cultures of the two species isolated from *M. omelianskii* (phase contrast). A, the methanogenic organism grown for four days in rumen fluid medium with 1:1 H_2-CO_2 gas phase; B, recombined culture of the methanogenic organism and the *S* organism grown in an ETY agar slant for four days; C, the *S* organism grown in an ETY agar slant for four days

lucent and up to 0.8 mm in diameter. Gas analysis indicated that these colonies produced a small amount of H_2 and no methane.

Six representative colonies from these tubes with 1:1 N_2-CO_2 gas phase and with 10^{-4} or 10^{-5} ml inoculum (Table 1) were picked and stabbed into the base of ETY agar slants containing CO_2 gas phase. All grew in 4 to 10 days and microscopic examination showed that all were straight to slightly curved rods, usually about $0.7-0.8$ μ wide by $2-4$ μ long (Fig. 1 C), and with rounded ends. They were sluggishly motile and were gram negative with some internal granules which sometimes tended to be gram positive. All showed only very light growth along the line of the stab and produced a small amount of H_2 but no methane even when incubated for one month. After two to four weeks of incubation, 3 to 5 μmoles of H_2 per ml of medium was recovered from the gas phase —about $0.07-0.11$ atm H_2 pressure in the gas phase. These bacteria will be referred to as S organisms.

Twelve well-isolated colonies were picked from roll tubes of rumen fluid-ethanol agar or ETY agar containing CO_2 gas phase in two experiments otherwise similar to that of Table 1, and all cultures were morphologically and culturally identical with the S organism.

Electron micrographs of one of the S strains showed cells to have one to about three peritrichous flagella which were extremely long and without a regular helicoidal shape (unpublished information of Kingsley Langenberg). Some were at least 20 times as long as the cell to which they were attached.

Further studies on one of the S strains showed it to be an anaerobe but not nearly as strict as the methanogenic bacterium. When stabbed into ETY agar deeps with the upper portion oxidized (resazurin indicator pink rather than colorless), growth was initiated in the reduced portion of the medium after about two days of incubation. After another day, growth was seen throughout the stab line but none occurred on the surface.

The S organism did not show visible growth in the ETY agar slant medium when ethanol was delected nor when the CO_2 gas phase was replaced with 1:1 H_2-CO_2. These results are in agreement with the results of the roll tube experiments and indicate that ethanol is used as energy source and that high partial pressures of H_2 are inhibitory.

Attempts to obtain good growth of the S organism in pure culture have so far been unsuccessful; however, growth is somewhat better in ETY liquid medium if the culture is slowly sparged with N_2-CO_2 to remove H_2 as it is produced. In one experiment with 4:1 N_2-CO_2, an O. D. of 0.11 was obtained in four days. A control which was not sparged produced barely visible growth (O. D. 0.01).

Analysis of the above cultures indicates that acetate is produced by the S organism. The sparged culture produced 28 μmoles per ml and the unsparged culture, 3 μmoles per ml, as measured by the acetokinase reaction (ROSE, 1955). Quantitative analyses of fermentation products produced have been delayed in the hope that better conditions for growth of the organism can be developed.

The tubes of the N_2-CO_2 medium (Table 1) inoculated with 10^{-3} ml of culture contained a very large number of very small colonies otherwise similar to those in tubes of higher dilution or to the diffuse colonies in the H_2-CO_2 medium but also contained a number of larger smooth colonies which were yellow to orange in color. These were lens-shaped when in the agar and entire, convex and opaque as surface colonies. Diameters varied from about one to three mm. Gas analyses showed that a large amount of methane was produced in these tubes, indicating ethanol fermentation. The large colonies appeared to contain both the methanogenic bacterium and the *S* organism. When picked and stabbed into ETY agar slants with CO_2 gas phase, good growth and large gas splits occurred within five days. Microscopic examination showed both organisms to be present and gas analyses showed the presence of a large amount of CH_4 and little or no H_2.

Table 2. *The effect of combination of methanogenic bacteria with the S organism on growth and gas formation in ET Y agar slant medium with CO₂ gas phase*

Inoculum[1]	Visible Growth[2]	H_2[3] μmoles	CH_4[3] μmoles
M	—	0	0
S	+	19	0
M + S	+ + +	0	198
M R	—	0	0
M R + S	+ + +	0	195

[1] *M* is the hydrogen grown methanogenic bacterium and *S* is the ethanol oxidizing bacterium isolated from the *M. omelianskii* culture. *MR* is *Methanobacterium ruminantium*.

[2] Growth was near maximum after 4—5 days of incubation.

[3] Gas analyses are means of duplicate tubes analyzed after 20 days of incubation.

Recombination Experiments

Because neither of the bacterial species isolated from the *M. omelianskii* culture grew well or produced methane in ethanol media with CO_2 gas phase, an experiment was designed to see if they would grow well if recombined. Results shown in Table 2 and Fig. 2 show that growth and methane formation were excellent in the ETY agar when the methanogenic isolate and *S* organism were recombined. Fig. 1 B shows the appearance of cells after recombination.

Results in Table 2 and Fig. 2 also show that when *Methanobacterium ruminantium*—a species that uses only formate or H_2 as electron donor in methane formation—is combined with the *S* culture very good growth and a large amount of methane are produced in the ETY agar medium.

Analyses of DNA base ratios by density gradient centrifugation by Dr. Manley Mandel, Texas Medical Center, Houston, show that two

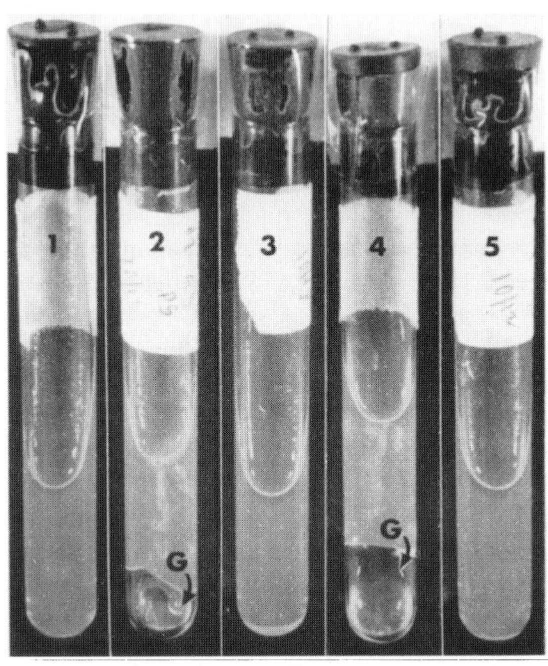

Fig. 2. Photographs of ETY agar slant cultures of the *S* organism and methanogenic species after five days of incubation in a recombination experiment. 1. *Methanobacterium ruminantium*; 2. *M. ruminantium* and *S* organism; 3. *S* organism; 4. methanogenic organism isolated from *M. omelianskii* and *S* organism; 5. methanogenic organism from *M. omelianskii*. *G* indicates gas splits. Tube 3 contained visible growth but too little to be seen in the photograph

kinds of DNA are present in *M. omelianskii* grown in ethanol-carbonate medium. Only one kind of DNA is present in the methanogenic isolate grown in rumen fluid medium with H_2-CO_2 gas phase. Details of these experiments will be published separately.

Discussion

The results of these studies show that *M. omelianskii* as usually cultured in ethanol-carbonate medium (Barker, 1940; Wolin *et al.*, 1964) consists of a symbiotic association of two species of bacteria, neither of which will grow well as pure cultures in ethanol-carbonate media even with complex sources of growth factors such as rumen fluid, trypticase and yeast extract added. One of these species, the *S* organism, oxidizes ethanol with production of H_2 and acetate. Its failure to grow well in ethanol media is at least partially explained by the fact that it is inhibited by the H_2 produced during growth. The other species, the methanogenic organism, utilizes H_2 but not ethanol as the source of electrons for growth and methane formation.

Concerning the nomenclature of these organisms, the specific name *Methanobacterium omelianskii* (BARKER, 1936, 1940) and the more recent name, *Methanobacillus omelianskii* (BARKER, 1956), do not appear to be valid since the present study shows that the description is based on two species rather than one (see rule 14a, International Code of Nomenclature of Bacteria). Further use of either of these names in reference to the methanogenic organism isolated from *M. omelianskii* would lead to considerable confusion. The specific names should be added to the lists of *nomina rejicienda* maintained by the International Committee on Nomenclature of Bacteria.

The methanogenic organism appears to belong to an as yet undescribed species. Among pure cultures so far described, it most closely resembles *Methanobacterium formicicum* (SCHNELLEN, 1947; MYLROIE and HUNGATE, 1954) but differs from this organism in colony type, in showing some definitely gram positive cells, in being somewhat larger in width and in its failure to utilize formic acid as an electron donor for methane formation. The organism will not be named until additional features can be determined and comparative studies carried out.

The *S* organism appears to be a new species of anaerobic bacterium but also will not be named at this time because of inadequate knowledge of its features. That the methanogenic bacterium utilizes H_2 but not ethanol and the *S* organism oxidizes ethanol suggests that the *S* organism also oxidizes other organic compounds shown by BARKER (1941) to be utilized by *M. omelianskii* as electron donors in methane formation. These include a number of simple primary and secondary alcohols in additional to ethanol but not acetate, propionate, n-butyrate, n-valerate, lactate, malate, succinate, glucose, glycerol, mannitol, or 2,3 butylene glycol.

BARKER (1940) detected spore-like bodies of unusually low heat resistance in *M. omelianskii*. We have not seen spores in either species but this might be due to poor conditions for spore formation.

The fact that 0.5 atm of H_2 is inhibitory to the *S* organism even though H_2 is produced during growth on ethanol is in agreement with experiments of JOHNS and BARKER (1960) on washed cells of *M. omelianskii*. It was shown that oxidation of ethanol in the absence of CO_2, i. e. reactions carried out under N_2 gas or in a vacuum, proceeded according to equation (3).

$$CH_3CH_2OH + H_2O \rightarrow CH_3COO^- + H^+ + 2\,H_2. \qquad (3)$$

However, 1 atm of H_2 pressure completely inhibited H_2 production. They calculated that oxidation of ethanol according to equation (3), at 25°C, pH 7, 1 atm H_2 pressure and 0.01 M acetate, involves an unfavorable free energy change of 1.5 kcal. However, with 0.01 atm H_2 the free energy change was —3.9 kcal and the reaction could proceed. In view of the present results, reaction 3 was undoubtedly carried out by cells of the *S* organism in the *M. omelianskii* culture. The inhibitory effect of H_2 on growth of the *S* organism is undoubtedly due to the inability of the organism to oxidize ethanol in the presence of high partial pressures of H_2. Thus, no energy source would be available for growth.

The inability of earlier workers to detect the presence of two species in cultures of *M. omelianskii* is probably due to a number of factors. The organisms are rather similar in morphology and both are highly restricted in substrates used as energy source. The methanogenic organism is probably limited to one source of electrons for growth and methane formation, i. e. H_2; and although Barker and others (Barker, 1943) showed that washed cells of *M. omelianskii* utilized H_2 for methane formation, it is probable that no attempt was made to maintain the organism on culture media with added H_2. The earlier workers used defined mineral culture media with ethanol and, sometimes, vitamins as the only organic compounds added. In studies now in progress, we have been able to maintain the methanogenic organism on a defined medium with H_2-CO_2 gas phase and including minerals, B vitamins, volatile fatty acids, cysteine and sulfide. Acetate appears to be highly stimulatory and "vitamin free" casitone is stimulatory. Thus, the use of complex culture media with and without H_2 gas were probably the more important factors involved in the detection of the two species.

Earlier work with pure or nearly pure cultures of methanogenic bacteria indicated that they are greatly restricted in substrates decomposed (Barker, 1956). Thus, only lower normal fatty acids containing from one to six carbons, the normal and iso alcohols containing from one to five carbons and the three gases, H_2, CO, and CO_2 were substrates. Essentially all of the information on normal and iso alcohols as substrates, with the exception of methanol, was based on studies of *M. omelianskii*. In view of the present results, it seems doubtful that methanogenic bacteria in nature attack alcohols other than methanol. One can speculate that fatty acids other than formate and acetate are anaerobically decomposed with H_2 production by nonmethanogenic bacteria similar to the *S* organism of the present study rather than by methanogenic bacteria. Well authenticated pure cultures of methanogenic bacteria which decompose acids such as propionate or butyrate have yet to be described.

Results to be published elsewhere show that extracts of H_2-CO_2 grown cells of the methanogenic bacterium isolated from *M. omelianskii* exhibit metabolic characteristics similar to those previously reported for extracts of *M. omelianskii* (WOLFE *et al.*, 1966) except that no ethanol dehydrogenase is detected.

Acknowledgements. We wish to thank Mr. FRANCIS ALTHAUS and Mr. KENNETH HOLMER for excellent assistance in various aspects of this study. Acetate and ethanol dehydrogenase determinations were done by Dr. ALAN E. JOYNER. The strain of *Methanobacterium ruminantium* was kindly furnished by Dr. P. H. SMITH, University of Florida, Gainesville.

The work was supported by grants from the United States Public Health Service (SW00045-02) and the United States Department of Agriculture (Hatch 35—331).

References

BARKER, H. A.: Studies upon the methane-producing bacteria. Arch. Mikrobiol. **7**, 420—438 (1936).
— Studies upon the methane fermentation. IV. The isolation and culture of *Methanobacterium omelianskii*. Antonie v. Leeuwenhoek. **6**, 201—220 (1940).
— Studies on the methane fermentation. V. Biochemical activities of *Methanobacterium omelianskii*. J. biol. Chem. **137**, 153—167 (1941).
— Studies on the methane fermentation. VI. The influence of carbon dioxide concentration on the rate of carbon dioxide reduction by molecular hydrogen. Proc. nat. Acad. Sci. (Wash.) **29**, 184—190 (1943).
BRYANT, M. P., and I. M. ROBINSON: An improved nonselective culture medium for ruminal bacteria and its use in determining diurnal variation in numbers of bacteria in the rumen. J. Dairy Sci. **44**, 1446—1456 (1961).
HEUKELEKIAN, H., and B. HEINEMANN: Studies on methane producing bacteria. I. Development of a method for enumeration. Sewage Works J. **11**, 426—435 (1939).
HUNGATE, R. E.: The anaerobic mesophilic cellulolytic bacteria. Bact. Rev. **14**, 1—49 (1950).
JOHNS, A. T., and H. A. BARKER: Methane formation; fermentation of ethanol in the absence of carbon dioxide by *Methanobacillus omelianskii*. J. Bact. **80**, 837—841 (1960).
MYLROIE, R. L., and R. E. HUNGATE: Experiments on the methane bacteria in sludge. Canad. J. Microbiol. **1**, 55—64 (1954).
ROSE, I. A.: Acetate kinase of bacteria (acetokinase), p. 591—595. In S. P. COLOWICK and N. O. KAPLAN (ed.): Methods in enzymology, Vol. 1. New York: Academic Press Inc. 1955.
SCHNELLEN, C. G. T. P.: Onderzoekingen over de methaangistung. Dissertation, Delft 1947.
WOLFE, R. S., E. A. WOLIN, M. J. WOLIN, A. M. ALLAM, and J. M. WOOD: Biochemistry of methane formation in *Methanobacillus omelianskii*, p. 162—169. In: Developments in industrial microbiology, Vol. 7. Washington, D.C.: American Institute of Biological Sciences 1966.
WOLIN, E. A., R. S. WOLFE, and M. J. WOLIN: Viologen dye inhibition of methane formation by *Methanobacillus omelianskii*. J. Bact. **87**, 993—998 (1964).

Dr. M. P. BRYANT
Dept. of Dairy Science
University of Illinois
Urbana, Illinois, U.S.A.

A Roll Tube Method for Cultivation of Strict Anaerobes

R. E. HUNGATE

The isolation and maintenance of obligately anaerobic microorganisms have been and still are difficult tasks. The removal of oxygen from a culture medium, gas phase, rubber tubing, glassware, and other equipment needed for culturing of anaerobic microorganisms requires a well-developed technique. Hungate, while studying obligately anaerobic bacteria of the rumen in 1947, developed a roll tube method with a thin agar layer under an anaerobic atmosphere. The technique was refined and improved after its original description. The paper selected here gives a good theoretical basis, as well as an outstanding description, of what is now known as the Hungate technique for cultivating anaerobic microorganisms. This method is used by any investigator of anaerobic microorganisms. With the Hungate technique and new improvements of it by other investigators, it has become possible to isolate and cultivate numerous new microorganisms including methanogens, acetogens, chemolithotrophs, photoautotrophs, extreme thermophiles, and also obligately anaerobic fungi. The work has led to much understanding of the diversity of anaerobic microbial life and of its importance in environmental settings, including the recycling of carbon and many other elements on Earth.

LARS G. LJUNGDAHL

Reprinted from J. R. Norris and D. W. Ribbon (ed.), *Methods in Microbiology*, vol. 3B, p. 117–132, copyright 1969, by permission of the publisher Academic Press.

CHAPTER IV

A Roll Tube Method for Cultivation of Strict Anaerobes

R. E. HUNGATE

Department of Bacteriology, University of California, Davis, California, U.S.A.

I. THEORETICAL CONSIDERATIONS

Categories such as micro-aerophilic, moderately anaerobic, strictly anaerobic, and extremely anaerobic lack precision in defining the degree of anaerobiosis. Oxidation–reduction potential can be a more precise index of anaerobiosis and can provide a long and continuous scale within which the range of oxygen relationships between the most aerobic and most anaerobic conditions can be compared.

Oxygen exerts its effect through chemical reaction with constituents of the cells or of the medium. Following this reaction these constituents are more oxidized and the oxygen atom is more reduced.

The degree of oxidation or reduction of a chemical system can be defined by the redox potential. Aerobes contain many systems at low redox potentials (i.e., the systems are anaerobic), but these are not subject to irreversible injury by a high potential and in the active cell are maintained at low potentials by continuous interaction with reducing systems. Some anaerobes are not killed by high potentials, but cannot grow until the medium is more reduced; others are killed, certain essential systems being irreversibly destroyed.

In nature, oxygen is the almost universal cause of high redox potential. The question has been raised, does the oxygen or the potential inhibit the anaerobes? Oxygen cannot exist apart from its potential, though similar potentials can be induced by chemicals other than oxygen. These other agents with high potentials will also affect reduced systems within the cell if they can react with them, either spontaneously or aided by enzymes. An oxidant not reacting with metabolic systems within the cell may affect an electrode inserted to measure the potential, but such a measurement provides little indication of the potential of cell systems not reacting with the electrode The present discussion refers to systems which are free to react with each other.

Oxygen oxidizes those biochemical systems in the cell which have an oxidation potential less than that of the oxygen.

The redox potential of oxygen is a function of the concentration of oxygen and its reduced form according to the equation—

$$E_{\text{oxygen}} = E_{\text{o of oxygen}} + \frac{RT}{nF} \ln \frac{\textit{conc. } O_2}{\textit{conc. reduced form of } O_2}$$

Since $n = 4$ in the reaction $O_2 \longrightarrow 2\,O^{2-}$, for each tenfold decrease in the concentration of oxygen there is a decrease of $0{\cdot}015$ volts in the oxygen potential, provided the concentration of the reduced product (assumed to be water) remains constant. The potential of the reaction $O_2 \rightarrow 2\,O^{2-}$ is $+0{\cdot}81$ V at pH $7{\cdot}0$ and at an oxygen concentration in equilibrium with one atmosphere pressure of O_2, or $0{\cdot}80$ V at the concentration in air.

Some anaerobes such as the methane-forming bacteria cannot initiate growth at potentials greater than $-0{\cdot}33$ V (Smith and Hungate, 1958). The oxygen concentration at $-0{\cdot}33$ V would be $10^{-(0{\cdot}80+0{\cdot}33)/(0{\cdot}015)} = 10^{-75}$ of the concentration of oxygen in the atmosphere.

The amount of oxygen in one litre of water at $30°C$ in equilibrium, with air at one atmosphere pressure is—

$$\frac{0{\cdot}02608 \times 1000 \times 0{\cdot}21}{22{\cdot}4} \ \text{mmoles/litre}$$

$$= 0{\cdot}245 \ \text{mmoles/litre}$$
$$= 2{\cdot}45 \times 10^{-4} \ \text{moles/litre}$$
$$= (2{\cdot}45 \times 10^{-4}) \times (6{\cdot}06 \times 10^{23}) \ \text{molecules/litre}$$

since 1 mole of oxygen contains $6{\cdot}06 \times 10^{23}$ molecules.

$$= 1{\cdot}48 \times 10^{19} \ \text{molecules/litre of water in equilibrium with}$$

air at 1 atm. pressure.

At a potential of $-0{\cdot}33$ V this becomes—

$$1{\cdot}48 \times 10^{19} \times 10^{-75} \ (\text{see above}).$$
$$= 1{\cdot}48 \times 10^{-56} \ \text{molecules/litre}.$$

This calculation strikingly illustrates: (a) that it is difficult to obtain low potentials for cultivation of strict anaerobes, (b) that the permissible concentration of oxygen in solution becomes a statistical function (10^{-55}

molecules/litre), rather than a finite number of molecules, (c) that it is impossible to obtain low potentials simply by removing oxygen, and, the corollary, (d) that to obtain the needed low potential some reduced system at a lower potential must be added. The concentration of the added reducing system should be minimal, to avoid toxicity and to simulate the concentrations in most natural environments.

Since a stable potential in a system is the expression of equilibrium or flux between all interacting oxidation–reduction systems, addition of an oxidizing agent will raise the potential, and additional reducing agent will be needed to lower it to the former level. But this leaves some of the originally reducing material in an oxidized form, and the greater the proportion of the oxidized form, the higher the potential in the system. The greater the concentration of oxidized materials in the system the greater the concentration of reducing agents needed to bring the system again to a fixed low potential. This is the reason why exposure to oxygen during preparation of media must be avoided, and the addition of reducing agent must be the terminal step in preparation, i.e., must occur just before inoculation.

Petri plates are unsuited for work with obligate anaerobes though their many advantages have led to numerous adaptations for anaerobic work. These have been successfully used with many bacteria only moderately sensitive to oxygen, but they are entirely unsuited for work with extreme anaerobes.

A roll tube method was developed in which agar medium was distributed as a thin layer over the internal surface of test tubes charged with an anaerobic atmosphere (Hungate, 1947) for the isolation of obligately anaerobic bacteria of the rumen. This technique has undergone numerous modifications and improvements (Hungate, 1966) since the original description (Hungate, 1950). Experience has shown a considerable superiority of these methods over most anaerobic methods in common use. It is the aim of this chapter to indicate some of the factors to be considered in developing stringently anaerobic techniques and to describe in detail the procedure and rationale of the roll tube method.

Since air is the primary source of oxygen, air must be excluded from cultures of delicate anaerobes. The low solubility of oxygen in aqueous solutions makes it generally more common in any gas phase than in contiguous aqueous solutions. Large gas spaces are therefore preferably avoided in strictly anaerobic procedures. A high ratio of liquid to gas volume is desirable.

In the roll tube method exposure of bacteria and culture medium to air is avoided by displacing the air in the culture vessel with an oxygen-free gas such as carbon dioxide, hydrogen, nitrogen, or mixtures of these gases. Carbon dioxide is the gas of choice because it is heavier than air, relatively cheap, and valuable in buffering. Any oxygen in the gas must be completely absorbed. Vessels are stoppered under conditions preventing access of air.

The cultures require no special incubators and can be removed and examined with no anaerobic precautions if kept stoppered. If opened, anaerobiosis can be continuously maintained during necessary manipulations, and the culture again closed without exposure to oxygen.

II. PROCEDURES

A. Avoidance of oxygen during media preparation and storage

1. *Removal of O_2 from gases*

A vertical column of heated copper (in the form of coarse copper filings) is satisfactory and economical for removing O_2 from gases. Various methods have been developed by many workers. In our experience it has been most satisfactory to pack the copper in a Pyrex glass column (Fig. 1) 25 mm inside diameter, narrowed at top and bottom to permit attachment of rubber tubing or ground glass connections. The column is heated electrically to about 350°C by a coil of nichrome wire wrapped around the column. The wattage needed will vary according to the extent of insulation. The gas stream enters at the bottom and leaves at the top of the column. The glass above the copper surface is not covered with asbestos. This permits visual inspection to determine whether the copper is reduced, i.e., with a bright copper colour rather than the dark colour of the CuO. Reduction is achieved by passing hydrogen gas through the column. To avoid an explosion, the tube must be filled with carbon dioxide or nitrogen to displace any oxygen before the hydrogen is introduced. Transition from one gas to another without entrance of any oxygen is easily accomplished by having a shortened Pasteur pipette (Fig. 1) introduced in the line prior to the copper column. The carbon dioxide or nitrogen is adjusted to flow at a slow rate and the conduit is opened by disconnecting at A as the rubber tube at B is closed by constricting it between thumb and forefinger and kept closed while the tube is disconnected. A rubber tube carrying a slow flow of hydrogen terminates also in a Pasteur pipette. This pipette is inserted in the disconnected open end of the rubber tube at A and air displaced by holding it within the open end a moment before the pipette is inserted tightly at A and the tube again opened at B. After the copper is reduced, the same procedure is used to change back to carbon dioxide without entrance of air.

The electric current in the nichrome coil is operated continuously. The expenditure of electricity and the heat evolved are modest if the insulation is sufficient.

In wiring the column, a thin asbestos sheet is first wrapped around the glass tube, extending from the base to the upper level of the copper. Plumbers' powdered asbestos is moistened and packed to the desired thickness around the asbestos sheet and nichrome coil. Each end of the nichrome wire can be attached firmly between two iron washers held between the nut and head of a $1\frac{1}{2}$ in. by $\frac{1}{8}$ in. iron bolt threaded throughout its length. The head and connections are buried in the asbestos, with the other end

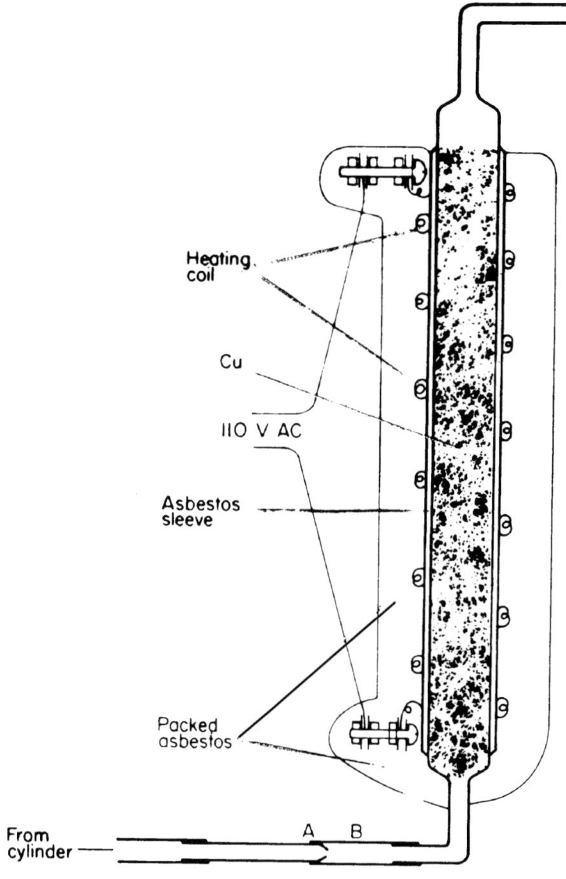

FIG. 1. Heated copper column for freeing gases from traces of oxygen. See text for explanation.

projecting at right angles from the asbestos surface. Two nuts with intervening washers are used to attach the electrical power supply leads to the external ends of the bolts, and additional asbestos can be added to cover and insulate these connections. The powdered asbestos dries and becomes sufficiently rigid to support the coil and leads. The glass column and asbestos at the base should be supported and the column also held near the top.

2. *Removal of O₂ from media and other solutions*

The easiest way to remove oxygen from a heat-stable solution is to boil it vigorously for about one minute. It can then be kept anaerobic by replacing the vapour with oxygen-free gas as the container cools. Agar to be included in the medium can be added prior to boiling.

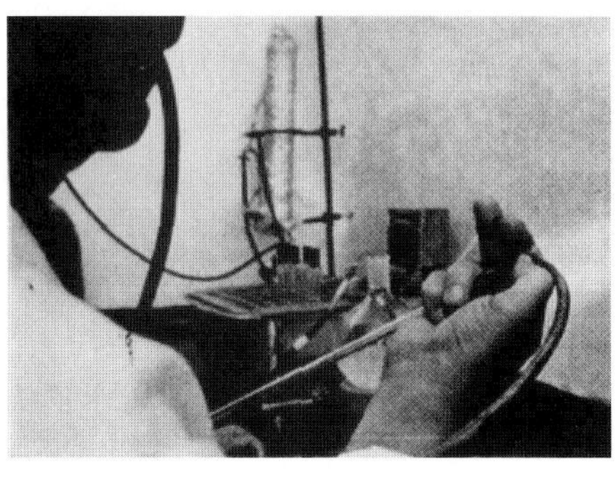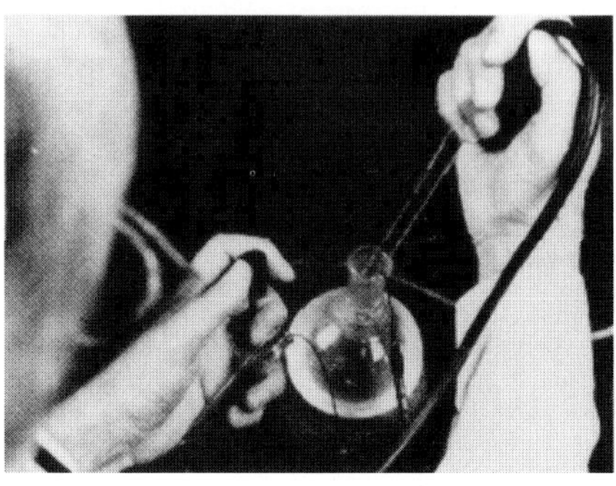

(a)   (b)

FIG. 2. (a) Method of holding pipetting tube to permit adjustment of level. (b) Method of gassing the tube during the pipetting process.

The tube delivering the gas into the vessel should extend down near to the liquid (Fig. 2b) but it is not necessary to bubble gas through the liquid if the latter is already oxygen-free. It is sufficient to exclude air.

If the solution is heat labile it can be freed of oxygen by bubbling oxygen-free gas through it but the rate of removal is slow unless the gas bubbles are very abundant. Oxygen escapes by diffusion through the gas-liquid interface, making the rate of removal a function of the interface area. *Thirty minutes to an hour* are usually required to rid a solution of most of the dissolved oxygen, but even more extended bubbling is necessary to equal the effect of boiling.

One of the best methods to ensure a low oxygen concentration in a medium is to allow sufficient growth of an aerobic micro-organism in the unsterilized stoppered tube to remove all the dissolved oxygen. The oxygen in the gas phase can also be absorbed if the tube is thoroughly shaken several times during a 30-minute incubation period prior to sterilization. This method is applicable only if the metabolic products or the bodies of the bacteria employed to remove the oxygen (which are killed later by autoclaving) do not interfere with the use of the culture.

As mentioned earlier, the concentration of dissolved oxygen is small in comparison to that in the equilibrated gas phase and in many instances the reducing agent can effectively dispose of the amounts in solution. As a precaution, however, it is much better to store ingredients such as sugar solutions, bicarbonate solutions, and other dissolved materials in closed containers from which air has been excluded. The reducing agent must always be kept under oxygen-free conditions.

3. *Dispensing anaerobic media*

It is most convenient to prepare media in large batches, transferring to the desired culture containers before or after autoclaving (Figs. 2b, 3a). Transfer prior to sterilization is more desirable because heat accelerates combination of O_2 with ingredients of the medium and removes some of the oxygen which almost always remains despite elaborate precautions.

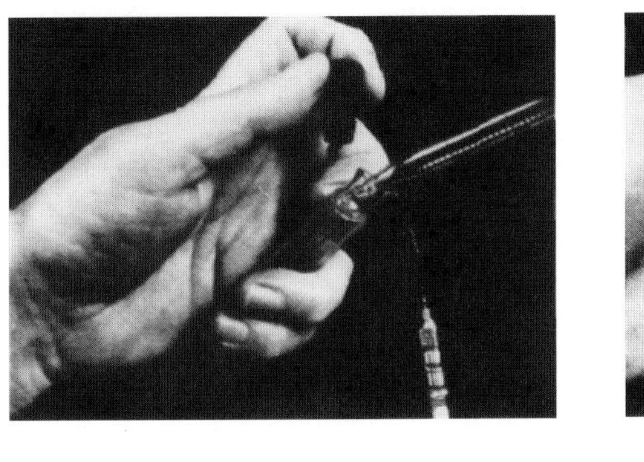

(a)   (b)

FIG. 3. (a) Disposition of tube, rubber stopper, pipette, and gassing needle during transfers with the pipette. (b) Withdrawal of the gassing needle as the rubber stopper is slipped into place.

Culture tubes are 16×150 mm, made of heat-resistant glass, with a thickened constricted neck taking a No. 00 rubber stopper (Fig. 4a).

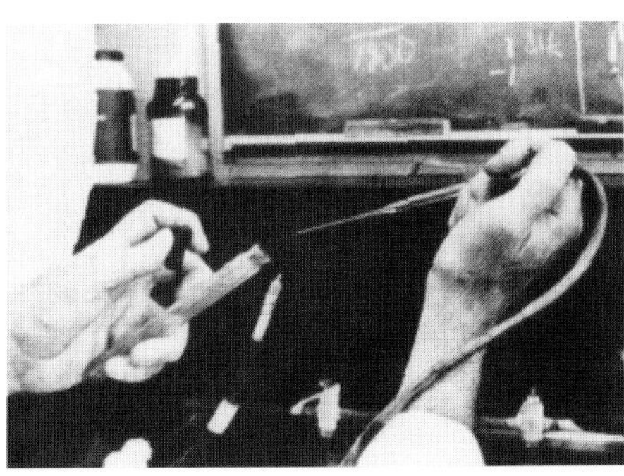
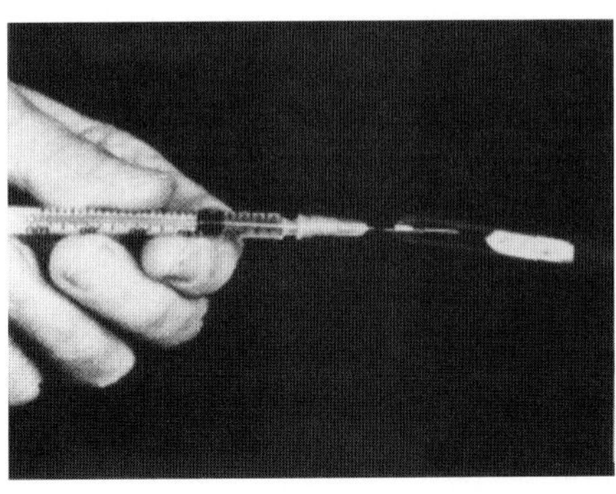

(a)   (b)

FIG. 4. (a) Method for holding tube, stopper, gassing needle, and bent capillary pipette for picking colonies. (b) Displacing the dead air space prior to use of a syringe.

An O_2-free gas stream (Fig. 2b) terminating in a sterile tube must flow into both the dispensing and the receiving vessel at a rate sufficient to exclude air. The rubber tube conducting each gas stream is attached distally to a conduit which can easily be flame sterilized and which is of a diameter small enough to permit partial insertion of the rubber-stopper closure while the conduit is still in the neck of the container. A 5-inch 20-gauge syringe needle locked to the cotton-filled barrel of a 1-ml Luer-Lok syringe serves as a suitable conduit (Fig. 3a). The flange at the open end of the barrel is cut off to allow attachment of the rubber tube. The syringe and needle are dry sterilized prior to initial use and thereafter the needle portion is flamed as necessary. Shorter needles are more convenient with shallow vessels, but it is important that the end of each needle be inserted well into the vessel to avoid a vortex at the neck which could draw in air. The distance below the open neck should be about five times the diameter of the neck.

The flow of gas should be sufficient for the purpose at hand but excess should be avoided, not only for economy but also because the gas is so often a source of oxygen. It is useful to verify the rate of gas flow either by testing in a beaker of water prior to flame sterilization, by noting its force in the flame as the needle is sterilized, or by inserting the sterile end momentarily into the liquid in the gassing container. To displace air from a 16×150 mm culture tube containing 5 to 10 ml of medium, a flow rate of about 5 ml of gas/second, continued for 5 seconds, is satisfactory. For simply excluding air a slower rate is adequate.

The medium is best transferred with a serological pipette to which a rubber mouth tube about 2 feet long is attached (Fig. 2b). Manipulation of the pipette by the mouth tube permits better exclusion of oxygen than do customary techniques. The pipette is inserted in the dispensing container, and O_2-free gas drawn through it before the liquid is sucked in. The rate at which gas is sucked into the pipette should not exceed the rate at which the gas enters the vessel. With CO_2, the taste discloses when the air has been displaced. The pipette is closed by squeezing the rubber tube between the thumb and the first joint of the forefinger (Fig. 2a). The pipette is held by the other three fingers. After a volume slightly in excess of that required has been drawn in, the quantity of solution in the closed pipette can be easily adjusted to the desired mark by changing the angle between thumb and forefinger. With practice, a culture tube, a gassing needle and the pipette can all be simultaneously and appropriately manipulated (Figs. 3a, b).

4. *Closing the vessel* (Fig. 3b)

This is done by inserting the stopper with the thumb and finger of one hand, holding the tube with the other three fingers, with the gassing needle still in place. The stopper should be loose enough to let the gassing needle be withdrawn with the other hand without moving the stopper. If the stopper is inserted too tightly, as the needle is pulled out the pressure against the needle may break the glass of culture tubes without a thickened neck. The stopper and vessel neck are held firmly with one hand so that friction of the needle as it is withdrawn does not change the position of the stopper in the neck. Gassing is continued for a moment, then the needle is quickly pulled out, as light but firm pressure with the other hand pushes the stopper into place, closing the tube. Both hands can then be used to seat the stopper more firmly. The stopper seats readily if a little of the medium is tipped up to moisten the inner end and a twisting motion is used to force the stopper into its final position with the basal solid portion just inside the narrowest part of the neck of the tube. The twisting motion gives less tube breakage and if a break occurs the fingers are less likely to be cut.

The removal of the needle should be rapid, as the end comes by the stopper. A slow removal can inject air into the vessel. Similarly, in opening an anaerobic vessel, the needle should be rapidly plunged to the optimal depth as soon as the stopper is removed. The stopper may first be loosened with both hands but not opened. Then it is pulled out with one hand as the gassing needle is slipped in with the other.

Butyl rubber stoppers are used to close the vessels (Hungate *et al.*, 1966). Butyl rubber is quite impermeable to oxygen and the permeability

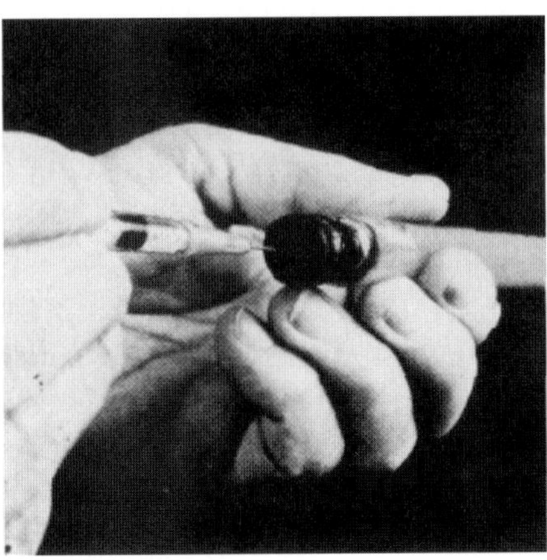
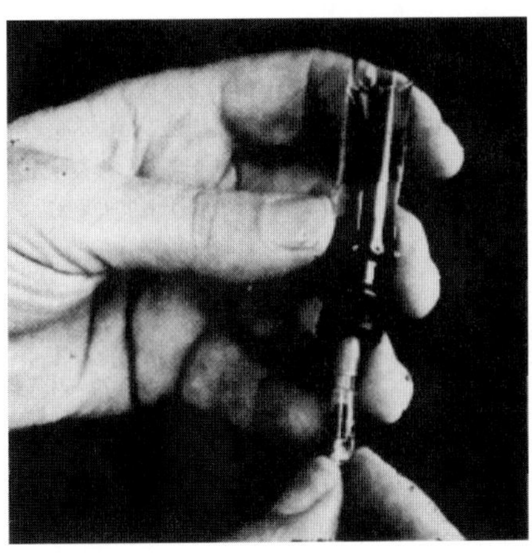

(a)   (b)

FIG. 5. (a) Inserting the needle into the recess in the rubber stopper. (b) Removal of sample from anaerobic tube into anaerobic syringe.

to other gases is less than with most types of rubber. Recessed (Fig. 5a) butyl rubber stoppers, No. 00, can be obtained from Arthur H. Thomas

Co., Philadelphia, U.S.A. Solid butyl-rubber stoppers can be obtained at reasonable price on special order from most manufacturers.

When asepsis is critical, it is well to place a cap of aluminium or other material over the top of culture tubes prior to sterilization, and to maintain it in place except when the tube is opened.

B. The nature of the culture medium

The composition of the medium is determined as for any other cultures, with the exception that the buffering capacity must be increased to take care of acid fermentation products, and reducing agents must be included. It should be clear that in designing the culture medium simulation of the natural habitat is often essential for success if organisms of unknown nutritional requirements are to be isolated directly from nature (Hungate, 1963). Ingredients of the medium can be added before heat sterilization except for reducing agents and heat-labile materials.

Inorganic salts are usually provided to balance the high concentration of sodium bicarbonate in the buffer. In practice it has been convenient to store a three-times concentrated balanced inorganic salts solution in two separate flasks; one, the A solution containing the acid phosphate plus the other salts—

Solution A

NaCl	6 g
KH_2PO_4	3 g
$(NH_4)_2SO_4$	3 g
$MgSO_4$	0·6 g
$CaCl_2$	0·6 g
Water	1 litre

Solution B contains 3 g K_2HPO_4 per litre. One part of A and one part of B together compose one third of the culture medium.

Carbon dioxide–bicarbonate is a more common buffer in nature than is phosphate. Concentrations up to one per cent are not usually toxic. Unless other buffers are abundant the pH of the medium is determined by the relative concentrations of carbon dioxide and bicarbonate. A solution of 0·5% sodium bicarbonate, equilibrated with an atmosphere of carbon dioxide, has a pH of about 6·7. Other desired acidities can be obtained by using carbon dioxide and bicarbonate concentrations calculated from the Henderson-Hasselbach equation—

$$pH = 6·52 + \log \frac{\text{molar conc. bicarb.}}{\text{molar conc. dissolved } CO_2}$$

Sodium bicarbonate in a 10% aqueous solution is sterilized by filtration because heating converts it to sodium carbonate, H_2O, and CO_2.

This raises the pressure within the culture medium container, increasing the risk of stoppers blowing out. A period of equilibration after sterilization is necessary to allow reconversion of the CO_2, H_2O and carbonate to bicarbonate. Conversion of bicarbonate to carbonate also increases the alkalinity of the medium and the extent to which salts of phosphate and sulphate precipitate. If no bicarbonate is added prior to autoclaving, and the vessels are filled with carbon dioxide, the salts do not precipitate.

In our laboratory, hydrogen (Mylroie and Hungate, 1954), sodium thioglycolate, cysteine, sodium sulphide, hydrogen sulphide and dithionite have been used as reducing agents. The hydrogen is used with a platinum catalyst. The sodium thioglycolate and cysteine can be heat sterilized and stored under carbon dioxide or other gases. A 1% solution of dithionite is sterile. The sodium sulphide must be stored under hydrogen or nitrogen since it is very alkaline, and carbon dioxide is rapidly absorbed, creating a vacuum which facilitates the entrance of air. Acidification to pH 6·7 converts sodium sulphide to hydrogen sulphide. The E_0' at pH 7 is $-0\cdot571$ for the reaction $H_2S \rightleftharpoons S + 2H^+ + 2e$. Sterile hydrogen sulphide gas can be drawn off in a syringe from a Kipp generator and is perhaps the best and most convenient reducing agent for stringent anaerobes, though in most cases also cysteine is needed. Final concentrations of reducing agents can be 0·02 to 0·05%; with higher concentrations inhibitory effects may be encountered, particularly with dithionite. For routine use we have employed chiefly a combination of 0·03% (w/v) cysteine hydrochloride and 0·25 ml of H_2S gas per 10 ml of medium.

The bicarbonate and the reducing solutions are added to the medium just before it is inoculated. Heat-labile nutrients can also be added just before inoculation.

The initial volume of the medium before autoclaving should be diminished by the volume of reagents to be added later. A final volume of 4·5 ml per tube containing 1·2% agar gives a usable thickness of agar after the tube is rolled, and is convenient for serial dilutions.

A low concentration (0·0001%) of resazurin is included in the culture medium to indicate oxidation–reduction potentials above $-0\cdot042$. Others (Bryant, 1963) have used phenosafranine, with a redox potential (E_0', pH 7·0 $=$ 0·252) better suited for registering low potentials. Benzyl viologen (E_0', pH 7·0 $= -0\cdot359$) has an appropriate potential but is toxic at the concentrations needed to give a detectable colour.

C. Sterilization

1. *Steam pressure sterilization*

Large batches of O_2-free medium can be sterilized in any container strong enough to withstand the pressure within the vessel as the autoclave cools. For quantities up to 5 litres, round-bottom boiling flasks can be used. The stopper must be securely wired in place. In many cases large quantities of medium can also be sterilized in open containers which are removed from the autoclave and gassed as soon as removal is safe. It is easier to maintain anaerobiosis in large rather than in small liquid volumes.

During sterilization, culture tubes containing medium can be held in a vertical rack with a lid which must not be removed until the tubes have cooled to a safe temperature. The test tube racks of the Labtool Specialties Co. of Ypsilanti, Michigan, U.S.A., can be used.

Stoppered tubes containing sterile medium can be stored for long periods without oxidation, but they may gradually oxidize due to slight permeability or leaks in the seal. Gradual diffusion of CO_2 or H_2 out of the container lowers the internal pressure and increases the chance of penetration of air.

2. *Filtration*

Filtration can be accomplished anaerobically by using filters designed for application of pressure rather than of vacuum. Air is first eliminated by passing O_2-free gas in either direction through the filter assembly and then gassing both inlet and outlet orifices during transfer of the medium to the filter. If the O_2-free gas source is under sufficient pressure it can be used to supply the pressure for filtration, but it is necessary to be sure that the gas conduit between the gas cylinder and the filter will withstand the pressure to be applied.

D. Obtaining pure cultures

1. *Inoculation*

The roll tube method cannot be used quantitatively for surface inoculation because of the vertical surface during incubation. It can be used to obtain growth of surface colonies by inoculating a liquid suspension of cells and rotating the tube to spread them over the surface. The excess liquid and cells drain to the base of the tube and do not affect the growth of cells left above on the agar surface, provided the tube is never tipped too far toward a horizontal position.

The roll tube method is best suited for inoculation just prior to solidification of the melted medium. Most of the colonies develop in the agar but as they increase in size many reach the gas interface and continue development as surface colonies. They may also reach the wall of the culture tube and grow as a film between agar and glass. The appearance of the colonies in each of these locations differs markedly from the appearance in the other locations and may give the inexperienced worker a false impression of impurity. Careful examination under a binocular dissecting microscope can reveal the exact location of colonies and assist the interpretation of their form.

2. Picking colonies

Colonies can be picked by touching with an inoculating needle and transferring the adhering cells quickly to the subculture or other preparation. This transfer exposes the cells to oxygen and since they are in a very thin layer on the needle, sensitive cells may be killed.

It is better to transfer some of the agar along with the cells. This may shield them and increase the viability. A platinum–iridium inoculating needle can be flattened at the end to form a microspade or spatula. The wire can be bent several mm from the flattened end to an angle which allows the flattened end to be pushed under a colony to lift it out. A little agar block containing the colony should first be cut out with the sharp edges of the microspade. A steady hand is needed during the cutting and also as the block is removed. If the freed block touches the agar or glass after it has been taken up it usually adheres and comes off the microspade. The microspade technique is useful and economical for routine transfers of pure colonies and others which are separated from contaminants by several mm.

Transfer with a Pasteur pipette is better for difficult isolations and for protecting the inoculum from air during transfer. The capillary of a sterile cotton-plugged Pasteur pipette is drawn out to a finer capillary at a distance from the main shank which will permit reaching the colony when the pipette is inserted into the tube (Fig. 4a). The diameter of the finer capillary should be about equal to that of the colony to be picked. This capillary is cut off about two inches from the main capillary shank and is bent at a right angle by momentarily placing it horizontally in a very small flame (the flame of a match is enough). With a piece of broken porcelain the capillary is scratched about 5 mm distal from the bend and is cleanly broken at the scratch. It is lightly flamed, i.e., enough to sterilize but not enough to melt the tip, and the plugged end attached to the rubber mouth tube. The capillary portion of the pipette is inserted into the culture tube alongside the gassing needle, and some O_2-free gas is sucked in. The tip is then placed on the agar surface over the colony to be picked, and forced down into the agar as gentle mouth suction is applied. The colony is drawn into the capillary together with a little agar and some liquid that squeezes out of the agar when it is broken.

The picked colony is rapidly introduced into the subculture tube and a little molten agar (cooled to 45°C) drawn in. This, with the colony, is then

extruded onto the dry wall of the culture tube and is drawn in and out of
the capillary tip to disperse the cells, but no gas is blown out of the capillary. A few gas bubbles do little damage, too many can introduce some
oxygen.

Additional molten agar is drawn up and ejected onto the inoculum on the
wall in order to avoid solidification. The pipette is withdrawn, the stopper
is held in place and inserted as the gassing needle is withdrawn, and the
tube is tilted back and forth a few times to mix the inoculum. Too rapid
tilting forms gas bubbles which may persist during solidification of the
agar and confuse the identification of colonies.

The inoculated agar is solidified in cold water either in a mechanical
tube roller (such as that supplied by Astell Laboratory Service Co.,
Catford, England) or by hand. Ice water is preferable for obtaining a firm
agar, and the speed of the mechanical roller holds the agar in an even film
until it is hard. If rolling is done by hand it is necessary to stop the motion
just before solidification occurs, since otherwise the forming gel is broken
and the resulting solidified agar is weak and ragged.

During incubation, water of syneresis drains down from within the
thin agar on the vertical walls of the tube. It does not usually disturb the
surface colonies but the accumulation of liquid in the base of the tube
makes it necessary to avoid spreading the bacteria in the basal liquid by
undue tilting of the tube during examination and handling.

3. *Dilution*

The concentration of colonies in the tube receiving the inoculum is
usually not low enough to permit picking of an isolated colony. It is necessary to inoculate several dilutions of the material. This dilution can be
directly in the tubes of molten agar at a temperature of about 45°C, or
special dilution medium can be made up and used for dilution prior to
inoculation into the culture tubes. Such dilution medium, if used, should
be prepared with all the anaerobic precautions described for the culture
media, and small volumes (0·1 ml) should be used for inoculation in order
to maintain agar concentrations in the culture tubes.

Once the inoculum has been placed in the first tube, further dilutions
are advantageously made by means of a hypodermic syringe. To permit
this, we use recessed No. 00 butyl rubber stoppers (Catalogue No. 8820-B
of the Arthur H. Thomas Co., Philadelphia, U.S.A.) or ordinary No. 00
butyl rubber stoppers in which a recess about 3 mm in diameter has been
bored with a high speed drill in the centre of the top, extending about
two-thirds of the length of the stopper. The 21-gauge needle easily pene-
trates the reduced thickness of rubber, and if locked to the syringe it does
not pull off as the needle is withdrawn. In non-quantitative transfers to

agar tubes, the dead space in the end of the syringe can be transferred and the same syringe used for all the dilutions. Usually four tubes serially inoculated in this way will give sufficiently well isolated colonies in one of the dilutions.

The syringe method is very rapid and convenient also for transfers of broth cultures (Fig. 5b), using the relatively inexpensive plastic syringes now available. Just before the syringe is used, the air in the dead space should be eliminated by inserting the needle in a stream of O_2-free gas (Fig. 4b) issuing from the orifice of a sterile Pasteur pipette with most of the capillary portion removed.

4. *Criteria of purity*

As with almost all aerobic bacteria growing on the surface of agar, colonies of different anaerobic bacterial species almost always differ in size, texture, colour, transparency, margin, shape, consistency, or other characteristics. Close examination under the dissecting microscope can usually detect contaminants. Keen observation is needed to avoid conclusions of impurity based on differences due to location of the colony in the agar. In the roll tube the proportion of colonies on the surface or between agar and glass is usually greater than in the similarly inoculated Petri dish because the agar layer is thinner.

The history of a culture is the best index of its purity. If all cultures derived from a single colony exhibit only colonies of one type, the number of colonies in the tubes of a dilution series decrease in accordance with the dilution, and there is no microscopic evidence of impurity, then the culture can be assumed to be pure.

When speed of isolation is critical, as in clinical examination of specimens, a single colony picked from a high dilution is often pure and can be subjected immediately to further tests, but its isolation should be continued to ensure that it actually was pure. With methanogenic, some cellulolytic, and some other anaerobic bacteria difficult to grow, it is as well to satisfy the above criteria of purity rigorously at an early stage of the investigation.

E. Storage of cultures

This has not been studied systematically and extensively for anaerobes grown with the roll-tube technique. The problem of preservation is no different from that of other culture methods, except that the stoppered tubes can be stored almost indefinitely without drying out or becoming oxidized. The same advantages apply also to storage of sterile uninoculated tubes of media.

Many strains of anaerobic bacteria remain viable for months if the culture is frozen quickly and held at the temperature of dry ice.

F. Applications of anaerobic methods to various media

Most media can be used to study anaerobes with little modification from aerobic methods. Reducing agents and some carbon dioxide and bicarbonate should usually be added. Some media, if prepared and inoculated with care for anaerobiosis, may be sufficiently reduced to start growth without adding reducing agents, particularly if inoculated heavily, but if nutritional requirements are under study it is well to subculture enough times to ensure that growth is not due to materials transferred with the inoculum. Many anaerobes require carbon dioxide. When acid production is to be measured it may be desirable to reduce the concentrations of carbon dioxide and sodium bicarbonate.

G. Other applications

The roll tube method may readily be adapted for the cultivation of micro-organisms requiring special gas atmospheres, such as hydrogen/oxygen/carbon dioxide mixtures or methane/air mixtures.

REFERENCES

Bryant, M. P. (1963). *J. Anim. Sci.*, **22**, 801–813.
Hungate, R. E. (1947). *J. Bact.*, **55**, 631–645.
Hungate, R. E. (1950). *Bact. Rev.*, **14**, 1–49.
Hungate, R. E. (1963). *In* "The Bacteria" (Ed. Gunsalus, J. C., and Stanier, R. Y.), Vol. IV, Chap. 3, pp. 95–119. Academic Press, New York.
Hungate, R. E. (1966). *In* "The Rumen and Its Microbes", pp. 23–30. Academic Press, New York.
Hungate, R. E. (1966). *J. Bact.*, **91**, 908–909.
Mylroie, R. L., and Hungate, R. E. (1954). *Can J. Microbiol.*, **1**, 55–64.
Smith, P. H., and Hungate, R. E. (1958). *J. Bact.*, **75**, 713–718.

Structure and Methylation of Coenzyme M (HSCH₂CH₂SO₃)

$(HSCH_2CH_2SO_3)$

C. D. TAYLOR AND R. S. WOLFE

The first description of the structure of a novel coenzyme from a methanogen, this work reinvigorated the search for new coenzymes in procaryotes. Since then, the structures of five more novel coenzymes have been described in methanogens. These include the deazaflavin coenzyme F420, the nickel tetrapyrrole coenzyme F430, tetrahydromethanopterin, methanofuran, and 7-mercaptoheptanoyl threonine phosphate (component B). Two of these, coenzyme F420 and tetrahydromethanopterin, were subsequently shown to be more widely distributed among the procaryotes. Similarly, novel coenzymes have been discovered in methanotrophic and other bacteria.

WILLIAM B. WHITMAN

Reprinted with permission from *Journal of Biological Chemistry* 249:4879–4885. Copyright © 1974. American Society of Biological Chemistry.

THE JOURNAL OF BIOLOGICAL CHEMISTRY
Vol. 249, No. 15, Issue of August 10, pp. 4879–4885, 1974
Printed in U.S.A.

Structure and Methylation of Coenzyme M ($HSCH_2CH_2SO_3$)*

(Received for publication, December 17, 1973)

CRAIG D. TAYLOR‡ AND RALPH S. WOLFE

From the Department of Microbiology, University of Illinois, Urbana, Illinois 61801

SUMMARY

Coenzyme M is a recently discovered cofactor which is involved in methyl transfer reactions in *Methanobacterium*. Information derived from infrared, proton NMR, and ultraviolet spectroscopy as well as from chemical tests and quantitative elemental analysis reveals that the coenzyme is 2,2'-dithiodiethanesulfonic acid. Verification of this structure resides in the comparison of authentic with chemically synthesized 2,2'-dithiodiethanesulfonic acid. Evidence indicates that an active form of this cofactor is 2-mercaptoethanesulfonic acid which is methylated producing 2-(methylthio)ethanesulfonic acid; this derivative is subsequently reductively demethylated, yielding methane.

Recent studies of methane formation have been devoted to methyltransfer reactions in *Methanobacterium* (1–3) and *Methanosarcina* (4–6). When cell extracts of *Methanobacterium* strain M. o. H. were subjected to anaerobic dialysis, they were resolved for a heat-stable organic cofactor, coenzyme M (CoM), which was required for methane formation from methylcobalamin (1). Two enzymic reactions were defined:

$$^{14}CH_3\text{-}B_{12} + CoM^1 \xrightarrow[\text{methyltransferase}]{\substack{\text{anaerobic conditions} \\ \text{electron donor}}} {}^{14}CH_2\text{-}CoM + B_{12r} \quad (1)$$

$$^{14}CH_3\text{-}CoM \xrightarrow[\text{methylreductase}]{\substack{\text{anaerobic conditions} \\ H_2,\ ATP,\ Mg^{2+}}} {}^{14}CH_4 + CoM \quad (2)$$

Reaction 2, catalyzed by methylreductase, was specifically inhibited by tripolyphosphate and measurement of the radioactivity trapped in [*methyl*-¹⁴C]CoM served as an assay for CoM (1). Subsequently, sodium borohydride was found to be an excellent electron donor for Reaction 1, and use of 100-fold purified methyltransferase eliminated methyl reductase activity (7). We present here the purification, structural identification,

and chemical synthesis of three biologically active forms of CoM. A preliminary report has appeared (8).

EXPERIMENTAL PROCEDURE

Materials—Methylcobalamin was chemically synthesized by the method of Müller and Müller (9); salts were removed by phenol extraction (10). The preparation was eluted from a water-equilibrated column of SP-Sephadex C-25 (ammonium form) with a 0 to 0.1 M linear, ammonium acetate gradient. Ammonium acetate was removed by lyophilization. [*methyl*-¹⁴C]Methylcobalamin was prepared as described by Wood *et al.* (11) and the preparation was purified by cation exchange chromatography as outlined above. The concentration of each preparation was determined spectrophotometrically (12). Methyl iodide and sodium 2-bromoethanesulfonate were purchased from Aldrich Chemical Co.; [¹⁴C]methyl iodide from Amersham-Searle; cyanocobalamin and sodium borohydride from Sigma Chemical Co.; deuterium oxide (99.8% isotopic purity) from Diaprep, Inc.; and sodium 3-trimethylsilyl-propionate-2,2,3,3-d₄ (TSP) from Merck, Sharp, and Dohme of Canada, Ltd. Methylmercaptan and argon were obtained from Union Carbide Corp.; QAE (quaternary aminoethyl)-Sephadex A-25, SP-Sephadex C-25, and Sephadex G-10 from Pharmacia Fine Chemicals, Inc.; and AG 50W-X4 (200 to 400 mesh) cation exchange resin (H⁺ form) from Bio-Rad Laboratories. *Methanobacterium* strain M. o. H. was mass cultured and cells were stored as previously described (1, 7).

Synthesis of Ammonium 2-Mercaptoethanesulfonate (HS-CoM)—Sodium 2-bromoethanesulfonate monohydrate (4.1 g) was dissolved in a 40-mole excess of concentrated ammonium hydroxide, and the solution was rapidly saturated with hydrogen sulfide gas. The mixture was stirred for 3 hours with a slow stream of hydrogen sulfide bubbling through the mixture. Precautions were taken to prevent entrance of air into the reaction mixture. The following steps were performed as quickly as possible to prevent autooxidation of the thiol. The ammonium hydroxide and volatile ammonium sulfide were removed by flash evaporation. The solid residue was dissolved in water and applied to a QAE-Sephadex A-25 (acetate form) column (5 × 20 cm) that had been equilibrated with water. HS-CoM was separated from the bromide salts by elution with a linear 0 to 3 M ammonium acetate gradient. The fractions which contained the thiol were located by reacting small samples with nitrous acid; HS-CoM was detected by the formation of the red *S*-nitroso derivative. The desired fractions were pooled and flash-evaporated to remove ammonium acetate. HS-CoM which may have oxidized to the corresponding disulfide was removed by precipitation with aqueous acetone. HS-CoM remained in solution. The acetone solution was removed by flash evaporation. HS-CoM was crystallized from methanol upon addition of diethylether and was recrystallized two times from the same solvent mixture. A 30% yield was obtained. Data from quantitative elemental analysis are provided in Table I.

Synthesis of Ammonium 2,2'-Dithiodiethanesulfonate—HS-CoM (0.5 g) was dissolved in 30 ml of 30% aqueous ammonium hydroxide, and the solution was bubbled with oxygen until the thiol could no longer be detected with nitrous acid. (S-CoM)₂ was crystallized from water upon addition of acetone. The compound was recrystallized two times from the same solvents. A 20% yield

* This work was supported by National Science Foundation Grants GB 27681 and GB 37826X.

‡ Predoctoral trainee, Public Health Service Training Grant GM-510. Present address, Woods Hole Oceanographic Institution, Woods Hole, Massachusetts 02543.

¹ The abbreviations used are: CoM, coenzyme M or HSCH₂CH₂SO₃; HS-CoM, 2-mercaptoethanesulfonic acid; (S-CoM)₂, 2,2'-dithiodiethanesulfonic acid; CH₃-S-CoM, 2-(methylthio)ethanesulfonic acid; (CH₃)₂-S-CoM, 2-(dimethylsulfonium)ethanesulfonate or 3-methyl-3-thioniabutane-1-sulfonate.

TABLE I

Per cent elemental composition of CoM derivatives

Samples were dried at 80–90° over phosphorous pentoxide.

Element	Authentic (S-CoM)$_2$ (NH$_4$$^+$ salt)		Synthetic (S-CoM)$_2$ (NH$_4$$^+$ salt)		Synthetic (CH$_3$-S-CoM) (NH$_4$$^+$ salt)		Synthetic HS-CoM (NH$_4$$^+$ salt)		Synthetic (CH$_3$)$_2$-S-CoM (internal salt)	
	Calculated	Found	Calculated	Found	Calculated	Found	Calculated	Found	Calculated	Found
Carbon	15.18	15.58	15.18	15.64	20.80	20.94	15.09	15.32	28.22	29.98
Hydrogen	5.10	5.09	5.10	4.98	6.40	6.25	5.70	5.68	5.92	5.90
Nitrogen	8.85	8.80	8.85	9.00	8.08	8.06	8.80	8.48	0.0	0.0
Sulfur	40.53	40.79	40.53	40.64	37.01	36.89	40.27	40.15	37.67	36.81

was obtained. Data from quantitative elemental analysis are presented in Table I. For certain studies the ammonium ions of synthetic (S-CoM)$_2$ were replaced by sodium ions by passage of an aqueous solution of the coenzyme through an SP-Sephadex C-25 column (sodium form) that had been equilibrated with water. Sodium (S-CoM)$_2$ was crystallized twice from aqueous acetone.

Synthesis of Ammonium 2-(Methylthio)ethanesulfonate—The procedure for the synthesis of CH$_3$-S-CoM was identical with the synthesis of HS-CoM except that methylmercaptan was used instead of hydrogen sulfide. CH$_3$-S-CoM was recrystallized three times from aqueous acetone. A 20% yield was obtained. Results of quantitative elemental analysis are shown in Table I.

A large scale enzymic methylation of HS-CoM was performed in a 125-ml double sidearm Erlenmeyer flask. The reaction mixture (25 ml) contained 1.25 mmoles of potassium phosphate buffer at pH 7.1, 154 μmoles of HS-CoM, 31 μmoles of [*methyl*-^{14}C]methylcobalamin (specific activity, 1020 cpm per nmole), 415 μmoles of unlabeled methylcobalamin, and 100-fold purified methyltransferase (1.6 mg of protein). The reaction flask was made anaerobic, and the reaction was initiated by tipping in [*methyl*-^{14}C]methylcobalamin from one of the sidearms. The reaction was allowed to proceed until all of the [*methyl*-^{14}C]methylcobalamin had reacted. The excess nonradioactive methylcobalamin was tipped in from the other sidearm, and the reaction was allowed to proceed until all of the HS-CoM was methylated. The reaction mixture was passed through a water-equilibrated Bio-Rad AG 50W-X4 (H$^+$ form) column (2.5 × 10 cm) to remove methylcobalamin and aquocobalamin. The radioactive effluent was concentrated by flash evaporation and eluted with water through a Sephadex G-10 column (2.5 × 90 cm). Ninety-eight per cent of the recovered radioactivity was contained within a single peak. The material from this peak was applied to a column (1.2 × 20 cm) of QAE-Sephadex A-25 (acetate form) that had been equilibrated with water; methylated CoM was eluted with a linear (0 to 1.5 M) ammonium acetate gradient. Ammonium acetate was removed by flash evaporation. The residue was dissolved in 4 ml of water, and methylated CoM was separated in a continuous electrophoresis apparatus as described below. The ammonium acetate buffer was removed, and the derivative was recrystallized twice from aqueous acetone.

Synthesis of 2-(Dimethylsulfonium)ethanesulfonate—2-(Methylthio)ethanesulfonate (ammonium salt) was dissolved in methanol at a concentration of 300 μmoles per ml. A 30-fold molar excess of methyl iodide was added, and the mixture was allowed to stand in the dark at room temperature overnight. (CH$_3$)$_2$-S-CoM crystallized as the internal salt from the reaction mixture. The crystalline residue was twice recrystallized from aqueous solution upon addition of methanol. A 73% yield was obtained. Results of quantitative elemental analysis are shown in Table I. 2-[*methyl*-^{14}C]Methylsulfonium ethanesulfonate was prepared by reacting 1.1 μmoles of ^{14}CH$_3$I (60 mCi per mmole) with 520 μmoles of CH$_3$-S-CoM for 30 min prior to addition of CH$_3$I.

Quantitative Assay of (S-CoM)$_2$ and HS-CoM—The enzymic assay for (S-CoM)$_2$ or HS-CoM was a modification of the method of McBride as described by Taylor and Wolfe (7). The upper limit of the assay was 300 nmoles for HS-CoM and 150 nmoles for (S-CoM)$_2$. A 100-fold purified preparation of methylcobalamin-CoM methyltransferase was employed and sodium borohydride was used as the electron donor when (S-CoM)$_2$ was assayed. For assay of column eluates, the procedure was the same as described above except that the reaction was initiated by injection of a freshly prepared anaerobic solution of [*methyl*-^{14}C]methylcobalamin and sodium borohydride to the reaction mixture which contained enzyme as well as (S-CoM)$_2$ (0.05 or 0.1 ml of the column eluate). Ammonium acetate did not interfere with the assay. The reaction was terminated after 20 to 30 min of incubation by addition of the cation exchange resin directly to the entire reaction mixture. The supernatant solution (0.5 ml) was counted for radioactivity.

Assay for Methane—A small scale modification of the method of McBride and Wolfe (1) was performed in a test tube (13 × 55 mm) which contained 12.5 μmoles of potassium phosphate buffer at pH 7.1, 1.2 μmoles of magnesium sulfate, 1.5 μmoles of (S-CoM)$_2$, 1.3 μmoles of ATP, 3.7 μmoles of methylcobalamin, dialyzed crude extract (2 to 4 mg protein), and water. When CH$_3$-S-CoM (3 μmoles) was used, methylcobalamin and (S-CoM)$_2$ were omitted. The volume of the reaction mixture was 0.25 ml. All constituents except enzyme were added to the tube which then was sealed with a serum stopper. A 24-gauge inlet hypodermic needle and a 22-gauge exit needle were inserted in the serum cap, and the tube was flushed for 10 min at a rate of 10 cc per min with hydrogen which had been scrubbed of oxygen by passage over heated copper wire. The gassing needles were removed, and the reaction was initiated by injecting enzyme. The enzyme preparation was stored prior to use in a separate tube under a hydrogen atmosphere. Each gas sample (100 μl) was withdrawn by syringe and analyzed for methane by gas chromatography as described previously (1).

Instruments—A Cary model 14 spectrophotometer was used for absorption spectra in the ultraviolet and visible ranges. A Beckman model DU spectrophotometer was used to follow the absorption of fractions at 260 and 400 nm. A Perkin-Elmer 521 infrared spectrometer, Varian HA-100 and A-60 NMR spectrometers, a Nuclear-Chicago Mark I liquid scintillation system, a Desaga-Brinkmann TLE system, and a Packard gas chromatograph were used for appropriate analyses.

RESULTS

Purification of CoM—Whole cells were suspended in distilled water (1:1, w/v) and stirred for 30 min at 80–90°. The cell residue from this extraction was subjected to a second extraction by the same procedure. The supernatant solution was lyophilized, ground into a fine powder, and stored in a sealed bottle at room temperature. CoM was extracted by stirring the powder in anhydrous methanol (1:10, w/v) for 30 to 40 min. The insoluble residue was removed by filtration. Two additional extractions were performed by this procedure. The clear red-orange supernatant solution was flash-evaporated to yield a red-brown oil-like residue. This fraction, obtained from the extraction of 1.5 to 2 kg of wet cells, was taken up in 100 ml of water and applied to a QAE-Sephadex A-25 column (5 × 30 cm) (acetate form) that had been equilibrated with water. The sample was eluted with 3 liters of a 0 to 3 M ammonium acetate linear gradient (Fig. 1). CoM activity was always located at two positions, *Peak A* and *Peak B*. *Peak B* co-eluted with coenzyme F$_{420}$, a fluorescent electron carrier found in high amounts

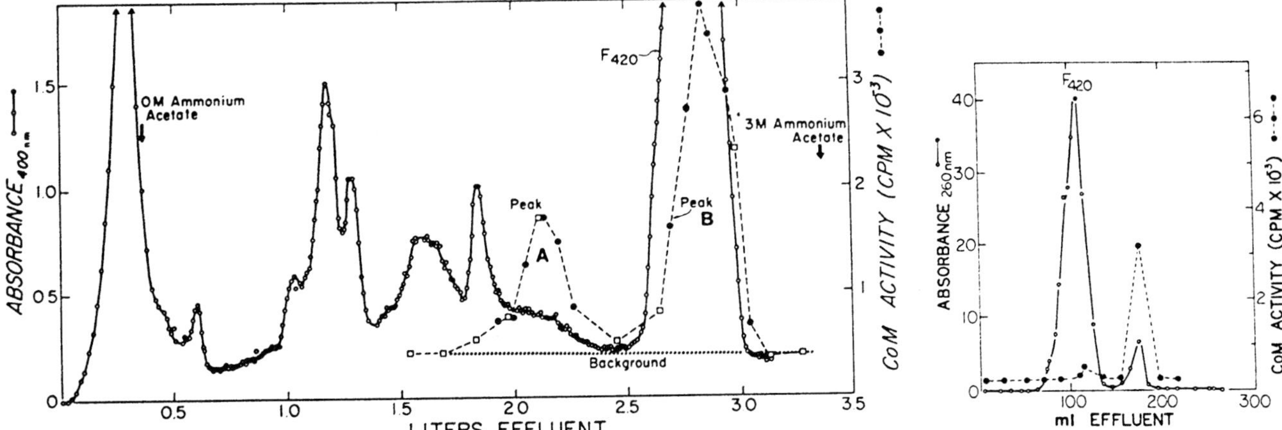

FIG. 1 (*left*). The elution profile of (S-CoM)₂ on QAE-Sephadex A-50. The elution profile was determined by directly measuring the absorbance of the fraction (13 ml) in each tube (18 × 150 nm) at 400 nm. (S-CoM)₂ activity was determined by assaying a 0.05-ml sample from each tube in a reaction mixture (0.25 ml) which contained 12.5 μmoles of potassium phosphate buffer at pH 7.1, 2.5 μmoles of sodium tripolyphosphate, 550 nmoles of sodium borohydride, 127 nmoles of [*methyl*-¹⁴C]methylcobalamin (specific activity, 4470 cpm per nmole), and 100-fold purified methylcobalamin-CoM methyltransferase (85 μg of protein). Reaction time, 15 min.

FIG. 2 (*right*). The elution profile from a Sephadex G-10 separation of the components of *Peak B*. The elution profile was determined by measuring the absorbance at 260 nm with a Beckman model DU spectrophotometer. Rectangular quartz cells with a 1-cm light path were used. (S-CoM)₂ activity was assayed as described in Fig. 1.

in *Methanobacterium* (13). From preparation to preparation, the per cent of the total activity found in *Peak A* varied in a reciprocal manner to that found in *Peak B*. Both derivatives possessed a strong negative charge, requiring 1.8 and 2.5 M ammonium acetate to elute *Peaks A* and *B*, respectively, from the column. Since *Peak B* contained from 70 to 88% of the total activity, this fraction was used for further purification. Ammonium acetate was removed by flash evaporation at 40°. The CoM-containing eluate from the equivalent of 4 kg of wet cells was dissolved in 8 to 10 ml of water and fractionated on a Sephadex G-10 column (2.5 × 90 cm) that had been equilibrated with water. The elution fluid was water. As shown in Fig. 2, CoM eluted in a single peak and was separated from much contaminating material that absorbed at 260 nm, the bulk of which was coenzyme F₄₂₀. The pooled active fractions from the G-10 eluate were reduced to 10 ml by flash evaporation. The sample was applied to a continuous electrophoresis apparatus, and separation was effected on a paper curtain at 400 volts (8 to 10 ma) in a pH 4, 0.02 M ammonium acetate buffer. CoM migrated 10 to 13 cm toward the positive pole while descending 7 cm and was separated from fluorescent compounds. The active fractions were flash-evaporated at 40° to remove the ammonium acetate buffer. The solid residue was taken up in a small amount of water and recrystallized three times from aqueous acetone, yielding colorless platelets with a melting point greater than 250°.

The recovery of CoM from this purification scheme was 65 mg of crystalline CoM per kg of wet packed cells. Accurate determination of the recovery from this scheme was not possible as inhibitors of the methyltransferase reaction were present in both the initial, lyophilized water extract and the methanol-soluble fraction. For example, the total activity found in *Peak B* of the anion exchange eluate (Table I) increased nearly 3-fold over that observed in the methanol-soluble residue.

Structure of CoM—Evidence obtained from infrared, proton NMR, and ultraviolet spectroscopy as well as from chemical tests and quantitative elemental analyses revealed that CoM as isolated is the ammonium salt of 2,2′-dithiodiethanesulfonic acid (S-CoM)₂. This compound possesses the following structure:

$$^+H_4N^-O_2SCH_2CH_2SSCH_2CH_2SO_3^-NH_4^+$$

Fig. 3 shows that the infrared spectra of authentic and synthetic (S-CoM)₂ are identical. Each spectrum is largely dominated by the absorptions of the sulfonate and ammonium ions. The strong absorptions at *Peaks d* and *e* (1170 and 1035 cm⁻¹) represent, respectively, the asymmetric and symmetric SO₃ stretching modes characteristic of sulfonic acid salts. The shoulders on the *left* of *Peak d* are probably the wag motion of the methylene groups attached to sulfur. The ammonium ion stretch and deformation modes appear at *a* and *c*, respectively (3200, 3060, and 1435 cm⁻¹); these absorptions disappear revealing the methylene stretch (2970 and 2930 cm⁻¹) and deformation (1420 cm⁻¹) modes when the ammonium ion is replaced by sodium ion as shown in the *inset* of Fig. 3. The fact that untreated aqueous solutions of both authentic and synthetic (S-CoM)₂ gave a positive reaction with Nessler's reagent provides additional chemical evidence for the presence of the ammonium ion. Evidence suggests that the peaks located at *g* (590 and 530 cm⁻¹) may involve the sulfonate group; Palmer (14), working with potassium dithionate, assigned the sharp peaks observed at 577 and 516 cm⁻¹ to be fundamental absorptions of the SO₃ deformation modes.

The absorptions observed with a variety of CoM derivatives which possess an intact sulfonate group as well as with sodium ethanesulfonate suggest that the peaks located at *f* and *g* involve the sulfonate moiety. All absorptions in this region as well as in the SO₃ stretch region were lost in the spectrum of ammonium 3,3′-dithiodipropionate where the sulfonate group is replaced by a carboxyl group. Data provided by Bellamy (15) and Colthup *et al.* (16) also were used to aid in peak assignments.

The proton NMR spectra of deuterium oxide solutions of authentic and synthetic (S-CoM)₂ are identical in all respects as shown in Fig. 4. There are no additional resonances downfield

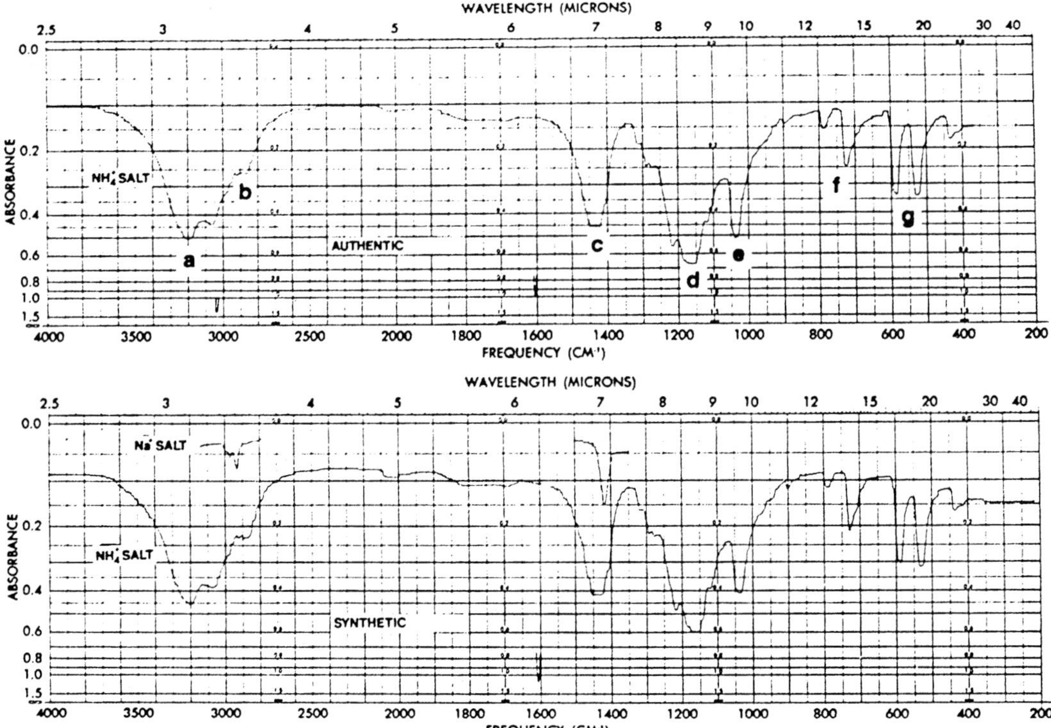

FIG. 3. Infrared spectra of authentic and synthetic (S-CoM)₂. One milligram of crystalline (S-CoM)₂ (ammonium salt) was mixed in 200 mg of anhydrous spectral grade potassium bromide. The mixture was heated over phosphorous pentoxide at 90° for 1 hour. Infrared spectra were obtained with a Perkin-Elmer 521 infrared spectrometer. Calibration peaks are located at 3027 and 1601 cm⁻¹. Pertinent absorption peaks of the sodium salt of (S-CoM)₂ are shown in the *inset*.

of 5 ppm. The observed symmetrical resonances at 3.28 and 3.11 ppm are of a 4-proton system of the type A₂B₂ characteristic of an aliphatic compound of the configuration XCH₂CH₂Y. The possibility of having an AA'BB' system of the configuration of X(CH₂CH₂)₂Y is eliminated on the basis of quantitative elemental analysis, and particularly the synthesis of (S-CoM)₂. The assignments made were based on observing the chemical shifts of the ethylene resonances of a variety of aliphatic sulfonate, sulfide, and disulfide compounds (17, 18). The presence of both a sulfonate and disulfide group in (S-CoM)₂ displaces the resonances of both methylene groups approximately 0.4 ppm downfield relative to the resonances observed in aliphatic compounds which contain only one of the functional groups of interest. Resonances of methylene groups next to the sulfonate functional group were in general located downfield of those methylene groups next to sulfide or disulfide moieties.

Ultraviolet spectra of authentic and synthetic (S-CoM)₂ exhibit absorption maxima at 191 to 193 and 245 nm. The extinction coefficients which were determined for the short wavelength band for both preparations ranged between 5800 and 6400 liters mole⁻¹ cm⁻¹. Because these measurements were made at the extreme lower end of the useful wavelength range of the Cary model 14 instrument, attention should be drawn only to the order of magnitude. The extinction coefficient at 245 nm, however, may be determined with accuracy and for synthetic (S-CoM)₂ was found to be 380 liters mole⁻¹ cm⁻¹. The absorption maxima and extinction coefficients for (S-CoM)₂ are close to those obtained for straight chain aliphatic disulfides. For example, diethyldisulfide possesses absorption maxima at 194 and 250 nm with extinction coefficients of 5500 and 380 liters

mole⁻¹ cm⁻¹, respectively (19). The elemental compositions of authentic and synthetic (S-CoM)₂ are shown in Table I; neither differs from that calculated by greater than 0.46%.

Authentic and synthetic (S-CoM)₂ were identical in their biological behavior. When 1.5 μmoles of synthetic (S-CoM)₂ were added to a standard reaction mixture which contained 4.9 μmoles of [*methyl*-¹⁴C]methylcobalamin and 100-fold purified methylcobalamin-CoM methyltransferase (8 μg of protein), a perfectly linear reaction rate was produced, with 400 nmoles of ¹⁴CH₃-S-CoM being produced in 10 min. The reaction rate of authentic (S-CoM)₂ was identical. In a separate experiment where the reaction was allowed to proceed to completion, the reaction mixture contained 52 μg of 100-fold purified methyltransferase and 2.5 μmoles of [*methyl*-¹⁴C]methylcobalamin. Two levels of synthetic (S-CoM)₂, 152 and 304 nmoles, and two levels of authentic (S-CoM)₂, 148 and 296 nmoles, were tested. For each mole of (S-CoM)₂ added 1.98 ± 0.10 moles of methyl groups were bound. The ratio remained within the above described limits when either 100-fold purified methyltransferase was used with sodium borohydride as the electron donor or when crude extracts were used with sodium borohydride or hydrogen as the electron source.

Reduction of (S-CoM)₂—The function of the electron requirement in the methyltransferase reaction was found to be that of reducing the disulfide bond of (S-CoM)₂ prior to methylation. HS-CoM functions as a methyl acceptor in the absence of an electron donor as shown by the *open* and *closed triangles* in Fig. 5. Comparison of the *open* and *closed circles* shows that methylation of (S-CoM)₂ requires an electron donor. There was a negligible chemical methylation of HS-CoM· from methyl-

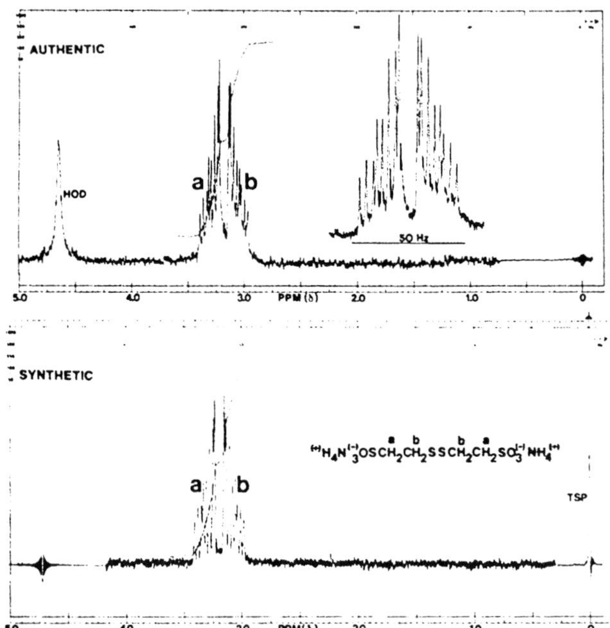

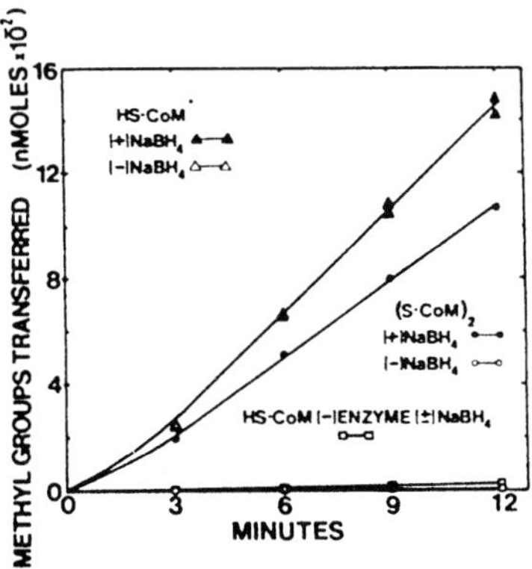

FIG. 5. Comparison of the reductive requirement of the methyl-cobalamin-CoM methyltransferase reaction when HS-CoM or (S-CoM)₂ was used as the methyl acceptor. The reaction mixtures (0.25 ml) contained 12.5 μmoles of potassium phosphate buffer at pH 7.1; where indicated, 1.50 μmoles of (S-CoM)₂ or 3.01 μmoles of HS-CoM; where indicated, 550 nmoles of sodium borohydride; 3.7 μmoles of [methyl-¹⁴C]methylcobalamin (specific activity, 1730 cpm per nmole); and 100-fold purified methylreductase (13 μg of protein).

FIG. 4. NMR spectra of authentic and synthetic (S-CoM)₂. The NMR spectra of (S-CoM)₂ in deuterium oxide (99.8% isotopic purity) were obtained with a Varian HA-100 NMR spectrometer. A NMR tube (2-mm inside diameter × 5-mm outside diameter × 17.7 mm) was used. The spectra were obtained under the following conditions (where the conditions are different, the value for authentic (S-CoM)₂ precedes that for synthetic (S-CoM)₂: concentration, 9 mg per 0.07 ml and 9 mg per 0.06 ml; temperature, 28°; frequency response, 2 and 20 Hz; radio frequency attenuator, 20 db; sweep time, 500 s; sweep width, 500 Hz (inset, 250 Hz); sweep offset, 0 and 472 Hz; spectrum amplitude, 10,000; lock signal, sodium 3-trimethylsilyl-propionate-2,2,3,3-d₄ (TSP) and hydrogendeuterium oxide (HOD); field milligauss (manual oscillator frequency), 0.1 mG; field milligauss (sweep frequency), 0.04 and 0.1 mG; reference compound, TSP.

cobalamin. HS-CoM was stoichiometrically methylated when the methyltransferase-catalyzed reaction was allowed to proceed to completion.

2-(Methylthio)ethanesulfonic Acid—The results presented above indicate that two methyl groups are bound for each mole of (S-CoM)₂. The methyl derivative of the coenzyme retained a strong negative charge, suggesting that methyl-CoM could be 2-(methylthio)ethanesulfonic acid.

Results presented in Fig. 6 show that the methylated coenzyme which was isolated from a large scale reaction mixture (see "Experimental Procedure") possesses an NMR spectrum identical with that of ammonium 2-(methylthio)ethanesulfonate which was chemically synthesized. The resonances of protons *a* and *b* appear, respectively, at 3.15 ppm (relative intensity, 1.8) and 2.88 ppm (relative intensity, 2.0). The large singlet labeled *c* at 2.17 ppm (relative intensity, 3.1) is typical of the proton resonances of a methyl group attached to sulfide. The small singlet at 2.24 ppm resulted from a contaminant in the solvent used for crystallization. The presence of the methyl group shifts resonances *a* and *b* upfield, with that of *b* being shifted to a greater extent. This lends additional support to the proton assignments made. No evidence was obtained to support the possibility that the methylated derivative of CoM is the methylsulfonate ester. Such a CoM derivative would be a neutral molecule and possess physical properties considerably

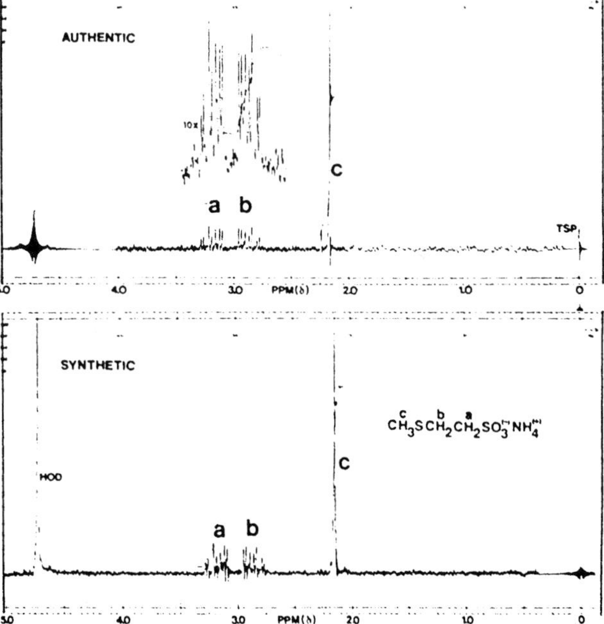

FIG. 6. Comparison of the NMR spectra of authentic and synthetic CH₃-S-CoM. The NMR spectra of CH₃-S-CoM in deuterium oxide (99.8% isotopic purity) were obtained with a Varian HA-100 NMR spectrometer. A NMR tube (2-mm inside diameter × 5-mm outside diameter × 17.7 mm) was used. The spectra were obtained under the following conditions (where the conditions are different, the value for the enzymically synthesized, authentic CH₃-S-CoM precedes that for synthetic CH₃-S-CoM): concentration, 3 mg per 0.05 ml and 10 mg per 0.07 ml; temperature, 28°; frequency response 1 and 5 Hz; radio frequency attenuator, 20 db; sweep time, 1000s and 500s, sweep width, 500 Hz; sweep offset, −473 and 0 Hz; spectrum amplitude, 10,000 (inset, 10 × gain), and 6,000; lock signal, sodium 3-trimethylsilyl-propionate-2,2,3,3-d₄ (TSP) and hydrogen deuterium oxide (HOD); field milligauss (manual oscillator frequency), 0.1 and 0.05 mG; field milligauss (sweep frequency), 0.05 and 0.1 mG; reference, compound, TSP.

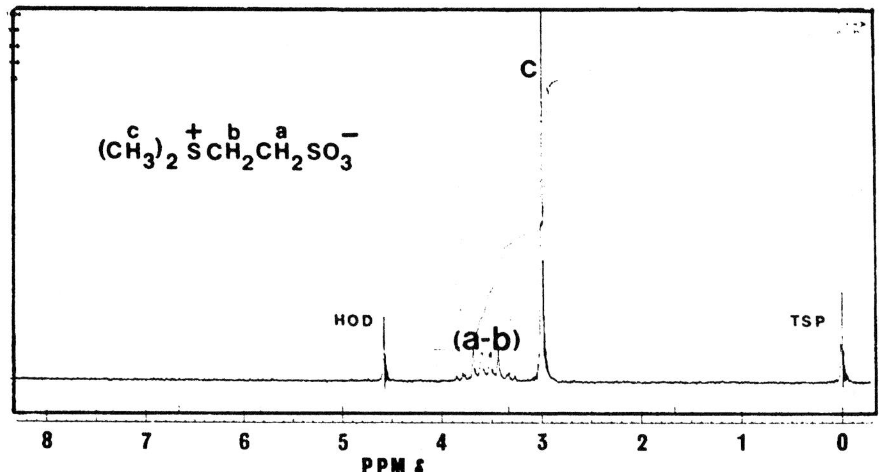

FIG. 7. NMR spectrum of (CH₃)₂-S-CoM. The spectrum was obtained in deuterium oxide (99.8% isotopic purity) with a Varian A-60 NMR spectrometer. An NMR sample tube (2-mm inside diameter × 5-mm outside diameter × 17.7 mm) was used. The spectrum was obtained under the following conditions: concentration of (CH₃)₂-S-CoM, 9 mg per 0.07 ml; temperature, 44°; filter band width 4 Hz; radio frequency field, 0.15 mG; sweep time 500 s; sweep width, 500 Hz; sweep offset, 0 Hz; spectrum amplitude, 32; integral amplitude, 50; reference compound, sodium 3-trimethylsilyl-propionate-2,2,3,3-d₄ (*TSP*). (*a-b*) indicates only that the cluster of resonances is from protons *a* and *b*; assignments have not been made. *HOD*, hydrogen deuterium oxide.

different than those observed. The methyl proton resonances of an aliphatic methylsulfonate ester lie very close to 3.9 ppm, a position very easily distinguishable from the resonances observed.

We examined the possibility that (S-CoM)₂ was methylated to yield 2-(dimethylsulfonium)ethanesulfonate ((CH₃)₂-S-CoM) which decomposed to CH₃-S-CoM prior to analysis. The NMR spectrum of chemically synthesized (¹⁴CH₃)₂-S-CoM is presented in Fig. 7. The spectrum in deuterium oxide revealed symmetrical ethylene proton resonances of the type A₂B₂ at 3.65 and 3.46 ppm (cumulative relative intensity, 4.0). A singlet corresponding to the methyl proton resonances was located at 3.00 ppm (relative intensity, 5.5). CH₃-S-CoM and (CH₃)₂-S-CoM were readily separable by anion exchange chromatography and thin layer electrophoresis. (CH₃)₂-S-CoM forms an internal salt between the sulfonium and sulfonate moieties. This imparts a charge considerably different from that of CH₃-S-CoM. (CH₃)₂-S-CoM was found to be completely stable under the conditions used for analysis. Reaction mixtures which contained methyltransferase were directly fractionated by anion exchange chromatography (Fig. 8A). When limiting amounts of [*methyl*-¹⁴C]methylcobalamin were allowed to react to completion, 99% of the radioactivity added was found in a single peak which possessed electrophoretic properties identical with that of chemically synthesized CH₃-S-CoM (Fig. 8B). No radioactivity above background was found in the void volume (V_0) where (CH₃)₂-S-CoM and methylcobalamin elute. Incubation of (¹⁴CH₃)₂-S-CoM in a methyltransferase reaction for 30 min at room temperature and for 30 min at 40° did not result in decomposition to ¹⁴CH₃-S-CoM.

Biological Activity of CH₃-S-CoM—Chemically synthesized and authentic CH₃-S-CoM exhibited identical biological activity. When 2.9 μmoles of authentic or synthetic CH₃-S-CoM were added to a reaction mixture (see "Experimental Procedure") which contained crude cell extract (3.3 mg of protein), a perfectly linear rate of methane formation was observed in each reaction vessel, the rates being identical; 22 nmoles of methane were produced in 25 min. In a separate experiment, the reaction was allowed to proceed to completion with the methyl group from 101 and 202 nmoles of synthetic CH₃-S-CoM being completely converted to methane. From 101 and 202 nmoles of authentic CH₃-S-CoM, 92 and 190 nmoles of methane, respectively, were detected.

As shown in Table II, (CH₃)₂-S-CoM was found to be completely inert as a methyl donor in the methylreductase-catalyzed reaction in incubation periods up to 70 min. No methane production was observed when (CH₃)₂-S-CoM and HS-CoM were present together in the reaction mixture; (CH₃)₂-S-CoM would not methylate HS-CoM to yield biologically active CH₃-S-CoM. Results from a separate experiment showed that CH₃-S-CoM could not be further methylated by methyltransferase.

DISCUSSION

The fact that CoM was isolated as the disulfide possibly reflects an artifact of isolation. Thiol derivatives such as HS-CoM are, under neutral or alkaline conditions, easily oxidized to disulfides. Anaerobic precautions were not taken when CoM was isolated. Early work on the isolation of CoA revealed that much of it was in the form of a disulfide (20). Mild hydrolysis of partially purified CoA released 2,2'-dithiodiethylamine. Disulfide formation may likely explain the presence of more than one active derivative of CoM in crude preparations. The component of *Peak A* in the QAE-Sephadex A-25 eluate may be a heterodisulfide possessing only one sulfonate moiety RSSCH₂CH₂SO₃⁻ (R-S-S-CoM). The sulfonate group will significantly govern the behavior of such a CoM derivative on an anion exchange column. Thus, (S-CoM)₂, possessing two sulfonate moieties would be expected to elute at a higher ionic strength than a derivative possessing only one sulfonate group. This is in fact the case. (S-CoM)₂ elutes from a QAE-Sephadex A-25 column at approximately 2.5 M ammonium acetate whereas CH₃-S-CoM and HS-CoM both elute at a concentration less than 2 M. The observation that the ratios of CoM activity in *Peaks A* and *B* vary in a reciprocal manner from preparation to preparation may simply reflect the proportions of (S-CoM)₂ and R-S-S-CoM present.

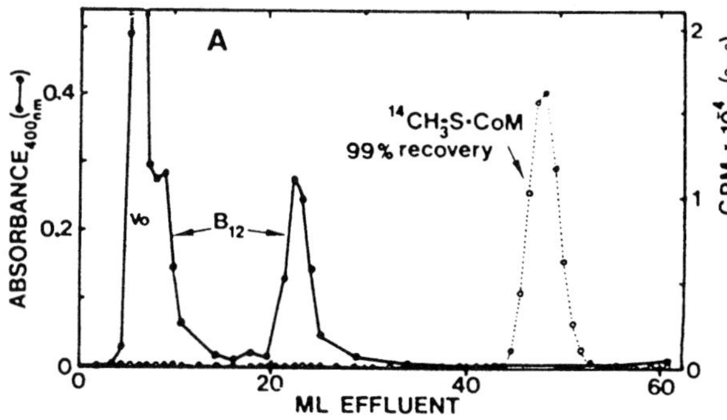

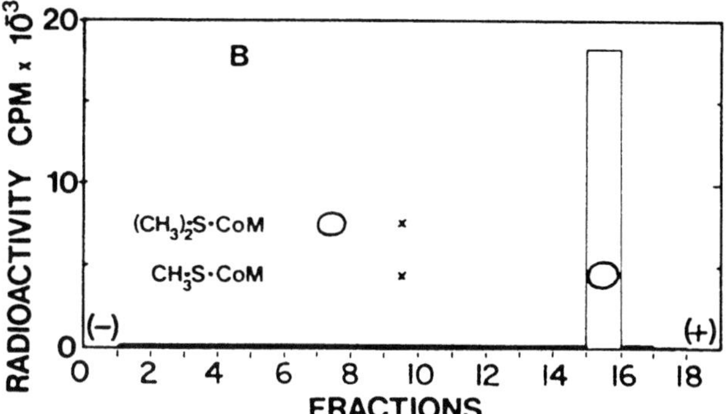

FIG. 8. *A*, anion exchange chromatography of a complete reaction mixture. The reaction mixture (0.50 ml) contained 25 μmoles of potassium phosphate buffer at pH 7.1, 4.2 μmoles of HS-CoM, and 1.75 μmoles of [*methyl*-^{14}C]methylcobalamin (specific activity, 1730 cpm per nmole). The reaction was initiated by injecting an anaerobic solution of enzyme (104 μg of protein) with a syringe. The reaction was allowed to proceed until all of the [*methyl*-^{14}C]methylcobalamin had reacted. A portion of the reaction mixture (0.35 ml) was applied to a QAE-Sephadex A-25 (acetate form) column (1.2 × 15 cm) that had been equilibrated with water. Elution was effected at 5.4 ml per hour with 200 ml of a 0 to 2 M ammonium acetate linear gradient. The profile shown was obtained by measuring the absorbance at 400 nm. $(CH_3)_2$-S-CoM if present would elute with the void volume, V_0. Cobalamin derivatives are indicated by B_{12}. The per cent recovery was determined by comparing the total counts applied to the column to that recovered in the radioactive peak. *B*, the material in the radioactive peak (Fig. 8*A*) was lyophilized to dryness and dissolved in 0.5 ml of water; 5 μl were applied to a Brinkmann MN 300 cellulose thin layer plate. Standard $(CH_3)_2$-S-CoM and CH_3-S-CoM also were applied. Electrophoresis was performed in 0.4% pyridine-0.8% acetic acid buffer at pH 4 (400 volts, 10 ma) for 30 min. The known samples were located by staining with iodine vapors. Radioactivity was determined by counting sections (1.0 × 1.5 cm) of the thin layer in Bray's solution. The origin is indicated by ×.

HS-CoM, 2-mercaptoethanesulfonate, has unique properties. It is one of the smallest and simplest of the known coenzymes, is highly acidic, and has an unusually high concentration of sulfur for its size. When the mercapto group is methylated, the coenzyme does not employ an "onium" ion as the active site of methyl transfer. This latter property stands in contrast to methyltransfer compounds such as *S*-adenosylmethionine, choline, betaine, dimethylthetin, or dimethyl propiothetin. The coenzyme has similarities to CoA and lipoic acid but so far appears to have a completely different role in biochemistry. It

Table II

Comparison of biological activity of CH_3-S-CoM and $(CH_3)_2$-S-CoM

Each reaction mixture (2.0 ml) contained 100 μmoles of potassium phosphate buffer at pH 7.1, 10 μmoles of magnesium sulfate (in the appropriate flask 4.9 μmoles of CH_3-S-CoM, 5.0 μmoles of HS-CoM, and 5.0 μmoles of $(CH_3)_2$-S-CoM), 10 μmoles of ATP, and dialyzed crude extract (33.8 mg of protein). The reaction rates of the positive control reactions were linear for 20 min.

Additions	Reaction rate (CH_4 produced per min)
	nmoles
No CoM derivative	2
HS-CoM	2
$(CH_3)_2$-S-CoM	2
HS-CoM + $(CH_3)_2$-S-CoM	2
CH_3-S-CoM	125
CH_3-S-CoM + $(CH_3)_2$-S-CoM	80

will be interesting to see if the coenzyme can handle a C_1 moiety more oxidized than a methyl group.

Acknowledgments—We thank Kenneth L. Rinehart for suggesting the use of 2-bromoethanesulfonic acid in the chemical synthesis of CoM and for helpful discussions on the structure of derivatives. We appreciate the interest of Robert L. Switzer in this investigation. The assistance of Victor Gabriel and Jerry Althaus in the mass culture of *Methanobacterium* is gratefully acknowledged. We thank the School of Chemical Sciences for use of their facilities in obtaining the NMR spectra, infrared spectra, and quantitative elemental analyses.

REFERENCES

1. McBride, B. C. & Wolfe, R. S. (1971) *Biochemistry* 10, 2317–2324
2. Wolfe, R. S. (1971) *Adv. Microbial. Physiol.* 6, 107–146
3. McBride, B. C. & Wolfe, R. S. (1971) *Adv. Chem. Ser.* 105, 11–22
4. Blaylock, B. A. & Stadtman, T. C. (1963) *Biochem. Biophys. Res. Commun.* 11, 34–38
5. Blaylock, B. A. & Stadtman, T. C. (1966) *Arch. Biochem. Biophys.* 116, 138–158
6. Blaylock, B. A. (1968) *Arch. Biochem. Biophys.* 124, 314–324
7. Taylor, C. D. & Wolfe, R. S. (1974) *J. Biol. Chem.* 249, 4886–4890
8. Taylor, C. D. & Wolfe, R. S. (1973) *Fed. Proc.* 32, 389
9. Müller, O. & Müller, G. (1962) *Biochem. Z.* 336, 299–313
10. Johnson, A. W., Mervyn, L., Shaw, N. & Smith, E. L. (1963) *J. Chem. Soc.* 4146–4156
11. Wood, J. M., Allam, A. M., Brill, W. J. & Wolfe, R. S. (1965) *J. Biol. Chem.* 240, 4564–4569
12. Hogenkamp, H. P. C. & Holmes, S. (1970) *Biochemistry* 9, 1886–1892
13. Cheeseman, P., Toms-Wood, A. & Wolfe, R. S. (1972) *J. Bacteriol.* 112, 527–531
14. Palmer, W. G. (1961) *J. Chem. Soc.*, 1552
15. Bellamy, L. J. (1958) *The Infrared Spectra of Complex Molecules*, John Wiley and Sons, Inc., New York
16. Colthup, N. B., Daly, L. H. & Wiberley, S. E. (1964) *Introduction to Infrared and Raman Spectroscopy*, Academic Press, New York
17. Bhacca, N. S., Johnson, L. F. & Schoolery, J. N. (1962) *NMR Spectra Catalog*, Vol. 1, Varian Associates, Palo Alto, Cal.
18. Bhacca, N. S., Hollis, D. P., Johnson, L. F. & Pier, E. A. (1963) *NMR Spectra Catalog*, Vol. 2, Varian Associates, Palo Alto, Cal.
19. Scott, A. I. (1964) *Interpretation of Ultraviolet Spectra of Natural Products*, Pergamon Press, New York
20. Challenger, F. (1959) *Aspects of the Organic Chemistry of Sulfur*, Butterworths, London

General and Applied Microbiology 263

Comparative Cataloging of 16S Ribosomal Ribonucleic Acid: Molecular Approach to Procaryotic Systematics

G. E. FOX, K. R. PECHMAN, AND C. R. WOESE

In this seminal work, George Fox and Carl Woese established that a comparison of oligonucleotides produced by enzymatic degradation of the 16S rRNA of various *Bacillus* spp. allowed the reconstruction of phylogenetic relationships. The application of this method to other procaryotes led to the creation of a unified phylogeny for all living organisms and the discovery of the Archaea. When nucleic acid sequencing became routine in the 1990s, this work became the foundation for a revolution in the study of microbial systematics, evolution, and ecology. Because these methods greatly simplified the task of identifying and classifying bacterial isolates, it was possible to study procaryotes within a comprehensive evolutionary framework.

WILLIAM B. WHITMAN

Reproduced from *International Journal of Systematic Bacteriology* 27:44-57. Copyright © 1977. American Society for Microbiology.

INTERNATIONAL JOURNAL OF SYSTEMATIC BACTERIOLOGY, Jan. 1977, p. 44–57

Vol. 27, No. 1

Printed in U.S.A.

Comparative Cataloging of 16S Ribosomal Ribonucleic Acid: Molecular Approach to Procaryotic Systematics

GEORGE E. FOX, KENNETH R. PECHMAN, AND CARL R. WOESE

Department of Genetics and Development, University of Illinois, Urbana, Illinois 61801

A taxonomy for *Bacillus subtilis*, *B. megaterium*, *B. cereus*, *B. pumilus*, *B. pasteurii*, *B. stearothermophilus*, and *Sporosarcina ureae* has been constructed from comparisons of T_1 ribonuclease digests of their respective 16S ribosomal ribonucleic acids. This molecular approach to systematics is shown to give results in essential agreement with traditional techniques for this group of organisms. In addition, the technique appears well suited for higher order classification, an area which has been difficult to approach with traditional techniques.

A prime objective of bacterial systematics certainly must be the establishment of a classification that spans the procaryote kingdom. However, classification by traditional techniques has been difficult, especially beyond the level of genera, reflecting the relative simplicity and/or antiquity of these organisms. It has been proposed that this problem might be resolved by integrating existing taxonomic studies where "suitable" overlap occurs (6). A direct experimental approach powerful enough to reveal the more distant relationships would certainly be more desirable, however.

Such an approach can be based on a molecular characterization of what Zuckerkandl and Pauling (24) call the "semantides"—informational macromolecules. Two considerations necessitate care. Not only are bacterial origins much more ancient than metazoan origins, but the genomes in the former case have a characteristically higher mutation rate (3). Consequently, one might expect an unmanageable degree of divergence in primary structures of any given semantide across the spectrum of procaryotes, again precluding higher order classification. Furthermore, bacterial genetic transfer mechanisms hold the potential for reticulate evolution (8), which could make for an unmanageable complexity.

It is herein contended that these difficulties can be avoided by primary structural characterization of ribosomal ribonucleic acids (rRNAs). The ribosome is clearly of very ancient origin and is necessarily ubiquitous. At least two of its RNA components (5S and 16S rRNA) apparently have functionally equivalent forms over a wide range of procaryotes, since both of these molecules can form functional chimera ribosomes in vitro (1, 9, 21). The primary structures of these rRNA molecules are sufficiently constrained that on the whole they have not changed rapidly in time. Moreover, they contain regions of both extreme conservation (4, 19) and hypervariablity (16, 19) so that both distant and close relationships can be examined. Since three distinct rRNA molecules exist, one can ultimately examine the extent to which a particular classification is dependent on the choice of RNA molecule. Finally the large number of ribosomal components (18), whose genes are not all necessarily contiguous (7), argues that the ribosomal system would not be readily transferred genetically from one organism to another.

In the present study the 16S rRNA molecule was selected since its $\approx 1,600$ nucleotides provide a more reliable classification than the small 5S rRNA (≈ 120 nucleotides), and experimentally it is more manageable than the larger 23S rRNA ($\approx 3,300$ nucleotides). Determination of 16S rRNA primary structure is sufficiently difficult at present that full sequence information cannot be used as a basis for phylogenetic study. However, a partial sequence characterization, in terms of a "comparative cataloging" approach, is both feasible and useful (11, 12, 20, 23). Herein this approach is applied to a group of bacilli to display the method and examine its potential for phylogenetic and general taxonomic classification.

In essence the method involves digesting the 16S rRNA from each organism under investigation with an endonuclease of restricted specificity and determining the sequence of the resulting oligonucleotides. This produces a unique catalog of digestion products from each organism. These catalogs are then intercompared by two essentially independent approaches. In the first, families of homologous digestion products are identified which allow direct examination of nucleotide replacements which have been introduced during the course of evolution. In the second, each individual digestion product is treated as a taxonomic character, with a dendrogram being produced by essentially normal taxonomic procedures.

MATERIALS AND METHODS

Strains. The strains included in this study were as follows: *Bacillus subtilis* 168; *Bacillus megaterium* KM; a *Bacillus cereus* strain obtained from P. Starr, University of Illinois; *Bacillus pumilus*, isolated and characterized by B. J. Lewis, University of Illinois, a characterization confirmed by R. Gordon, Rutgers University (personal communication); *Bacillus pasteurii*, ATCC 11859; *Bacillus stearothermophilus* strain 10 (10); and *Sporosarcina ureae*, ATCC 6473.

Growth, labeling, and RNA isolation. *B. subtilis, B. megaterium, B. pumilus,* and *B. cereus* were generally grown aerobically at 37°C in a yeast extract-peptone medium (pH adjusted to 7.0) and labeled with $^{32}PO_4$ in a dephosphorylated version of that medium (15). *B. pasteurii* and *S. ureae* were grown and labeled in the same medium with 0.3 M urea (final concentration) added and the pH adjusted to 8.5 (11). *B. stearothermophilus* was grown in a peptone medium at 60°C (10).

For labeling RNA, $^{32}PO_4$ (New England Nuclear Corp.) was added to each culture (10 to 30 cm³) in early log phase at a final concentration of 0.5 to 1.0 mCi/ml. After three to four divisions, cells were collected by centrifugation, washed, and opened by passage through a French pressure cell at 15,000 lb/in². The rRNA was then phenol-extracted and the 16S rRNA was separated by polyacrylamide gel electrophoresis, with final purification obtained by passage over a Whatman CF-11 cellulose column (K. J. Pechman, Ph.D. thesis, University of Illinois, Urbana, Ill., 1975).

Determination of oligonucleotide catalog. Each purified 16S rRNA was digested with T_1 ribonuclease, and the resulting oligonucleotides were separated and sequenced by the two-dimensional paper electrophoresis technique of Sanger and co-workers (13) with certain modifications introduced in this laboratory (17, 18a).

In brief, the analysis involves electrophoresis of the digest on cellulose acetate at pH 3.5 (in the presence of 7 M urea), followed by transfer of the resulting oligonucleotide pattern to diethylaminoethyl-cellulose paper, with subsequent orthogonal electrophoresis in one of two buffer systems (17, 18a). The resulting oligonucleotide "fingerprint" comprises a pattern of "isopleths." Each isopleth grouping has a characteristic uridylic acid (U) content per oligonucleotide, i.e., the U = 0 isopleth, the U = 1 isopleth, and so on. Within each isopleth, oligonucleotides are displayed by order of size, and within each size "isocline" they are separated to some extent on the basis of composition. Figure 1 is a representative fingerprint of this type.

Sequences of oligonucleotides on this fingerprint, the primary pattern, are determined (where not obvious from their positions) by various "secondary" and "tertiary" enzymatic digestions which produce a sufficient number and kind of recognizable products that the sequence of the original oligonucleotide can be deduced therefrom (17, 18a).

Data analysis. To identify families of related oligomers, a systematic search was conducted with the aid of an IBM 360/75 electronic computer. In this procedure all nonidentical sequences in any two catalogs, A and B, were examined. Every sequence in A was individually aligned so as to maximize similarity with each sequence in B. A matrix resulted wherein each row represented a particular sequence in catalog A and each column represented one in catalog B. Every entry in the matrix thus represented the number of residues in common between two specific sequences when they were aligned in the most favorable way. Next, each row in the matrix was scanned for an absolute maximum, and if, when identified, that entry was also an absolute maximum in its respective column, the two sequences were presumed to be corresponding related sequences. Finally, the column and row corresponding to the two newly identified homologous sequences were deleted and the process was continued.

This procedure was most powerful when applied to a group of related organisms, since the various binary comparisons tended to bear upon one another. For example, corresponding oligonucleotides from two distantly related species were often undetectable because of "noise" (usually tie scores in a column or row). However, if any one of the alternative possibilities was unequivocally identified in a comparison involving a third organsim, more closely related to one of the original organisms, then it was justifiable to define this relationship to be the correct one in the more distant comparison as well. Such subjective aspects of the analysis required considerable care during application if reliable results were to be obtained.

In a reasonably compact group of organisms, such as the bacilli, this approach identified a significant portion, but not all, of the corresponding related sequences between any two catalogs. In general then, families of related oligomers were identified, many of which had representatives from each organism under study. These families could be used in their entirety to produce individual binary comparisons. However, it alleviated any question as to a proper normalization procedure if only complete families were utilized so that the individual binary comparisons could all be based on counts of the variations at equivalent nucleotide positions. This information could then be used, in principle at least, to calculate a true phylogenetic tree.

Alternatively, an appropriate association coefficient could be defined as follows: $S_{AB} = 2N_{AB}/(N_A + N_B)$, where N_A = total number of residues represented by oligomers of at least length L in catalog A, N_B = total number of residues represented by oligomers of at least length L in catalog B, and N_{AB} = total number of residues represented by all the coincident oligomers between the two catalogs, A and B, of at least length L. For a reasonably compact group

FIG. 1. *Representative 16S rRNA oligonucleotide fingerprint.*

TABLE 1. *Oligonucleotide catalogs*[a]

Oligonucleotide	B. subtilis	B. pumilus	B. megaterium	B. cereus	B. pasteurii	S. ureae	B. stearothermophilus	S. inulinus
5-mers								
CCCCG	0	0	0	0	0	0	1	−
CCCAG	1	1	1	1	1	1	1	−
CCACG	1	1–2	1	1[b]	1	?[c]	0	+
(C, C, AC)G	0	0	0	0	0	1	0	−
CACCG	0	0	0	1	0	?	0	+
ACCCG	1	0	0	1	1	1	0	−
CAACG	2	1–2	2	2	1	1	1	+
ACACG	1	1	1	1	1	1	1	+
AACCG	0	1	0	0	0	0	0	−
ACAAG	0	0	0	0	0	0	1	+
AACAG	1	1	0	0	0	0	0	−
AAAAG	0	0	1	1	0	0	0	−
UCCCG	1	1	1	1	1	1	1	+
CCCUG	0	0	0	0	1	0	0	−
CUCAG	1	1	1	1	1	1	1	+
CUACG	0	1	0	0	0	0	0	−
ACCUG	1	1	2	1	2–1	2	1[b]	+
AUCCG	0	0	0	1	0	0	0	−
CUAAG	1	1	2	1	1	1	1[b]	−
UACAG	0	0	0	1	1–2	1	0	−
UACCG	1	?	0	1/2	1	1	1	−
UAACG	1	0	1	1	1	0	1	+
CAUAG	0	0	1	0	0	0	0	+
CAAUG	1	1	1	1	0	1	1	−
AUCAG	1	1	1	1	1	1	1	+
ACAUG	0	0	0	0	0	1	0	−
AAUCG	1	1	1	1	1	1	1	−
AACUG	0	0	1	0	0	1	0	−
UAAAG	2	2–1	1	1	2	1–2	1	+ +
AUAAG	0	0	0	0	0	0	1	−
AAUAG	1	0	0	0	0	0	0	−
AAAUG	1	1	1	1	1	1	1	+ +
CCUUG	1	1	1	1	0	0	2–1	+
CUCUG	1	1	0	0	1	1	2–1	+
(UC)UCG	1	0	0	0	0	0	0	−
CUUCG	?	0	0	1	0	0	0	−
UCAUG	0	0	0	0	0	1	0	−
CUAUG	0	0	0	0	1	0	0	−
UAUCG	1	1–2	1	1	1	1	0	+
ACUUG	0	0	0	1	0	1	1	−
UUAAG	2	1	1	1–2	1	1	1	+
UAAUG	0	1	0	0	1	1	0	+
UAUAG	0	0	0	0	1	0	0	−
AUUAG	1	1	2	1	1	1	1	+
AUAUG	0	0	0	1	0	0	0	−
AAUUG	1	1	1	1	1	1	1	+ +
UUUCG	0	1	1[b]	0	0	0	0	+
UCUUG	1	1	1	1	2	2	1	+
CUUUG	0	0	0	0	0	0	1	−
UUUAG	0	1	0	0	0	0	0	−
AUUUG	0	0	0	0	0	0	1	−

TABLE 1—*Continued*

Oligonucleotide	*B. subtilis*	*B. pumilus*	*B. megaterium*	*B. cereus*	*B. pasteurii*	*S. ureae*	*B. stearothermophilus*	*S. inulinus*
6-mers								
CCCCCG	1	1	0	0	0	0	0	−
CCCACG	0	0	1	1	1	1	?	−
(C, C, C, AC)G	0	0	0	0	0	0	1	−
AACCCG	0	0	0	0	0	0	1	−
CACAAG	1	1	1	1	1	1	1	+
CAAACG	0	0	0	0	1	0	0	−
AAACCG	1	1	1	1	1	1	1	+
AACAAG	1	1	1	1	1	1	0	−
UCCACG	1	1	1	1	0	1	1	−
UAACCG	0	0	1	0	0	0	0	−
CUAACG	1	1	1	0	0	1	1	−
CCUAAG	0	0	1	1	0	0	0	−
CAACUG	0	0	0	0	0	0	1	−
ACACUG	0	0	0	1	0	1	0	−
UAAACG	1	1	1	1	1	1	1	+
AUACAG	0	0	0	0	0	1	0	−
AAUACG	1	1	1	1	1	1	1	−
AAACUG	1	1	1	1	1	1	1	+
AAUAAG	0	0	0	1	0	0	0	−
UUCCCG	1	1	1	1	1	1	1	+
CCUUCG	0	0	0	0	0	0	1	−
UCCAUG	0	0	0	0	1	0	0	+
UCACUG	0	0	0	0	1	1	0	−
AUUCCG	0	0	0	0	0	1	0	−
AUCCUG	1	1	1	1	1	1	1	+
UUCAAG	0	1	0	0	0	0	0	−
CAUUAG	1	1	1	1	1	1	1	+
UAAUCG	1	1	1	1	1	1	1	+
UAACUG	1	1	1	1	1	1	0	−
AACUUG	1	0	1	0	0	0	0	−
AUUAAG	0	0	0	1	0	0	0	−
UUUCCG	1	1	1	1	1	1	0	−
CUUUCG	0	0	?	1	0	0	0	−
(CU)UUCG	0	0	1	0	0	0	0	−
CCUUUG	0	0	0	0	0	0	1	−
UCAUUG	1	1	1	1	1	1	1	+
UUUUCG	1	1	0	0	1	1	0	+
CUUUUG	0	1[b]	0	0	0	0	1	−
7-mers								
CAAACAG	0	0	1	1[b]	1	1	1	+
AACAAAG	1	1	0	0	0	0	0	−
CACUCCG	1	1	1	1	1	1	1	−
CAACUCG	1	1	1	1	1	1	1	+
CAACCUG	0	0	1	0	1	1	1	+
UAACACG	1	1	1	1	1	1	1	+
UACAAAG	0	0	1	1	0	0	1	+
UAAAAAG	0	0	0	0	0	1[b]	0	−

TABLE 1—*Continued*

Oligonucleotide	*B. subtilis*	*B. pumilus*	*B. megaterium*	*B. cereus*	*B. pasteurii*	*S. ureae*	*B. stearothermophilus*	*S. inulinus*
CUCCCUG	1	0	0	0	0	0	0	−
CCUCUAG	0	0	0	0	0	0	1	−
CUCUCAG	0	0	0	0	0	0	1	−
UUCCCAG	0	0	0	0	1	1	0	−
CACCUUG	0	1	0	1	0	0	0	−
CAUUCAG	1	0	?	0	1	0	1	−
UAACCUG	1	1	0	1	0	0	0	−
ACCUUAG	0	0	0	0	1	0	0	−
AACUCUG	0	0	0	0	1	0	0	−
CAUUAAG	1	1	1	1	1	1	0	+
UAAUACG	1	1	1	1	1	1	1	+
AAACUUG	0	1	0	0	1	0	0	−
CUCUCUG	0	0	0	0	0	0	1	+
UUCUCAG	1	1	1	1	0	0	0	−
UCACUUG	0	0	0	0	0	0	1	−
UUAUCCG	0	0	1	1	0	0	0	−
UACCUUG	1	0	1	0	1	1	0	−
CAUUUAG	0	1	1	0	0	0	0	−
AUCUUAG	1	1	1	1	0	0	0	+
CUUUCUG	0	0	0	0	1	1	0	−
(U$_5$, C)G	0	0	1	0	0	0	0	−
8-mers								
CCAACCCG	0	0	0	0	1	0	0	−
CAAACCCG	0	0	0	0	0	1	0	−
ACAACCCG	0	0	0	0	0	0	1	−
ACAAACCG	1[b]	1	1[b]	1	1	1[b]	0	+
AACACCAG	1	1	1[b]	1[b]	1	1	1	−
CUCAACCG	1	1	1	1[b]	0	1	1	+
CCACACUG	1	1	1	1	1	1	1	−
ACAUCCCG	0	0	0	0	1	0	0	−
CCCCUUAG	1[b]	1	0	0	1	1	0	−
CCCUUCAG	0	0	0	0	0	1	0	−
CCUACAUG	0	0	1[b]	1	0	0	0	−
CUUAACCG	0	0	0	0	1	0	0	−
CACUCUAG	0	0	0	0	0	0	1	−
AUACCCUG	1	1	1	1	1	1	1	+
CUACAAUG	1	1	1[b]	1	1	1	1	+
CCCUUUAG	0	0	1[b]	1	0	0	0	−
ACUCUCUG	1	1	?	0	0	0	0	−
A$_{1-0}$CUCUCUG	0	0	1	0	0	0	0	−
UUCCUUCG	0	0	0	0	1	0	0	−
CUCUUAUG	0	0	1	1	0	0	0	−
ACUUCUG	0	0	0	1	0	0	0	−
AAUUAUUG	1	1	1	1	1	1	1	+
9-mers								
CAACCCUCG	0	0	0	0	0	0	1	−
CUCACCAAG	1[b]	1	1[b]	1	0	1	1	−
CUACACACG	1	1	1	1	1	1	1	+
UACACACCG	1	1	1	1	1	1	1	+
UAACACCCG	1	1	1[b]	1	1	1	0	+

TABLE 1—*Continued*

Oligonucleotide	*B. subtilis*	*B. pumilus*	*B. megaterium*	*B. cereus*	*B. pasteurii*	*S. ureae*	*B. stearothermophilus*	*S. inulinus*
ACAUCCCAG	0	0	0	0	0	1	0	−
ACCAAAUCG	0	0	0	0	0	0	1	−
AUAACACCG	0	0	0	0	0	0	1	−
CAACCCUUG	1	1	1[b]	1	1[b]	0	0	+
UACCUCACG	0	0	0	0	0	0	1	−
ACUCCUACG	1	1	1	1[b]	1	1	1	−
CUUACCAAG	0	0	0	0	1	0	0	−
CUAACUACG	1	1	1	1	1	1	1	+
CUAAUACCG	1	1	1	1	1	1	1	+
CACUCUAAG	1[b]	1	1	1	1	0	0	+
AUAACUCCG	1	1	0	1	1	1	1	−
AAUUCCACG	1	1	1	1	1	1	1	+
AAUAAUCAG	0	0	0	0	1[b]	0	0	−
UCCCCUUCG	1	0	0	0	0	0	0	−
C(C$_{1-0}$CU)CUUAG	0	0	1	0	0	0	0	−
CCCCUUAUG	1	1	1	1	1	1	1	+
CAUCCUCUG	0	0	1[b]	0	0	0	0	−
CACUCUAUG	0	0	0	0	0	1[b]	0	−
UCCCUUAAG	0	0	0	0	0	0	1	−
AAUCUUCCG	1[b]	1	1[b]	1	0	0	1	−
AUAACUUCG	0	0	1[b]	0	0	0	0	+
AAUCUCAUG	0	0	0	1	0	0	0	−
UUCCCUUCG	0	0	0	0	0	0	1[b]	+
CCUUUUAAG	0	0	0	0	1[b]	1[b]	0	−
UCACUUAUG	0	0	0	1	0	0	0	−
UUUCUUAAG	1	1	1[b]	1	0	0	0	−
UUUAAUUCG	1	1	1	1	1	1	1	+
≥10-mers								
ACAACCCAAG	0	0	0	0	0	0	1	−
ACAACCCUAG	0	1	0	1	0	0	0	−
AAACUCAAAG	1	1	1	1	1	1	1	+
ACAUCCCCUG	0	0	0	0	0	0	1	−
A(C,A)ACUCUAG	0	0	1	0	0	0	0	−
ACAAUCCUAG	1	0	0	0	0	0	0	−
UAAAACUCUG	0	0	1	1	0	0	0	−
AAAUUCAAAG	0	0	0	1	0	0	0	−
ACAUCCUCUG	1[b]	1	1[b]	1	0	0	0	+
UCACUUACAG	0	1	0	0	0	0	0	−
CUUCCCUUCG	0	0	1[b]	1[b]	0	0	0	−
UUU(CU)CUUUG	0	0	0	0	1	0	0	−
UACCUAACCAG	1[b]	1	1	1	0	0	0	−
AACCUUACCAG	1	1	1	1	1	1	1	+
CCUAAUACAUG	1	1	1[b]	1	1	1	1	+
CCAUCAUUCAG	0	0	0	0	0	1	0	−
UACCUUACCAG	0	0	0	0	0	1[b]	0	−
CCAUCAUUAAG	0	0	0	1	0	0	0	−
(C~$_5$, U$_4$)AG	0	0	0	0	0	1	0	−
UACCUCAUUAG	0	0	0	0	1	0	0	−
CUUUCCCUUCG	0	1	0	0	0	0	0	−
UAACCCUUUUG	0	0	0	0	1	0	0	−

Table 1—*Continued*

Oligonucleotide	*B. subtilis*	*B. pumilus*	*B. megaterium*	*B. cereus*	*B. pasteurii*	*S. ureae*	*B. stearothermophilus*	*S. inulinus*
UAACCCUUAG	0	0	0	0	0	1	0	
UAACCUUUAUG	0	1	0	0	0	0	0	−
UAAC$_{1-2}$CUUUUAG	1	0	0	0	0	0	0	−
AUAACAUUUUG	0	0	0	1[b]	0	0	0	−
UAAC(CU)UUUUG	0	0	0	1	0	0	0	−
AAUCCCAAAAAG	0	0	0	0	0	0	1	−
CAACCCUUAAUG	0	0	0	0	0	1[b]	0	−
AAUCUUCCACAAUG	0	0	0	0	1	1	0	−
UCACACCCUUUAG	0	0	0	0	0	0	1[b]	−
UCAAAUCAUCAUG	1	1	1	1	*[e]	1	1	+
(AU, C$_x$, U$_y$, CUUA, CUA)AG	0	0	1	0	0	0	0	−
(U$_3$, C$\sim_4$)UCUAUG	0	0	1	0	0	0	0	−
AUUUC$_{0-1}$UCCCUUCG	0	0	0	0	0	1	0	−
CCAAUCCCACAAAAUCG	0	0	0	0	0	1	0	−
UUCAAAC$_{0-1}$CAUAAAAG	1[b]	0	0	0	0	0	0	−
(AUAA, UCCA)U(CU, CUU)CG	0	0	0	0	0	1	0	−
(C, C, AAU, C, C, C, AC, AAAU, C, U)G	1	0	0	0	0	0	0	−
CUAAUCCCAUAAAACCG	0	0	0	0	1	0	0	−
CCAAUCCCAUAAAUCUG	0	1	0	0	0	0	0	−
CUAAUCUCAUAAAACCG	0	0	0	1[b]	0	0	0	−
AUAUACCUUUCCCUUCG	0	0	0	0	1	0	0	−
Modified oligomers[d]								
AAĠ	1	1	1	1	1	1	1	+
ȦAG	1	1	1	1	1	1	1	+
ĠCCG	1	1	1	1	1	1	1	+
CĊCCG	1	1	1	1	1	1	1	+
ĊAACG	1	1	1	1	?	1	1	+
U̇AACAAG	1	1	1	1	1	1	1	+
UCAĊACCACG	1	1	1[b]	1	1	1	1	+
UĊAAAUCAUCAUG	0	0	0	0	1	0	0	−
Terminii								
AUCACCUCCUUUU$_{OH}$	1[b]	1[b]	1[b]	1[b]	1[b]	1[b]	1[b]	+
pUCUUAUG	0	0	0	0	0	1[b]	0	−
pUUUAUCG	1	0	0	0	0	0	0	−
pUAUUAUG	0	0	0	0	1	0	0	−
pCUUUUUG	0	0	0	0	0	0	0	+
pUUCUUUG	0	0	0	0	0	0	1	−
pUUUUUCG	0	1	1	0	0	0	0	−
pUUUAUUG	0	0	0	1	0	0	0	−

[a] Symbols: 1/0, the mole fraction of each oligonucleotide is indicated for each organism except *S. inulinus*; +/−, the presence or absence of each oligonucleotide in *S. inulinus* is indicated (the complete catalog for this organism will be presented elsewhere); 2–1/1–2, the oligonucleotide in question may be present in multiple copies (the first number indicates the more likely quantitation).

[b] The sequence given is probable but not certain.

[c] Question mark indicates that presence or absence of the oligonucleotide in question is not determined.

[d] Dot indicates base which is post-transcriptionally modified.

[e] Asterisk indicates that the identical oligonucleotide, but containing a post-transcriptionally modified base, is found in *B. pasteurii*.

such as the bacilli examined here, $L = 5$. The similarity coefficient, S_{AB}, was calculated for each individual binary comparison and the resulting similarity matrix was used to produce a dendrogram in one of the usual ways, in this study single-linkage clustering.

RESULTS

Table 1 lists the oligonucleotide catalogs for the various species of the genus *Bacillus* (11; Pechman, thesis; 18a). Table 1 also records whether any given sequence was present or absent in the catalog of *Sporolactobacillus inulinus*, a closely related but presumably "non-*Bacillus*" organism. In Table 2 the 52 sequences of universal occurrence within the *Bacillus* species examined here are shown. Table 3 summarizes the 39 families of corresponding related oligomers that were identified.

Single-linkage clustering utilizing the association coefficient, S_{AB}, defined the dendrogram shown in Fig. 2. *Escherichia coli* (17) is here included to provide perspective. The individual binary comparisons upon which this dendrogram was based are indicated in Table 4.

The closest relationship found was between *B. subtilis* and *B. pumilus*. *B. megaterium* and *B. cereus* were also found to cluster as were *B. pasteurii* and *S. ureae*. Assuming this tree can be attributed phylogenetic significance, the *B. megaterium*-*B. cereus* divergence from the common *Bacillus* ancestor appeared to have

been considerably more recent than that of the *B. pasteurii*-*S. ureae* couple. Hence, the seven organisms screened could be crudely thought of as defining three groups of species of *Bacillus* — one group containing *B. subtilis*, *B. pumilus*, *B. megaterium*, and *B. cereus*; a second here represented only by *B. stearothermophilus*; and a third containing *B. pasteurii* and *S. ureae*. That all of these groups are indeed best considered divisions within the genus *Bacillus* rather than individual genera was indicated by the fact that they are each more closely related to the others than any one is to the outside "reference" organism—*Sporolactobacillus inulinus*.

Binary comparison of the individual organisms over the entire set of 27 complete families allowed a direct determination of the nucleotide replacements which were introduced during the evolution of the portions of the 16S rRNA molecule represented by these families. These numbers are displayed in Table 4. In the present case the complete families represented 239 nucleotide positions, whereas the 52 universal sequences in Table 1 represented another 364 positions. The combined total of 603 nucleotide positions is the vast majority of the total number of positions accessible by the method used herein (i.e., the average *Bacillus* catalog given here contains 746 nucleotides in digestion products of size five and larger) and is a significant portion, $\approx 37\%$, of the entire 16S rRNA sequence. Since ancestral sequences can be postulated for each family, it is in principle possible to calculate a phylogenetic tree from these families (Fox and Woese, unpublished data). Such a tree is topologically very similar to that obtained with the association coefficient employed here.

In any event, many of the families reaffirm the branching points suggested in the tree generated by the association coefficeint S_{AB}. In particular, families 1, 5, 14, 15, 22, and 32 indicate *Sporolactobacillus* to be outside the main group; families 3, 7, 11, 24, and 34 indicate the close relationship between *B. subtilis* and *B. pumilus*; families 8, 19, 20, 21, 23, and 35 emphasize the interrelatedness among *B. subtilis*, *B. pumulis*, *B. megaterium* and *B. cereus*; whereas numbers 8, 11, 19, 21, 24, 30, and 38 reflect the group comprising *B. pasteurii* and *S. ureae*. In addition the families allow some insight into the evolutionary process and how it has affected the RNAs. The incorporated mutations appear in some sense nonrandom; transition mutations appear far more readily than do transversions. Likewise, insertions or deletions of single nucleotides appear to be rare.

TABLE 2. *Universal oligomers*

AAĠ	AAACCG	ACUCCUACG
ȦȦG	UAAACG	CUAACUACG
ĠCCG	AAUACG	CUAAUACCG
CĊCCG	AAACUG	AAUUCCACG
ĊAACG[a]	UUCCCG	CCCCUUAUG
U̇AACAAG	AUCCUG	UUUAAUUCG
UCAĊACCACG	CAUUAG	
	UAAUCG	AAACUCAAAG
CCCȦG	UCAUUG	
CAACG		AACCUUACCAG
ACACG	CACUCCG	CCUAAUACAUG
UCCCG	CAACUCG	
CUCAG	UAACACG	UCAAAUCAUCAUG[b]
ACCUG	UAAUACG	AUCACCUCCUUUU$_{OH}$
CUAAG		
AUCAG	AACACCAG	
AAUCG	CCACACUG	
UAAAG	AUACCCUG	
AAAUG	CUACAAUG	
UUAAG	AAUUAUUG	
AUUAG		
AAUUG	CUACACACG	
UCUUG	UACACACCG	
CACAAG		

[a] Presence or absence not determined in *B. pasteurii*.

[b] This oligomer is post-transcriptionally modified in *B. pasteurii* —e.g., UĊAAAUCAUCAUG.

Table 3. *Oligonucleotide families*[a]

Families		*B. subtilis*	*B. pumilus*	*B. megaterium*	*B. cereus*	*B. pasteurii*	*S. ureae*	*B. stearothermophilus*	*S. inulinus*
Complete families									
1.	AAUACG	1	1	1	1	1	1	1	−
	AAUCCG	0	0	0	0	0	0	0	+
2.	UAACUG	1	1	1	1	1	1	0	−
	UUACUG	0	0	0	0	0	0	0	+
	CAACUG	0	0	0	0	0	0	1	−
3.	CAAACAG	0	0	1	1	1	1	1	+
	AACAG[b]	1	1	0	0	0	0	0	−
4.	AACAAG	1	1	1	1	1	1	0	−
	ACAAG[b]	0	0	0	0	0	0	1	+
5.	CACUCCG	1	1	1	1	1	1	1	−
	CAUUCCG	0	0	0	0	0	0	0	+
6.	UCCACG	1	1	1	1	0	1	1	−
	UCCAUG	0	0	0	0	1	0	0	+
7.	CCCCCG	1	1	0	0	0	0	0	−
	CCC<u>AC</u>G	0	0	1	1	1	1?[c]	1	−
8.	UUCUCAG	1	1	1	1	0	0	0	−
	UUCCCAG	0	0	0	0	1	1	0	−
	CUCUCAG	0	0	0	0	0	0	1	−
9.	UAACCUG	1	1	0	1	0	0	0	−
	CAACCUG	0	0	1	0	1	1	1	+
10.	pUAUUAUG	0	0	0	0	1	0	0	−
	pUCUUAUG	0	0	0	0	0	1	0	−
	pUUUAUUG	0	0	0	1	0	0	0	−
	pUUUAUCG	1	0	0	0	0	0	0	−
	pUUUUUCG	0	1	1	0	0	0	0	−
	pUUCUUUG	0	0	0	0	0	0	1	−
	pCUUUUUG	0	0	0	0	0	0	0	+
11.	ACUUUCUG	0	0	0	1	0	0	0	−
	<u>A</u>CUCUCUG	1	1	1?	0	0	0	0	−
	<u>C</u>UCUCUG	0	0	0	0	0	0	1	+
	CUUUCUG	0	0	0	0	1	1	0	−
12.	ACAAACCG	1	1	1	1	1	1	0	+
	ACAACCCG	0	0	0	0	0	0	1	−
13.	CUCAACCG	1	1	1	1	0	1	1	+
	CUUAACCG	0	0	0	0	1	0	0	−
14.	AACACCAG	1	1	1	1	1	1	1	−
	AAUACCAG	0	0	0	0	0	0	0	+
15.	CCACACUG	1	1	1	1	1	1	1	−
	CCACAUUG	0	0	0	0	0	0	0	+

TABLE 3—*Continued*

Families		B. subtilis	B. pumilus	B. megaterium	B. cereus	B. pasteurii	S. ureae	B. stearothermophilus	S. inulinus
16.	AUAACUCCG	1	1	0	1	1	1	1	−
	AUAACUUCG	0	0	1?	0	0	0	0	+
17.	CUCACCAAG	1	1	1	1	0	1	1	−
	CUUACCAAG	0	0	0	0	1	0	0	−
	CCCACCAAG	0	0	0	0	0	0	0	+
18.	CACUCUAAG	1	1	1	1	1	0	0	+
	CACUCUAUG	0	0	0	0	0	1	0	−
	CACUCUAG	0	0	0	0	0	0	1	−
19.	UUUCUUAAG	1	1	1	1	0	0	0	−
	CUUCUUAAG	0	0	0	0	0	0	0	+
	UCCCUUAAG	0	0	0	0	0	0	1	−
	CCUUUUAAG	0	0	0	0	1	1	0	−
20.	UACCUAACCAG	1	1	1	1	0	0	0	−
	UACCUCAUUAG	0	0	0	0	1	0	0	−
	UACCUUACCAG	0	0	0	0	0	1	0	−
	UACCUCACG	0	0	0	0	0	0	1	−
21.	AAUCUUCCG	1	1	1	1	0	0	1	−
	AAUCUUCCACAAUG	0	0	0	0	1	1	0	−
22.	CCCAG/ACUCCUACG	1	1	1	1	1	1	1	−
	CCCAAACUCCUACG	0	0	0	0	0	0	0	+
23.	ACAUCCUCUG	1	1	1	1	0	0	0	+
	ACAUCCCCUG	0	0	0	0	0	0	1	−
	ACAUCCOG	0	0	0	0	1	0	0	−
	ACAUCCCAG	0	0	0	0	0	1	0	−
24.	CCCCUUAG	1	1	0	0	1	1	0	−
	CCCUUUAG	0	0	1	1	0	0	0	−
	UCCAACCCCUUAG[b]	0	0	0	0	0	0	0	+
	UCACACCCUUUAG[b]	0	0	0	0	0	0	1	−
25.	UCCCCUUCG	1	0	0	0	0	0	0	−
	UUCCCUUCG	0	0	0	0	0	0	1	+
	CUUCCCUUCG	0	0	1	1	0	0	0	−
	AUUUCUCCCUUCG (0–1)	0	0	0	0	0	1	0	−
	CUUUCCCUUCG	0	1	0	0	0	0	0	−
	AUAUACCUUUCCCUUCG	0	0	0	0	1	0	0	−
26.	CAACCCUUG	1	1	1	1	1	0	0	+
	CAACCCUCG	0	0	0	0	0	0	1	−
	CAACCCUUAAUG	0	0	0	0	0	1	0	−
27.	UAAAACUCUG	0	0	1	1	0	0	0	−
	UAAAG/CUCUG	1	1	0	0	1	1	1	+
Incomplete families									
28.[d]	CAUUCAG	1	0	0	0	1	0	1	−
	CAUUUAG	0	1	1	0	0	0	0	−

Table 3—*Continued*

Families		B. subtilis	B. pumilus	B. megaterium	B. cereus	B. pasteurii	S. ureae	B. stearothermophilus	S. inulinus
29.[d]	CCAUCAUUAAG	0	0	0	1	0	0	0	−
	CCAUCAUUCAG	0	0	0	0	0	1	0	−
	CCAUCACUUACAG	0	0	0	0	0	0	0	+
30.	UACCUUG	1	0	1	0	1	1	0	−
	CACCUUG	0	1	0	1	0	0	0	−
31.	CUAACG	1	1	1	0	0	1	1	−
	CAAACG	0	0	0	0	1	0	0	−
32.	UUUCCG	1	1	1	1	1	1	0	−
	UAUCCG	0	0	0	0	0	0	0	+
33.	CUAAUCUCAUAAAACCG	0	0	1	0	0	0	0	−
	CCAAUCCCAUAAAUCUG	0	0	0	1	0	0	0	−
	CCAAUCCCACAAAUCUG	1?	0	0	0	0	0	0	−
	CUAAUCCCAUAAAACCG	0	0	0	0	1	0	0	−
	CCAAUCCCACAAAAUCG	0	0	0	0	0	1	0	−
	CCAAUCCCAUAAAG	0	0	0	0	0	0	0	+
	AAUCCCAAAAG	0	0	0	0	0	0	1	−
34.	AACAAAG	1	1	0	0	0	0	0	−
	UACAAAG	0	0	1	1	0	0	1	+
	UAAAAAG[b]	0	0	0	0	0	1	0	−
35.	AUCUUAG	1	1	1	1	0	0	0	+
	ACCUUAG	0	0	0	0	1	0	0	
36.	ACAAUCCUAG	1	0	0	0	0	0	0	−
	ACAACUCUAG	0	0	1?	0	0	0	0	−
	ACAACCCUAG	0	1	0	1	0	0	0	−
	ACAACCCAAG	0	0	0	0	0	0	1	−
37.	UAACCUUUUAG	1	0	0	0	0	0	0	−
	UAACCUUUUUG	0	0	0	1?	0	0	0	−
	UAACCUUUAUG	0	1	0	0	0	0	0	−
	AACCUUUAUG	0	0	0	0	0	0	0	+
	UAACCCUUAG	0	0	0	0	0	1	0	−
	UAACCCUUUUG	0	0	0	0	1	0	0	−
38.	CCAACCCG	0	0	0	0	1	0	0	−
	CAAACCCG	0	0	0	0	0	1	0	−
39.	UCACUUAUG	0	0	0	1	0	0	0	−
	UCACUUACAG	0	1	0	0	0	0	0	−

[a] The notation here is similar to that in Table 1, except that tentative sequences are not indicated.
[b] Membership in family is tentative.
[c] 1?, The underlined portion of the sequence indicated is not well determined in this organism.
[d] These families may be tentatively combined to form one complete family.

DISCUSSION

The limited group of species of the genus *Bacillus* examined here serves to illustrate the potential of the comparative cataloging approach in the study of bacterial systematics. To be genuinely useful, the method must not only give reliable results in general, but it must be effective where traditional techniques have not proven adequate.

Since many of the organisms examined here have been well characterized by more conventional methods, relationships herein defined can be compared to existing classifications. The arrangement of strains originally suggested by Smith et al. (14) and recently updated by Gordon et al. (5) was based on the most extensive study available. They classify the genus *Bacillus* into three major groups, based primarily on the shape of the spores and the swelling of the sporangium by the spore. Of the organisms examined here, *B. megaterium*, *B. cereus*, *B. subtilis*, and *B. pumilus* occur in their group 1, *B. stearothermophilus* occurs in group 2, and *B. pasteurii* occurs in group 3. Among the group 1 organisms, Gordon et al. (5) further suggested that *B. subtilis* and *B. pumilus* were very closely related, whereas *B. megaterium* and *B. cereus* were closer to each other than either was to any other species. These same conclusions also emerged from the present approach. In addition, our results provided strong evidence that *S. ureae*, whose classification has traditionally been difficult, should not only be classified as a member of the genus *Bacillus*, but should be placed in group 3 along with *B. pasteurii* (11).

Comparative cataloging of 16S rRNA offers several advantages over traditional methods. (i) The identical procedure can be applied to any genus, resulting in similarity coefficients which are directly comparable. Hence, data from a diversity of organisms can be readily and objectively integrated, ultimately permitting quantitative definition of terms such as "species," "genus," and "family." (ii) Exceedingly ancient divergences should be detectable, allowing identification of relationships throughout the procaryotic kingdom. That this is indeed the case has already been demonstrated by the succesful application of the comparative cataloging approach to the elucidation

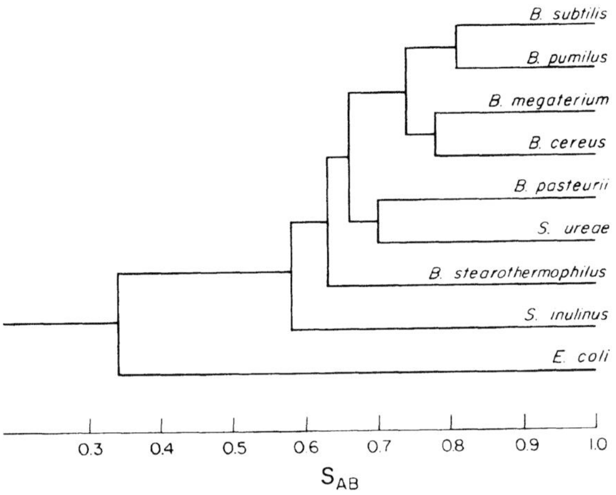

FIG. 2. *Dendrogram for strains indicated derived by single linkage clustering using a Dice type similarity coefficient, S_{AB}.*

TABLE 4. *Number of oligonucleotide differences—complete families*

Strain	B. subtilis	B. pumilus	B. megaterium	B. cereus	B. pasteurii	S. ureae	B. stearothermophilus	S. inulinus
B. subtilis		0.81	0.73	0.73	0.65	0.66	0.63	0.56
B. pumilus	3		0.74	0.73	0.64	0.65	0.63	0.57
B. megaterium	9	7		0.78	0.62	0.63	0.61	0.58
B. cereus	8	8	5		0.61	0.62	0.59	0.54
B. pasteurii	24	21	22	21		0.70	0.56	0.54
S. ureae	22	18	21	20	12		0.62	0.50
B. stearothermophilus	21	20	18	19	24	23		0.53

Similarity coefficient (S_{AB}) with $L = 6$

of the procaryotic nature of the chloroplast (2, 22).

Comparative cataloging is not, however, a panacea. Compared to traditional techniques, it is relatively expensive and time consuming, and it requires considerable specialized expertise. Thus it is appropriate to view comparative cataloging of 16S rRNA as an adjunct to and not a replacement for the more usual approaches.

For phylogenetic purposes it would be highly desirable to define a similarity coefficient that relates in a linear fashion the actual number of base changes between two 16S rRNA molecules. S_{AB} as defined herein is not such a coefficient. A similarity coefficient, P_{AB}, which has this property, was previously proposed (18a, 20). However, the derivation of this coefficient assumes that the probability of a mutation being accepted is the same at all positions in the 16S rRNA. This has been shown to be a nonvalid assumption (19), and therefore P_{AB} is only likely to be a useful approximation of true homology for closely related organisms. Thus, until these matters receive further attention, S_{AB}, which is readily calculated and cannot be misconstrued, seems to be the association coefficient of choice.

ACKNOWLEDGMENTS

This study was supported by NASA grant NSG-7044 and Public Health Service grant AI-6457 from the National Institute of Allergy and Infectious Diseases to C. R. W. We thank L. B. Zablen and C. D. Pribula for technical assistance in obtaining data on *Sporolactobacillus inulinus* 16S rRNA.

REPRINT REQUESTS

Address reprint requests to: Dr. C. R. Woese, Department of Genetics and Development, University of Illinois, Urbana, Ill. 61801.

LITERATURE CITED

1. Bellemare, G., R. Vigne, and B. Jordan. 1973. Interaction between *Escherichia coli* ribosomal proteins and 5S RNA molecules: recognition of prokaryotic 5S RNAs and rejection of eukaryotic 5S RNA. Biochimie 55:29–35.
2. Bonen, L., and W. F. Doolittle. 1975. On the prokaryotic nature of red algal chloroplasts. Proc. Natl. Acad. Sci. U.S.A. 72:2310–2314.
3. Drake, J. W. 1974. The role of mutation in microbial evolution, p. 41–58. *In* M. J. Carlile and J. J. Skehel (ed.), 24th Symp. Soc. Gen. Microbiol. Cambridge University Press, London.
4. Fox, G. E., and C. R. Woese. 1975. The architecture of 5S rRNA and its relation to function. J. Mol. Evol. 6:61–76.
5. Gordon, R. E., W. C. Haynes, and C. Hor-Nay Pang. 1973. The genus *Bacillus*. Agriculture Handbook #427, U.S. Department of Agriculture, Washington, D.C.
6. Hill, L. R. 1975. Interlocking numerical taxonomies. Int. J. Syst. Bacteriol. 25:245–251.
7. Jaskunas, S. R., M. Nomura, and J. Davies. 1974. Genetics of bacterial ribosomes, p. 333–368. *In* M. Nomura, A. Tissières, and P. Lengyel (ed.), Ribosomes. Cold Spring Harbor Laboratory, Cold Spring Harbor, New York.
8. Jones, D., and P. H. A. Sneath. 1970. Genetic transfer and bacterial taxonomy. Bacteriol. Rev. 34:40–81.
9. Nomura, M., P. Traub, and H. Bechmann. 1968. Hybrid 30S ribosomal particles reconstituted from components of different bacterial origins. Nature (London) 219:793–799.
10. Pace, B., and L. L. Campbell. 1971. Homology of ribosomal ribonucleic acid of diverse bacterial species with *Escherichia coli* and *Bacillus stearothermophilus*. J. Bacteriol. 107:543–547.
11. Pechman, K. J., B. J. Lewis, and C. R. Woese. 1976. Phylogenetic status of *Sporosarcina ureae*. Int. J. Syst. Bacteriol. 26:305–310.
12. Pechmann, K. J., and C. Woese. 1972. Characterization of the primary structural homology between the 16S ribosomal RNAs of *Escherichia coli* and *Bacillus megaterium* by oligomer cataloging. J. Mol. Evol. 1:230–240.
13. Sanger, F., G. G. Brownlee, and B. G. Barrell. 1965. A two dimensional fractionation procedure for radioactive nucleotides. J. Mol. Biol. 13:373–398.
14. Smith, N. R., R. E. Gordon, and F. E. Clark. 1952. Aerobic spore forming bacteria. Agriculture Monograph No. 16. U.S. Department of Agriculture, Washington, D.C.
15. Sogin, M. L., K. J. Pechman, L. Zablen, B. J. Lewis, and C. R. Woese. 1972. Observations on the post-transcriptionally modified nucleotides in the 16S ribosomal ribonucleic acid. J. Bacteriol. 112:13–16.
16. Sogin, S. J., M. L. Sogin, and C. R. Woese. 1972. Phylogenetic measurement in procaryotes by primary structural characterization. J. Mol. Evol. 1:173–184.
17. Uchida, T., L. Bonen, H. W. Schaup, B. J. Lewis, L. Zablen, and C. Woese. 1974. The use of ribonuclease U_2 in RNA sequnce determination: some corrections in the catalog of oligomers produced by ribonuclease T_1 digestion of *Escherichia coli* 16S ribosomal RNA. J. Mol. Evol. 3:63–77.
18. Wittman, H. G. 1976. Structure, function, and evolution of ribosomes. Eur. J. Biochem. 61:1–13.
18a. Woese, C., M. Sogin, D. Stahl, B. J. Lewis, and L. Bonen. 1976. A comparison of the 16S ribosomal RNAs from mesophilic and thermophilic *Bacilli*: some modifications in the Sanger method for RNA sequencing. J. Mol. Evol. 7:197–213.
19. Woese, C. R., G. E. Fox, L. Zablen, T. Uchida, L. Bonen, K. Pechman, B. J. Lewis, and D. Stahl. 1975. Conservation of primary structure in 16S ribosomal RNA. Nature (London) 254:83–86.
20. Woese, C. R., M. L. Sogin, and L. A. Sutton. 1974. Procaryote phylogeny I: concerning the relatedness of *Aerobacter aerogenes* to *Escherichia coli*. J. Mol. Evol. 3:293–299.
21. Wrede, P., and V. A. Erdmann. 1973. Activities of *B. stearothermophilus* 50S ribosomes reconstituted with prokaryotic and eucaryotic 5S RNA. FEBS Lett. 33:315–319.
22. Zablen, L. B., M. S. Kissil, C. R. Woese, and D. E. Buetow. 1975. Phylogenetic origin of the chloroplast and prokaryotic nature of its ribosomal RNA. Proc. Natl. Acad. Sci. U.S.A. 72:2418–2422.
23. Zablen, L., L. Bonen, R. Meyer, and C. R. Woese. 1975. The phylogenetic status of *Pasteurella pestis*. J. Mol. Evol. 4:347–358.
24. Zuckerkandl, E., and L. Pauling. 1965. Molecules as documents of evolutionary history. J. Theoret. Biol. 8:357–366.

Retrieval of Concentrated and Undecompressed Microbial Populations from the Deep Sea

H. W. JANNASCH AND C. O. WIRSEN

Holger Jannasch, with many colleagues, left a legacy of knowledge and observation regarding microorganisms and their activities in the oceans. His work demonstrated many important aspects related to microbial growth in seawater at low substrate concentrations, the presence of several types of psychrophilic bacteria and barophilic bacteria, and the astounding microbial activities around deep-sea hydrothermal vents. These discoveries, which showed that microbial sulfur oxidation supports dense animal communities in the absence of light, were a result, according to Jannasch, of close collaboration between biologists, geochemists, and geophysicists. The growing understanding of the many aspects of marine microbial activities did not come easily nor without the development of devices for conducting the research. This article by Jannasch and Wirsen describes a device used for sampling microbial populations at depths of 6,000 m, continued subsampling, and studying the natural population without decompression. Our understanding of the global importance of microbial activities of the oceans clearly will guide us in solving many of our environmental problems, including pollution and greenhouse gas effects.

LARS G. LJUNGDAHL

APPLIED AND ENVIRONMENTAL MICROBIOLOGY, Mar. 1977, p. 642–646

Vol. 33, No. 3

Printed in U.S.A.

Retrieval of Concentrated and Undecompressed Microbial Populations from the Deep Sea[1]

H. W. JANNASCH* AND C. O. WIRSEN

Department of Biology, Woods Hole Oceanographic Institution, Woods Hole, Massachusetts 02543

Received for publication 16 September 1976

A device for sampling at depths of up to 6,000 m is described in which 3 liters of seawater is concentrated over a Nucleopore filter to about 13 ml and retrieved under in situ pressure and temperature. Subsamples can be withdrawn into transfer units that are equipped with individual gas accumulators for preventing loss of pressure during prolonged periods of storage. Transfer of samples or sample portions into sterile medium contained in pre-pressurized incubation vessels and continued subsampling therefrom permit time course experiments for the study of natural populations of deep-sea microorganisms in the absence of decompression. A test experiment with a water sample from a depth of 2,600 m supplemented with radioactively labeled Casamino Acids showed reduced rates of substrate incorporation and respiration as compared with data from a decompressed control. The barotolerance observed in this study was characterized by reduced, rather than equal, activities recorded at elevated pressures as compared with 1-atm controls.

The need for studying deep-sea microorganisms in the absence of decompression has been discussed with emphasis on the possible pressure sensitivity of potentially pressure-adapted organisms by ZoBell (5) and Jannasch et al. (3). In the latter publication, we reported on data obtained by using two pressure-retaining samplers for the retrieval of water from depths of up to 6,000 m and subsequently as incubation vessels. The capability to add soluble substrates and to withdraw subsamples from the 1-liter culture chamber permitted direct experimental studies on undecompressed microbial populations from the deep sea.

A limitation of this approach has been the fact that each sampler/incubation vessel could be used for only one sample per cruise. Apart from the slow accumulation of data to a level of statistical significance, possible operational failures confer a high risk with respect to invested cruise time. This situation would be vastly improved if undecompressed samples could be taken in larger numbers and from different areas, stored, and transferred to the incubation vessels later. Such an approach would keep the original sampler/incubation vessels available for continuous laboratory use. It would be necessary, however, to work with smaller sample volumes, making sample concentration desirable. This, in turn, would render it possible to study aliquots of identical

[1] Contribution no. 3844 of the Woods Hole Oceanographic Institution.

samples, a requirement for a comparative investigation of various effects of decompression on microbial populations. Finally, anticipated attempts to isolate pure cultures under pressure will depend on concentrated samples for the streaking procedure.

These definite advantages led us to construct the pressure-retaining filter sampler and storage/transfer units described in this paper. After the appropriate laboratory and pressure tests, samplings at sea and transfers to the incubation vessels have been successful, and data are reported below. Four storage/transfer units have been constructed at a cost considerably below that of one sampler/incubation vessel.

MATERIALS AND METHODS

Filter sampler. The filter sampler (Fig. 1b and 2) is lowered on the hydrowire and triggered (arrows in Fig. 1b) at two pivoted mechanisms (S) simultaneously by one messenger at the desired depth: first, opening a lid (A), which covers the sterilized channel for sample intake (a small amount of seawater entering the intake channel during descent due to compression is kept from contaminating the sample by a membrane filter inserted in the face of the lid) and, second, operating a toggle valve (F) to start the flow of seawater into the sampler.

A slow rate of filling is set by the snubbing device (J) located below the membrane, which prevents shear forces from being exerted on the sample as it passes through the 0.2 μm-pore-size Nucleopore filter (I). It has been calculated that 45 min is required for complete filling of the sampler at a depth of 6,000 m. During filling, the free-floating piston (K), ini-

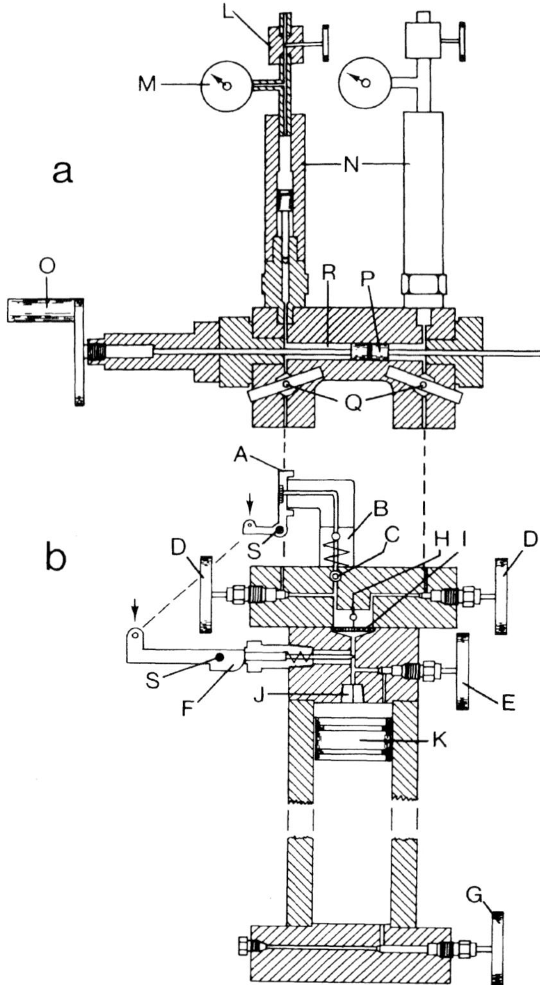

FIG. 1. (a) Transfer unit. Abbreviations: L, Gas precharge valve; M, pressure gauge; N, gas accumulator with differential piston; O, crank; P, movable piston; Q, transfer connecting valves; R, transfer chamber. (b) Filter sampler. Abbreviations: A, Intake lid; B, check valve; C, hand valve; D, connecting valves; E, bypass valve; F, toggle valve; G, gas precharge valve; H, magnetic stirring bar and scraper; I, Nucleopore filter on support; J, flow rate snubbing orifice; K, free-floating piston; S, pivot.

the filter sampler are the subject of a detailed engineering study by K. W. Doherty (unpublished data). According to his figures, retrieval of the 3-liter filter sampler from a depth of 6,000 m in the absence of a gas accumulator will result in a pressure drop of 24 atm due to expansion of the steel housing. Under the same conditions, a 1-ml leak will result in a 5.6-atm pressure loss. This value increases to a 17-atm loss if the sample is only 1 liter, as in the cultivation vessels. In other words, the smaller the volume of the sample, the more important is the need for a gas accumulator.

The check valve (B) retains the internal pressure of the sampler during ascent. Upon retrieval, the hand valve (C) is closed. The intake nozzle, as well as the check valve, are then removed before attachment of the storage/transfer unit. The pressure retained in the sampler, and thereby the actual depth of sampling, is measured by attaching a pressure gauge. If the sampler is brought to a cold room (ca. 3°C) immediately after retrieval, no substantial temperature change in the 13-ml sample above the filter will occur. Simple foam insulation is possible, but has not been found to be necessary. The filter sampler can be used to a depth of 6,000 m (600 atm), while a fourfold margin of safety is maintained.

A magnetic stirrer placed on top of the sampler will turn a stirring bar (H) contained in the sample chamber over the filter. Attached to the bar is a membrane-type scraper that gently removes attached bacteria from the filter surface. This procedure was shown to increase the relative number of viable suspended bacteria (plated at 1 atm) by a factor of approximately 3 compared with procedures that use a stirring bar without a scraper.

After sample aliquots have been removed from the filter chamber, buffered glutaraldehyde is injected into the chamber by using a transfer unit for fixation of the remaining cells on the membrane in the undecompressed state for scanning electron microscopy. (The results will be published separately.) The top section of the sampler is then washed and sterilized again by autoclaving and readied for the next deployment by precharging the lower section with nitrogen gas from a scuba tank.

Storage/transfer units. The transfer units (Fig. 1a) are, in principle, similar to those used for the removal of samples from the original sampler/incubation vessels (2, 3). Since, however, they are also intended to be used for prolonged storage of subsamples, individual gas accumulators (N) are attached to the transfer chamber (R, maximally 13 ml). The different surface areas of the floating pistons in the gas accumulators result in a hydraulic differential of a factor of 6 between the pressure measured by the gauge (M) and that in the transfer chamber (R).

After flame and alcohol sterilization of the sampler surface and filling of the connecting channels with sterile seawater, the autoclaved transfer unit is attached by eight bolts to the top of the filter sampler (Fig. 2). The piston (Fig. 1, P) has been moved to one end of the transfer chamber, which is filled with sterile seawater. After the two gas accumulators are precharged (one-sixth of the sample pressure), the connecting valves (Fig. 1, D and

tially held in its upper position by a nitrogen gas precharge in the lower part of the sampler, moves downward until inside and outside pressure have equilibrated. The sample filtrate is then contained above and the compressed gas below the piston. This gas accumulator is essential in compensating for losses of pressure caused by small volume changes, which occur during attachment of pressure gauges and transfer units, and expansion of the vessel, compression of sealing material, etc. The gas precharge (maximum of 25 atm for sampling at 6,000 m) is calculated to allow for free play of the piston at full pressure.

Calculations of some technical characteristics of

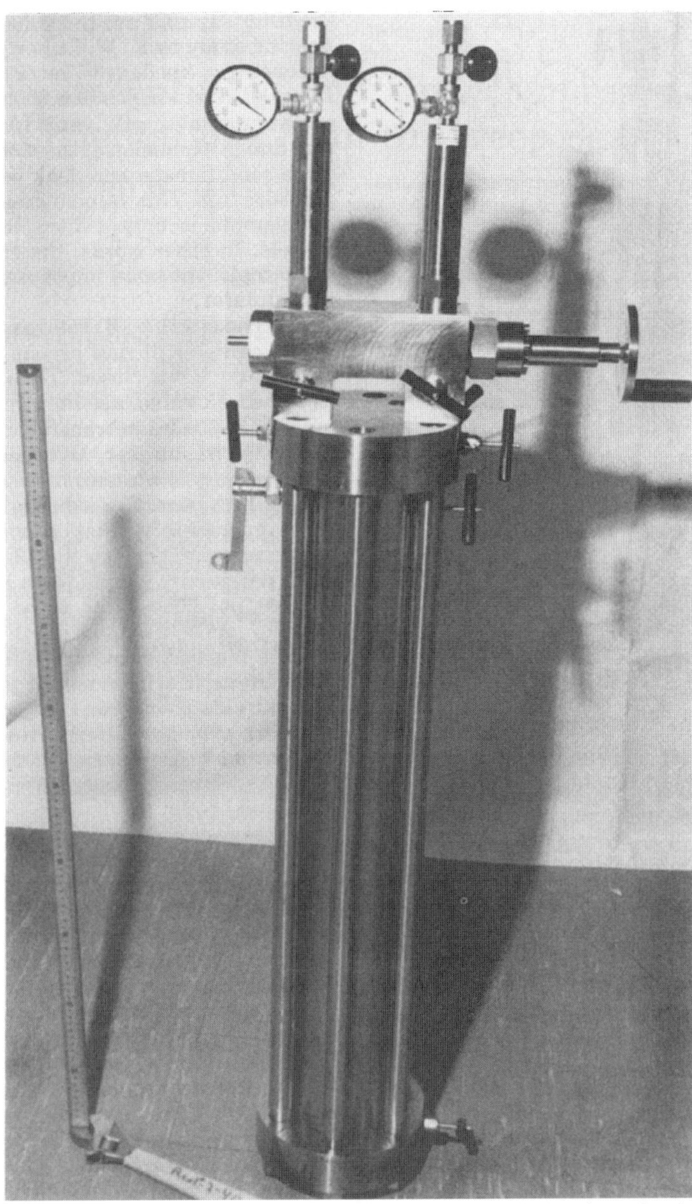

FIG. 2. *Filter sampler with attached transfer unit (scale = 1 m).*

Q) are opened, and the inside pressure of the filter sampler is transmitted to the transfer chamber. When the piston is moved by the crank (O), the whole 13-ml portion, or any fraction of it, is removed from the filter chamber while sterile seawater is flushed in through the opposite channel with an unavoidable, but slight, mixing. The connecting valves are then closed, and the transfer unit detached and stored at 3°C. Tests at 1 atm with a pure culture of a known organism have shown an average removal of 65% of the cells during transfer.

A number of transfer units containing samples from different locations or depths, or aliquots of the same sample, are kept in cold storage and brought to the laboratory for inoculation into the pre-pressur-ized incubation vessels described in detail by Jannasch et al. (2). The top plates of the incubation vessels are designed similarly to that of the filter sampler, and the inoculation procedure is essentially the same as that of sample removal, but in reverse. By this procedure, the organisms contained in 3 liters of seawater are initially concentrated to a volume of 13 ml, removed (at about 65% recovery) in the undecompressed state to a storage/transfer unit, and then reinoculated back into a pre-pressurized incubation vessel containing 1 liter of seawater with a specific substrate added.

Sampling for the specific test experiment discussed here took place in July 1975 (*Chain* cruise no. 124) at the depth of 2,600 m near the newly established Deep Ocean Station no. 2, 3,640 m) at 38°19′N, 69°41′W. The sample in the transfer unit was held in storage at 3.0°C for 45 days before transfer into the incubation vessel. The latter contained 1 liter of sterilized seawater with 5.0 mg of Casamino Acids supplemented with uniformly ^{14}C-labeled amino acids and was pressurized to the in situ pressure of 260 atm. The decompressed seawater for the control experiment was taken from a sample obtained at the same depth (2,600 m) and time with a sterile Niskin sampler. After storage for 45 days at 1 atm and 3°C, this control sample was inoculated with labeled substrate and incubated in a stoppered 1-liter Erlenmeyer flask. When samples were removed, an equal volume of sterile seawater was introduced to simulate the sampling procedure done with the pressurized culture vessel. Both the decompressed and the undecompressed samples were incubated with stirring at the in situ temperature of 3.0°C.

Analyses. A mixture of uniformly ^{14}C-labeled amino acids (I.C.N. Corp., specific activity 1.1 mCi/mg) was added to unlabeled Casamino Acids carrier solution. The final concentration in the cultures was 5.0 μg of Casamino Acids per ml, with an activity of about 0.005 μCi/ml of labeled amino acids. A 10-ml portion of each subsample, taken after various intervals of incubation, from the decompressed and undecompressed samples, was filtered through a 0.22-μm-pore-size membrane filter and washed with 2 volumes of chilled seawater to determine the amount of substrate incorporated in cell material and pools. Duplicate 0.3-ml portions of the subsamples were used for measuring $^{14}CO_2$ production (respiration) after the procedure of Wirsen and Jannasch (4). The term "total substrate utilization" refers to the sum of incorporated and respired labeled material, since remaining substrate and labeled dissolved intermediates were not measured. The radioactivity was counted using a scintillation spectrometer (Intertechnique SL-20). Quenching was corrected for by the channels ratio method.

RESULTS

The data given in Fig. 3 demonstrate an increased metabolic activity of the natural microbial population in the decompressed deep-water sample. In the undecompressed sample incubated at the in situ pressure of 260 atm, the total substrate utilization was less than half of that of the 1-atm control. Special tests on the effects of sample concentration and storage are underway.

If separated into incorporation and respiration of Casamino Acids carbon, the values were 9 and 19%, respectively, of the total substrate utilized in the undecompressed sample, as compared with 21 and 47%, respectively, in the decompressed control after 8 days. After 2 weeks of incubation of the undecompressed sample, incorporation, as well as respiration, data reached a plateau, indicating no further significant activity during an additional 3

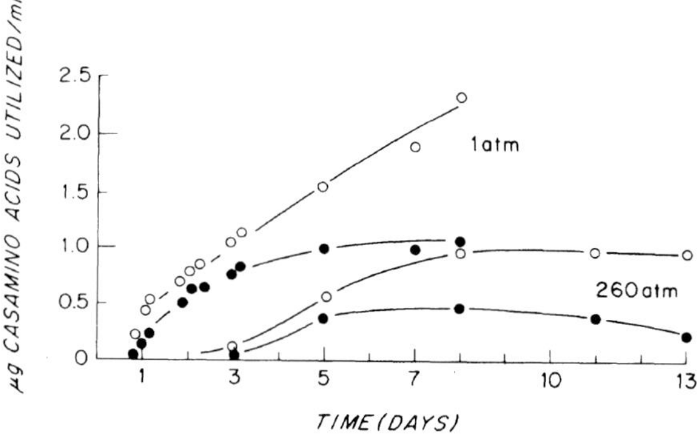

FIG. 3. *Incorporation* (●) *and respiration* (○) *of* ^{14}*C-labeled Casamino Acids (initial concentration, 5 μg/ml) in a water sample taken at a depth of 2,600 m at the North Atlantic Station "DOS 2" (38°19′N, 69°41′W) during incubation at in situ pressure (undecompressed) and at 1 atm, both at in situ temperature (3.0°C).*

weeks of measurements. When the pressure was released and the temperature raised to 22°C, at the end of the 5-week incubation period, the activity resumed after a brief lag.

DISCUSSION

The results confirm earlier data obtained by using the incubation vessels as samplers as originally intended (3). The approach described in this paper has been demonstrated to be practical and to offer substantial advantages in terms of economy and time-effort efficiency in the procurement of data.

Populations metabolizing at equal or reduced rates in the presence of elevated hydrostatic pressures as compared with 1-atm controls are defined here as barotolerant. The type of barotolerance observed in this as well as in our earlier studies (1, 3) was invariably characterized by reduced activities. It must be noted, however, that these rates of substrate conversion at in situ pressure are still significant as compared with the amounts of organic substrates that normally reach these depths.

A barophilic response, i.e., an increased activity at elevated pressure as compared with 1 atm, has not been found. This does not exclude the possible existence of barophilic bacterial species, since their relatively high metabolic activity at in situ pressure may be masked by that of a predominant population of nonbarophilic organisms. In principle, the same holds for the detection of decompression-sensitive organisms in natural populations.

To identify and to study these traits of microbial metabolism at high hydrostatic pressure, a pure-culture approach is indispensable. One of the purposes of the filter sampler system described in this paper is to provide concentrated and undecompressed cell suspensions for streaking on solidified media. Preliminary studies in this direction (C. D. Taylor and H. W. Jannasch, Abstr. Annu. Meet. Am. Soc. Microbiol. 1976 N42, p. 177) have shown that growth is not affected in a helium-oxygen atmosphere at 400 atm as compared with a hydrostatic pressure control, indicating that a streaking procedure is possible.

ACKNOWLEDGMENTS

We gratefully acknowledge the engineering and laboratory assistance of K. W. Doherty and S. J. Molyneaux.

This work was supported by the research grants DES75-15017 and OCE75-21278 from the National Science Foundation.

LITERATURE CITED

1. Jannasch, H. W., and C. O. Wirsen. 1973. Deep sea microorganisms: *In situ* response to nutrient enrichment. Science 180:641–643.
2. Jannasch, H. W., C. O. Wirsen, and C. L. Winget. 1973. A bacteriological pressure-retaining deep-sea sampler and culture vessel. Deep-Sea Res. 20:661–664.
3. Jannasch, H. W., C. O. Wirsen, and C. D. Taylor. 1976. Studies on undecompressed microbial populations from the deep sea. Appl. Environ. Microbiol. 32:360–367.
4. Wirsen, C. O., and H. W. Jannasch. 1974. Microbial transformations of some ^{14}C-labeled substrates in coastal water and sediment. Microb. Ecol. 1:25–37.
5. ZoBell, C. E. 1968. Bacterial life in the deep sea. Bull. Misaki Mar. Biol. Inst. Kyoto Univ. 12:77–96.

Whole-Genome Random Sequencing and Assembly of Haemophilus influenzae Rd

R. D. FLEISCHMANN, M. D. ADAMS, O. WHITE, ET AL.

This paper heralded a new chapter in microbiology. It was the first report of a complete sequence of a genome from a free-living organism, and it marked a new level of understanding of microbial life. It was followed by the sequencing of several other bacterial genomes. Microbiology has entered what now is called the genomic era. Having the genomic information in hand fundamentally changes the scope of questions that can be asked within every facet of microbiology, including evolution, genetics, physiology, enzymology, and biotechnology. The availability of this database, including the genome, increases the validity of and confirms work in other branches of microbiology.

JUDY WALL

Whole-Genome Random Sequencing and Assembly of *Haemophilus influenzae* Rd

Robert D. Fleischmann, Mark D. Adams, Owen White, Rebecca A. Clayton,
Ewen F. Kirkness, Anthony R. Kerlavage, Carol J. Bult, Jean-Francois Tomb,
Brian A. Dougherty, Joseph M. Merrick, Keith McKenney, Granger Sutton,
Will FitzHugh, Chris Fields,* Jeannine D. Gocayne, John Scott, Robert Shirley,
Li-Ing Liu, Anna Glodek, Jenny M. Kelley, Janice F. Weidman, Cheryl A. Phillips,
Tracy Spriggs, Eva Hedblom, Matthew D. Cotton, Teresa R. Utterback,
Michael C. Hanna, David T. Nguyen, Deborah M. Saudek, Rhonda C. Brandon,
Leah D. Fine, Janice L. Fritchman, Joyce L. Fuhrmann, N. S. M. Geoghagen,
Cheryl L. Gnehm, Lisa A. McDonald, Keith V. Small, Claire M. Fraser,
Hamilton O. Smith, J. Craig Venter†

An approach for genome analysis based on sequencing and assembly of unselected pieces of DNA from the whole chromosome has been applied to obtain the complete nucleotide sequence (1,830,137 base pairs) of the genome from the bacterium *Haemophilus influenzae* Rd. This approach eliminates the need for initial mapping efforts and is therefore applicable to the vast array of microbial species for which genome maps are unavailable. The *H. influenzae* Rd genome sequence (Genome Sequence DataBase accession number L42023) represents the only complete genome sequence from a free-living organism.

A prerequisite to understanding the complete biology of an organism is the determination of its entire genome sequence. Several viral and organellar genomes have been completely sequenced. Bacteriophage φX174 [5386 base pairs (bp)] was the first to be sequenced, by Fred Sanger and colleagues in 1977 (*1*). Sanger *et al.* were also the first to use strategy based on random (unselected) pieces of DNA, completing the genome sequence of bacteriophage λ (48,502 bp) with cloned restriction enzyme fragments (*1*). Subsequently, the 229-kb genome of cytomegalovirus (CMV) (*2*), the 192-kb genome of vaccinia (*3*), and the 187-kb mitochondrial and 121-kb chloroplast genomes of *Marchantia polymorpha* (*4*) have been sequenced. The 186-kb genome of variola (smallpox) was the first to be completely sequenced with automated technology (*5*).

At the present time, there are active genome projects for many organisms, including *Drosophila melanogaster* (*6*), *Escherichia coli* (*7*), *Saccharomyces cerevisiae* (*8*), *Bacillus subtilis* (*9*), *Caenorhabditis elegans* (*10*), and

Homo sapiens (*11*). These projects, as well as viral genome sequencing, have been based primarily on the sequencing of clones usually derived from extensively mapped restriction fragments, or λ or cosmid clones. Despite advances in DNA sequencing technology (*12*) the sequencing of genomes has not progressed beyond clones on the order of the size of λ (~40 kb). This has been primarily because of the lack of sufficient computational approaches that would enable the efficient assembly of a large number (tens of thousands) of independent, random sequences into a single assembly.

The computational methods developed to create assemblies from hundreds of thousands of 300- to 500-bp complementary DNA (cDNA) sequences (*13*) led us to test the hypothesis that segments of DNA several megabases in size, including entire microbial chromosomes, could be sequenced rapidly, accurately, and cost-effectively by applying a shotgun sequencing strategy to whole genomes. With this strategy, a single random DNA fragment library may be prepared, and the ends of a sufficient number of randomly selected fragments may be sequenced and assembled to produce the complete genome. We chose the free-living organism *Haemophilus influenzae* Rd as a pilot project because its genome size (1.8 Mb) is typical among bacteria, its G+C base composition (38 percent) is close to that of human, and a physical clone map did not exist.

Haemophilus influenzae is a small, nonmotile, Gram-negative bacterium whose only

natural host is human. Six *H. influenzae* serotype strains (a through f) have been identified on the basis of immunologically distinct capsular polysaccharide antigens. Non-typeable strains also exist and are distinguished by their lack of detectable capsular polysaccharide. They are commensal residents of the upper respiratory mucosa of children and adults and cause otitis media and respiratory tract infections, mostly in children. More serious invasive infection is caused almost exclusively by type b strains, with meningitis producing neurological sequelae in up to 50 percent of affected children. A vaccine based on the type b capsular antigen is now available and has dramatically reduced the incidence of the disease in Europe and North America.

Genome sequencing. The strategy for a shotgun approach to whole genome sequencing is outlined in Table 1. The theory follows from the Lander and Waterman (*14*) application of the equation for the Poisson distribution. The probability that a base is not sequenced is $P_\mathrm{o} = e^{-m}$, where m is the sequence coverage. Thus after 1.83 Mb of sequence has been randomly generated for the *H. influenzae* genome ($m = 1$, 1 × coverage), $P_\mathrm{o} = e^{-1} = 0.37$ and approximately 37 percent of the genome is unsequenced. Fivefold coverage (approximately 9500 clones sequenced from both insert ends and an average sequence read length of 460 bp) yields $P_\mathrm{o} = e^{-5} = 0.0067$, or 0.67 percent unsequenced. If L is genome length and n is the number of random sequence segments done, the total gap length is Le^{-m}, and the average gap size is L/n. Fivefold coverage would leave about 128 gaps averaging about 100 bp in size.

To approximate the random model during actual sequencing, procedures for library construction (*15*) and cloning (*16*) were developed. Genomic DNA from *H. influenzae* Rd strain KW20 (*17*) was mechanically sheared, digested with BAL 31 nuclease to produce blunt ends, and size-fractionated by agarose gel electrophoresis. Mechanical shearing maximizes the randomness of the DNA fragments. Fragments between 1.6 and 2.0 kb in size were excised and recovered. This narrow range was chosen to minimize variation in growth of clones. In addition, we chose this maximum size to minimize the number of complete genes that might be present in a single fragment, and thus might be lost as a result of expression of deleterious gene products. These fragments were ligated to Sma I–cut, phosphatase-treated pUC18 vector, and the ligated products were fractionated on an agarose gel. The linear vector plus insert band was excised and recovered. The ends of the linear recombinant molecules were repaired with T4 polymerase, and the molecules were then ligated into circles. This two-

J.-F. Tomb, B. A. Dougherty, and H. O. Smith are with the Johns Hopkins University School of Medicine, Baltimore, MD 21205, USA. J. M. Merrick is with the State University of New York, Department of Microbiology, Buffalo, NY, 14214, USA. K. McKenney is with the National Institute for Standards and Technology, Gaithersburg, MD 20878, USA. All other authors are with The Institute for Genomic Research (TIGR), Gaithersburg, MD, 20878, USA. The address for TIGR as of 9 September 1995 is 9712 Medical Center Drive, Rockville, MD 20850, USA.

*Present address: The National Center for Genome Resources, Santa Fe, NM, 87505, USA.
†To whom correspondence should be addressed.

stage procedure resulted in a collection of single-insert plasmid recombinants with minimal contamination from double-insert chimeras (<1 percent) or free vector (<3 percent). Because deviation from randomness is most likely to occur during cloning, *E. coli* host cells deficient in all recombination and restriction functions (*18*) were used to prevent rearrangements, deletions, and loss of clones by restriction. Transformed cells were plated directly on antibiotic diffusion plates (*16*) to avoid the usual broth recovery phase that would have allowed multiplication and selection of the most rapidly growing cells and could lead to deviation from randomness. All colonies were used for template preparation regardless of size. Only clones lost because of expression of deleterious gene products would be deleted from the library, resulting in a slight increase in gap number over that expected.

To evaluate the quality of the *H. influenzae* library, sequence data were obtained from ~4000 templates by means of the M13-21 primer. Sequence fragments were assembled with the AUTOASSEMBLER software [Applied Biosystems division of Perkin-Elmer (AB)] after obtaining 1300, 1800, 2500, 3200, and 3800 sequence fragments, and the number of unique assembled base pairs was determined. The data obtained from the assembly of up to 3800 sequence fragments were consistent with a Poisson distribution of fragments with an average "read" length of 460 bp for a genome of 1.9×10^6 bp, indicating that the library was essentially random.

Plasmid DNA templates that were double-stranded and of high quality (19,687) were prepared by a method developed in collaboration with Advanced Genetic Technology Corporation (*19*). Plasmids were prepared in a 96-well format for all stages of DNA preparation from bacterial growth through final DNA purification. Template concentration was determined with Hoechst dye and a Millipore Cytofluor 2350. DNA concentrations were not adjusted, but low-yielding templates (<30 ng/μl) were identified where possible and not sequenced. Templates were also prepared from two *H. influenzae* λ genomic libraries (*20*). An amplified library was constructed in vector λ GEM-12 and an unamplified library was constructed in λ DASH II. Both libraries contained inserts in the size range of 15 to 20 kb. Liquid lysates (10 ml) were prepared from selected plaques and templates were prepared on an anion-exchange resin (Qiagen). Sequencing reactions were carried out on plasmid templates by means of a Catalyst LabStation (AB) and PRISM Ready Reaction Dye Primer Cycle Sequencing Kits (AB) for the M13 forward (M13-21) and the M13 reverse (M13RP1)

primers (*21*). Dye terminator sequencing reactions were carried out on the λ templates on a Perkin-Elmer 9600 Thermocycler with the Applied Biosystems Prism Ready Reaction Dye Terminator Cycle Sequencing Kits. We used T7 and SP6 primers to sequence the ends of the inserts from the λ GEM-12 library and T7 and T3 primers to sequence the ends of the inserts from the λ DASH II library. Sequencing reactions (28,643) were performed by eight individuals using an average of 14 AB 373 DNA Sequencers per day over a 3-month period. All sequencing reactions were analyzed with the Stretch modification of the AB 373 sequencer. These sequencers were modified to include a heat plate and the height of the laser was reduced. With standard gel plates the "well-to-read" length was increased to 34 cm when standard sequencing plates were used and to 48 cm when 60-cm plates were used. The sequencing reactions in this project were analyzed primarily with a 34-cm well-to-read distance. The overall sequencing success rate was 84 percent for M13-21 sequences, 83 percent for M13RP1 sequences, and 65 percent for dye-terminator reactions. The average usable read length was 485 bp for M13-21 sequences, 444 bp for M13RP1 sequences, and 375 bp for dye-terminator reactions. The high-throughput sequencing phase of the project is summarized in Table 2.

We balanced the desirability of sequencing templates from both ends, in terms of ordering of contigs and reducing the cost of lower total number of templates, against shorter read lengths for sequencing reactions performed with the M13RP1 primer compared to the M13-21 primer. Approximately one-half of the templates were sequenced from both ends. Altogether, 9297 M13RP1 sequencing reactions were done. Random reverse sequencing reactions were done on the basis of successful forward se-

quencing reactions. Some M13RP1 sequences were obtained in a semidirected fashion; for example, M13-21 sequences pointing outward at the ends of contigs were chosen for M13RP1 sequencing in an effort to specifically order contigs. The semidirected strategy was effective, and clone-based ordering formed an integral part of assembly and gap closure.

In the course of our research on expressed sequence tags (ESTs), we developed a laboratory information management system for a large-scale sequencing laboratory (*22*). The system was designed to automate data flow wherever possible and to reduce user error. It has at its core a series of databases developed with the Sybase relational data management system. The databases store and correlate all information collected during the entire operation from template preparation to final analysis. Although the system was originally designed for EST projects, many of its features were applicable or easily modified for a genomic sequencing project. Because the raw output of the AB 373 sequencers is collected on a Macintosh system and our data management system is based on a Unix system, it was necessary to design and implement multiuser, client-server applications that allow the raw data as well as analysis results to flow seamlessly into the database with a minimum of user effort. To process data collected by the AB 3735, sequence files were first analyzed with FACTURA, an AB program that runs on the Macintosh and is designed for automatic vector sequence removal and end-trimming of sequence files. The Macintosh program ESP, written at The Institute for Genomic Research (TIGR), loaded the feature data extracted from sequence files by FACTURA to the Unix-based *H. influenzae* relational database. Assembly was accom-

Table 1. Whole-genome sequencing strategy.

Stage	Description
Random small insert and large insert library construction	Shear genomic DNA randomly to ~2 kb and 15 to 20 kb, respectively
Library plating	Verify random nature of library and maximize random selection of small insert and large insert clones for template production
High-throughput DNA sequencing	Sequence sufficient number of sequence fragments from both ends for 6× coverage
Assembly	Assemble random sequence fragments and identify repeat regions
Gap closure	
Physical gaps	Order all contigs (fingerprints, peptide links, λ clones, PCR) and provide templates for closure
Sequence gaps	Complete the genome sequence by primer walking
Editing	Inspect the sequence visually and resolve sequence ambiguities, including frameshifts
Annotation	Identify and describe all predicted coding regions (putative identifications, starts and stops, role assignments, operons, regulatory regions)

plished by first retrieving a specified set of sequence files and their associated features by means of STP, another TIGR program, which is an X-windows graphical interface that retrieves sequences from the database with user-defined queries.

TIGR ASSEMBLER is the software component that enabled us to assemble the *H. influenzae* genome. It simultaneously clusters and assembles fragments of the genome. In order to obtain the speed necessary to assemble more than 10^4 fragments, the algorithm builds a table of all 10-bp oligonucleotide subsequences to generate a list of potential sequence fragment overlaps. When TIGR ASSEMBLER is used, a single fragment begins the initial contig; to extend the contig, a candidate fragment is chosen with the best overlap based on oligonucleotide content. The current contig and candidate fragment are aligned by a modified version of the Smith-Waterman (23) algorithm, which provides for optimal gapped alignments. The contig is extended by the fragment only if strict criteria for the quality of the match are met. The match criteria include the minimum length of overlap, the maximum length of an unmatched end, and the minimum percentage match. The algorithm automatically lowers these criteria in regions of minimal coverage and raises them in regions with a possible repetitive element. The number of potential overlaps for each fragment determines which fragments are likely to fall into repetitive elements. Fragments representing the boundaries of repetitive elements and potentially chimeric fragments are often rejected on the basis of partial mismatches at the needs of alignments and excluded from the contig.

TIGR ASSEMBLER was designed to take advantage of clone size information coupled with sequence information from both ends of each template. It enforces the constraint that sequence fragments from two ends of the same template point toward one another in the contig and are located within a certain range of base pairs (definable for each clone on the basis of the insert length or the clone size range for a given library). In order for the assembly process to be successful it was essential that the sequence data be of the highest quality and that sequence fragment lengths be sufficient to span most small repeats. Less than 13 percent of our random sequence fragments were smaller than 400 bp after vector removal and end trimming. Assembly of 24,304 sequence fragments of *H. influenzae* required 30 hours of central processing unit time with the use of one processor on a SPARCenter 2000 containing 512 Mb of RAM. This process resulted in approximately 210 contigs. Because of the high stringency of the TIGR ASSEMBLER, all contigs were searched against each other with GRASTA, which is a modified version of the program FASTA (24). In this way, additional overlaps that enabled compression of the data set into 140 contigs were detected. The location of each fragment in the contigs and extensive information about the consensus sequence itself were loaded into the *H. influenzae* relational database.

After assembly, the relative positions of the 140 contigs were unknown. The program ASM_ALIGN, developed at TIGR, identified clones whose forward and reverse sequencing reactions indicated that they were in different contigs and ordered and displayed these relationships. With this program, the 140 contigs were placed into 42 groups totaling 42 physical gaps (no template DNA for the region) and 98 sequence gaps (template available for gap closure).

Four integrated strategies were developed to order contigs separated by physical gaps. Oligonucleotide primers were designed and synthesized from the end of each contig group. These primers were then available for use in one or more of the strategies outlined below:

1) DNA hybridization (Southern) analysis was done to develop a "fingerprint" for a subset of 72 of the above oligonucleotides. This procedure was based on the supposition that labeled oligonucleotides homologous to the ends of adjacent contigs should hybridize to common DNA restriction fragments, and thus share a similar or identical hybridization pattern or fingerprint (25). Adjacent contigs identified in this manner were targeted for specific PCR reactions.

2) Peptide links were made by searching each contig end with BLASTX (26) against a peptide database. If the ends of two contigs matched the same database sequence appropriately, then the two contigs were tentatively considered to be adjacent.

Table 2. Summary of features of whole-genome sequencing of *H. influenzae* Rd.

Description	Number
Double-stranded templates	19,687
Forward-sequencing reactions (M13-21 primer)	19,346
Successful (%)	16,240 (84)
Average edited read length (bp)	485
Reverse sequencing reactions (M13RP1 primer)	9,297
Successful (%)	7,744 (83)
Average edited read length (bp)	444
Sequence fragments in random assembly	24,304
Total base pairs	11,631,485
Contigs	140
Physical gap closure	42
PCR	37
Southern analysis	15
λ clones	23
Peptide links	2
Terminator sequencing reactions*	3,530
Successful (%)	2,404 (68)
Average edited read length (bp)	375
Genome size (bp)	1,830,137
G+C content (%)	38
rRNA operons	6
rrnA, rrnC, rrnD (spacer region) (bp)	723
rrnB, rrnE, rrnF (spacer region) (bp)	478
tRNA genes identified	54
Number of predicted coding regions	1,743
Unassigned role (%)	736 (42)
No database match	389
Match hypothetical proteins	347
Assigned role (%)	1,007 (58)
Amino acid metabolism	68 (6.8)
Biosynthesis of cofactors, prosthetic groups, and carriers	54 (5.4)
Cell envelope	84 (8.3)
Cellular processes	53 (5.3)
Central intermediary metabolism	30 (3.0)
Energy metabolism	105 (10.4)
Fatty acid and phospholipid metabolism	25 (2.5)
Purines, pyrimidines, nucleosides and nucleotides	53 (5.3)
Regulatory functions	64 (6.3)
Replication	87 (8.6)
Transcription	27 (2.7)
Translation	141 (14.0)
Transport and binding proteins	123 (12.2)
Other	93 (9.2)

*Includes gap closure, walks on rRNA repeats, random end-sequencing of λ clones for assembly confirmation, and alternative reactions for ambiguity resolution.

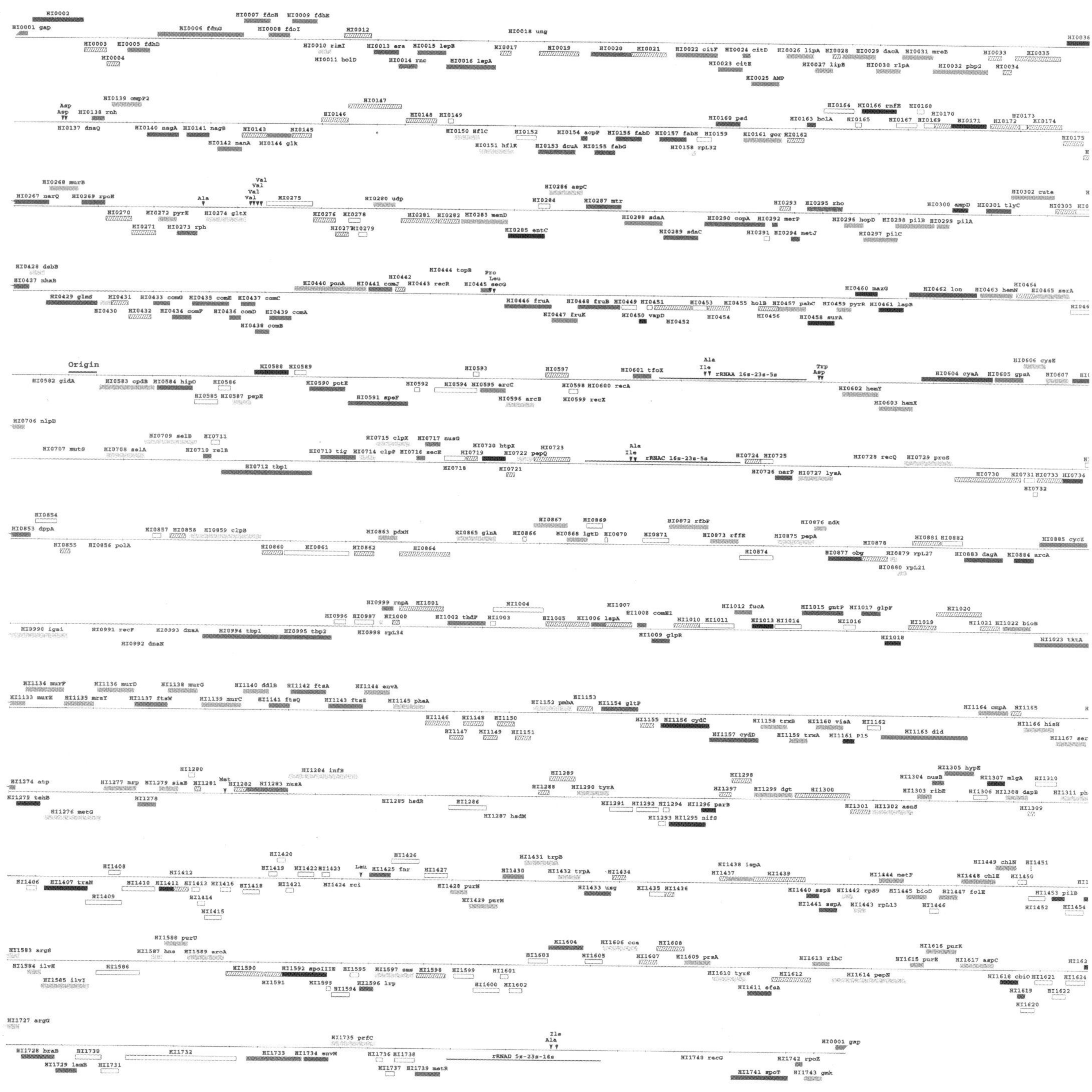

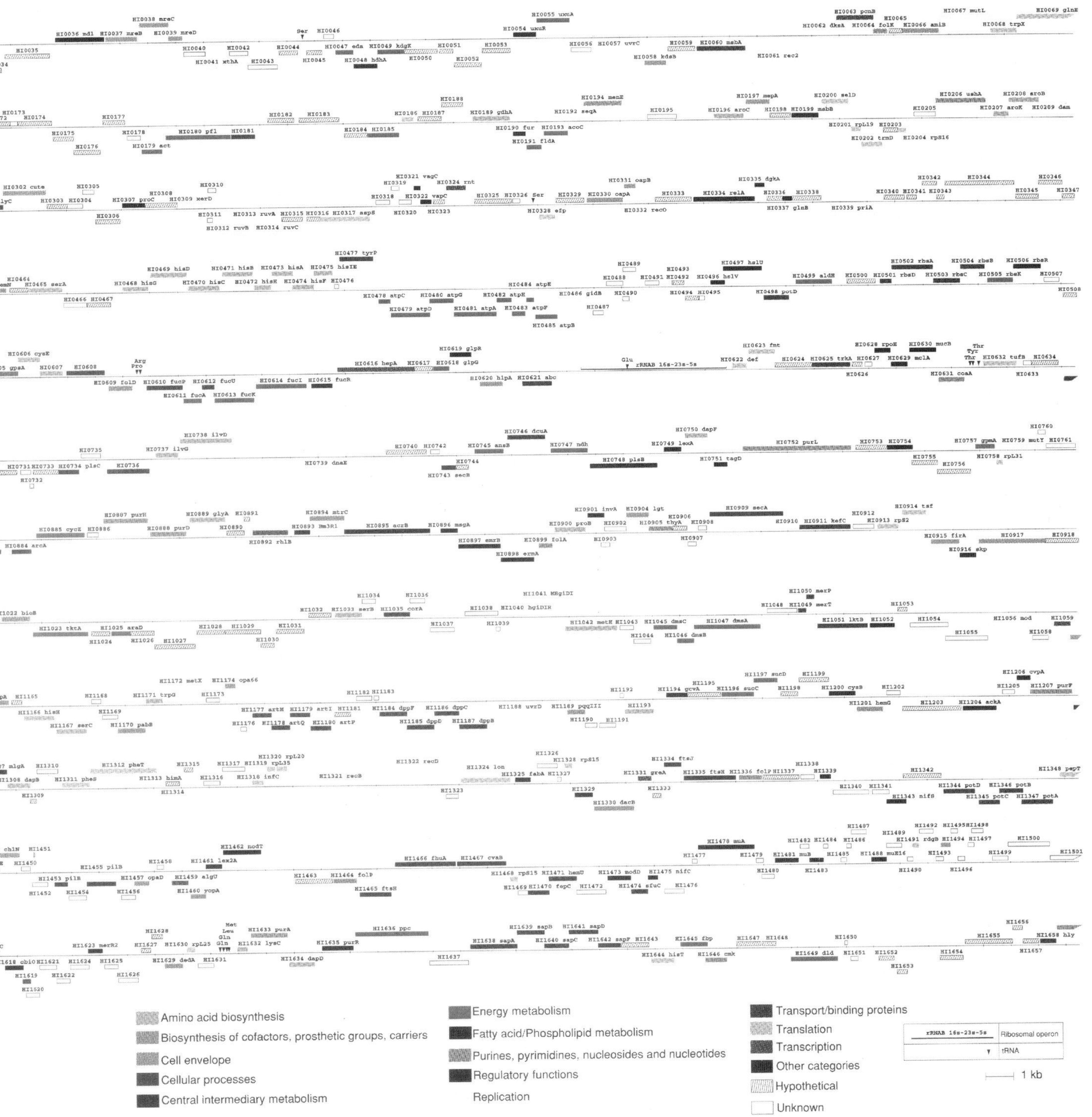

Amino acid biosynthesis
Biosynthesis of cofactors, prosthetic groups, carriers
Cell envelope
Cellular processes
Central intermediary metabolism
Energy metabolism
Fatty acid/Phospholipid metabolism
Purines, pyrimidines, nucleosides and nucleotides
Regulatory functions
Replication
Transport/binding proteins
Translation
Transcription
Other categories
Hypothetical
Unknown
rRNAB 16s-23s-5s   Ribosomal operon
tRNA
1 kb

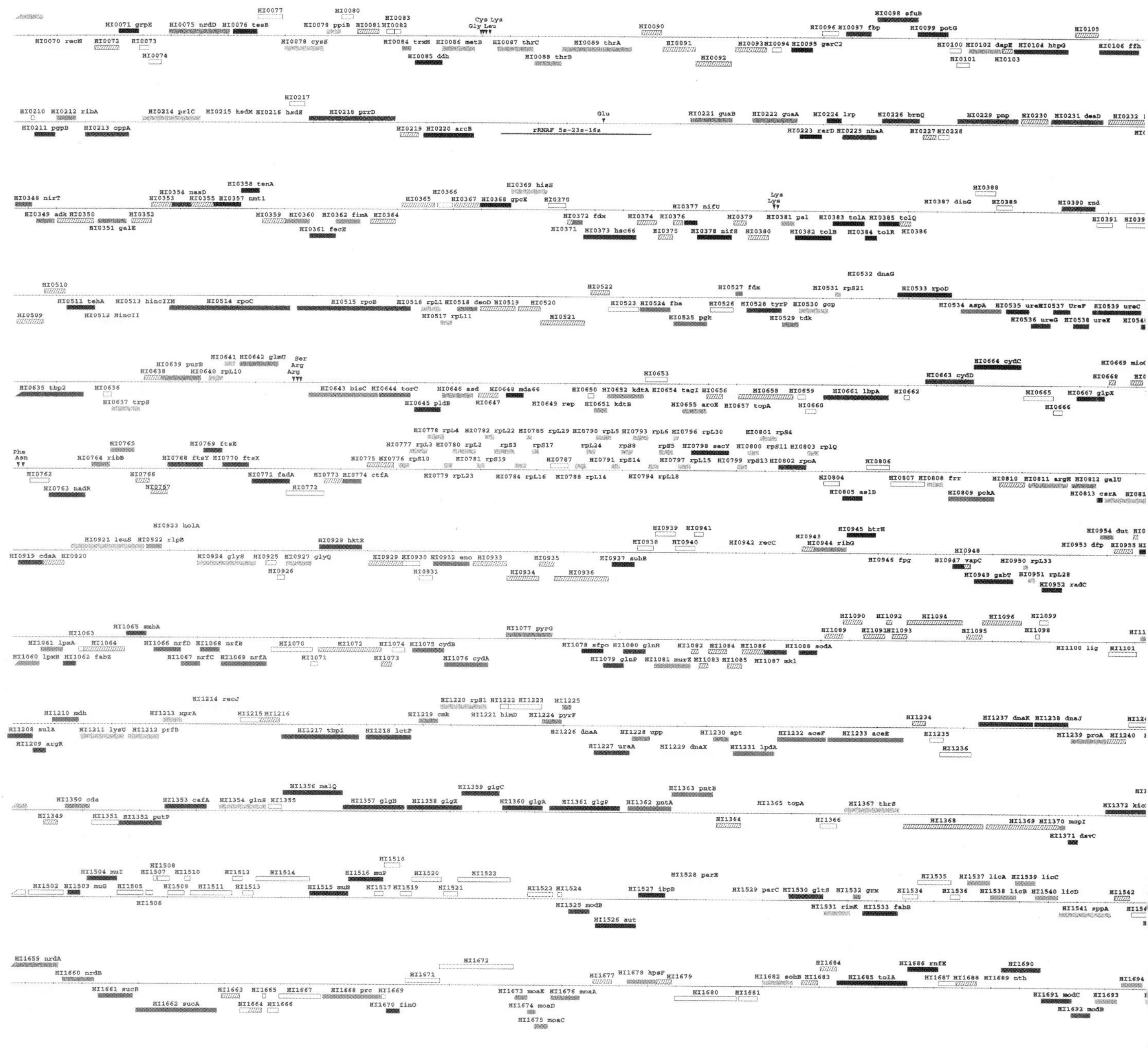

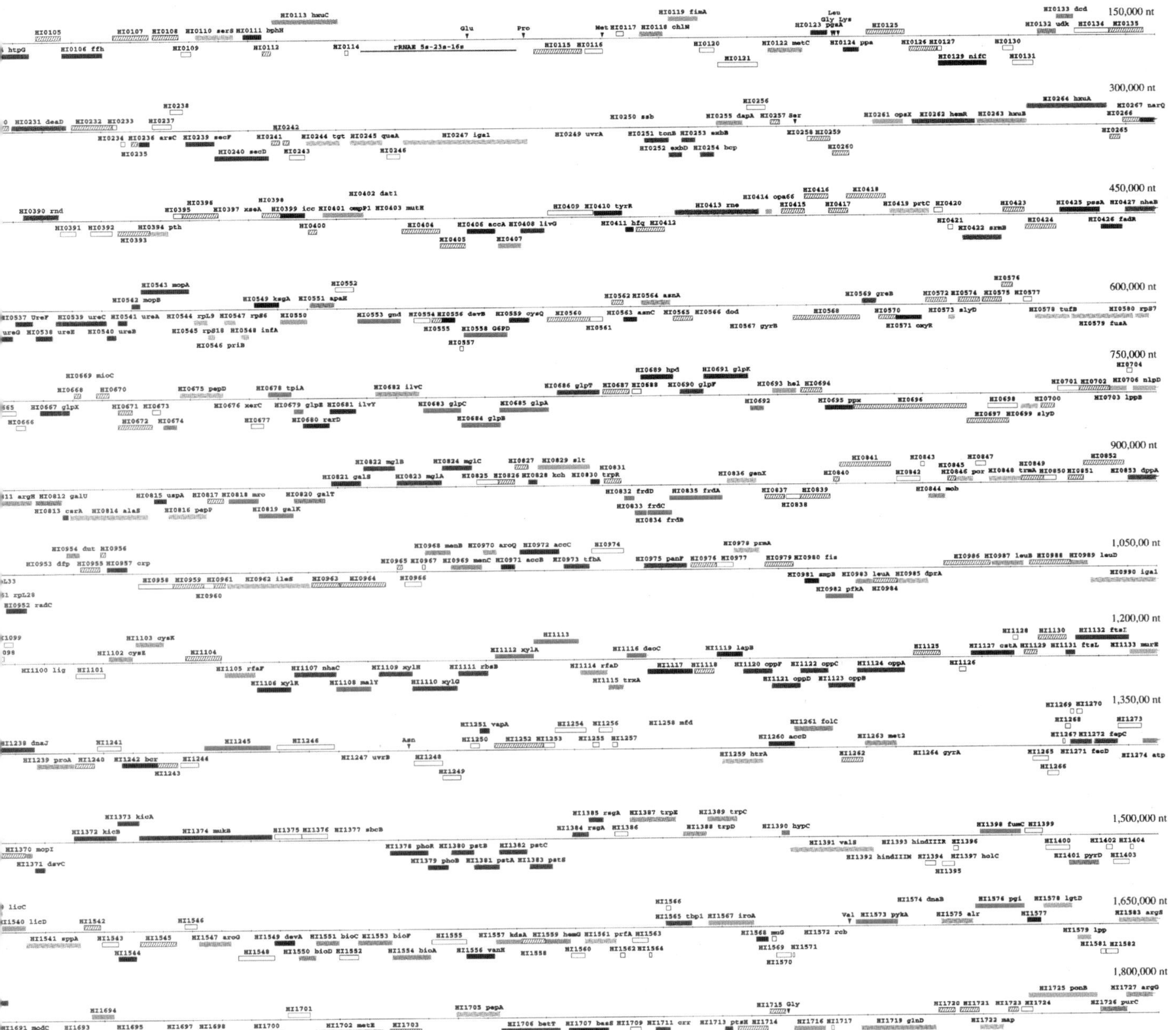

150,000 nt
HI0105
HI0106 ffh
HI0107 HI0108 HI0110 serS HI0111 hphN
HI0113 hxuC
HI0119 fimA
HI0123 pgsA
Leu
Gly Lys
HI0132 udk
HI0133 dcd
HI0125
htpG
HI0109
HI0112
Glu
Pro
Met HI0117 HI0118 chlN
HI0114 rRNAB 5s-23s-16s
HI0115 HI0116
HI0120
HI0122 metC
HI0124 ppa
HI0126 HI0127
HI0130
HI0134 HI0135
HI0121
HI0131
HI0129 nifC
300,000 nt
HI0231 deaD HI0232 HI0233
HI0238
HI0237
HI0256
HI0250 ssb
HI0255 dapA HI0257 Ser
HI0261 opaX HI0262 hemR HI0263 hxuB
HI0264 hxuA
HI0267 narQ
HI0234 HI0236 arsC HI0239 secF
HI0242
HI0244 tgt HI0245 quaA
HI0247 lgal
HI0249 uvrA
HI0251 tonB HI0253 exbB
HI0258 HI0259
HI0266
HI0235
HI0240 secD
HI0241
HI0243
HI0346
HI0252 exbD HI0254 bcp
HI0260
HI0265
450,000 nt
HI0390 rnd
HI0395
HI0396
HI0398
HI0402 dat1
HI0409 HI0410 tyrR
HI0413 rne
HI0414 opa66
HI0416
HI0419
HI0423
HI0425 pssA HI0427 nhaB
HI0391 HI0392 HI0394 pth
HI0397 xseA HI0399 icc HI0401 ompP1 HI0403 mutH
HI0400
HI0406 accA HI0408 ilvG
HI0411 hfq HI0412
HI0415
HI0417
HI0419 prtC HI0420
HI0421 HI0422 srmB
HI0424
HI0426 fadR
HI0393
HI0404
HI0405
HI0407
600,000 nt
HI0537 UreF HI0539 ureC HI0541 ureA HI0544 rpL9 HI0547 rpS6
HI0543 mopA
HI0552
HI0562 HI0564 asnA
HI0569 greB
HI0572 HI0574 HI0575 HI0577
HI0576
HI0578 tufB
HI0580 rpS7
ureG HI0538 ureE HI0540 ureB
HI0542 mopB
HI0549 ksgA HI0551 apaH
HI0550
HI0553 gnd HI0554 HI0556 devB HI0559 cysQ
HI0560
HI0561 asmC HI0565 HI0566 dod
HI0568
HI0570
HI0573 slyD
HI0579 fusA
HI0545 rpS18 HI0548 infA
HI0555 HI0558 G6PD
HI0561
HI0567 gyrB
HI0571 oxyR
HI0546 priB
HI0557
750,000 nt
HI0704
HI0669 mioC
HI0675 pepD
HI0678 tpiA
HI0682 ilvC
HI0686 glpT
HI0687 HI0688 HI0690 glpF
HI0693 hel HI0694
HI0701 HI0702 HI0706 nlpD
HI0668
HI0670
HI0676 xerC HI0679 glpR HI0681 ilvY
HI0683 glpC
HI0685 glpA
HI0692
HI0695 ppx
HI0696
HI0698
HI0700
HI0703 lppB
HI0667 glpX
HI0671 HI0673
HI0684 glpB
HI0697 HI0699 slyD
HI0666
HI0672 HI0674
HI0677
HI0680 narD
900,000 nt
HI0822 mglB
HI0824 mglC
HI0827 HI0829 slt
HI0841
HI0843
HI0847
HI0849
HI0852
HI0821 galS
HI0823 mglA
HI0825 HI0826 HI0828 kch HI0830 trpB
HI0831
HI0836 genX
HI0840
HI0845
HI0846 por HI0848 trmA HI0850 HI0851
HI0853 dppA
argH HI0812 galU
HI0815 uspA HI0817 HI0818 aroE HI0820 galT
HI0832 frdD
HI0835 frdA
HI0837 HI0839
HI0844 moh
HI0813 carA HI0814 alaS
HI0816 pepP
HI0819 galK
HI0833 frdC
HI0838
HI0834 frdB
1,050,000 nt
HI0954 dut HI0956
HI0968 menB HI0970 aroQ HI0972 accC
HI0976
HI0978 prmA
HI0986 HI0987 leuB HI0988 HI0989 leuD
HI0953 dfp HI0955 HI0957 crp
HI0965 HI0967 HI0969 menC HI0971 accB
HI0973 tfbA
HI0975 panF HI0976 HI0977
HI0979 HI0980 fis
HI0990 lgal
rpL33
HI0958 HI0959 HI0961
HI0962 ileS
HI0963 HI0964
HI0966
HI0981 ampB HI0983 leuA HI0985 dprA
rpS1 rpL28
HI0960
HI0982 pfkA HI0984
HI0952 radC
1,200,000 nt
HI1099
HI1103 cysK
HI1113
HI1116 deoC
HI1119 lapB
HI1128 HI1130
HI1132 ftsI
HI1098
HI1102 cysE
HI1104
HI1112 xylA
HI1117 HI1118
HI1120 oppF HI1122 oppC
HI1124 oppA
HI1137 cstA HI1129 HI1131 ftsL HI1133 murE
HI1100 lig HI1101
HI1105 rfaF
HI1107 nhaC
HI1109 xylH
HI1111 rbsB
HI1114 rfaD
HI1121 oppD HI1123 oppB
HI1126
HI1106 xylR
HI1108 malY
HI1110 xylG
HI1115 trxA
HI1269 HI1270
1,350,000 nt
HI1238 dnaJ
HI1241
HI1245
HI1246
HI1251 vapA
HI1254 HI1256
HI1258 mfd
HI1261 folC
HI1268
HI1273
HI1239 proA HI1240 HI1242 ber
HI1244
Asn
HI1250
HI1252 HI1253
HI1255 HI1257
HI1260 accD
HI1263 met2
HI1247 HI1272 fspC
HI1239 proA HI1240 HI1242 ber
HI1244
HI1247 uvrB
HI1248
HI1259 btrA
HI1262
HI1264 gyrA
HI1265 HI1271 fecD
HI1274 atp
HI1243
HI1249
HI1266
1,500,000 nt
HI1373 kicA
HI1385 rsgA HI1387 trpE
HI1389 trpC
HI1372 kicB
HI1374 mukB
HI1375 HI1376 HI1377 sbcB
HI1384 rsgA HI1386
HI1388 trpD
HI1390 hypC
HI1398 fumC HI1399
HI1370 mopI
HI1378 phoR HI1380 pstB HI1382 pstC
HI1391 valS
HI1393 hindIIIR HI1396
HI1400
HI1402 HI1404
HI1371 dsvC
HI1379 phoB HI1381 pstA HI1383 pstS
HI1392 hindIIIM HI1394 HI1397 holC
HI1401 pyrD HI1403
HI1395
1,650,000 nt
licC
HI1566
HI1574 dnaB
HI1576 pgi
HI1578 lgtD
HI1565 tbp1 HI1567 iroA
Val HI1573 pykA
HI1575 elr
HI1577
HI1583 argS
HI1540 licD HI1542
HI1546
HI1541 zppA HI1543
HI1545 HI1547 aroG HI1549 devA HI1551 bioC HI1553 bioF
HI1555 HI1557 kdsA HI1559 hemG HI1561 prfA HI1563
HI1568 mnG
HI1572 rcb
HI1579 lpp
HI1544
HI1548 HI1550 bioD HI1552
HI1554 bioA HI1556 vanH
HI1558 HI1560 HI1562 HI1564
HI1569 HI1571
HI1581 HI1582
HI1570
1,800,000 nt
HI1694
HI1701
HI1705 pepA
HI1715 Gly
HI1720 HI1721 HI1723 HI1724
HI1725 ponB
HI1727 argG
HI1691 modC HI1693 HI1695 HI1697 HI1698
HI1700
HI1702 metE HI1703
HI1706 betT HI1707 basS HI1709 HI1711 crr HI1713 ptsH HI1714
HI1716 HI1717
HI1719 glnD
HI1722 map
HI1726 purC
HI1692 modB HI1696
HI1699
HI1704
HI1708 basR HI1710 HI1712 ptsI
HI1718

The Genome of *Haemophilus influenzae* Rd

Figure 2. Gene map of the *H. influenzae* Rd genome. Predicted coding regions are shown on each strand. The rRNA and tRNA genes are shown as lines and triangles, respectively. Genes are color-coded by role category as described in the Figure key. Gene identification numbers correspond to those in Table 3. Where possible, three-letter designations are also provided. In the region containing ribosomal proteins HI0782-HI0796 some identification numbers have been omitted because of space limitations. Predicted coding regions with similarity to database sequences designated as hypothetical coding regions are represented as white, cross-hatched rectangles. Predicted coding regions that have no database match are represented as white, unfilled rectangles.

Table 3. Identification of *H. influenzae* genes. Gene identification numbers are listed with the prefix HI in Fig. 3. Each identified gene is listed in its role category [adapted from Riley (*36*)]. The percentage of similarity (Sim) of the best match to the NRBP (as described in the text) is also shown. The amino acid substitution matrix used in the BLAZE analysis is BLOSUM60. An expanded version of this table with additional match information, including species, is available via World Wide Web (URL: http://www.tigr.org/). Abbreviations used: Ac, acetyl; ATase, aminotransferase; BP, binding protein; biosyn, biosynthesis; CoA, coenzyme A; DCase, decarboxylase; DHase, dehydrogenase; DMSO, dimethyl sulfoxide; f-Met, formylmethionine; G3PD, glyceraldehyde-3-phosphate dehydrogenase; GABA, γ-aminobutyric acid; GlcNAc, *N*-acetylglucosamine; LOS, Lipooligosaccharide; lpp, lipoprotein; MTase, methyltransferase; MurNAc, *N*-acetylmuramyl; P, phosphate; prt, protein; PRTase, phosphoribosyltransferase; RDase, reductase; SAM, *S*-adenosylmethionine; Sase, synthase-synthetase; sub, subunit; Tase, transferase. The following hypothetical proteins were matched from the other species as indicated (percent similarity in parentheses after gene identification number): *Alcaligenes eutrophus*: 1053(52); *Anabaena variabilis*: 1349(54); *Bacillus subtilis*: 0115(53), 0259(54), 0355(61), 0404(47), 0415(69), 0416(63), 0417(66), 0454(64), 0456(56), 0522(54), 0687(49), 0775(54), 0959(50), 1083(53), 1203(63), 1627(59), 1647(81), 1648(65), 1654(64); Bacteriophage P22: 1412(54); *Buchnera aphidicola*: 1199(65); *Campylobacter jejuni*: 0560(71); *Chromatium vinosum*: 0105(75); *Clostridium acetobutylicum*: 0773(72); *Clostridium kluyveri*: 0976(48); *Clostridium perfringens*: 0143(58); *Coxiella burnetii*: 1590(74), 1591(50); *Erwinia carotovora*: 1436(72); *Escherichia coli*: 0003(52), 0012(67), 0017(91), 0028(68), 0033(90), 0034(84), 0035(79), 0044(80), 0045(67), 0050(70), 0051(50), 0052(56), 0053(56), 0059(72), 0065(75), 0072(65), 0081(71), 0091(72), 0092(49), 0093(59), 0103(71), 0107(54), 0108(65), 0125(88), 0126(87), 0135(68), 0145(69), 0146(58), 0147(61), 0148(62), 0162(47), 0172(67), 0174(84), 0175(70), 0176(87), 0182(60), 0183(66), 0184(73), 0187(58), 0188(81), 0198(75), 0203(86), 0227(51), 0230(71), 0232(69), 0235(80), 0241(82), 0242(50), 0258(95), 0257(76), 0265(77), 0266(83), 0270(80), 0271(73), 0276(70), 0281(76), 0282(59), 0293(61), 0303(81), 0306(70), 0308(58), 0315(87), 0316(68), 0329(79), 0336(91), 0338(68), 0340(72), 0341(84), 0342(60), 0343(67), 0344(85), 0345(82), 0346(77), 0347(67), 0364(55), 0365(86), 0367(48), 0371(84), 0374(64), 0375(62), 0376(75), 0379(57), 0380(58), 0386(76); 0393(93), 0396(54), 0398(72), 0400(65), 0409(69), 0412(85), 0418(68), 0423(67), 0424(66), 0431(76), 0432(68), 0442(93), 0452(73), 0464(78), 0467(80), 0493(64), 0494(69), 0500(63), 0508(82), 0509(69), 0510(74), 0519(71), 0520(59), 0521(58), 0562(83), 0565(63), 0568(71), 0570(80), 0572(70), 0574(63), 0575(80), 0576(65), 0597(57), 0617(54), 0624(72), 0626(81), 0634(78), 0638(68), 0647(64), 0656(74), 0658(56), 0668(76), 0670(83), 0671(87), 0696(54), 0697(64), 0700(77), 0702(71), 0719(86), 0721(78), 0723(73), 0724(64), 0730(65), 0733(55), 0744(70), 0755(61), 0756(60), 0766(87), 0767(72), 0810(74), 0817(68), 0826(70), 0827(86), 0831(77), 0837(74), 0839(69), 0840(72), 0841(66), 0849(75), 0851(71), 0852(66), 0855(75), 0858(68), 0860(86), 0862(81), 0864(92), 0878(77), 0881(81), 0890(69), 0891(79), 0906(71), 0918(81), 0929(58), 0933(71), 0934(52), 0935(63), 0936(64), 0943(83), 0948(67), 0955(72), 0956(73), 0963(67), 0965(81), 0979(79), 0984(79), 0986(81), 0988(85), 1000(80), 1001(75), 1005(61), 1007(86), 1010(53), 1019(65), 1020(65), 1021(71), 1024(67), 1026(85), 1027(72), 1028(77), 1029(83), 1030(62), 1031(87), 1032(79), 1064(57), 1072(57), 1073(62), 1082(67), 1084(61), 1085(76), 1086(89), 1089(70), 1090(82), 1091(76), 1092(73), 1093(72), 1094(81), 1095(79), 1096(64), 1104(53), 1118(84), 1125(87), 1129(77), 1130(80), 1146(80), 1147(68), 1148(88), 1149(73), 1150(59), 1151(81), 1153(84), 1155(79), 1165(87), 1181(68), 1195(76), 1198(85), 1216(73), 1234(80), 1240(77), 1243(74), 1252(93), 1262(61), 1280(71), 1282(74), 1288(84), 1289(74), 1297(67), 1298(69), 1300(58), 1301(82), 1309(67), 1314(70), 1315(66), 1333(79), 1337(84), 1342(57), 1364(56), 1368(53), 1369(44), 1437(72), 1463(84), 1542(61), 1545(80), 1558(62), 1598(58), 1608(76), 1612(72), 1628(61), 1643(70), 1652(68), 1653(88), 1655(56), 1656(69), 1657(65), 1664(50), 1677(72), 1679(69), 1703(74), 1704(73), 1714(78), 1715(86), 1721(71), 1723(92); *Klebsiella pneumoniae*: 0021(63); *Lactobacillus johnsonii*: 0112(54), 1720(55); *Lactococcus lactis*: 0555(69); *Mycobacterium leprae*: 0004(62), 0019(62), 0136(58), 0260(56), 0694(54), 0740(56), 0920(57), 1663(55); *Mycoplasma hyopneumoniae*: 1281(71); *Pasteurella haemolytica*: 0219(92); *Pseudomonas aeruginosa*: 0090(68), 0177(56); *Rhodobacter capsulatus*: 0170(62), 0672(59), 1439(65), 1683(75), 1684(60), 1688(58); *Salmonella typhimurium*: 0405(51), 0964(67), 1434(76), 1607(51); *Shigella flexneri*: 0277(52); *Streptococcus parasanguis*: 0359(65); *Synechococcus* sp.: 0961(70); *Vibrio parahaemolyticus*: 0323(87), 0325(75); *Vibrio* sp.: 0333(70); *Yersinia enterocolitica*: 0753(69).

HI#	Identification	%Sim

Amino acid biosynthesis

Aromatic amino acid family
HI#	Identification	%Sim
0970	3-dehydroquinase (aroQ)	83
0208	3-dehydroquinate Sase (aroB)	77
0472	amidotransferase (hisH)	70
1387	anthranilate Sase component I (trpE)	73
1388	anthranilate Sase component II (trpD)	74
1389	anthranilate isomerase (trpC)	75
1171	anthranilate Sase Gln amidotransferase (trpG)	59
0468	ATP PRTase (hisG)	82
1290	chorismate mutase (tyrA)	77
1145	chorismate mutase-prephenate dehydratase (pheA)	75
0196	chorismate Sase (aroC)	88
1547	DAHP Sase (aroG)	84
0607	dehydroquinase shikimate DHase	48
1589	enolpyruvylshikimatephosphateSyn (aroA)	98
1166	Gln amidotransferase (hisH)	61
0469	histidinol dehydrogenase (hisD)	78
0474	hisF cyclase (hisF)	91
0470	histidinol-P ATase (hisC)	77
0471	imidazoleglycerol-P dehydratase (hisB)	81
0475	phosphoribosyl-AMP cyclohydrolase (hisIE)	77
0473	phosphoribosylformimino-5-aminoimidazole caarboximde ribotide isomerase (hisA)	77
0655	shikimate Sase (aroK)	70
0207	shikimic acid kinase I (aroK)	88
1432	Trp Sase α chain (trpA)	73
1431	Trp Sase β chain (trpB)	90

Aspartate family
HI#	Identification	%Sim
0564	Asn Sase A (asnA)	77
0286	Asp ATase (aspC)	54
1617	Asp ATase (aspC)	84
0646	Asp-semialdehyde DHase (asd)	85
1632	aspartokinase III (lysC)	73
0089	aspartokinase-homoserine DHase (thrA)	77
1042	B12-dependent homocysteine-N5-methyltetrahydrofolate transmethylase (metH)	70
0122	β-cystathionase (metC)	84
0086	cystathionine γ-Sase (metB)	62
1308	dehydrodipicolinate RDase (dapB)	83
0727	diaminopimelate DCase (lysA)	79
0750	diaminopimelate epimerase (dapF)	86
0255	dihydrodipicolinate Sase (dapA)	80
1263	homoserine acetyltransferase (met2)	57
0088	homoserine kinase (thrB)	81
0102	succinyl-diaminopimelate desuccinylase (dapE)	80
1634	tetrahydrodipicolinate N-succinyltransferase (dapD)	99
1702	tetrahydropteroyltriglutamate MTase (metE)	68
0087	Thr Sase (thrC)	81

Branched chain family
HI#	Identification	%Sim
0989	3-isopropylmalate dehydratase (leuD)	86
0987	3-isopropylmalate DHase (leuB)	80
0737	acetohydroxy acid Sase II (ilvG)	80
1585	acetolactate Sase III large chain (ilvI)	84
1584	acetolactate Sase III small chain (ilvH)	85
1193	branched-chain amino acid transaminase	49
0738	dihydroxyacid dehydrase (ilvD)	90
0983	α isopropylmalate Sase (leuA)	100
0682	ketol acid reductoisomerase (ilvC)	90

Glutamate family
HI#	Identification	%Sim
0811	argininosuccinate lyase (argH)	84
1727	argininosuccinate Sase (argG)	87
0900	γ-glutamyl kinase (proB)	84
1239	γ-glutamyl-P RDase (proA)	79
0865	Gln Sase (glnA)	86
0189	Glu DHase (gdhA)	84
0596	ornithine carbamoyltransferase (arcB)	91
1719	uridylyl Tase (glnD)	68

Pyruvate family
HI#	Identification	%Sim
1575	Ala racemase, biosynthetic (alr)	75

Serine family
HI#	Identification	%Sim
1102	Cys Sase (cysZ)	76
1103	Cys Sase (cysK)	84
0465	phosphoglycerate DHase (serA)	84
1167	phosphoserine ATase (serC)	72
1033	phosphoserine phosphatase (serB)	70
0606	Ser acetyltransferase (cysE)	88
0889	Ser hydroxymethyltransferase (glyA)	94

Biosynthesis of cofactors, prosthetic groups, and carriers

Biotin
HI#	Identification	%Sim
1554	7,8-diamino-pelargonic acid ATase (bioA)	74
1553	7-keto-8-aminopelargonic acid Sase (bioF)	56
1551	biotin synthesis prt (bioC)	47
0643	biotin sulfoxide RDase (bisC)	72
1022	biotin biosyn (bioB)	78
1550	dethiobiotin Sase (bioD)	60
1445	dethiobiotin Sase (bioD)	62

Folic acid
HI#	Identification	%Sim
1444	5,10-methylenetetrahydrofolate RDase (metF)	83
0609	5,10-methylenetetrahydrofolate DHase (folD)	82
0064	7,8-dihydro-6-hydroxymethylpterin-pyrophosphokinase (folK)	78
0457	aminodeoxychorismate lyase (pabC)	67
1629	dedA	55
0899	dehydrofolate RDase, type I (folA)	68
1336	dihydropteroate Sase (folP)	71
1464	dihydropteroate Sase (folP)	71
1261	folylpolyglutamate Sase (folC)	68
1447	GTP cyclohydrolase I (folE)	79
1170	p-aminobenzoate Sase (pabB)	54

Heme and porphyrin
HI#	Identification	%Sim
1160	ferrochelatase (visA)	69
0113	heme utilization prt (hxuC)	46
0263	heme-hemopexin utilization (hxuB)	99
0463	oxygen-independent coproporphyrinogen III oxidase (hemN)	52
0602	protoporphyrinogen oxidase homolog	64
1201	protoporphyrinogen oxidase (hemG)	57
1559	protoporphyrinogen oxidase (hemG)	73
0603	uroporphyrinogen III methylase (hemX)	60

Lipoate
HI#	Identification	%Sim
0026	lipoate biosyn prt A (lipA)	84
0027	lipoate biosyn prt B (lipB)	84

Menaquinone and ubiquinone
HI#	Identification	%Sim
0283	2-succinyl-6-hydroxy-2,4-cyclohexadiene-1-carboxylate Sase (menD)	64
0969	4-(2'-carboxyphenyl)-4-oxybutyric acid Sase (menC)	74
1189	coenzyme PQQ synthesis prt III (pqqIII)	49
0968	dihydroxynaphthoic acid Sase (menB)	95
1438	farnesyldiphosphate Sase (ispA)	73
0194	O-succinylbenzoate-CoA Sase (menE)	67

Molybdopterin
HI#	Identification	%Sim
1676	molybdenum biosyn prt A (moaA)	78
1675	molybdenum biosyn prt C (moaC)	89
1370	molybdenum-pterin-BP (mopI)	74
1448	molybdopterin biosyn prt (chlE)	73
0118	molybdopterin biosyn prt (chlN)	78
1449	molybdopterin biosyn prt (chlN)	78
1674	molybdopterin converting factor, sub 1 (moaD)	79
1673	molybdopterin converting factor, sub 2 (moaE)	76
0844	molybdopterin-dinucleotide biosyn prt (mob)	62

Pantothenate
HI#	Identification	%Sim
0953	pantothenate metabolism flavoprotein (dfp)	77
0631	pantothenate kinase (coaA)	78

Pyridoxine
HI#	Identification	%Sim
0863	pyridoxamine phosphate oxidase (pdxH)	73

Riboflavin
HI#	Identification	%Sim
0764	3,4-dihydroxy-2-butanone 4-P Sase (ribB)	83
0212	GTP cyclohydrolase II (ribA)	81
0944	riboflavin biosyn prt (ribG)	76
1613	riboflavin Sase α chain (ribC)	82
1303	riboflavin Sase β chain (ribE)	90

Thioredoxin, glutaredoxin, and glutathione
HI#	Identification	%Sim
0161	glutathione RDase (gor)	85
1115	thioredoxin (trxA)	59
1159	thioredoxin (trxA)	62
0084	thioredoxin m (trxM)	79

Cell envelope

Membranes, lipoproteins, and porins
HI#	Identification	%Sim
1579	15 kD peptidoglycan-assoc lpp (lpp)	95
0620	28 kD membrane prt (hlpA)	100
0302	apolipoprotein N-acyltransferase (cute)	64
0407	hydrophobic membrane prt	61
0360	hydrophobic membrane prt	67
1567	iron-regulated outer membrane prt A (iroA)	51
0693	lpp (hel)	100
0706	lpp (nlpD)	65
0703	lpp B (lppB)	90
0894	membrane fusion prt (mtrC)	54
0401	outer membrane prt P1 (ompP1)	97
0139	outer membrane prt P2 (ompP2)	98
1164	outer membrane prt P5 (ompA)	96
0904	prolipoprotein diacylglyceryl Tase (lgt)	80
0030	rare lpp A (rlpA)	58
0922	rare lpp B (rlpB)	62

Murein sacculus and peptidoglycan
HI#	Identification	%Sim
1140	D-Ala-D-Ala ligase (ddlB)	76
1330	D-alanyl-D-Ala carboxypeptidase (dacB)	68
1138	GlcNAc transferase (murG)	76
1494	MurNAc-L-Ala amidase	62
0066	N-acetylmuramoyl-L-Ala amidase (amiB)	77
0440	penicillin-BP (ponA)	100
1725	penicillin-BP 1B (ponB)	67
0032	penicillin-BP 2 (pbp2)	74
1668	penicillin-BP 3 (prc)	70
0029	penicillin-BP 5 (dacA)	68
0197	penicillin-insensitive murein endopeptidase (mepA)	67
0381	peptidoglycan-assoc outer membrane lpp (pal)	100
1135	phospho-N-acetylmuramoyl-pentapeptide-Tase E (mraY)	89
0031	rod shape-determining prt (mreB)	81
0037	rod shape-determining prt (mreB)	90
0038	rod shape-determining prt (mreC)	74
0039	rod shape-determining prt (mreD)	72
0829	soluble lytic murein transglycosylase (slt)	59
1081	UDP-GlcNAc enolpyruvyl Tase (murZ)	85
1139	UDP-MurNAc-Ala ligase (murC)	82
1136	UDP-MurNAc-Ala-D-Glu ligase (murD)	74
1134	UDP-MurNAc-pentapeptide Sase (murF)	68
1133	UDP-MurNAc-tripeptide Sase (murE)	73
0268	UDP-NAc-enolpyruvoylglucosamine RDase (murB)	76

Surface polysaccharides, lipopolysaccharides and antigens
HI#	Identification	%Sim
1557	2-dehydro-3-deoxyphosphooctonate aldolase (kdsA)	92
0652	3-deoxy-D-manno-octulosonic-acid Tase (kdtA)	70
1105	ADP-heptose-lps heptosyltransferase II (rfaF)	79
1114	ADP-L-glycero-D-mannoheptose-6-epimerase (rfaD)	88
0058	CTP:CMP-3-deoxy-D-manno-octulosonate-cytidylyl-transferase (kdsB)	82
0868	glycosyl Tase (lgtD)	55
1578	glycosyl Tase (lgtD)	64
1678	kpsF prt (kpsF)	71
1537	lic-1 operon prt (licA)	100
1538	lic-1 operon prt (licB)	99
1539	lic-1 operon prt (licC)	99
1540	lic-1 operon prt (licD)	94
1060	lipid A disaccharide Sase (lpxB)	77
0765	LOS biosyn prt	60
0550	LOS biosyn prt	99
0651	lipopolysaccharide core biosyn prt (kdtB)	76
1700	lsg locus prt 1	100
0867	lsg locus prt 1	83
1699	lsg locus prt 2	99
1698	lsg locus prt 3	97
1697	lsg locus prt 4	98
1696	lsg locus prt 5	98
1695	lsg locus prt 6	99
1694	lsg locus prt 7	98
1693	lsg locus prt 8	99
0261	lipopolysaccharide biosyn prt (opsX)	57
1716	rfe prt	77
1144	UDP-3-O-acyl GlcNAc deacetylase (envA)	88
0915	UDP-3-O-(R-3-hydroxymyristoyl)-glucosamine N-acetyltransferase (firA)	91
1061	UDP-GlcNAc acetyltransferase (lpxA)	79
0873	UDP-GlcNAc epimerase (rffE)	79
0872	undecaprenyl-P Gal-P Tase (rfbP)	75

Surface structures
HI#	Identification	%Sim
0119	adhesin B precursor (fimA)	48
0362	adhesin B precursor (fimA)	62
0330	cell envelope prt (oapA)	100
0331	opacity assoc prt (oapB)	99
1174	opacity prt (opa66)	59
0414	opacity prt (opa66)	91
1457	opacity prt (opaD)	56
1460	outer membrane adhesin (yopA)	62
0299	pilin biogenesis prt (pilA)	52
0298	pilin biogenesis prt (pilB)	65
0297	pilin biogenesis prt (pilC)	57
0917	protective surface antigen D15	99

Cellular processes

Cell division
HI#	Identification	%Sim
0769	cell division ATP-BP (ftsE)	78
1208	cell division inhibitor (sulA)	56
1142	cell division prt (ftsA)	74
1335	cell division prt (ftsH)	88
1465	cell division prt (ftsH)	88
1334	cell division prt (ftsJ)	90
1131	cell division prt (ftsL)	60
1141	cell division prt (ftsQ)	58
1137	cell division prt (ftsW)	75
0768	cell division prt (ftsY)	81
1143	cell division prt (ftsZ)	83
1374	cell division prt (mukB)	77
1353	cytoplasmic axial filament prt (cafA)	86
0770	cell division membrane prt (ftsX)	70
1065	mukB suppressor prt (smbA)	90
1132	penicillin-BP 3 (ftsI)	71

Cell killing
HI#	Identification	%Sim
0301	hemolysin (tlyC)	58
1658	hemolysin, 21 kD (hly)	72
1373	killing prt (kicA)	84
1372	killing prt suppressor (kicB)	83
1051	leukotoxin secretion ATP-BP (lktB)	55

Chaperones
HI#	Identification	%Sim
0373	heat shock cognate prt 66 (hsc66)	82
1238	heat shock prt (dnaJ)	83
1237	heat shock prt 70 (dnaK)	88
0104	heat shock prt C62.5 (htpG)	88
0543	heat shock prt groEL (mopA)	95
0542	heat shock prt groES (mopB)	95

Detoxification
HI#	Identification	%Sim
0928	catalase (hktE)	99
1088	superoxide dismutase (sodA)	100
1002	thiophene and furan oxidation prt (thdF)	85

Protein and peptide secretion
HI#	Identification	%Sim
1467	colicin V secretion ATP-BP (cvaB)	56
0016	GTP-binding membrane prt (lepA)	91
1006	lpp signal peptidase (lspA)	72
1642	peptide transport system ATP-BP (sapF)	71
0716	preprotein translocase (secE)	62
0798	preprotein translocase (secY)	87
0240	protein-export membrane prt (secD)	77
0239	protein-export membrane prt (secF)	73
0445	protein-export membrane prt (secG)	81
0743	protein-export prt (secB)	81
0909	preprotein translocase sub (secA)	82
0015	signal peptidase I (lepB)	65
0106	signal recognition particle prt 54 (ffh)	91
0713	trigger factor (tig)	80
0296	type 4 prepilin-like prt specific leader peptidase (hopD)	49

Transformation
HI#	Identification	%Sim
1008	competence locus E (comE1)	70
0601	tfoX	100
0439	transformation prt (comA)	100
0438	transformation prt (comB)	100
0437	transformation prt (comC)	100
0436	transformation prt (comD)	100
0435	transformation prt (comE)	100
0434	transformation prt (comF)	100

Central intermediary metabolism

Amino sugars
HI#	Identification	%Sim
0140	GlcNAc-6-P deacetylase (nagA)	72
0429	Gln amidotransferase (glmS)	84
0141	glucosamine-6-P deaminase (nagB)	88

Degradation of polysaccharides
HI#	Identification	%Sim
1356	amylomaltase (malQ)	62

Other
HI#	Identification	%Sim
0048	7-α-hydroxysteroid DHase (hdhA)	55
1204	acetate kinase (ackA)	84
0949	GABA transaminase (gabT)	56
0111	glutathione Tase (bphH)	57
0691	glycerol kinase (glpK)	89
0584	hippuricase (hipO)	50
0541	urease (ureA)	76
0539	urease α sub (urea amidohydrolase) (ureC)	82
0537	urease accessory prt (UreF)	55
0538	urease prt (ureE)	57
0536	urease prt (ureG)	87
0535	urease prt (ureH)	54
0540	urease prt (ureB)	77

Phosphorus compounds
HI#	Identification	%Sim
0695	exopolyphosphatase (ppx)	77
0124	inorganic PPase (ppa)	50
0645	lysophospholipase L2 (pldB)	53

Polyamine biosynthesis
HI#	Identification	%Sim
0099	nucleotide-BP (potG)	67
0591	ornithine DCase (speF)	80

Polysaccharides - (cytoplasmic)
HI#	Identification	%Sim
1357	1,4-α-glucan branching enzyme (glgB)	80
1361	α-glucan phosphorylase (glgP)	79
1359	ADP-glucose Sase (glgC)	74
1358	glycogen operon prt (glgX)	68
1360	glycogen Sase (glgA)	71

Sulfur metabolism
HI#	Identification	%Sim
0805	arylsulfatase regulatory prt (aslB)	67
1371	desulfoviridin γ sub (dsvC)	58
0559	sulfite synthesis pathway prt (cysQ)	56

Energy metabolism

Aerobic
HI#	Identification	%Sim
1163	D-lactate DHase (dld)	48
1649	D-lactate DHase (dld)	78
0605	glycerol-3-P DHase (gpsA)	81
0747	NDH DHase (ndh)	75

Amino acids and amines
HI#	Identification	%Sim
0534	aspartase (aspA)	89
0595	carbamate kinase (arcC)	88
0745	L-asparaginase II (ansB)	81
0288	L-Ser deaminase (sdaA)	83

Anaerobic
HI#	Identification	%Sim
1047	anaerobic DMSO RDase A (dmsA)	86
1046	anaerobic DMSO RDase B (dmsB)	85
1045	anaerobic DMSO RDase C (dmsC)	55
0644	cytochrome C-type prt (torC)	55
0348	denitrification system component (nirT)	72
0009	formate DHase pathway prt (fdhE)	72
0006	formate DHase (fdnG)	79
0005	formate DHase-N affector (fdhD)	71
0008	formate DHase-O γ sub (fdoI)	72
0007	formate DHase-O, β sub (fdoH)	86
1069	formate-dependent nitrite RDase (nrfA)	75
1068	formate-dependent nitrite RDase (nrfB)	68
1067	formate-dependent nitrite RDase prt Fe-S centers (nrfC)	81
1066	formate-dependent nitrite RDase transmembrane prt (nrfD)	68
0833	fumarate RDase (frdC)	72
0832	fumarate RDase 13 kD hydrophobic prt (frdD)	77
0835	fumarate RDase, flavoprotein sub (frdA)	87
0834	fumarate RDase, iron-sulfur prt (frdB)	85
0685	G3PD, sub A (glpA)	83
0684	G3PD, sub B (glpB)	60
0683	G3PD, sub C (glpC)	76
0679	glpE	63
0618	glpG	65
1390	hydrogenase isoenzymes formation prt (hypC)	72

ATP-proton motive force interconversion
HI#	Identification	%Sim
0484	ATP Sase C chain (atpE)	82
0485	ATP Sase F0 α sub (atpB)	78

HI#	Identification	%Sim
0483	ATP Sase F0 β sub (atpF)	79
0481	ATP Sase F1 α sub (atpA)	95
0479	ATP Sase F1 β sub (atpD)	96
0482	ATP Sase F1 δ sub (atpH)	78
0478	ATP Sase F1 ε sub (atpC)	76
0480	ATP Sase F1 γ sub (atpG)	83
1274	ATP Sase sub 3 region prt (atp)	50

Electron transport

HI#	Identification	%Sim
0885	C-type cytochrome biogenesis prt (copper tolerance) (cycZ)	68
1076	cytochrome oxidase d sub I (cydA)	82
1075	cytochrome oxidase d sub II (cydB)	78
0527	ferredoxin (fdx)	77
0372	ferredoxin (fdx)	84
0191	flavodoxin (fldA)	91
1362	NAD(P) transhydrogenase sub α (pntA)	84
1363	NAD(P) transhydrogenase sub β (pntB)	88
1278	NAD(P)H-flavin oxidoreductase	55

Entner-Doudoroff

HI#	Identification	%Sim
0047	2-keto-3-deoxy-6-phosphogluconate aldolase (eda)	63
0049	2-keto-3-deoxy-D-gluconate kinase (kdgK)	64

Fermentation

HI#	Identification	%Sim
0499	aldehyde DHase (aldH)	62
0774	butyrate-acetoacetate CoA-Tase sub A (ctfA)	75
0185	glutathione-dependent formaldehyde DHase (gd-faldH)	78
1305	hydrogenase gene region (hypE)	48
1636	phosphoenolpyruvate carboxylase (ppc)	80
0180	pyruvate formate-lyase (pfl)	93
0179	pyruvate formate-lyase activating enzyme (act)	85
1430	short chain alcohol DHase	69

Gluconeogenesis

HI#	Identification	%Sim
1645	fructose-1,6-bisphosphatase (fbp)	84
0809	phosphoenolpyruvate carboxykinase (pckA)	83

Glycolysis

HI#	Identification	%Sim
0447	1-phosphofructokinase (fruK)	74
0982	6-phosphofructokinase (pfkA)	84
0932	enolase (eno)	79
0524	fructose-bisphosphate aldolase (fba)	86
1576	glucose-6-P isomerase (pgi)	89
0001	G3PD (gap)	90
0525	phosphoglycerate kinase (pgk)	91
0757	phosphoglyceromutase (gpmA)	75
1573	pyruvate kinase type II (pykA)	87
0678	triosephosphate isomerase (tpiA)	81

Pentose phosphate pathway

HI#	Identification	%Sim
0553	6-phosphogluconate DHase (gnd)	71
0558	glucose-6-P 1-DHase (G6PD)	65
1023	transketolase 1 (tktA)	88

Pyruvate dehydrogenase

HI#	Identification	%Sim
1232	dihydrolipoamide acetyltransferase (aceF)	82
0193	dihydrolipoamide acetyltransferase (acoC)	49
1231	lipoamide DHase (lpdA)	82
1233	pyruvate DHase (aceE)	84

Sugars

HI#	Identification	%Sim
0818	aldose 1-epimerase precursor (mro)	55
0055	D-mannonate hydrolase (uxuA)	86
1116	deoxyribose aldolase (deoC)	69
0613	fucokinase (fucK)	65
1012	fuculose-1-P aldolase (fucA)	52
0611	fuculose-1-P aldolase (fucA)	81
0819	galactokinase (galK)	99
0144	glucose kinase (glk)	53
0614	L-fucose isomerase (fucI)	85
1025	L-ribulose-P 4-epimerase (araD)	82
1108	mal inducer biosyn blocker (malY)	52
0142	N-acetylneuraminate lyase (nanA)	61
0505	ribokinase (rbsK)	75
1112	xylose isomerase (xylA)	87
1113	xylulose kinase	50

TCA cycle

HI#	Identification	%Sim
1662	2-oxoglutarate DHase (sucA)	81
0025	acetate:SH-citrate lyase ligase (AMP)	68
0022	citrate lyase α chain (citF)	86
0023	citrate lyase β chain (citE)	81
0024	citrate lyase γ chain (citD)	72
1661	dihydrolipoamide succinyltransferase (sucB)	84
1398	fumarate hydratase (fumC)	74
1210	malate DHase (mdh)	85
1245	malic acid enzyme	68
1197	succinyl-CoA Sase α sub (sucD)	92
1196	succinyl-CoA Sase β sub (sucC)	80

Fatty acid and phospholipid metabolism

HI#	Identification	%Sim
1062	(3R)-hydroxymyristol acyl carrier prt dehydrase (fabZ)	85
0734	1-acyl-glycerol-3-P acyltransferase (plsC)	78
0155	3-ketoacyl-acyl carrier prt RDase (fabG)	85
0771	Ac-CoA acetyltransferase (fadA)	80
0406	Ac-CoA carboxylase (accA)	88
0154	acyl carrier prt (acpP)	91
0076	acyl-CoA thioesterase II (tesB)	73
1533	β-ketoacyl-ACP Sase I (fabB)	84
0157	β-ketoacyl-acyl carrier prt Sase III (fabH)	80
0971	biotin carboxyl carrier prt (accB)	83
0972	biotin carboxylase (accC)	91
0919	CDP-diglyceride Sase (cdsA)	67
1325	D-3-hydroxydecanoyl-(acyl carrier-prt) dehydratase (fabA)	92
0335	diacylglycerol kinase (dgkA)	72
0426	fatty acid metabolism prt (fadR)	68
0748	glycerol-3-P acyltransferase (plsB)	76
0002	long chain fatty acid CoA ligase	53
0156	malonyl CoA acyl carrier prt transacylase (fabD)	82
0211	phosphatidylglycerophosphate phosphatase B (pgpB)	60
0123	phosphatidylglycerophosphate Sase (pgsA)	83
0160	phosphatidylserine DCase proenzyme (psd)	76
0425	phosphatidylserine Sase (pssA)	71
0689	prt D (hpd)	99
1734	short chain alcohol DHase homolog (envM)	85
1433	USG-1 prt (usg)	54

Purines, pyrimidines, nucleosides, and nucleotides

2'-Deoxyribonucleotide metabolism

HI#	Identification	%Sim
0075	anaerobic ribonucleoside-triphosphate RDase (nrdD)	88
0133	deoxycytidine triphosphate deaminase (dcd)	87
0954	deoxyuridinetriphosphatase (dut)	91
1532	glutaredoxin (grx)	80
1660	ribonucleoside diphosphate RDase β2 sub (nrdB)	93
1659	ribonucleoside-diphosphate RDase 1	92
1158	thioredoxin RDase (trxB)	86
0905	thymidylate Sase (thyA)	55

Nucleotide and nucleoside interconversions

HI#	Identification	%Sim
1077	CTP Sase (pyrG)	90
1299	dGTP triphosphohydrolase (dgt)	58
0132	uridine kinase (udk)	85

Purine ribonucleotide biosynthesis

HI#	Identification	%Sim
1616	5'-phosphoribosyl-5-amino-4-imidazole carboxylase II (purK)	72
1429	5'-phosphoribosyl-5-aminoimidazole Sase (purM)	87
1743	5'-guanylate kinase (gmk)	82
0349	adenylate kinase (adk)	100
0639	adenylosuccinate lyase (purB)	88
1633	adenylosuccinate Sase (purA)	87
1207	amidoPRTase (purF)	84
0752	formylglycineamide ribonucleotide Sase (purL)	82
1588	formyltetrahydrofolate hydrolase (purU)	85
0222	GMP Sase (guaA)	88
0221	inosine-5'-monophosphate DHase (guaB)	81
0876	nucleoside diphosphate kinase (ndk)	74
0888	phosphoribosylamine-Gly ligase (purD)	85
0887	phosphoribosylaminoimidazole carboxamide formyltransferase (purH)	87
1615	phosphoribosylaminoimidazole carboxylase catalytic sub (purE)	97
1428	phosphoribosylglycinamide formyltransferase (purN)	71
1609	phosphoribosylpyrophosphate Sase (prsA)	91
1726	SAICAR Sase (purC)	55

Pyrimidine ribonucleotide biosynthesis

HI#	Identification	%Sim
1401	dihydroorotate DHase (pyrD)	77
0272	orotate PRTase (pyrE)	84
1225	orotidine-5'-monophosphate DCase	83
1224	orotidine-5'-monophosphate DCase (pyrF)	79
0459	uracil PRTase (pyrR)	74

Salvage of nucleosides and nucleotides

HI#	Identification	%Sim
0583	2',3'-cyclic nucleotide 2'-phosphodiesterase (cpdB)	78
1230	adenine PRTase (apt)	83
0551	adenosine tetraphosphatase (apaH)	73
1350	cytidine deaminase (cda)	77
1646	cytidylate kinase (cmk)	77
1219	cytidylate kinase (cmk)	79
0518	purine-nucleoside phosphorylase (deoD)	90
1277	putative ATPase (mrp)	79
0529	thymidine kinase (tdk)	82
1228	uracil PRTase (upp)	94
0280	uridine phosphorylase (udp)	85
0674	xanthine-guanine PRTase	88
0692	xanthine-guanine PRTase	88

Sugar-nucleotide biosynthesis and conversions

HI#	Identification	%Sim
0206	5'-nucleotidase (ushA)	55
1279	CMP-NeuNAc Sase (siaB)	64
0820	Gal-1-P uridylyltransferase (galT)	100
0812	Glc-P uridylyltransferase (galU)	86
0351	UDP-Glc 4-epimerase (galE)	83
0642	UDP-GlcNAc pyrophosphorylase (glmU)	83

Regulatory functions

HI#	Identification	%Sim
0604	adenylate cyclase (cyaA)	100
0884	aerobic respiration control prt (arcA)	88
0220	aerobic respiration control sensor prt (arcB)	70
1052	araC-like transcription regulator	48
1209	Arg repressor prt (argR)	81
0236	arsC prt (arsC)	57
0462	ATP-dependent proteinase (lon)	88
0334	ATP:GTP 3'-pyrophosphotransferase (relA)	80
1127	carbon starvation prt (cstA)	54
0813	carbon storage regulator (csrA)	91
0957	cyclic AMP receptor (crp)	100
1200	cys regulon transcriptional activator (cysB)	79
0190	ferric uptake regulation prt (fur)	75
1453	fimbrial transcription regulation repressor (pilB)	53
1455	fimbrial transcription regulation repressor (pilB)	73
1260	folylpolyglutamate-dihydrofolate Sase expression regulator (accD)	83
1425	fumarate (and nitrate) reduction regulatory prt (fnr)	89
0821	galactose operon repressor (galS)	99
0754	glucokinase regulator	56
1194	Gly cleavage system transcriptional activator (gcvA)	69
1009	glycerol-3-P regulon repressor (glpR)	50
0619	glycerol-3-P regulon repressor (glpR)	77
0013	GTP-BP (era)	87
0877	GTP-BP (obg)	71
0571	hydrogen peroxide-inducible activator (oxyR)	86
0615	L-fucose operon activator (fucR)	56
0399	lacZ expression regulator (icc)	71
0224	Leu responsive regulatory prt (lrp)	53
1596	Leu responsive regulatory prt (lrp)	87
0749	LexA repressor (lexA)	85
1461	lipooligosaccharide prt (lex2A)	67
1611	maltose regulatory prt sfs1 (sfsA)	71
0294	metF aporepressor (metJ)	93
1473	molybdenum transport system (modD)	52
0199	msbB	67
0763	nadAB transcriptional regulator (nadR)	75
0710	negative regulator of translation (relB)	48
0629	negative rpo regulator (mclA)	63
0267	nitrate sensor prt (narQ)	63
0726	nitrate, nitrite response regulator prt (narP)	79
0337	nitrogen regulatory prt P-II (glnB)	94
1741	penta-P guanosine-3'-pyrophosphohydrolase (spoT)	77
1378	phosphate regulon sensor prt (phoR)	67
1379	phosphate regulon transcriptional regulatory prt (phoB)	72
1635	purine nucleotide synthesis repressor prt (purR)	74
0163	putative murein gene regulator (bolA)	66
0506	rbs repressor (rbsR)	71
0563	regulatory prt (asnC)	81
0893	repressor for cytochrome P450 (Bm3R1)	51
0269	RNA polymerase sigma-32 factor (rpoH)	87
0533	RNA polymerase sigma-70 factor (rpoD)	81
0628	RNA polymerase sigma-E factor (rpoE)	88
1707	sensor prt for basR (basS)	56
1440	stringent starvation prt (sspB)	81
1441	stringent starvation prt A (sspA)	87
1739	trans-activator of metE and metH (metR)	61
0358	transcription activator (tenA)	48
0681	transcriptional activator prt (ilvY)	70
1708	transcriptional regulatory prt (basR)	60
0410	transcriptional regulatory prt (tyrR)	67
0830	Trp repressor (trpR)	67
0054	uxu operon regulator (uxuR)	72
1106	xylose operon regluatory prt (xylR)	75

Replication

Degradation of DNA

HI#	Identification	%Sim
1689	endonuclease III (nth)	92
0249	excinuclease ABC sub A (uvrA)	91
1247	excinuclease ABC sub B (uvrB)	88
0057	excinuclease ABC sub C (uvrC)	80
1377	exodeoxyribonuclease I (sbcB)	75
1321	exodeoxyribonuclease V (recB)	58
0942	exodeoxyribonuclease V (recC)	61
1322	exodeoxyribonuclease V (recD)	59
0041	exonuclease III (xthA)	84
0397	exonuclease VII, large sub (xseA)	74
1214	single-stranded DNA-specific exonuclease (recJ)	77

DNA replication, restriction, modification, recombination, and repair

HI#	Identification	%Sim
0759	A/G-specific adenine glycosylase (mutY)	75
1226	chromosomal replication initiator (dnaA)	75
0993	chromosomal replication initiator (dnaA)	80
0314	crossover junction endodeoxyribonuclease (ruvC)	88
0209	DNA adenine methylase (dam)	71
1264	DNA gyrase, sub A (gyrA)	85
0567	DNA gyrase, sub B (gyrB)	86
0728	DNA helicase (recQ)	78
1188	DNA helicase II (uvrD)	98
1100	DNA ligase (lig)	80
0654	DNA 3-methyladenine glycosidase I (tagI)	76
0403	DNA mismatch repair prt (mutH)	81
0067	DNA mismatch repair prt (mutL)	67
0707	DNA mismatch repair prt (mutS)	84
0856	DNA polymerase I (polA)	77
0992	DNA polymerase III β sub (dnaN)	80
0923	DNA polymerase III δ sub (holA)	62
0455	DNA polymerase III δ' sub (holB)	57
0137	DNA polymerase III ε sub (dnaQ)	76
0739	DNA polymerase III α chain (dnaE)	86
1397	DNA polymerase III χ sub (holC)	99
0011	DNA polymerase III psi sub (holD)	59
0532	DNA primase (dnaG)	74
1740	DNA recombinase (recG)	80
0070	DNA repair prt (recN)	67
0657	DNA topoisomerase I (topA)	55
0566	dod	93
0062	dosage-dependent dnaK suppressor prt (dksA)	84
0946	formamidopyrimidine-DNA glycosylase (fpg)	75
0582	glucose-inhibited division prt (gidA)	87
0486	glucose-inhibited division prt (gidB)	78
0980	Hin recombinational enhancer BP (fis)	93
0512	HincII endonuclease (HincII)	98
1392	HindIII modification MTase (hindIIIM)	99
1393	HindIII restriction endonuclease (hindIIIR)	100
0313	Holliday junction DNA helicase (ruvA)	80
0312	Holliday junction DNA helicase (ruvB)	90
0676	integrase-recombinase prt (xerC)	74
0309	integrase-recombinase prt (xerD)	85
1313	integration host factor α sub (himA)	83
1221	integration host factor β sub (IHF-β) (himD)	77
0402	methylated-DNA--prt-Cys MTase (dat1)	62
0669	mioC	72
1041	modification methylase HgiDI (MHgiDI)	70
0513	modification methylase HincII (hincIIM)	99
0910	mutator mutT	72
0192	negative modulator of initiation of replication (seqA)	72
0546	primosomal prt n precursor (priB)	100
0339	primosomal prt replication factor (priA)	70
0387	probable ATP-dependent helicase (dinG)	51
0991	DNA, ATP-BP (recF)	76
0332	DNA repair prt (recO)	77
0600	recombinase (recA)	100
0061	recombination prt (rec2)	100
0443	recR prt (recR)	88
0599	regulatory prt (recX)	50
0649	rep helicase (rep)	83
1229	replication prt (dnaX)	70
1574	replicative DNA helicase (dnaB)	83
1040	restriction enzyme (hgiDIR)	64
1172	SAM Sase 2 (metX)	92
1424	shufflon-specific DNA recombinase (rci)	56
0250	single-stranded DNA BP (ssb)	98
1572	site-specific recombinase (rcb)	57
1365	topoisomerase I (topA)	84
0444	topoisomerase III (topB)	79
1529	topoisomerase IV sub A (parC)	85
1528	topoisomerase IV sub B (parE)	89
1258	transcription-repair coupling factor (mfd)	83
0216	type I restriction enzyme ECOK1 specificity prt (hsdS)	59
1287	type I restriction enzyme ECOR124/3 I M (hsdM)	54
0215	type I restriction enzyme ECOR124/3 I M (hsdM)	89
1285	type I restriction enzyme ECOR124/3 R (hsdR)	53
1056	type III restriction-modification ECOP15 enzyme (mod)	56
0018	uracil DNA glycosylase (ung)	80

Transcription

Degradation of RNA

HI#	Identification	%Sim
0218	anticodon nuclease masking-agent (prrD)	86
1733	exoribonuclease II	68
0390	ribonuclease D (rnd)	65
0413	ribonuclease E (rne)	72
0138	ribonuclease H (rnh)	76
1059	ribonuclease HII	83
0014	ribonuclease III (rnc)	80
0273	ribonuclease PH (rph)	88
0999	RNase P (rnpA)	81
0324	RNase T (rnt)	81

RNA synthesis, modification, and DNA transcription

HI#	Identification	%Sim
0616	ATP-dependent helicase (hepA)	74
0231	ATP-dependent RNA helicase (deaD)	79
0892	ATP-dependent RNA helicase (rhlB)	84
0422	ATP-dependent RNA helicase (srmB)	61
0802	DNA-directed RNA polymerase α chain (rpoA)	97
0515	DNA-directed RNA polymerase β chain (rpoB)	92
0514	DNA-directed RNA polymerase β' chain (rpoC)	91
1304	N utilization substance prt B (nusB)	71
0063	plasmid copy number control prt (pcnB)	73
0229	polynucleotide phosphorylase (pnp)	87
1742	RNA polymerase omega sub (rpoZ)	76
1459	sigma factor (algU)	49
0717	transcription antitermination prt (nusG)	68
1331	transcription elongation factor (greA)	90
0569	transcription elongation factor (greB)	79
1283	transcription factor (nusA)	95
0295	transcription termination factor rho (rho)	95

Translation

Amino acyl tRNA synthetases and tRNA modification

HI#	Identification	%Sim
0814	Ala-tRNA Sase (alaS)	83
1583	Arg-tRNA Sase (argS)	84
1302	Asn-tRNA Sase (asnS)	91
0317	Asp-tRNA Sase (aspS)	85
0708	Cys-tRNA selenium Tase (selA)	76
0078	Cys-tRNA Sase (cysS)	87
1354	Gln-tRNA Sase (glnS)	84
0274	Glu-tRNA Sase (gltX)	84
0927	Gly-tRNA Sase α chain (glyQ)	95
0924	Gly-tRNA Sase β chain (glyS)	82

HI#	Identification	%Sim
0369	His-tRNA Sase (hisS)	79
0962	Ile-tRNA Sase (ileS)	78
0921	Leu-tRNA Sase (leuS)	82
1211	Lys-tRNA Sase (lysU)	84
0836	Lys-tRNA Sase analog (genX)	78
0623	Met-tRNA formyltransferase (fmt)	77
1276	Met-tRNA Sase (metG)	83
0394	peptidyl-tRNA hydrolase (pth)	81
1311	Phe-tRNA Sase α sub (pheS)	82
1312	Phe-tRNA Sase β sub (pheT)	80
0729	Pro-tRNA Sase (proS)	87
1644	pseudouridylate Sase I (hisT)	83
0245	queuosine biosyn prt (queA)	86
0200	selenium metabolism prt (selD)	80
0110	Ser-tRNA Sase (serS)	86
1367	Thr-tRNA Sase (thrS)	86
0202	tRNA (guanine-N1)-MTase (trmD)	93
0848	tRNA (U-5-)-MTase (trmA)	86
0068	tRNA δ(2)-isopentenylpyrophosphate Tase (trpX)	87
1606	tRNA nucleotidyltransferase (cca)	73
0244	tRNA-guanine transglycosylase (tgt)	91
0637	Trp-tRNA Sase (trpS)	86
1610	Tyr-tRNA Sase (tyrS)	73
1391	Val-tRNA Sase (valS)	83

Degradation of proteins, peptides, and glycopeptides

HI#	Identification	%Sim
0875	aminopeptidase A (pepA)	58
1705	aminopeptidase a/i (pepA)	78
1614	aminopeptidase N (pepN)	76
0816	aminopeptidase P (pepP)	88
0714	ATP-dependent clp protease (clpP)	84
1597	ATP-dependent protease (sms)	92
0715	ATP-dependent protease ATPase sub (clpX)	83
0859	ATP-dependent protease ATP-binding sub (clpB)	89
0419	collagenase (prtC)	53
0150	HflC	78
0990	IgA1 protease (iga1)	100
0247	IgA1 protease (iga1)	57
1324	lon protease (lon)	47
0214	oligopeptidase A (prlC)	85
0675	peptidase D (pepD)	72
0587	peptidase E (pepE)	60
1348	peptidase T (pepT)	71
1259	periplasmic Ser protease Do (htrA)	74
0722	Pro dipeptidase (pepQ)	70
1682	protease (sohB)	74
1541	protease IV (sppA)	64
0151	protease for λ cII repressor (hflK)	73
0530	sialoglycoprotease (gcp)	92

Nucleoproteins

HI#	Identification	%Sim
0186	DNA-BP	64
1491	DNA-BP (rdgB)	61
1587	DNA-BP H-NS (hns)	65
0430	DNA-BP HU-α	87

Protein modification and translation factors

HI#	Identification	%Sim
0846	disulfide oxidoreductase (por)	100
0985	DNA processing chain A (dprA)	60
0914	elongation factor EF-Ts (tsf)	85
0578	elongation factor EF-Tu (tufB)	96
0632	elongation factor EF-Tu (tufB)	96
0579	elongation factor G (fusA)	92
0328	elongation factor P (efp)	86
0622	f-Met deformylase (def)	80
0069	Glu-ammonia-ligase adenylyltransferase (glnE)	70
0548	initiation factor IF-1 (infA)	99
1284	initiation factor IF-2 (infB)	85
1318	initiation factor IF-3 (infC)	95
1152	maturation of antibiotic MccB17 (pmbA)	79
1722	Met aminopeptidase (map)	80
0428	oxido-RDase (dsbB)	69
1561	peptide chain release factor 1 (prfA)	88
1212	peptide chain release factor 2 (prfB)	94
1735	peptide chain release factor 3 (prfC)	93
0079	peptidyl-prolyl cis-trans isomerase B (ppiB)	84
0808	ribosome releasing factor (frr)	85
0573	rotamase, peptidyl prolyl cis-trans isomerase (slyD)	73
0699	rotamase, peptidyl prolyl cis-trans isomerase (slyD)	79
0709	translation factor (selB)	65
1213	thiol:disulfide interchange prt (xprA)	67

Ribosomal proteins: sthesis and modification

HI#	Identification	%Sim
0516	ribosomal prt L1 (rpL1)	93
0640	ribosomal prt L10 (rpL10)	89
0517	ribosomal prt L11 (rpL11)	94
0978	ribosomal prt L11 MTase (prmA)	83
1443	ribosomal prt L13 (rpL13)	96
0788	ribosomal prt L14 (rpL14)	98
0797	ribosomal prt L15 (rpL15)	91
0784	ribosomal prt L16 (rpL16)	96
0803	ribosomal prt L17 (rplQ)	92
0794	ribosomal prt L18 (rpL18)	91
0201	ribosomal prt L19 (rpL19)	98
0780	ribosomal prt L2 (rpL2)	93
1320	ribosomal prt L20 (rpL20)	97
0880	ribosomal prt L21 (rpL21)	86
0782	ribosomal prt L22 (rpL22)	97
0779	ribosomal prt L23 (rpL23)	83
0789	ribosomal prt L24 (rpL24)	86
1630	ribosomal prt L25 (rpL25)	77
0879	ribosomal prt L27 (rpL27)	92
0951	ribosomal prt L28 (rpL28)	95
0785	ribosomal prt L29 (rpL29)	87
0777	ribosomal prt L3 (rpL3)	92
0796	ribosomal prt L30 (rpL30)	86
0758	ribosomal prt L31 (rpL31)	86
0158	ribosomal prt L32 (rpL32)	86
0950	ribosomal prt L33 (rpL33)	91
0998	ribosomal prt L34 (rpL34)	93
1319	ribosomal prt L35 (rpL35)	84
0778	ribosomal prt L4 (rpL4)	93
0790	ribosomal prt L5 (rpL5)	96
0793	ribosomal prt L6 (rpL6)	90
0641	ribosomal prt L7/L12 (rpL7/L12)	92
0544	ribosomal prt L9 (rpL9)	86
1220	ribosomal prt S1 (rpS1)	89
0776	ribosomal prt S10 (rpS10)	99
0800	ribosomal prt S11 (rpS11)	96
0799	ribosomal prt S13 (rpS13)	93
0791	ribosomal prt S14 (rpS14)	95
1328	ribosomal prt S15 (rpS15)	87
1468	ribosomal prt S15 (rpS15)	87
0204	ribosomal prt S16 (rpS16)	85
0786	ribosomal prt S17 (rpS17)	94
0545	ribosomal prt S18 (rpS18)	95
0781	ribosomal prt S19 (rpS19)	98
0913	ribosomal prt S2 (rpS2)	89
0531	ribosomal prt S21 (rpS21)	87
0783	ribosomal prt S3 (rpS3)	93
0801	ribosomal prt S4 (rpS4)	95
0795	ribosomal prt S5 (rpS5)	96
0547	ribosomal prt S6 (rpS6)	87
1531	ribosomal prt S6 modification prt (rimK)	69
0580	ribosomal prt S7 (rpS7)	94
0792	ribosomal prt S8 (rpS8)	91
1442	ribosomal prt S9 (rpS9)	98
0010	ribosomal-prt-Ala acetyltransferase (riml)	73
0581	streptomycin resistance prt (strA)	100

Transport and binding proteins

Amino acids, peptides and amines

HI#	Identification	%Sim
1177	Arg permease (artM)	80
1178	Arg permease (artQ)	78
1179	Arg-BP (artl)	73
1180	Arg transport ATP-BP artP (artP)	83
0253	biopolymer transport prt (exbB)	99
0252	biopolymer transport prt (exbD)	55
1728	branched chain AA transport system II (braB)	50
0883	D-Ala permease (dagA)	65
1187	dipeptide permease (dppB)	79
1186	dipeptide permease (dppC)	83
1185	dipeptide transport ATP-BP (dppD)	84
1184	dipeptide transport ATP-BP (dppF)	87
1079	Gln permease (glnP)	59
1080	Gln-BP (glnH)	48
1530	Glu permease (gltS)	73
0408	Leu-specific transport prt (livG)	55
0226	LIV-II transport system (brnQ)	60
0213	oligopeptide-BP (oppA)	53
1124	oligopeptide-BP (oppA)	69
1123	oligopeptide permease (oppB)	61
1122	oligopeptide permease (oppC)	87
1121	oligopeptide permease ATP-BP (oppD)	85
1120	oligopeptide permease ATP-BP (oppF)	84
1638	peptide permease (sapA)	64
1639	peptide permease (sapB)	64
1640	peptide permease (sapC)	60
1641	peptide permease ATP-BP (sapD)	80
1154	proton Glu symport prt (gltP)	54
0590	putrescine permease (potE)	88
0289	Ser transporter (sdaC)	78
1346	spermidine-putrescine permease (potB)	84
1345	spermidine-putrescine permease (potC)	89
1347	spermidine-putrescine permease ATP-BP (potA)	83
1344	spermidine-putrescine-BP (potD)	72
0498	spermidine-putrescine-BP (potD)	75
0287	Trp-specific permease (mtr)	73
0528	Tyr-specific transport prt (tyrP)	65
0477	Tyr-specific transport prt (tyrP)	68

Anions

HI#	Identification	%Sim
1691	hydrophilic membrane-bound prt (modC)	75
1692	hydrophobic membrane-bound prt (modB)	85
1381	integral membrane prt (pstA)	78
0354	nitrate transporter ATPase component (nasD)	58
1380	peripheral membrane prt B (pstB)	87
1382	peripheral membrane prt C (pstC)	79
1383	periplasmic phosphate-BP (pstS)	68
1604	phosphate permease	60

Carbohydrates, organic alcohols, and acids

HI#	Identification	%Sim
0020	2-oxoglutarate/malate translocator	60
0153	Asp transport prt (dcuA)	70
0746	Asp transport prt (dcuA)	70
1110	D-xylose transport ATP-BP (xylG)	86
1111	D-xylose-BP (rbsB)	88
1712	enzyme I (ptsI)	84
0181	formate transporter	73
0448	fructose permease IIA/FPR component (fruB)	68
0446	fructose permease IIBC component (fruA)	72
0612	fucose operon prt (fucU)	80
1711	Glc phosphotransferase enzyme III (crr)	83
1017	glycerol uptake facilitator prt (glpF)	55
0690	glycerol uptake facilitator prt (glpF)	87
1015	gluconate permease (gntP)	56
0686	glycerol-3-phosphatase transporter (glpT)	79
0502	high affinity ribose transport prt (rbsA)	85
0503	high affinity ribose transport prt (rbsC)	86
0501	high affinity ribose transport prt (rbsD)	78
0610	L-fucose permease (fucP)	58
1218	L-lactate permease (lctP)	54
1729	lactam utilization prt (lamB)	60
0823	methylgalactoside permease ATP-BP (mglA)	85
0822	methylgalactoside-BP (mglB)	81
0824	methylgalactoside permease (mglC)	90
1690	Na+ and Cl- dependent GABA transporter	53
0736	Na+-dependent noradrenaline transporter	54
0504	periplasmic ribose-BP (rbsB)	87
1713	phosphohistidinoprotein-hexose phosphotransferase (ptsH)	88
0828	potassium channel homolog (kch)	80
1109	ribose permease (xylH)	84

Cations

HI#	Identification	%Sim
0254	bacterioferritin comigratory prt (bcp)	80
0251	energy transducer (tonB)	98
1272	ferric enterobactin transport ATP-BP (fepC)	51
1470	ferric enterobactin transport ATP-BP (fepC)	55
1466	ferrichrome-iron receptor (fhuA)	49
1385	ferritin like prt (rsgA)	74
1384	ferritin like prt (rsgA)	79
1271	iron(III) dicitrate permease (fecD)	61
0361	iron(III) dicitrate transport ATP-BP (fecE)	56
1035	magnesium and cobalt transport prt (corA)	85
0097	major ferric iron-BP precursor (fbp)	82
1049	mercury transport prt (merT)	54
1050	mercury scavenger prt (merP)	46
0292	mercury scavenger prt (merP)	67
1525	molybdate-BP (modB)	43
0427	Na+,H+ antiporter (nhaB)	87
1107	Na+,H+ antiporter (nhaC)	62
0225	Na+,H+ antiporter 1 (nhaA)	75
0098	periplasmic-BP-dependent iron transport (sfuB)	59
1474	periplasmic-BP-dependent iron transport (sfuC)	58
0911	potassium efflux system (kefC)	66
0290	potassium, copper-transporting ATPase A (copA)	64
1352	sodium, Pro symporter (putP)	79
0625	TRK system potassium uptake prt (trkA)	83

Nucleosides, purines and pyrimidines

HI#	Identification	%Sim
1087	ribonucleotide transport ATP-BP (mkl)	61
1227	uracil permease (uraA)	62

Other

HI#	Identification	%Sim
0621	ATP-BP (abc)	87
0060	ATP-dependent translocator (msbA)	100
1619	cystic fibrosis transmembrane conductance regulator	61
0853	heme-binding lpp (dppA)	99
0264	heme-hemopexin-BP (hxuA)	89
1471	hemin permease (hemU)	63
0262	hemin receptor precursor (hemR)	46
1706	high-affinity choline transport prt (betT)	62
0661	lactoferrin-BP (lbpA)	48
0608	Na+, sulfate cotransporter	86
0975	pantothenate permease (panF)	78
0973	transferrin-BP (tfbA)	48
0712	transferrin-BP 1 (tbp1)	49
1565	transferrin-BP 1 (tbp1)	59
0994	transferrin-BP 1 (tbp1)	69
1217	transferrin-BP 1 (tbp1)	80
0635	transferrin-BP 1 (tbp2)	52
0995	transferrin-BP 2 (tbp2)	55
0663	transport ATP-BP (cydD)	54
1157	transport ATP-BP (cydD)	73

Other categories

Adaptations and atypical conditions

HI#	Identification	%Sim
1526	autotrophic growth prt (aut)	61
0071	heat shock prt B253 (grpE)	66
0720	heat shock prt (htpX)	82
1527	heat shock prt B (ibpB)	71
0945	htrA-like prt (htrH)	73
0901	invasion prt (invA)	61
1544	NAD(P)H:menadione oxidoreductase	55
0458	survival prt (surA)	58
0815	universal stress prt (uspA)	87
1251	virulence assoc prt A (vapA)	58
0322	virulence assoc prt C (vapC)	57
0947	virulence assoc prt C (vapC)	61
0450	virulence assoc prt D (vapD)	67
1307	virulence plasmid prt (mlgA)	56
0321	virulence plasmid prt (vagC)	58

Colicin-related functions

HI#	Identification	%Sim
0382	colicin tolerance prt (tolB)	78
1206	colicin V production prt (cvpA)	79
0384	inner membrane prt (tolR)	79
0385	inner membrane prt (tolQ)	83
1685	outer membrane integrity prt (tolA)	48
0383	outer membrane integrity prt (tolA)	57

Drug and analog sensitivity

HI#	Identification	%Sim
0895	acriflavine resistance prt (acrB)	55
0300	ampD signalling prt (ampD)	75
1242	bicyclomycin resistance prt (bcr)	69
1623	mercury resistance regulatory prt (merR2)	58
0648	modulator of drug activity (mda66)	75
0897	multidrug resistance prt (emrB)	85
0898	multidrug resistance prt (ermA)	66
0036	multidrug resistance prt (mdl)	51
1462	nodulation prt T (nodT)	46
0549	rRNA (adenosine-N6,N6-)-dimethyltransferase (ksgA)	81
0511	tellurite resistance prt (tehA)	62
1275	tellurite resistance prt (tehB)	71

Phage-related functions and prophages

HI#	Identification	%Sim
1488	E16 prt (muE16)	53
1503	G prt (muG)	52
1568	G prt (muG)	54
1483	gam prt	74
0411	host factor-I (HF-I) (hfq)	97
1504	I prt (mul)	55
1481	MuB prt (muB)	70
1515	N prt (muN)	52
1516	P prt (muP)	61
1411	terminase sub 1	52
1478	transposase A (muA)	60

Radiation sensitivity

HI#	Identification	%Sim
0952	DNA repair prt (radC)	72

Transposon-related functions

HI#	Identification	%Sim
1577	IS1016-V6	61
1329	IS1016-V6	75
1018	IS1016-V6	94

Other

HI#	Identification	%Sim
1161	15 kD prt (P15)	68
0085	2-hydroxyacid dehydrogenase (ddh)	73
0460	β-lactamase regulatory prt (mazG)	73
0223	chloramphenicol-sensitive prt (rarD)	53
0680	chloramphenicol-sensitive prt (rarD)	55
1670	conjugative transfer co-repressor (finO)	52
0307	δ-1-pyrroline-5-carboxylate RDase (proC)	60
1549	heterocyst maturation prt (devA)	66
1339	embryonic abundant prt, group 3	89
0916	export factor homolog (skp)	76
0937	extragenic suppressor (suhB)	80
0667	glp regulon prt (glpX)	83
1013	glyoxylate-induced prt	58
0497	heat shock prt (hslU)	90
0496	heat shock prt (hslV)	89
1117	ilv-related prt	77
0285	isochorismate Sase (entC)	49
1618	membrane assoc ATPase (cbiO)	53
0461	membrane prt (lapB)	56
1119	membrane prt (lapB)	80
0630	mucoid status locus prt (mucB)	52
0588	N-carbamyl-L-amino acid amidohydrolase	59
1295	nitrogen fixation prt (nifS)	56
1343	nitrogen fixation prt (nifS)	59
0378	nitrogen fixation prt (nifS)	67
0377	nitrogen fixation prt (nifU)	74
0166	nitrogen fixation prt (rnfE)	48
1686	nitrogen fixation prt (rnfE)	59
0129	nitrogenase C (nifC)	53
1475	nitrogenase C (nifC)	60
1296	partitioning system prt (parB)	68
0171	phenolhydroxylase	57
0368	prt E (gpcE)	94
0556	putative glucose-6-P DHase isozyme (devB)	52
0981	small prt (smpB)	91
1592	spoIIIE prt (spoIIIE)	75
0095	spore germination and vegetative growth prt (gerC2)	55
0896	suppressor prt (msgA)	56
1078	surfactin (sfpo)	78
0357	thiamine-repressed prt (nmt1)	55
0751	toxR regulon (tagD)	64
1407	traN	62
0664	transport ATP-BP (cydC)	52
1156	transport ATP-BP (cydC)	70
1556	vanamycin-resistance prt (vanH)	57

3) The two λ libraries constructed from *H. influenzae* genomic DNA were probed with oligonucleotides designed from the ends of contig groups (27). The positive plaques were then used to prepare templates, and the sequence was determined from each end of the λ clone insert. These sequence fragments were searched with GRASTA against a database of all contigs. Two contigs that matched the sequence from the opposite ends of the same λ clone were ordered. The λ clone then provided the template for closure of the sequence gap between the adjacent contigs.

4) To confirm the order of contigs found by the other approaches and establish the order of the remaining contigs, we performed amplifications by polymerase chain reaction (PCR), both standard and long range (XL) (28). Although a PCR reaction was done for essentially every combination of physical gap ends, techniques such as DNA fingerprinting, database matching, and the probing of large insert clones were particularly valuable in ordering contigs adjacent to each other and reducing the number of combinatorial PCRs necessary to achieve complete gap closure. Use of these strategies to an even greater extent in future genome projects will increase the overall efficiency of complete genome closure. In the program ASM_ALIGN Southern analysis data, identification of peptide links, forward and reverse sequence data from λ clones, and PCR data are used to establish the relative order of the contigs separated by physical gaps. The number of physical gaps ordered and closed by each of these techniques is summarized in Table 2.

Lambda clones were a central feature for completion of the genome sequence and assembly. It was probable that some fragments of the *H. influenzae* genome would be nonclonable in a high copy plasmid because they would produce deleterious proteins in the *E. coli* host cells. Lytic λ clones would provide DNA for these segments because such genes would not inhibit plaque production. Furthermore, sequence information from the ends of 15- to 20-kb clones is particularly suitable for gap closure and providing general confirmation of genome assembly. Because of their size, they would be likely to span any physical gap. Approximately 100 random plaques were picked from the amplified λ library, templates were prepared, and sequence information was obtained from each end. These sequences were searched (GRASTA) against the contigs and linked in the database to their appropriate contig, thus providing a scaffolding of λ clones that contributed additional support to the accuracy of the genome assembly (Fig. 1). In addition to confirmation of the contig structure, the λ clones provided closure for 23 physical gaps.

Approximately 78 percent of the genome was covered by λ clones.

The λ clones were particularly useful for solving repeat structures. All repeat structures identified in the genome were small enough to be spanned by a single clone from the random insert library, except for the six ribosomal RNA (rRNA) operons and one repeat (two copies) that was 5340 bp in length. The ability to distinguish and assemble the six rRNA operons of *H. influenzae* (each containing in order 16S, 23S, and 5S subunit genes) was a test of our overall strategy to sequence and assemble a complex genome that might contain a significant number of repeat regions. The high degree of sequence similarity and the length of the six operons caused the assembly process to cluster all the underlying sequences into a few indistinguishable contigs. To determine the correct placement of the operons in the sequence, unique sequences were identified at the 5S ends. Oligonucleotide primers were designed from these six flanking regions and used to probe the two λ libraries. For five of the six rRNA operons at least one positive plaque was identified that completely spanned the rRNA operon and contained uniquely identifying flanking sequence at the 16S and 5S ends. These plaques provided the templates for obtaining the sequence for these rRNA operons. For *rrnA* a plaque was identified that contained the particular 5S end and terminated in the 16S end. The 16S end of *rrnA* was obtained by PCR from *H. influenzae* Rd genomic DNA.

An additional confirmation of the global structure of the assembled circular genome was obtained by comparing a computer-

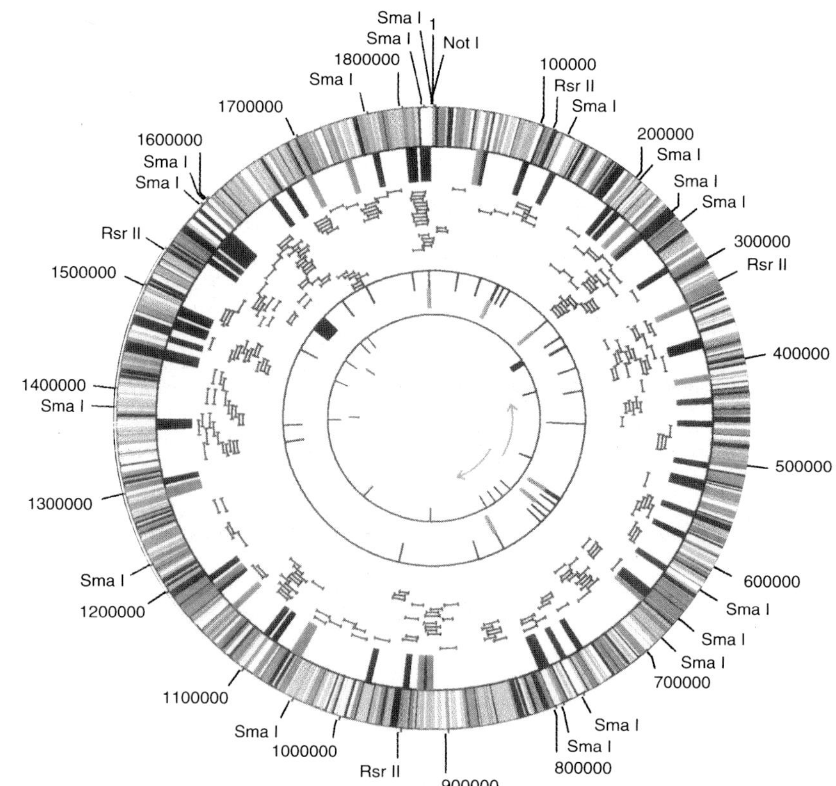

Fig. 1. A circular representation of the *H. influenzae* Rd chromosome illustrating the location of each predicted coding region containing a database match as well as selected global features of the genome. Outer perimeter: The location of the unique Not I restriction site (designated as nucleotide 1), the Rsr II sites, and the Sma I sites. Outer concentric circle: Coding regions for which a gene identification was made. Each coding region location is classified as to role according to the color code in Fig. 2. Second concentric circle: Regions of high G+C content (>42 percent, red; >40 percent, blue) and high A+T content (>66 percent, black; >64 percent, green). Third concentric circle: Coverage by λ clones (blue). More than 300 λ clones were sequenced from each end to confirm the overall structure of the genome and identify the six ribosomal operons. Fourth concentric circle: The locations of the six ribosomal operons (green), the tRNAs (black) and the cryptic mu-like prophage (blue). Fifth concentric circle: Simple tandem repeats. The locations of the following repeats are shown: CTGGCT, GTCT, ATT, AATGGC, TTGA, TTGG, TTTA, TTATC ,TGAC, TCGTC, AACC, TTGC, CAAT, CCAA. The putative origin of replication is illustrated by the outward pointing arrows (green) originating near base 603,000. Two potential termination sequences are shown near the opposite midpoint of the circle (red).

generated restriction map based on the assembled sequence for the endonucleases Apa I, Sma I, and Rsr II with the predicted physical map of Lee *et al.* (*29*). The restriction fragments from the sequence-derived map matched those from the physical map in size and relative order (Fig. 1).

At the same time that the final gap filling process occurred, each contig was edited visually by reassembling overlapping 10-kb sections of contigs by means of the AB AUTOASSEMBLER and the Fast Data Finder hardware. AUTOASSEMBLER provides a graphical interface to electropherogram data for editing. The electropherogram data was used to assign the most likely base at each position. Where a discrepancy could not be resolved or a clear assignment made, the automatic base calls were initially left unchanged. Individual sequence changes were written to the electropherogram files and a program was designed (CRASH) to maintain the synchrony of sequence data between the *H. influenzae* database and the electropherogram files. After the editing, contigs were reassembled with TIGR ASSEMBLER prior to annotation.

Potential frameshifts identified in the course of annotating the genome were saved as reports in the database. These frameshifts were used to indicate areas of the sequence that might require further editing or sequencing. Frameshifts were not corrected for cases in which clear electropherogram data disagreed with a frameshift. Frameshift editing was done with TIGR EDITOR. This program was developed as a collaborative effort between TIGR and AB and is a modification of the AB AUTOASSEMBLER. TIGR EDITOR can download contigs from the database and thus provides a graphical interface to the electropherogram for the purpose of editing data associated with the aligned sequence file output of TIGR ASSEMBLER. The program maintains synchrony between the electropherogram files on the Macintosh system and the sequence data in the *H. influenzae* database on the Unix system. TIGR EDITOR is now our primary tool for sequence viewing and editing for the purpose of genome assembly.

The final assembly of the *H. influenzae* genome with the TIGR ASSEMBLER was precluded by the rRNA and other repeat regions, and was accomplished by means of COMB_ASM (a program written at TIGR) that splices together contigs on the basis of short sequence overlaps.

Throughout the project, we paid particular attention to the accuracy of the sequence generated and included various quality control measures. In particular, we constructed random small and large insert libraries (as described above), used strict criteria for excluding any single sequence in which more than 3 percent of the nucleo-

tides could not be identified with certainty, determined that there was no vector contamination in each sequence, and rejected chimeric sequences from the assembly process. The most important measure of the sequence accuracy is the correct assembly of the 1.8-Mb genome. Any deviation from inclusion of only high-quality sequences would have resulted in an inability to assemble the final genome. In addition, the use of the large insert λ clones confirmed the accuracy of the final assembly. Our finding that the restriction map of the *H. influenzae* Rd genome based on our sequence data is in complete agreement with that previously published (*29*) further confirms the accuracy of the assembly.

As a consequence of our shotgun approach, we reached an average of more than sixfold redundancy across the genome, although there are some regions in which the coverage is lower. The criteria that we used to define overall sequence quality and completion were as follows: (i) The sequence should have less than 1 percent single sequence coverage. Because *H. influenzae* is a genome rich in AT pairs, it is possible to obtain a highly accurate sequence with single-pass coverage. However, any regions with single sequence coverage that contained ambiguities were again sequenced with an alternative sequencing chemistry. (ii) Areas with more than single sequence coverage that contained ambiguities or G–C compressions were also sequenced again with an alternative sequencing chemistry. The combination of sequence redundancy together with the application of an alternative sequencing chemistry in areas with ambiguities is, we believe at least as accurate, if not more so, than double-stranded coverage. By these criteria we have reduced the number of nucleotide ambiguities [International Union of Biochemistry (IUB) codes] in the sequence to less than 1 in 19,000. The same approaches used to resolve ambiguities were also applied to areas where apparent frameshifts were indicated. Sixty potential frameshifts were identified by comparison to entries in peptide databases. Although some of these potential frameshifts are undoubtedly real, others may reflect the hundreds of frameshifts present in GenBank sequences from public databases (*30*). They may also represent biologically significant phenomena such as insertions or deletions in insertion elements, or in tandem repeats often associated with virulence genes (*31*).

We also considered comparison of our sequence to existing GenBank *H. influenzae* Rd sequences as a method for evaluating sequence accuracy as reported for yeast chromosome VIII (*32*). Unlike yeast, only a limited number of *H. influenzae* sequences are in GenBank (38 *H. influenzae* Rd accessions) and these are not necessarily of high

accuracy. The results of such a comparison show that our sequence is 99.67 percent identical overall to those GenBank sequences annotated as *H. influenzae* Rd. Two problems were apparent with this type of comparison. Sequences could differ because of strain variation, which is poorly annotated in the GenBank entries. It is also difficult to evaluate the significance of differences as the accuracy of the GenBank entries was impossible to assess. We compared GenBank accession M86702 (*strA* resistance gene) to our sequence and found the identity to be 94.7 percent over 545 bp. There are 24 single base pair mismatches relative to our sequence as well as an insertion and a deletion. Comparison of our sequence to GenBank accession L23824 (adenylate cyclase) shows a 99.7 percent match over 2960 bp. There are nine single base pair mismatches and one insertion. In this case the mismatches all fall in the noncoding flanking regions. While we cannot speak to the accuracy of these GenBank sequences, we are very confident of our sequences in these regions because of the 3× to 9× coverage with high-quality sequence data. Thus, a comparison of our sequence to sequences in GenBank annotated as *H. influenzae* Rd is not a meaningful way to evaluate the accuracy of the sequence.

Although it is extremely difficult to assess sequence accuracy, we wanted to provide an approximation of accuracy based on frequency of shifts in open reading frames, unresolved ambiguities, overall quality of raw data, and fold coverage. We estimate our error rate to be between 1 base in 5000 and 1 base in 10,000.

We also attempted to estimate the cost of the complete sequencing of the genome. Reagent and labor costs for construction of small insert and λ libraries, template preparation and sequencing, gap closure, sequence confirmation, annotation, and preparation for publication were summed and divided by the genome length. Sequencing projects that require up front mapping should include the cost of construction of the clone maps for sequencing. Not included were costs associated with development of technology and software that will be used for future sequencing projects. The estimated direct cost was 48 cents per finished base pair. Because of the techniques developed during this project any future genomes of this size should cost less.

Data and software availability. The *H. influenzae* genome sequence has been deposited in the Genome Sequence DataBase (GSDB) with the accession number L42023 and is termed version 1.0. The nucleotide sequence and peptide translation of each predicted coding region with identified start and stop codons have also been accessioned by GSDB. We consider annotation, accuracy checking, and error resolution to be ongoing tasks. As outlined above, there are predicted coding regions with potential frameshift errors in the sequence. As these are resolved, they will be deposited with GSDB. We also expect the annotation of the sequence to increase over time and be updated in GSDB.

Additional data are available on our World Wide Web site (http://www.tigr.org). An expanded version of Table 3 has links to the database accessions that were used to identify the predicted coding regions, additional sequence similarity data, and coordinates of the predicted coding regions. The alignments between the predicted coding regions and the database sequences are also available. The data can also be queried by gene identification number, putative identification, matching accession, and role. The entire sequence and the sequences of all predicted coding regions and their translations, including those having frameshifts, are also available. This Web site will be maintained as an up-to-date source of *H. influenzae* genome sequence data, and we encourage the scientific community to forward their results for inclusion (with proper attribution) at this site.

The software developed at TIGR that is described in the article is still under development. However, TIGR will work with other genome centers to make its software available upon request.

Genome analysis. We have attempted to predict all of the coding regions and identify genes, transfer RNAs (tRNAs) and rRNAs, as well as other features of the DNA sequence (such as repeats, regulatory sites, replication origin sites, and nucleotide composition), with the realization that biochemical and biological conformation of many of these will be an ongoing task. We include a description of some of the most obvious sequence features.

The *H. influenzae* Rd genome is a circular chromosome of 1,830,137 bp. The overall G+C nucleotide content is approximately 38 percent (A, 31 percent; C, 19 percent; G, 19 percent; T, 31 percent). The G+C content of the genome was examined with several window lengths to look for global structural features. With a window of 5000 bp, the G+C content is relatively even except for seven large regions rich in G+C and several regions rich in A+T (Fig. 1). The G+C–rich regions correspond to six rRNA operons and a cryptic mu-like prophage. Genes for several proteins similar to proteins encoded by bacteriophage mu are located at approximately position 1.56 to 1.59 Mbp of the genome. This area of the genome has a markedly higher G+C content than average for *H. influenzae* (~50 percent G+C compared to ~38 percent for

the rest of the genome).

The minimal origin of replication (*oriC*) in *E. coli* is a 245-bp region defined by three copies of a 13-bp repeat at one end (sites for initial DNA unwinding) and four copies of a 9-bp repeat (sites for DnaA binding, the first step in replication) at the other (*33*). An approximately 280-bp sequence containing structures similar to the three 13-bp and four 9-bp repeats defines the putative origin of replication in *H. influenzae* Rd. This region lies between sets of ribosomal operons *rrnF*, *rrnE*, *rrnD* and *rrnA*, *rrnB*, *rrnC*. These two groups of ribosomal operons are transcribed in opposite directions and the placement of the origin is consistent with their polarity for transcription. Termination of *E. coli* replication is marked by two 23-bp termination sequences located ~100 kb on either side of the midway point at which the two replication forks meet. Two potential termination sequences sharing a 10-bp core sequence with the *E. coli* termination sequence were identified in *H. influenzae*. These two regions are offset approximately 100 kb from a point approximately 180° opposite of the proposed origin of *H. influenzae* replication.

Six rRNA operons were identified. Each contains three subunits and a variable spacer region in the order: 16S subunit—spacer region—23S subunit—5S subunit. The subunit lengths are 1539, 2653, and 116 bp, respectively. The G+C content of the three ribosomal subunits (50 percent) is higher than that of the genome as a whole. The G+C content of the spacer region (38 percent) is consistent with the remainder of the genome. The nucleotide sequence of the three rRNA subunits is completely identical in all six ribosomal operons. The rRNA operons can be grouped into two classes based on the spacer region between the 16S and 23S sequences. The shorter of the two spacer regions is 478 bp (*rrnb*, *rrnE*, and *rrnF*) and contains the gene for tRNA^{Glu}. The longer spacer is 723 bp (*rrnA*, *rrnC*, and *rrnD*) and contains the genes for tRNA^{Ile} and tRNA^{Ala}. The two sets of spacer regions are also completely identical across each group of three operons. Other tRNA genes are present at the 16S and 5S ends of two of the rRNA operons. The genes for tRNA^{Arg}, tRNA^{His}, and tRNA^{Pro} are located at the 16S end of *rrnE* while the genes for tRNA^{Trp} and tRNA^{Asp} are located at the 5S end of *rrnA*.

The predicted coding regions were initially defined by evaluating their coding potential with the program GENEMARK (*34*) based on codon frequency matrices derived from 122 *H. influenzae* coding sequences in GenBank. The predicted coding region sequences (plus 300 bp of flanking sequence) were used in searches against a database of nonredundant bacterial proteins

(NRBP) created specifically for the annotation. Redundancy was removed from NRBP at two stages. All DNA coding sequences were extracted from GenBank (release 85), and sequences from the same species were searched against each other. Sequences having more than 97 percent identity over regions longer than 100 nucleotides were combined. In addition, the sequences were translated and used in protein comparisons with all sequences in Swiss-Prot (release 30). Sequences belonging to the same species and having more than 98 percent similarity over 33 amino acids were combined. NRBP is composed of 21,445 sequences extracted from 23,751 GenBank sequences and 11,183 Swiss-Prot sequences from 1099 different species.

A total of 1743 predicted coding regions was identified. Searches of the predicted coding regions for *H. influenzae* were performed against NRBP with BLAZE (*35*) run on a Maspar MP-2 massively parallel computer with 4096 microprocessors. BLAZE translates the query DNA sequence in the three plus-strand reading frames and identifies the protein sequences that match the query. The protein-protein matches were aligned with PRAZE, a modified Smith-Waterman (*23*) algorithm. In cases where insertions or deletions in the DNA sequence produced a potential frameshift, the alignment algorithm started with protein regions of maximum similarity and extended the alignment to the same database match in alternative frames by means of the 300-bp flanking region. Unidentified predicted coding regions and the remaining intergenic sequences were searched against a dataset of all available peptide sequences from Swiss-Prot, the Protein Information Resource (PIR), and GenBank. Identification of operon structures is expected to be facilitated by experimental determination of promoter and termination sites.

Each putatively identified *H. influenzae* gene was assigned to one of 102 biological role categories adapted from Riley (*36*). Assignments were made by linking the protein sequence of the predicted coding regions with the Swiss-Prot sequences in the Riley database. Of the 1743 predicted coding regions, 736 have no role assignment. Of these, no database match was found for 389, while 347 matched "hypothetical proteins" in the database. Role assignments were made for 1007 of the predicted coding regions. Each of the 102 role categories was grouped into one of 14 broader role categories (Table 2). A compilation of all the predicted coding regions, their identifiers, a three-letter gene identifier, and percent similarity are presented in Table 3 (foldout). An annotated complete genome map of *H. influenzae* Rd is presented in Fig. 2 (foldout). The map places each predicted coding region on the *H. influenzae* chromosome, indicates its direction of transcription and color codes its role assignment. Role assignments are also represented in Fig. 1.

A survey of the genes and their chromosomal organization in *H. influenzae* Rd makes possible a description of the metabolic processes *H. influenzae* requires for survival as a free-living organism, the nutritional requirements for its growth in the laboratory, and the characteristics that make it different from other organisms specifically as they relate to its pathogenicity and virulence. The genome would be expected to have complete complements of certain classes of genes known to be essential for life. For example, there is a one-to-one correspondence of published *E. coli* ribosomal protein sequences to potential homologs in the *H. influenzae* database. Likewise, as shown in Table 3, an aminoacyl tRNA synthetase is present in the genome for each amino acid. Finally, the location of tRNA genes was mapped onto the genome. There are 54 identified tRNA genes, including representatives of all 20 amino acids.

In order to survive as a free-living organism, *H. influenzae* must produce energy in the form of ATP via fermentation or electron transport. As a facultative anaerobe, *H. influenzae* Rd is known to ferment glucose, fructose, galactose, ribose, xylose, and fucose (*37*). As indicated by the genes identified in Table 3, transport systems are available for the uptake of these sugars by the phosphoenolypyruvate-phosphotransferase system (PTS), and by non-PTS mechanisms. Genes that specify the common phosphate-carriers enzyme I and Hpr (*ptsI* and *ptsH*) of the PTS system were identified as well as the glucose-specific *crr* gene. We have not, however, identified the gene-encoding, membrane-bound, glucose-specific enzyme II. The latter enzyme is required for transport of glucose by the PTS system. A complete PTS system for fructose was identified.

Genes encoding the complete glycolytic pathway and for the production of fermentative end products were identified. Also identified were genes encoding functional anaerobic electron transport systems that depend on inorganic electron acceptors such as nitrates, nitrites, and dimethyl sulfoxide. Genes encoding three enzymes of the tricarboxylic acid (TCA) cycle appear to be absent from the genome. Citrate synthase, isocitrate dehydrogenase, and aconitase were not found by searching the predicted coding regions or by using the *E. coli* enzymes as peptide queries against the entire genome in translation. This provides an explanation for the large amount of glutamate (1 g/liter) that is required in defined culture media (*38*). Glutamate can be directed into the TCA cycle by conversion to α-ketoglutarate by glutamate dehydrogenase. In the absence of a complete TCA cycle, glutamate presumably serves as the source of carbon for biosynthesis of amino acids from precursors that branch from the TCA cycle. Functional electron transport systems that depend on oxygen as a terminal electron acceptor are available for the production of adenosine triphosphate.

Previously unanswered questions regarding pathogenicity and virulence can be addressed by examining certain classes of genes such as adhesins and the lipo-oligosaccharide biogenesis genes. Moxon and coworkers (*31*) have obtained evidence that a number of these virulence-related genes contain tandem tetramer repeats that undergo frequent addition and deletion of one or more repeat units during replication such that the reading frame of the gene is changed and its expression thereby altered. It is now possible, by means of the complete genome sequence, to locate all such tandem repeat tracts (Fig. 2) and to begin to determine their roles in phase variation of such potential virulence genes.

Haemophilus influenzae Rd has a highly efficient, DNA transformation system. The DNA uptake sequence site, 5′ AAGTGCGGT, present in multiple copies in the genome, is necessary for efficient DNA uptake (*39*). It is now possible to locate all of these sites and describe their distribution with respect to genic and intergenic regions (*40*). Fifteen genes involved in transformation have already been described and sequenced (*41*). Six of the genes, *comA* to *comF*, comprise an operon that is under positive control by a 22-bp, palindromic, competence regulatory element (CRE) located approximately one helix turn upstream of the promoter. It is now feasible to locate additional copies of CRE in the genome and discover potential transformation genes under CRE control (*42*). In addition, other global regulatory elements may be discovered with an ease not previously possible.

One well-described system for gene regulation in bacteria is the "two-component" system composed of a sensor molecule that detects an environmental signal and a regulator molecule that is phosphorylated by the activated form of the sensor. The regulator protein is generally a transcription factor that, when activated by the sensor, turns on or off expression of a specific set of genes. It has been estimated that *E. coli* harbors 40 sensor-regulator pairs (*43*). The *H. influenzae* genome was searched with representative proteins from each family of sensor and regulator proteins with TBLASTN and TFASTA. Four sensor and five regulator proteins were identified with similarity to proteins from other species (Table 4). There appears to be a corresponding sensor for each regulator protein except CpxR. Searches with the CpxA protein from *E. coli* identified three of the four sensors listed in Table 4, but no additional significant matches were found. It is possible that the sequence similarity is low enough to be undetectable with TFASTA. All of the regulator proteins present fall into the OmpR subclass (*43*). No representatives of the NtrC class of regulators were found. This class of proteins interacts directly with the sigma-54 subunit of RNA polymerase, which is absent from *H. influenzae*, and which plays a major role in the regulation of a large number of operons in *E. coli* and other enterobacteria. The absence of the Ntr network in *H. influenzae* suggests significant differences in the regulatory processes between these two groups of organisms.

Some of the most interesting questions that can be answered by a complete genome sequence relate to the genes or pathways that are absent. The nonpathogenic *H. influenzae* Rd strain varies significantly from the pathogenic serotype b strains. Many of the differences between these two strains appear in factors affecting infectivity. For example, we have found that the eight genes that make up the fimbrial gene cluster (*44*) involved in adhesion of bacteria to host cells are absent in the Rd strain. The *pepN* and *purE* genes, which flank the fimbrial cluster in *H. influenzae* type b strains,

Table 4. Two-component systems in *H. influenzae* Rd. ID, identity; Sim, similarity.

Identification number	Location	Best match*	Id (%)	Sim (%)	Length (bp)
		Sensors			
HI0220	239,378	*arcB*	39.5	63.9	200
HI0267	299,541	*narQ*	38.1	68.0	562
HI1707	1,781,143	*basS*	27.7	51.5	250
HI1378	1,475,017	*phoR*	38.1	61.6	280
		Regulators			
HI0726	777,934	*narP*	59.3	77.0	209
HI0837	887,011	*cpxR*	51.9	73.0	229
HI0884	936,624	*arcA*	77.2	87.8	236
HI1379	1,475,502	*phoB*	52.9	71.4	228
HI1708	1,781,799	*basR*	43.5	59.3	219

*In all cases, the best match was to a gene of *E. coli*.

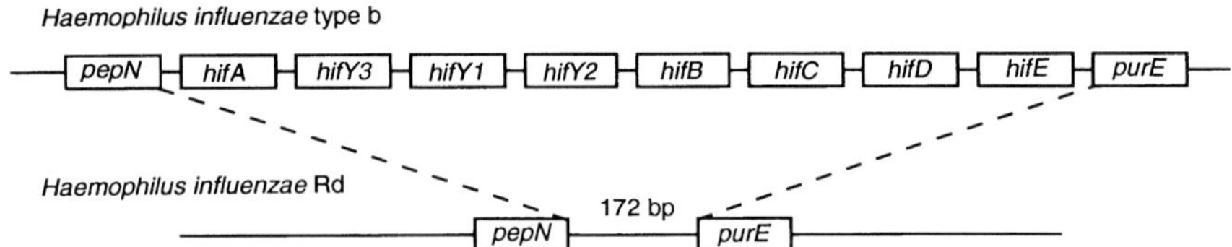

Fig. 3. A comparison of the region of the *H. influenzae* chromosome containing the eight genes of the fimbrial gene cluster present in *H. influenzae* type b and the same region in *H. influenzae* Rd. The region is flanked by *pepN* and *purE* in both organisms. However, in the noninfectious Rd strain the eight genes of the fimbrial gene cluster have been excised. A 172-bp spacer region is located in this region in the Rd strain and continues to be flanked by the *pepN* and *purE* genes.

are adjacent to one another in the Rd strain (Fig. 3), suggesting that the entire fimbrial cluster was excised.

On a broader level, we determined which *E. coli* proteins are not in *H. influenzae* by taking advantage of a nonredundant set of protein-coding genes from *E. coli*, namely the University of Wisconsin Genome Project contigs in GenBank: 1216 predicted protein sequences from GenBank accessions D10483, L10328, U00006, U00039, U14003, and U18997 (45). The minimum threshold for matches was set so that even weak matches would be scored as positive, thereby giving a minimal estimate of the *E. coli* genes not present in *H. influenzae*. We used TBLASTN to search each of the *E. coli* proteins against the complete genome. All BLAST scores greater than 100 were considered matches. Altogether 627 *E. coli* proteins matched at least one region of the *H. influenzae* genome and 589 proteins did not. The 589 nonmatching proteins were examined and found to contain a disproportionate number of hypothetical proteins from *E. coli*. Sixty-eight percent of the identified *E. coli* proteins were matched by an *H. influenzae* sequence whereas only 38 percent of the hypothetical proteins were matched. Proteins are anno-

tated as hypothetical on the basis of a lack of matches with any other known proteins (45). At least two potential explanations can be offered for the overrepresentation of hypothetical proteins among those without matches: (i) some of the hypothetical proteins are not, in fact, translated (at least in the annotated frame), or (ii) these are *E. coli*–specific proteins that are unlikely to be found in any species except those most closely related to *E. coli*, for example, *Salmonella typhimurium*.

A total of 389 predicted coding regions did not display significant similarity with a six-frame translation of GenBank release 87. These unidentified coding regions were compared to one another with FASTA. Two previously unidentified gene families were identified. Two predicted coding regions without database matches (HI0589 and HI0850) share 75 percent identity over almost their entire lengths (139 and 143 amino acid residues respectively). A second pair of predicted coding regions (HI1555 and HI1548) encode proteins that share 30 percent identity over almost their entire lengths (394 and 417 amino acids respectively). These similarities suggest that there may be previously unidentified gene families present in these regions.

Another analysis that can be applied to the unidentified coding regions is hydropathy analysis, which indicates the patterns of potential membrane-spanning domains that are often conserved between members of receptor and transporter gene families, even in the absence of significant amino acid identity. The five best examples of unidentified predicted coding regions that display potential transmembrane domains with a periodic pattern that is characteristic of membrane-bound channel proteins are shown in Fig. 4. Such information can be used to focus on specific aspects of cellular function that are affected by targeted deletion or mutation of these genes.

We have learned some important lessons concerning overall strategy from the *H. influenzae* sequencing project that should reduce the effort required for future bacterial genome sequencing projects. For example, the small insert library and the large insert library should be constructed and end-sequenced concurrently. It is essential that the sequence fragments used for the assembly are of the highest quality. The sequences should be rigorously checked for vector contamination. Although it is important that sequence read lengths be long enough to span most small repeats, they must also be highly accurate. Our raw sequence data contained on average less than 1.5 percent uncertainties. The use of high quality individual sequence fragments and a rigorous assembly algorithm essentially eliminated difficulty with achieving closure. The success of whole genome shotgun sequencing offers the potential to accelerate research in a number of areas. Comparative genomics could be advanced by the availability of an increased number of complete genomes from a variety of prokaryotes and eukaryotes. Knowledge of the complete genomes of pathogenic organisms could lead to new vaccines. Information obtained from the genomes of particular organisms could have industrial applications. Finally, this strategy has potential to facilitate the sequencing of the human genome.

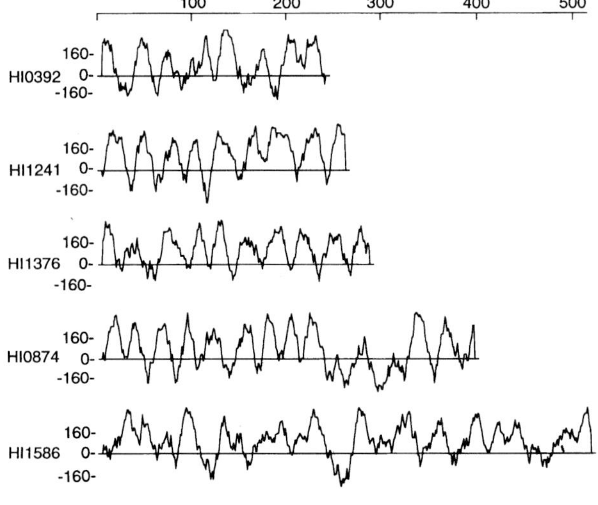

Fig. 4. Hydrophobicity analysis of five potential channel proteins. The amino acid sequences of five predicted coding regions that do not display similarity with known peptide sequences (GenBank release 87), each exhibit multiple hydrophobic domains that are characteristic of channel-forming proteins. The predicted coding region sequences were analyzed by the Kyte-Doolittle algorithm (46) (with a range of 11 residues) with the GENEWORKS software package (Intelligenetics).

REFERENCES AND NOTES

1. F. Sanger *et al.*, *Nature* **246**, 687 (1977); F. Sanger, A. R. Coulson, G. F. Hong, D. F. Hill, G. B. Petersen, *J. Mol. Biol.* **162**, 729 (1982).
2. A. T. Bankier *et al.*, *DNA Seq.* **2**, 1 (1991).
3. S. J. Goebel *et al.*, *Virology* **179**, 247 (1990).
4. K. Oda *et al.*, *J. Mol. Biol.* **223**, 1 (1992); K. Ohyama *et al.*, *Nature* **322**, 572 (1986).
5. R. F. Massung *et al.*, *Nature* **366**, 748 (1993).
6. D. L. Hartl and M. J. Palazzolo, *Genome Research in Molecular Medicine and Virology*, K. W. Adolph, Ed. (Academic Press, Orlando, FL, 1993), pp. 115–129.
7. H. J. Sofia *et al.*, *Nucleic Acids Res.* **22**, 2576 (1994).
8. J. Levy, *Yeast* **10**, 1689 (1994).
9. P. Glaser *et al.*, *Mol. Microbiol.* **10**, 371 (1993).
10. J. Sulston *et al.*, *Nature* **356**, 37 (1992).
11. W. F. Bodmer, *Rev. Invest. Clin.* (suppl., pp. 3–5) (1994).
12. M. D. Adams, C. Fields, J. C. Venter, Eds. *Automat-*

ed DNA Sequencing and Analysis (Academic Press, San Diego, CA, 1994).

13. M. D. Adams *et al.*, *Science* **252**, 1651 (1991); M. D. Adams *et al.*, *Nature* **355**, 632 (1992); M. D. Adams *et al.*, *ibid.*, in press.

14. E. S. Lander and M. S. Waterman, *Genomics* **2**, 231 (1988).

15. *Haemophilus influenzae* Rd KW20 DNA was prepared by extraction with phenol. A mixture (3.3 ml) containing 600 μg of DNA, 300 mM sodium acetate, 10 mM tris-HCl, 1 mM Na-EDTA, and 30 percent glycerol was sonicated (Branson Model 450 Sonicator) at the lowest energy setting for 1 minute at 0°C with a 3-mm probe. The DNA was precipitated in ethanol and redissolved in 500 μl of tris-EDTA (TE) buffer to create blunt ends; a 100-μl portion was digested for 10 minutes at 30°C in 200 μl of BAL 31 buffer with 5 units of BAL 31 nuclease (New England BioLabs). The DNA was extracted with phenol, precipitated in ethanol, redissolved in 100 μl of TE buffer, and fractionated on a 1.0 percent low melting agarose gel. A fraction (1.6 to 2.0 kb) was excised, extracted with phenol, and redissolved in 20 μl of TE buffer. A two-step ligation procedure was used to produce a plasmid library in which 97 percent of the recombinants contained inserts, of which >99 percent were single inserts. The first ligation mixture (50 μl) contained 2 μg of DNA fragments, 2 μg of Sma I + bacterial alkaline phosphatase pUC18 DNA (Pharmacia), and 10 units of T4 ligase (Gibco/BRL), and incubation was at 14°C for 4 hours. After extraction with phenol and ethanol precipitation, the DNA was dissolved in 20 μl of TE buffer and separated by electrophoresis on a 1.0 percent low melting agarose gel. A ladder of ethidium bromide–stained linearized DNA bands, identified by size as insert (i), vector (v), v+i, v+2i, v+3i, and so on, was visualized by 360-nm ultraviolet light, and the v+i DNA was excised and recovered in 20 μl of TE. The v+i DNA was blunt-ended by T4 polymerase treatment for 5 minutes at 37°C in a reaction mixture (50 μl) containing the linearized v+i fragments four deoxynucleotide triphosphates (dNTPs) (500 μM each) and 9 units of T4 polymerase (New England BioLabs) under buffer conditions recommended by the supplier. After phenol extraction and ethanol precipitation, the repaired v+i linear pieces were dissolved in 20 μl of TE. The final ligation to produce circles was carried out in a 50-μl reaction containing 5 μl of v+i DNA and 5 units of T4 ligase at 14°C overnight. The reaction mixture was heated for 10 minutes at 70°C and stored at −20°C.

16. A 100-μl portion of Epicurian Coli SURE 2 Supercompetent Cells (Stratagene 200152) was thawed on ice and transferred to a chilled Falcon 2059 tube on ice. A 1.7-μl volume of 1.42 M β-mercaptoethanol was added to the cells to a final concentration of 25 mM. Cells were incubated on ice for 10 minutes. A 1-μl sample of the final ligation mix was added to the cells and incubated on ice for 30 minutes. The cells were heat-treated for 30 seconds at 42°C and placed back on ice for 2 minutes. The outgrowth period in liquid culture was omitted to minimize the preferential growth of any given transformed cell. Instead, the transformed cells were plated directly on a nutrient rich SOB plate containing a 5-ml bottom layer of SOB agar (1.5 percent SOB agar consisted of 20 g of tryptone, 5 g of yeast extract, 0.5 g of NaCl, and 1.5 percent Difco agar/liter). The 5-ml bottom layer was supplemented with 0.4 ml of ampicillin (50 mg/ml) per 100 ml of SOB agar. The 15-ml top layer of SOB agar was supplemented with 1 ml of X-gal (2 percent), 1 ml of MgCl$_2$ (1 M), and 1 ml of MgSO$_4$ (1 M) per 100 ml of SOB agar. The 15-ml top layer was poured just before plating. Our titer was approximately 100 colonies per 10-μl aliquot of transformation.

17. K. W. Wilcox and H. O. Smith, *J. Bact.* **122**, 443 (1975).

18. A. Greener, *Strategies* **3**, 5 (1990).

19. T. R. Utterback *et al.*, in preparation.

20. For the unamplified λ library, *H. influenzae* Rd KW20 DNA (>100 kb) was partially digested in a reaction mixture (200 μl) containing 50 μg of DNA, 1× Sau3A

I buffer, and 20 units of Sau3A I for 6 minutes at 23°C. The digested DNA was extracted with phenol and fractionated on a 0.5 percent low melting agarose gel at 2 V/cm for 7 hours. Fragments from 15 to 25 kb were excised and recovered in a final volume of 6 μl. We used 1 μl of fragments with 1 μl of DASHII vector (Stratagene) in the recommended ligation reaction. One microliter of the ligation mixture was used per packaging reaction as recommended in the protocol with the Gigapack II XL Packaging Extract (Stratagene, 227711). Phage were plated directly without amplification from the packaging mixture (after dilution with 500 μl of recommended SM buffer and treatment with chloroform). [SM buffer contains (per liter) 5.8 g of NaCl, 2 g of MgSO$_4$ · H$_2$O, 50 ml of 1 M tris-HCl, pH7.5, and 5 ml of a 2 percent solution of gelatin.] The yield was about 2.5 × 10^3 plaque-forming units (PFU) per microliter. The amplified library was prepared essentially as above except the λ GEM-12 vector was used. After packaging, about 3.5 × 10^4 PFU were plated on the restrictive NM539 host. The lysate was harvested in 2 ml of SM buffer and stored frozen in 7 percent dimethyl sulfoxide. The phage titer was approximately 1 × 10^9 PFU/ml.

21. M. D. Adams, *et al.*, *Nature* **368**, 474 (1994).

22. A. R. Kerlavage *et al.*, *Proceedings of the Twenty-Sixth Annual Hawaii International Conference on System Science* (IEEE Computer Society Press, Washington, DC, 1993), p. 585; A. R. Kerlavage *et al.*, *IEEE Computers in Medicine and Biology* (IEEE, Computer Society Press, Washington, DC, in press).

23. M. S. Waterman, *Methods Enzymol.* **164**, 765 (1988).

24. W. Pearson and D. Lipman, *Proc. Natl. Acad. Sci. U.S.A.* **85**, 2444 (1988).

25. Oligonucleotides were labeled by combining 50 pmol of each 20-mer and 250 mCi of [γ-^{32}P] adenosine triphosphate and T4 polynucleotide kinase. The labeled oligonucleotides were purified with Sephadex G-25 superfine (Pharmacia). A portion containing 10^7 counts per minute of each was used in a Southern hybridization analysis of *H. influenzae* Rd chromosomal DNA digested with one frequently cleaving endonuclease (Ase I) and five less-frequent ones (Bgl II, Eco RI, Pst I, Xba I, and Pvu II). The DNA from each digest was fractionated on a 0.7 percent agarose gel and transferred to nylon (Nytran Plus) membranes (Schleicher & Schuell). Hybridization was carried out for 16 hours at 40°C. To remove nonspecific signals, we sequentially washed each blot at room temperature with increasingly stringent conditions up to 0.1× saline sodium citrate and 0.5 percent SDS. Blots were exposed to a PhosphorImager cassette (Molecular Dynamics) for several hours; hybridization patterns were compared visually.

26. S. Altschul *et al.*, *J. Mol. Biol.* **215**, 403 (1990).

27. E. F. Kirkness *et al.*, *Genomics* **10**, 985 (1991).

28. Standard amplification by polymerase chain reaction (PCR) was performed in the following manner. Each reaction (57 μl) contained a 37-μl mixture of 16.5 μl of H$_2$O, 3 μl of 25 mM MgCl$_2$, 8 μl of a dNTP mix (1.25 mM each dNTP), 4.5 μl of 10× PCR core buffer II (Perkin-Elmer N808-0009), and 25 ng of *H. influenzae* Rd KW20 genomic DNA. The appropriate two primers (4 μl, 3.2 pmol/μl) were added to each reaction. A preliminary incubation (hotstart) was performed at 95°C for 5 minutes followed by a 75°C hold. During the holding period, Amplitaq DNA polymerase (Perkin-Elmer N801-0060, 0.3 μl in 4.3 μl of H$_2$O, 0.5 μl of 10× PCR core buffer II) was added to each reaction. The PCR profile was 25 cycles of 94°C for 45 seconds, then denature; 55°C for 1 minute, then aneal; 72°C for 3 minutes, then extension. All reactions were performed in a 96-well format on a Perkin-Elmer GeneAmp PCR System 9600. Long-range PCR was performed as follows: Each reaction contained a 35.2-μl mixture of 12.0 μl of H$_2$O, 2.2 μl of 25 mM magnesium acetate, 4 μl of a dNTP mixture (200 μM final concentration), 12.0 μl of 3.3× PCR buffer, and 25 ng of *H. influenzae* Rd KW20 genomic DNA. The appropriate two primers (5 μl, 3.2 pmol/μl) were added to each reaction. A preliminary incubation (hot start) was performed at

94°C for 1 minute. Then r*Tth* polymerase (Perkin-Elmer N808-0180) (4 units per reaction) in 2.8 μl of 3.3× PCR buffer II was added to each reaction. The PCR profile was 18 cycles of 94°C for 15 seconds, denature; 62°C for 8 minutes, anneal and extend followed by 12 cycles 94°C for 15 seconds, denature; 62°C for 8 minutes (increase 15 per cycle), anneal and extend; and 72°C for 10 minutes, final extension. All reactions were done in a 96-well format on a Perkin-Elmer GeneAmp PCR System 9600.

29. J. J. Lee, H. O. Smith, R. R. Redfield, *J. Bacteriol.* **171**, 3016 (1989).

30. J. M. Claverie, *J. Mol. Biol.* **234**, 1140 (1993).

31. J. N. Weiser *et al.*, *Cell* **59**, 657 (1989).

32. M. Johnston *et al.*, *Science* **265**, 2077 (1994).

33. B. Lewin, Ed., *Genes V* (Oxford Univ. Press, New York, 1994), chaps. 18 and 19.

34. M. Borodovsky and J. McIninch, *Comp. Chem.* **17**, 123 (1993). In the GeneMark program second-order phased Markov chain models were used; it was trained on 188,572 bp of protein coding sequence and 33,118 bp of noncoding sequence as annotated in GenBank *H. influenzae* entries. It was shown that the second-order program is the most accurate given the size of the training set. The accuracy level was assessed by a cross-validation procedure with a set of 96-bp nonoverlapping fragments derived from the same sets of sequences. With the use of a threshold of 0.5, coding fragments were identified correctly in 91.2 percent of the cases; noncoding fragments were identified correctly in 93.3 percent of the cases.

35. D. Brutlag *et al.*, *ibid.*, p. 203. The BLOSUM 60–amino acid substitution matrix was used in all protein-protein comparisons [S. Henikoff and J. G. Henikoff, *Proc. Natl. Acad. Sci. U.S.A.* **89**, 10915 (1992)].

36. M. Riley, *Microbiol. Rev.* **57**, 862 (1993).

37. I. R. Dorocicz *et al.*, *J. Bacteriol.* **175**, 7142 (1993); B. Dougherty, unpublished results.

38. R. D. Klein and G. H. Luginbuhl, *J. Gen. Microbiol.* **113**, 409 (1979).

39. D. B. Danner *et al.*, *Gene* **11**, 311 (1980); D. B. Danner *et al.*, *Proc. Natl. Acad. Sci. U.S.A.* **79**, 2393 (1982); M. E. Kahn and H. O. Smith, *J. Membr. Biol.* **138**, 155 (1984).

40. H. O. Smith *et al.*, *Science* **269**, 538 (1995).

41. R. R. Redfield, *J. Bacteriol.* **173**, 5612 (1991); M. S. Chandler, *Proc. Natl. Acad. Sci. U.S.A.* **89**, 1616 (1992); R. Barouki and H. O. Smith, *J. Bacteriol.* **163**, 629 (1985); J.-F. Tomb, H. El-Haji, H. O. Smith, *Gene* **104**, 1 (1991); J.-F. Tomb, *Proc. Natl. Acad. Sci. U.S.A.* **89**, 10252 (1992).

42. J.-F. Tomb, unpublished results.

43. L. M. Albright, E. Huala, F. M. Ausubel, *Annu. Rev. Genet.* **23**, 311 (1989); J. S. Parkinson and E. C. Kofoid, *Am. Rev. Genet.* **26**, 71 (1992).

44. M. S. vanHam, L. vanAlphen, F. R. Mooi, J. P. Van-Pattern, *Mol. Microbiol.* **13**, 673 (1994).

45. T. Yura *et al.*, *Nucleic Acids Res.* **20**, 3305 (1992); V. Burland *et al.*, *Genomics* **16**, 551 (1993).

46. J. Kyte and R. F. Doolittle, *J. Mol. Biol.* **157**, 105 (1982).

47. Supported in part by a core grant from Human Genome Sciences and an American Cancer Society grant (NP-838C) (to H.O.S.). Reagents for sequencing reactions and the synthesis of the oligonucleotides were a gift from the Applied Biosystems Division of Perkin-Elmer. We thank T. Burcham of Applied Biosystems for his contribution in the development of the TIGR EDITOR software; M. Riley, Marine Biological Laboratory, Woods Hole, for making her *E. coli* database available; M. Borodovsky and W. Hayes, School of Biology, Georgia Institute of Technology for providing and tuning the GeneMark software for use with *H. influenzae*; and J. Kelley, T. Dixon, and V. Sapiro for their excellent computer system support. H.O.S. is an American Cancer Society research professor.

16 May 1995; accepted 28 June 1995

General and Applied Microbiology 303

SECTION IV
Molecular Biology and Physiology

Molecular Biology and Physiology

The last half of the 20th century witnessed an extraordinary expansion of our knowledge of virtually every aspect of microbial molecular biology and physiology. Consequently, today's practicing microbiologists believe that most biological problems are amenable to experimental attack. Given the task of our group, selecting 10 classic scientific contributions in this area during this period, we relied mostly on our individual recollections and made the somewhat arbitrary decision to emphasize papers that presented conceptual advances rather than technological breakthroughs. However, we recognize the importance of the many impressive technological achievements. We also decided that it would be inappropriate to predict which papers published in the last 20 years will someday be considered classics. We apologize to our many fellow scientists whose outstanding contributions also deserve to be cited. In this preface we introduce each classic paper in the context of what was known and not known at the time the paper was published. Members of the selection committee were Jonathan Beckwith, Ph.D., Jeffrey H. Miller, Ph.D., Lucy Shapiro, Ph.D., Thomas J. Silhavy, Ph.D., and Charles Yanofsky, Ph.D. (Chair). A. Dale Kaiser, Ph.D., and Norman R. Pace, Ph.D., were invited to contribute individual paper introductions.

Genetic Control of Biochemical Reactions in Neurospora

G. W. BEADLE AND E. L. TATUM

One gene, one enzyme. What a simple concept, how obvious! This fundamental relationship between gene and gene product, hinted at by several previous investigators, did not become firmly established until George W. Beadle and Edward L. Tatum performed their pioneering analyses with *Neurospora crassa*. Prior to their proposing the one-gene–one-enzyme hypothesis, in the early 1940s, there was little understanding of how genetic material determined the characteristics of every living organism. One need only recall how little was known about genes and proteins prior to their pathfinding investigations, to appreciate the significance of their impressive contribution. The chemical nature of genetic material had not yet been established, and it was not yet proven that proteins (polypeptides) consist of linear sequences of amino acids. Not until the early 1950s did the initial findings of Avery, MacLeod, and McCarty (see Section II), and the more significant observations of Hershey and Chase (see Section V), convince the scientific community that genetic material was most likely DNA. And it was not until the mid-to-late 1950s, some years after Watson and Crick presented their elegant structure of DNA, that our view of genetic material became fixed forever. The first complete amino acid sequence of a protein, insulin, was presented in the early 1950s, by Fred Sanger and co-workers (A. P. Ryle et al., *Biochem. J.* **60**:541–556, 1955). Only when these essential findings were in place could the significance of the one-gene–one-enzyme concept be fully appreciated. Beadle points this out in a retrospective article written in 1966 for the book "Phage and the Origins of Molecular Biology." He states that at the Cold Spring Harbor Symposium of 1951, "The number whose faith in one-gene–one-enzyme remained steadfast could be counted on the fingers of one hand—with a couple of fingers left over."

How did Beadle and Tatum come to make this momentous discovery? Did their scientific backgrounds, interests, and prior experience direct them along the right path? As we so often note in reviewing progress in science, great contributions stem from insight gained from ongoing investigations on a related subject or question. This was certainly true for Beadle, and for Tatum. Prior to initiating his *Neurospora* studies, Beadle had spent several years collaborating with Boris Ephrussi in an attempt to determine the sequential steps in an eye pigment pathway in *Drosophila*. Similarly, before he joined Beadle in these eye pigment studies, Tatum's research focussed on identifying growth factors required by different species of bacteria. What could be more logical than for these two investigators to combine their interests and expertise, and ask the basic unanswered question: do mutations altering a specific gene, mutations that change the characteristics of an organism, do so by altering the structure and enzymatic activity of a single, corresponding enzyme? If this relationship was correct, they argued, couldn't we use a haploid organism with a simple life cycle, an organism that grows on a chemically-defined medium, to isolate nutrient-requiring mutants that prove the one-gene–one-enzyme relationship?

The rest is history. Many of us consider this contribution by Beadle and Tatum to be one of the most important in the 20th century!

CHARLES YANOFSKY

GENETIC CONTROL OF BIOCHEMICAL REACTIONS IN NEUROSPORA*

By G. W. BEADLE AND E. L. TATUM

BIOLOGICAL DEPARTMENT, STANFORD UNIVERSITY

Communicated October 8, 1941

From the standpoint of physiological genetics the development and functioning of an organism consist essentially of an integrated system of chemical reactions controlled in some manner by genes. It is entirely tenable to suppose that these genes which are themselves a part of the system, control or regulate specific reactions in the system either by acting directly as enzymes or by determining the specificities of enzymes.[1] Since the components of such a system are likely to be interrelated in complex ways, and since the synthesis of the parts of individual genes are presumably dependent on the functioning of other genes, it would appear that there must exist orders of directness of gene control ranging from simple one-to-one relations to relations of great complexity. In investigating the rôles of genes, the physiological geneticist usually attempts to determine the physiological and biochemical bases of already known hereditary traits. This approach, as made in the study of anthocyanin pigments in plants,[2] the fermentation of sugars by yeasts[3] and a number of other instances,[4] has established that many biochemical reactions are in fact controlled in specific ways by specific genes. Furthermore, investigations of this type tend to support the assumption that gene and enzyme specificities are of the same order.[5] There are, however, a number of limitations inherent in this approach. Perhaps the most serious of these is that the investigator must in general confine himself to a study of non-lethal heritable characters. Such characters are likely to involve more or less non-essential so-called "terminal" reactions.[5] The selection of these for genetic study was perhaps responsible for the now rapidly disappearing belief that genes are concerned only with the control of "superficial" characters. A second difficulty, not unrelated to the first, is that the standard approach to the problem implies the use of characters with visible manifestations. Many such characters involve morphological variations, and these are likely to be based on systems of biochemical reactions so complex as to make analysis exceedingly difficult.

Considerations such as those just outlined have led us to investigate

TABLE 1

GROWTH OF PYRIDOXINLESS STRAIN OF *N. sitophila* ON LIQUID MEDIUM CONTAINING INORGANIC SALTS,[9] 1% SUCROSE, AND 0.004 MICROGRAM BIOTIN PER CC. TEMPERATURE 25°C. GROWTH PERIOD, 6 DAYS FROM INOCULATION WITH CONIDIA

MICROGRAMS B_6 PER 25 CC. MEDIUM	STRAIN	DRY WEIGHT MYCELIA, MG.
0	Normal	76.7
0	Pyridoxinless	1.0
0.01	"	4.2
0.03	"	5.7
0.1	"	13.7
0.3	"	25.5
1.0	"	81.1
3.0	"	81.1
10.0	"	65.4
30.0	"	82.4

the general problem of the genetic control of developmental and metabolic reactions by reversing the ordinary procedure and, instead of attempting to work out the chemical bases of known genetic characters, to set out to determine if and how genes control known biochemical reactions. The ascomycete *Neurospora* offers many advantages for such an approach and is well suited to genetic studies.[6] Accordingly, our program has been built around this organism. The procedure is based on the assumption that x-ray treatment will induce mutations in genes concerned with the control of known specific chemical reactions. If the organism must be able to carry out a certain chemical reaction to survive on a given medium, a mutant unable to do this will obviously be lethal on this medium. Such a mutant can be maintained and studied, however, if it will grow on a medium to which has been added the essential product of the genetically blocked reaction. The experimental procedure based on this reasoning can best be illustrated by considering a hypothetical example. Normal strains of *Neurospora crassa* are able to use sucrose as a carbon source, and are therefore able to carry out the specific and enzymatically controlled

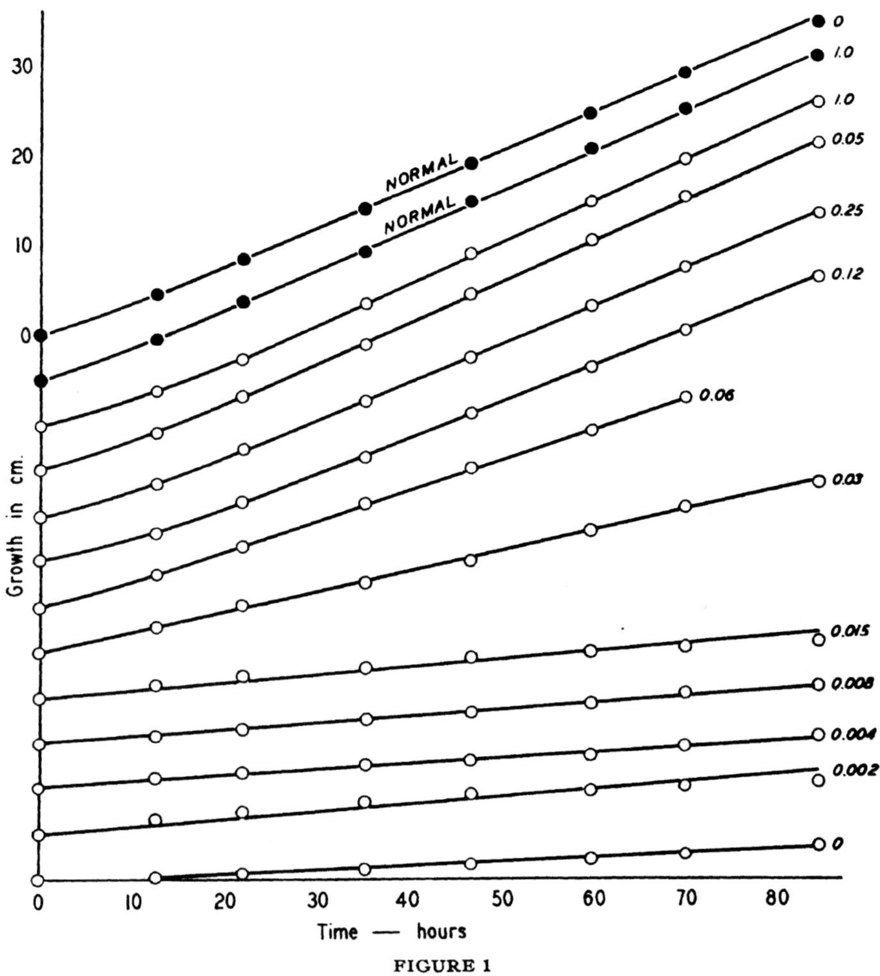

FIGURE 1

Growth of normal (top two curves) and pyridoxinless (remaining curves) strains of *Neurospora sitophila* in horizontal tubes. The scale on the ordinate is shifted a fixed amount for each successive curve in the series. The figures at the right of each curve indicate concentration of pyridoxine (B_6) in micrograms per 25 cc. medium.

reaction involved in the hydrolysis of this sugar. Assuming this reaction to be genetically controlled, it should be possible to induce a gene to mutate to a condition such that the organism could no longer carry out sucrose hydrolysis. A strain carrying this mutant would then be unable to grow on a medium containing sucrose as a sole carbon source but should be able to grow on a medium containing some other normally utilizable carbon source. In other words, it should be possible to establish and maintain such a mutant strain on a medium containing glucose and detect its inability to utilize sucrose by transferring it to a sucrose medium.

Essentially similar procedures can be developed for a great many metabolic processes. For example, ability to synthesize growth factors (vitamins), amino acids and other essential substances should be lost through gene mutation if our assumptions are correct. Theoretically, any such metabolic deficiency can be "by-passed" if the substance lacking can be supplied in the medium and can pass cell walls and protoplasmic membranes.

In terms of specific experimental practice, we have devised a procedure in which x-rayed single-spore cultures are established on a so-called "complete" medium, i.e., one containing as many of the normally synthesized constituents of the organism as is practicable. Subsequently these are tested by transferring them to a "minimal" medium, i.e., one requiring the organism to carry on all the essential syntheses of which it is capable. In practice the complete medium is made up of agar, inorganic salts, malt extract, yeast extract and glucose. The minimal medium contains agar (optional), inorganic salts and biotin, and a disaccharide, fat or more complex carbon source. Biotin, the one growth factor that wild type *Neurospora* strains cannot synthesize,[7] is supplied in the form of a commercial concentrate containing 100 micrograms of biotin per cc.[8] Any loss of ability to synthesize an essential substance present in the complete medium and absent in the minimal medium is indicated by a strain growing on the first and failing to grow on the second medium. Such strains are then tested in a systematic manner to determine what substance or substances they are unable to synthesize. These subsequent tests include attempts to grow mutant strains on the minimal medium with (1) known vitamins added, (2) amino acids added or (3) glucose substituted for the more complex carbon source of the minimal medium.

Single ascospore strains are individually derived from perithecia of *N. crassa* and *N. sitophila* x-rayed prior to meiosis. Among approximately 2000 such strains, three mutants have been found that grow essentially normally on the complete medium and scarcely at all on the minimal medium with sucrose as the carbon source. One of these strains (*N. sitophila*) proved to be unable to synthesize vitamin B_6 (pyridoxine). A second strain (*N. sitophila*) turned out to be unable to synthesize vitamin B_1 (thiamine). Additional tests show that this strain is able to synthesize the pyrimidine half of the B_1 molecule but not the thiazole half. If thiazole alone is added to the minimal medium, the strain grows essentially normally. A third strain (*N. crassa*) has been found to be unable to synthesize para-aminobenzoic acid.   This mutant strain appears to be entirely normal when grown on the minimal medium to which *p*-aminobenzoic acid has been added.   Only in the case of the "pyridoxinless" strain has an analysis of the inheritance of the induced metabolic defect been investigated.   For this reason detailed accounts of the thiamine-deficient and *p*-aminobenzoic acid-deficient strains will be deferred.

Qualitative studies indicate clearly that the pyridoxinless mutant, grown on a medium containing one microgram or more of synthetic vitamin B_6 hydrochloride per 25 cc. of medium, closely approaches in rate and characteristics of growth normal strains grown on a similar medium with

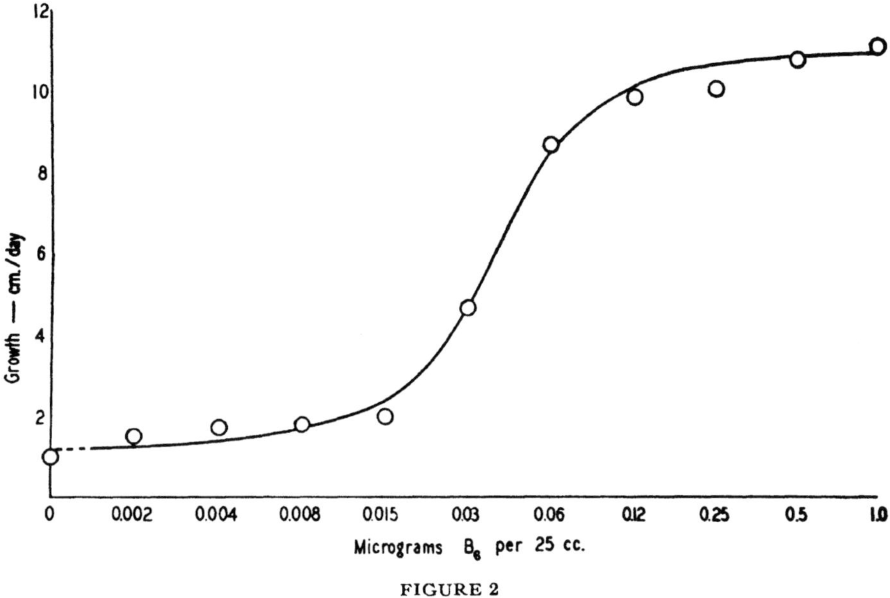

FIGURE 2

The relation between growth rate (cm./day) and vitamin B_6 concentration.

no B_6.   Lower concentrations of B_6 give intermediate growth rates.   A preliminary investigation of the quantitative dependence of growth of the mutant on vitamin B_6 in the medium gave the results summarized in table 1.   Additional experiments have given results essentially similar but in only approximate quantitative agreement with those of table 1. It is clear that additional study of the details of culture conditions is necessary before rate of weight increase of this mutant can be used as an accurate assay for vitamin B_6.

It has been found that the progression of the frontier of mycelia of *Neurospora* along a horizontal glass culture tube half filled with an agar medium provides a convenient method of investigating the quantitative effects of growth factors. Tubes of about 13 mm. inside diameter and about 40 cm. in length are used. Segments of about 5 cm. at the two ends are turned up at an angle of about 45°. Agar medium is poured in so as to fill the tube about half full and is allowed to set with the main segment of the tube in a horizontal position. The turned up ends of the tube are stoppered with cotton plugs. Inoculations are made at one end of the agar surface and the position of the advancing front recorded at convenient intervals. The frontier formed by the advancing mycelia is remarkably well defined, and there is no difficulty in determining its position to within a millimeter or less. Progression along such tubes is strictly linear with time and the rate is independent of tube length (up to 1.5 meters). The rate is not changed by reducing the inside tube diameter to 9 mm., or by

TABLE 2

RESULTS OF CLASSIFYING SINGLE ASCOSPORE CULTURES FROM THE CROSS OF PYRI-
DOXINLESS AND NORMAL *N. sitophila*

ASCUS NUMBER	1	2	3	4	5	6	7	8
17	—	*pdx*	*pdx*	*pdx*	*N*	*N*	*N*	—
18	—	—	*N*	*N*	—	—	*pdx*	*pdx*
19	—	*pdx*	—	—	—	—	—	*N*
20	—	—	*N*	—	—	—	—	*pdx*
22	—	—	*N*	—	—	—	—	—
23	—	*	*	*	*N*	*N*	*pdx*	*pdx*
24	*N*	*N*	*N*	*N*	*pdx*	*pdx*	*pdx*	*pdx*

N, normal growth on B_6-free medium.   *pdx*, slight growth on B_6-free medium.   Failure of ascospore germination indicated by dash.

 * Spores 2, 3 and 4 isolated but positions confused.   Of these, two germinated and both proved to be mutants.

sealing one or both ends. It therefore appears that gas diffusion is in no way limiting in such tubes.

The results of growing the pyridoxinless strain in horizontal tubes in which the agar medium contained varying amounts of B_6 are shown graphically in figures 1 and 2. Rate of progression is clearly a function of vitamin B_6 concentration in the medium.[10] It is likewise evident that there is no significant difference in rate between the mutant supplied with B_6 and the normal strain growing on a medium without this vitamin. These results are consistent with the assumption that the primary physiological difference between pyridoxinless and normal strains is the inability of the former to carry out the synthesis of vitamin B_6. There is certainly more than one step in this synthesis and accordingly the gene differential involved is presumably concerned with only one specific step in the biosynthesis of vitamin B_6.

In order to ascertain the inheritance of the pyridoxinless character, crosses between normal and mutant strains were made. The techniques for hybridization and ascospore isolation have been worked out and described by Dodge, and by Lindegren.[6] The ascospores from 24 asci of the cross were isolated and their positions in the asci recorded. For some unknown reason, most of these failed to germinate. From seven asci, however, one or more spores germinated. These were grown on a medium containing glucose, malt-extract and yeast extract, and in this they all grew normally. The normal and mutant cultures were differentiated by growing them on a B_6 deficient medium. On this medium the mutant cultures grew very little, while the non-mutant ones grew normally. The results are summarized in table 2. It is clear from these rather limited data that this inability to synthesize vitamins B_6 is transmitted as it should be if it were differentiated from normal by a single gene.

The preliminary results summarized above appear to us to indicate that the approach outlined may offer considerable promise as a method of learning more about how genes regulate development and function. For example, it should be possible, by finding a number of mutants unable to carry out a particular step in a given synthesis, to determine whether only one gene is ordinarily concerned with the immediate regulation of a given specific chemical reaction.

It is evident, from the standpoints of biochemistry and physiology, that the method outlined is of value as a technique for discovering additional substances of physiological significance. Since the complete medium used can be made up with yeast extract or with an extract of normal *Neurospora*, it is evident that if, through mutation, there is lost the ability to synthesize an essential substance, a test strain is thereby made available for use in isolating the substance. It may, of course, be a substance not previously known to be essential for the growth of any organism. Thus we may expect to discover new vitamins, and in the same way, it should be possible to discover additional essential amino acids if such exist. We have, in fact, found a mutant strain that is able to grow on a medium containing Difco yeast extract but unable to grow on any of the synthetic media we have so far tested. Evidently some growth factor present in yeast and as yet unknown to us is essential for *Neurospora*.

Summary.—A procedure is outlined by which, using *Neurospora*, one can discover and maintain x-ray induced mutant strains which are characterized by their inability to carry out specific biochemical processes.

Following this method, three mutant strains have been established. In one of these the ability to synthesize vitamin B_6 has been wholly or largely lost. In a second the ability to synthesize the thiazole half of the vitamin B_1 molecule is absent, and in the third para-aminobenzoic acid is not synthesized.   It is therefore clear that all of these substances are essential growth factors for *Neurospora*.[11]

Growth of the pyridoxinless mutant (a mutant unable to synthesize vitamin B_6) is a function of the B_6 content of the medium on which it is grown.   A method is described for measuring the growth by following linear progression of the mycelia along a horizontal tube half filled with an agar medium.

Inability to synthesize vitamin B_6 is apparently differentiated by a single gene from the ability of the organism to elaborate this essential growth substance.

NOTE: Since the manuscript of this paper was sent to press it has been established that inability to synthesize both thiazole and *p*-aminobenzoic acid are also inherited as though differentiated from normal by single genes.

* Work supported in part by a grant from the Rockefeller Foundation.   The authors are indebted to Doctors B. O. Dodge, C. C. Lindegren and W. S. Malloch for stocks and for advice on techniques, and to Miss Caryl Parker for technical assistance.

[1] The possibility that genes may act through the mediation of enzymes has been suggested by several authors.   See Troland, I. T., *Amer. Nat.*, **51**, 321–350 (1917); Wright, S., *Genetics*, **12**, 530–569 (1927); and Haldane, J. B. S., in *Perspectives in Biochemistry*, Cambridge Univ. Press, pp. 1–10 (1937), for discussions and references.

[2] Onslow, Scott-Moncrieff and others, see review by Lawrence, W. J. C., and Price, J. R., *Biol. Rev.*, **15**, 35–58 (1940).

[3] Winge, O., and Laustsen, O., *Compt. rend. Lab. Carlsberg, Serie physiol.*, **22**, 337–352 (1939).

[4] See Goldschmidt, R., *Physiological Genetics*, McGraw-Hill, pp. 1–375 (1939), and Beadle, G. W., and Tatum, E. L., *Amer. Nat.*, **75**, 107–116 (1941) for discussion and references.

[5] See Sturtevant, A. H., and Beadle, G. W., *An Introduction to Genetics*, Saunders, pp. 1–391 (1931), and Beadle, G. W., and Tatum, E. L., loc. cit., footnote 4.

[6] Dodge, B. O., *Jour. Agric. Res.*, **35**, 289–305 (1927), and Lindegren, C. C., *Bull. Torrey Bot. Club*, **59**, 85–102 (1932).

[7] In so far as we have carried them, our investigations on the vitamin requirements of *Neurospora* corroborate those of Butler, E. T., Robbins, W. J., and Dodge, B. O., *Science*, **94**, 262–263 (1941).

[8] The biotin concentrate used was obtained from the S. M. A. Corporation, Chagrin Falls, Ohio.

[9] Throughout our work with *Neurospora*, we have used as a salt mixture the one designated number 3 by Fries, N., *Symbolae Bot. Upsalienses*, Vol. 3, No. 2, 1–188 (1938). This has the following composition: NH_4 tartrate, 5 g.; NH_4NO_3, 1 g.; KH_2PO_4, 1 g.; $MgSO_4 \cdot 7H_2O$, 0.5 g.; NaCl, 0.1 g.; $CaCl_2$, 0.1 g.; $FeCl_3$, 10 drops 1% solution; H_2O, 1 l.   The tartrate cannot be used as a carbon source by *Neurospora*.

[10] It is planned to investigate further the possibility of using the growth of *Neurospora* strains in the described tubes as a basis of vitamin assay, but it should be emphasized that such additional investigation is essential in order to determine the reproducibility and reliability of the method.

[11] The inference that the three vitamins mentioned are essential for the growth of normal strains is supported by the fact that an extract of the normal strain will serve as a source of vitamin for each of the mutant strains.

Mutations of Bacteria from Virus Sensitivity to Virus Resistance

S. E. Luria and M. Delbrück

After the introduction of pure culture techniques for the study of bacteria in the late 19th century, it became accepted that bacteria, like higher organisms, have inherited qualities. A single bacterium isolated on the surface of agar produces a clone of descendants, all of which resemble the original parent in virtually every respect. However, it also became clear that variants appear in otherwise pure cultures and that these variants breed true. What was the origin of such variants?

Despite the widespread recognition of the role of chance mutations in the genetic variation of higher organisms, bacteriologists were reluctant to extend this concept to bacteria. Bacterial cultures appeared to be plastic. Exposed to adverse environments, cultures quickly gave rise to genetically stable types adapted to the new conditions. The majority of bacteriologists in the early 20th century believed that environment directly induced some or all of the cells in the population to become stably adapted. Obviously the much faster rate of growth of bacteria than of higher organisms contributed to the appearance of plasticity. Various mechanisms other than mutation were proposed for these variations, including shifts in chemical equilibria by Hinshelwood (A. C. R. Dean and C. Hinshelwood, p. 21–45, *in* E. F. Davies and R. Davies [ed.], *Adaptation in Microorganisms,* Cambridge University Press, Cambridge, U.K., 1953). The absence of sexual differentiation and of a chromosome visible by light microscopy made it easier to accept a different mode of inheritance for bacteria. The phenomenon of induced enzyme biosynthesis, which had not been clarified at that time, lent circumstantial weight to the plasticity view.

Against this background, the first clear-cut evidence for the occurrence of spontaneous mutation in bacteria was provided by Luria and Delbrück in 1943. They demonstrated that, when a sensitive culture is plated in the presence of an excess of bacteriophage T1, colonies arise from phage-resistant mutants, which were present in the inoculum before exposure to the phage. Luria and Delbrück used fluctuations in the fraction of resistant mutants across a series of cultures to argue that the phage-resistant mutants predated selection by the phage. The Luria-Delbrück paper took a solid first step toward showing that bacteria had genes like plants and animals.

Luria and Delbrück were an effective team. They had agreed to study the problem of "secondary growth" (of the phage-resistant mutants) in April of 1941. However, the problem proved harder than expected. Early attempts to devise critical experiments were unsuccessful during Luria's visit to Delbrück at Vanderbilt University in the fall term of 1942. However, Luria continued working on the problem at Indiana University, where he was an Assistant Professor. Luria, in Delbruck's words, ". . . early in 1943 hit on a satisfactory method of deciding the issue. This method was worked out, the theory by me here at Vanderbilt, the experimental technique by Dr. Luria at Indiana." Delbruck and Luria then went on to create a whole new school of phage genetic research.

A. Dale Kaiser

Reprinted from *Genetics* 28:491–511. Copyright © 1943 by Genetics Society of America.

MUTATIONS OF BACTERIA FROM VIRUS SENSITIVITY
TO VIRUS RESISTANCE[1,2]

S. E. LURIA[3] AND M. DELBRÜCK

Indiana University, Bloomington, Indiana, and
Vanderbilt University, Nashville, Tennessee

Received May 29, 1943

INTRODUCTION

WHEN a pure bacterial culture is attacked by a bacterial virus, the culture will clear after a few hours due to destruction of the sensitive cells by the virus. However, after further incubation for a few hours, or sometimes days, the culture will often become turbid again, due to the growth of a bacterial variant which is resistant to the action of the virus. This variant can be isolated and freed from the virus and will in many cases retain its resistance to the action of the virus even if subcultured through many generations in the absence of the virus. While the sensitive strain adsorbed the virus readily, the resistant variant will generally not show any affinity to it.

The resistant bacterial variants appear readily in cultures grown from a single cell. They were, therefore, certainly not present when the culture was started. Their resistance is generally rather specific. It does not extend to viruses that are found to differ by other criteria from the strain in whose presence the resistant culture developed. The variant may differ from the original strain in morphological or metabolic characteristics, or in serological type or in colony type. Most often, however, no such correlated changes are apparent, and the variant may be distinguished from the original strain only by its resistance to the inciting strain of virus.

The nature of these variants and the manner in which they originate have been discussed by many authors, and numerous attempts have been made to correlate the phenomenon with other instances of bacterial variation.

The net effect of the addition of virus consists of the appearance of a variant strain, characterized by a new stable character—namely, resistance to the inciting virus. The situation has often been expressed by saying that bacterial viruses are powerful "dissociating agents." While this expression summarizes adequately the net effect, it must not be taken to imply anything about the mechanism by which the result is brought about. A moment's reflection will show that there are greatly differing mechanisms which might produce the same end result.

D'HERELLE (1926) and many other investigators believed that the virus by direct action induced the resistant variants. GRATIA (1921), BURNET (1929), and others, on the other hand, believed that the resistant bacterial variants are produced by mutation in the culture prior to the addition of virus. The

[1] Theory by M. D., experiments by S. E. L.
[2] Aided by grants from the DAZIAN FOUNDATION FOR MEDICAL RESEARCH and from the ROCKEFELLER FOUNDATION.
[3] Fellow of the GUGGENHEIM FOUNDATION.

virus merely brings the variants into prominence by eliminating all sensitive bacteria.

Neither of these views seems to have been rigorously proved in any single instance. BURNET'S (1929) work on isolations of colonies, morphologically distinguishable prior to the addition of virus, which proved resistant to the virus comes nearest to this goal. His results appear to support the mutation hypothesis for colony variants. It may seem peculiar that this simple and important question should not have been settled long ago, but a close analysis of the problem in hand will show that a decision can only be reached by a more subtle quantitative study than has hitherto been applied in this field of research.

Let us begin by restating the basic experimental finding.

A bacterial culture is grown from a single cell. At a certain moment the culture is plated with virus in excess. Upon incubation, one finds that a very small fraction of the bacteria survived the attack of the virus, as indicated by the development of a small number of resistant colonies, consisting of bacteria which do not even adsorb the virus.

Let us focus our attention on the first generation of the resistant variant—that is, on those bacteria which survive immediately after the virus has been added. These survivors we may call the "original variants." We know that these bacteria and their offspring are resistant to the virus. We may formulate three alternative hypotheses regarding them.

a. *Hypothesis of mutation to immunity.* The original variants were resistant before the virus was added, and, like their offspring, did not even adsorb it. On this hypothesis the virus did not interact at all with the original variants, the origin of which must be ascribed to "mutations" that occur quite independently of the virus. Naming such hereditary changes "mutations" of course does not imply a detailed similarity with any of the classes of mutations that have been analyzed in terms of genes for higher organisms. The similarity may be merely a formal one.

b. *Hypothesis of acquired immunity.* The original variants interacted with the virus, but survived the attack. We may then inquire into the predisposing cause which effected the survival of these bacteria in contradistinction to the succumbing ones. The predisposing cause may be hereditary or random. Accordingly we arrive at two alternative hypotheses—namely,

b_1. *Hypothesis of acquired immunity of hereditarily predisposed individuals.* The original variants originated by mutations occurring independently of the presence of virus. When the virus is added, the variants will interact with it, but they will survive the interaction, just as there may be families which are hereditarily predisposed to survive an otherwise fatal virus infection. Since we know that the offspring of the original variants do not adsorb the virus, we must further assume that the infection caused this additional hereditary change.

b_2. *Hypothesis of acquired immunity—hereditary after infection.* The original variants are predisposed to survival by random physiological variations in size, age, etc. of the bacteria, or maybe even by random variations in the point of attack of the virus on the bacterium. After survival of such random individuals, however, we must assume that their offspring are hereditarily immune, since they do not even adsorb the virus.

These alternative hypotheses may be grouped by first considering the origin of the hereditary difference. Do the original variants trace back to mutations which occur independently of the virus, such that these bacteria belong to a few clones, or do they represent a random sample of the entire bacterial population? The first alternative may then be subdivided further, according to whether the original variants do or do not interact with the virus. Disregarding for the moment this subdivision, we may formulate two hypotheses:

1. *First hypothesis (mutation)*: There is a finite probability for any bacterium to mutate during its life time from "sensitive" to "resistant." Every offspring of such a mutant will be resistant, unless reverse mutation occurs. The term "resistant" means here that the bacterium will not be killed if exposed to virus, and the possibility of its interaction with virus is left open.

2. *Second hypothesis (acquired hereditary immunity)*: There is a small finite probability for any bacterium to survive an attack by the virus. Survival of an infection confers immunity not only to the individual but also to its offspring. The probability of survival in the first instance does not run in clones. If we find that a bacterium survives an attack, we cannot from this information infer that close relatives of it, other than descendants, are likely to survive the attack.

The last statement contains the essential difference between the two hypotheses. On the mutation hypothesis, the mutation to resistance may occur any time prior to the addition of virus. The culture therefore will contain "clones of resistant bacteria" of various sizes, whereas on the hypothesis of acquired immunity the bacteria which survive an attack by the virus will be a random sample of the culture.

For the discussion of the experimental possibility of distinction between these two hypotheses, it is important to keep in mind that the offspring of a *tested* bacterium which survives is resistant on either hypothesis. Repeated tests on a bacterium at different times, or on a bacterium and on its offspring, could therefore give no information of help in deciding the present issue. Thus, one has to resort to less direct methods. Two main differences may be derived from the hypotheses:

First, if the individual cells of a very large number of microcolonies, each containing only a few bacteria, were examined for resistance, a pronounced correlation between the types found in a single colony would be expected on the mutation hypothesis, while a random distribution of resistants would be expected on the hypothesis of acquired hereditary immunity. This experiment, however, is not practicable, both on account of the difficulty of manipulation and on account of the small proportion of resistant bacteria.

Second, on the hypothesis of resistance due to mutation, the proportion of resistant bacteria should increase with time, in a growing culture, as new mutants constantly add to their ranks.

In contrast to this increase in the proportion of resistants on the mutation hypothesis, a constant proportion of resistants may be expected on the hypothesis of acquired hereditary immunity, as long as the physiological conditions of the culture do not change. To test this point, accurate determinations of the proportion of resistant bacteria in a growing culture and in successive subcultures are required. In the attempt to determine accurately the proportion of resistant bacteria, great variations of the proportions were found, and results did not seem to be reproducible from day to day.

Eventually, it was realized that these fluctuations were not due to any uncontrolled conditions of our experiments, but that, on the contrary, large fluctuations are a necessary consequence of the mutation hypothesis and that the quantitative study of the fluctuations may serve to test the hypothesis.

The present paper will be concerned with the theoretical analysis of the probability distribution of the number of resistant bacteria to be expected on either hypothesis and with experiments from which this distribution may be inferred.

While the theory is here applied to a very special case, it will be apparent that the problem is a general one, encountered in any case of mutation in uniparental populations. It is the belief of the authors that the quantitative study of bacterial variation, which until now has made such little progress, has been hampered by the apparent lack of reproducibility of results, which, as we shall show, lies in the very nature of the problem and is an essential element for its analysis. It is our hope that this study may encourage the resumption of quantitative work on other problems of bacterial variation.

THEORY

The aim of the theory is the analysis of the probability distributions of the number of resistant bacteria to be expected on the hypothesis of acquired immunity and on the hypothesis of mutation.

The basic assumption of the hypothesis of acquired hereditary immunity is the assumption of a fixed small chance for each bacterium to survive an attack by the virus. In this case we may therefore expect a binomial distribution of the number of resistant bacteria, or, in cases where the chance of survival is small, a Poisson distribution.

The basic assumption of the mutation hypothesis is the assumption of a fixed small chance *per time unit* for each bacterium to undergo a mutation to resistance. The assumption of a fixed chance per time unit is reasonable only for bacteria in an identical state. Actually the chance may vary in some manner during the life cycle of each bacterium and may also vary when the physiological conditions of the culture vary, particularly when growth slows down on account of crowding of the culture. With regard to the first of these variations, the assumed chance represents the average chance per time unit, averaged over the life cycle of a bacterium. With regard to the second variation, it seems reasonable to assume that the chance is proportional to the growth rate of the bacteria. We will then obtain the same results as on the simple assumption of a fixed chance per time unit, *if we agree to measure time in units of division cycles of the bacteria,* or any proportional unit.

We shall choose as time unit the average division time of the bacteria, divided by ln 2, so that the number N_t of bacteria in a growing culture as function of time t follows the equations

$$(1) \qquad dN_t/dt = N_t, \quad \text{and} \quad N_t = N_0 e^t.$$

We may then define the chance of mutation for each bacterium during the time element dt as

$$(2) \qquad adt,$$

so that a is the chance of mutation per bacterium per time unit, or the "mutation rate."

If a bacterium is capable of different mutations, each of which results in resistance, the mutation rate here considered will be the sum of the mutation rates associated with each of the different mutations.

The number dm of mutations which occur in a growing culture during a time interval dt is then equal to this chance (2) multiplied by the number of bacteria,[4] or

$$(3) \qquad dm = adtN_t,$$

and from this equation the number m of mutations which occur during any finite time interval may be found by integration to be

$$(4) \qquad m = a(N_t - N_0)$$

or, in words, to be equal to the chance of mutation per bacterium per time unit multiplied by the increase in the number of bacteria.

The bacteria which mutate during any time element dt form a random sample of the bacteria present at that time. For small mutation rates, their number will therefore be distributed according to Poisson's law. Since the mutations occuring in different time intervals are quite independent from each other, the distribution of all mutations will also be according to Poisson's law.

This prediction cannot be verified directly, because what we observe, when we count the number of resistant bacteria in a culture, is not the number of mutations which have occurred, but the number of resistant bacteria which have arisen by multiplication of those which mutated, the amount of multiplication depending on how far back the mutation occurred.

If, however, the premise of the mutation hypothesis can be proved by other means, the prediction of a Poisson distribution of the number of mutations

---

[4] We assume that the number of resistant bacteria is at all times small in comparison with the total number of bacteria. If this condition is not fulfilled, the total number of bacteria in this equation has to be replaced by the number of sensitive bacteria. The subsequent theoretical developments will then become a little more complicated. For the case studied in the experimental part of this paper the condition is fulfilled.

may be used to determine the mutation rate. It is only necessary to determine the fraction of cultures showing no mutation in a large series of similar cultures. This fraction p_0, according to theory, should be:

$$(5) \qquad\qquad\qquad p_0 = e^{-m}.$$

From this equation the average number m of mutations may be calculated, and hence the mutation rate a from equation (4).

Let us now turn to the discussion of the distribution of the number of resistant bacteria.

The average number of resistant bacteria is easily obtained by noting that this number increases on two accounts—namely, first on account of new mutations, second on account of the growth of resistant bacteria from previous mutations. During a time element dt the increase on the first account will be, by equation (3): $adtN_t$. N_t, the number of bacteria present at time t, is given by equation (1). The increase on the second account will depend on the growth rate of the resistant bacteria. In the simple case, which we shall treat here, this growth rate is the same as that of the sensitive bacteria, and the increment on this account is $\rho\, dt$, where ρ is the average number of resistant bacteria present at time t. We have then as the total rate of increase of the average number of resistant bacteria $d\rho/dt = aN_t + \rho$ and upon integration

$$(6) \qquad\qquad\qquad \rho = taN_t$$

if we assume that at time zero the culture contained no resistant bacteria.

It will be seen that the average number of resistant bacteria increases more rapidly than the total number of bacteria. Indeed the fraction of resistant bacteria in the culture increases proportionally to time. This, as pointed out in the introduction, is a distinguishing feature of the mutation hypothesis but unfortunately, as will be seen in the sequel, is not susceptible to experimental verification due to statistical fluctuations.

The resistant bacteria in any culture may be grouped, for the purpose of this analysis, into clones, taking together all those which derive from the same mutation. We may say that the culture contains clones of various age and size, calling "age" of a clone the time since its parent mutation occurred and "size" of a clone the number of bacteria in a clone at the time of observation. It is clear that size and age of a clone determine each other. If, in particular, we make the simplifying hypothesis that the resistant bacteria grow as fast as the normal sensitive strain, the relation between size and age will be expressed by equation (1), with appropriate meaning given to the symbols.

The relation implies that the size of a clone increases exponentially with its age. On the other hand, the frequency with which clones of different ages may be encountered in any culture must decrease exponentially with age, according to equations (3) and (1).

Combining these two results—namely, that clone size increases exponentially with clone age and that frequency of clones of different age decreases exponentially with clone age—we see that the two factors cancel when the *average number of bacteria belonging to clones of one age group* is considered. In other words, at the time of observation we shall have, *on the average,* as many resistant bacteria stemming from mutations which occurred during the first generation after the culture was started as stemming from mutations which occurred during the last generation before observation, or during any other single generation.

On the other hand, for small mutation rates it is very improbable that any mutation will occur during the early generations of a single or of a limited number of experimental cultures. It follows that the average number of resistant bacteria derived from a limited number of experimental cultures will, probably, be considerably smaller than the theoretical value given by equation (6), and, improbably, the experimental value will be much larger than the theoretical value. The situation is similar to the operation of a (fair) slot machine, where the average return from a limited number of plays is probably considerably less than the input, and improbably, when the jackpot is hit, the return is much bigger than the input.

This result characterizes the distribution of the number of resistant bacteria as a distribution with a long and significant tail of rare cases of high numbers of resistant bacteria, and therefore as *a distribution with an abnormally high variance.* This variance will be calculated below.

For such distributions the averages derived from limited numbers of samples yield very poor estimates of the true averages. Somewhat better estimates of the averages may in such cases be obtained by omitting, in the calculation of the theoretical averages, the contribution to these averages of those events which probably will not occur in any of our limited number of samples. We may do this, in the integration leading to equation (6), by putting the lower limit of integration not at time zero, when the cultures were started, but at a certain time t_0, prior to which mutations were not likely to occur in any of our experimental cultures. We then obtain as a *likely average* r of the number of resistant bacteria in a limited number of samples, instead of equation (6),

$$(6a) \qquad\qquad r = (t - t_0)aN_t.$$

It now remains to choose an appropriate value for the time interval $t - t_0$.

For this purpose we return to equation (4), in which it was stated that the average number of mutations which occur in a culture is equal to the mutation rate multiplied by the increase of the number of bacteria. Let us then choose t_0 such that up to that time just one mutation occurred, on the average, in a group of C similar cultures, or

$$1 = aC(N_{t_0} - N_0).$$

In this equation we may neglect N_0, the number of bacteria in each inoculum, in comparison with N_{t_0}, the number of bacteria in each culture at the critical time t_0. We may also express N_{t_0} in terms of N_t, the number of bacteria at the time of observation, applying equation (1):

$$N_{t_0} = N_t e^{-(t - t_0)}.$$

We thus obtain

$$(7) \qquad\qquad t - t_0 = \ln(N_t Ca).$$

Equations (6a) and (7) may be combined to eliminate $t - t_0$ and to yield a relation between the observable quantities r and N_t on the one hand and the mutation rate a on the other hand, to be determined by this equation:

$$(8) \qquad\qquad r = aN_t \ln (N_t Ca).$$

This simple transcendental equation determining a may be solved by any standard numerical method. In figure 1, the relation between r and aN_t is plotted for several values of C.

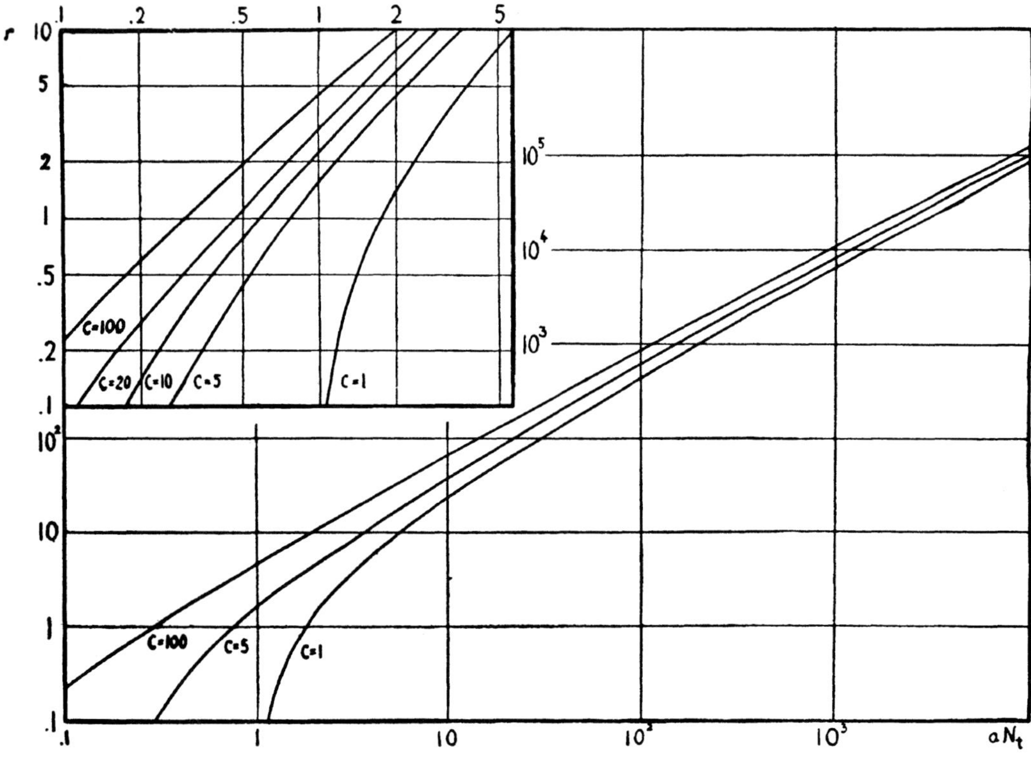

FIGURE 1.—The value of aN_t as a function of r for various values of C. The upper left hand part of the figure gives the curves for low values of aN_t and of r on a larger scale. See text.

Estimates of a obtained from equation (8) will be too high if in any of the experimental cultures a mutation happened to occur prior to time t_0. From the definition of t_0 it will be seen that this can be expected to happen in little more than half of the cases.

While we have thus obtained a relation permitting an estimate of the mutation rate from the observation of a limited number of cultures, this relation is in no way a test of the correctness of the underlying assumptions and, in particular, is not a test of the mutation hypothesis itself. In order to find such tests of the correctness of the assumption we must derive further quantitative relations concerning the distribution of the number of resistant bacteria and compare them with experimental results.

Since we have seen that the mutation hypothesis, in contrast to the hypothesis of acquired immunity, predicts a distribution of the number of resistant bacteria with a long tail of high numbers of resistant bacteria, the determination of the *variance* of the distribution should be helpful in differentiating between the two hypotheses. We may here again determine first the true variance—that is, the variance of the complete distribution—and second the likely variance in a limited number of cultures, by omitting those cases which are not likely to occur in a limited number of cultures.

The variance may be calculated in a simple manner by considering separately the variances of the partial distributions of resistant bacteria, each partial distribution comprising the resistant bacteria belonging to clones of one age group. The distribution of the total number of resistant bacteria is the resultant of the superposition of these independent partial distributions.

Each partial distribution is due to the mutations which occurred during a certain time interval $d\tau$, extending from $(t-\tau)$ to $(t-\tau+d\tau)$. The average number of mutations which occurred during this interval is, according to equation (3),

$$(9) \qquad dm = aN_\tau d\tau = aN_t e^{-\tau} d\tau.$$

These mutations will be distributed according to Poisson's law, so that the variance of each of these distributions is equal to the mean of the distribution. We are however not interested in the distribution of the number of mutations but in the distribution of the number of resistant bacteria which stem from these mutations at the time of observation—that is, after the time interval τ. Each original mutant has then grown into a clone of size e^τ. The distribution of the resistant bacteria stemming from mutations occurred in the time interval $d\tau$ has therefore an average value which is e^τ times greater than the average number of mutations, and a variance which is $e^{2\tau}$ times greater than the variance of the number of mutations. Thus we find for the average number of resistant bacteria:

$$d\rho = aN_t d\tau,$$

and for the variance of this number

$$\mathrm{var}_{d\rho} = aN_t e^\tau d\tau.$$

From this variance of the partial distribution, the variance of the distribution of all resistant bacteria may be found simply by integrating over the appropriate time interval—that is, either from time t to time o (τ from o to t), if the true variance is wanted, or from time t to time t_0 (τ from o to $t-t_0$), if the likely variance in a limited number of cultures is wanted. In the first case we obtain:

$$(10) \qquad \mathrm{var}_\rho = aN_t(e^t - 1).$$

In the second case we obtain:

$$(10a) \qquad \mathrm{var}_\tau = aN_t[e^{(t-t_0)} - 1].$$

Substituting here the previously found value of $(t - t_0)$ and neglecting the second term in the brackets, we obtain:

$$(11) \qquad var_r = Ca^2 N_t^2.$$

Comparing this value of the likely variance with the value of the likely average, from equation (8), we see that the ratio of the standard deviation to the average is:

$$(12) \qquad \sqrt{var_r}/r = \sqrt{C}/\ln(N_t Ca).$$

It is seen that this ratio depends on the logarithm of the mutation rate and will consequently be only a little smaller for mutation rates many thousand times greater than those considered in the experiments reported in this paper.

In the beginning of this theoretical discussion we pointed out that the hypothesis of acquired immunity leads to the prediction of a distribution of the number of resistant bacteria according to Poisson's law, and therefore to the prediction of a variance equal to the average. On the other hand, if we compare the average, equation (8), with the variance, equation (11), (not, as above, with the square root of the variance), we obtain

$$(12a) \qquad var_r = r N_t Ca/\ln(N_t Ca).$$

Equation (12a) shows that the likely ratio between variance and average is much greater than unity on the hypothesis of mutation, if $(N_t Ca)$, the total number of mutations which occurred in our cultures, is large compared to unity.[5]

It is possible to carry the analysis still further and to evaluate the higher moments of the distribution function of the number of resistant bacteria, or even the distribution function itself. The moments are comparatively easy to obtain, while the calculation of the distribution function involves considerable

---

[5] In some of the experiments reported in the present paper we did not determine the total number of resistant bacteria in each culture, but the number contained in a small sample from each culture. In these cases the variance of the distribution of the number of resistant bacteria will be slightly increased by the sampling error. The proper procedure is here first to find the average number of resistant bacteria per culture by multiplying the average per sample by the ratio

$$(13) \qquad \frac{\text{volume of culture}}{\text{volume of sample}};$$

second, to evaluate the mutation rate with the help of equation (8); third, to figure the likely variance for the cultures by equation (11); fourth, to divide this variance by the square of the ratio (13) to obtain that part of the variance in the samples which is due to the chance distribution of the mutations. The experimental variance should be greater than this value, on account of the sampling variance. The sampling variance is in all our cases only a small correction to the total variance, and it is sufficient to use its upper limit, that of the Poisson distribution, in our calculations. Consequently, when comparing the experimental with the calculated values, we first subtract from the experimental value the sampling variance, which we take to be equal to the average number of resistant bacteria.

mathematical difficulties. An approximation to the beginning of the distribution function—that is, to its values for small numbers of resistant bacteria—may be obtained by grouping mutations according to the bacterial generation during which they occurred. For instance, the probability of obtaining seven resistant bacteria may be broken down into the sum of the following alternative events: (a) seven mutations during the last generation; (b) three mutations during the last generation and two mutations one generation back; (c) three mutations during the last generation and one mutation two generations back; (d) one mutation during the last generation and three mutations one generation back; (e) one mutation during the last generation, one mutation one generation back and one mutation two generations back.

The probability of each of these events depends only on the mutation rate and on the final number of bacteria.

The grouping of mutations according to the bacterial generation during which they occurred, and the assumption that the bacteria increase in simple geometric progression, simplify the calculation sufficiently to permit numerical computation. On the other hand, the classes with two, four, eight, etc., mutants are artificially favored by this procedure, so that a somewhat uneven distribution results, with too high values for two, four, eight, etc., resistant bacteria (see fig. 2).

MATERIAL AND METHODS

The material used for our experimental study consisted of a bacterial virus α and of its host, *Escherichia coli* B (DELBRÜCK and LURIA 1942). Secondary cultures after apparently complete lysis of B by virus α show up within a few hours from the time of clearing. They consist of cells which are resistant to the action of virus α, but sensitive to a series of other viruses active on B. The resistant cells breed true and can be established easily as pure cultures. No trace of virus could be found in any pure culture of the resistant bacteria studied in this paper. The resistant strains are therefore to be considered as non-lysogenic.

Tests were made to see whether the resistance to virus α was a stable character of the resistant strains. In the first place, it was found that virus α is not appreciably adsorbed by any of the resistant strains. In the second place, when a certain amount of virus α is mixed with a growing culture of a resistant strain, no measurable increase of the titer of virus α occurs over a period of several hours. This is a very sensitive test for the occurrence of sensitive bacteria, and its negative result for all resistant strains shows that reversion to sensitivity must be a very rare event.

Morphologically at least two types of colonies of resistant bacteria may be distinguished. The first type of colony is similar to the type produced by the sensitive strain both in size and in the character of the surface and of the edge. The second type of colony is much smaller and translucent. The difference in colony type is maintained in subcultures. Microscopically the bacteria from these two types of colonies are indistinguishable. They also do not differ from each other or from the sensitive strain in their fermentation reactions on common sugars and in the characteristics of their growth curves in nutrient broth. In particular, the lag periods, the division times during the logarithmic phase of growth, and the maximum titers attained are identical for the sensitive strain and for the two variants. Both variants, therefore, fulfill the requirements for the applicability of the theory developed above.

In the presentation of our experimental results we have lumped the counts of the two types of colonies together, because: (1) theoretically, this is equivalent to summing the corresponding mutation rates; (2) experimentally, we are not certain whether each of these types does not actually comprise a diversity of variants; (3) experimentally, no correlation appeared to exist between the occurrence of these variants, which shows the independence of the causes of their occurrence.

Cultures of B were grown either in nutrient broth (containing .5 percent NaCl) or in an asparagin-glucose synthetic medium. In the latter, the division time during the logarithmic phase of growth was 35 minutes, as compared with 19 minutes in broth. In synthetic medium, the acidity increased during the time of incubation from pH 7 to pH 5.

In cultures of strain B, between 10^{-8} and 10^{-5} of the bacteria are found usually to give colonies resistant to the action of virus α when samples of such cultures are plated with large amounts of virus. In order to be reasonably certain that the resistant bacteria found in the test had not been introduced into the test culture with the initial inoculum, the test cultures were always started with very small inocula, containing between 50 and 500 bacteria from a growing culture. Thus any resistant bacterium found at the moment of testing (when the culture contains between 10^8 and 5×10^9 bacteria/cc) must be an offspring of one of the sensitive bacteria of the inoculum.

All platings were made on nutrient agar plates. The plating experiments for counting the number of resistant bacteria in a liquid culture of the sensitive strain were done by plating either a portion or the entire culture with a large amount of virus α. The virus was plated first, and spread over the entire surface of the agar. A few minutes later the bacterial suspension to be tested was spread over the central part of the plate, leaving a margin of at least one centimeter. Thus all bacteria were surrounded by large numbers of virus particles.

Microscopic examination of plates seeded in this manner showed that lysis takes place very quickly; only bacteria which at the time of plating were in the process of division may sometimes complete the division. The resistant colonies which appear after incubation are therefore due to resistant bacterial cells present at the time of plating.

The total number of bacteria present in the culture to be tested was determined by colony counts in the usual manner.

The resistant colonies of the large type appear after 12–16 hours of incubation, the colonies of the small type appear after 18–24 hours, and never reach half the size of the former ones. Counts were usually made after 24 and 48 hours.

EXPERIMENTAL

A Test of the Reliability of the Plating Method

In our experiments we wanted to study the fluctuations of the numbers of resistant bacteria found in cultures of sensitive bacteria. It was therefore necessary to show first that the method of testing did not involve any unrecognized variables, which caused the number of resistant colonies to vary from plate to plate or from sample to sample.

Therefore, parallel platings were made using a series of samples from the same bacterial culture. If our plating method is reliable, fluctuations should in this arrangement be due to random sampling only, and the variance from a series of such samples should be equal to the mean.

Table 1 gives the results of three such experiments. It will be seen that in

TABLE 1

The number of resistant bacteria in different samples from the same culture.

SAMPLE NO.	EXP. NO. 10a RESISTANT COLONIES	EXP. NO. 11a RESISTANT COLONIES	EXP. NO. 3 RESISTANT COLONIES
1	14	46	4
2	15	56	2
3	13	52	2
4	21	48	1
5	15	65	5
6	14	44	2
7	26	49	4
8	16	51	2
9	20	56	4
10	13	47	7
mean	16.7	51.4	3.3
variance	15	27	3.8
χ^2	9	5.3	12
P	.4	.8	.2

all three cases variance and mean agree as well as may be expected. There is therefore no reason to assume that the method of sampling or plating introduces any fluctuations into our results besides the sampling error.

Fluctuations of the Number of Resistant Bacteria in Samples from a Series of Similar Cultures

As pointed out in the introduction and in the theoretical part, the hypothesis of acquired immunity and the hypothesis of mutation lead to radically different predictions regarding the distribution of the number of resistant bacteria in a series of similar cultures. The hypothesis of acquired immunity predicts a variance equal to the average, as in sampling, while the mutation hypothesis predicts a much greater variance.

Series of five to 100 cultures were set up in parallel with small equal inocula, and were grown until maximum titer was reached. Three kinds of cultures were used—namely: (1) 10.0 cc aerated broth cultures; (2) .2 cc broth cultures; (3) .2 cc synthetic medium cultures.

The results of all tests for the number of resistant bacteria are summarized in table 2 and table 3.

TABLE 2

The number of resistant bacteria in series of similar cultures.

EXPERIMENT NO.	1	10	11	15	16	17	21a	21b
Number of cultures	9	8	10	10	20	12	19	5
Volume of cultures, cc	10.0	10.0	10.0	10.0	.2*	.2*	.2	10.0
Volume of samples, cc	.05	.05	.05	.05	.08	.08	.05	.05
Culture No.								
1	10	29	30	6	1	1	0	38
2	18	41	10	5	0	0	0	28
3	125	17	40	10	3	0	0	35
4	10	20	45	8	0	7	0	107
5	14	31	183	24	0	0	8	13
6	27	30	12	13	5	303	1	
7	3	7	173	165	0	0	0	
8	17	17	23	15	5	0	1	
9	17		57	6	0	3	0	
10			51	10	6	48	15	
11					107	1	0	
12					0	4	0	
13					0		19	
14					0		0	
15					1		0	
16					0		17	
17					0		11	
18					64		0	
19					0		0	
20					35			
Average per sample	26.8	23.8	62	26.2	11.35	30	3.8	48.2
Variance (corrected for sampling)	1217	84	3498	2178	694	6620	40.8	1171
Average per culture	5360	4760	12400	5240	28.4	75	15.1	8440
Bacteria per culture	3.4×10^{10}	4×10^{10}	4×10^{10}	2.9×10^{10}	5.6×10^{8}	5×10^{8}	1.1×10^{8}	3.2×10^{10}
Mutation rate	1.8×10^{-8}	1.4×10^{-8}	4.1×10^{-8}	2.1×10^{-8}	1.1×10^{-8}	3.0×10^{-8}	3.3×10^{-8}	3.0×10^{-8}
Standard deviation { exp.	1.3	.39	.95	1.8	2.3	2.7	1.7	.71
Average { calc.	.35	.33	.33	.37	.94	.67	1.04	.26

* Cultures in synthetic medium.

It will be seen that in every experiment the fluctuation of the numbers of resistant bacteria is tremendously higher than could be accounted for by the sampling errors, in striking contrast to the results of plating from the same culture (see table 1) and in conflict with the expectations from the hypothesis of acquired immunity.

We want to see next whether these results fit the expectations from the hypothesis of mutation. We must therefore compare the experimental results with the relations developed in the theoretical part, keeping in mind that the theory contains several simplifying assumptions.

First we can compare, according to equation (12), the experimental and the calculated values of the ratio between the standard deviation and the average of the numbers of resistant bacteria. These ratios are included in tables 2 and 3. It is seen that the experimental and theoretical values are reasonably close.

However, in all but one case the experimental ratio is greater than the value calculated from the theory—that is, the variability is even greater than predicted.

TABLE 3

Distribution of the numbers of resistant bacteria in series of similar cultures.

EXPERIMENT NO.	22		23	
Number of cultures	100		87	
Volume of cultures, cc	.2*		.2*	
Volume of samples, cc	.05		.2	

	Resistant bacteria	Number of cultures	Resistant bacteria	Number of cultures
	0	57	0	29
	1	20	1	17
	2	5	2	4
	3	2	3.	3
	4	3	4	3
	5	1	5	2
	6– 10	7	6– 10	5
	11– 20	2	11– 20	6
	21– 50	2	21– 50	7
	51– 100	0	51– 100	5
	101– 200	0	101– 200	2
	201– 500	0	201– 500	4
	501–1000	1	501–1000	0

	22	23
Average per sample	10.12	28.6
Variance (corrected for sampling)	6270	6431
Average per culture	40.48	28.6
Bacteria per culture	2.8×10^8	2.4×10^8
Mutation rate	2.3×10^{-8}	2.37×10^{-8}
Standard deviation ⌠exp.	7.8	2.8
Average ⌡calc.	1.5	1.5

* Cultures in synthetic medium.

A part of this discrepancy may be accounted for by the fact that the time t_0, mutations occurring prior to which were disregarded by the theory, was chosen in such a manner that on the average one mutation would occur prior to time t_0. This mutation, if it occurs, will of course tend to increase the variance, and in some of the experiments the high value of the experimental variance can be traced directly to one exceptional culture in which a mutation had evidently occurred several generations prior to time t_0. Unfortunately, there is no general criterion by which one might eliminate such cultures from the statistical analysis, because, in a culture with an exceptionally high count of resistant bacteria, these do not necessarily stem from one exceptionally early mutation, but may also be due to an exceptionally large number of mutations after time t_0.

There may also be other reasons why the observed variances are higher than the expected ones. First of all, the simplifying assumption that the mutation rate per bacterial generation is independent of the physiological state of the bacteria may be too simple. If the mutation rate is higher for actively growing bacteria than for bacteria near the saturation limit of the cultures, early mutations and big clone sizes will be favored, and therefore higher variations of the numbers of resistant bacteria can be expected. Second, the assumption of a sudden transition from sensitivity to resistance may also be too simple. It is conceivable that the character "resistance to virus" may not fully develop in the bacterial cell in which the mutation occurs, but only in its offspring, after one or more generations. However, if this were the case, cultures with only one or two resistant bacteria should be relatively rare. The last experiment listed in table 3, in which the entire cultures were plated, shows a rather high proportion of cultures with only one resistant bacterium. This seems to show that the

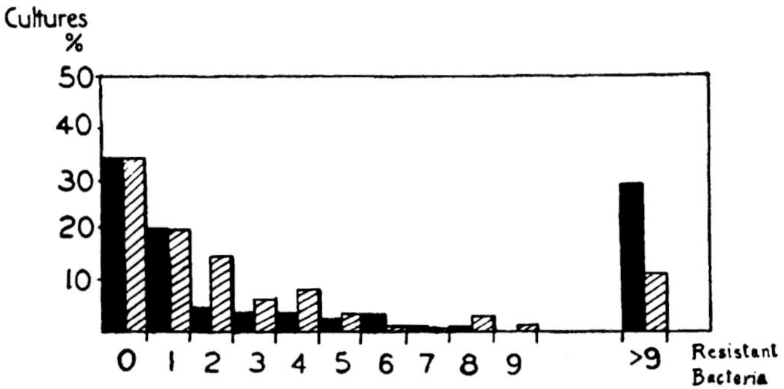

FIGURE 2.—Experimental (Experiment No. 23) and calculated distributions of the numbers of resistant bacteria in a series of similar cultures. Solid columns: experimental. Cross-hatched columns: calculated.

character "resistance to virus" in general does come to expression in the bacterial cell in which the corresponding mutation occurred, as assumed by the theory.

Another way of comparing the experimental results with the theory is to compare the experimental distribution of resistant bacteria with the approximate distribution calculated by the method outlined at the end of the theoretical part. The theoretical distribution has to be calculated from the average number of mutations per culture given by equation (5). Only experiments where the whole culture is tested can therefore be used for such a comparison. This method tests the fitting of the expectations for small numbers of resistant bacteria, in contrast to the comparison of the standard deviations, which involves predominantly the cultures with high numbers of resistant bacteria.

Figure 2 shows the experimental and calculated distributions for Experiment No. 23; the cultures with more than nine resistant bacteria are lumped together in one class, since the distribution has not been calculated for values higher than nine.

It is seen that the fitting for small values is satisfactory. In particular, the number of cultures with one resistant bacterium very closely fits the expectation. The classes with two, four, eight, etc., resistant bacteria are bound to be favored in the theoretical distribution, as explained in the theoretical part.

The results shown in figure 2 also confirm the assumption that the discrepancy between experimental and calculated standard deviations must be due to an excess of cultures with large numbers of resistant bacteria.

Summing up the evidence, we may say that the experiments show clearly that the resistant bacteria appear in similar cultures not as random samples but in groups of varying sizes, indicating a correlating cause for such grouping, and that the assumption of genetic relatedness of the bacteria of such groups offers the simplest explanation for them.

Mutation Rate

As pointed out in the theoretical part of this paper, mutation rates may be estimated from the experiments by two essentially different methods. The first method makes use of the fact that the number of mutations in a series of similar cultures should be distributed in accordance with Poisson's law; the average number of mutations per culture is calculated from the proportion of cultures containing no resistant bacteria at the moment of the test, according to equation (5).

There are two technical difficulties involved in the application of this method. In the first place, rather large numbers of cultures have to be handled and conditions have to be chosen so that the proportion of resistant bacteria is neither too small nor too large. In the second place, the entire cultures have to be tested, which means, in our method of testing, that cultures of rather small volume have to be used and great care must be taken to plate as nearly as possible the entire culture.

Experiment No. 23 (see table 3) permits an estimate of the mutation rate by this method. Out of 87 cultures, no resistant bacteria were found in 29 cultures, a proportion of .33. From equation (5) we calculate therefore that the average number of mutations per culture in this experiment was 1.10. Since the total number of bacteria per culture was 2.4×10^8, we obtain as the mutation rate, from equation (4),

$$a = .47 \times 10^{-8} \text{ mutations per bacterium per time unit}$$

$$= .32 \times 10^{-8} \text{ mutations per bacterium per division cycle.}$$

This calculation makes use exclusively of the proportion of cultures containing no resistant bacteria. It is therefore inefficient in its use of the information gathered in the experiment.

The second method makes use of the average number of resistant bacteria per culture. The relation of this average number with the mutation rate was discussed in the theoretical part of this paper and was found to be expressed by equation (8). The mutation rates calculated by this method for each experiment are collected in table 4.

TABLE 4

Values of mutation rate from different experiments.

EXPERIMENT NO.	NUMBER OF CULTURES	VOLUME OF CULTURES	MUTATION RATE
		cc	*Mutations per bacterium per time unit*
1	9	10.0	1.8×10^{-8}
10	8	10.0	1.4×10^{-8}
11	10	10.0	4.1×10^{-8}
15	10	10.0	2.1×10^{-8}
16	20	.2*	1.1×10^{-8}
17	12	.2*	3.0×10^{-8}
21a	19	.2	3.3×10^{-8}
21b	5	10.0	3.0×10^{-8}
22	100	.2*	2.3×10^{-8}
23	87	.2*	2.4×10^{-8}
Average			2.45×10^{-8}

* Cultures in synthetic medium.

It will be seen that the values of the mutation rate obtained by the second method are all higher than the value found by the first method. This discrepancy may be traced back to the same cause as the discrepancy between the calculated and observed values of the standard deviation of the numbers of resistant bacteria. This, we found, was due to an excess of early mutations, giving rise to big clones of resistant bacteria. These big clones do not affect the mutation rate calculated by the first method, but they do affect the results of the second method, which is based on the average number of resistant bacteria.

One sees in table 4 that the mutation rate calculated by the second method does not vary greatly from experiment to experiment. In particular, it will be noted that there is no significant difference between the values obtained from cultures in broth and from cultures in synthetic medium, notwithstanding the considerable difference of metabolic activity and of growth rate of the bacteria in these two media. This shows that the simple assumption of a fixed small chance of mutation per physiological time unit is vindicated by the results. It may also be noted in table 4 that there is no significant difference between the mutation rates obtained from 10 cc cultures and those obtained from .2 cc cultures, or between the experiments with many and those with few cultures. The variability of the value of the mutation rate seems to be solely due to the peculiar probability distribution of the number of resistant bacteria in series of similar cultures predicted by the mutation theory.

At this point an experiment may be mentioned by which it was desired to find out whether or not mutations occur in a culture after the bacteria have ceased growing. A culture was grown to saturation and was then tested repeatedly for resistant bacteria and for total number of bacteria over several days. The proportion of resistant bacteria did not change, even when the sensitive bacteria began to die, showing that the resistant bacteria have the same death rate in aging cultures as the sensitive bacteria.

DISCUSSION

We consider the above results as proof that in our case the resistance to virus is due to a heritable change of the bacterial cell which occurs independently of the action of the virus. It remains to be seen whether or not this is the general rule. There is reason to suspect that the mechanism is more complex in cases where the resistant culture develops only several days after lysis of the sensitive bacteria.

The proportion of mutant organisms in a culture and the mutation rate are far smaller in our case than in other studied cases of heritable bacterial variation. The possibility of investigation of such rare mutations is in our case merely the result of the method of detecting the mutant organisms. In other cases, the variants are detected by changes in the colony type which is produced by the mutant organism, either in the pigmentation or in the character of the surface or the edge of the colony. Often, colonies of intermediate character occur, and it is difficult to decide whether they are mixed colonies or stem from bacteria with intermediate character. This is particularly true of cases where the mutation rate is high and where reverse mutation occurs. Fairly high mutation rates, however, are a prerequisite of any study of colony variants, since the number of colonies that can be examined is limited by practical reasons.

The study of mutations causing virus resistance is free of these difficulties. The segregation of the mutant from the normal organisms occurs in the one-cell stage by elimination of the normal individuals, and the character of the colony which develops from a mutant organism is of secondary importance. Owing to the total elimination of the normal individuals, the number of organisms which may be examined is very much higher than for any other method; more than 10^8 bacteria may be tested on a single plate. Since the mutations to virus resistance are often associated with other significant characters, the method may well assume importance with regard to the general problems of bacterial variation.

It must not be supposed that the peculiar statistical difficulties encountered in our case are restricted to cases of very low mutation rates. The essential condition for the occurrence of the peculiar distribution studied in the theoretical part of this paper is the following: *the initial number of bacteria in a culture must be so small that the number of mutations which occur during the first division cycle of the bacteria is a small number.* This will always be true, however great the mutation rate, if one studies cultures containing initially a small number of organisms.

In a series of very interesting studies of the color variants of *Serratia marcescens*, BUNTING (1940a, 1940b, 1942; BUNTING and INGRAHAM 1942) succeeded to some extent in obviating the statistical difficulties by always using inocula of about 100,000 bacteria. In some of her cases this number was sufficiently high to result in numerous mutations during the first division cycle of the bacteria. In other cases the number was apparently not high enough, since the author reports troublesome variations of the fractions of variants in successive subcultures. In those cases where the size of the inocula was high enough, the author succeeded in deriving reproducible values for the mutation rates from the study of single cultures, followed through numerous subcultures. In these cases it is sufficient to apply the equations of the theory referring to the *average* numbers of mutants as a function of time. It is clear, however, that this method is applicable only in cases of mutation rates of at least 10^{-4} per bacterium per division cycle.

In our case, as in many others, the virus resistant variants do not exhibit any striking correlated physiological changes. There is therefore little opportunity for an inquiry into the nature of the physiological changes responsible for the resistance to virus. Since the offspring of the mutant bacteria, when isolated after the test, are unable to synthesize the surface elements to which the virus is specifically adsorbed in the sensitive strain, one might suppose that this loss is a direct effect of the mutation. However, it is also conceivable that the loss occurs upon contact with virus, since it is detected only after such contact (hypothesis b_1). In some of the cases studied by BURNET (1929), where the mutational change to resistance is correlated with a change of phase, from smooth to rough or vice versa, the change of the surface structure must be a direct result of the mutation, since the mutant colonies may be picked up prior to the resistance test and, when tested, exhibit the typical change of affinity of the surface structure. These findings make it more probable that the loss of surface affinity to virus is a direct effect of the mutation.

The alteration of specific surface structures due to genetic change is a phenomenon of the widest occurrence. The genetic factors determining the antigenic properties of erythrocytes are well known. There is evidence (WEBSTER 1937; HOLMES 1938; STEVENSON, SCHULTZ, and CLARK 1939) that resistance or sensitivity to virus in plants and animals is correlated with, or even dependent on, genetic changes, possibly affecting the antigenic make-up of the cellular surface. The proof that resistance to a bacterial virus may be traced to a specific genetic change may assume importance, therefore, with regard to the general problems of virus sensitivity and virus resistance.

SUMMARY

The distribution of the numbers of virus resistant bacteria in series of similar cultures of a virus-sensitive strain has been analyzed theoretically on the basis of two current hypotheses concerning the origin of the resistant bacteria.

The distribution has been studied experimentally and has been found to conform with the conclusions drawn from the hypothesis that the resistant bacteria arise by mutations of sensitive cells independently of the action of virus.

The mutation rate has been determined experimentally.

LITERATURE CITED

BUNTING, M. I., 1940a A description of some color variants produced by *Serratia marcescens*, strain 274. J. Bact. **40**: 57–68.

1940b The production of stable populations of color variants of *Serratia marcescens* #274 in rapidly growing cultures. J. Bact. **40**: 69–81.

1942 Factors affecting the distribution of color variants in aging broth cultures of *Serratia marcescens* #274. J. Bact. **43**: 593–606.

BUNTING, M. I., and L. J. INGRAHAM, 1942 The distribution of color variants in aging broth cultures of *Serratia marcescens* #274. J. Bact. **43**: 585–591.

BURNET, F. M., 1929 Smooth-rough variation in bacteria in its relation to bacteriophage. J. Path. Bact. **32**: 15–42.

DELBRÜCK, M., and S. E. LURIA, 1942 Interference between bacterial viruses. I. Arch. Biochem. **I**: 111–141.

GRATIA, A., 1921 Studies on the d'Herelle phenomenon. J. Exp. Med. **34**: 115–131.

D'HERELLE, F., 1926 The Bacteriophage and Its Behavior. Baltimore: Williams and Wilkins.

HOLMES, F. O., 1938 Inheritance of resistance to tobacco-mosaic disease in tobacco. Phytopathology **28**: 553–561.

STEVENSON, F. J., E. S. SCHULTZ, and C. F. CLARK, 1939 Inheritance of immunity from virus X (latent mosaic) in the potato. Phytopathology **29**: 362–365.

WEBSTER, L. T., 1937 Inheritance of resistance of mice to enteric bacterial and neurotropic virus infections. J. Exp. Med. **65**: 261–286.

The Elementary Units of Heredity
S. BENZER

Seymour Benzer's work changed our notion of the concept of the gene, by demonstrating that the gene had a fine structure consisting of a linear array of subelements. At the time Benzer began his classic work, the concept of the gene was different from what it is today. Genes were thought to be indivisible and to be the smallest units of recombination, mutation, and function. Genes could have different allelic states, but these alleles represented the whole gene, not parts of it. In one sense, genes were thought of as beads on a necklace, the necklace being the chromosome. This picture of the gene proved to be at odds with the physical structure of DNA elucidated by Watson and Crick in 1953, which revealed the physical structure of the gene to consist of a sequence of nucleotides. Each nucleotide should be able to mutate and should also be the smallest unit of recombination. Benzer's work bridged the gap between the classic view of the gene as an indivisible unit and the physical structure of DNA. By exploiting the rII system of phage T4 and refining high-resolution genetic selection, Benzer was able to show that the subelements in the gene could mutate and recombine with one another. The smallest unit of mutation and recombination was now shown to be on the order of only a few nucleotides or less, based mainly on genetic analysis. Benzer also refined the *cis-trans* test for use with phage T4, and he defined the cistron as a unit of gene function, a term and concept that were used for many years.

Benzer developed the use of deletions in genetic crosses, which opened up the entire field of fine structure analysis. It laid the groundwork for his subsequent studies, which defined the concept of mutational hot spots, and profoundly influenced work on mutagenesis for a generation.

JEFFREY H. MILLER

Reprinted from *The Chemical Basis of Heredity*, p. 70–93. Copyright © 1957, by Johns Hopkins University Press.

THE ELEMENTARY UNITS OF HEREDITY*

Seymour Benzer

*Biophysical Laboratory, Purdue University,
Lafayette, Indiana*

Introduction

The techniques of genetic experiments have developed to a point where a highly detailed view of the hereditary material is attainable. By the use of selective procedures in recombination studies with certain organisms, notably fungi (14), bacteria (2), and viruses (1), it is now feasible to "resolve" detail on the molecular level. In fact, the amount of observable detail is so enormous as to make an exhaustive study a real challenge.

A remarkable feature of genetic fine structure studies has been the ability to construct (by recombination experiments) genetic maps which remain one-dimensional down to the smallest levels. The molecular substance (DNA) constituting the hereditary material in bacteria and bacterial viruses is also one-dimensional in character. It is therefore tempting to seek a relation between the linear genetic map and its molecular counterpart which would make it possible to convert "genetic length" (measured in terms of recombination frequencies) to molecular length (measured in terms of nucleotide units).

The classical "gene," which served at once as the unit of genetic recombination, of mutation, and of function, is no longer adequate. These units require separate definition. A lucid discussion of this problem has been given by Pontecorvo (13).

The unit of recombination will be defined as the smallest element in the one-dimensional array that is interchangeable (but not divisible) by genetic recombination. One such element will be referred to as a

* This research has been supported by grants from the American Cancer Society, upon recommendation of the Committee on Growth of the National Research Council, and from the National Science Foundation.

70

"recon." The unit of mutation, the "muton," will be defined as the smallest element that, when altered, can give rise to a mutant form of the organism. A unit of function is more difficult to define. It depends upon what level of function is meant. For example, in speaking of a single function, one may be referring to an ensemble of enzymatic steps leading to *one* particular physiological end-effect, or of the synthesis of *one* of the enzymes involved, or of the specification of *one* peptide chain in one of the enzymes, or even of the specification of *one* critical amino acid.

A functional unit can be defined genetically, independent of biochemical information, by means of the elegant *cis-trans* comparison devised by Lewis (12). This test is used to tell whether two mutants, having apparently similar defects, are indeed defective in the same way. For the *trans* test, both mutant genomes are inserted in the same cell (e.g., in heterocaryon form, or, in the case of a bacterial virus, the equivalent obtained by infecting a bacterium with virus particles of both mutant types). If the resultant phenotype is defective, the mutants are said to be non-complementary, i.e., defective in the same "function." As a control, the same genetic material is inserted in the *cis* configuration, i.e., as the genomes from one double mutant and one non-mutant. The *cis* configuration usually produces a non-defective phenotype (or a close approximation to it). It turns out that a group of non-complementary mutants falls within a limited segment of the genetic map. Such a map segment, corresponding to a function which is unitary as defined by the *cis-trans* test applied to the heterocaryon, will be referred to as a "cistron."

The experiments to be described in this paper represent an attempt to place limits on the sizes of these three genetic units in the case of a specific region of the hereditary material of the bacterial virus T4. A group of "*rII*" mutants of T4 has particularly favorable properties for this kind of analysis. Mutants are easily isolated. Recombinants can be detected, even in extremely low frequency, by a selective technique. The system is sufficiently sensitive to permit extension of genetic mapping down to the molecular (nucleotide) level, so that the recon and muton become accessible to measurement. The *rII* mutants are defective in the sense of being unable to multiply in cells of a certain host bacterium (although they do infect and kill the cell). The *cis-trans* test can therefore be readily applied.

GENETIC MAPS

Method of Construction.

The construction of a genetic map of an organism starts with the selection of a standard ("wild") type. From the progeny of the wild type, mutant forms can be isolated on the basis of some heritable difference. When two mutants are crossed, there is a possibility that a wild-type organism will be formed as a result of recombination of genetic material. The reciprocal recombinant, containing both mutational alterations, also occurs. The proportion of progeny constituting such recombinant types is characteristic of the particular mutants used. The results of crosses involving a group of mutants can be plotted on a one-dimensional diagram where each mutant is represented by a point. The interval between two points signifies the proportion of recombinants occurring in a cross between the two corresponding mutants. Usually, it is not possible to construct a single map for all the mutants of an organism; instead the mutants must be broken up into "linkage groups." A linear map may be constructed within each linkage group, but the mutant characters assigned to different linkage groups assort randomly among the progeny. The number of linkage groups, in some cases, has been shown to correspond to the number of visible chromosomes.

The procedure for constructing a genetic map for a bacterial virus is much the same (9). A genetically uniform population of a mutant can readily be grown from a single individual. Two mutants are crossed by infecting a susceptible bacterium with both types and examining the resulting virus progeny for recombinant types. Virus T4 has been mapped in some detail (3, 1), and behaves as a haploid organism with a single linkage group (18).

Relativity of Genetic Maps.

A genetic map is an image composed of individual points. Each point represents a mutation which has been localized with respect to other mutations by recombination experiments. The image thus obtained is a highly colored representation of the hereditary material. Alterations in the hereditary material will lead to noticeable mutations only if they affect some phenotypic characteristic to a visible degree. Innocuous changes may pass unnoticed, leaving their corresponding regions on the map blank. At the other extreme, alterations having a

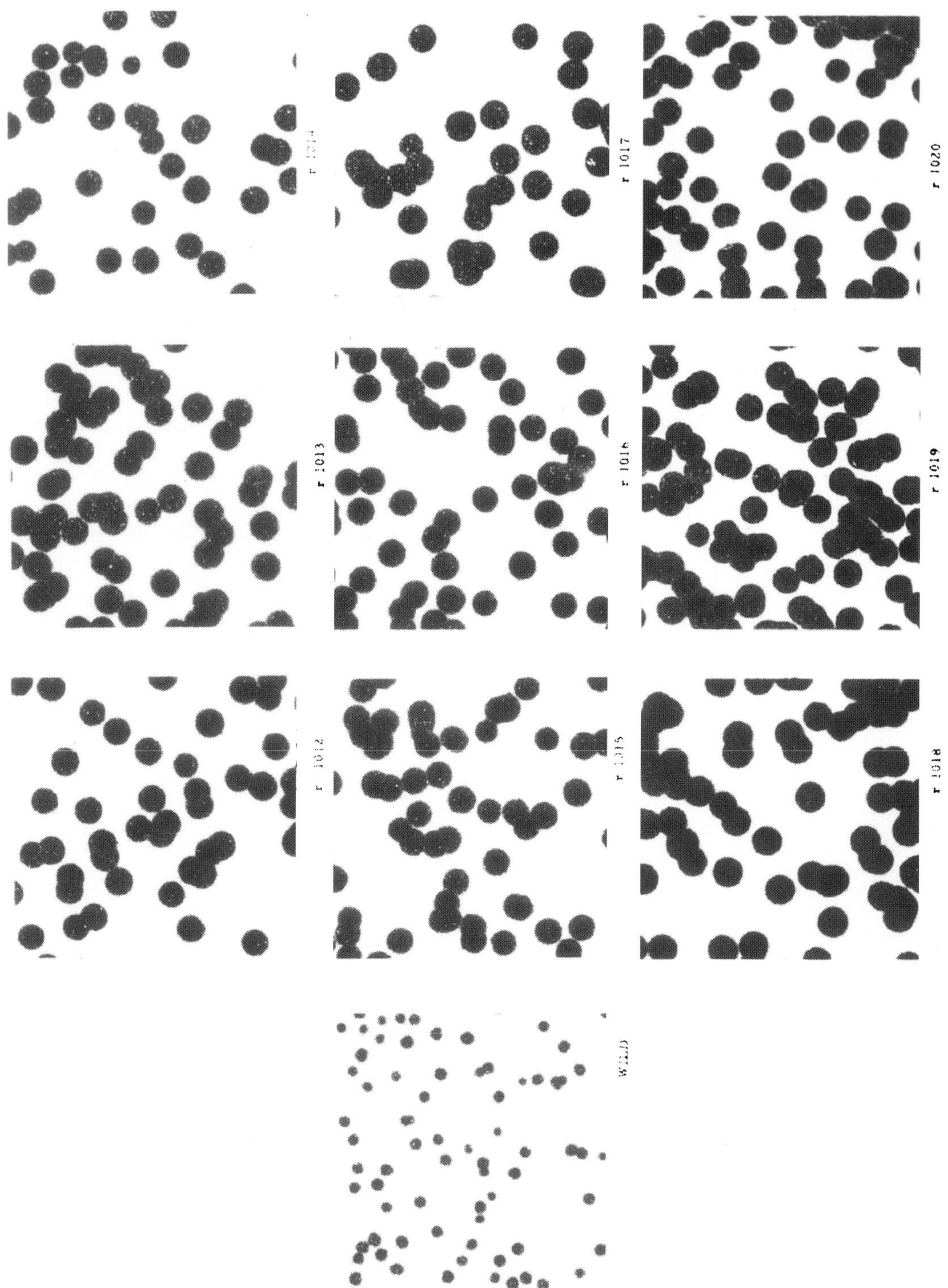

Fig. 1. Photographs of plaques formed on *E. coli* B by T4 "wild-type" and nine independently arising *r* mutants.

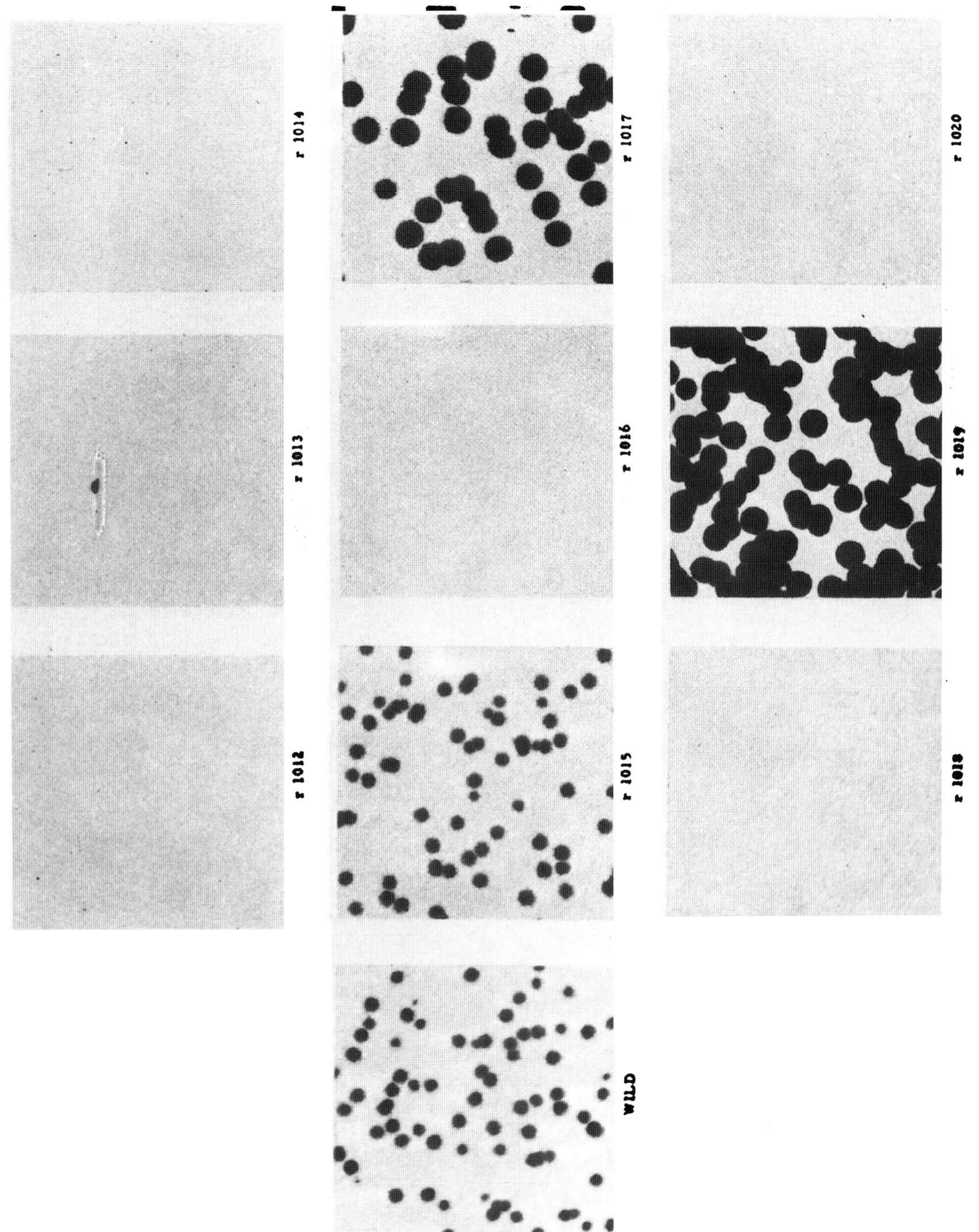

Fig. 2. The same mutants as used in Fig. 1, plated on *E. coli* K.

lethal effect will also be missed (in a haploid organism). The map represents, therefore, only cases which fall between these extremes under the conditions of observation. By varying these conditions, a given mutational event may be shifted from one of these categories (innocuous, noticeable, or lethal) to another, thereby appearing on, or disappearing from the map.

This effect may be illustrated by the "*r*" mutants of bacterial virus T4. Wild-type T4 produces small, fuzzy plaques on *Escherichia coli* B (Fig. 1). From plaques of wild-type T4, *r*-type mutants can be isolated which produce a different sort of plaque. Fig. 1 shows the plaques of nine *r* mutants, each isolated from a different plaque of the wild type in order to assure independent origin. The similarity of plaque type of these *r*'s on B disappears when they are plated on another host strain, *E. coli* K (a lysogenic K12 strain (10) carrying phage lambda), as shown in Fig. 2. Here, they split into three groups: two mutants form *r*-type plaques, one forms wild-type plaques, and the remaining six do not register. Thus, with B as host, all three types of mutation lead to visible effects, while with K as host, the effects may be visible, innocuous, or lethal.

When the same set of mutants is plated on a third strain, *E coli* S (K12S, a non-lysogenic derivative (10) of K12) or BB (a "Berkeley" derivative (17) of B) the pattern of plaque morphology is different from that on either B or K (Table 1).

TABLE 1

PLAQUE MORPHOLOGY OF T4 STRAINS (ISOLATED IN B) PLATED ON VARIOUS HOSTS

PHAGE STRAIN	BACTERIAL HOST STRAIN		
	B	S	K
wild	wild	wild	wild
r I	*r*	*r*	*r*
r II	*r*	wild	—
r III	*r*	wild	wild

If a genetic map is constructed for these mutants, using B as host, the three groups fall into different map regions, as indicated in Fig. 3. On strain S, the *rII* and *rIII* types of mutation are innocuous. Thus, if S had been used as host in the isolation of *r* mutants, only the *rI* region would have appeared on the map. On K as host, the *rIII* mutation is innocuous, and the *rII* mutation is (usually) lethal, so that only the

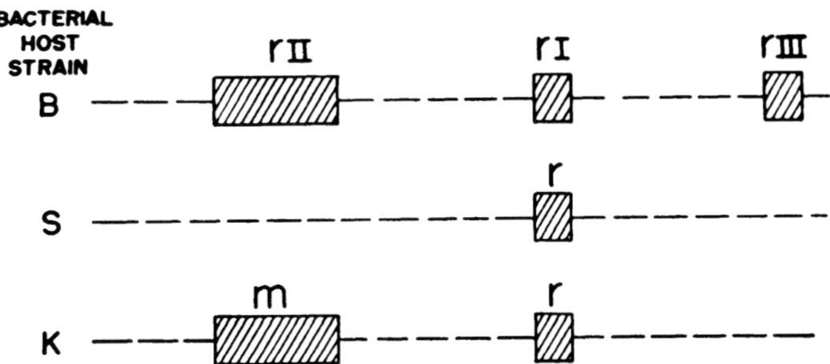

FIG. 3. Dependence of the genetic map of T4 upon the choice of host. Three regions of the map are shown as they probably would appear if *E. coli* strains B, S, or K were used as the host.

rI region would appear. Actually, a few *rII* mutants are able to multiply somewhat on K, producing visible tiny plaques. If K were used as host in the isolation and testing of mutants from wild-type T4, these mutants could be noticed and would probably be designated by some other name, perhaps "minute." The map would then appear as in the bottom row of Fig. 3. The distribution of points on the map within this "minute" region would be very different from those for the *rII* region using B as the host.

The appearance of a genetic map also depends on the choice of the standard type, which is, after all, arbitrary. For example, suppose an *r* form were taken as the standard type and non-*r* mutants were isolated from it. Then a completely different map would result. An example of this is to be found in the work of Franklin and Streisinger (5) on the $h \rightarrow h+$ mutation in T2, as compared with that of Hershey and Davidson (7) on the $h+ \rightarrow h$ mutation.

Another way in which the picture is weighted is by local variations in the stability of the genetic material. Certain types of structural alterations may occur more frequently than others. Thus, a perfectly stable genetic element (i.e., one which never errs during replication) would not be represented by any point on the map.

Determination of the Sizes of the Hereditary Units by Mapping.

Determination of the recon requires "running the map into the ground" (Delbrück's expression), that is, isolation and mapping of so large a linear density of mutants that their distances apart diminish to the point of being comparable to the indivisible unit. With a finite set of mutants, only an upper limit can be set upon the recon, which must be smaller than (or equal to) the smallest non-zero interval observed between pairs of mutants.

To determine the length of map involved in a mutational alteration, a group of three closely linked mutants is needed. Since map distances are (approximately) additive, a calculation of the "length" of the central mutation can be attempted (15) from the discrepancy observed between the longest distance and the sum of the two shorter ones, as shown in Fig. 4. The upper limit to the size of the muton would be

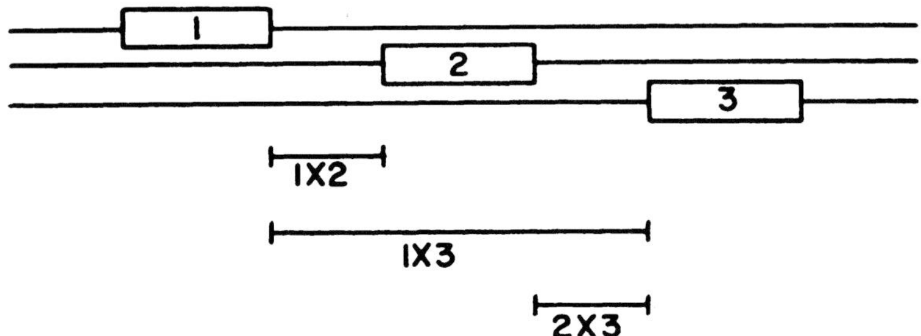

Fig. 4. Method for determining the "length" of a mutation. The discrepancy between the long distance and the sum of the two short distances measures the length of the central mutation.

the smallest discrepancy observed by this method, which can be determined accurately only if the three mutants are very closely linked. It should be noted that since the degree to which the genetic structure can be sliced by recombination experiments is limited by the size of recon, the size of the muton will register as zero by this method if it is equal to or smaller than one recon. A second method for determining the muton size is by the maximum number of mutations, separable by recombination, that can be packed into a definite length of the map.

For the cistron size, only a *lower* limit can be set with a finite group of mutants. The cistron must be at least as large as the distance between the most distant pair within it. Its boundaries become more sharply defined the larger the number of points which are shown to lie inside them.

Thus, the determination of the sizes of all three units requires the isolation and crossing of large numbers of mutants. The magnitude of this undertaking increases with the square of the number of mutants, since to cross n mutants in all possible pairs requires $n(n-1)/2$ crosses, or approximately $n^2/2$. Fortunately, however, the project can be shortened considerably by means of a trick.

The Method of Overlapping "Deletions."

Certain *rII* mutants are anomalous in the sense that they cannot be represented as *points* on the map. The anomalous mutants give no detectable wild recombinants with any of several other mutants which *do* give wild recombinants with each other. An anomalous mutant can be represented (Fig. 5) as covering a segment of the map. Reversion

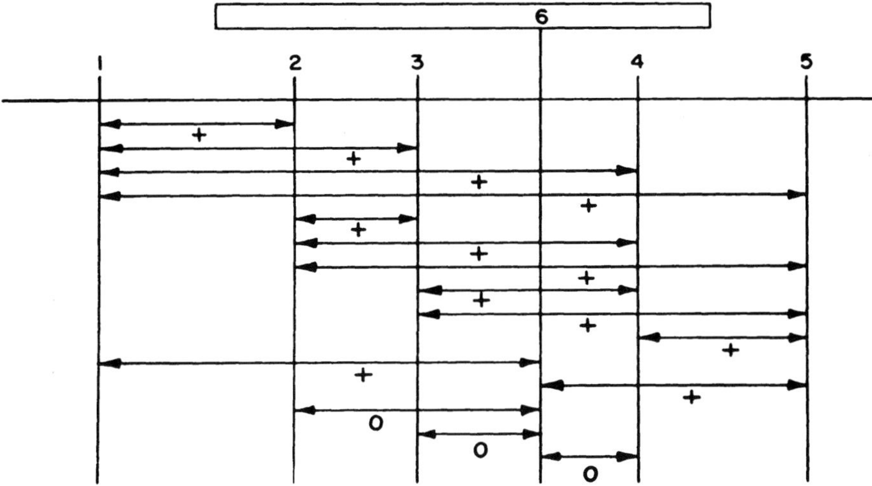

Fig. 5. Illustration of the behavior of an "anomalous" mutant. Mutant no. 6 is anomalous with respect to the segment of the map indicated by the bar; it fails to give wild recombinants with mutants (nos. 2, 3, and 4) located within that segment. A + signifies production, and 0 lack of production, of wild recombinants in a cross.

of such an *rII* mutant has never been observed; also, no mutant which does revert has been found to have this anomalous character. The properties of an anomalous mutant can be explained as owing to the deletion (i.e. loss) of a segment of hereditary material corresponding to the map span covered. However, anomalous behavior and stability against reversion are not sufficient to establish that a deletion has occurred. Similar properties could be expected of a double mutant when crossed with either of two different single mutants located at the same points. An inversion also would show the same behavior. However, the occurrence of a deletion seems to be the only reasonable explanation in the cases of several of the *rII* mutants, since they fail to give recombinants with any of three or more (in one case as many as 20) well-separated mutants.

Whether a given mutation belongs in the region covered by a deletion can be determined by the appropriate cross. If wild recombinants are produced, the mutant must have a map position *outside* the region of the deletion. This eliminates the need to cross that mutant with any of the mutants whose map positions lie *within* the region of the deletion. The problem of mapping a large number of mutants is greatly simplified by this system of "divide and conquer." The mutants can first be classified into groups that fall into different regions on the basis of crosses with mutants of the deletion-type. Further crossing in all possible pairs is then necessary only within each group.

Suppose that three deletions occur in overlapping configuration, as shown in Fig. 6A. Fig. 6B represents the results that would be obtained

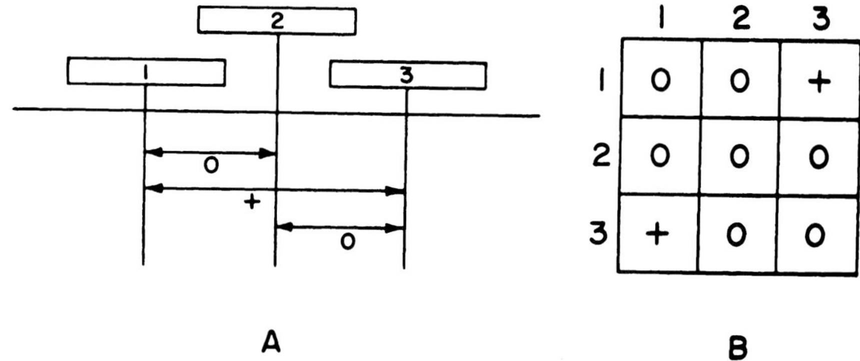

A   B

FIG. 6. The method of overlapping deletions. Three mutants are shown, each differing from wild type in the deletion of a portion of the genetic material. Mutants no. 1 and no. 3 can recombine with each other to produce wild type, but neither of them can produce wild recombinants when crossed to mutant no. 2. The matrix B represents the results obtained by crossing three such mutants in pairs and testing for wild recombinants; the results uniquely determine the order of the mutations on the map A.

in crosses of pairs of these three mutants. A diagonal element (representing a cross of a mutant with itself) is, of course, zero, since no wild recombinants can be produced. An overlap is reflected by the pattern of non-diagonal zeros. These results would establish a unique *order* of the deletions (without resort to the three-factor crosses that would ordinarily be necessary). With a sufficient number and appropriate distribution of deletions, one could hope to order a large length of map. The reader will note an analogy (not altogether without significance!) to the technique used by Sanger (16) to order the amino acids in a polypeptide chain by means of overlapping peptide segments.

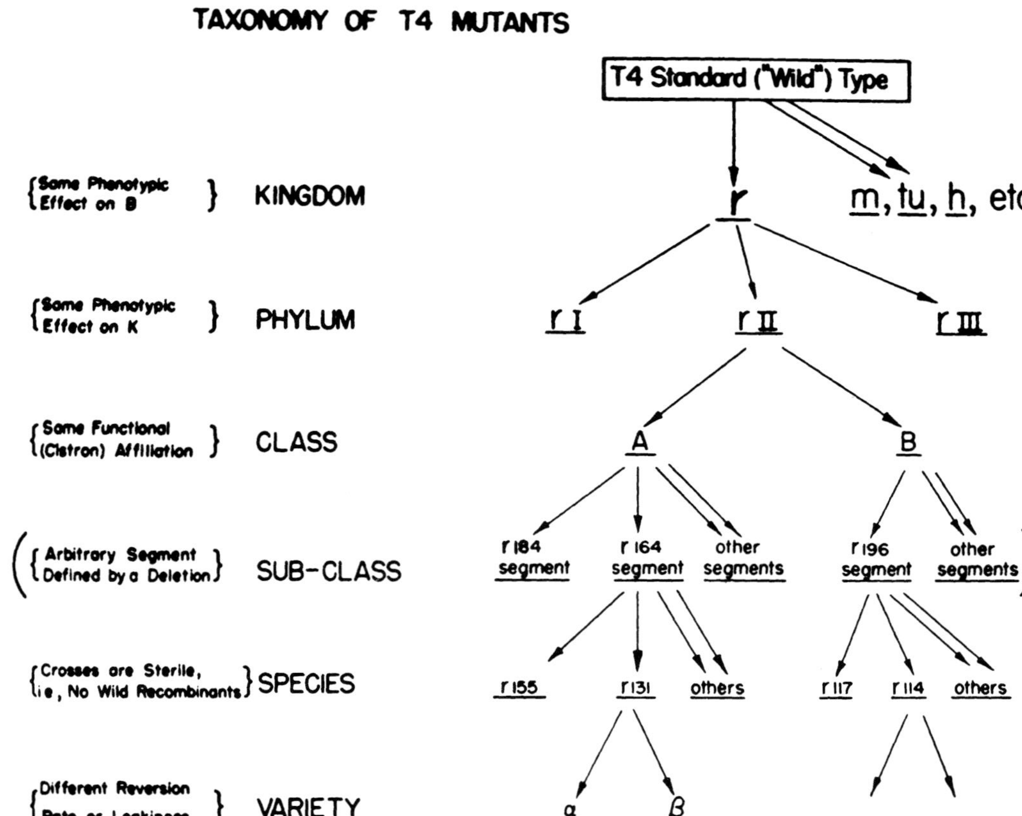

Fig. 7. Classification scheme for *r* mutants of T4.

Mapping the *rII* Region of T4

Taxonomy of r Mutants.

In classifying the mutants of T4, classical terminology may be conveniently used for the taxonomic scheme shown in Fig. 7. Mutants of the *r* "kingdom," isolated on B, can be separated into three "phyla" by testing on K. For the present purposes, our attention will be limited to mutants of the *rII* phylum, which are inactive on K.

A pair of *rII* mutants may be subjected to the *cis-trans* test. The *cis* configuration (mixed infection of K with double mutant and wild-type particles) is active, since the presence of a wild particle in the cell enables both types to multiply. The *trans* configuration (mixed infection of K with the two single mutants) may be active or inactive. If inactive, the two mutants are placed in the same "class." Since the members of a class fail to complement each other, they can be considered as belonging to a single functional group. On the basis of this test, the *rII* mutants divide into two clear-cut classes. The map positions of the mutants in each class have been found to be restricted to separate map segments, the A and B "cistrons."

Arbitrary sub-classes can be chosen (from among the available deletions) for convenience in mapping: mutants falling within the map region encompassed by a particular deletion form a sub-class.

Reverting mutants are considered as of different "species" if crosses between them yield wild recombinants. Among a group of mutants which have not yielded to resolution by recombination tests, "varieties" can in some cases be distinguished by other criteria (reversion rate or degree of ability to grow on K).

Procedures in the Classification of r Mutants of T4.

(1) Isolation of mutants

Each mutant is isolated from a separate plaque of wild-type T4 (plated on B) and freed from contaminating wild-type particles by replating. Stocks of mutants are prepared by growth on S (to avoid the selective advantage which wild type revertants would have on B). Mutants are numbered in the order of isolation, starting with 101 to avoid confusion with mutants previously isolated by others.

(2) Spot test on K

In this first test of a new *r* mutant, 10^8 particles are plated on K and then the plate is spotted with one drop (10^6 particles) of *r164* (a mutant having a "deletion" in an A cistron) and one drop of *r196* (a mutant having a deletion located in the B cistron). Typical examples of the results of this test are shown in Fig. 8. If the new mutant belongs either to the *rI* or the *rII* phylum, the plating bacteria will be completely lysed (except for a background of colonies formed by mutants of K which are resistant to T4), as typified by mutant *X* in Fig. 8.

Mutant *Y* in Fig. 8 is typical of a stable *rII* mutant. The background shows no plaques, indicating that the proportion of revertants in the stock is less than 10^{-8}. The spot of *r196* is completely clear, in contrast to the *r164* spot. This massive lysis is caused by the ability of mutant *Y* and *r196* to complement each other for growth on K. From this result, it may be concluded that mutant *Y* belongs to the A class. Within the *r164* spot, however, some plaques may be seen. These are due to wild recombinants arising from *r164* and mutant *Z* (by virtue of very feeble growth of *rII* mutants on K). Therefore, mutant *Y* is not in the subclass defined by *r164*.

The third test plate is typical of a reverting *rII* mutant. The stock contains a fraction 10^{-6} of wild-type particles which produce the plaques seen in the background. It is evident from the spot tests that the mutant *Z* belongs in the B class, and appears to lie within the *r196* subclass.

(3) Spot test on a mixture of K and B cells

Once the class of a new mutant is known, it can be tested on a single plate against several mutants of the same class. For this purpose, the sensitivity of the test may be increased enormously by the addition of some B cells (about one part in a hundred) to the K used for plating. The additional growth possible for the mutants on B cells enhances their opportunity to produce wild recombinants. This test gives a positive response down to the level of around 0.01 per cent recombination. A negative result does not, of course, eliminate the possibility that recombination occurs with a lower frequency.

(4) Preliminary crosses

A semiquantitative measure of recombination frequency may be obtained by mixedly infecting B with two mutants and plating the infected cells on K. B cells which liberate one or more wild-type particles can produce plaques. This method is convenient for preliminary testing for recombination in the range from 0.0001 to 0.1 per cent.

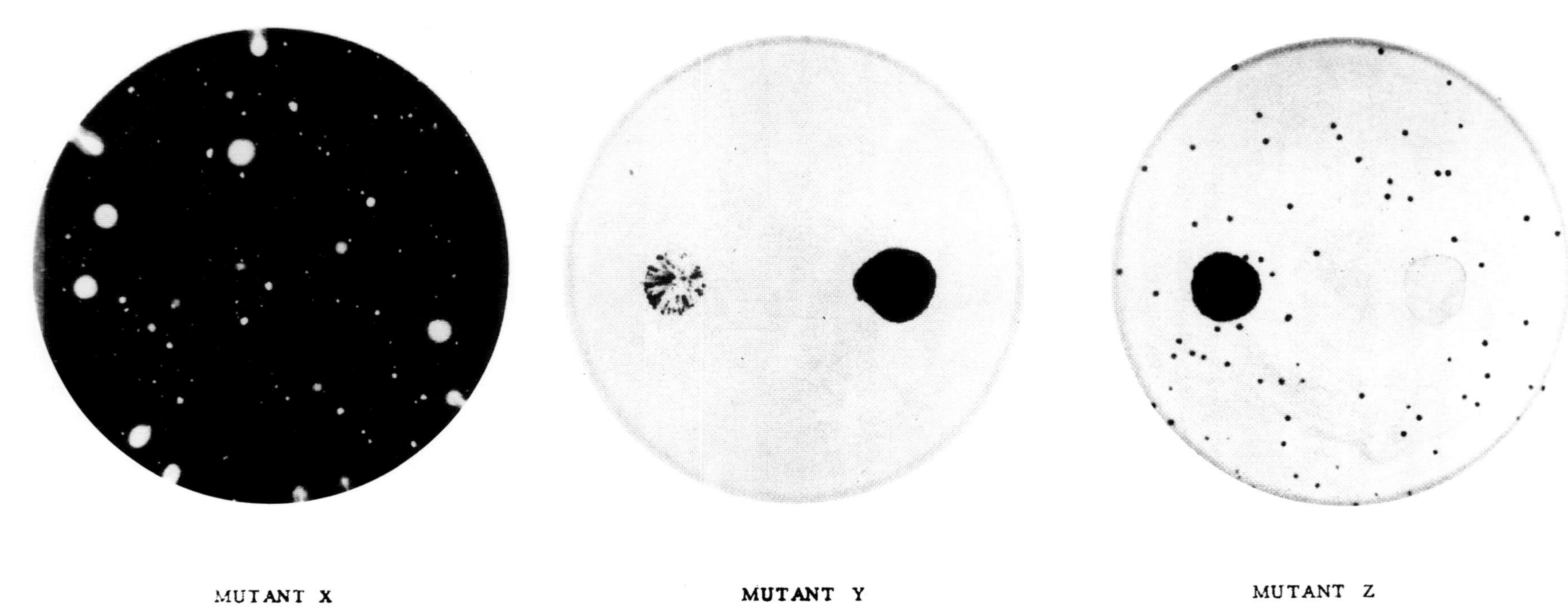

Fɪɢ. 8. Spot test used in classification of r mutants. To test a mutant, 10^8 mutant particles are plated on bacterial strain K and the plate is spotted with one drop of *r164* (left) and one drop of *r196* (right). Mutants *X*, *Y*, and *Z* illustrate typical results. Mutant *X* is of the *rI* phylum. Mutant *Y* is a stable mutant of the *rII* phylum and the A class, but is not in the *r164* sub-class. Mutant *Z* is a reverting mutant (wild-type plaques in background) of the *rII* phylum, B class, and *r196* subclass.

With higher frequencies of recombination, approaching the point where a large fraction of the mixedly infected cells liberate recombinants, saturation sets in.

(5) Standard crosses

Standard measurements of recombination frequency are made in conventional crosses. B cells are infected with an average of three particles per cell of each phage. The infected cells are allowed to burst in a liquid medium, and the progeny are plated on K and on B to determine the proportion of wild-type particles. The reciprocal recombinant (double *rII* mutant) does not, in general, produce plaques on K, but since the two recombinant types are produced in statistically equal numbers (9), the proportion of recombinants in the progeny can be taken as twice the ratio of plaques on K to plaques on B (corrected for the relative efficiency of plating of wild type on these two strains, which is close to unity).

(6) Reversion rates

The reversion rate of a mutant is reflected in the proportion of wild-type particles present in a stock. This value is an important characteristic of each mutant, varying over an enormous range for different mutants. It may be less than 10^{-8} for "stable" (i.e., non-reverting) mutants or as high as several per cent. (For one exceedingly unstable mutant the proportion of revertants averages 70 per cent, even in stocks derived from individual mutant particles.) The precision with which a mutant can be localized on the map is inversely related to its reversion rate; only relatively stable mutants are useful for mapping. In the experiments here reported, it has been assumed that the reversion rate of a mutant is not altered during a cross; the reversion contribution is subtracted from the observed percentage of wild particles in the progeny. In most cases, this correction is negligible.

(7) "Leakiness" of *rII* mutants

rII mutants differ greatly in their ability to grow on K cells. A sensitive measure of this ability can be obtained by infecting K cells and plating them on B. Any K cell that liberates one or more virus particles can give rise to a plaque. The fraction of infected cells yielding virus progeny, which is a characteristic property of each mutant (when measured under fixed conditions), may vary from almost 100 per cent down to less than one per cent for different mutants.

Leakiness has the effect of limiting the sensitivity of K as a tool for selection of wild recombinants, thereby hampering the mapping of very leaky mutants.

TABLE 2

CLASSIFICATION OF AN UNSELECTED GROUP OF 241 *r* MUTANTS OF T4

The number of mutants in each classification is given in parentheses. An asterisk indicates that reversion of the mutant to wild type has not been detected. A few mutants (indicated as "not determined") could not be further classified due to excessively high reversion rate or leakiness.

Kingdom	Phylum	Class	Subclass	Species	Variety
r mutants (241)	*rI* (96)		*r164** (27)	1 sp. (20)	1 var. (11) 1 var. (9)
				1 sp. (2)	
				4 sp. (1 ea.)	
			*r184** (5)	1 sp. (2)	1 var. (1) 1 var. (1)*
				1 sp. (2)*	1 var. (1)
			*r221** (4)	1 sp. (3)	1 var. (1) 1 var. (1)
	rII (134)	IIA (73)	*r47** (4)	1 sp. (2)*	
				1 sp. (1)*	
			*r197** (2)	1 sp. (1)	
			others (29)	1 sp. (3)*	
				1 sp. (2)	
				1 sp. (2)	1 var. (1) 1 var. (1)
				1 sp. (2)	
				1 sp. (2)	
				1 sp. (2)	1 var. (1) 1 var. (1)
				1 sp. (2)	
				14 sp. (1 ea.)	
			not determined (1)		
		IIB (60)	*r196** (34)	1 sp. (21)	1 var. (19) 1 var. (2)
				1 sp. (9)	1 var. (5) 1 var. (4)
				3 sp. (1 ea.)	
			*r187** (6)	1 sp. (3)*	1 var. (1) 1 var. (1)
				1 sp. (2)	
			others (16)	1 sp. (3)	1 var. (2) 1 var. (1)
				1 sp. (3)	
				1 sp. (2)	
				8 sp. (1 ea.)	
			not determined (4)		
	rIII (11)				
		not determined (1)			

Classification of a Set of 241 r Mutants.

A set of *r* mutants was isolated, using B as host, and given numbers from *r101* to *r338*; the mutants *r47*, *r48*, and *r51*, isolated by Doermann (3), were added to this set, making a total of 241 *r* mutants. These were analyzed according to methods already described.

The results are shown in Table 2. Of these mutants, 134 fell into the *rII* phylum. Each of these (with the exception of one very leaky mutant) could be assigned unambiguously to either of two classes on the basis of the test for complementary action of pairs of mutants for growth on K. Mutants within each class were crossed with stable mutants of the same class; those giving no detectable wild recombinants with a particular stable mutant were assigned to the same subclass. Mutants of each subclass were crossed in all pairs. When two or more mutants were found to be of the same species (i.e., showed, in a "preliminary" type cross, recombination of less than about 0.001 per cent, or less than the uncertainty level set by the reversion rate, whichever was greater), one was used to represent the species in further crosses. Those mutants not falling into any of the subclasses defined by the available stable mutants were crossed with each other in pairs. By these procedures, the classification was carried to the species level for the entire set of mutants, except for six highly revertible or leaky mutants whose subclass was not established.

Several of the species showed evidence of splitting into varieties distinguishable by reversion rate or degree of leakiness. Some mutant varieties recurred frequently (e.g., 19, 11, 9 times). These recurrences were far outside the expectation for a Poisson distribution, and are indicative of local variations of mutability. The fact that many species were represented by only one occurrence suggests that many other species remain to be found.

The 33 species found in the A cistron and the 18 species found in the B cistron are sufficient to define reasonably well the limits of each cistron. The minimum size of a cistron in recombination units is determined by the maximum amount of recombination observed in standard crosses between pairs of mutants within it. On the basis of the standard crosses performed so far, this value is about 4 per cent recombination for the A cistron, and 2 per cent for the B cistron.

Study of 923 r Mutants.

While the study of the foregoing 241 mutants yielded a good idea of the sizes and complexity of the A and B cistrons, it fell short of "saturating" the map sufficiently to provide the close clusters of mutants required for the determination of the sizes of the recon and muton. To this end, it was decided to isolate many more mutants. By confining attention to those falling into the *r164* subclass, a more exhaustive study could be made of a selected portion of the map.

In a group of **923** *r* mutants (*r101* through *r1020*, plus Doermann's three), 149 were found to belong to the *r164* subclass. Four of those were stable. The remaining 145 mutants separated into the 11 species shown in Fig. 9. One of the species accounted for 123 of the mutants! As shown in Table 3, this species included three varieties as distinguished by their reversion rates; two were of roughly equal abundance, while the third occurred only once.

Results of standard crosses between the mutants of the *r164* subclass are presented in Fig. 10. The smallest recombination distance, setting an upper limit to the size of the recon, is around 0.02 per cent (between

TABLE 3

REVERSION DATA FOR MUTANTS OF THE *r131* SPECIES

The "reversion index" is the proportion of wild-type particles in a lysate prepared from a few mutant particles (to avoid introduction of any revertants present in the original stock) using S as host. The measurement is subject to large fluctuations due to the clonal growth of the revertants formed Therefore, four separate lysates are made for each mutant. Parentheses indicate extreme fluctuations. An asterisk indicates a background of tiny plaques (smaller than those produced by wild type) when the lysate is plated on K.

In addition to the examples listed in the table, 67 other mutants of this species are also of variety α, as judged by the proportion of revertants (from 0.2×10^{-6} to 4.0×10^{-6}) in single lysates; 45 additional mutants apparently are of variety β (having values from 300×10^{-6} to 4.000×10^{-6}).

The mutants of variety α give less than 0.001 per cent recombination with *r274*. For the more highly revertible mutants of variety β, this limit can only be set at less than 0.02 per cent recombination with *r274*. The mutant *r973*, of variety γ. gives less than 0.005 per cent (the limit set by background on K) recombination with *r274*.

	Mutant	Reversion Index (units of 10^{-6})			
variety α	*r200*	0.47	0.91	0.17	0.25
	r220	0.55	0.41	0.25	0.21
	r274	0.24	0.29	0.61	(10.)
	r930	0.17	0.66	0.19	0.21
	r1012	0.58	0.27	0.22	0.15
variety β	*r245*	420.	490.	490.	1120.
	r353	540.	530.	3900.	(500,000.)
	r376	520.	360.	1240.	450.
	r510	2000.	640.	460.	860.
	r888	610.	530.	250.	570.
variety γ	*r973*	0.01*	0.005*	0.005*	0.01*

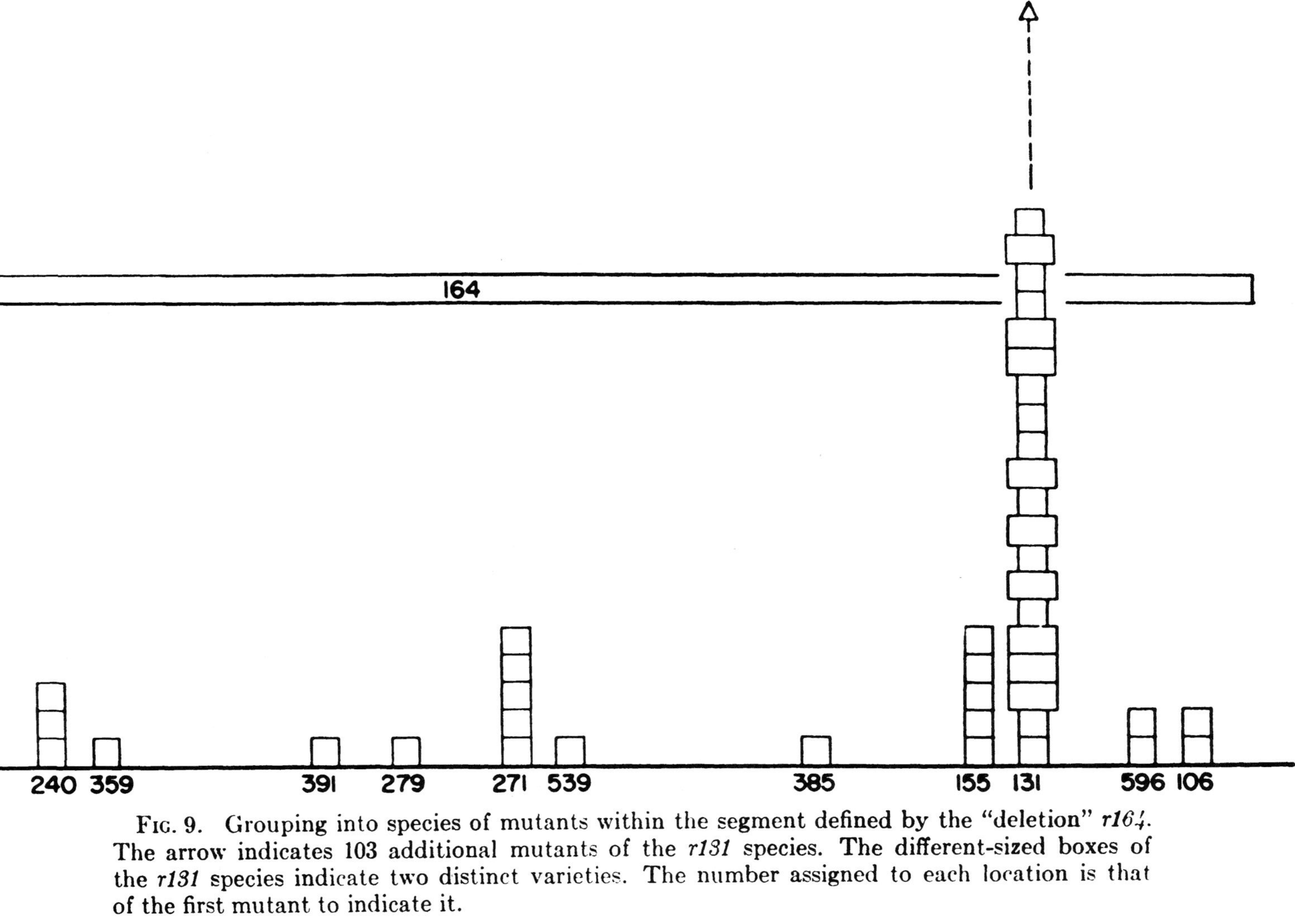

Fig. 9. Grouping into species of mutants within the segment defined by the "deletion" *r164*. The arrow indicates 103 additional mutants of the *r131* species. The different-sized boxes of the *r131* species indicate two distinct varieties. The number assigned to each location is that of the first mutant to indicate it.

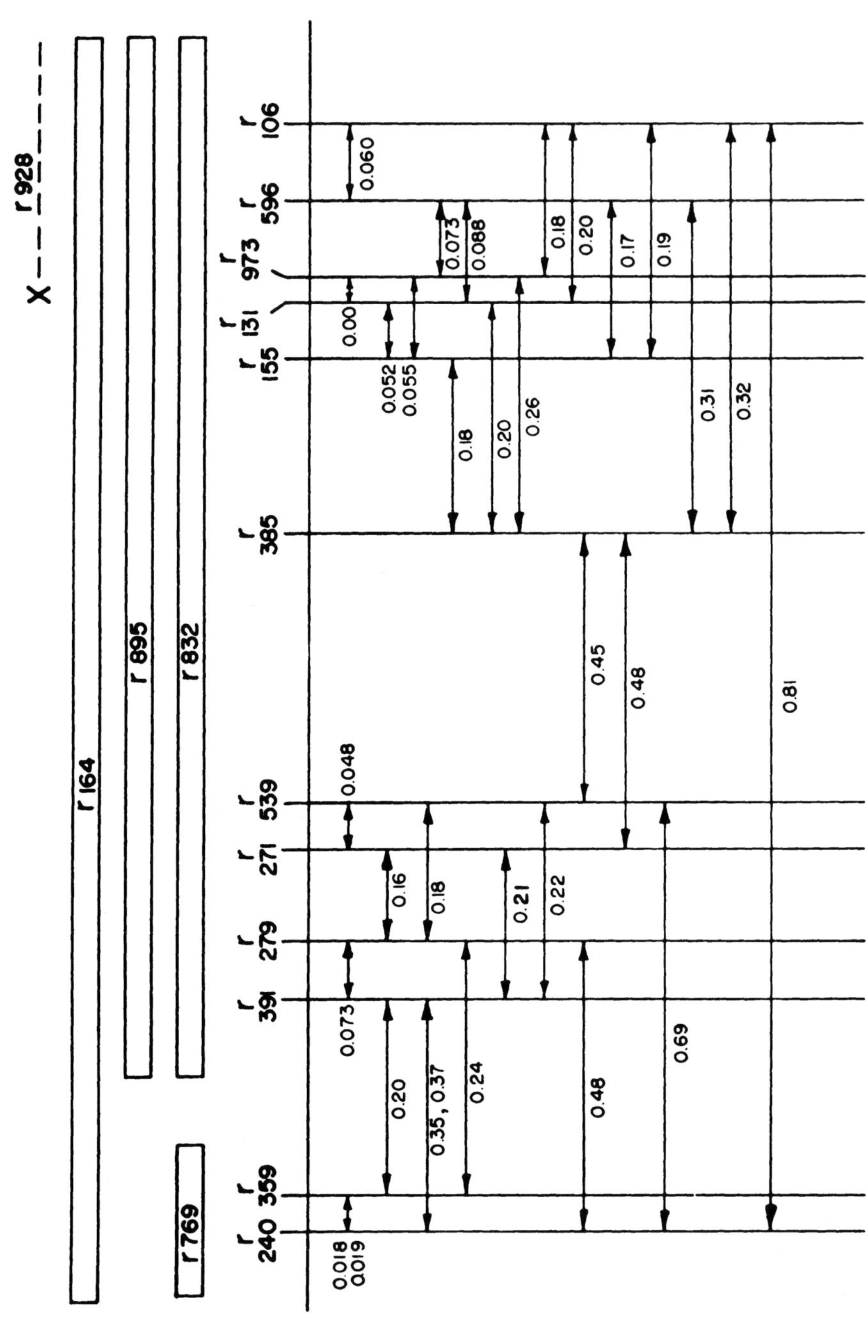
X
r 928
r 164
r 895
r 832
r 769
r 106
r 596
r 973
r 131
r 155
r 385
r 539
r 271
r 279
r 391
r 359
r 240
0.060
0.073
0.088
0.18
0.20
0.17
0.19
0.00
0.052
0.055
0.18
0.20
0.26
0.31
0.32
0.45
0.48
0.81
0.048
0.16
0.18
0.21
0.22
0.073
0.20
0.35, 0.37
0.24
0.48
0.69
0.018
0.019

r240 and *r359*). Since only one interval has this value, the possibility of smaller values is not ruled out.

One procedure for measuring the size of the muton requires a group of three closely linked mutants in order to compare the long distance with the sum of the two shorter ones. If the central mutation has an appreciable size, there should be a discrepancy between these values. There are eight cases in Fig. 10 for which the distances have been measured for three adjacent mutants: 240-359-391, 359-391-279, 391-279-271, 279-271-539, 271-539-385, 385-155-131, 155-131-596, 131-596-106. The discrepancies for these groups are, reading from left to right, + 0.14, — 0.03, — 0.02, — 0.03, — 0.02, — 0.03, + 0.03, and + 0.05 per cent recombination. The average of these values is + 0.01, with an average deviation of ± 0.05. Since each measurement of recombination frequency is subject to experimental error of the order of 20 per cent of its magnitude, these determinations of mutation size (each derived from three measurements) are uncertain to plus or minus about 0.05 per cent recombination. Therefore, the latter is the smallest upper limit than can be set upon the size of the muton by these data.

Another measure of muton size can be attempted by finding the number of species that can exist within a given length of the map. As shown in Fig. 10, a map length of 0.8 per cent recombination includes 9 separable mutant species, or no more than 0.09 per cent per species. Both this determination and the previous one suffer uncertainty due to imperfect additivity of map distances ("negative interference"— see Discussion).

Stable Mutants.

Since the problem of mapping large numbers of mutants is greatly facilitated by the use of "deletions," particular attention has been paid to non-reverting mutants, in the hope of obtaining a complete set of

---

Fɪɢ. 10. Map of the mutants in the *r164* segment. The numbers give the percentage of recombination observed in standard crosses between pairs of mutants. The arrangement on this map is that suggested by these recombination values; it has not yet been verified by three-point tests. Stable mutants are represented as bars above the axis; the span of the bar covers those mutants with which the stable mutant produces no detectable wild recombinants. The stable mutant *r928* appears to be a double mutant having one mutation at the highly mutable *r131* location and a second mutation at a point in the B cistron. Mutants *r131* and *r973* are separated on the map so that the data for each can be indicated. Some of the data here given differ from (and supersede) previously published data based upon unconventional crosses which turned out to be incorrect.

overlapping deletions. Among the series of 923 *r* mutants, 72 stable *rII* mutants were found, 47 in the A cistron and 24 in the B cistron. One mutant *(r928)* was exceptional: it failed to complement mutants of either the A or B cistrons and therefore belongs to *both* classes.

The stable mutants of the B cistron have been crossed (by spot tests on K plus B) in all possible pairs. The results are shown in Fig. 11.

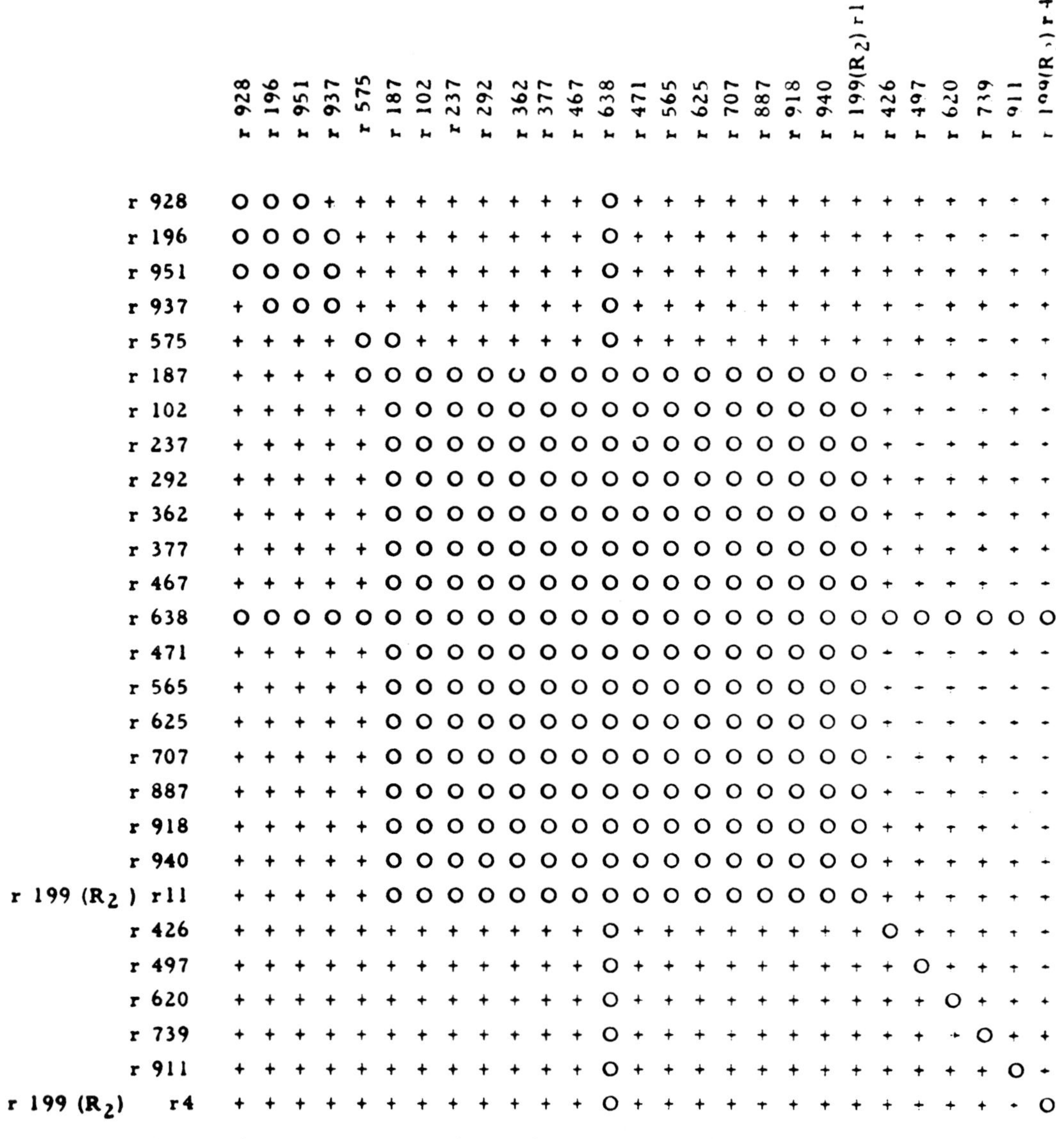

	r 928	r 196	r 951	r 937	r 575	r 187	r 102	r 237	r 292	r 362	r 377	r 467	r 638	r 471	r 565	r 625	r 707	r 887	r 918	r 940	r 199(R_2) r11	r 426	r 497	r 620	r 739	r 911	r 199(R_2) r4
r 928	O	O	O	+	+	+	+	+	+	+	+	+	O	+	+	+	+	+	+	+	+	+	+	+	+	+	+
r 196	O	O	O	O	+	+	+	+	+	+	+	+	O	+	+	+	+	+	+	+	+	+	+	+	+	+	+
r 951	O	O	O	O	+	+	+	+	+	+	+	+	O	+	+	+	+	+	+	+	+	+	+	+	+	+	+
r 937	+	O	O	O	+	+	+	+	+	+	+	+	O	+	+	+	+	+	+	+	+	+	+	+	+	+	+
r 575	+	+	+	+	O	O	+	+	+	+	+	+	O	+	+	+	+	+	+	+	+	+	+	+	+	+	+
r 187	+	+	+	+	O	O	O	O	O	O	O	O	O	O	O	O	O	O	O	O	O	+	+	+	+	+	+
r 102	+	+	+	+	+	O	O	O	O	O	O	O	O	O	O	O	O	O	O	O	O	+	+	+	+	+	+
r 237	+	+	+	+	+	O	O	O	O	O	O	O	O	O	O	O	O	O	O	O	O	+	+	+	+	+	+
r 292	+	+	+	+	+	O	O	O	O	O	O	O	O	O	O	O	O	O	O	O	O	+	+	+	+	+	+
r 362	+	+	+	+	+	O	O	O	O	O	O	O	O	O	O	O	O	O	O	O	O	+	+	+	+	+	+
r 377	+	+	+	+	+	O	O	O	O	O	O	O	O	O	O	O	O	O	O	O	O	+	+	+	+	+	+
r 467	+	+	+	+	+	O	O	O	O	O	O	O	O	O	O	O	O	O	O	O	O	+	+	+	+	+	+
r 638	O	O	O	O	O	O	O	O	O	O	O	O	O	O	O	O	O	O	O	O	O	O	O	O	O	O	O
r 471	+	+	+	+	+	O	O	O	O	O	O	O	O	O	O	O	O	O	O	O	O	+	+	+	+	+	+
r 565	+	+	+	+	+	O	O	O	O	O	O	O	O	O	O	O	O	O	O	O	O	+	+	+	+	+	+
r 625	+	+	+	+	+	O	O	O	O	O	O	O	O	O	O	O	O	O	O	O	O	+	+	+	+	+	+
r 707	+	+	+	+	+	O	O	O	O	O	O	O	O	O	O	O	O	O	O	O	O	+	+	+	+	+	+
r 887	+	+	+	+	+	O	O	O	O	O	O	O	O	O	O	O	O	O	O	O	O	+	+	+	+	+	+
r 918	+	+	+	+	+	O	O	O	O	O	O	O	O	O	O	O	O	O	O	O	O	+	+	+	+	+	+
r 940	+	+	+	+	+	O	O	O	O	O	O	O	O	O	O	O	O	O	O	O	O	+	+	+	+	+	+
r 199 (R_2) r11	+	+	+	+	+	O	O	O	O	O	O	O	O	O	O	O	O	O	O	O	O	+	+	+	+	+	+
r 426	+	+	+	+	+	+	+	+	+	+	+	+	O	+	+	+	+	+	+	+	+	O	+	+	+	+	+
r 497	+	+	+	+	+	+	+	+	+	+	+	+	O	+	+	+	+	+	+	+	+	+	O	+	+	+	+
r 620	+	+	+	+	+	+	+	+	+	+	+	+	O	+	+	+	+	+	+	+	+	+	+	O	+	+	+
r 739	+	+	+	+	+	+	+	+	+	+	+	+	O	+	+	+	+	+	+	+	+	+	+	+	O	+	+
r 911	+	+	+	+	+	+	+	+	+	+	+	+	O	+	+	+	+	+	+	+	+	+	+	+	+	O	+
r 199 (R_2) r4	+	+	+	+	+	+	+	+	+	+	+	+	O	+	+	+	+	+	+	+	+	+	+	+	+	+	O

Fɪɢ. 11. Recombination matrix for stable mutants of the B class. A + indicates production of wild recombinants (around 0.01 per cent recombination or more would be detected) in the cross between the indicated pair of mutants (by spot test on K plus B). All diagonal elements (self-crosses) are zero; non-diagonal zeros indicate overlaps. Two of the mutants are derived, not from the original wild type, but from a revertant of *r199*.

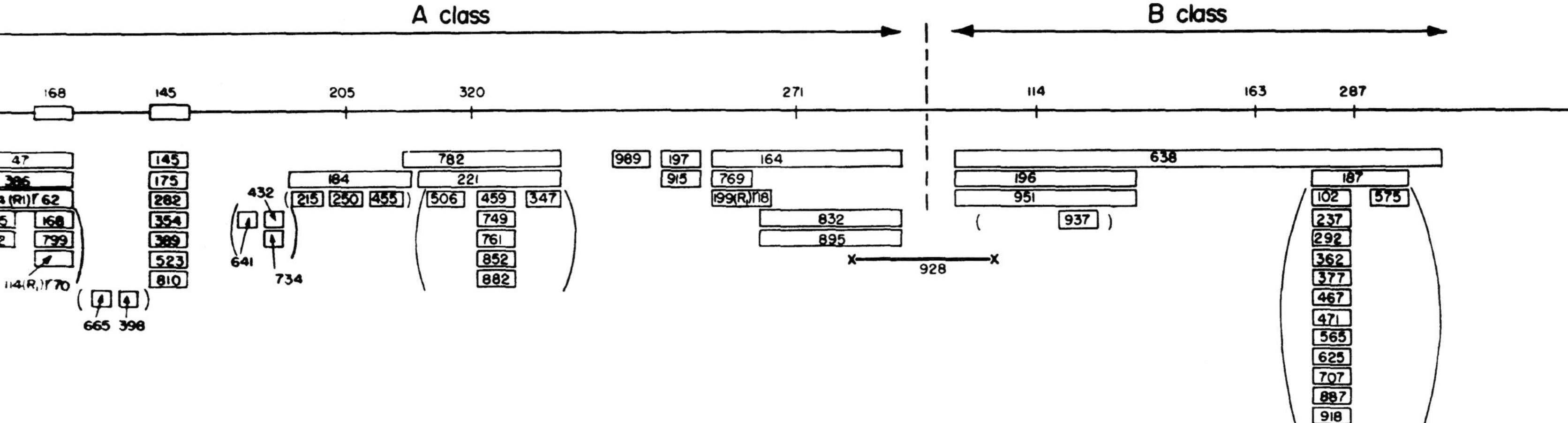

FIG. 12. Preliminary locations of stable *rII* mutants. Mutants producing no wild recombinants with each other are drawn in overlapping configuration. Pairs which produce small amounts are placed near each other. Since there remain some gaps, the order shown depends upon that established by Doermann (4) for the mutants shown on the axis. The scale is somewhat distorted in order to show the overlap relationships clearly. Brackets indicate groups, the internal order of which is not established. Ten stable mutants of the A class and six of the B class were not sufficiently close to any others to permit them to be placed on the map. (A class equals A cistron; B class equals B cistron.)

Overlapping relationships are indicated by non-diagonal zeros. Fig. 12 shows the genetic map representation of these results together with the results derived from an analysis (as yet incomplete) of the stable mutants of the A cistron. Unfortunately, gaps still remain in the map.

The mutant *r638* is of particular note. No B class mutant has been found that gives wild recombinants with it, so that it appears to be due to deletion of the entire B cistron. In spite of this gross defect, it is capable of normal reproduction on *E. coli* strains B and S.

In order to characterize a stable mutant of a "deletion" type, it is necessary to show that it gives no wild recombinants with at least three other mutants that do give recombination with each other (to exclude the possibilities that it is a double mutant or has an inversion). This criterion cannot be applied unless a suitable set of three mutants is available. Only some of the stable mutants (164, 184, 221, 196, 782, 638, 832, 895, 951) have as yet been shown to satisfy this criterion.

Stable mutations tend to occur "all over the map." However, as in the case of reverting mutants, certain localities show a strikingly high recurrence tendency, as illustrated by *r102*, et al., and by *r145*, et al.

DISCUSSION

Relation of Genetic Length to Molecular Length.

We would like to relate the genetic map, an abstract construction representing the results of recombination experiments, to a material structure. The most promising candidate in a T4 particle is its DNA component, which appears to carry the hereditary information (6). DNA also has a linear geometry (20). The problem, then, is to derive a relation between genetic map distance (a probability measurement) and molecular distance. In order to have a unit of molecular distance which is invariant to changes in molecular configuration, the interval between two points along the (paired) DNA structure will be expressed in nucleotide (pair) units, which are more meaningful for our purposes than, say, Ångstrom units.

Unfortunately, present information is inadequate to permit a very accurate calculation to be made of map distance in terms of nucleotide units. First, it is not known whether the probability of recombination is constant (per unit of molecular length) along the entire genetic structure. Second, there is the question of what portion of the total DNA of a T4 particle constitutes hereditary material (for a discussion of this problem, see the paper by Delbrück and Stent in this volume). The result of Levinthal's elegant experiment (11) suggests a value of 40 per cent. Since the total DNA content of a T4 particle is 4×10^5 nucleotides (8), it would seem that the hereditary information of T4 is carried in 1.6×10^5 nucleotides. We do not know, however, whether the information exists in one or in many copies. If there is just one copy, and if it has the paired structure of the model of Watson and Crick (20), the total length of hereditary material should be 8×10^4 nucleotide pairs.

There are difficulties on the genetic side as well. The total length of the genetic map is not well established. The determination of this length requires a number of genetic markers sufficient to define the ends of the map. It also requires a favorable distribution of markers in order that the intervals between them can be summated; if the distance between two markers is sufficiently large, the frequency of recombination between them approaches that for unlinked markers and therefore loses its value as a measure of the linkage distance. Unfortunately, the linkage data presently available for T4 leave much to be desired. The experiments of Streisinger (18) indicate that the map of T4 consists of a single linkage group. Adding up the intervals between markers (corrected for successive rounds of mating according to the theory of Visconti and Delbrück, 19) leads to a total value of the order of 200 per cent recombination units. This estimate is very rough, since the number of available markers upon which it is based is small.

A further difficulty arises from the fact that the map distances measured in standard crosses are not quite additive: a large distance tends to be less than the sum of its component smaller distances. For distances of the order of 10 per cent recombination units and more, the deviations from additivity, referred to as "negative interference," can be accounted for by the Visconti-Delbrück considerations. However, a "negative interference" effect, not accountable for by their theory, persists, and apparently gets worse, at very small distances (4). This presents a serious obstacle for our purposes, since we are interested in knowing what fraction of the total map is represented by a small distance. According to preliminary data on this point, summation of the smallest available distances between *rII* mutants yields a total length for the *rII* region which is several fold greater than that found for crosses involving distant *rII* markers. If the total T4 map length could be obtained by a similar summation of small distances, the indications are that it might be of the order of 800 per cent recombination units in length.

Thus, there are plenty of uncertainties involved in relating the genetic map quantitatively to the DNA structure. The best we can do at present is to make a rough estimate based upon the following assumptions: (1) the genetic information of T4 is carried in one copy consisting of a DNA thread 80,000 nucleotide pairs long; (2) the genetic map has a total length of about 800 per cent recombination units; (3) the probability of recombination per unit molecular length is uniform. According to these assumptions, the ratio of recombination probability (at small distances) to molecular distance would be 800 per cent recombination divided by 80,000 nucleotide pairs, or 0.01 per cent recombination per nucleotide pair. That is to say, if two mutants, having mutations one nucleotide pair apart, are crossed, the proportion of recombinants in the progeny should be 0.01 per cent. This estimate is greater, by a factor of ten, than one made a year ago, in which it was assumed that all the DNA was genetic material and that the effect of negative interference was negligible. It should become possible to improve this calculation as more information becomes available.

The estimate indicates that the level of genetic fine structure which has been reached in these experiments is not far removed from that of the individual nucleotides. Furthermore, the estimate is useful in that it defines an "absolute zero" for recombination probabilities: if a cross between two (single) T4 mutants does not give at least 0.01 per cent recombination, the locations of the two mutations probably are not separated by even one nucleotide pair.

Molecular Sizes of the Genetic Units.

Recon: The smallest non-zero recombination value so far observed among the *rII* mutants of T4 is around 0.02 per cent recombination. If the estimate of 0.01 per cent recombination per nucleotide pair should prove to be correct, the size of the recon would be limited to no more than two nucleotide pairs.

Muton: Evidently, among the stable mutants, mutations may involve varied lengths of the map. The muton is defined as the *smallest* element, alteration of which can be effective in causing a mutation. In the case of reverting mutants, it has not been possible, so far, to demonstrate any appreciable mutation size greater than around 0.05 per cent recombination. This would indicate that alteration of very few nucleotides (no more than five, according to the present estimate) is capable of causing a visible mutation.

Cistron: A cistron turns out to be a very sophisticated structure. The function to which it corresponds can be impaired by mutation at many different locations. In the study of **241** *r* mutants, **33** species were found to be located in the A cistron, of which 6 were in the *r164* subclass. In extending the survey to **923** *r* mutants, the number of known species in the *r164* subclass was doubled. Consequently, it may be expected that about 60 A cistron species will be found—when the analysis of the **923** mutants is completed. Since many species are represented by only one occurrence, implying that many more are yet to be found, it seems safe to conclude that in the A cistron alone there are over a hundred "sensitive" points, i.e., locations at which a mutational event leads to an observable phenotypic effect. Just as in the case of the entire genetic map of an organism, the portrait of a cistron is weighted by considerations of which alterations are effectual. It should be fascinating to try to translate the "topography" within a cistron into that of a physiologically active structure, such as a polypeptide chain folded to form an enzyme.

REFERENCES

1. Benzer, S., *Proc. Natl. Acad. Sci. U. S.*, **41**, 344 (1955).
2. Demerec, M., Blomstrand, I., and Demerec, Z. E., *Proc. Natl. Acad. Sci. U. S.*, **41**, 359 (1955).
3. Doermann, A. H., and Hill, M. B., *Genetics*, **38**, 79 (1953).
4. Doermann, A. H., and collaborators, pers. commun.
5. Franklin, N., and Streisinger, G., pers. commun.
6. Hershey, A. D., and Chase, M., *J. Gen. Physiol.*, **36**, 39 (1952).
7. Hershey, A. D., and Davidson, H., *Genetics*, **36**, 667 (1951).
8. Hershey, A. D., Dixon, J., and Chase, M., *J. Gen. Physiol.*, **36**, 777 (1953).
9. Hershey, A. D., and Rotman, R., *Genetics*, **34**, 44 (1949).
10. Lederberg, E. M., and Lederberg, J., *Genetics*, **38**, 51 (1953).
11. Levinthal, C., *Proc. Natl. Acad. Sci. U S.*, **42**, 394 (1956)
12. Lewis, E. B., *Cold Spring Harbor Symposia Quant. Biol.*, **16**, 159 (1951).
13. Pontecorvo, G., *Advances in Enzymol.*, **13**, 121 (1952).
14. Pritchard, R. H., *Heredity,* **9**, 343 (1955).
15. Roper, J. A., *Nature*, **166**, 956 (1950).
16. Sanger, F., *Advances in Protein Chem.*, **7**, 1 (1952).
17. Stent, G. S., pers. commun.
18. Streisinger, G., pers. commun.
19. Visconti, N., and Delbrück, M., *Genetics*, **38**, 5 (1953).
20. Watson, J. D., and Crick, F. H. C., *Cold Spring Harbor Symposia Quant. Biol.*, **18**, 123 (1953).

The Genetic Control and Cytoplasmic Expression of "Inducibility" in the Synthesis of β-Galactosidase by E. coli

A. B. PARDEE, F. JACOB, AND J. MONOD

The PaJaMo paper—as it came to be called in abbreviation of the authors' names—represents the culmination of a series of findings made at the Institut Pasteur that progressively illuminated how the expression of genes could be controlled. Monod and coworkers had shown that induced proteins such as β-galactosidase were made de novo rather than reshaped from preexisting proteins. They also isolated *Escherichia coli* mutants (*lacI⁻*) that were constitutive for β-galactosidase synthesis, demonstrating genetic control over inducibility. The PaJaMo paper itself, proposing a mechanism for genetic regulation with broad explanatory power, was a flash of clarity in the murky sea of vaguely outlined theories that had been proposed up to this time. The concreteness and the simplicity of the repressor model and the mode of analysis suddenly turned the intractable problem of gene regulation into one that could be readily studied by the classical genetic approach of dominance-recessiveness analysis.

Most research in the field of gene regulation today can trace its origins to the concept proposed in this paper of a regulatory protein interacting with a site on DNA to control the expression of adjacent genes. These studies also contributed in a major way to the development of the concept of messenger RNA (mRNA). The rapidity with which the transferred *lac* genes were expressed in the conjugation experiments, along with other studies on induction of the *lac* system, necessitated the postulation of a rapidly synthesized, unstable intermediate in gene expression. François Jacob himself was one of the contributors to the first experiments demonstrating the existence of mRNA.

Finally, the experiments described in the PaJaMo paper and the way in which they were used to explain genetic regulation in general reflect a style of science that reached its height at the Institut Pasteur during this period. This style is reflected in recollections by Arthur Pardee (A. B. Pardee, A tribute to Jacques Monod, p. 109–116, *in* A. Lwoff and A. Ullmann [ed.], *Origins of Molecular Biology*, Academic Press, Inc., New York, N.Y., 1979)," who stated that Monod ". . . looked upon science like a finished painting. Perfection consists of doing just enough, not one stroke too many or one too few."

JONATHAN BECKWITH

The Genetic Control and Cytoplasmic Expression of "Inducibility" in the Synthesis of β-galactosidase by *E. Coli*[†]

ARTHUR B. PARDEE[‡], FRANÇOIS JACOB AND JACQUES MONOD

Institut Pasteur, Paris and University of California, Berkeley, California, U.S.A.

(*Received 16 March 1959*)

A number of extremely closely linked mutations have been found to affect the synthesis of β-galactosidase in *E. coli*. Some of these (z mutations) are expressed by loss of the capacity to synthesize active enzyme. Others (i mutations) allow the enzyme to be synthesized constitutively instead of inducibly as in the wild type. The study of galactosidase synthesis in heteromerozygotes of *E. coli* indicates that the z and i mutations belong to different cistrons. Moreover the constitutive allele of the i cistron is recessive over the inducible allele. The kinetics of expression of the i^+ (inducible) character suggest that the i gene controls the synthesis of a specific substance which represses the synthesis of β-galactosidase. The constitutive state results from loss of the capacity to synthesize active repressor.

1. Introduction

Any hypothesis on the mechanism of enzyme induction implies an interpretation of the difference between "inducible" and "constitutive" systems. Conversely, since specific, one-step mutations are known, in some cases, to convert a typical inducible into a fully constitutive system, an analysis of the genetic nature and of the biochemical effects of such a mutation should lead to an interpretation of the control mechanisms involved in induction. This is the subject of the present paper.

It should be recalled that the metabolism of lactose and other β-galactosides by intact *E. coli* requires the sequential participation of two distinct factors:

(1) The galactoside-permease, responsible for allowing the entrance of galactosides into the cell.

(2) The intracellular β-galactosidase, responsible for the hydrolysis of β-galactosides.

Both the permease and the hydrolase are inducible in wild type *E. coli*. Three main types of mutations have been found to affect this sequential system:

(1) $z^+ \rightarrow z^-$: loss of the capacity to synthesize β-galactosidase;

(2) $y^+ \rightarrow y^-$: loss of the capacity to synthesize galactoside-permease;

(3) $i^+ \rightarrow i^-$: conversion from the inducible (i^+) to the constitutive (i^-) state.

The $i^+ \rightarrow i^-$ mutation always affects *both* the permease and the hydrolase. All these mutations are extremely closely linked: so far all independent occurences of each of these types have turned out to be located in the "*Lac*" region of the *E. coli* K 12 chromosome. However, the mutations appear to be *independent* since all the different phenotypes resulting from combinations of the different alleles are observed (Rickenberg, Cohen, Buttin & Monod, 1955; Cohen & Monod, 1957; Cohn, 1957).

† This work has been aided by a grant from the Jane Coffin Childs Memorial Fund.
‡ Senior Postdoctoral Fellow of the National Science Foundation (1957–58).

165

It should also be recalled that conjugation in *E. coli* involves the injection of a chromosome from a ♂ (Hfr) into a ♀ (F⁻) cell, and results generally in the formation of an incomplete zygote (merozygote) (Wollman, Jacob & Hayes, 1956). Recombination between ♂ and ♀ chromosome segments does not take place until about 60 to 90 min after injection; moreover segregation of recombinants from heteromerozygotes occurs only after several hours, thus allowing ample time for experimentation.

In order to study the interaction of these factors, their expression in the cytoplasm and their dominance relationships, we have developed a technique which allows one to determine the kinetics of β-galactosidase synthesis in merozygotes of *E. coli*, formed by conjugation of ♂ (Hfr) and ♀ (F⁻) cells carrying different alleles of the factors z, y and i (Pardee, Jacob & Monod, 1958). Before discussing the results obtained with this technique, we shall summarize some preliminary observations on the genetic structure of the "*Lac*" region in *E. coli* K 12.

2. Materials and Methods

(a) *Bacterial strains*

A ♂ (Hfr) strain (no. 4,000) of *E. coli* K 12 was used in most experiments. It was derived from strain 58,161 F⁺, and was selected for early injection of the "*Lac*" marker (Jacob & Wollman, 1957). This strain is streptomycin sensitive (Sˢ), requires methionine for growth and carries the phage λ. A second Hfr strain (no. 3,000), isolated by Hayes (1953), was used in some experiments. This strain is Sˢ, requires vitamin B_1, and does not carry λ prophage. Other Hfr strains carrying mutations for galactosidase (z), inducibility-constitutivity (i), and permease (y) were isolated from the Hayes strain after u.v. irradiation. These markers were also put into ♀ (F⁻) strains, by appropriate matings and selection of the desired recombinants.

A synthetic medium (M 63) was commonly used. It contained per liter: 13·6 g KH_2PO_4, 2·0 g $(NH_4)_2SO_4$, 0·2 g $MgSO_4$. $7H_2O$, 0·5 mg $FeSO_4$. $7 H_2O$, 2·0 g glycerol, and KOH to make pH 7·0. If amino-acids were required, they were added at a concentration of 10 mg/l. of the L-form. For mating experiments, the above stock medium was adjusted to pH 6·3 and vitamin B_1 (0·5 mg/l.) was added prior to use. Aspartate (0·1 mg/ml.) was generally added at the time of mating, according to Fisher (1957).

(b) *Mating experiments*

The desired volume of fresh medium was inoculated with an overnight culture (grown in the same medium) to an initial density of approximately 2×10^7 bacteria/ml. This culture was aerated by shaking at 37°C in a water bath. Turbidity was measured from time to time; and when the density reached 1 to 2×10^8 bacteria/ml., the experiment was started. Usually small volumes of ♂ and ♀ bacteria were mixed in a large Erlenmeyer flask, with the ♀ strain in excess (e.g. 3 ml. ♂ plus 7 ml. ♀ in a 300 ml. flask). The mixed bacteria were agitated very gently so that the motion of the liquid was barely perceptible. From time to time samples were removed for enzyme assay and plating on selective media, usually lactose-B_1-streptomycin agar, for measurement of recombinants. Under these conditions, in a mating of ♂ z^+Sm^s by ♀ z^-Sm^r, up to 20 % of the ♂ population formed z^+Sm^r recombinants (as tested by selection on lactose-streptomycin agar). More often 5 to 10 % recombinants were found.

Streptomycin (Sm)† was used in many mating experiments, to block enzyme synthesis by z^+Sm^s ♂ cells. Controls showed that the synthesis of β-galactosidase was blocked in these strains immediately upon addition of 1 mg/ml. of Sm. Incorporation of ³⁵S from ³⁵SO₄⁻ as well as increase of turbidity were also suppressed by this treatment. This concentration of Sm had no effect on Sm-resistant (Sm^r) mutants. In some experiments, virulent phage (T6) was used to kill the ♂ cells, thus preventing remating.

† The following abbreviations are used in this paper:

Sm = streptomycin	ONPG = *o*-nitrophenyl-β-D-galactoside
IPTG = *iso*propyl-thio-β-D-galactoside	TMG = methyl-thio-β-D-galactoside

It should be noted that if streptomycin was added initially, it significantly reduced the number of recombinants (e.g., 75 % fewer colonies were formed on lactose-B₁-streptomycin plates after 80 min mating in the presence of 1 mg/ml. streptomycin) relative to mating in the absence of streptomycin; but the antibiotic had little effect on enzyme formation by zygotes if added at the commencement of the experiment or after the z^+ locus had been injected.

When galactosidase synthesis had to be induced in zygotes, *isopropyl-thio-β-D-galactoside* (IPTG) was used at 10^{-3}M, a concentration at which this inducer is known to be active even in the absence of permease (Rickenberg *et al.*, 1956).

(c) *Recombination studies*

The blender technique of Wollman & Jacob (1955) was used to determine the times of penetration of markers into the zygotes. It should be noted that this treatment reduces enzyme-forming capacity in zygotes by 30 to 60 %. Recombinant colonies, selected on appropriate selective media, were restreaked on the selector medium and replica plating was used to determine unselected characters. Tests for galactosidase synthesis (with or without induction) were performed on maltose-synthetic agar plates with or without IPTG, using filter paper impregnated with ONPG, according to Cohen-Bazire & Jolit (1953).

Transductions were performed with phage 363, according to Jacob (1955).

(d) *β-galactosidase assay*

For this enzyme assay, 1 ml. aliquots of culture were pipeted into tubes containing 1 drop of toluene. The tubes were shaken vigorously and were incubated for 30 min at 37°C. They were then brought to 28°C; 0·2 ml. of a solution of M/75 *o*-nitrophenyl-β-D-galactoside in M/4 sodium phosphate (pH 7·0) was added, and the tubes were incubated a measured time, until the desired intensity of color had developed. The reaction was halted by addition of 0·5 ml. of 1 M-Na₂CO₃, and the optical density was measured at 420 mμ with the Beckman spectrophotometer. A correction for turbidity could be made by multiplying the optical density at 550 mμ by 1·65 and subtracting this value from the density at 420 mμ. One unit of enzyme is defined as producing 1 mμ-mole *o*-nitrophenol/minute at 28°C, pH 7·0. The units of enzyme in the sample can be calculated from the fact that 1 mμ-mole/ml. *o*-nitrophenol has an optical density of 0·0075 under the above conditions (using 10 mm light-path).

(e) *Chemicals*

o-nitrophenyl-β-D-galactoside (ONPG), methyl-thio-β-D-galactoside (TMG) and *isopropyl-thio-β-D-galactoside* (IPTG) were synthesized at the Institut Pasteur by Dr. D. Türk. Other chemicals were commercial products.

3. Genetic Structure of the " Lac " Region

Figure 1 presents the structure of the *"Lac"* region, as it can be sketched from the data available at present. This complex locus, as established long ago by Lederberg (1947) and confirmed by the blender experiments of Wollman & Jacob (1955), lies at about equal distances from the classical markers *TL* and *Gal*. The closest known markers are *Proline* (left) and *Adenine* (or *T6*) (right). As shown in the map, the several (about 10) occurrences of the y^- mutation all lie together probably at the left of the segment, while the different z^- mutations and the i^- mutant are packed together at the other end. No attempt has been made to establish the order of individual y^- mutations. The order of the z^- mutations relative to each other and to the i^- marker is unambiguously established, as shown, except for the z_U^- mutation, whose position is largely undetermined. Several independent occurrences of the i^- mutation have been isolated. They all appear to be closely linked to the i_3^- marker, but they have not been mapped, for lack of adequate methods of selection i^+ recombinants. The evidence for this structure is briefly as follows:

(1) The frequency of recombination between z and y mutations is very low:

M

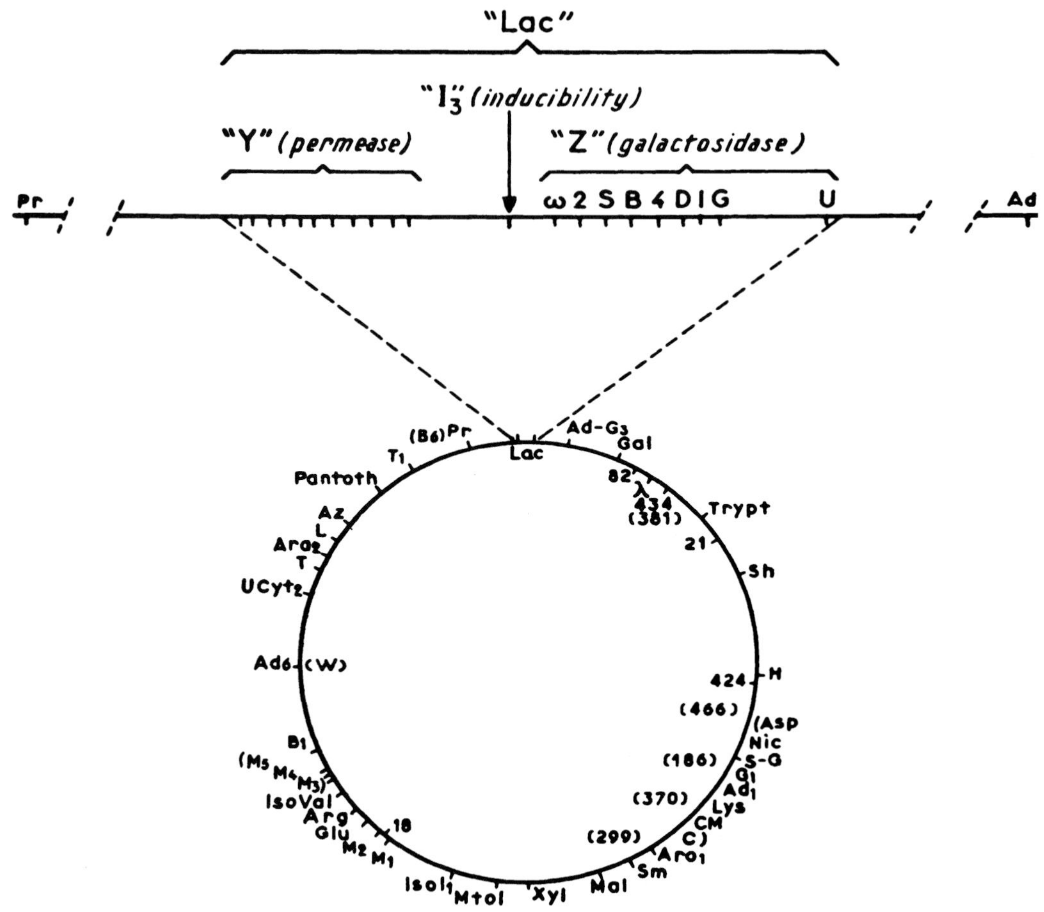

FIG. 1. Fine structure of the "*Lac*" segment.

The "*Lac*" segment is shown enlarged and positioned with respect to the rest of the *E. coli* K 12 linkage group for which the circular model (Jacob & Wollman, 1958) has been adopted.

roughly 1/100th of the frequency of recombination between TL and *Gal*. The frequency of recombination between individual z markers is about one order of magnitude lower.

(2) When y^+z^+ recombinants are selected (by growth on lactose-agar) in crosses of the type:

$$y^+i^-z^- \times y^-i^+z^+$$

the i^+ marker remains associated with z^+ 85 % of the time.

(3) The frequency of cotransduction of i with z (selecting for z^+ alone) is very high ($> 90 \%$), while the frequency for i and y is also high, although definitely lower (about 70 %). (These data are somewhat ambiguous, because of the heterogeneity of the clones resulting from a transduction.)

(4) The selection of z^+ recombinants in crosses involving different z^- mutants, and i as unselected marker, invariably results in about 90 % of the progeny being either i^- or i^+, depending on the particular z^- mutants used. Assuming this result to be due to the position (left or right) of i with respect to the z group:

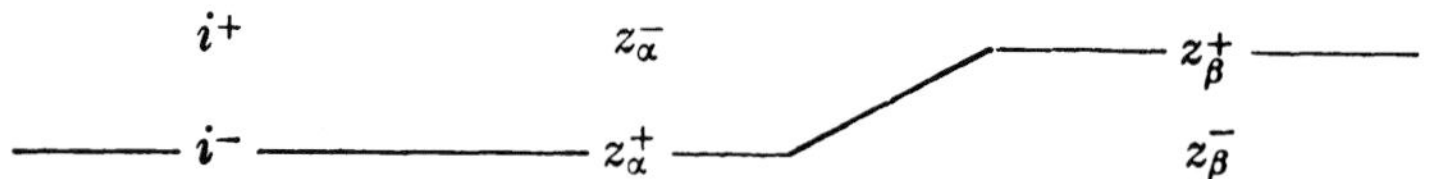

a linear order can be established, without contradictions, for the eight markers shown. This however leaves an ambiguity as to whether i lies *between* the y and the z groups, or outside.

Let us emphasize that this sketch of the *Lac* region is preliminary and very incomplete, and that the results concerning the relationships of certain markers are not understood. For instance, the z_U^- marker recombines rather freely with all the other mutants shown (both y and z) yet, by cotransduction tests, it is closely linked to i (25 % cotransduction). It should also be mentioned that certain of the z^- mutants (z_w^- ; z_2^- ; z_G^-) have apparently lost the capacity to synthesize *both* the galactosidase *and* the permease. Yet these mutations do not seem to be deletions. We shall not attempt, here, to interpret this finding, since we shall center our attention on the interaction between the i marker and the z region.†

A question which should now be considered is whether we may regard the z region as possessing the specific structural information concerning the galactosidase molecule. The fact that so far all the independent mutations resulting in loss of the capacity to synthesize galactosidase were located in this region might not constitute sufficient evidence‡. However, it has been found by Perrin, Bussard & Monod (1959, in preparation) that several of the z^- mutants synthesize, instead of active galactosidase, an antigenically identical, or closely allied, protein. Moreover several of these mutant proteins are different from one another by antigenic and other tests. These findings appear to prove that the z region indeed corresponds to the "structural" genetic unit for β-galactosidase.

4. β-Galactosidase Synthesis by Heteromerozygotes

(a) *Preliminary experiments*

The feasibility and significance of experiments on the expression and interaction of the z, y and i factors depended primarily on whether *E. coli* merozygotes are physiologically able to synthesize significant amounts of enzyme very soon after mating. It was equally important to determine whether the mating involved any cytoplasmic mixing. These questions were investigated in a series of preliminary experiments.

Since the physical separation of *E. coli* zygotes from unmated or exconjugant parent cells cannot be achieved at present, test conditions must be set up, such that the zygotes only, but not the parents, can synthesize the enzyme. This is obtained when the following mating:

$$\text{♂ } z^+y^+i^+Sm^s \times \text{ ♀ } z_s^-y^+i^+Sm^r \tag{A}$$

is performed in the presence of inducer (IPTG) and of 1 mg/ml. of streptomycin. The ♀ lack the z^+ factor; the ♂ are inhibited by streptomycin (cf. Methods); the zygotes are not, because they inherit their cytoplasm from the ♀ cells (see below

† Interaction of i with the y region is of course equally interesting, but since determinations of activity are much less sensitive with the galactoside-permease than with the galactosidase, we have used the latter almost exclusively.

‡ In addition to the mutants shown on Fig. 1, 20 other galactosidase-negative mutants, as yet unmapped, have been found to belong to the same segment by contransduction tests. None was found outside. Lederberg *et al.* (1951), however, have isolated some lactose-"non-fermenting" mutants (as tested on EMB-lactose agar) which are located at other points on the *E. coli* chromosome. In our hands, one of these mutants (Lac_3^-) formed normal amounts of both galactosidase and galactoside-permease (although it did form white colonies on EMB-lactose). Another one (Lac_7^-) formed reduced, but significant, amounts of both. A third (Lac_8) which is a galactosidase-negative, appears to belong to the "*Lac*" segment, by cotransduction tests.

pages 170 and 171), and because the type of ♂ used transfers the *Sm*[s] gene to only a very small percentage of the cells. Under these conditions, enzyme is formed in the mated population with a time course and in amounts showing that the synthesis can be due only to zygotes having received the z^+ factor. Figure 2 shows the

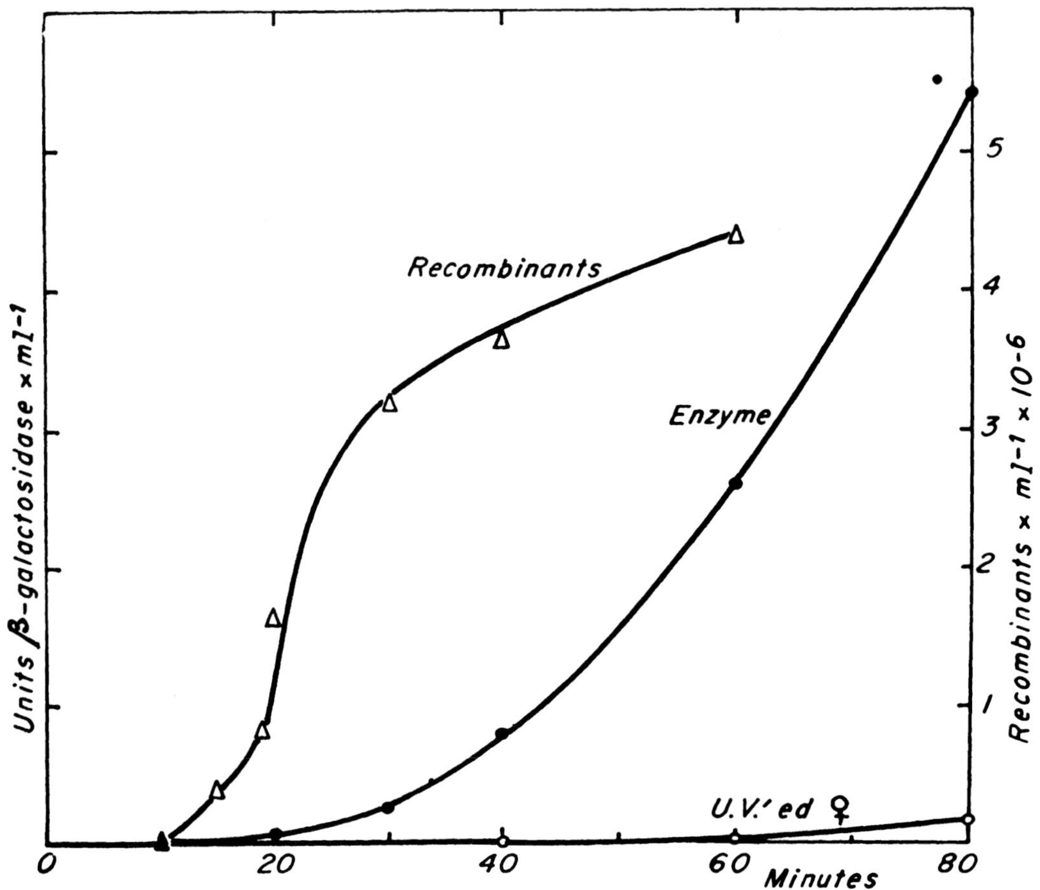

Fig. 2. Enzyme formation and appearance of recombinants in mating A.
Mating in presence of streptomycin (1 mg/ml.) and IPTG (10⁻³M). A control with u.v.-treated ♀ cells (0·01 % survival) is shown. Recombinants (z^+Sm^r) selected by plating on Sm-lactose agar after blending (separate experiment with the same ♂ culture).

kinetics of galactosidase accumulation, compared with the appearance of z^+Sm^r recombinants, determined on aliquots of the same population (cf. Methods). The latter curve corresponds, as shown by Wollman & Jacob (1955), to the distribution of times of penetration of z^+ genes in the zygote population. It will be remarked that enzyme synthesis commences just within a few minutes after the first z^+ genes enter into zygotes. Assuming that the number of zygotes having received a z^+ gene is 4 to 5 times the number of recovered z^+Sm^r recombinants, and taking into account the fact that normal cells are on the average trinucleate (i.e., have three z^+ genes), the rate of enzyme synthesis per injected z^+ appears nearly normal.

This rapid expression of the z^+ factor poses the problem whether cytoplasmic constituents are injected from the ♂ into the zygote. This already appeared unlikely from the previous observations of Jacob & Wollman (1956). We reasoned that if there occurred any significant cytoplasmic mixing, such a mixing should allow the

♂ cells to feed the ♀ cells with any small metabolites which the ♂ had and the ♀ lacked. This condition is obtained in the following mating:

$$♂ \; z^+Sm^s \; maltose^+ \times \; ♀ \; z^-Sm^r \; maltose^-$$

if it is performed in presence of maltose as sole carbon source, using a ♂ which virtually does not inject the *maltose*^+ gene. It results in a very strong inhibition of enzyme synthesis (and recombinant formation) showing that the ♂ cannot effectively

TABLE 1

Enzyme formation in nutritionally deficient zygotes

Deficiency	Rate of enzyme formation †		Mean % inhibition	Mean % inhibition of recombinant formation
	Control	Deficient		
Carbon source ‡	1·6	0·4	73	75
	0·66	0·20		
Arginine §	0·28	0·02	96	65
	0·36	0·01		

† Units of enzyme × hr⁻¹.

‡ ♂ z^+Sm^s *maltose*^+ × ♀ z^-Sm^r *maltose*^- mated in presence of inducer and Sm, with glycerol plus maltose (control) or maltose as sole carbon source.

§ ♂ z^+Sm^s Arg^+ × ♀ z^-Sm^r Arg^- mated in presence of inducer and Sm with and without arginine (10 μg/ml.).

feed the ♀. An even stronger effect is observed when the ♀ requires arginine, the ♂ not, and mating takes place in absence of arginine (again on condition that the Ar^+ gene is not injected by the ♂) (Table 1). These observations indicate that even small molecules do not readily pass from the ♂ into the ♀ cell during conjugation.†

It therefore appears that cytoplasmic fusion or mixing does not occur to an extent which might allow cross-feeding. That the contribution of the ♂ is exclusively genetic, and does not involve cytoplasmic constituents of a nature, or in amounts, significant for our purposes, is however only proved by the results of the opposite matings, which we shall consider in the next section.

(b) *Expression and interaction of the alleles of the z and i factors*

We should first consider which of the alleles of the *z* factors are dominant, and whether they all belong to a single cistron. Experiments of the type described above (mating A) were performed with each of the eight z^- mutants, used as ♀ cells, receiving a z^+ from the ♂. Enzyme was synthesized to similar extents in all cases, showing that the z^- mutants in question were all recessive. Each of the mutants was also mated (as ♂) to a z^- ♀. No enzyme was synthesized by any of these double recessive heterozygotes where the mutations were in the *trans* position

† However such leakage may occur when the concentration of a compound is exceptionally high in the ♂. This happens when a ♂ with the constitution $z^-i^-y^+$ is used in the presence of lactose. The constitutive permease then may concentrate lactose up to 20 % of the cells' dry-weight (Cohen & Monod, 1957). Adequate tests have shown that this lactose does flow from the ♂ into a permease-less ♀ during conjugation.

$$\frac{z_\alpha^+ \; z_\beta^-}{z_\alpha^- \; z_\beta^+}$$

showing that all the (tested) z^- mutants belong to the same cistron as defined **by** Benzer (1957).

The next and most critical problem is whether the z and i factors also belong to the same unit of function (gene or cistron) or not. Let us recall that cells with the constitution z^+i^+ synthesize enzyme in presence of inducer only, while z^+i^- cells synthesize enzyme without induction, and z^-i^+ or z^-i^- cells do not synthesize enzyme under any condition. The extremely close linkage of z and i mutations suggests that they may belong to the same unit. If this were so, they would not be able to interact through the cytoplasm, but could act together only when in *cis* position within the same genetic unit. The heterozygote, z^+i^+/z^-i^- would then be expected not to synthesize galactosidase constitutively.

In order to test this expectation, the following mating:

$$\text{♂} \; z^+i^+ \; \times \; \text{♀} \; z_2^- i_3^- \qquad\qquad \textbf{(B)}$$

was performed *in absence of inducer*. The ♂ cannot synthesize enzyme, because they are i^+. The ♀ cannot because they are z^-. The zygotes however do synthesize enzyme (Fig. 3): during the first hour following mating the synthesis is, if anything, even more rapid and vigorous than when both parents are i^+ and inducer is used, as in mating (A).

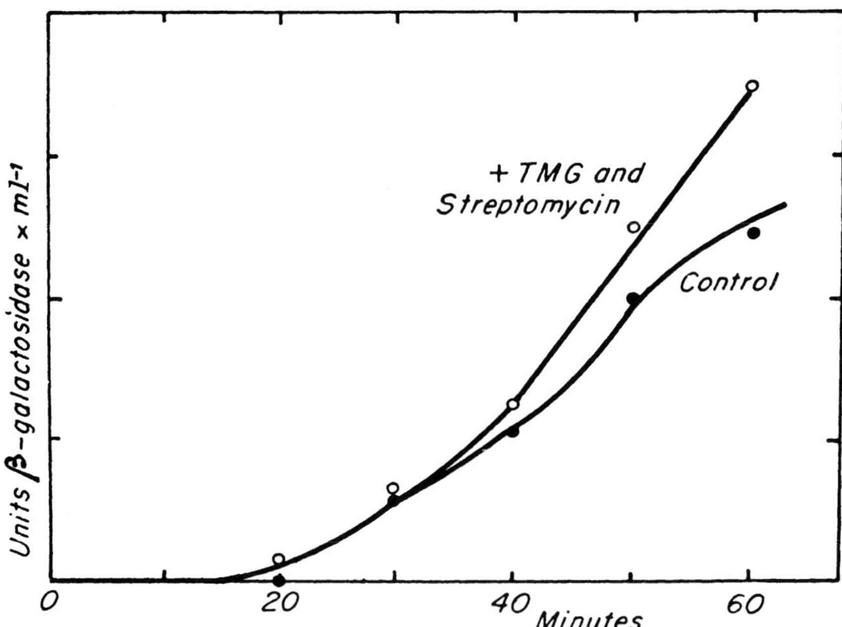

FIG. 3. Enzyme formation during first hour in mating B.
Mating under usual conditions. To an aliquot streptomycin (0·8 mg/ml.) was added at 20 minutes, and TMG at 25 minutes, to allow comparison of synthesis with and without inducer.

Such a mating therefore allows immediate and complete interaction of the z^+ from the ♂ with the i^- from the ♀. The possibility that the interaction depends upon actual recombination yielding z^+i^- in *cis* configuration is excluded because: (a) the synthesis begins virtually immediately after injection whereas genetic recombination is known (Jacob & Wollman, 1958) not to occur until 60 to 90 min after injection; (b) the factors z and i are so closely linked that recombination is an exceedingly rare event (less than 10^{-4} of the zygotes) while the rate of enzyme synthesis is of an order indicating that most or all of the zygotes participate.

The possibility should also be considered that, rather than taking place through the cytoplasm, the interaction requires actual *pairing* of the homologous chromosome segments. This is excluded by the fact that the following mating:

$$\male\ z_2^- i_3^- \times \female\ z^+ i^+ \tag{C}$$

when performed in the *absence* of inducer, yields no trace of enzyme, at any time after mixing, although conjugation and chromosome injection occur normally as shown by adequate controls involving other markers. The zygotes obtained in matings B and C are genetically identical, except that the wild type alleles ($z^+ i^+$) are in relative excess (about 3 to 1) in (B), while the mutant alleles are in similar excess in (C). This quantitative difference cannot account for the absolute contrast of the results of the reciprocal matings, one allowing vigorous constitutive synthesis, the other none at all. This can only be attributed to the fact that the cytoplasm of the zygote is entirely furnished by the $\female$ cell, with no significant contribution from the $\male$. Therefore the $i^- \rightarrow z^+$ interaction must be considered to take place through the cytoplasm.

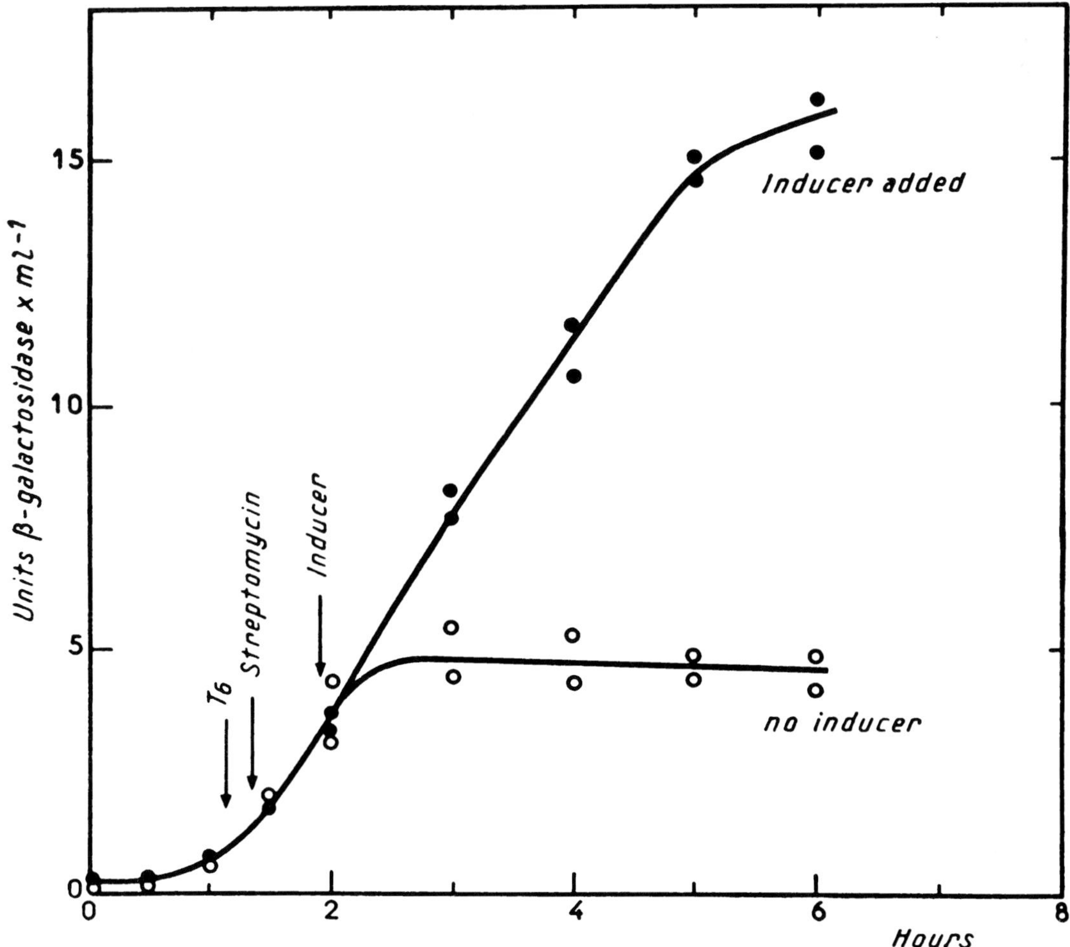

Fig. 4. Enzyme formation in mating D.
    Mating performed under usual conditions in quadruplicate in absence of inducer. At times indicated, a suspension of phage T6 (20ϕ/B final concentration) and streptomycin (1 mg/ml.) were added to all of the cultures and TMG (2×10^{-3}M) was added to two of them (black circles) while the other two (white circles) received no addition.

This result may also be expressed by saying that the i factor sends out a cytoplasmic message which is picked up by the z gene, or gene products. Postulating, as we must, that this message is borne by a specific compound synthesized under the control of the i gene, we may further assume that one of the alleles of the i gene provokes the synthesis of the message, while the other one is inactive in this respect. If these assumptions are adequate, one of the alleles should be absolutely dominant over the other, but the dominance should become expressed only gradually when the cytoplasm of the zygotes came from the recessive parent, while it should be expressed immediately when the cytoplasm came from the dominant parent.

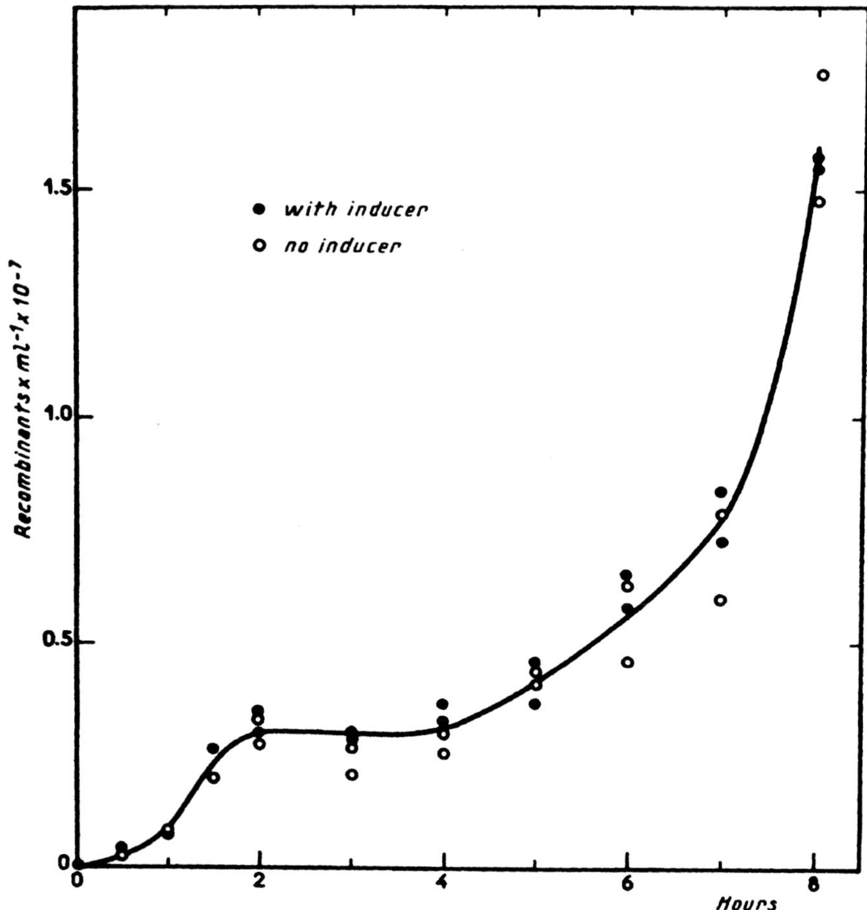

Fig. 5. Recombinant appearance in mating D.

Formation of z^+Sm^r recombinants tested by plating aliquots of the four cultures used in the experiment above (Fig. 4) on lactose-Sm agar. Portions of the culture were diluted 1000-fold and shaken vigorously at 100 minutes to prevent further mating. The increase up to the second hour is due to increasing numbers of zygotes. The increase after the fourth hour is due to *multiplication* of segregants (Wollman, Jacob & Hayes, 1956).

The fact that in matings of type (C) no enzyme is synthesized, even several hours after mating, means that the constitutive (i^-) allele from the ♂ is never expressed. This suggests that the dominant allele is the inducible (i^+). If so, the i^+ should eventually become expressed in matings of type (B)—i.e., the zygotes, initially constitutive (since their cytoplasm comes from the i^- parent), should eventually become inducible. To test this prediction, the following mating was performed:

$$\text{♂ } z^+i^+Sm^s\,T6^s \times \text{♀ } z_2^-i_3^-Sm^r\,T6^r \qquad\qquad \textbf{(D)}$$

and the synthesis of enzyme, in the absence and in the presence of inducer, **was** followed over several hours (in order to block induction of the ♂ and remating, **a** mixture of streptomycin and T6 phage was used). Figure 4 shows that, in the **absence** of inducer, enzyme synthesis stops about 90 min (or earlier) after entry of the z^+i^+ genes into the ♀ cells. When inducer is added at this stage, enzyme synthesis is resumed, showing that the initially constitutive z^+i^+/z^-i^- zygotes have **not been** inactivated, but have become inducible.

It should be asked whether this conversion to inducibility, rather than **occurring** in the heterozygotes, might not correspond to the segregation of homozygous $z^+i^+Sm^rT6^r$ recombinants with concomitant disappearance of the **heterozygotes.** This is excluded because the earliest homozygous recombinants only **appear 2 hr** after the time when constitutive synthesis ceases† (Fig. 5).

From these observations we may conclude that the constitutive (i^-) **allele** is inactive, while the i^+ is dominant, provoking the synthesis of a **substance** responsible specifically for the inducible behaviour of the galactosidase **enzyme-** forming-center.

5. Discussion and Conclusions

(1) The conclusions which can be directly drawn from the evidence **presented** above may be summarized as follows:

The synthesis of β-galactosidase and galactoside-permease in *E. coli* is **controlled** by three extremely closely linked genes (cistrons), *z*, *i* and *y*. The *z* gene **determines,** in part at least, the structure of the galactosidase protein molecule. The *y* gene **pro-** bably does the same for the permease molecule, but there is no evidence **on this** point. The *i* gene in its active form controls the synthesis of a product which, **when** present in the cytoplasm, prevents the synthesis of β-galactosidase and **galactoside-** permease, unless inducer is added externally (inducible behaviour). When the *i* **gene-** product is absent or inactive as a result of mutation within the gene, no **external** inducer is required for β-galactosidase and galactoside-permease synthesis **(constitu-** tive behaviour). The *i* gene product is very highly specific, having no effect on **any** other known system.

(2) While proving that the interaction of the *i* and *z* factors involves a **specific** cytoplasmic messenger, the data presented here do not, by themselves, give **any** indication as to the mode of action of this compound. Two alternative models of **this** action should be considered.

According to one, which we shall call the "inducer" model, the activity of **the** galactosidase-forming system‡ requires the presence of an inducer, both in **the** constitutive and in the inducible organism. Such an inducer (a galactoside) is **syn-** thesized by *both* types of organisms. The i^+ gene controls the synthesis of **an enzyme** which destroys or inactivates the inducer: hence the requirement for **external** inducer in the wild type. The i^- mutation inactivates the gene (or its product, **the** enzyme) allowing accumulation of endogenous inducer. This model accounts for **the** dominance of inducibility over constitutivity, and for the kinetics of conversion of the zygotes.

† It may also be recalled that, according to Anderson & Maze (1957), heterozygosis prevails for many generations in the descendants of *E. coli* zygotes.

‡ By this term we designate the system of all cellular constituents *specifically* involved in galactosidase synthesis. This includes the *z* gene and its cytoplasmic products.

According to the other, or "repressor", model the activity of the galactosidase-forming system is inhibited in the wild type by a specific "repressor" (probably also involving a galactosidic residue) synthesized under the control of the i^+ gene. The inducer is required only in the wild-type as an *antagonist* of the repressor. In the constitutive (i^-), the repressor is not formed, or is inactive, hence the requirement for an inducer disappears. This model accounts equally well for the dominance of i^+ and for the kinetic relationships.

(3) The "repressor" hypothesis might appear strictly *ad hoc* and arbitrary were it not also suggested by other facts which should be briefly recalled. That the synthesis of certain constitutive enzyme systems may be specifically inhibited by certain products (or even substrates) of their action, was first observed in 1953 by Monod & Cohen-Bazire working with constitutive galactosidase (of *E. coli*) (1953*a*) or with tryptophan-synthetase (of *A. aerogenes*) (1953*b*), and by Wijesundera & Woods (1953), and Cohn, Cohen & Monod (1953) independently working with the methionine-synthase complex of *E. coli*. It was suggested at that time that this remarkable inhibitory effect could be due to the displacement of an internally-synthesized inducer, responsible for constitutive synthesis, and it was pointed out that such a mechanism could account, in part at least, for the proper adjustment of cellular syntheses (Cohn & Monod, 1953; Monod, 1955). During the past two or three years, several new examples of this effect have been observed and studied in some detail by Vogel (1957), Yates & Pardee (1957), Gorini & Maas (1957). It now appears to be a general rule, for bacteria, that the formation of sequential enzyme systems involved in the synthesis of essential metabolites is *inhibited* by their end product. The convenient term "repression" was coined by Vogel to distinguish this effect from another, equally general, phenomenon: the control of enzyme *activity* by end products of metabolism.

(4) The facts which demonstrate the existence and wide occurrence of repression effects justify the basic assumptions of the repressor model. They do not allow a choice between the two models. Further considerations make the repressor model appear much more adequate:

(a) The repressor model is simpler since it does not require an independent inducer-synthesizing system.

(b) It predicts that constitutive mutants should, as a rule, synthesize more enzyme than induced wild-type. This appears to be the case for such different systems as galactosidase, amylomaltase (Cohen-Bazire & Jolit, 1953), glucuronidase (Stoeber, 1959, unpublished data), galactokinase of *E. coli* and penicillinase of *B. cereus* (Kogut, Pollock & Tridgell, 1956).

(c) The inducer model, if generalized, implies that internally synthesized inducers (Buttin, unpublished) operate in all constitutive systems. This assumption, first suggested as an interpretation of repression effects, has not been vindicated in recent work on repressible biosynthetic systems (Vogel, 1957; Gorini & Maas, 1957; Yates & Pardee, 1957). In contrast, the synthesis of numerous inducible systems has been known for many years (Dienert, 1900; Stephenson & Yudkin, 1936; Monod, 1942) to be inhibited by glucose and other carbohydrates. The recent work of Neidhardt & Magasanik (1957) has shown this glucose effect to be comparable to a non-specific repression and these authors have suggested that glucose acts as a preferential metabolic source of internally synthesized repressors. If this is so, and if our repressor model is correct, the conversion of glucose into specific galactosidase-repressor should be blocked in the constitutives. Accordingly the galactosidase-forming system of the mutant should be largely insensitive to the glucose effect while other inducible systems should retain their sensitivity. That this is precisely the case (Cohn & Monod, 1953) is a very strong argument in favor of the repressor model.

(5) If adopted and confirmed with other systems, the repressor model may lead to a generalizable picture of the regulation of protein syntheses; according to this scheme, the basic mechanism common to all protein-synthesizing systems would be inhibition by specific repressors formed under the control of particular genes, and antagonized, in some cases, by inducers. Although the wide occurrence of repression effects is certain, the situation revealed with the present system, namely a genetic "complex" comprising, besides the "structural" genes (z, y) a repressor-making gene (i) whose function is to block or regulate the expression of the neighboring genes is, so far, unique for enzyme systems. But the formal analogy between this situation and that which is known to exist in the control of immunity and zygotic induction of temperate bacteriophage is so complete as to suggest that the basic mechanism might be essentially the same. It should be recalled that according to Jacob & Wollman (1956), when a chromosome from a λ-lysogenic $\male$ of $E.\ coli$ is injected into a non-lysogenic $\female$, the process of vegetative phage development is started, which involves as an essential, probably as a primary, step the synthesis of specific proteins. When the reverse mating ($\male$ non-lysogenic $\times$ $\female$ λ-lysogenic) is performed, zygotic induction does not occur; nor does vegetative phage develop when such zygotes are superinfected with λ particles. The λ-lysogenic cell is therefore immune against manifestations of prophage or phage potentialities, *and the immunity is expressed in the cytoplasm* (Jacob, 1958–59). Moreover the immunity is strictly specific, since it does not extend to other, even closely related, phages. The formation, under the control of a phage gene, of a specific repressor, able to block synthesis of proteins determined by other genes of the phage, would account for these findings.

(6) Implicit in the repressor model are two critical questions, which for lack of evidence we have avoided discussing, but which should be explicitly stated in conclusion. These questions are:

(a) What is the chemical nature of the repressor? Should it be considered a primary or a secondary product of the gene?

(b) Does the repressor act at the level of the gene itself, or at the level of the cytoplasmic gene-product (enzyme-forming system)?

We are much indebted to Professor Leo Szilard for illuminating discussions during this work and to Mme M. Beljanski, Mme M. Jolit and Mr. R. Barrand for assistance in certain experiments.

REFERENCES

Anderson, T. F. & Maze, R. (1957). *Ann. Inst. Pasteur*, **93**, 194.
Benzer, S. (1957). "The elementary units of heredity", in *The Chemical Basis of Heredity*, ed. by W. McElroy & B. Glass, p. 70. Baltimore: John-Hopkins Press.
Cohen, G. N. & Monod, J. (1957). *Bact. Rev.* **21**, 169.
Cohen-Bazire, G. & Jolit, M. (1953). *Ann. Inst. Pasteur*, **84**, 937.
Cohn, M. (1957). *Bact. Rev.* **21**, 140.
Cohn, M., Cohen, G. N. & Monod, J. (1953). *C.R. Acad. Sci., Paris*, **236**, 746.
Cohn, M. & Monod, J. (1953). In *Adaptation in Microorganisms*, p. 132. Cambridge: University Press.
Dienert, F. (1900). *Ann. Inst. Pasteur*, **14**, 139.
Fisher, K. W. (1957). *J. Gen. Microbiol.* **16**, 120.

Gorini, L. & Maas, W. K. (1957). *Biochim. biophys. Acta*, **25**, 208.
Hayes, W. (1953). *Cold Spr. Harb. Symp. Quant. Biol.* **18**, 75.
Jacob, F. (1955). *Virology*, **1**, 207.
Jacob, F. (1958–59). Harvey Lectures, Series **54**, in the press.
Jacob, F. & Wollman, E (1956). *Ann. Inst. Pasteur*, **91**, 486.
Jacob, F. & Wollman, E. (1957). *C.R. Acad. Sci., Paris*, **244**, 1840.
Jacob, F. & Wollman, E. (1958). *Symp. Soc. Exp. Biol.* **12**, 75. Cambridge: University
     Press.
Kogut, M., Pollock, M. R. & Tridgell, E. J. (1956). *Biochem. J.* **62**, 391.
Lederberg, J. (1947). *Genetics*, **32**, 505.
Lederberg, J., Lederberg, E. M., Zinder, N. D. & Lively, E. R. (1951). *Cold Spr. Harb.
     Symp. Quant. Biol.* **16**, 413.
Monod, J. (1942). *Recherches sur la croissance des cultures bactériennes*. Paris: Herman Edit.
Monod, J. (1955). *Exp. Ann. Biochim. Méd., série* **17**, 195. Paris: Masson & Cie Edit.
Monod, J. & Cohen-Bazire, G. (1953*a*). *C.R. Acad. Sci., Paris*, **236**, 417.
Monod, J. & Cohen-Bazire, G. (1953*b*). *C.R. Acad. Sci., Paris*, **236**, 530.
Neidhardt, F. C. & Magasanik, B. (1957). *J. Bact.* **73**, 253.
Pardee, A. B., Jacob, F. & Monod, J. (1958). *C.R. Acad. Sci., Paris*, **246**, 3125.
Rickenberg, H. V., Cohen, G. N., Buttin, G. & Monod, J. (1956). *Ann. Inst. Pasteur*, **91**,
     829.
Stephenson, M. & Yudkin, J. (1936). *Biochem. J.* **30**, 506.
Vogel, H. J. (1957). In *The Chemical Basis of Heredity*, ed. by W. D. McElroy & B. Glass,
     p. 276. Baltimore: Johns Hopkins Press.
Wijesundera, S. & Woods, D. D. (1953). *Biochem. J.* **55**, viii.
Wollman, E. & Jacob, F. (1955). *C.R. Acad. Sci., Paris*, **240**, 2449.
Wollman, E., Jacob, F. & Hayes, W. (1956). *Cold Spr. Harb. Symp. Quant. Biol.* **21**, 141.
Yates, R. A. & Pardee, A. B. (1957). *J. Biol. Chem.* **227**, 677.

General Nature of the Genetic Code for Proteins

F. H. C. CRICK, L. BARNETT, S. BRENNER, AND R. J. WATTS-TOBIN

Once the genetic material had been identified as DNA and its structure had been determined, biologists faced the problem of how one could arrange the sequence of the four DNA bases to code for the enormous variety of protein sequences. Persuasive answers to many questions about the genetic code came from the elegant experiments described in this paper. This study remains one of the most telling testimonials to the power of genetics, inspiring many budding young molecular biologists at the time to approach fundamental biological problems with genetics. Note that the conclusions were arrived at without any knowledge of the nature of the bacteriophage T4 *rIIA* and *B* gene products. Nevertheless, molecular biologists were largely convinced by this purely genetic paper that the code was triplet, that it was not overlapping, that the sequence of codons was read from a fixed starting point, and that the code was probably degenerate. Although many of these conclusions were thought likely to be true anyway, these experiments pretty much put an end to debates on the subject.

This paper is also inspirational as an example of the scientific discovery that results from the classical case of the "prepared mind." In *The Eighth Day of Creation* (Cold Spring Harbor Press, Plainview, N.Y., 1996), H. F. Judson pointed out that Crick initiated these experiments to test a "loopy code" theory, according to which base-pairing in mRNA played a role in coding. But, after characterization of a large number of acridine-induced *rII* mutations and their suppressors, it became clear that the mutations did not fit the theory. Then, recognizing the potential implications of the properties of these mutations, the Cambridge group carried out phage crosses with various combinations of *rII* mutations to "determine" how many bases composed a codon. As reported by Judson, late one evening after Leslie Barnett and Francis Crick had examined the crucial plates for phage plaques, Crick turned to Barnett and told her, "We're the only two who know it's a triplet code!"

JONATHAN BECKWITH

Reprinted by permission from *Nature* 192:1227–1232. Copyright © 1961. Macmillan Magazines Ltd.

GENERAL NATURE OF THE GENETIC CODE FOR PROTEINS

By Dr. F. H. C. CRICK, F.R.S., LESLIE BARNETT, Dr. S. BRENNER
and Dr. R. J. WATTS-TOBIN

Medical Research Council Unit for Molecular Biology,
Cavendish Laboratory, Cambridge

THERE is now a mass of indirect evidence which suggests that the amino-acid sequence along the polypeptide chain of a protein is determined by the sequence of the bases along some particular part of the nucleic acid of the genetic material. Since there are twenty common amino-acids found throughout Nature, but only four common bases, it has often been surmised that the sequence of the four bases is in some way a code for the sequence of the amino-acids. In this article we report genetic experiments which, together with the work of others, suggest that the genetic code is of the following general type:

(a) A group of three bases (or, less likely, a multiple of three bases) codes one amino-acid.

(b) The code is not of the overlapping type (see Fig. 1).

(c) The sequence of the bases is read from a fixed starting point. This determines how the long sequences of bases are to be correctly read off as triplets. There are no special 'commas' to show how to select the right triplets. If the starting point is displaced by one base, then the reading into triplets is displaced, and thus becomes incorrect.

(d) The code is probably 'degenerate'; that is, in general, one particular amino-acid can be coded by one of several triplets of bases.

The Reading of the Code

The evidence that the genetic code is not overlapping (see Fig. 1) does not come from our work, but from that of Wittmann[1] and of Tsugita and Fraenkel-Conrat[2] on the mutants of tobacco mosaic virus produced by nitrous acid. In an overlapping triplet code, an alteration to one base will in general change three adjacent amino-acids in the polypeptide chain. Their work on the alterations produced in the protein of the virus show that usually only one amino-acid at a time is changed as a result of treating the ribonucleic acid (RNA) of the virus with nitrous acid. In the rarer cases where two amino-acids are altered (owing presumably to two separate deaminations by the nitrous acid on one piece of RNA), the altered amino-acids are not in adjacent positions in the polypeptide chain.

Brenner[3] had previously shown that, if the code were universal (that is, the same throughout Nature), then all overlapping triplet codes were impossible. Moreover, all the abnormal human hæmoglobins studied in detail[4] show only single amino-acid changes. The newer experimental results essentially rule out all simple codes of the overlapping type.

If the code is not overlapping, then there must be some arrangement to show how to select the correct triplets (or quadruplets, or whatever it may be) along the continuous sequence of bases. One obvious suggestion is that, say, every fourth base is a 'comma'. Another idea is that certain triplets make 'sense', whereas others make 'nonsense', as in the comma-free codes of Crick, Griffith and Orgel[5]. Alternatively, the correct choice may be made by starting at a fixed point and working along the sequence of bases three (or four, or whatever) at a time. It is this possibility which we now favour.

Experimental Results

Our genetic experiments have been carried out on the *B* cistron of the r_{II} region of the bacteriophage *T*4, which attacks strains of *Escherichia coli*. This is the system so brilliantly exploited by Benzer[6],[7]. The r_{II} region consists of two adjacent genes, or 'cistrons', called cistron *A* and cistron *B*. The wild-type phage will grow on both *E. coli B* (here called *B*) and on *E. coli K12* (λ) (here called *K*), but a phage which has lost the function of either gene will not grow on *K*. Such a phage produces an *r* plaque on *B*. Many point mutations of the genes are known which behave in this way. Deletions of part of the region are also found. Other mutations, known as 'leaky', show partial function; that is, they will grow on *K* but their plaque-type on *B* is not truly wild. We report here our work on the mutant *P* 13 (now re-named *FC* 0) in the *B*1 segment of the *B* cistron. This mutant was originally produced by the action of proflavin[8].

We[9] have previously argued that acridines such as proflavin act as mutagens because they add or delete a base or bases. The most striking evidence in favour of this is that mutants produced by acridines are seldom 'leaky'; they are almost always completely lacking in the function of the gene. Since our note was published, experimental data from two sources have been added to our previous evidence: (1) we have examined a set of 126 r_{II} mutants made with acridine yellow; of these only 6 are leaky (typically about half the mutants made with base analogues are leaky); (2) Streisinger[10] has found that whereas mutants of the lysozyme of phage *T*4 produced by base-analogues are usually leaky, all lysozyme mutants produced by proflavin are negative, that is, the function is completely lacking.

If an acridine mutant is produced by, say, adding a base, it should revert to 'wild-type' by deleting a base. Our work on revertants of *FC* 0 shows that it usually

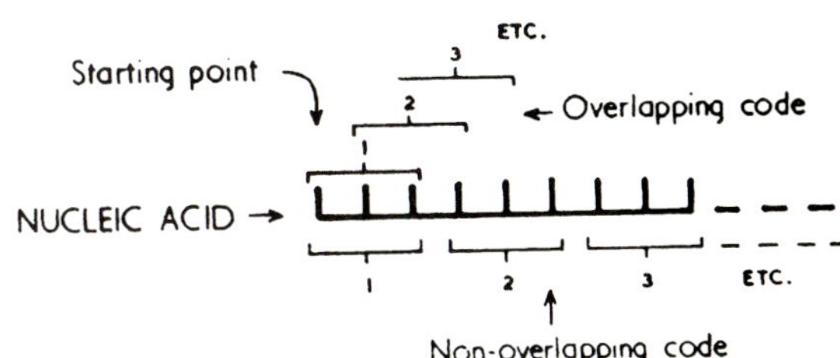

Fig. 1. To show the difference between an overlapping code and a non-overlapping code. The short vertical lines represent the bases of the nucleic acid. The case illustrated is for a triplet code

reverts not by reversing the original mutation but by producing a second mutation at a nearby point on the genetic map. That is, by a 'suppressor' in the same gene. In one case (or possibly two cases) it may have reverted back to true wild, but in at least 18 other cases the 'wild type' produced was really a double mutant with a 'wild' phenotype. Other workers[11] have found a similar phenomenon with r_{II} mutants, and Jinks[12] has made a detailed analysis of suppressors in the h_{III} gene.

The genetic map of these 18 suppressors of FC 0 is shown in Fig. 2, line a. It will be seen that they all fall in the B1 segment of the gene, though not all of them are very close to FC 0. They scatter over a region about, say, one-tenth the size of the B cistron. Not all are at different sites. We have found eight sites in all, but most of them fall into or near two close clusters of sites.

In all cases the suppressor was a non-leaky r. That is, it gave an r plaque on B and would not grow on K. This is the phenotype shown by a complete deletion of the gene, and shows that the function is lacking. The only possible exception was one case where the suppressor appeared to back-mutate so fast that we could not study it.

Each suppressor, as we have said, fails to grow on K. Reversion of each can therefore be studied by the same procedure used for FC 0. In a few cases these mutants apparently revert to the original wild-type, but usually they revert by forming a double mutant. Fig. 2, lines b–g, shows the mutants pro-

duced as suppressors of these suppressors. Again all these new suppressors are non-leaky r mutants, and all map within the B1 segment for one site in the B2 segment.

Once again we have repeated the process on two of the new suppressors, with the same general results, as shown in Fig. 2, lines i and j.

All these mutants, except the original FC 0, occurred spontaneously. We have, however, produced one set (as suppressors of FC 7) using acridine yellow as a mutagen. The spectrum of suppressors we get (see Fig. 2, line h) is crudely similar to the spontaneous spectrum, and all the mutants are non-leaky r's. We have also tested a (small) selection of all our mutants and shown that their reversion-rates are increased by acridine yellow.

Thus in all we have about eighty independent r mutants, all suppressors of FC 0, or suppressors of suppressors, or suppressors of suppressors of suppressors. They all fall within a limited region of the gene and they are all non-leaky r mutants.

The double mutants (which contain a mutation plus its suppressor) which plate on K have a variety of plaque types on B. Some are indistinguishable from wild, some can be distinguished from wild with difficulty, while others are easily distinguishable and produce plaques rather like r.

We have checked in a few cases that the phenomenon is quite distinct from 'complementation', since the two mutants which separately are phenotypically r, and together are wild or pseudo-wild,

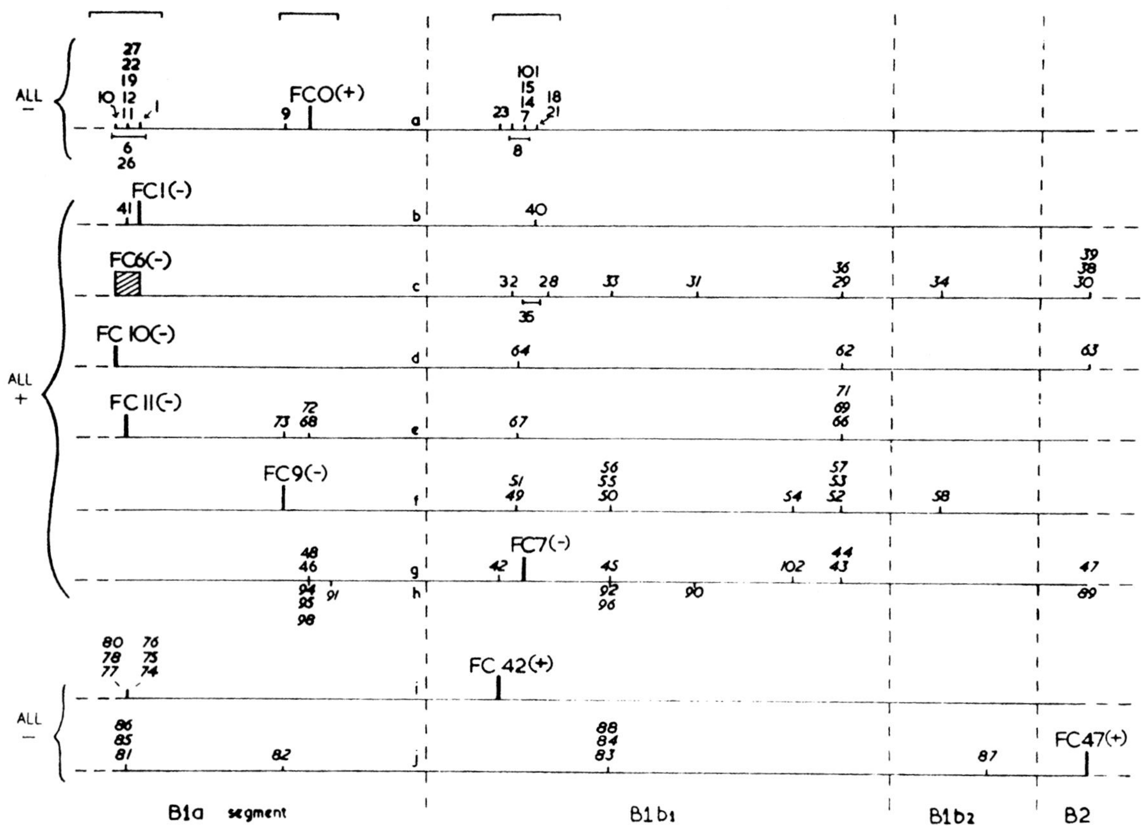

Fig. 2. A tentative map—only very roughly to scale—of the left-hand end of the B cistron, showing the position of the FC family of mutants. The order of sites within the regions covered by brackets (at the top of the figure) is not known. Mutants in italics have only been located approximately. Each line represents the suppressors picked up from one mutant, namely, that marked on the line in bold figures

must be put together in the same piece of genetic material. A simultaneous infection of K by the two mutants in separate viruses will not do.

The Explanation in Outline

Our explanation of all these facts is based on the theory set out at the beginning of this article. Although we have no direct evidence that the B cistron produces a polypeptide chain (probably through an RNA intermediate), in what follows we shall assume this to be so. To fix ideas, we imagine that the string of nucleotide bases is read, triplet by triplet, from a starting point on the left of the B cistron. We now suppose that, for example, the mutant FC 0 was produced by the insertion of an additional base in the wild-type sequence. Then this addition of a base at the FC 0 site will mean that the reading of all the triplets to the right of FC 0 will be shifted along one base, and will therefore be incorrect. Thus the amino-acid sequence of the protein which the B cistron is presumed to produce will be completely altered from that point onwards. This explains why the function of the gene is lacking. To simplify the explanation, we now postulate that a suppressor of FC 0 (for example, FC 1) is formed by deleting a base. Thus when the FC 1 mutation is present by itself, all triplets to the right of FC 1 will be read incorrectly and thus the function will be absent. However, when both mutations are present in the same piece of DNA, as in the pseudo-wild double mutant FC (0 + 1), then although the reading of triplets between FC 0 and FC 1 will be altered, the original reading will be restored to the rest of the gene. This could explain why such double mutants do not always have a true wild phenotype but are often pseudo-wild, since on our theory a small length of their amino-acid sequence is different from that of the wild-type.

For convenience we have designated our original mutant FC 0 by the symbol + (this choice is a pure convention at this stage) which we have so far considered as the addition of a single base. The suppressors of FC 0 have therefore been designated −. The suppressors of those suppressors have in the same way been labelled as +, and the suppressors of these last sets have again been labelled − (see Fig. 2).

Double Mutants

We can now ask: What is the character of any double mutant we like to form by putting together in the same gene any pair of mutants from our set of about eighty? Obviously, in some cases we already know the answer, since some combinations of a + with a − were formed in order to isolate the mutants. But, by definition, no pair consisting of one + with another + has been obtained in this way, and there are many combinations of + with − not so far tested.

Now our theory clearly predicts that all combinations of the type + with + (or − with −) should give an r phenotype and not plate on K. We have put together 14 such pairs of mutants in the cases listed in Table 1 and found this prediction confirmed.

Table 1. DOUBLE MUTANTS HAVING THE r PHENOTYPE

− With −	+ With +	
FC (1 + 21)	FC (0 + 58)	FC (40 + 57)
FC (23 + 21)	FC (0 + 38)	FC (40 + 58)
FC (1 + 23)	FC (0 + 40)	FC (40 + 55)
FC (1 + 9)	FC (0 + 55)	FC (40 + 54)
	FC (0 + 54)	FC (40 + 38)

At first sight one would expect that all combinations of the type (+ with −) would be wild or pseudo-wild, but the situation is a little more intricate than that, and must be considered more closely. This springs from the obvious fact that if the code is made of triplets, any long sequence of bases can be read correctly in one way, but incorrectly (by starting at the wrong point) in two different ways, depending whether the 'reading frame' is shifted one place to the right or one place to the left.

If we symbolize a shift, by one place, of the reading frame in one direction by → and in the opposite direction by ←, then we can establish the convention that our + is always at the head of the arrow, and our − at the tail. This is illustrated in Fig. 3.

We must now ask: Why do our suppressors not extend over the whole of the gene? The simplest postulate to make is that the shift of the reading frame produces some triplets the reading of which is 'unacceptable'; for example, they may be 'nonsense', or stand for 'end the chain', or be unacceptable in some other way due to the complications of protein structure. This means that a suppressor of, say, FC 0 must be within a region such that no 'unacceptable' triplet is produced by the shift in the reading frame between FC 0 and its suppressor. But, clearly, since for any sequence there are *two* possible misreadings, we might expect that the 'unacceptable' triplets produced by a → shift would occur in different places on the map from those produced by a ← shift.

Examination of the spectra of suppressors (in each case putting in the arrows → or ←) suggests that while the → shift is acceptable anywhere within our region (though not outside it) the shift ←, starting from points near FC 0, is acceptable over only a more limited stretch. This is shown in Fig. 4. Somewhere in the left part of our region, between FC 0 or FC 9 and the FC 1 group, there must be one or more unacceptable triplets when a ← shift is made; similarly for

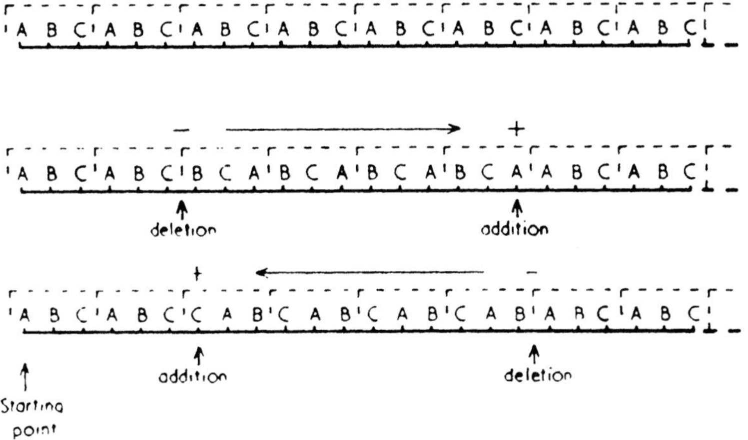

Fig. 3. To show that our convention for arrows is consistent. The letters A, B and C each represent a different base of the nucleic acid. For simplicity a repeating sequence of bases, ABC, is shown. (This would code for a polypeptide for which every amino-acid was the same.) A triplet code is assumed. The dotted lines represent the imaginary 'reading frame' implying that the sequence is read in sets of three starting on the left

the region to the right of the *FC* 21 cluster. Thus we predict that a combination of a + with a − will be wild or pseudo-wild if it involves a → shift, but that such pairs involving a ← shift will be phenotypically *r* if the arrow crosses one or more of the forbidden places, since then an unacceptable triplet will be produced.

Table 2. Double Mutants of the Type (+ with −)

	FC 41	*FC* 0	*FC* 40	*FC* 42	*FC* 58*	*FC* 63	*FC* 38
FC 1	*W*	*W*	*W*		*W*		*W*
FC 86		*W*	*W*	*W*	*W*	*W*	
FC 9	*r*	*W*	*W*	*W*	*W*		*W*
FC 82	*r*		*W*	*W*	*W*	*W*	
FC 21	*r*	*W*			*W*		*W*
FC 88	*r*	*r*			*W*	*W*	
FC 87	*r*	*r*	*r*	*r*			*W*

W, wild or pseudo-wild phenotype; *W*, wild or pseudo-wild combination used to isolate the suppressor; *r*, *r* phenotype.
* Double mutants formed with *FC* 58 (or with *FC* 34) give sharp plaques on *K*.

We have tested this prediction in the 28 cases shown in Table 2. We expected 19 of these to be wild, or pseudo-wild, and 9 of them to have the *r* phenotype. In all cases our prediction was correct. We regard this as a striking confirmation of our theory. It may be of interest that the theory was constructed before these particular experimental results were obtained.

Rigorous Statement of the Theory

So far we have spoken as if the evidence supported a triplet code, but this was simply for illustration. Exactly the same results would be obtained if the code operated with groups of, say, 5 bases. Moreover, our symbols + and − must not be taken to mean literally the addition or subtraction of a single base.

It is easy to see that our symbolism is more exactly as follows:

$$+ \text{ represents } +m, \text{ modulo } n$$
$$- \text{ represents } -m, \text{ modulo } n$$

where n (a positive integer) is the coding ratio (that is, the number of bases which code one amino-acid) and m is any integral number of bases, positive or negative.

It can also be seen that our choice of reading direction is arbitrary, and that the same results (to a first approximation) would be obtained in whichever direction the genetic material was read, that is, whether the starting point is on the right or the left of the gene, as conventionally drawn.

Triple Mutants and the Coding Ratio

The somewhat abstract description given above is necessary for generality, but fortunately we have convincing evidence that the coding ratio is in fact 3 or a multiple of 3.

This we have obtained by constructing triple mutants of the form (+ with + with +) or (− with − with −). One must be careful not to make shifts

Table 3. Triple Mutants having a Wild or Pseudo-wild Phenotype

FC (0 + 40 + 38)
FC (0 + 40 + 58)
FC (0 + 40 + 57)
FC (0 + 40 + 54)
FC (0 + 40 + 55)
FC (1 + 21 + 23)

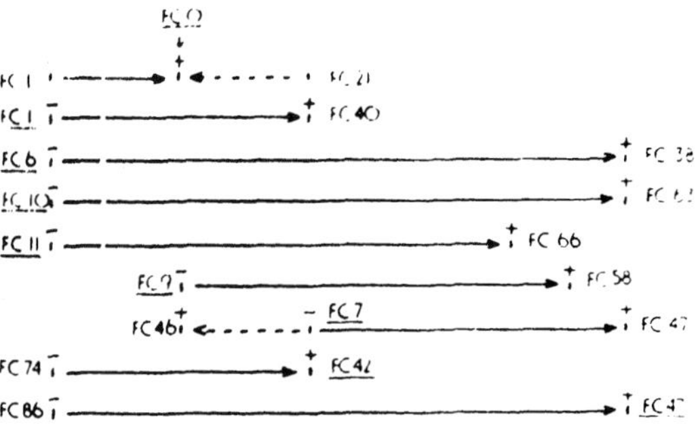

Fig. 4. A simplified version of the genetic map of Fig. 2. Each line corresponds to the suppressor from one mutant, here underlined. The arrows show the range over which suppressors have so far been found, the extreme mutants being named on the map. Arrows to the right are shown solid, arrows to the left dotted

across the 'unacceptable' regions for the ← shifts, but these we can avoid by a proper choice of mutants.

We have so far examined the six cases listed in Table 3 and in all cases the triples are wild or pseudo-wild.

The rather striking nature of this result can be seen by considering one of them, for example, the triple (*FC* 0 with *FC* 40 with *FC* 38). These three mutants are, by themselves, all of like type (+). We can say this not merely from the way in which they were obtained, but because each of them, when combined with our mutant *FC* 9 (−), gives the wild, or pseudo-wild phenotype. However, either singly or together in pairs they have an *r* phenotype, and will not grow on *K*. That is, the function of the gene is absent. Nevertheless, the combination of all three in the same gene partly restores the function and produces a pseudo-wild phage which grows on *K*.

This is exactly what one would expect, in favourable cases, if the coding ratio were 3 or a multiple of 3.

Our ability to find the coding ratio thus depends on the fact that, in at least one of our composite mutants which are 'wild', at least one amino-acid must be added to or deleted from the polypeptide chain without disturbing the function of the gene-product too greatly.

This is a very fortunate situation. The fact that we can make these changes and can study so large a region probably comes about because this part of the protein is not essential for its function. That this is so has already been suggested by Champe and Benzer[13] in their work on complementation in the r_{II} region. By a special test (combined infection on *K*, followed by plating on *B*) it is possible to examine the function of the *A* cistron and the *B* cistron separately. A particular deletion, 1589 (see Fig. 5) covers the right-hand end of the *A* cistron and part of the left-hand end of the *B* cistron. Although 1589 abolishes the *A* function, they showed that it allows the *B* function to be expressed to a considerable extent. The region of the *B* cistron deleted by 1589 is that into which all our *FC* mutants fall.

Joining two Genes Together

We have used this deletion to re-inforce our idea that the sequence is read in groups from a fixed starting point. Normally, an alteration confined to the *A* cistron (be it a deletion, an acridine mutant, or any other mutant) does not prevent the expression of the *B* cistron. Conversely, no alteration within the *B* cistron prevents the function of the *A* cistron. This implies that there may be a region between the two cistrons which separates them and allows their functions to be expressed individually.

We argued that the deletion 1589 will have lost this separating region and that therefore the two (partly damaged) cistrons should have been joined together. Experiments show this to be the case, for now an alteration to the left-hand end of the A cistron, if combined with deletion 1589, can prevent the B function from appearing. This is shown in Fig. 5. Either the mutant P43 or X142 (both of which revert strongly with acridines) will prevent the B function when the two cistrons are joined, although both of these mutants are in the A cistron. This is also true of X142 S1, a suppressor of X142 (Fig. 5, case b). However, the double mutant (X142 with X142 S1), of the type (+ with −), which by itself is pseudo-wild, still has the B function when combined with 1589 (Fig. 5, case c). We have also tested in this way the 10 deletions listed by Benzer[7], which fall wholly to the left of 1589. Of these, three (386, 168 and 221) prevent the B function (Fig. 5, case f), whereas the other seven show it (Fig. 5, case e). We surmise that each of these seven has lost a number of bases which is a multiple of 3. There are theoretical reasons for expecting that deletions may not be random in length, but will more often have lost a number of bases equal to an integral multiple of the coding ratio.

It would not surprise us if it were eventually shown that deletion 1589 produces a protein which consists of part of the protein from the A cistron and part of that from the B cistron, joined together in the same polypeptide chain, and having to some extent the function of the undamaged B protein.

Is the Coding Ratio 3 or 6 ?

It remains to show that the coding ratio is probably 3, rather than a multiple of 3. Previous rather rough extimates[10,14] of the coding ratio (which are admittedly very unreliable) might suggest that the coding ratio is not far from 6. This would imply, on our theory, that the alteration in FC 0 was not to one base, but to two bases (or, more correctly, to an even number of bases).

We have some additional evidence which suggests that this is unlikely. First, in our set of 126 mutants produced by acridine yellow (referred to earlier) we have four independent mutants which fall at or

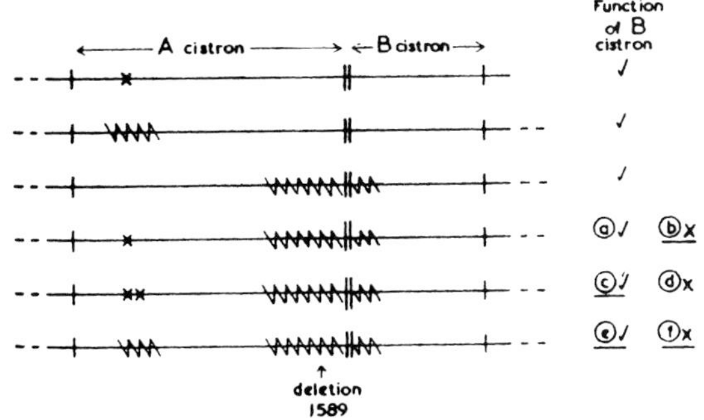

Fig. 5. Summary of the results with deletion 1589. The first two lines show that without 1589 a mutation or a deletion in the A cistron does not prevent the B cistron from functioning. Deletion 1589 (line 3) also allows the B cistron to function. The other cases, in some of which an alteration in the A cistron prevents the function of the B cistron (when 1589 is also present), are discussed in the text. They have been labelled (a), (b), etc., for convenience of reference, although cases (a) and (d) are not discussed in this paper. ✓ implies function; × implies no function

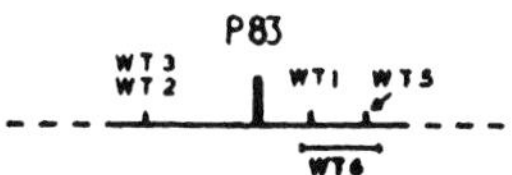

Fig. 6. Genetic map of P 83 and its suppressors, WT 1, etc. The region falls within segment B9a near the right-hand end of the B cistron. It is not yet known which way round the map is in relation to the other figures

close to the FC 9 site. By a suitable choice of partners, we have been able to show that two are + and two are −. Secondly, we have two mutants (X146 and X225), produced by hydrazine[15], which fall on or near the site FC 30. These we have been able to show are both of type −.

Thus unless both acridines and hydrazine usually delete (or add) an even number of bases, this evidence supports a coding ratio of 3. However, as the action of these mutagens is not understood in detail, we cannot be certain that the coding ratio is not 6, although 3 seems more likely.

We have preliminary results which show that other acridine mutants often revert by means of close suppressors, but it is too sketchy to report here. A tentative map of some suppressors of P 83, a mutant at the other end of the B cistron, in segment B 9a, is shown in Fig. 6. They occur within a shorter region than the suppressors of FC 0, covering a distance of about one-twentieth of the B cistron. The double mutant WT (2 + 5) has the r phenotype, as expected.

Is the Code Degenerate?

If the code is a triplet code, there are 64 ($4 \times 4 \times 4$) possible triplets. Our results suggest that it is unlikely that only 20 of these represent the 20 amino-acids and that the remaining 44 are nonsense. If this were the case, the region over which suppressors of the FC 0 family occur (perhaps a quarter of the B cistron) should be very much smaller than we observe, since a shift of frame should then, by chance, produce a nonsense reading at a much closer distance. This argument depends on the size of the protein which we have assumed the B cistron to produce. We do not know this, but the length of the cistron suggests that the protein may contain about 200 amino-acids. Thus the code is probably 'degenerate', that is, in general more than one triplet codes for each amino-acid. It is well known that if this were so, one could also account for the major dilemma of the coding problem, namely, that while the base composition of the DNA can be very different in different micro-organisms, the amino-acid composition of their proteins only changes by a moderate amount[16]. However, exactly how many triplets code amino-acids and how many have other functions we are unable to say.

Future Developments

Our theory leads to one very clear prediction. Suppose one could examine the amino-acid sequence of the 'pseudo-wild' protein produced by one of our double mutants of the (+ with −) type. Conventional theory suggests that since the gene is only altered in two places, only two amino-acids would be changed. Our theory, on the other hand, predicts that a string of amino-acids would be altered, covering the region of the polypeptide chain corresponding to the region on the gene between the two mutants. A good protein on which to test this hypothesis is the lysozyme of the phage, at present being studied chemically by Dreyer[17] and genetically by Streisinger[10].

At the recent Biochemical Congress at Moscow, the audience of Symposium I was startled by the announcement of Nirenberg that he and Matthaei[18] had produced polyphenylalanine (that is, a polypeptide all the residues of which are phenylalanine) by adding polyuridylic acid (that is, an RNA the bases of which are all uracil) to a cell-free system which can synthesize protein. This implies that a sequence of uracils codes for phenylalanine, and our work suggests that it is probably a triplet of uracils.

It is possible by various devices, either chemical or enzymatic, to synthesize polyribonucleotides with defined or partly defined sequences. If these, too, will produce specific polypeptides, the coding problem is wide open for experimental attack, and in fact many laboratories, including our own, are already working on the problem. If the coding ratio is indeed 3, as our results suggest, and if the code is the same throughout Nature, then the genetic code may well be solved within a year.

We thank Dr. Alice Orgel for certain mutants and for the use of data from her thesis, Dr. Leslie Orgel for many useful discussions, and Dr. Seymour Benzer for supplying us with certain deletions. We are particularly grateful to Prof. C. F. A. Pantin for allowing us to use a room in the Zoological Museum, Cambridge, in which the bulk of this work was done.

[1] Wittman, H. G., Symp. 1, Fifth Intern. Cong. Biochem., 1961, for refs. (in the press).

[2] Tsugita, A., and Fraenkel-Conrat, H., *Proc. U.S. Nat. Acad. Sci.*, **46**, 636 (1960); *J. Mol. Biol.* (in the press)

[3] Brenner, S., *Proc. U.S. Nat. Acad. Sci.*, **43**, 687 (1957).

[4] For refs. see Watson, H. C., and Kendrew, J. C., *Nature*, **190**, 670 (1961).

[5] Crick, F. H. C., Griffith, J. S., and Orgel, L. E., *Proc. U.S. Nat. Acad. Sci.*, **43**, 416 (1957).

[6] Benzer, S., *Proc. U.S. Nat. Acad. Sci.*, **45**, 1607 (1959), for refs. to earlier papers.

[7] Benzer, S., *Proc. U.S. Nat. Acad. Sci.*, **47**, 403 (1961); see his Fig. 3.

[8] Brenner, S., Benzer, S., and Barnett, L., *Nature*, **182**, 983 (1958).

[9] Brenner, S., Barnett, L., Crick, F. H. C., and Orgel, A., *J. Mol. Biol.*, **3**, 121 (1961).

[10] Streisinger, G. (personal communication and in the press).

[11] Feynman, R. P.; Benzer, S.; Freese, E. (all personal communications).

[12] Jinks, J. L., *Heredity*, **16**, 153, 241 (1961).

[13] Champe, S., and Benzer, S. (personal communication and in preparation).

[14] Jacob, F., and Wollman, E. L., *Sexuality and the Genetics of Bacteria* (Academic Press, New York, 1961). Levinthal, C. (personal communication).

[15] Orgel, A., and Brenner, S. (in preparation).

[16] Sueoka, N. *Cold Spring Harb. Symp. Quant. Biol.* (in the press).

[17] Dreyer, W. J., Symp. 1, Fifth Intern. Cong. Biochem., 1961 (in the press).

[18] Nirenberg, M. W., and Matthaei, J. H., *Proc. U.S. Nat. Acad. Sci.*, **47**, 1588 (1961).

On the Colinearity of Gene Structure and Protein Structure

C. YANOFSKY, B. C. CARLTON, J. R. GUEST, D. R. HELINSKI, AND U. HENNING

In the late 1950s, much thought and research were focused on deciphering the genetic code, determining how proteins are synthesized, and explaining how genetic information is translated into proteins. Whether or not the linear nucleotide sequence of a gene corresponds to the linear amino acid sequence of a polypeptide, "colinearity" was of major concern. But how could this relationship be examined experimentally? Specific genes could not be isolated, much less sequenced. Similarly, detecting single amino acid changes in a set of mutant proteins seemed beyond the realm of possibility. Benzer's studies in the late 1950s established that a fine-structure genetic map could suffice as a representation of the nucleotide sequence of a gene. Further, with the development of the peptide "fingerprinting" technique by Vernon Ingram in 1958, there was the hope that single amino acid changes could be detected in some proteins (V. E. Ingram, *Biochim. Biophys. Acta* **28:**539–545, 1958).

Colinearity of gene structure and protein structure was demonstrated in the early 1960s by my group and by Sydney Brenner and his co-workers (A. S. Sarabhai, A. O. W. Stretton, S. Brenner, and A. Bolle, *Nature* **201:**13–17, 1964). Interestingly, the types of mutants examined and the experimental approaches used by the two groups were quite different. Our studies dealt with missense mutants and determination of the single amino acid changes in the corresponding protein. Brenner's group examined amber nonsense mutants and demonstrated changes in the length of a prematurely terminated protein. The unique feature of the protein we selected for study, the tryptophan synthetase α subunit of *Escherichia coli,* was its role as a component of an enzyme complex. Mutant missense α subunits could be readily recognized because they activated the β subunit of the enzyme. The positions of amino acid changes in purified missense proteins were then related to a fine structure genetic map of the *trpA* gene. Sydney Brenner's group solved colinearity by using amber nonsense mutants altered in the head protein of bacteriophage T4. They exploited the finding that, during the late stages of T4 infection, >50% of the protein synthesized by the host is head protein. Head proteins produced by amber mutants were isotopically labeled and then digested with trypsin or chymotrypsin, and the labeled peptides were separated by electrophoresis. Scoring the presence or absence of each peptide allowed the linear ordering of the terminated proteins with respect to the genetic map. We were all delighted to find that nature had not played one of its masterful tricks. Imagine our surprise when mRNA splicing was discovered!

CHARLES YANOFSKY

Reprinted from *Proceedings of the National Academy of Sciences USA* 51:266–272. Copyright © 1964, by permission of the authors.

Reprinted from the PROCEEDINGS OF THE NATIONAL ACADEMY OF SCIENCES
Vol. 51, No. 2, pp. 266–272. February, 1964.

ON THE COLINEARITY OF GENE STRUCTURE AND PROTEIN STRUCTURE*

BY C. YANOFSKY, B. C. CARLTON,† J. R. GUEST,‡ D. R. HELINSKI,†
AND U. HENNING†

DEPARTMENT OF BIOLOGICAL SCIENCES, STANFORD UNIVERSITY

Communicated by Victor Twitty, December 18, 1963

The pioneering studies of Beadle and Tatum with *Neurospora crassa*[1] led to the concept that there is a 1:1 relationship between gene and enzyme. Subsequent studies on the structure of proteins[2] and genetic material[3] permitted a restatement of this relationship in molecular terms;[4] the linear sequence of nucleotides in a gene specifies the linear sequence of amino acids in a protein.

Several years ago studies were initiated with the A gene–A protein system of the tryptophan synthetase of *Escherichia coli* with the intention of examining this concept of a colinear relationship between gene structure and protein structure. A large number of mutant strains which produced altered A proteins were isolated, and genetic and protein primary structure studies were performed with these strains to locate the positions of the alterations within the A gene and the A protein.[5–10] It was hoped that with information of this type it would be possible to determine whether a genetic map and the primary structure of the corresponding protein were colinear. Recently, Kaiser[11] has demonstrated the correspondence of the genetic map with the sequence of blocks of nucleotides in DNA. Thus if a colinear relationship could be established between a genetic map and the primary structure of a protein, it would be reasonable to conclude that this relationship extends to the nucleotide sequence corresponding to the genetic map.

In previous reports on studies with the tryptophan synthetase A protein, conclusive evidence was presented for the colinearity of a segment of the A gene and a segment of the A protein.[12, 13] The present communication deals with more extensive data with 16 mutants with mutational alterations in one segment of the A gene and the A protein.

Materials and Methods.—Mutant strains: Of the A-protein mutants examined in detail in this paper, strains A23, A27, A28, A36, A46, A58, A78, A90, A94, A95, A169, A178, and A187 were isolated following ultraviolet irradiation of the K-12 wild-type strain of *E. coli*, and strain A223 was isolated following treatment of the wild-type strain with ethylmethanesulfonate. Mutant A446 (previously designated PR8)[14] and mutant A487 were initially isolated as spontaneous second-site reversions and subsequently were separated from the original A mutants with which they had been associated. Strains $anth_1^-$ and $anth_2^-$ are blocked prior to anthranilic acid in the tryptophan pathway and respond to anthranilic acid, indole, or tryptophan. V_1^R and V_1^R $tryp^-$ deletion mutants were isolated by treatment of T1-sensitive populations of the various mutants with phage T1h$^+$. All of the V_1^R mutations mentioned are very closely linked to the A gene, and the V_1^R $tryp^-$ deletions include the V_1^R locus and some segment of the A gene.[15] A stock of unrestricted T1 phage (uT1)[16] was kindly supplied by J. R. Christensen.

Protein studies: The altered A proteins were isolated and examined for primary structure changes as described previously.[6, 7, 9] The ordering of the tryptic peptides mentioned in the paper will be described in detail elsewhere.[17, 18]

Genetic studies: Recombination experiments were performed with the temperate-transducing phage P1kc.[19] Recombination distances between A mutants were obtained by determining the frequency of appearance of $tryp^+$ transductants. Transduction from $his^- \rightarrow his^+$ was scored in each experiment for internal reference, and the ratio of $tryp^+/his^+$ transductants calculated.[20]

Each value was halved to correct for the difference in relative frequency of transduction in the *his* and *tryp* regions.[20] In transduction experiments with leaky mutants (A169, A223, A446, and A487) the plating medium was supplemented with 0.1 µg/ml DL-5-methyltryptophan to suppress growth of the leaky mutants. This supplement has little or no effect on the growth of wild-type recombinants and does not appear to affect the recombination values obtained with nonleaky mutants. In spite of the presence of 5-methyltryptophan, it was frequently difficult to score recombination in experiments with leaky mutants. Anthranilic acid requirement was scored either by picking and streaking or by replication to appropriate test media. Resistance to phage T1 (V_1^R) was scored by picking, streaking, and spot testing with phage uT1. As shown by Drexler and Christensen,[16] P1 lysogeny does not prevent the multiplication of uT1.

Results and Discussion.—Relative order of mutational alterations in the A gene: The genetic map based on recombination frequencies, deletion mapping, and three-point crosses is shown in Figure 1.

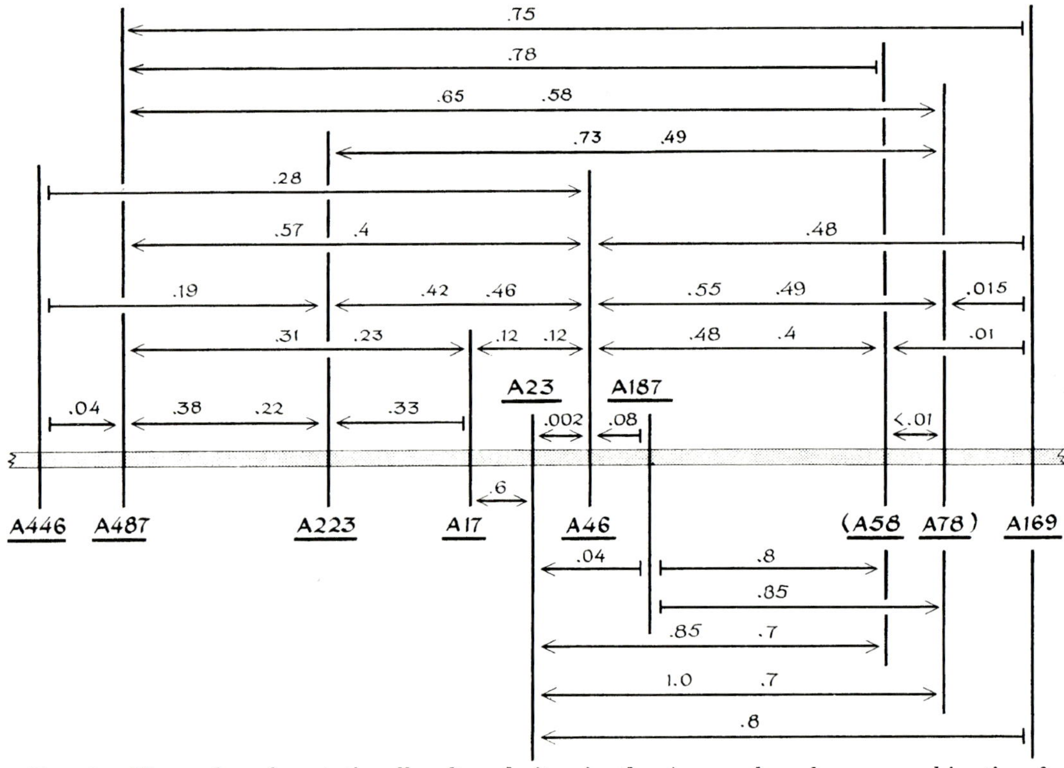

FIG. 1.—The order of mutationally altered sites in the A gene based on recombination frequencies, deletion mapping, and three-point genetic tests. The head of each arrow points to the recipient in each transduction cross. If two values are given, the cross was performed in both directions. In such cases each value is placed near the recipient in the cross.

Recombination frequency data: Recombination frequency data alone establishes close linkage of three groups of mutants: (1) A446 and A487; (2) A46, A23, and A187; and (3) A58, A78, and A169. The recombination values obtained also suggest that the order of these groups is as indicated in the figure, but they do not permit ordering within any of the groups.[21] In most cases recombination experiments were performed in both directions, i.e., each strain served as donor and recipient, and fair agreement was observed between the two values. However, it is evident from the data in Figure 1 that there were some exceptions. The exceptions generally involved leaky mutants, and it is possible that the scoring difficulties encountered with these strains were responsible.

Two of the mutants examined, A23 and A187, gave recombination values considerably higher than those obtained with mutant A46. Other mutants which were independent isolates resembling A23 (A27, A28, A36) and A46 (A95, A178) exhibited the same recombination behavior as the strain they resembled. Extensive mapping experiments with other A mutants and the strains mentioned above indicate that the A23 and A187 recombination values are probably exceptionally high rather than the A46 values exceptionally low. Since the mutational alterations in strains of the A23 and A46 type were probably single nucleotide changes,[22] it would appear that differences of single nucleotide pairs between donor DNA and recipient DNA can influence the frequency of recombinational events. An alternative explanation, that all A23 and A187 double mutants are prototrophic, seems very unlikely on the basis of numerous tests with recombinants from crosses involving A23 and A187.

Recombinants were not obtained in crosses of A23 × A27, A28, or A36; A46 × A95 or A178; or A58 × A90 or A94. On the basis of these tests and the primary structure studies to be described, it was concluded that each of these groups consists of members which arose by repeat identical mutations at the same site. Mutants A58 and A78 are clearly different, but they do not give recombinants at the 0.01 per cent level.

Deletion mapping: Information on the order of the mutational alterations in the relevant mutant strains was obtained in transduction experiments with a series of $V_1{}^R$-*tryp*$^-$ deletion mutants (Table 1). These latter strains had regions of the A gene deleted, including, in each case, the V_1 locus which is situated at one end of the A gene.[15] Thus all the deletions extended into the A gene from the same side, and are overlapping, but may have different end points in the A gene. The recovery of tryptophan-independent recombinants from a transduction cross between any A mutant and a deletion mutant would indicate that the mutationally altered site in the A mutant was outside the region of the A gene that was missing in the deletion mutant. Recombination values were determined in transduction crosses with some of the deletion mutants to approximate the relative distance from a mutationally altered site in an A mutant to the end point of a deletion.

It is clear from the crosses with T$^-$70 and T$^-$689 that the altered sites in mutants A23, A46, A187, A58, A78, and A169—but not those in A446, A487, and A223—are within the region of the A gene that is missing in these deletion mutants. The rela-

TABLE 1

RECOMBINATION TESTS WITH VARIOUS TRYP$^-$ DELETION MUTANTS

| | *his*$^-V_1{}^R$ *tryp*$^-$ Deletion Mutant Recipient | | | | | |
Donor	T$^-$201	T$^-$5	T$^-$70	T$^-$689	T$^-$211	T$^-$226
A446	0	0	0.08‡	0.08‡		
A487	0	0	0.19*	0.5		
A223	0	0	0.17	0.06	+	+
A23	0	0	0	0	1.3	+
A46	0	0	0	0	0.86	0.77
A187	0	0	0		1.3	+
A58	0	0	0	0	0.09	0.1
A78	0	0	0		0.1	0.1
A169			0		0.1	0.1
A38	+	+	+	+	+	+

+, 0 = recombinants or no recombinants, respectively, in qualitative transduction experiments.
* = each recombination value represents the uncorrected observed ratio of *tryp*$^+$ to *his*$^+$ transductants.
‡ A446 gives unusually low recombination values in all experiments.

tive order of the altered sites in mutants A446, A487, and A223 could not be established by quantitative transduction experiments with the same deletion mutants. The quantitative transduction experiments with deletion mutants T^--211 and T^--226 divide the other mutants into two groups: (A46, A23, and A187) and (A58, A78, and A169). The combined results presented suggest the following order of altered sites: A38......(A446, A487, A223)......(A46, A23, A187)......(A58, A78, A169)......V_1.

Three-point genetic tests: Genetic tests employing outside markers were performed to determine relative order within each group of closely linked mutants. These tests are summarized in Table 2. The results obtained support the conclusions concerning group order that were arrived at by the previous methods and indicate relative orders within each closely linked group. Crosses 1–7 establish the order anth......A34......A446......A487......A223......A46, crosses 8 and 9 the order anth ......A46......A78......A169, and crosses 10 and 11 the order anth......(A23, A46) ......A187. The order anth......A23-A46 had tentatively been assigned on the basis of other data.[22]   Crosses 12–15 confirm orders established by other methods. The combined genetic analyses with the mutants examined suggest the sequence of mutational alterations shown in Figure 1—A446-A487-A223-A23-A46-A187-(A58, A78)-A169.

TABLE 2
Outside-Marker Ordering of Mutationally Altered Sites

Transduction cross	Nonselective markers	Recombinants detected*		$anth^+$ (%)	Order
(1) 34 $\rightarrow anth_2^--223$	$anth_2$	72 $anth^+$;	332 $anth^-$	18	$anth$-34-223
(2) 46 $\rightarrow anth_2^--223$	$anth_2$	180 $anth^+$;	196 $anth^-$	48	$anth$-223-46
(3) 446 $\rightarrow anth_2^--223$	$anth_2$	14 $anth^+$;	112 $anth^-$	11	$anth$-446-223
(4) 487 $\rightarrow anth_2^--223$	$anth_2$	29 $anth^+$;	118 $anth^-$	20	$anth$-487-223
(5) 34 $\rightarrow anth_2^--487$	$anth_2$	86 $anth^+$;	368 $anth^-$	19	$anth$-34-487
(6) 46 $\rightarrow anth_2^--487$	$anth_2$	27 $anth^+$;	28 $anth^-$	49	$anth$-487-46
(7) 446 $\rightarrow anth_2^--487$	$anth_2$	7 $anth^+$;	31 $anth^-$	23	$anth$-446-487
(8) 46 $\rightarrow anth_2^--78$	$anth_2$	52 $anth^+$;	220 anth$^-$	19	$anth$-46-78
(9) 169 $\rightarrow anth_2^--78$	$anth_2$	22 $anth^+$;	24 $anth^-$	48	$anth$-78-169
(10) 187 $\rightarrow anth_1^--23$	$anth_1$	9 $anth^-$;	42 $anth^+$	83	$anth$-23-187
(11) 187 $\rightarrow anth_1^--46$	$anth_1$	14 $anth^-$;	116 $anth^+$	89	$anth$-46-187
(12) 46 $V_1^R \rightarrow anth_1^--58 V_1^S$	$anth_1$; V_1	22 $anth^- V_1^R$;	1 $anth^+ V_1^R$	4	$anth$-46-58-V_1
(13) 58 $V_1^R \rightarrow anth_1^--46 V_1^S$	$anth_1$; V_1	10 $anth^+ V_1^S$;	1 $anth^+ V_1^R$; 73 4 $anth^- V_1^S$		$anth$-46-58-V_1
(14) 169 $V_1^R \rightarrow anth_1^--46 V_1^S$	$anth_1$; V_1	6 $anth^+ V_1^S$;	1 $anth^- V_1^S$ 90		$anth$-46-169-V_1
(15) 446 $\rightarrow anth_1^--46$	$anth_1$	81 $anth^-$; 20 $anth^-$-446-46;	22 $anth^+$ 65 $anth^+$ 446$^-$-46	21 76	$anth$-446-46

Order of markers: $anth$-A34 . . . . (A446, etc.) - - - V_1^R.
* In crosses with the outside marker $anth_2$ the percentage of $anth^+$ recombinants is approximately 20% if the order is $\dfrac{+ \quad x}{anth^- \quad y}$ and approximately 50% if the order of x and y is reversed. With the marker $anth_1$ different values are obtained but the order of x and y relative to $anth$ can be clearly established. The explanation for the different values obtained with the two outside markers is not known.

Primary Structure Studies.—Amino acid substitutions have been detected in primary structure studies with each of the 16 A mutants.[6, 7, 14, 17, 18]   The substitutions observed and the peptides in which they are present are shown in Figure 2. The conclusions from the genetics studies are also included in the figure (i.e., order of alterations and approximation of recombination distances between alterations). It is apparent that mutants with alterations extremely close to one another in the A gene have amino acid substitutions close to one another in the A protein. Primary structure studies with the A protein[17, 18] have established the linear sequence

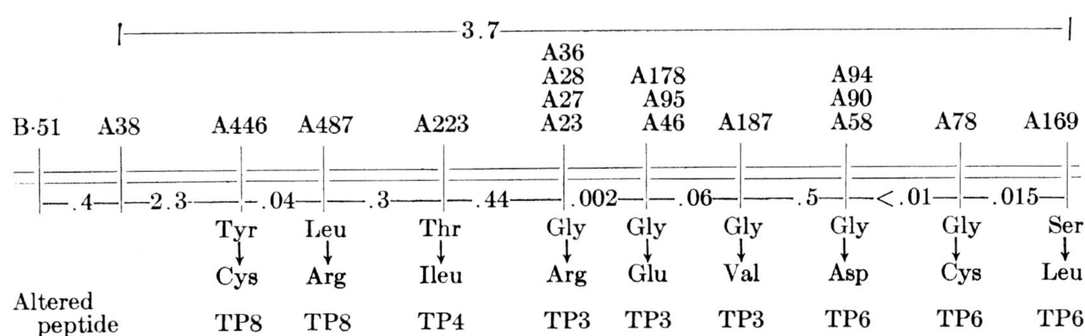

Fig. 2.—Amino acid substitutions in the A proteins of various mutants.[6, 7, 14, 17, 18] The A58, A78, A90, A94, and A169 substitutions will be described in detail elsewhere.[18]

of peptides TP-11, TP-8, TP-4, TP-18, TP-3, and TP-6, constituting a 75-residue segment of the A protein. This sequence and the sequence of the amino acids of most of these peptides are presented in Figure 3. This segment accounts for approximately one fourth of the residues in the A protein and is not at the amino or car-

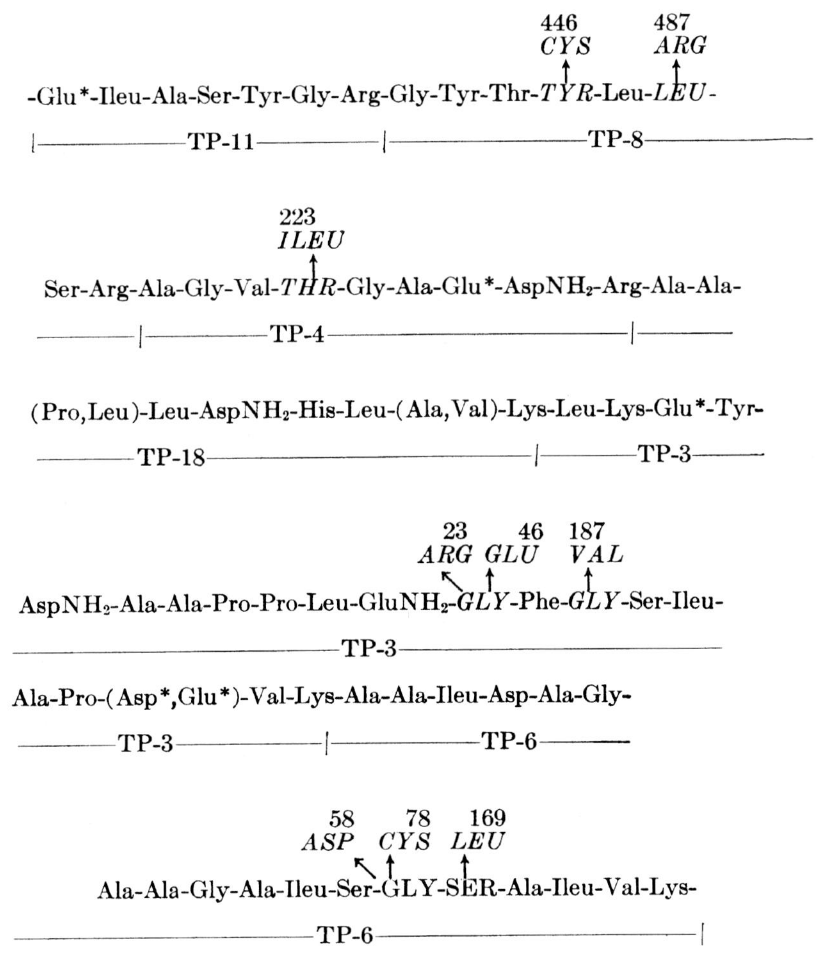

Fig. 3.—Peptide and partial amino acid sequence of a segment of the A protein. The positions of amino acid substitutions in the A proteins of the various mutants are indicated. Based on the studies by Carlton and Yanofsky[17] and Guest and Yanofsky.[18] (* = not known whether present as acid or amide.)

boxyl end of the protein.   The positions of the amino acid replacements in the A proteins of the various mutants are also shown.   It is clear that the positions of the amino acid replacements in the segment of the A protein are in the same relative order as the order of the mutationally altered sites of the corresponding mutants in the A gene.   These findings convincingly demonstrate a colinear relationship between gene structure and protein structure.

The relationship between the map and residue distances observed is also of interest and can be seen most clearly from the representative values summarized in Table 3.   Although the map distance/residue distance ratio varies between 0.01 and 0.05, in most cases this value is approximately 0.02.   It would appear, therefore, that distances on the genetic map are representative of distances between amino acid residues in the corresponding protein.

TABLE 3

RELATIONSHIP BETWEEN MAP DISTANCES AND RESIDUE DISTANCES

Mutant pair	Map distance	Amino acid residue distance	$\dfrac{\text{Map distance}}{\text{Residue distance}}$
A58-A78	<0.01	0	—
A46-A23	0.002	0	—
A58-A169	0.01	1	0.01
A78-A169	0.015	1	0.015
A46-A187	0.08	2	0.04
A23-A187	0.04	2	0.02
A446-A487	0.04	2	0.02
A487-A223	0.3	6	0.05
A446-A223	0.19	8	0.02
A46-A58	0.44	23	0.02
A46-A78	0.52	23	0.02
A23-A58	0.78	23	0.03
A23-A78	0.85	23	0.04
A46-A169	0.48	24	0.02
A23-A169	0.8	24	0.03
A223-A46	0.44	28	0.02
A487-A46	0.48	34	0.01
A446-A46	0.28	36	0 01
A78-A223	0.61	51	0.01
A169-A487	0.75	58	0.01

Summary.—The concept of colinearity of gene structure and protein structure was examined with 16 mutants with alterations in one segment of the A gene and the A protein of tryptophan synthetase.   The results obtained demonstrate a linear correspondence between the two structures and further show that genetic recombination values are representative of the distances between amino acid residues in the corresponding protein.

Note added in proof: Dr. Francis Crick has recently forwarded a manuscript which deals with a study of colinearity in another system (Sarabhai *et al.*, *Nature*, in press).

The authors are indebted to Virginia Horn for performing the genetic analyses described in this paper.   They are also indebted to Deanna Thorpe, Patricia Schroeder, Donald Vinicor, and John Horan for their excellent technical assistance.

* This investigation was supported by grants from the National Science Foundation and the U.S. Public Health Service.

† Present address: B. C. Carlton, Department of Biology, Yale University, New Haven, Connecticut; D. R. Helinski, Department of Biology, Princeton University, Princeton, New Jersey; U. Henning, Max-Planck Institut für Zellchemie, Munich, Germany.

‡ Guinness Research Fellow, on leave from Department of Biochemistry, Oxford, England.

[1] Beadle, G. W., and E. L. Tatum, these Proceedings, **27**, 499 (1941).

[2] Sanger, F., and H. Tuppy, *Biochem. J.*, **49**, 481 (1951).

[3] Watson, J. D., and F. H. C. Crick, in *Viruses*, Cold Spring Harbor Symposia on Quantitative Biology, vol. 18 (1953), p. 123.

[4] Crick, F. H. C., *Symp. Soc. Exptl. Biol.*, **12**, 138 (1958).

[5] Yanofsky, C., D. R. Helinski, and B. D. Maling, in *Cellular Regulatory Mechanisms*, Cold Spring Harbor Symposia on Quantitative Biology, vol. 26 (1961), p. 11.

[6] Helinski, D. R., and C. Yanofsky, these Proceedings, **48**, 173 (1962).

[7] Henning, U., and C. Yanofsky, these Proceedings, **48**, 183 (1962).

[8] *Ibid.*, 1497 (1962).

[9] Helinski, D. R., and C. Yanofsky, *Biochim. Biophys. Acta*, **63**, 10 (1962).

[10] Carlton, B. C., and C. Yanofsky, *J. Biol. Chem.*, **238**, 2390 (1963).

[11] Kaiser, A. D., *J. Mol. Biol.*, **4**, 275 (1962).

[12] Yanofsky, C., in *Synthesis and Structure of Macromolecules*, Cold Spring Harbor Symposia on Quantitative Biology, vol. 28 (1963), in press.

[13] Yanofsky, C., in *The Bacteria*, ed. I. C. Gunsalus and R. Y. Stanier (Academic Press), vol. 5, in press.

[14] Helinski, D. R., and C. Yanofsky, *J. Biol. Chem.*, **238**, 1043 (1963).

[15] Somerville, R., and C. Yanofsky, unpublished observations.

[16] Drexler, H., and J. R. Christensen, *Virology*, **13**, 31 (1961).

[17] Carlton, B. C., and C. Yanofsky, in preparation.

[18] Guest, J., and C. Yanofsky, in preparation.

[19] Lennox, E. S., *Virology*, **1**, 190 (1955).

[20] Yanofsky, C., and E. S. Lennox, *Virology*, **8**, 425 (1959).

[21] On the basis of preliminary two-point mapping data with mutant A2, it earlier had tentatively been concluded that the alterations in mutants A58, A78, A90, and A94 were to the left of the alterations in A17 and A46.[5] The more extensive data presented in Figure 1 and the deletion and three-point data indicate that the correct order is as shown in Figure 1.

[22] Yanofsky, C., in *Synthesis and Structure of Macromolecules*, Cold Spring Harbor Symposia on Quantitative Biology, vol. 28 (1963), in press.

Positive Control of Enzyme Synthesis by Gene C in the L-Arabinose System

E. ENGLESBERG, J. IRR, J. POWER, AND N. LEE

Negative control—the repressor-operator model for genetic regulation—held sway for many years as an account of the control of gene expression in general. The success of this account owes much to its simplicity, to the extent to which it offered explanation for a host of biological phenomena, and to its impressive articulation by Monod and Jacob. Many assumed that the problem of genetic regulation had been "solved." Yet, researchers often found themselves uncomfortably forcing data on their systems into a repressor-operator mold paradigm.

Years before the emergence of any other challenge to the dominance of the negative-control mechanism, Englesberg and his coworkers systematically accumulated evidence for a different mode of gene regulation: positive control. Using as a standard the genetic approaches of the Pasteur group—constitutive mutations and dominance-recessiveness relationships—they deduced the existence of an activator protein required for the expression of the genes determining arabinose metabolism in *Escherichia coli*. Despite their extensive studies documenting this positive control mechanism, however, the paradigmatic power of the negative-control model prevented rapid acceptance of a second genetic regulatory mechanism. The resistance to Englesberg's proposals may be further explained by the apparently complicated nature of the *ara* system, in contrast to the simple mode of control of *lac*; the *araC* gene product appeared to function as both an activator and a repressor.

By the late 1960s, however, the sheer weight of the evidence in the *ara* system and the appearance of other systems that shared its features established positive control as a bona fide mechanism. The floodgates were opened. Once an alternative to negative control became acceptable, discovery of numerous other regulatory mechanisms followed. Ironically, it is now positive control—control by activators—that appears, in fact, to have the broadest explanatory power, being the most commonly used mechanism in eucaryotic systems.

JONATHAN BECKWITH

JOURNAL OF BACTERIOLOGY, Oct., 1965
Copyright © 1965 American Society for Microbiology

Vol. 90, No. 4
Printed in U.S.A.

Positive Control of Enzyme Synthesis by Gene C in the L-Arabinose System

ELLIS ENGLESBERG,[1] JOSEPH IRR,[1] JOSEPH POWER,[1] AND NANCY LEE[1]

Department of Biology, University of Pittsburgh, Pittsburgh, Pennsylvania

Received for publication 13 May 1965

ABSTRACT

ENGLESBERG, ELLIS (University of Pittsburgh, Pittsburgh, Pa.), JOSEPH IRR, JOSEPH POWER, AND NANCY LEE. Positive control of enzyme synthesis by gene C in the L-arabinose system. J. Bacteriol **90**:946–957. 1965.—The L-arabinose gene complex consists of genes D, A, B, and C, linked in that order between the markers *thr* and *leu*, and an unlinked gene E. Genes D, A, B, and E are the structural genes for three inducible enzymes and permease, respectively. Gene C, with two mutant alleles, C^- and C^c, is the regulatory gene exhibiting positive and negative control. C^- mutants are deficient and C^c mutants are constitutive for all three enzymes and permease. Complementation analysis, employing sexual merozygotes ($A^-C^+ \times A^+C^-$), with six different C^- mutants, demonstrates that C^- is recessive to C^+ (positive control). A total of 61 C^c mutants, isolated as clones resistant to D-fucose inhibition, are linked to the *leu ara* region of the chromosome, and the 22 C^c mutants that were analyzed in detail mapped within the C gene among the C^- mutant sites. C^c mutants produce various but coordinate levels of the two enzymes measured, and permease. Complementation analysis ($A^-C^c \times A^+C^-$, $A^-C^c \times A^+C^+$) shows that C^c is dominant to C^- (positive control) and recessive to C^+ (negative control). Deletion mutants that extend into the C gene are L-arabinose permease-negative, thus supporting the positive regulatory role of the C gene. The name "activator gene" is proposed for genes of the C type to accentuate their positive role in gene expression. A working model consistent with these results is presented.

The C gene has been shown to be a new type of regulatory gene exhibiting positive control of the expression of genes A, B, D, and E, the structural genes for L-arabinose isomerase, L-ribulokinase, L-ribulose-5-phosphate 4-epimerase, and L-arabinose permease, respectively, in the L-arabinose gene complex (Gross and Englesberg, 1959; Englesberg and Kileen, 1959; Englesberg, 1961; Englesberg et al., 1962; Helling and Weinberg, 1963; Isaacson and Englesberg, 1964; Novotny and Englesberg, 1964). (Although no modified epimerase or permease protein has been detected, we are assuming that genes D and E are the structural genes for L-ribulose-5-phosphate 4-epimerase and L-arabinose permease, respectively, on the basis of physiological and enzymatic characteristics of the mutants concerned.) The C gene is linked to genes D, A, and B in the order DABC between the markers for a threonine (*thr*) and a leucine (*leu*) requirement. Gene E is unlinked to this area of the coli chromosome by transduction analysis, but is linked to *ser*-A and *arg*-1 (Fig. 1) (Isaacson and Englesberg, 1964; Englesberg and Lieberman, *unpublished data*).

L-Arabinose-negative mutations in the C gene (C^-) have a pleiotropic effect, leading to the inability to induce the synthesis of the enzymes listed above (Englesberg and Killeen, 1959; Englesberg, 1961) and the L-arabinose permease (Novotny and Englesberg, 1964). Complementation analysis has shown that the C gene (as defined by the C^- alleles) is separate and distinct from genes A, B, and D. The C^- allele, analyzed in mutant C^-19, was found to be recessive to C^+ (Helling and Weinberg, 1963). This evidence, together with the finding that the C gene controls the expression of the unlinked L-arabinose permease gene, E gene (Novotny and Englesberg, 1964), suggested that the C gene produced a cytoplasmic product essential for the expression of the structural genes in the L-arabinose gene complex.

This is an example of positive control of gene expression and is in sharp contrast to the negative or repressor control as has been primarily elaborated in the β-galactosidase system (Jacob and Monod, 1961b).

[1] Present address: Department of Biological Sciences, University of California, Santa Barbara.

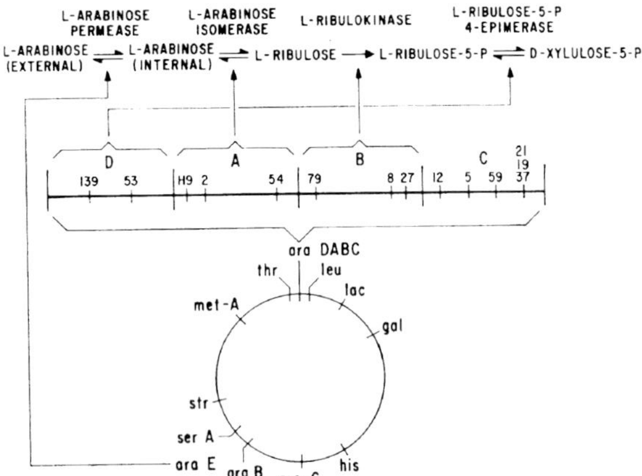

Fig. 1. *L-Arabinose gene-enzyme complex.*

In this paper, we shall show that complementation analysis with other C⁻ mutants confirms the above findings. In addition, we shall describe the isolation of constitutive mutants (Cᶜ) in this system, all of which map within the C gene among the C⁻ mutant sites (Englesberg, Irr, and Power, 1964). The results of an analysis of the interaction of Cᶜ with C⁻ and C⁺ in the synthesis of L-arabinose isomerase are in agreement with what would be expected from the previous findings, but indicate, in addition, that the C gene exerts a negative as well as a positive control over gene expression. Since a regulator gene has been defined as producing a substance that only inhibits the expression of a structural gene (Jacob and Monod, 1961a), we propose the name "activator gene" for the C gene to accentuate its positive role in this transfer of information from structural genes to protein.

MATERIALS AND METHODS

Nomenclature. Genes A, B, C, D, and E are defined as follows: A, L-arabinose isomerase gene; B, L-ribulokinase gene; C, regulatory gene (activator type); D, L-ribulose-5-phosphate 4-epimerase gene; E, L-arabinose permease gene.

A⁺, B⁺, D⁺, E⁺ or A, B, D, E indicates ability to produce the respective enzyme due to wild-type configuration of the gene involved.

A⁻, B⁻, D⁻, E⁻, usually followed by a number (B⁻8), indicates inability to produce the wild-type enzyme due to mutation in the respective structural gene. The number indicates the mutational site involved.

C⁺, C⁻, and Cᶜ are genetic markers for inducibility (wild type), inability to induce, and constitutive formation, respectively, of the first three metabolic enzymes and permease in L-arabinose metabolism. C⁻ and Cᶜ are usually followed by numbers (Cᶜ5) indicating the mutational site in the C gene.

Hfr indicates male type (Wollman, Jacob, and Hayes, 1956); F⁻, female type; *str*ˢ and *str*ʳ, genetic markers for sensitivity and resistance to streptomycin; *ara*⁺ and *ara*⁻, ability and inability to utilize L-arabinose as sole carbon and energy source for growth; *leu*, leucine; *thr*, threonine; and a minus superscript indicates inability to grow in the absence of the particular amino acid.

Strains of bacteria. Mutants derived from *Escherichia coli* B/r are: *ara*A⁻2, *ara*B⁻8, *ara*C⁻12, *ara*C⁻5 (Gross and Englesberg, 1959; Englesberg, 1961; Lee and Englesberg, 1962); *ara*D⁻139, *ara*D⁻53 (Englesberg et al., 1962); *ara*A⁻H9, *ara*A⁻54, *ara*B⁻79, *ara*B⁻27, *ara*C⁻37, *ara*C⁻47, *ara*C⁻59 (Cribbs and Englesberg, 1964); *ara*Cᶜ1 to *ara*Cᶜ50, constitutive mutants isolated from B/r *leu*⁻; *ara*D⁻53,204 and *ara*D⁻53,201 are deletion mutants to be described in Results. Various competent females (F⁻) were constructed from these strains as described below (see *Complementation*). The bacterial strains were constructed in *thr leu* and in either *thr leu*⁻ or *thr*⁻ *leu*⁻ genetic backgrounds (Gross and Englesberg, 1959).

E. coli B/r (hs) produces a heat-stable L-ribulokinase; *ara*Cᶜ60 to *ara*Cᶜ70 are constitutive mutants isolated from B/r (hs).

E. coli K-12, Hfr KH600 *leu*⁻ *str*ˢ, has the entire threonine, arabinose, leucine region replaced by that of *E. coli* B/r (Helling and Weinberg, 1963), and was employed in the construction of males with the appropriate arabinose markers as described below (see *Complementation*).

Transduction. Phage P1bt lysates (Boyer, Englesberg, and Weinberg, 1962), were used for transduction (Gross and Englesberg, 1959).

Media. The following media were used: mineral base with the carbon and energy sources and amino acids added as indicated (Gross and Englesberg, 1959); L-broth and EMB (Gross and Englesberg, 1959); nutrient agar and nutrient broth (Difco).

Enzyme assays. L-Arabinose isomerase and isocitric dehydrogenase were determined by the methods of Cribbs and Englesberg (1964), and L-ribulokinase by the method of Lee and Englesberg (1963).

Complementation. Appropriate arabinose mutations were transferred from *E. coli* B/r, where they were originally isolated, by transduction into KH600 males, and genotypes of the males were then verified by progeny tests (Helling and Weinberg, 1963). The females in these experiments were prepared from *E. coli* B/r by crossing into this strain the region of the K-12 chromosome linked to *thr* that controls K-12 modification and restriction (Boyer, 1964).

The actual procedures for complementation analysis were essentially those of Helling and Weinberg (1963), except that the male to female ratio of 1:1 was used and mating was performed on a slow-shaking water bath at 37 C.

In each cross, the female was *str*ʳ and the male *str*ˢ. After 50 min of mating, streptomycin was added (to 0.5 mg/ml). In some experiments, where indicated, L-arabinose was added after 10 min at a final concentration of 0.4%. Samples were then removed at stated times, and extracts were prepared for L-arabinose isomerase assays. The amount of protein present in the mating mixture was determined by turbidity measurements in a previously calibrated Klett-Summerson colorimeter [1 Klett unit (42 filter) = 1.18 μg/ml of protein (Lowry et al., 1951)]. The amount of protein per milliliter was multiplied by the specific activity of the cell-free extract to obtain units of enzyme present per milliliter of culture. Controls of males and females alone were maintained under the same conditions. The units of isomerase per milliliter in a mixture of the same composition of male and female as in the mating mixture were calculated from the separate control cultures and given as the control values in the results.

We assayed for isomerase activity in these crosses because the assay procedure for this enzyme was the simplest. Since genes BAD act as a genetic unit of coordinate expression, similar results would probably have been obtained if kinase (Helling and Weinberg, 1963; Lee and Englesberg, 1963) and epimerase (Lee and Englesberg, 1963) activities had been determined. The coordination of kinase and isomerase activities in the constitutive mutants, as presented in this paper, further supports this conclusion.

Cell suspension for permease assay and for preparation of cell-free extracts. L-Arabinose permease, L-arabinose isomerase, and L-ribulokinase, and, as controls, glucose permease and isocitric dehydrogenase, were determined on the same culture with each of 22 constitutive mutants in the following manner. Overnight cultures of each of the constitutive mutants grown in a mineral glycerol (0.1%) liquid medium with aeration were used to inoculate 200 ml of mineral glycerol (0.2%) medium contained in 500-ml Erlenmeyer flasks to approximately 6 × 10⁷ cells per milliliter. The cultures were aerated by agitation at 37 C and removed to an ice bath during the exponential phase of growth when the turbidity reached approximately 5.4 × 10⁸ cells per milliliter. A 15-ml amount of each culture was removed for permease assay, and the remainder was employed for the preparation of cell-free extracts (Englesberg, 1961). The 15-ml sample was centrifuged in the cold, and the cells were suspended in 7.5 ml of suspension fluid (KH₂PO₄ − K₂HPO₄, pH 7.0, 1.0%; MgSO₄·7H₂O, 0.01%), recentrifuged, and resuspended in the suspension fluid at a turbidity equivalent to 156 μg (dry weight) of bacteria per ml. This suspension was employed for both permease assays.

Since a few of the constitutive cultures are unstable (there appears to be a high selection pressure in some cases for C^+ and C^-), each culture employed in the assays for permease and metabolic enzymes was streaked on nutrient agar, and the colonies appearing on this medium were assayed by use of the filter-paper technique for their ability to utilize L-arabinose constitutively. A maximum of 9% "reversion" from C^c was found

with C^c8, and appropriate corrections were made in the assays.

Detection of constitutive mutants by a filter-paper assay. Whatman no. 2 filter paper (diameter, 8.7 cm) was impregnated with 1 ml of a solution containing 2% L-arabinose, 0.0025 M MnCl₂, and 1.0% streptomycin, and was air-dried. The streptomycin was added to prevent induction during the assay. The surface of a nutrient agar plate containing colonies to be tested was partially dried by exposure to flowing air at 37 C for 1 hr. The impregnated filter paper was placed over the nutrient agar and pressed down lightly to insure contact with the individual colonies. The lids were placed back on the dishes, and the dishes were incubated for 4 hr at 37 C. The filter paper was then removed and air-dried. To detect L-arabinose utilization by the individual colonies, the filter paper was dipped briefly into a developing solution containing equal quantities of 1 M NaOH in methanol and 2% triphenyltetrazolium chloride in methanol. The paper was hung in an oven at 100 C and removed as soon as a uniform red color developed. The position of the individual colonies can be readily noted due to the adherence of a portion of the colony on the filter paper. Nonconstitutive variants, unable to utilize L-arabinose under these conditions, yielded a red color similar to the background, whereas constitutive colonies left a white spot, indicating complete utilization of the L-arabinose. As a control, each agar plate assayed in this manner was previously inoculated at a defined area with a C^+ and C^c culture.

L-Arabinose permease. D-Xylose-*1*-C^{14} accumulation was employed as a measure of the L-arabinose permease (Novotny and Englesberg, 1964). The membrane-filter technique was used (Hoffee and Englesberg, 1962). A 2-ml amount of a previously washed and standardized cell suspension brought to 25 C was added to 0.5 ml of a reaction mixture consisting of the following at the given final concentrations: D-xylose-*1*-C^{14}, 10⁻³ M, specific activity, 1.75 × 10⁵ counts per min per μmole; chloramphenicol, 50 μg/ml; and glycerol, 0.1%. The reaction mixture was in a test tube at approximately an 18° angle from the horizontal in a New Brunswick shaking water bath at 25 C. Incubation was continued for 10 min to insure that the cells had reached the steady state of accumulation, and two 1-ml samples were removed and filtered onto each of two membrane filters. Each filter was washed with 1 ml of the suspending fluid (as previously described) maintained at 25 C. The membrane filters were air-dried, mounted with Vaseline on planchets, and counted in a Nuclear-Chicago Micromil thin-window Geiger counter. At least 1,000 counts were determined for each sample. Counts due to diffusion, based upon cells containing 80% water, were subtracted. Results of duplicate assays were averaged and expressed as micromoles of D-xylose accumulated per gram (dry weight) of cells.

Glucose permease. The glucose permease was assayed by use of α-methyl glucoside-C^{14} (Englesberg, Watson, and Hoffee, 1961). The procedure was similar to that described for the L-arabinose permease, except that α-methyl glucoside-C^{14} at 10^{-3} M (final concentration), specific activity, 8.1×10^4 counts per min per μmole, was employed in place of D-xylose.

RESULTS

Isolation of constitutive mutants (C^c). D-Fucose inhibits the growth of *E. coli* B/r in mineral L-arabinose medium, and selection occurs for mutants that are constitutive for the three metabolic enzymes and permease involved in the initial steps in L-arabinose metabolism. Beginning with a *leu⁻* parental strain, we isolated 50 independent L-arabinose constitutive mutants (C^c1 to C^c50) from late growth in mineral L-arabinose (0.04%)-D-fucose (0.08%) (MAF)-leucine liquid medium, in the following manner. An overnight nutrient broth culture of *E. coli* B/r *leu⁻* was diluted into fresh nutrient broth to yield 10^3 viable cells per milliliter. The broth suspension was divided into 50 separate 1-ml portions and was incubated overnight. Fifty tubes containing 2.5 ml of MAF-leucine were inoculated with 0.05 ml of these overnight cultures; the tubes were incubated at 37 C until they became turbid (2 to 7 days), and the contents were streaked onto nutrient agar. Colonies that appeared were tested by picking and streaking onto MAF-leucine agar containing 0.1% L-arabinose and 0.2% D-fucose. One colony from each of the original cultures that grew on MAF-leucine agar was further purified on nutrient agar; it was then transferred to nutrient agar slants and deeps for storage after verifying that each of the isolates were *leu⁻* and resistant to fucose inhibition.

We also isolated 11 independent constitutive mutants (C^c60 to C^c70) directly from colonies that appeared on MAF agar plates. These plates were originally smeared with approximately 10^8 *E. coli* B/r (hs) cells from nutrient broth cultures started from small inocula as described above.

Mapping of the constitutive mutants. By use of the linkage of leucine to L-arabinose (approximately 50% cotransduction of leucine and any arabinose negative markers in that area), it was possible to determine whether the C^c mutant sites map within this L-arabinose region of the chromosome. Phage grown on the wild-type B/r was crossed to each of the 50 C^c *leu⁻* (C^c1 *leu⁻* to C^c50 *leu⁻*) mutants. We selected for *leu⁺* and scored each of the *leu⁺* transductants appearing on mineral glucose plates for C^c by replica plating onto MAF, upon which C^+ fails to produce visible growth. Among the *leu⁺* transductants analyzed in each cross, we found approximately equal

numbers of C^+ and C^c recombinants, indicating that each of the C^c mutant sites analyzed is loosely linked with *leu* and, therefore, may map within the arabinose region of the chromosome located next to *leu*.

A more definitive mapping of 22 C^c mutant sites (C^c1 to C^c11, C^c60 to C^c70) was performed with reference to C^- sites distributed throughout the C gene (C^-5, C^-12, C^-59, C^-37), and a B mutant site (B^-8) closely linked to the C gene. (Mapping of the C^c sites in this manner was indicated on the basis of results of preliminary experiments which showed that the C^c sites tested were more closely linked to C^- mutant sites than to mutant sites in the other arabinose genes.)

The order of the C^- and the B^-8 mutant sites had been previously established by three-factor crosses (Gross and Englesberg, 1959; Cribbs and Englesberg, 1964). The relative distances between the four reference C^- mutant sites and B^-8 were determined by the per cent of *ara⁺* to *leu⁺* recombinants per unit volume of the transduction mixture, as determined by plating on mineral arabinose-threonine-leucine plates (selection for *ara⁺*) and on mineral glucose-threonine plates (selection for *leu⁺*). The distances given (Fig. 3) are the averages of the distances determined by reciprocal crosses. The following is a typical cross: phage *ara*C^-12 × bacterium *ara*C^-5 *thr⁻ leu⁻*; phage *ara*C^-5 × bacterium *ara*C^-12 *thr⁻ leu⁻*.

The position of the C^c mutant sites within the C gene was ascertained by determining the recombination frequencies between each of the C^c mutant sites and some or all of the five *ara⁻* sites (Fig. 2), in the following manner: C^c1 to C^c11, originally all *leu⁻*, were converted to *leu⁺* by transduction with phage grown on the *ara⁺* prototroph, and nonlysogenic *leu⁺* constitutive transductants were isolated. Phage were grown on each of the *leu⁺* constitutive mutants and crossed to C^-12, C^-5, and C^-59, or C^-37, *leu⁻*

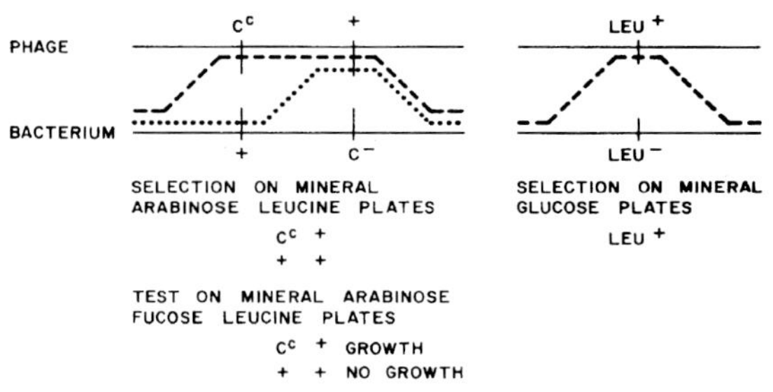

$$\frac{+ \ +}{LEU} \times 100 = \text{RECOMBINATION FREQUENCY OR RELATIVE DISTANCE}$$

FIG. 2. *Mapping of constitutive mutants.*

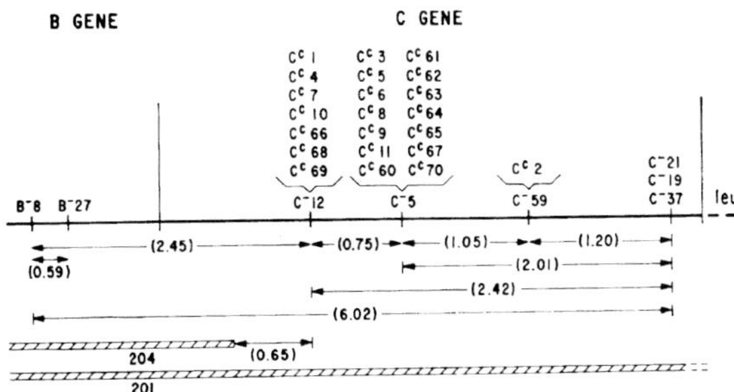

FIG. 3. *Map of the C gene. The figures in paren-thesis represent the average of reciprocal distance measurements (see Results). The order of B⁻8, C⁻12, C⁻5, C⁻19, C⁻21 (Gross and Englesberg, 1959) and B⁻27 (Cribbs and Englesberg, 1964) has been established by reciprocal three-factor crosses. The distance between B⁻8 and B⁻27 was established by Cribbs and Englesberg (1964). C⁻59 has been er-roneously placed to the right of the cluster of mutants C⁻19, C⁻21, C⁻37, etc. (Cribbs and Englesberg, 1964). Reciprocal three-factor crosses and distance measurements that we have performed place C⁻59 unambiguously to the left of C⁻37 and to the right of C⁻5. C⁻37 is placed at the same site as C⁻19 and C⁻21, since it fails to yield any recombinants with these two mutants.*

mutants. Crosses were made with B⁻8 when recombination frequencies indicated that a Cᶜ mutant site might be to the left of C⁻12. Selection was made for *leu⁺* on mineral glucose agar and for *ara⁺* on mineral L-arabinose leucine agar. The *ara⁺* recombinants were of two types: wild type (+ +) or constitutive (Cᶜ +). To determine the recombination frequency between a Cᶜ site and a C⁻ or B⁻ site, it is necessary to know the number of wild-type recombinants. This was determined by testing each of the *ara⁺* recombinants for ability to grow on MAF agar. Constitutives grow, whereas wild type is inhibited on this medium. The per cent wild type/ *leu⁺* recombinants gives the recombination fre-quency or relative distance between the mutant sites involved. We settled for scoring a minimum of 800 *ara⁺* recombinants for each cross because of the difficulties of the mapping procedure. There is no positive selection for wild type (inducibility), and Cᶜ sites are closely linked to the C⁻ sites. Since recombination frequencies were low, and it was not possible to perform reciprocal crosses, which are crucial for any fine structure mapping, the constitutive mutant sites were merely grouped next to the C⁻ mutant sites with which they gave the smallest recombination frequencies (Fig. 3). Of the 22 constitutive mutants, 18 mapped between the two distal C⁻ mutant sites. Cᶜ8 and Cᶜ70 failed to give any wild-type re-combinants with C⁻5, and, based upon the num-

ber of *ara⁺* recombinants assayed, we estimated that Cᶜ8 and Cᶜ70 are less than 0.02 and 0.1 recombination units away from C⁻5, respectively. The great majority of the Cᶜ mutants mapped

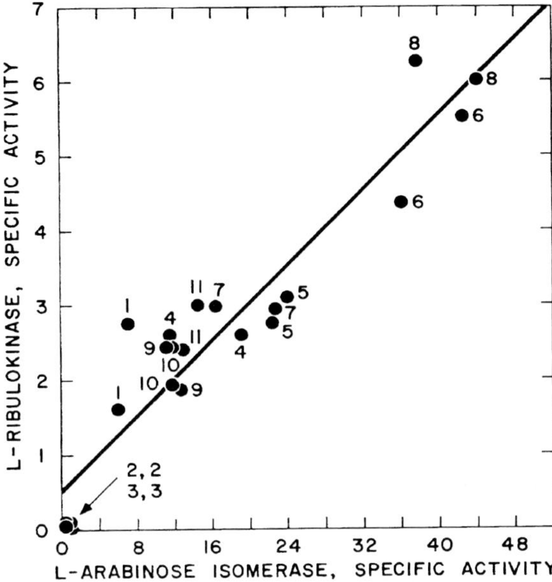

FIG. 4. *L-Arabinose isomerase versus L-ribu-lokinase. The numbers denote particular constitutive mutants isolated from Escherichia coli B/r. Assays are from duplicate extracts prepared from the same cell suspension. The straight line was determined by the method of least squares; data from the wild type were not included. Specific activity = micromoles of product formed per hour per milligram of protein.*

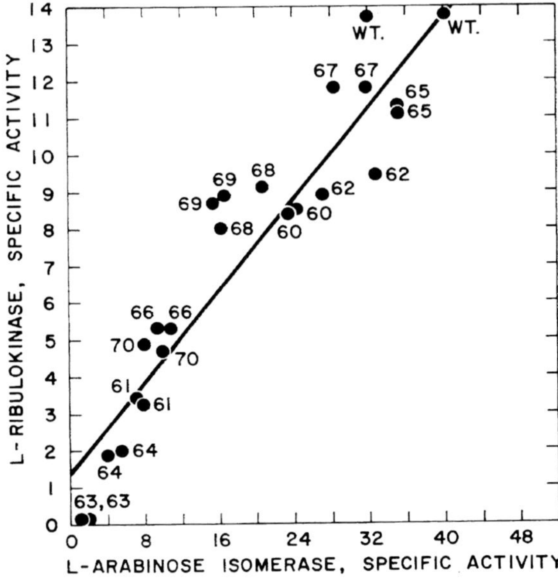

FIG. 5. *L-Arabinose isomerase versus L-ribu-lokinase. As in Fig. 4, except constitutive mutants were derived from Escherichia coli B/r (hs).*

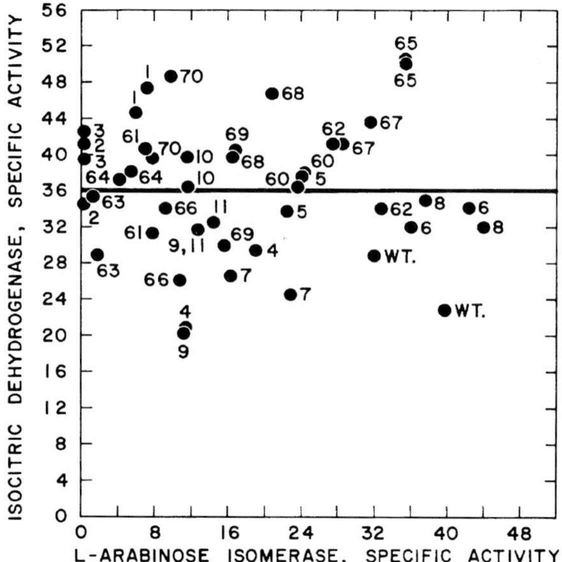

Fig. 6. *L-Arabinose isomerase versus isocitric dehydrogenase (as in Fig. 4 and 5).*

close to C^-5. On the basis of additive distances from each of the C^- markers, four C^c mutant sites that map in the vicinity of C^-12 may map to the left of C^-12. C^c4 was perhaps the farthest to the left of C^-12 by a distance of 0.47. Since the distance between the closest B mutant site (B^-27) and C^-12 was 1.85 recombination units, and on the basis of phenotypic characteristics, we can, with some justification, assume that these four mutants also map within the confines of the C gene.

Enzyme and permease levels in mutants resistant to D-fucose inhibition. Mineral glycerol grown cells in exponential phase were assayed for L-arabinose isomerase, L-ribulokinase, and L-arabinose permease, and, as controls, the same preparations were assayed for glucose permease and isocitric dehydrogenase. The C^c mutants have wide differences in constitutive levels of isomerase, kinase, and permease (approximately a 100-fold difference between the highest and lowest constitutive level of isomerase and kinase and approximately a 16-fold difference between the highest and lowest constitutive level of L-arabinose permease). A plot of constitutive L-arabinose isomerase versus L-arabinose permease and L-ribulokinase specific activities indicates that they are coordinate (Fig. 4, 5, 6, 7, and 8). Since genes BAD act as a genetic unit of coordinate expression (Lee and Englesberg, 1963), we presume that epimerase activity would also vary coordinately.

Isocitric dehydrogenase and glucose permease levels varied at a maximum of 2.5, and there

was no coordination between these levels and L-arabinose isomerase levels.

The coordination between L-arabinose isomerase and L-arabinose permease is interesting, since the isomerase gene (gene A) and the permease gene (gene E) are unlinked. This finding demonstrates that coordination of enzyme activities alone cannot be taken as evidence for linkage of the genes concerned, or for the production of a multicistronic message from a series of linked genes. The slope of the lines representing isomerase versus kinase activity of the C^c mutants 1 to 11 and 60 to 70 are noticeably different. This difference is, no doubt, due to the heat stability of the kinase in C^c60 to C^c70 mutants, since each of these mutants was isolated from a parent strain having a heat-stable kinase. This difference in kinase, probably due to a difference in the B gene, obviously does not interfere with coordinate control between isomerase and kinase exercised by the C gene.

Complementation analysis ($A^-C^+ \times A^+C^-$). It has been previously demonstrated that C^+ is dominant to C^- (Helling and Weinberg, 1963). These experiments were performed in a K-12 cross with just one C^- mutant, C^-19. This mutation

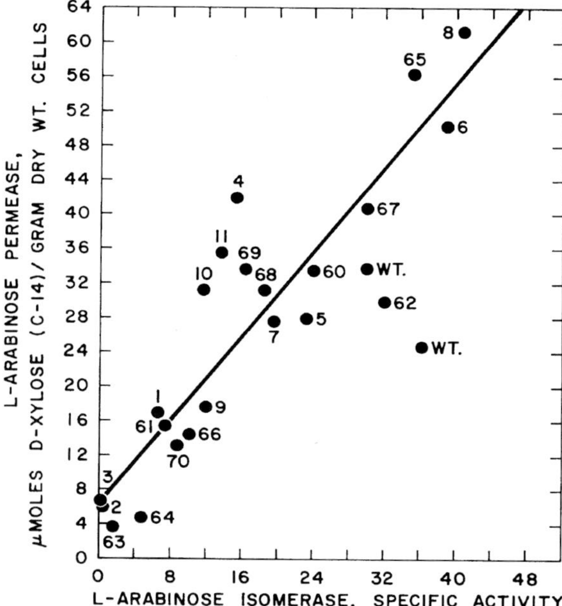

Fig. 7. *L-Arabinose isomerase versus L-arabinose permease. The numbers denote particular constitutive mutants isolated from Escherichia coli B/r and B/r (hs). Points represent the average of permease assays from duplicate samples of the permease reaction mixture and the average of kinase activities of the duplicate extracts. The straight line was determined by the method of least squares; data from the wild type were not included.*

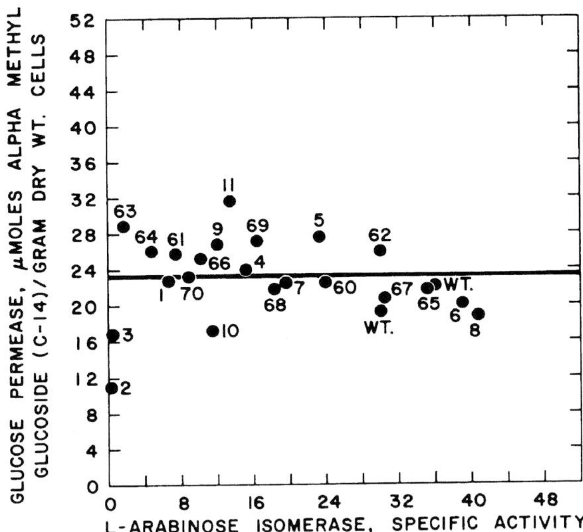

Fig. 8. *L-Arabinose isomerase versus glucose permease (as in Fig. 7).*

was initially isolated in B/r and transferred into K-12. The following experiments were conducted with five additional C⁻ mutants, this time with F⁻ B/r females (*see* Materials and Methods). In each case, merozygotes formed between *ara*A⁻ 2 males, and the C⁻ females produced substantially more L-arabinose isomerase than the controls (Table 1). Thus, all C⁻ mutants tested behave in the same manner as C⁻19: C⁺ is dominant to C⁻. The isomerase structural gene on the same strand as the *ara*C⁻ mutations is functional if an *ara*C⁺ gene is present in the same cell.

In a second cross (A⁻C⁻ × A⁺C⁻), merozygotes from A⁻C⁻ males and C⁻ B/r females, in the absence of the inducer L-arabinose, produced 6- to 30-fold more L-arabinose isomerase than the respective controls (Table 2). The results of these two types of crosses, A⁻C⁺ × A⁺C⁻ and A⁻C⁻ × A⁺C⁻, are compatible and point to the same conclusion. The C gene produces a product required for the expression of the isomerase structural gene. The isomerase structural gene on the same strand as the *ara*C⁻ mutations is functional and can be activated if an *ara*C⁺ gene plus L-arabinose is present, or, in the case of a constitutive mutant, the presence of the C⁻ allele alone is sufficient. The results are exactly what one would have anticipated if control was purely positive.

In a third cross (A⁻C⁻ × A⁺C⁺), merozygotes from A⁻C⁻ males and A⁺C⁺ B/r females, in the absence of L-arabinose, failed in every case to yield enzyme activity significantly higher than determined in the unmated controls. This is the first evidence for the production of a repressor substance by the C gene.

Deletion mutants. If the C gene does produce an activating substance, i.e., if it does exert positive control over the expression of genes A, B, D, and E, a deletion in the C gene should produce an *ara*⁻ phenotype. On the other hand, if control is only negative, a deletion should produce a constitutive phenotype. Deletions were isolated in the following manner. Growth of *ara*D⁻53 is inhibited by L-arabinose, and it is a simple matter to isolate hundreds of spontaneous L-arabinose-resistant mutants on EMB-L-arabinose agar plates. The great majority of these

TABLE 1. *L-Arabinose isomerase production by merozygotes formed with* araA⁻2 *males (Hfr) and different* araC⁻ *females (F⁻)**

C⁻ marker in F⁻ parent	Units of L-arabinose isomerase/ml	
	Mating	Control
*ara*C⁻12	3.4	0.7
*ara*C⁻5	1.5	<0.1
*ara*C⁻37	1.8	0.1
*ara*C⁻47	2.2	0.8
*ara*C⁻59	0.8	0.1

* The Hfr in each case was KH600 *ara*A⁻2 *str*ˢ, and the females were B/r *ara*C⁻ *str*ʳ (see Materials and Methods). The inducer L-arabinose was added 1 hr after the initiation of mating, and extracts were prepared from samples after 2 additional hr of incubation. Units of isomerase = micromoles of ribulose produced per hour.

TABLE 2. *L-Arabinose isomerase production by merozygotes formed from* araA⁻ C⁻ *males (Hfr) and* araC⁻ *or* araC⁺ *females (F⁻)**

Hfr	F⁻	L-Arabinose isomerase activity/ml	
		Mating	Control
*ara*A⁻2 C⁻8	*ara*C⁻37	1.0	<0.1
*ara*A⁻H9 C⁻11	*ara*C⁻3	0.6	0.1
*ara*A⁻2 C⁻60	*ara*C⁻5	1.8	0.06
*ara*A⁻2 C⁻62	*ara*C⁻5	1.7	0.06
*ara*A⁻2 C⁻65	*ara*C⁻5	1.7	0.06
*ara*A⁻2 C⁻8	C⁺ (wild type)	0.2	0.1
*ara*A⁻2 C⁻8	C⁺ (wild type)	0.1	0.1
*ara*A⁻2 C⁻60	C⁺ (wild type)	0.08	0.12
*ara*A⁻2 C⁻62	C⁺ (wild type)	0.07	0.12
*ara*A⁻2 C⁻65	C⁺ (wild type)	0.01	0.12

* The Hfr in each case was KH600 *ara*A⁻C⁻ *str*ˢ, and the females were B/r *ara*C⁻ *str*ʳ or B/r *ara*C⁺ *str*ʳ. No L-arabinose was added in these experiments. Extracts were prepared from cells 3 hr after the initiation of mating. See Table 1 for further details.

mutants are double L-arabinose-negative. They contain the original $araD^-53$ mutant site and, in addition, another mutation in the A, B, or C genes (Englesberg et al., 1962). A group of 200 L-arabinose-resistant mutants of thr^- $araD^-53$ leu^-, isolated on EMB-L-arabinose plates, were screened for deletions in the A, B, and C genes by crossing each of them with phage grown previously on $araA^-H9$, $araA^-54$, $araB^-87$, $araB^-8$, $araC^-12$, and $araC^-37$. Two deletion mutants were isolated that are of interest to this study. $AraD^-53,201$ failed to yield any ara^+ recombinants with any of the phage employed, although leu^+ recombinants were plentiful. Ara^+ recombinants were isolated with crosses with $araD^-139$; thus, it appears that the deletion extends from the $araC^-37$ mutant site in the C gene through the B and A genes and possibly further into the D gene. A second deletion, $araD^-53,204$, gave recombinants with both $araC^-37$ and $araC^-12$, but failed to yield recombinants with B or A mutants. Crosses with $araD^-53,204$ and $araC^-12$, in which the per cent of ara^+ recombinants to leu^+ recombinants were scored, placed the right end of the 204 deletion 0.65 map units to the left of C^-12, which may place it within the C gene.

Enzymatic analysis of cell-free extracts of both deletion mutants prepared from cells grown in a casein hydrolysate-L-arabinose medium (Cribbs and Englesberg, 1964) showed them to be devoid of L-arabinose isomerase and L-ribulokinase activity, as would be expected, since both deletions included genes A and B. However, both deletion mutants were also deficient in L-arabinose permease determined by gene E, which is not included in the deletion.

DISCUSSION

Based upon the analysis of the β-galactoside gene-enzyme complex (Jacob and Monod, 1961a, b; Pardee, Jacob, and Monod, 1959), it is generally inferred that the transfer of all information from deoxyribonucleic acid (DNA) to the final synthesis of enzyme molecules is under negative (repressor) control, with no further specific genetic element intervening in the regulation (Willson et al., 1964). This system of control, as initially conceived, was composed of essentially three elements: one or more structural genes, and an operator (O) and a regulator gene (i). The concept of the operator was derived from an analysis of two types of mutants in this system: Lac O° (operator-negative) and Lac O^c (operator-constitutive). Both types were originally thought to map at the end of the z gene, the β-galactosidase structural gene, and exert a pleiotropic effect. O° mutants are phenotypically β-galacto-sidase-, β-galactoside permease-, and β-galactoside transacetylase-negative. O^c mutants produce all three enzymes constitutively. In complementation studies, both types of mutation exert a dominant effect on the control of the adjacent structural genes (z, y, Ac) for the above enzymes (*cis* dominant), but exert no effect on the same genes in another strand of DNA (*trans* recessive). Some revertants of O° mutants, closely linked to the original O° site, produced an altered β-galactosidase molecule (Jacob and Monod, 1961b). From these results, together with evidence from complementation studies with i gene mutants, Jacob and Monod (1961b) proposed that the end of the z gene, the operator region, was the site for repressor attachment and for initiation of messenger ribonucleic acid (mRNA) synthesis. [In a recent paper, Jacob, Ullman, and Monod (1964) have separated these two functions. The site for repressor attachment (operator region) is now thought to be outside the z cistron and is defined solely by O^c mutants, whereas the site for initiation of transcription, now called the promoter area (P), remains part of this gene.]

According to the model proposed by Jacob and Monod (1961a, b), the O° and O^c phenotypes were the result of changes in the DNA in the operator region, preventing transcription and repressor attachment, repectively. Attardi et al. (1963) were not able to detect any Lac mRNA synthesis by Lac O° mutants, thereby offering support for this hypothesis. However, the finding of O°-like mutations in the middle of various structural genes which exert a pleiotropic negative effect in a polar direction (dual effect or polarity mutants; Englesberg, 1961; Franklin and Luria, 1961; Jacob and Monod, 1961b; Lee and Englesberg, 1963; Anies and Hartman, 1963), together with the finding that it is possible to obtain unlinked suppressors to Lac O° mutations (Beckwith, 1963; Orias and Gartner, 1964; Schwartz, 1964), indicates that the effect of O° mutations may be at the level of messenger reading rather than at the level of transcription. This leaves only O^c mutants to define the operator region.

The regulator gene (i) was defined on the basis of an analysis of two types of mutations: i$^-$ (constitutive) and i^s (super-repressed). The former are constitutive, whereas the latter are negative for the three enzymes in the system. In complementation analysis, i$^-$ is recessive to i$^+$, (Pardee et al., 1959; Jacob and Monod, 1961b), and i^s is dominant to i$^-$ and i$^+$ (Jacob and Monod, 1961b; Willson et al., 1964). On the basis of these experiments, negative control by a repressor substance was formulated. According to this hypothesis, the i gene in the wild-type configuration produces a repressor. The repressor has two sites of attachment. The inducer reacts with the repressor at one site and, by an allosteric effect, turns it into an inactive substance. In the absence of the inducer, the other site of the repressor is free to react with the operator region. It is this reaction that prevents transcription. The i^s mutants produce a repressor which has lost the site for inducer attachment, and thereby cannot be neutralized. Constitutive i^- mutants fail to produce repressor or *any biologically active entity*. In the absence of an active repressor, the transfer of information flows freely from genes z, y, and Ac. Is this the only type of control of gene expression found in microorganisms?

Similar types of experiments which, in the β-galactosidase gene complex, indicated a negative control system have led to defining a system that exhibits both a negative and positive control in the L-arabinose gene complex.

The regulator gene in the L-arabinose system, gene C, is adjacent to, but functionally distinct from, gene B. Mutations in gene C are of two types: C^- (phenotypically kinase-, isomerase-, epimerase-, and permease-negative) and C^c (constitutive for all three enzymes and permease).

Is this C region the operator region of the L-arabinose gene complex? Are C^- and C^c mutants, which define this region, similar to Lac O^o and Lac O^c mutants in the β-galactosidase system?

The fact that C^- mutants are deficient in L-arabinose permease by itself would rule out the possibility that the C^- mutants are similar to Lac O^o mutants, since the L-arabinose permease gene is unlinked to the C region of the chromosome. Whether Lac O^o mutants, as originally defined by Jacob and Monod (1961*b*), really exist, or whether they are really polarity mutants, control by these mutants probably rests either upon continuous transcription or translation in a polar fashion of an intact DNA or mRNA sequence. The fact that the E gene for the L-arabinose permease is approximately one-half the total length of the chromosome away from the C gene makes this an impossibility. The C gene would then have to control not only L-arabinose metabolism but something like one-half of the metabolic activities of the cell.

Complementation analysis with temporary merozygotes furnishes further evidence that the C region is not the operator region of the L-arabinose gene complex. Six C^- mutants defining the C region were shown to be recessive to C^+. In particular, in the cross $A^-C^+ \times A^+C^-$, the temporary merozygotes produce L-arabinose isomerase, demonstrating that C^- is comple-

mented by mutants in the A gene and, therefore, that the isomerase gene on the strand of DNA containing the C^- mutation is activated by C^+. C^+ must be producing some diffusible product required to activate the C^- strand, whereas the operator region in the β-galactosidase system apparently does not produce such a product. If this mutation were of an O^o type, no isomerase would be produced in this cross.

C^c mutants are constitutive for the inducible enzymes and permease in L-arabinose metabolism. These mutants differ from Lac O^c mutants. C^c is dominant to C^-. In the cross $A^-C^c \times A^+C^-$, L-arabinose isomerase is produced constitutively, indicating that C^c produces a product which is able to activate the C^- strand in the absence of L-arabinose. If C^c were the same type of allele as Lac O^c, no L-arabinose isomerase activity would be produced, since O^c is only *cis* dominant. We must, therefore, conclude that the C region is not the operator region of the L-arabinose system.

Is the C gene a regulatory gene of the i type? In the merozygotes produced by the cross $A^-C^c \times A^+C^+$, no L-arabinose isomerase is formed in the absence of L-arabinose; C^+ (inducibility) is dominant to C^c. In this respect, the C gene is similar to the i gene in the β-galactosidase system, but here the similarity ceases. No i^s-like mutants have been uncovered in the C gene, and no C^--like mutants have been found in the i gene. C^- is distinct from i^s since C^- is recessive to C^+ and C^c, whereas i^s is dominant to i^+ and i^-. Using analogous reasoning, as employed in deriving the negative control model, we arrive at the following. The dominance of C^+ over C^c, in the absence of inducer, indicates that the C^+ gene produces a repressor. However, the fact that both C^+ and C^c are dominant over C^- indicates that C^- does not produce a repressor; therefore, C^c, in the absence of L-arabinose, and C^+, in the presence of L-arabinose, produce a cytoplasmic substance which is required to activate the transfer of information from the L-arabinose structural genes to protein. We shall call this substance "activator." This is an example of positive control—a switching-on mechanism that is necessary for gene expression. Although by an elaborate series of *ad hoc* postulates it may be possible to explain these results on the basis that the C gene is solely a repressor gene, the experiments with the deletion mutants offer further evidence supporting the activator function of the C gene. A deletion in the i gene and in general in any regulatory gene of a negative control system leads to constitutive enzyme synthesis. The deletions in the C gene lead to failure of expression of the unlinked E gene.

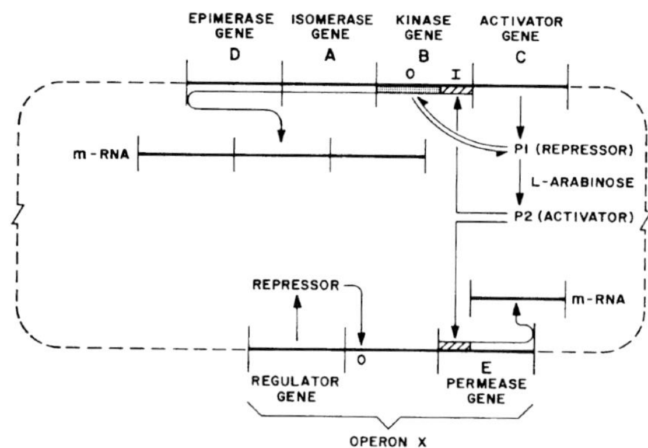

Fig. 9. *Model for regulation in the L-arabinose system.*

The R1 gene in the alkaline phosphatase system (Garen and Echols, 1962a, b) bears a resemblance to the C gene in that both exert positive control over gene expression. That such control exists in the alkaline phosphatase system is based mainly upon the finding that phosphatase-negative mutants (C mutants) that may map within the R1 gene are recessive to wild type. In addition, under conditions of nonrepression (low concentration of phosphate), most constitutive mutants (R1⁻) produce less phosphatase than the parent strain, and this trait is recessive to wild type. There are, however, certain differences between the R1 and C genes. Although negative mutants in both genes are recessive to wild type, many of the mutants in the R1 gene are dominant to R1⁻ (constitutive), whereas all negative mutants in the C gene (C⁻) so far tested are recessive to Cᶜ. In addition the R1 gene acts in conjunction with another regulatory gene (R2), whereas the C gene appears, so far, to be the only regulatory gene (if we exclude the B gene) in the L-arabinose system.

We shall consider one model (Fig. 9) to explain the results presented and to guide us in future experiments. We propose that the C gene produces a product P1, the repressor, presumably a protein. P1 has two active sites: one for attachment to an operator region in the B gene and the other for L-arabinose. By an allosteric transition, L-arabinose converts P1 to P2, the activator. This system is so far somewhat similar to that proposed for the β-galactosidase system. It differs in that instead of the repressor being converted into an inert substance, we propose that, in the L-arabinose system, it is converted to a *biologically necessary* entity, P2. P2 activates the BAD region and the E gene for transcription at initiation sites of attachment (I), which we propose are at the beginning of the B and E genes. (There is

little, if any, evidence at present to suggest where P1 and P2 act in the transfer of information from gene to protein. However, for the sake of simplicity, the model shows all the control at the transcription level.) C⁻ alleles fail to produce a substance that has any repressor function or that can be converted into an activator by L-arabinose. Thus, C⁺ is dominant to C⁻ (in the presence of L-arabinose). In merozygotes A⁻C⁺ × A⁺C⁻, in the presence of L-arabinose, the C product, P1, is converted to P2, which activates the strand containing the C⁻ allele. Cᶜ mutant sites, all of which map within the C gene, alter the amino acid sequence of the C product so that altered proteins (P3, P4, P5, etc.) are formed by the different constitutive mutants. These proteins may function partially or fully as activators in the absence of L-arabinose. (Some may still have some weak repressor capabilities.) In merozygotes between Cᶜ and C⁻ (A⁻Cᶜ × A⁺C⁻), in the absence of L-arabinose, C⁻ does not produce repressor or activator, whereas Cᶜ produces activator and, therefore, activates the BAD and E region for protein synthesis. In merozygotes A⁻Cᶜ × A⁺C⁺, in the absence of L-arabinose, C⁺ produces repressor and Cᶜ produces activator, but no L-arabinose enzymes are produced. Although such merozygotes were produced with hyperconstitutives, and enzyme assays were conducted for periods up to 3 hr after mating, no significant enzyme activity was detected in the merozygotes over that of the controls. Therefore, we propose that P1 and P2 or P3, etc., act at different sites and that only L-arabinose can neutralize the function of repressor, P1.

Since the different constitutive mutants produce various but coordinate levels of isomerase, kinase (and probably epimerase), and permease, we propose that the initiation sites in the B and E genes must be genetically similar to respond in this manner.

This model also provides an explanation for the dual effects of mutations in the B gene. Mutations in this kinase structural gene affect a coordinate increase or decrease in inducible levels of isomerase, epimerase, and kinase protein. Evidence indicates that mutation in the A gene has no effect on levels of epimerase activity. Thus, modulation as an explanation for the polarity mutants in the histidine system (Ames and Hartman, 1963) may not apply in this case. To explain our results, we further propose that most of the B gene, except that portion adjacent to the C gene (I or initiator region), serves as the site for P1 attachment, i.e., the operator. One can explain the dual effect on the basis that changes in base sequence in the O region of the B gene affect its affinity for P1, thereby modifying the equilibrium between the components of this system. B gene–O–P1 $\rightleftharpoons$ P1 $\rightleftharpoons$ P1 + L-arabinose $\rightarrow$ P2. In some cases, mutation leads to a looser, and in other cases to a tighter, attachment to B. The rate at which P1 is removed from the B gene may determine the rate at which the message BAD is formed. Of course, since the permease gene is unlinked to BAD, its rate of synthesis is not so affected by mutation in the B gene (Novotny and Englesberg, 1964).

What is the biological significance of this control system that combines both a negative and a positive controlling element? We propose that such a system is required in coordinate control involving genes in different operon systems. In the L-arabinose system, the permease gene is unlinked to the other structural genes, and the C gene is apparently adjacent to gene B in the BAD genetic unit of coordinate expression. Since there is selective advantage in the bacteria for adjacent linkage of genes controlling a particular biochemical pathway, we have assumed that the L-arabinose permease gene is part of another control system or operon (operon X), and that positive control, in the form of P2, is required to remove the E gene from the control exerted by this other system. Similarly, in the CBAD area of the chromosome, message synthesis of the C gene must occur and stop at B in the absence of activator. We propose that P2 also functions to activate the BAD strand.

Why is the repressor needed at all? The activator-repressor system may function as a double lock. Activator may be similar to a master key, perhaps being less specific and thereby controlling a number of regions on the chromosome, whereas P1 is the specific element removable only by the specific inducer L-arabinose.

It should be realized that this proposed model may have no more than a limited conceptual utility. However, it makes several predictions that can be experimentally verified. For instance, it proposes that gene E is not subject to control by repressor P1; therefore, in merozygotes of $E^- C^c \times E^+ C^+$, constitutive permease synthesis should occur. It also predicts that permease may be induced by another inducer besides L-arabinose. Further, it predicts the following types of mutants that so far we have not obtained: C^- *cis* and *trans* dominant—a mutation producing a repressor that cannot be converted into an activator; I^-—a mutation in the initiator area which fails to respond to activator and is *cis* dominant for isomerase, kinase, and epimerase, but still produces an inducible permease.

ACKNOWLEDGMENTS

We wish to thank Evelyn Lieberman and Linda McAfee for their able technical assistance.

This investigation was supported by a research grant from the National Science Foundation, by Public Health Service grant GM10165 from the Division of General Medical Sciences, and by a contract between the Office of Naval Research and the University of Pittsburgh. The work of J. Power was carried out during tenure of a Predoctoral Research Fellowship, National Institutes of Health, U.S. Public Health Service.

LITERATURE CITED

AMES, B., AND P. HARTMAN. 1963. The histidine operon. Cold Spring Harbor Symp. Quant. Biol. **28**:349–356.

ATTARDI, G., S. NAONO, J. ROUVIÈRE, F. JACOB, AND F. GROS. 1963. Production of messenger RNA and regulation of protein synthesis. Cold Spring Harbor Symp. Quant. Biol. **28**:363–372.

BECKWITH J. 1963. Restoration of operon activity by suppressor. Biochim. Biophys. Acta **76**:162–164.

BOYER, H. 1964. Genetic control of restriction and modification in *Escherichia coli*. J. Bacteriol. **88**:1652–1660.

BOYER, H., E. ENGLESBERG, AND R. WEINBERG. 1962. Direct selection of L-arabinose negative mutants of *Escherichia coli* B/r. Genetics **47**:417–425.

CRIBBS, R., AND E. ENGLESBERG. 1964. L-Arabinose negative mutants of the L-ribulokinase structural gene affecting the levels of L-arabinose isomerase in *Escherichia coli*. Genetics **49**:95–108.

ENGLESBERG, E. 1961. Enzymatic characterization of 17 L-arabinose negative mutants of *Escherichia coli*. J. Bacteriol. **81**:996–1006.

ENGLESBERG, E., R. L. ANDERSON, R. WEINBERG, N. LEE, P. HOFFEE, G. HUTTENHAUER, AND H. BOYER. 1962. L-Arabinose-sensitive, L-ribulose 5-phosphate 4-epimerase-deficient mutants of *Escherichia coli*. J. Bacteriol. **84**:137–146.

ENGLESBERG, E., J. IRR, AND J. POWER. 1964. Evidence for a new type of regulatory gene in the L-arabinose gene enzyme complex. Bacteriol Proc., p. 19.

ENGLESBERG, E., AND N. L. KILLEEN. 1959. Mixed gene loci controlling L-arabinose isomerase and L-ribulokinase activities in *Escherichia coli* B/r. Genetics **44**:508–509.

ENGLESBERG, E., J. WATSON, AND P. HOFFEE. 1961. The glucose effect and the relationship between glucose permease, acid phosphatase and glucose resistance. Cold Spring Harbor Symp. Quant. Biol. **26**:261–276.

FRANKLIN, N. C., AND S. E. LURIA. 1961. Transduction by bacteriophage P1 and the properties of the lac genetic region in *E. coli* and *S. dysenteriae*. Virology **15**:299–311.

GAREN, A., AND H. ECHOLS 1962a. Properties of

two regulating genes for alkaline phosphatase. J. Bacteriol. **83**:297–300

GAREN, A., AND H. ECHOLS. 1962b. Genetic control of induction of alkaline phosphatase synthesis in *E. coli*. Proc. Natl. Acad. Sci. U.S. **48**:1398–1402.

GROSS, J., AND E. ENGLESBERG. 1959. Determination of the order of mutational sites governing L-arabinose utilization in *Escherichia coli* B/r by transduction with phage P1bt. Virology **9**:314–331.

HELLING, R. B., AND R. WEINBERG. 1963. Complementation studies of arabinose genes in *Escherichia coli*. Genetics **48**:1397–1410.

HOFFEE, P., AND E. ENGLESBERG. 1962. Effect of metabolic activity on the glucose permease of bacterial cells. Proc. Natl. Acad. Sci. U.S. **48**:1759–1765.

ISAACSON, D., AND E. ENGLESBERG. 1964. L-Arabinose inhibition of growth and isolation of L-arabinose permeaseless, cryptic mutants of *Escherichia coli* B/r. Bacteriol Proc., p. 113–114.

JACOB, F., AND J. MONOD, 1961a. Genetic regulatory mechanisms in the synthesis of proteins. J. Mol. Biol. **3**:318–356.

JACOB F., AND J. MONOD. 1961b. On the regulation of gene activity. Cold Spring Harbor Symp. Quant. Biol. **26**:193–211.

JACOB, F., A. ULLMAN, AND J. MONOD. 1964. Le promoteur, élément genétíque nécessaire à l'expression d'un opéron. Compt. Rend. **258**:3125–3128.

LEE, N., AND E. ENGLESBERG. 1962. Dual effects of structural genes in *Escherichia coli*. Proc. Natl. Acad. Sci. U.S. **48**:335–348.

LEE, N., AND E. ENGLESBERG. 1963. Coordinate variations in induced syntheses of enzymes associated with mutations in a structural gene. Proc. Natl. Acad. Sci. U.S. **50**:696–702.

LOWRY, O. H., N. J. ROSEBROUGH, A. L. FARR, AND R. J. RANDALL. 1951. Protein measurement with the Folin phenol reagent. J. Biol. Chem. **193**:265–275.

NOVOTNY, C., AND E. ENGLESBERG. 1964. The L-arabinose permease of *Escherichia coli*, B/r. Proc. Intern. Congr. Biochem., 6th, New York **3**:46.

ORIAS, E., AND T. K. GARTNER. 1964. Suppression of a class of rII mutants of T4 by a suppressor of a lac "operator negative" mutation. Proc. Natl. Acad. Sci. U.S. **52**:859–864.

PARDEE, A. B., F. JACOB, AND J. MONOD. 1959. The genetic control and cytoplasmic expression of "inducibility" in the synthesis of β-galactosidase by *E. coli*. J. Mol. Biol. **1**:165–178.

SCHWARTZ, N. 1964. Suppression of a *lac O^o* mutation in *Escherichia coli*. J. Bacteriol. **88**:996–1001.

WILLSON, C., D. PERRIN, M. COHN, F. JACOB, AND J. MONOD. 1964. Non-inducible mutants of the regulator gene in the "lactose" system of *Escherichia coli*. J. Mol. Biol. **8**:582–592.

WOLLMAN, E. L., F. JACOB, AND W. HAYES. 1956. Conjugation and genetic recombination in *Escherichia coli*. Cold Spring Harbor Symp. Quant. Biol. **21**:141–162.

Transposition of the Lac Region of Escherichia coli. I. Inversion of the Lac Operon and Transduction of Lac by φ80

J. R. BECKWITH AND E. R. SIGNER

The power of recombinant DNA technology stems from the ability it provides to create novel DNA joints; this method uses enzymes to join unrelated DNA fragments in vitro. For example, two novel DNA joints are created when a gene is cloned in vitro onto a small plasmid vector. Transposition and illegitimate recombination also create novel DNA joints. Accordingly, when harnessed properly by genetic selection, these reactions provide an equally powerful experimental tool for the clever bacterial geneticist. Beckwith and Signer first demonstrated this with the isolation of φ80*lac*. In effect, they cloned the *lacZ* gene with toothpicks.

In 1966 Beckwith and Signer were postdoctoral students in the lab of François Jacob. At that time there were two genetic methods for cloning genes from *Escherichia coli*, and both were severely limited. Certain genes could be incorporated into specialized transducing phage; for example, the *gal* and *bio* genes could be incorporated into phage λ and the *trp* and *supF* (Su3) genes could be incorporated into φ80. However, this method was limited to genes that were adjacent to phage attachment sites in the bacterial chromosome. A more general method involved the isolation of F' factors, but these episomes were extremely large and practically impossible to manipulate biochemically.

Beckwith and Signer combined these methods in ingenious fashion. Using an F'*lac* strain, which carried a temperature-sensitive mutation that prevented episome replication, and a strain in which the chromosomal *lac* region had been deleted, François Cuzin and François Jacob had shown that Hfr strains could be isolated by selecting for Lac$^+$ at 42°C. Under these conditions the episome is lost (Lac$^-$) unless it integrates into the chromosome, thus forming an Hfr derivative. Influenced by the work on λ integration by Allan Campbell (*Adv. Genet.* **11**:101–145, 1962), Beckwith and Signer reasoned correctly that F'*lac* integration, if it occurred within a gene, would "split" that gene and thus destroy its function. This impressive insight occurred years before transposons and insertion elements were discovered in bacteria, at a time when spontaneous insertion mutations were incomprehensible.

As it turns out, the T1*rec* gene, now called *tonB*, is located adjacent to the attachment site for φ80 (*att*$_{80}$). Mutations in *tonB* confer resistance to several different phages and colicins. Accordingly, "directed transposition" of *lac* to a chromosomal location near *att*$_{80}$ could be accomplished by using the Cuzin and Jacob strain by simultaneous selection for Lac$^+$ and phage and colicin resistance at 42°C. By using a strain that answered this selection (Hfr EC15), φ80*lac* could be isolated in the same manner as φ80*trp* from wild-type φ80 lysogens.

The implications of this work were profound. Chromosomes could be redesigned at will, and genes could be moved from one replicon to another. With specialized transducing phage, genetically well-characterized genes such as *lac* could be studied by biochemical and biophysical methods. In many ways this represents the dawn of genetic engineering. Indeed, the search for new and better ways to make transducing phage led ultimately to modern recombinant DNA technology.

THOMAS J. SILHAVY

Reprinted from *Journal of Molecular Biology* 19:254–265. Copyright © 1966, by permission of the publisher, Academic Press.

J. Mol. Biol. (1966) **19**, 254–265

Transposition of the *Lac* Region of *Escherichia coli*

I. Inversion of the *Lac* Operon and Transduction of *Lac* by $\phi80$

Jonathan R. Beckwith† and Ethan R. Signer‡

Service de Genetique Microbienne, Institut Pasteur
Paris, France

(*Received 11 March 1966*)

We describe a technique for isolating transpositions of *lac* to specific regions of the *Escherichia coli* chromosome. From two strains, EC8 and EC15, in which *lac* is transposed, we have isolated $\phi80lac$ transducing particles which also carry a part of the sex factor, F. We have shown that in EC8 and EC15 the *lac* operons are inverted relative to one another. In addition, we discuss and present evidence in support of a model for the origin of such Hfr's.

1. Introduction

Gene transposition in bacteria is a rare event (see, for example, Jacob & Wollman, 1961). Recently however, methods of selecting transpositions of chromosomal genes carried by the F-factor in *Escherichia coli* have been described (Cuzin & Jacob, 1964; Scaife & Pekhov, 1964; Scaife, 1966).

In this paper we describe experiments in which the transposition of the *lac*§ region to different sites on the *E. coli* chromosome suggests new ways of studying the expression of the *lac* operon. We find that it is possible in certain cases to obtain transpositions of the *lac* genes to a region of the chromosome chosen in advance. This technique makes it feasible to isolate Hfr's the origins of which lie in regions where they may be useful, and to move the *lac* region and F-factor into positions on the chromosome where different aspects of their functions may be studied. In particular, we are able to obtain specialized transduction of both the *lac* region and a portion of the F-factor by the phage $\phi80$ in some of the transposed strains.

Transduction of *lac* by $\phi80$ has permitted several novel experimental approaches to *lac* operon expression and to the nature of specialized transduction. In further communications we shall describe deletions joining the *lac* and *try* operons; studies on the expression of the *lac* operon after induction of, or infection by, the transducing phage; properties of transducing phages; and experiments on the determination of specificity in lysogeny.

† Present address: Department of Bacteriology and Immunology, Harvard Medical School, Boston, Mass., U.S.A.
‡ Present address: Department of Biology, Massachusetts Institute of Technology, Cambridge, Mass., U.S.A.
§ Abbreviations used: *lac*, utilization of lactose; *pro, try, pur, pyr*, requirement for proline, tryptophan, purine, pyrimidine; *T1rec, T6rec*, synthesis of receptor for phage T1, phage T6; $T1^r$, $T1,5^r$, resistance to phage T1, phages T1 and T5; att_{80}, attachment site for phage $\phi80$; *Su*, suppressor; Sm^s, Sm^r, sensitive and resistant to streptomycin; *col* colicin; HFT, high-frequency transducing lysate; $\phi80lac$, $\phi80$ transducing phage carrying the *lac* region.

The technique used may, of course, be extended to markers other than *lac* in suitable systems (Scaife & Pekhov, 1964; Scaife, 1966).

2. Materials and Methods

Methods are those of Signer, Beckwith & Brenner (1965) for bacterial genetics and of Signer (1966) for $\phi 80$.

Media are those of Signer *et al.* (1965), except that minimal medium was medium 63 of Pardee, Jacob & Monod (1959). We recall that, on LZ agar, isolated lac^+ colonies are white, and isolated lac^- colonies are red.

In addition to strains described previously (Signer *et al.*, 1965), we have used the following *lac* mutants:

deletion: X74, deletion of the entire *lac* operon, from Dr F. Jacob;

X111, deletion of *proB* and *lac*, from Dr F. Jacob;

z^-: X90, *ochre* mutant (Newton, Beckwith, Zipser & Brenner, 1965);

y^-: NG328, *amber* mutant, from Dr D. Zipser;

NG707, *amber* mutant, from Dr D. Zipser;

R, from Dr F. Jacob;

U8202, from Dr J. Scaife.

Strain EC0 is $(proB\ lac)_{X111}^-Sm^s$ carrying $F_{TS-114}lac^+$, which is an F-linked episome unable to replicate autonomously at high temperature (Cuzin & Jacob, 1964). Temperature-resistant lac^+ Hfr's were isolated by plating EC0 on LZ agar at 42°C and picking white colonies.

Strain EZ0 is $lac_{X74}^-Sm^s$ carrying $F\text{-}lac\ z^-y^+\text{-}purE\text{-}d25\,(F'\text{-}ad\text{-}z^-y^+)$; the episomal $lac\ z^-y^+$ is linked to *purE* by a deletion which places the y gene under purine control (Jacob, Ullman & Monod, 1965).

F_{def} is a defective F^+ episome (Signer & Beckwith, unpublished experiments) isolated in $F^-\ (proB\ lac)_{X111}^-Sm^s$ after crossing with Hfr P10 by Dr F. Cuzin.

$Flac^+pro\ C^+$ was isolated by Dr F. Jacob.

Male-specific phage F2 (Loeb & Zinder, 1961) and female-specific phage ϕII (Cuzin, 1965) were obtained from Dr F. Jacob.

Colicin lysates were prepared from a strain carrying F *col V col B* (P. Fredericq, personal communication; see Signer *et al.*, 1965) and $\phi 80$. Induction of $\phi 80$ led to lysates having high colicin titre as judged from killing of $T1,5^r$ strains. $\phi 80$-virulent (Signer, 1966) was added to the lysates for mutant selection (see text).

3. Results

Jacob, Brenner & Cuzin (1963) have isolated an $F\text{-}lac$ episome the replication of which is temperature-sensitive ($F_{TS}\text{-}lac$). In strains carrying this episome and a lac^- mutation on the host chromosome, they selected for integration of $F_{TS}\text{-}lac$ into the chromosome by selecting rare derivatives which remained lac^+ at high temperature. Such derivatives were Hfr donors as a result of integration of the episome into the *lac* region. When the host chromosome carried an extensive deletion of the *lac* region, the episome integrated not in the *lac* region, but at several different chromosomal sites, resulting in transposition of the *lac* genes (Cuzin & Jacob, 1964).

We have isolated a series of Hfr's (to be described in a subsequent paper) using the technique of Cuzin & Jacob (1964). To isolate temperature-resistant lac^+ Hfr's, we have used EC0, a strain carrying an $F_{TS}\text{-}lac$ (114) and a deletion, X111, of the *lac* and *proB* markers (Cuzin & Jacob, 1964). In one Hfr, EC8, $F_{TS}\text{-}lac$ has integrated between the attachment site for $\phi 80$ (att_{80}) and an *ochre* suppressor locus, Su_o (Signer *et al.*, 1965). EC8 transfers Su_c^-, but not att_{80} as an early marker (Table 1 and Fig. 1). The *lac* region is transferred as the terminal marker, indicating the order, Su_o^- F-factor origin-*lac*-att_{80}. We have used EC8 in experiments described below.

TABLE 1

Transfer of markers by EC8 and EC15 in 60-min blendor experiments

Time of mating (min)	Number of recombinants/ml. ($\times 10^{-3}$)						
	EC8	EC15	EC8	EC8	EC15	EC8	EC15
	purB		Su_o^-	att_{80}		try	
0	0	0	0	–	–	0	0
60	300	0	300	0	0	0	500

Transfer of Su_o^- was determined in a cross of EC8 with a strain carrying a lac^- *ochre* mutant (lac^-_2) (Brenner & Beckwith, 1965) and Su_c^+. The strain is phenotypically lac^+. Su_o^- (lac^-) colonies were scored on LZ agar (Signer *et al.*, 1965). To determine transfer of att_{80}, both EC8, lysogenic for $\lambda h80$ (Signer, 1964), and EC15 were crossed with an F$^-$ strain which was lysogenic for the temperature-inducible phage $\lambda i_{857}h80$ and which was purB$^-$ and try^-. The transfer of att_{80} was determined by examining for the loss of temperature sensitivity among 100 purB$^+$ recombinants (EC8) and 50 try^+ recombinants (EC15). In the case of EC8, the cross was also allowed to proceed for 6 hr, after which we found that 12 of 25 try^+ recombinants were temperature-insensitive, indicating transfer of att_{80} as a terminal marker with try.

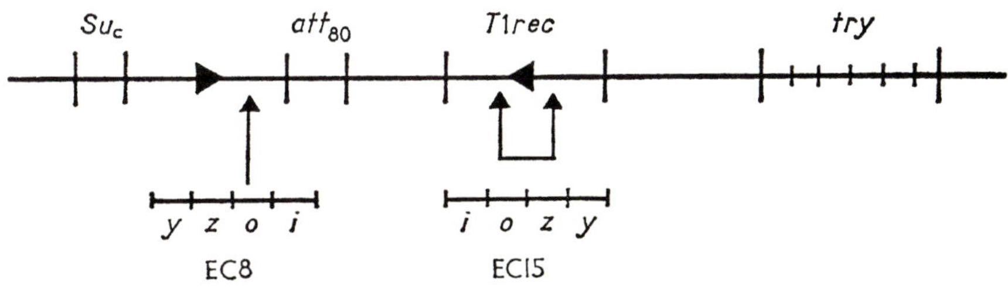

FIG. 1. The orientations of Hfr EC8 and Hfr EC15. See Signer *et al.* (1965) for a detailed map of this region of the chromosome. The double arrow for the *lac* region of EC15 indicates that the position of *lac* relative to the Hfr origin is not known. *Lac* is normally located far from this region.

(a) *The integration of* F$_{TS}$-lac *into genes*

If the integration of F_{TS}-*lac* into the chromosome is due to recombination between episome and chromosome in regions of weak homology, it may be possible, occasionally, that this homology lies within a gene on the chromosome. If the integration of the F_{TS}-*lac* into this gene occurs by a single reciprocal cross-over (Campbell, 1962), the gene would be split by the cross-over and the two resultant fragments of the gene separated by the length of the F_{TS}-*lac*. The gene should thus be inactivated. We have isolated three different Hfr's in which the F_{TS}-*lac* appears to have integrated into a gene resulting in the loss of gene function. The first, Hfr EC11, is an Hfr derived from EC0, which grows on nutrient but not on synthetic medium. The integration of the F_{TS}-*lac* appears to have resulted in auxotrophy for a requirement, as yet uncharacterized.

In the other two examples, in order to isolate strains in which the F_{TS}-*lac* is integrated into a specific locus, we have simultaneously selected for temperature-independent *lac⁺* and resistance to a bacteriophage. In this way we have isolated a *T6ʳ*, *lac⁺* temperature-resistant Hfr, EC102, which transfers its chromosome as though it were in or very close to the *T6rec* gene near *pur*E (Jacob & Beckwith, unpublished results). In order to obtain another Hfr in which *lac* is integrated near ϕ80, as in EC8, we selected for integration of the F_{TS}-*lac* into the *T1rec* locus adjacent to att_{80}. A culture of EC0 was spread on LZ agar at 42°C, together with a lysate containing ϕ80-virulent (Signer, 1964), colicin V and colicin B; mutants resistant to all three of these agents should arise only at the *T1rec* locus adjacent to att_{80} (Gratia, 1964). Of seven colonies which appeared, one (EC15) was *lac⁺*. Since the normal frequency of integration of F_{TS}-*lac* in EC0 is approximately 10^{-4}, it appears likely that a single event is responsible for the mutation to *T1ʳ* and the integration of F_{TS}-*lac*. EC15 is an Hfr which transfers its chromosome with an origin very close to *T1rec* (Fig. 1). It donates *try* but neither att_{80} nor *pur*B as an early marker (Fig. 2 and Table 1). These results indicate that the *T1ʳ* phenotype is caused by the integration of F_{TS}-*lac* into *T1rec*. Due to the high frequency of reversion of this strain to the autonomous F_{TS}-*lac* condition, we have not been able to determine whether *lac* is donated as the first or last marker.

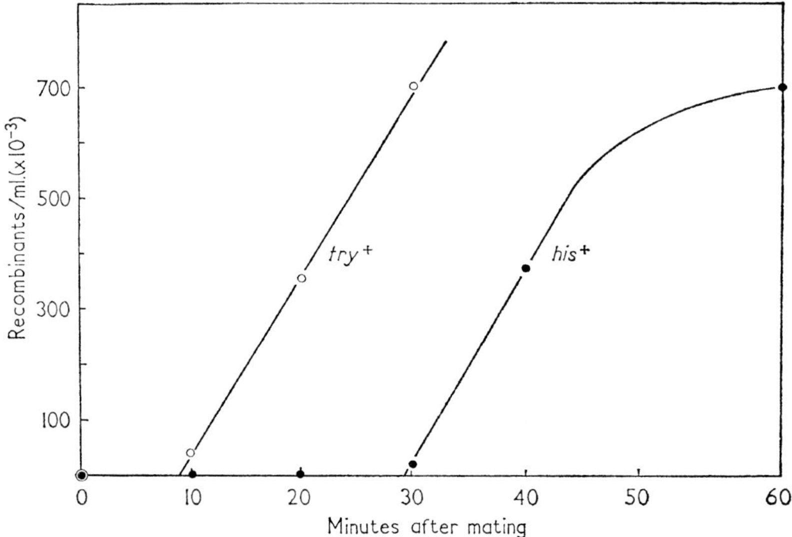

FIG. 2. Kinetics of marker transfer in interrupted mating experiments. Donor EC15, recipient *try⁻his⁻*.

(b) *Specialized transduction of* lac *by* ϕ80

We have isolated two Hfr's, EC8 and EC15, in which *lac* is located very close to att_{80}, and we should like to know whether ϕ80 can transduce *lac* from lysogens of these strains. This seems possible, since in both strains F_{TS}-*lac* is integrated between ϕ80 and a transducible marker: Su_{c} (Signer, 1966) in the case of EC8, and *try* (Matsushiro, 1963) in the case of EC15. Table 2 shows that *lac* is indeed transducible by ϕ80 in both cases. Since ϕ80 never transduces *lac* from wild-type strains (Matsushiro, 1963), this strongly indicates that specialized transduction of a marker is a consequence only of its location next to the prophage on the donor chromosome (for a discussion of this point, see Signer, 1966). Transduction of *lac* by ϕ80 closely resembles that of other markers by ϕ80 (Matsushiro, 1963; Signer, 1966) and of *gal* by λ (Morse, Leder-

Table 2

Low-frequency transduction by $\phi 80$ induced from lysogens of EC8

$\phi 80$ induced from lysogen of	Transductants per active phage for		
	lac^+	try^+	$lac^+ try^+$
EC8	8×10^{-7}	3×10^{-7}	$< 1 \times 10^{-10}$
EC15	3×10^{-9}	see text	$< 1 \times 10^{-10}$

Since EC15 is $T1^r$ its lysogen was made by phenotypic mixing (Signer, 1966); use of the host range mutant $\phi 80h$ (obtained from Dr B. Hall) gave similar frequencies. The transduction recipient was Hfr $lac^-_{X74} try^-_{X25}$; similar frequencies were obtained with lac^- point mutants and with F^- strains.

berg & Lederberg, 1956). Further details of the system will be described elsewhere (Signer & Beckwith, manuscript in preparation).

We have also tried to determine the effect of the integrated *Flac* in EC8 and EC15 on low-frequency transduction of other markers by $\phi 80$. Signer (1966) has shown that the transduction frequency of a marker is increased when the marker is moved closer to the prophage by deletion of intervening genes. Analogously, we should expect that integration of F_{TS}-*lac* between a given marker and att_{80} would decrease the transduction frequency of that marker by $\phi 80$. That is, we expect the transduction of Su_c to be reduced in EC8, and that of *try* to be reduced in EC15 (see Fig. 1). However, since both strains are Su_c^- we could not score transduction of this marker. In the case of EC8, tranduction of *try* occurs at the normal frequency (Table 2), as expected from the fact that it is on the opposite side of att_{80} from the integrated F_{TS}-*lac*. In the case of EC15, the high reversion to autonomous F_{TS}-*lac* meant that any *try* transductants observed might have originated from cells which had already lost F_{TS}-*lac* from the chromosome before induction of the prophage. Nevertheless, the absence of simultaneous tranduction of *try* and *lac* (Table 2) is consistent with our expectation.

(c) *Specialized transduction of the F-factor by $\phi 80$*

In order to see whether $\phi 80$ can also transduce the F-factor from EC8 and EC15, we have screened one $\phi 80$ *lac* transducing phage derived from each strain. The following results show that both transducing phages carry a portion of the F-factor in addition to the *lac* genes. The F-factor carries loci responsible for sensitivity to the male-specific phage F2 (Loeb & Zinder, 1961) and resistance to the female-specific phage ϕII (Cuzin, 1965). Standard male ($F2^s \phi II^r$) and female ($F2^r \phi II^s$) strains, and recombinant strains $F2^s \phi II^s$ and $F2^r \phi II^r$, do not change their response to F2 or ϕII, demonstrating that neither $\phi 80 lac$ carries these two loci. Standard female strains carrying either $\phi 80 lac$ do not donate markers to other females in mating. However, lysogenization with either $\phi 80 lac$ at att_{80} modifies the mating behaviour of a strain carrying a defective episomal F-factor ($F2^s \phi II^s$), as shown in Table 3. The presence of $\phi 80 lac$ increases the transfer of certain markers several hundredfold: in each case the transducing phage confers upon the strain an origin and a direction corresponding to those of the strain from which it was derived, EC8 or EC15 (compare with Table 1 and Fig. 2). This might be due to complementation by the defective F-factor of an origin present in $\phi 80lac$, or to chromosome mobilization (Scaife & Gross, 1963) by the defective F-factor through some other portion of F present in the transducing phage. Since replacement of the defective F-factor by an active *Flac pro C* results in only a slight increase in transfer frequency (Table 3), the former explanation would require that the complementing function be much more active in *cis* than in *trans* position. In any event, it is clear that both transducing phages carry in addition to the *lac* genes some but not all of the F-factor.

TABLE 3

Transfer in 60-min blendor experiments

| Experiment | Episomal F-factor | $\phi 80lac$ derived from | \multicolumn{4}{c}{Transfer frequency ($\times 10^5$) per donor cell for} |
			purB	(*lac*)	tryC	pyrF
1	—	—	<0.1	<0.1	<0.1	<0.1
	F_{def}	—	5	<0.1	4	8
	F_{def}	EC8	100	13	11	14
	F_{def}	EC15	5	5	130	160
2	F_{def}	EC8	13	<0.1		2
	Flac$^+$*pro C*$^+$	EC8	32	32,000		<0.3
	F_{def}	EC15	<0.3	<0.3		16
	Flac$^+$*pro C*$^+$	EC15	<0.6	16,000		28

Donors: F$^-$ (*proB lac*)$_{\text{X111}}$$^-$ carrying F_{def} or *Flac pro C* and $\phi 80lac$ derived from EC8 or EC15 located at att_{80}. Recipient: F$^-$ *lac* z_2^- *pur*B *try*C$^-$ *pyr*F$^-$ S^r ($\phi 80$). Similar results are obtained when the donors are additionally lysogenic for $\phi 80$.

(d) *The opposite orientation of* lac *in EC8 and EC15*

The Hfr's, EC8 and EC15, transfer their chromosomes in opposite directions. It seems likely that this is due to the insertion of the F-factor into the chromosomes of these two strains in opposite orientations. Since the autonomous $F_{\text{TS}}lac$ is presumed to be a single molecule of DNA, it is then possible that integration has resulted in the *lac* operon as well being in opposite orientations in EC8 and EC15, as indicated in Fig. 1. We shall now show that this is the case. In Fig. 1 the *lac* operator region is closest to att_{80} and the *y* gene farthest from it in both Hfr's. If the $\phi 80$ transducing particles obtained in these strains arise by the single reciprocal cross-over event suggested by Campbell (1962), then those particles which carry a marker near att_{80} may not always carry a more distant marker. However, a transducing particle which carries one of the more distant markers of a group would always carry the markers between att_{80} and that marker. These predictions of the Campbell model have been confirmed for $\phi 80$ by the studies of Matsushiro (1963) and Signer (1966), and for λ by Adler & Templeton (1963). We shall assume that these hold true as well for transduction of *lac* by $\phi 80$. In order to determine the orientation of the *lac* operons in these strains, we have studied the kinds of $\phi 80lac$ transducing particles that can be obtained from these strains.

Since we propose that in both Hfr's EC8 and EC15 the z gene is closer to att_{80} than the y gene, we predict that from these strains occasional $\phi80$ transducing particles could be obtained which carry an intact z gene and only part of the y gene. In contrast, it should not be possible to obtain from these strains $\phi80lac\,z^-y^+$ transducing particles. We have devised a simple method for screening for such transducing phages which are z^+y^-. The principle of this isolation is to obtain lac^+ transductants in a recipient strain which carries a conditionally expressed y gene. In such a strain, those transductants which carry $\phi80lac\,z^+y^+$ will be lac^+ whether or not the y gene of the recipient is expressed; in contrast, those transductants which carry $\phi80lac\,z^+y^-$ will be lac^+ only when the y gene of the recipient is expressed. In this way, the two types of transducing phages can be distinguished. Jacob $et\ al.$ (1965) have isolated strains in which the lac permease gene is connected to the purE operon and is thus under purine control. Permease production is, therefore, repressed in the presence of adenine. The hybrid operon, which has a partially deleted z gene, is carried on an F' episome. One of these strains, EZ0, was used to screen for $\phi80lac\,z^+y^-$ transducing phages.

Using low-frequency transducing lysates made from lysogens of EC8 and EC15, transductants of EZ0, which were lac^+ in the absence of adenine, were isolated on lactose–synthetic agar plates. Under these conditions, the permease of EZ0 is made. The lac^+ transductants were then partially purified by picking on to the same selective plates and then replicated on to lactose synthetic medium with and without adenine. From lysates of $both$ strains, we have found lac^+ transductants of EZ0 which are lac^- in the presence of adenine (9/383 from EC8 and 4/181 from EC15). These transductants were then further purified and lysates of high-frequency transducing titre (HFT) were made. All HFT lysates were tested on a series of lac^- strains to confirm their z^+y^- character. None of the lysates could transduce to lac^+ a strain carrying a complete deletion of the lac region, X74, but all could transduce a z^-y^+ strain, X90. Furthermore, different HFT transducing phages were shown by their recombination patterns with various y^- mutants to carry different sized deletions of the y end of the operon (Fig. 3). Thus we have obtained $\phi80lac\,z^+y^-$ phages from both EC8 and EC15. We have also screened for $\phi80lac\,z^-y^+$ transducing particles by selecting for y^+ transductants of a lac-deletion strain, X74, and then scoring the z character. It is possible to select only for y function by selecting for growth on melibiose as sole

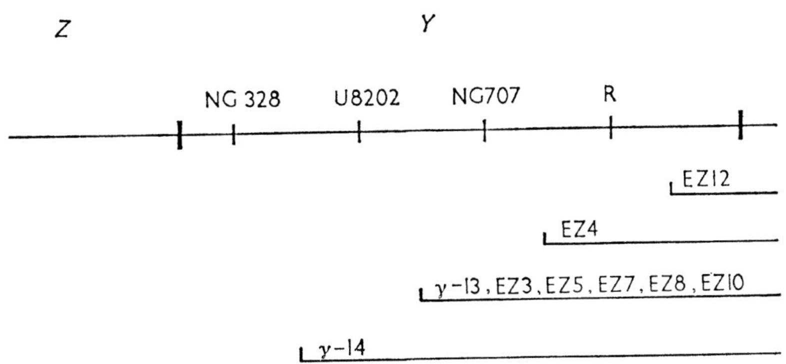

Fig. 3. Extent of y deletions in $\phi80lac\,z^+y^-$ inferred from recombination as determined by spotting HFT lysates on streaks of the appropriate lac mutants. EZ7, 8 and 10 are derived ultimately from EC15 and the rest from EC8.

carbon source (Pardee, 1957; Beckwith, 1963). We have not obtained $\phi80lac\ z^-y^+$ transducing particles from either EC8 (0/412) or EC15 (0/360). These results provide substantial evidence that the orientation of the *lac* operons relative to att_{80} in EC8 and EC15 is as indicated in Fig. 1.

(e) *Further evidence for inversion*

The Campbell model predicts that the orientation of a transduced gene located in a transducing phage lysogenized at the phage attachment site be the same as in the strain from which it was derived. We have already shown above that the direction of marker transfer conferred upon an appropriate strain by $\phi80lac$ is the same as that of the Hfr from which it was derived. Therefore, in cases where $\phi80lac\ z^+y^-$ has integrated at att_{80}, the partial *lac* operons should have the same orientation as in the original Hfr's. The *lac* operon in transductants of EZ0 deriving from EC8 should be oriented in the opposite direction from those deriving from EC15. The following experiment demonstrates that this is in fact the case. In these tranductants of EZ0 there exists a small degree of genetic homology between the *lac* regions of the F' episome and of the integrated phage (Fig. 4). It should be possible to isolate, from these

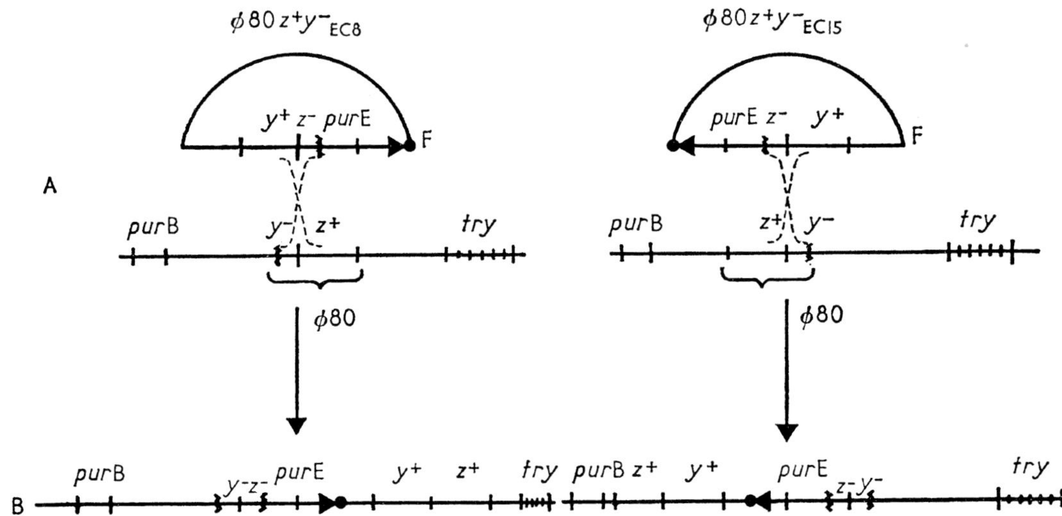

FIG. 4. Integration of F'-ad-z^-y^+ into 2 chromosomes lysogenic for $\phi80lac\ z^+y^-$ phages with opposite orientations. Part A shows synapsis between the homologous regions of the F'-ad-lac episome (the closed line) and the chromosome (straight line) carrying either of the integrated transducing phages. The dotted lines represent a reciprocal cross-over event within the region of homology. Part B shows the products of these cross-overs, Hfr's the direction of transfer of which depends upon the orientation of the chromosomal *lac* region in part A.

strains, recombinants in which the permease is no longer under the control of adenine and which are, therefore, lac^+ even in the presence of adenine. When the recombination event is due to a single reciprocal cross-over in the *lac* region, the integration of the episome into the chromosome should result in an Hfr in which the direction of chromosome transfer is dependent upon the orientation of the *lac* operon in the transducing phage (Fig. 4). We have isolated recombinants which are lac^+ in the presence of adenine from three transductants of EZ0, two of which derive from EC15 and one from EC8. These recombinants are Hfr's with direction of chromosome

 J. R. BECKWITH AND E. R. SIGNER

transfer the same as the Hfr from which the transducing phage is derived (Table 4 and Fig. 5).

TABLE 4

Transfer of markers by Hfr's derived from the integration of F'-ad-z^-y^+ in strains lysogenic for $\phi 80\ z^+y^-_{EC8}$ and $\phi 80\ z^+y^-_{EC15}$ in 60-min blendor experiments

| | Number of recombinants/ml. ($\times 10^{-3}$) | | | |
| | *pur*B$^+$ | | *try*$^+$ | |
	0 min	60 min	0 min	60 min
EZ103$_{EC8}$	0	400	0	2
EZ104$_{EC15}$	0	5	0	100
EZ105$_{EC15}$	0	4	0	150

Strain EZ103 was isolated from a derivative of EZ0, which is lysogenic for a $\phi 80\ z^+y^-$ phage derived originally from EC8. EZ104 and EZ105 are similar strains, except that the transducing phages are derived originally from EC15. All 3 strains were picked up as rare *lac*$^+$ colonies on LZ agar, where most of the colonies were *lac*$^-$ due to the repression of the permease by the adenine in the nutrient medium. The Hfr's were crossed with a recipient F$^-$ strain, *pur*B$^-$, *try*$^-$ and lysogenic for $\phi 80$ to avoid the complication of zygotic induction.

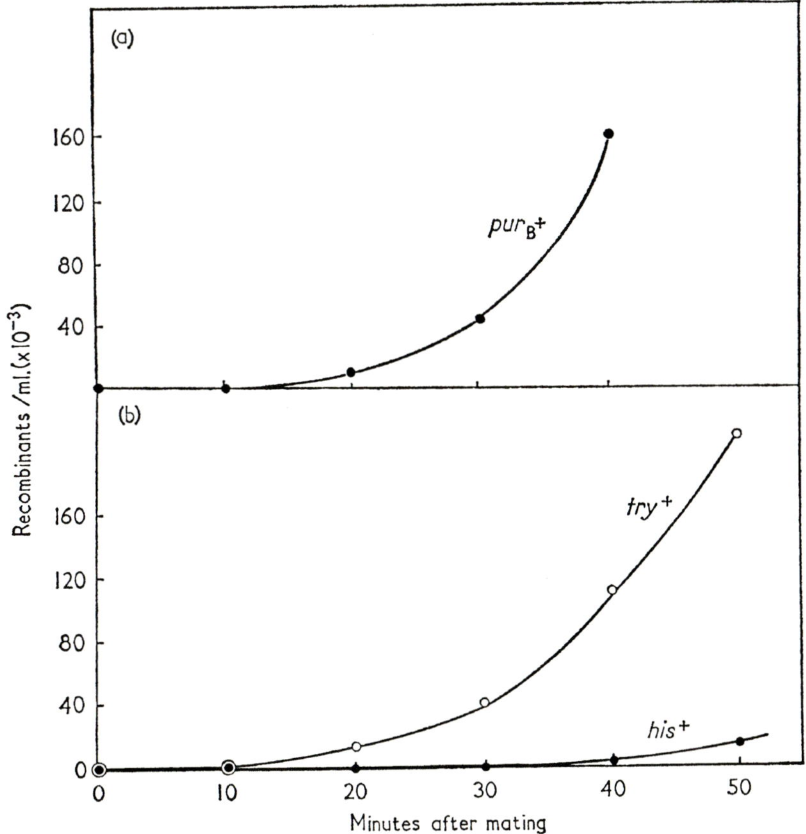

Fig. 5. Kinetics of marker transfer in interrupted mating experiments. (a) Donor EZ103, recipient *pur*B$^-$. (b) Donor EZ104, recipient *try*$^-$*his*$^-$.

4. Discussion

(a) *Isolation of Hfr's*

The evidence presented here strongly suggests that the Hfr EC15 arose from the integration of the F_{TS}-*lac* into the *T1rec* gene, since the origin of the Hfr is in the very short region (Signer *et al.*, 1965) between att_{80} and *try*, where the *T1rec* gene lies, and since this Hfr occurred at a high frequency amongst $T1^r$ derivatives of the parental strain. The inactivation of the *T1rec* gene would result from its splitting into two fragments as a consequence of the integration of the F_{TS}-*lac*. We cannot exclude the possibility that integration of the episome in close proximity to a gene could inactivate it; however, no examples of this type of position effect have been described in *E. coli*. We propose the same type of event to account for the auxotrophic Hfr, EC11, and for the $T6^r$ temperature-resistant *lac*$^+$ Hfr, EC102.

(b) *A model for inversion of* lac *resulting from integration*

The Hfr's EC8 and EC15 donate their chromosomes in opposite directions. One explanation for the formation of Hfr's which donate in opposite directions is presented in Fig. 6. Evidence for the role of homology in the integration of episomes into the chromosome has already been presented by other workers (Scaife & Gross, 1963; Jacob *et al.*, 1963). These workers found that integration of *F-lac* into the chromosome

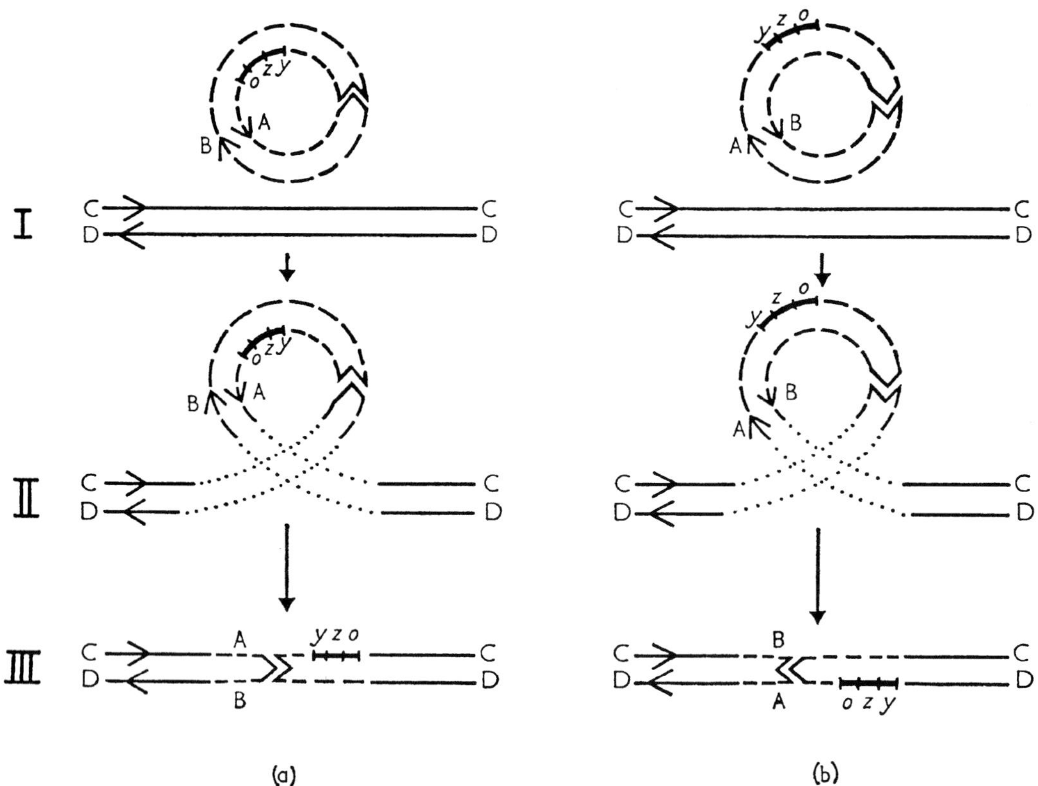

FIG. 6. Origin of Hfr's with opposite polarities (see text for details and discussion). The circular broken lines represent the *F-lac* episomes, and the continuous straight lines the chromosome. The arrows on complementary strands indicate their antiparallel nature. The directions of transfer of the episomes and of the resultant Hfr's are indicated by the dovetailing segments.

In part II, the dotted lines represent the cross-over between episome and chromosome.

   J. R. BECKWITH AND E. R. SIGNER

of *point lac* mutants occurred preferentially in the *lac* region. Cuzin & Jacob (1964) proposed that the integration which is found at other sites of the chromosomes of strains carrying large deletions, such as that discussed in this paper, is due to the presence at such points of regions of weak homology with the episome. We, therefore, present the following detailed model. We assume that the integration of the F_{TS}-*lac* episome into the chromosome by a single reciprocal cross-over results in the formation of an Hfr (Campbell, 1962). Both episome and chromosome are DNA double helices. We assume that the direction of transfer determined by the F-factor in bacterial crosses is constant with respect to the orientation of its (antiparallel) strands. We also assume that only one strand of the operon carries the information for synthesis of the *lac* proteins.

The stages of integration of the episome involve first the pairing between homologous regions of the episome and chromosome. If the homology is to lead to integration, we must define homology to mean that not only are there similar nucleotide sequences, but also the chemical directions of these sequences are the same. In case (a) (Fig. 6) a region of strand A of the episome is homologous with one on strand C of the chromosome, and similarly on the complementary strands B and D (I). A single reciprocal cross-over in this region (II) will lead to an Hfr (III) with the polarity of transfer indicated in Fig. 6(a). Strand A of the episome, including the *lac* informational strand, becomes continuous with strand C of the chromosome, as does strand B with strand D (III).

In contrast, in case (b) (Fig. 6), in a different region of the chromosome, the homology may be between nucleotide sequences on strand A of the episome and strand D of the chromosome, and similarly between the complementary strands B and C (I). For the integration of episome into chromosome as a result of this homology, the episome must be inverted relative to the chromosome compared to case (a), in order that the homologous strands do, indeed, have the same chemical direction. The result of subsequent integration of the episome (II) is an Hfr (III) the polarity of transfer of which is opposite to that derived in case (a). Now strand A of the episome, again including the *lac* informational strand, is continuous with strand D of the chromosome, and strand B with strand C. Thus, in the two cases, not only do the two Hfr's donate in opposite directions, but also the *lac* operons in the two strains are opposite in orientation and furthermore are on different strands of the host chromosome DNA duplex.

According to the Campbell model, integration results in the presence at each end of the integrated episome of a complete copy of the homology region through which pairing and recombination have taken place. However, if this homology region is only a short segment of a functional gene, then clearly integration of the episome will lead to the separation of the gene into two inactive fragments, in spite of the duplication of the homology segment at each end of the integrated episome. This will account for the formation of EC11 and EC15.

(c) *Experimental verification of inversion*

We have been able to test the prediction that in two Hfr's with opposite polarities of transfer derived from an $F_{TS}{}^{-}$*lac*, the *lac* operons will be oriented in opposite directions. From our information on the Hfr origins of EC8 and EC15, we know that the *lac* operons are on opposite sides of att_{80} on the chromosomes of the two Hfr's (Fig. 1). Since we have been able to obtain $\phi 80lac\ z^{+}y^{-}$ transducing phages from *both* Hfr's, it seems very likely that the order of the *lac* genes relative to att_{80} is as indicated in Fig. 1. This conclusion is based on the evidence, from other systems, that specialized transducing phages can transduce a near marker without necessarily transducing a more distant marker, but not *vice versa*.

Further evidence supporting the scheme presented in Fig. 6 comes from studies on the Hfr's derived from recombination between the *lac* regions of the $F\text{-}ad\text{-}z^-y^+$ episome and different $\phi 80lac\ z^+y^-$ phages integrated in the chromosome. When Hfr's are obtained in this way from two such strains which carry $\phi 80lac\ z^+y^-$ phages having *lac* regions in opposite orientations, the direction of transfer of the resultant Hfr's corresponds to the *lac* orientation.

We have thus shown that in EC8 and EC15 the *lac* operons are pointing in opposite directions. It follows from the model presented in Fig. 6, that the informational *lac* strand relative to the rest of the chromosome is different in the two Hfr's. Studies on the expression of *lac* operon on different strands indicate that it functions equally well on both strands. These studies will be presented and discussed in a subsequent paper (Beckwith & Signer, in preparation).

We thank Dr Francois Jacob and Dr Francois Cuzin for many useful discussions during the course of this work and Dr Jacob and Dr John Scaife for invaluable suggestions in the preparation of the manuscript. We are grateful to Dr Jacob for the use of his laboratory. We were both Fellows of the Jane Coffin Childs Memorial Fund for Medical Research. This investigation was supported in part by grants from the National Science Foundation and the Délégation Générale à la Recherche Scientifique et Technique to Dr F. Jacob.

REFERENCES

Adler, J. & Templeton, B. (1963). *J. Mol. Biol.* **7**, 710.
Beckwith, J. R. (1963). *Biochim. biophys. Acta*, **76**, 162.
Brenner, S. & Beckwith, J. R. (1965). *J. Mol. Biol.* **13**, 629.
Campbell, A. (1962). *Advanc. Genetics*, **11**, 101.
Cuzin, F. (1965). *C.R. Acad. Sci. Paris*, **260**, 6482.
Cuzin, F. & Jacob, F. (1964). *C.R. Acad. Sci., Paris*, **258**, 1350.
Gratia, J. P. (1964). *Ann. Inst. Pasteur*, **107**, 132.
Jacob, F., Brenner, S. & Cuzin, F. (1963). *Cold Spr. Harb. Symp. Quant. Biol.* **28**, 329.
Jacob, F., Ullman, A. & Monod, J. (1965). *J. Mol. Biol.* **13**, 704.
Jacob, F. & Wollman, E. L. (1961). In *Sexuality and the Genetics of Bacteria*, pp. 164–167. New York: Academic Press.
Loeb, T. & Zinder, N. D. (1961). *Proc. Nat. Acad. Sci., Wash.* **47**, 282.
Matsushiro, A. (1963). *Virology*, **19**, 475.
Morse, M. L., Lederberg, E. & Lederberg, J. (1956). *Genetics*, **41**, 758.
Newton, W. A., Beckwith, J. R., Zipser, D. & Brenner, S. (1965). *J. Mol. Biol.* **14**, 290.
Pardee, A. B. (1957). *J. Bact.* **73**, 376.
Pardee, A. B., Jacob, F. & Monod, J. (1959). *J. Mol. Biol.* **1**, 165.
Scaife, J. (1966). *Genet. Res., Camb.* in the press.
Scaife, J. & Gross, J. (1963). *Genet. Res., Camb.* **4**, 328.
Scaife, J. & Pekhov, A. (1964). *Genet. Res., Camb.* **5**, 495.
Signer, E. R. (1964). *Virology*, **22**, 650.
Signer, E. R. (1966). *J. Mol. Biol.* **15**, 243.
Signer, E. R., Beckwith, J. R. & Brenner, S. (1965). *J. Mol. Biol.* **14**, 153.

Chemoreceptors in Bacteria

J. Adler

For a hundred years it was known that motile bacteria are attracted to a variety of small organic molecules. However, few scientists were interested in bacterial chemotaxis, probably because they were unwilling to believe that these lowly organisms possessed any capability for information processing or could exhibit even simple forms of behavior. Despite evidence to the contrary, it was generally assumed that chemotaxis and metabolism were hopelessly entwined. Bacteria simply congregated where the food was; after all, that was where growth rates were fastest. Julius Adler broke this prejudice.

Undaunted by peer pressure, Adler set out to uncover the molecular basis for bacterial chemotaxis and, in particular, to test rigorously the perceived connection between this phenomenon and metabolism. First he modified a method developed by Pfeffer in the 1880s to permit a quantitative analysis of chemotaxis with *Escherichia coli*, an experimentally tractable organism. Basically this method involves inserting a capillary containing an attractant solution into a suspension of bacteria and then counting the cells that swim into the tube after a defined incubation period. Legend has it that he searched the sewers of Madison, Wis., to find an intelligent strain of *E. coli*. Domesticated strains, which are used to a life of luxury, had become either stupid or paralyzed.

The paper is written in a beautifully clear, Socratic style; questions are posed and answers are provided. With this quantitative assay, Adler presented five lines of evidence demonstrating that bacteria have chemoreceptors for attractants: (i) some metabolites fail to attract, (ii) some attractants cannot be metabolized, (iii) attractants can be detected even when cells are flooded with metabolites, (iv) competition is observed with structurally related attractants, and (v) mutants defective in chemotaxis can still metabolize the molecule in question. Moreover, using attractant competition and mutant analysis, he went on to identify at least five different chemoreceptors. Appropriately enough, the paper ends with a section entitled "Implications for neurobiology and behavioral biology."

Adler's elegantly simple experiments demonstrated that bacteria such as *E. coli* can sense and process environmental information with surprising sophistication. Now many scientists were "attracted" to chemotaxis, and the field grew exponentially. What is remarkable is the diversity of these scientific converts. They include mathematicians and physicists, biochemists and structural biologists, geneticists and molecular biologists, and neurobiologists. Despite the fact that the components of *E. coli's* "brain" have been identified and analyzed in great detail, important questions remain, including the basis for the large range of ligand sensitivity and the mechanisms of signal amplification and adaptation. Because these questions are fundamental to any sensory system, it is likely that bacterial chemotaxis will remain at the forefront of this important research field. Julius Adler spawned an enormously productive enterprise.

Thomas J. Silhavy

Chemoreceptors in Bacteria

Studies of chemotaxis reveal systems that detect attractants independently of their metabolism.

Julius Adler

Motile bacteria are attracted to a variety of chemicals—a phenomenon called chemotaxis [for a review, see (1)]. Although chemotaxis by bacteria has been recognized since the end of the 19th century, thanks to the pioneering work of Engelmann, Pfeffer, and other biologists, the mechanisms involved are still almost entirely unknown. How do bacteria detect the attractants? How is this sensed information translated into action; that is, how are the flagella directed? This article deals primarily with the first question.

To learn about the detection mechanism that bacteria use in chemotaxis, it is important first to know *what* is being detected. One possibility is that the attractants themselves are detected. In that case, extensive metabolism of the attractants would not be necessary for chemotaxis. There is another possibility: the attractants themselves are not detected but, instead, some metabolite of the attractants is detected (for example, the pyruvate inside the cell); or the energy produced from the attractants, perhaps in the form of adenosine triphosphate, is detected. In these cases, metabolism of the attractants would be necessary for chemotaxis. The idea that bacteria sense the energy produced from the attractants has, in fact, gained wide acceptance for explaining chemotaxis (and also phototaxis) (2).

To try to determine which of these possibilities is correct, experiments were carried out with *Escherichia coli* bacteria, which had previously been demonstrated to exhibit chemotaxis toward various organic nutrients (*3*). The results show that extensive metabolism of the attractants is not required, or sufficient, for chemotaxis. Instead, the attractants themselves are detected.

The systems that bacteria use to detect chemicals without metabolizing them are here called "chemoreceptors." Efforts to identify the chemoreceptors are described.

A Quantitative Method for Studying Chemotaxis

In the 1880's Pfeffer (*4*) demonstrated chemotaxis by exposing a suspension of motile bacteria to a solution of an attractant in a capillary tube and then observing microscopically that the bacteria accumulated first at the mouth of the capillary (Fig. 1) and later inside. A modification of this method, which permits quantitative study of chemotaxis, is here described briefly (*5*).

Wild-type *Escherichia coli* K12, strain W3110, was used, except where otherwise indicated. A capillary tube containing a solution of attractant was pushed into a suspension of bacteria on a slide (*6*). After incubation at 30°C (*7*) for 60 minutes, the capillary was taken out of the bacterial suspension and washed to remove bacteria adhering to the outside. The number of bacteria inside the capillary was then measured by plating the contents of the capillary and counting colonies the next day. The error is $\pm$ 15 percent.

A typical result for glucose (*8*) at various concentrations is shown in Fig. 2. From such a dose-response curve— or, better, from a double log plot—one can estimate a threshold concentration for accumulation inside the capillary, in this case about $4 \times 10^{-7}M$. (The threshold is actually lower than this, since the glucose is being used up.) At the highest concentrations, so much attractant diffuses out that the bacteria which have accumulated outside the capillary do not enter in the time allowed. The peak concentration varies with time of incubation, rate of use of the attractant, and other factors (*5*).

Results similar to that shown in Fig. 2 were obtained for other attractants— for example, galactose, ribose, aspartate, and serine (*8*).

Are the attractants themselves detected, or is it something that results

The author is a professor in the departments of biochemistry and genetics at the University of Wisconsin, Madison.

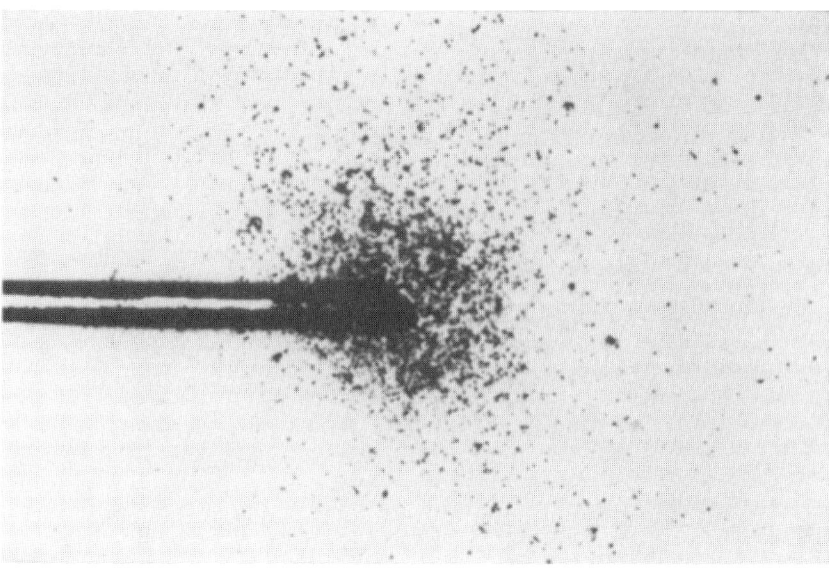

Fig. 1. Photomicrograph showing attraction of *Escherichia coli* bacteria to aspartate. The capillary tube (diameter, $\sim$ 25 microns) contained aspartate at a concentration of $2 \times 10^{-3}M$. [Photomicrograph by Scott W. Ramsey; dark-field photography]

Table 1. The ability of various metabolizable chemicals to attract *Escherichia coli*.

Attractant	Threshold molarity	Chemotaxis* Maximum response† Molarity	No. of bacteria attracted	Doubling time for growth (hours)
Galactose	4×10^{-7}	10^{-3}	125,000	2.6
Galactonate	10^{-1}	10^{-1}	5,000	2.0
Glucose	4×10^{-7}	10^{-3}	187,000	1.2
Gluconate	$> 10^{-1}$	(No response)		1.1
Glucuronate	$> 10^{-1}$	(No response)		1.1
Glycerol	$> 10^{-1}$	(No response)		1.8
α-Ketoglutarate	$> 10^{-1}$	(No response)		2.5
Succinate	10^{-2}	10^{-1}	8,000	2.0
Fumarate	$> 10^{-1}$	(No response)		2.0
Malate	10^{-1}	10^{-1}	5,000	1.7
Pyruvate	$> 10^{-1}$	(No response)		3.0

* The chemotaxis studies were carried out for 1 hour with wild-type (W3110) bacteria grown on each chemical ($0.025M$) as sole source of carbon and energy, in a medium described elsewhere (*45*).
† "Maximum response" refers to the number of bacteria attracted into a capillary tube in 1 hour at the peak concentration of attractant. The peak concentration was determined from a dose-response curve for concentrations between $10^{-6}M$ and $10^{-1}M$ (as in Fig. 2) for each chemical. A background value (the value obtained when there is no attractant in the capillary tube) of about 3000 bacteria has been subtracted (see *46*).

Table 2. The ability of L-aspartate and L-serine and of some of their products to attract *Escherichia coli*.

Attractant	Threshold molarity	Chemotaxis* Maximum response Molarity	No. of bacteria attracted	Oxygen uptake† (μl/hr per 10^{-9} cells)
Aspartate	6×10^{-8}	3×10^{-3}	330,000	12
Serine	2×10^{-7}	10^{-3}	194,000	25
Succinate	10^{-2}	10^{-1}	43,000	61
Fumarate	10^{-1}	10^{-1}	3,000	30
Malate	10^{-1}	10^{-1}	3,000	44
Oxalacetate	$> 10^{-1}$	(No response)		5
Pyruvate	$> 10^{-1}$	(No response)		61

* The bacteria were grown on glycerol as sole source of carbon and energy. Otherwise the conditions were as described for Table 1.   † Oxygen uptake was measured in chemotaxis medium (see *6*) at 30°C.

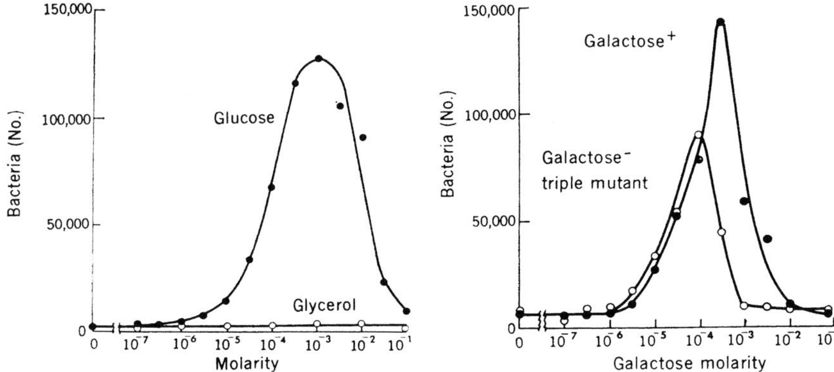

Fig. 2 (left). Graph showing chemotaxis toward glucose (solid circles) but not toward glycerol (open circles). Wild-type (strain W3110) bacteria, grown on glycerol, were used, and the experiment lasted 1 hour. Fig. 3 (right). Graph showing chemotaxis toward galactose by a wild-type strain, W3110 (solid circles), and by a galactose⁻ mutant, W4690 (open circles), which has a mutation in each of the genes for three enzymes needed in the metabolism of galactose. Both strains of bacteria were grown on mannose (see *30*). The experiment lasted 1 hour.

from the metabolism of the attractants —an intermediate, or the energy produced—that is detected? The following five approaches lead to the conclusion that chemotaxis is not a consequence of the metabolism of the attractants but, rather, that the attractants themselves are detected.

Inert Metabolizable Chemicals

1) *Some chemicals that are extensively metabolized fail to attract bacteria.* This result makes it clear that metabolism of a chemical and energy production from it are not sufficient to make a chemical an attractant.

Table 1 shows that among a number of chemicals that are readily metabolized and yield energy, as judged by the ability of *Escherichia coli* to grow on them in the absence of any other carbon and energy source, there are some that fail to attract the bacteria or that attract them very weakly. (The weak response of some of the citric acid cycle compounds could result from their structural resemblance to aspartate, and this resemblance might permit slight detection by an aspartate receptor.) The case of glycerol is shown in more detail in Fig. 2. The inability of glycerol to attract bacteria had already been shown by Pfeffer in 1888 (*1, 4*). The failure of some of the chemicals of Table 1 to attract bacteria was evidently not attributable to an inhibition of chemotaxis, since the presence of each of these chemicals in combination with chemicals that are attractants did not prevent the response of bacteria to the attractants, as docu-

mented below for the case of pyruvate and succinate.

Among the chemicals that are extensively metabolized but fail to attract bacteria (or attract them very weakly) are some that are the first products in the metabolism of the attractants aspartate and serine. First products should also attract, if bacteria detect metabolites of the attractants, or energy produced from the attractants.

In *Escherichia coli*, aspartate is known to be converted to fumarate, a reaction catalyzed by aspartase (*9*). Some aspartate may also be converted to oxalacetate by oxidation or transamination. The resulting fumarate and oxalacetate would give rise to succinate, malate, and pyruvate by way of the

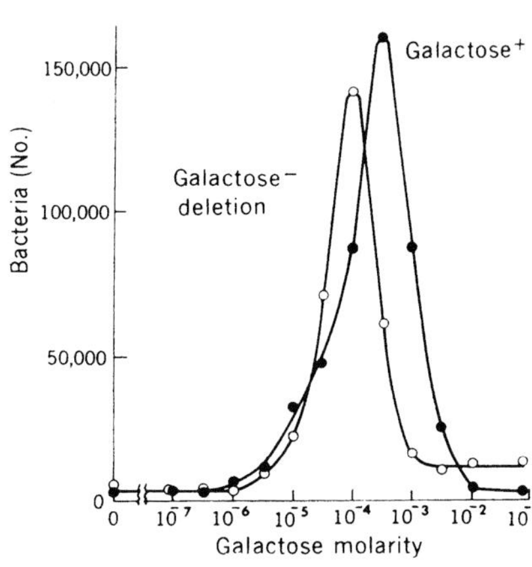

Fig. 4. Graph showing chemotaxis toward galactose by a wild-type strain, W3110 (solid circles), and by galactose⁻ mutant, SU742 (open circles) which has the genes for the three enzymes of galactose metabolism deleted. Both strains of bacteria were grown on mannose (see *30*). The incubation time was 1 hour.

citric acid cycle. Table 2 shows that cells that are strongly attracted to aspartate are not attracted to any of these intermediates or are attracted to them very weakly; these cells are able to use the intermediates readily, as one finds by measuring the rates of oxidation under the same conditions and by the same cells that are used in the chemotaxis studies. (The inertness of most of these chemicals in chemotaxis is shown in Table 1, but in those chemotaxis studies the bacteria are not strictly comparable to one another since they are first grown on the chemical in question with that chemical as the sole source of carbon and energy before being tested for chemotaxis.)

One of the prominent routes of L-serine metabolism in *Escherichia coli* is conversion to pyruvate by L-serine deaminase (*10*). Table 2 shows that the bacteria are attracted strongly to L-serine but not at all to pyruvate, though they oxidize pyruvate readily.

Pyruvate, oxalacetate, malate, fumarate, and succinate are, of course, also intermediates in the metabolism of glucose, galactose, and ribose, which are good attractants.

Nonmetabolizable Attractants

2) *Some chemicals that are essentially nonmetabolizable attract bacteria.* It has now been found that mutant bacteria that have lost the ability to metabolize an attractant are still attracted to it, and that bacteria are attracted to largely nonmetabolizable analogs of attractants.

2a) *Mutant bacteria that have lost the ability to metabolize a chemical are attracted to it.* An *Escherichia coli* mutant, W4690, which lacks three enzymatic activities essential for the metabolism of galactose (galactokinase, galactose-1-phosphate uridyltransferase, and uridine diphosphogalactose-4-epimerase) because of a point mutation in each of the genes for these three enzymes (*11*) and another *E. coli* mutant, SU742, in which these three genes are deleted altogether (*12*) are both strongly attracted to galactose, as compared to wild-type bacteria (Figs. 3 and 4). The response peak occurs at a somewhat higher concentration for the wild-type bacteria than for the mutants; this may reflect the fact that the wild-type bacteria consume the galactose and in this way alter the gradient. The galactose used in these experiments had been purified (*13*) to remove any contaminating attractants, such as glucose.

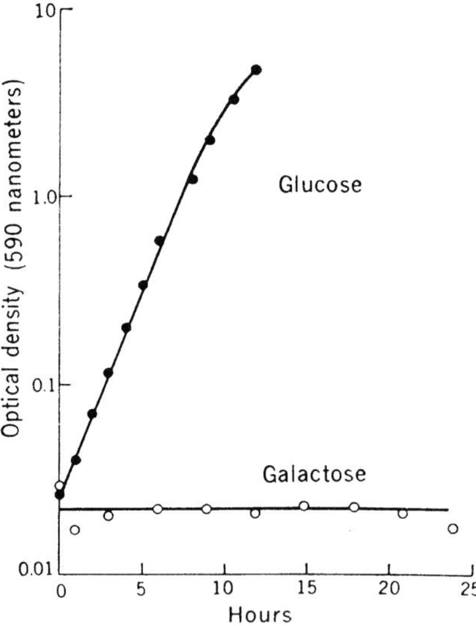

Fig. 5. Graph showing lack of growth on galactose by the galactose⁻ mutant W4690. Bacteria that had been grown on mannose were washed and inoculated into growth medium (*45*) containing galactose (open circles) or glucose (solid circles) at concentration of 0.05*M*, and then shaken at 35°C.

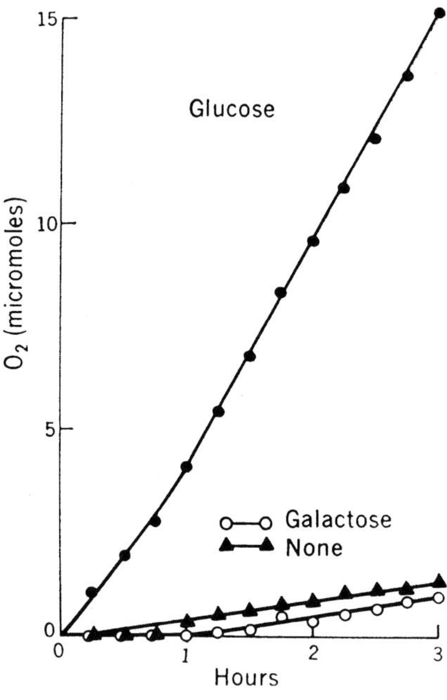

Fig. 6. Graph showing lack of oxidation of galactose by the galactose⁻ mutant W4690 as measured by oxygen uptake. The bacteria (5×10^8 per milliliter) were suspended in a Warburg flask containing 3 milliliters of chemotaxis medium (see *6*) to which galactose (open circles) or glucose (solid circles) was added at a concentration of $1.3 \times 10^{-3}M$, or to which no substrate was added (triangles), and the uptake of oxygen was measured in a Warburg respirometer at 30°C. The results are expressed as micromoles of O_2 taken up per flask. The bacteria had been grown on mannose.

The following evidence shows that mutant W4690 does not metabolize galactose. It failed to grow on galactose, though it grew normally on glucose (Fig. 5), and there was no detectable oxygen uptake on galactose, though there was on glucose (Fig. 6). Measurement of C^{14}-labeled carbon dioxide released from uniformly labeled C^{14}-galactose (*14*) showed that the mutant's production of carbon dioxide is 99.9 percent blocked relative to that of a wild-type strain (Fig. 7). Assays for the three enzymes of galactose metabolism were negative (*11*). Chromatography of an incubation mixture of a heavy suspension of bacteria in a medium containing radioactive galactose (Fig. 8) showed that the galactose was unused, and that slight amounts of only one, or perhaps two, products could be detected. The radioactivity near the origin may be galactose-6-phosphate, known to be produced from galactose and phosphoenolpyruvate (*15, 16*) in mutants that do not metabolize galactose (galactose⁻ mutants) (*17*). That reaction, however, could not be required for chemotaxis, since mutants unable to carry out this reaction showed normal taxis toward galactose.

The deletion mutant SU742 also failed to grow on galactose or to take up any detectable oxygen on galactose, and its production of C^{14}-labeled carbon dioxide from uniformly labeled C^{14}-galactose was 99.5 percent blocked relative to that of a wild-type strain.

An *Escherichia coli* mutant, DF2000, defective in its ability to metabolize glucose because of mutations in the genes for phosphoglucose isomerase and glucose-6-phosphate dehydrogenase (*18*) was attracted to glucose as strongly as its wild-type parent, strain K10, was (Fig. 9). The mutant failed to grow on glucose, and both its oxygen uptake on glucose and its production of C^{14}-labeled carbon dioxide from uniformly labeled C^{14}-glucose (*19*) were 97 percent blocked, relative to the wild-type parent. Being unable to metabolize glucose, the mutant is also unable to metabolize galactose; as expected, the mutant was attracted to galactose as strongly as its parent was.

Analogs That Attract

2b) *Some essentially nonmetabolizable analogs of metabolizable chemicals attract bacteria.* D-Fucose (6-deoxy-D-galactose) is a galactose analog that is not a source of carbon and energy for growth (*20*). Nevertheless the bacteria are attracted to it. In Fig. 10 the responses of bacteria to D-fucose and D-galactose are compared. It may be seen that D-fucose is an effective attractant, though its threshold concentration for chemotaxis is higher than that of D-galactose, as might be expected for an analog. The D-fucose had been purified (*13*) to remove metabolizable impurities such as galactose or glucose. (L-Fucose, while an excellent source of carbon and energy for growth, is inert as an attractant.)

The evidence that D-fucose is essentially nonmetabolizable may be summarized as follows. Buttin has reported (*20*) that D-fucose does not support the growth of *Escherichia coli* (this is demonstrated in Fig. 11), that it is not detectably phosphorylated by galactokinase or by a crude extract of *E. coli*, and that it is not consumed by these bacteria at an appreciable rate. Figure 12 shows that D-fucose, unlike D-galactose, is not detectably oxidized. Chro-

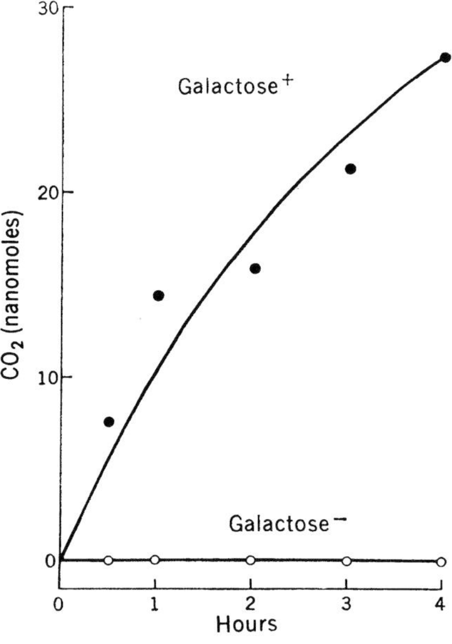

Fig. 7. Graph showing lack of oxidation of galactose by the galactose⁻ mutant W4690 as measured by production of C^{14}-labeled carbon dioxide. Mutant bacteria (open circles) or wild-type bacteria, W3110 (solid circles), each at a concentration of 4×10^8 bacteria per milliliter, were suspended in 1 milliliter of chemotaxis medium containing uniformly labeled C^{14}-galactose (*14*) ($1 \times 10^{-4}M$; 5×10^5 counts per minute per milliliter). Then portions (0.2 milliliter) were incubated at 30°C for various times in flasks equipped with a center well (*47*) for trapping C^{14}-labeled carbon dioxide in ethanolamine and ethylene glycol monomethyl ether (1 : 1) (*48*); a count was made of the trapped carbon dioxide (*48*). The results are expressed as nanomoles of carbon dioxide per flask. The bacteria had been grown on mannose.

Table 3. Effect of added metabolizable chemical on chemotaxis.

Attractant	Number of bacteria attracted			
	No added metabolizable chemical	Pyruvate $(3 \times 10^{-3}M)$	Succinate $(3 \times 10^{-3}M)$	Glucose $(3 \times 10^{-3}M)$
*Wild-type strain**				
None	3,000	4,000	3,000	3,000
Galactose $(10^{-3}M)$	83,000	69,000	71,000	
Fucose $(10^{-2}M)$	168,000	149,000	152,000	
Glucose $(10^{-3}M)$	138,000	109,000	86,000	
Aspartate $(10^{-3}M)$	537,000	548,000	752,000	503,000
Serine $(10^{-3}M)$	382,000	299,000	418,000	494,000
Galactose⁻ strain†				
None	10,000	9,000	6,000	
Galactose $(10^{-4}M)$	67,000	128,000	68,000	
Glucose⁻ strain‡				
None	5,500	3,000	4,000	
Glucose $(10^{-4}M)$	17,000	72,000	87,000	

* The wild-type strain, W3110, was grown on galactose as sole source of carbon and energy. † The galactose⁻ strain, W4690, was grown on mannose (see *30*); the same result was obtained with the galactose⁻ deletion strain, SU742. ‡ The glucose⁻ strain, DF2000, was grown on mannose. All incubations were carried out for 1 hour at 30°C.

matography of an incubation mixture of a heavy suspension of *E. coli* with radioactive D-fucose has revealed that some metabolism does occur (*21*). Five products were formed by wild-type *E. coli*, and three of these were formed by a mutant of *E. coli*, W4690, that lacks the enzymes of galactose metabolism but is attracted to D-fucose as strongly as wild-type bacteria are. In wild-type *E. coli*, D-fucose is converted into products at only 0.5 to 1.0 percent of the rate at which D-galactose metabolism occurs.

Three glucose analogs which failed to support growth and which were oxidized at less than 5 percent of the rate for glucose are also attractants: 2-deoxyglucose, α-methyl glucoside, and L-sorbose. It had been known that these analogs do not undergo extensive conversion (*22*). The thresholds for chemotaxis were $3 \times 10^{-4}M$, $10^{-3}M$, and $3 \times 10^{-4}M$, respectively, and the maximum responses were 125,000 bacteria at $10^{-1}M$, 210,000 at $10^{-1}M$, and 123,000 at $3 \times 12^{-2}M$, respectively. The three analogs were purified before use in order to remove glucose or other contaminants (*13*).

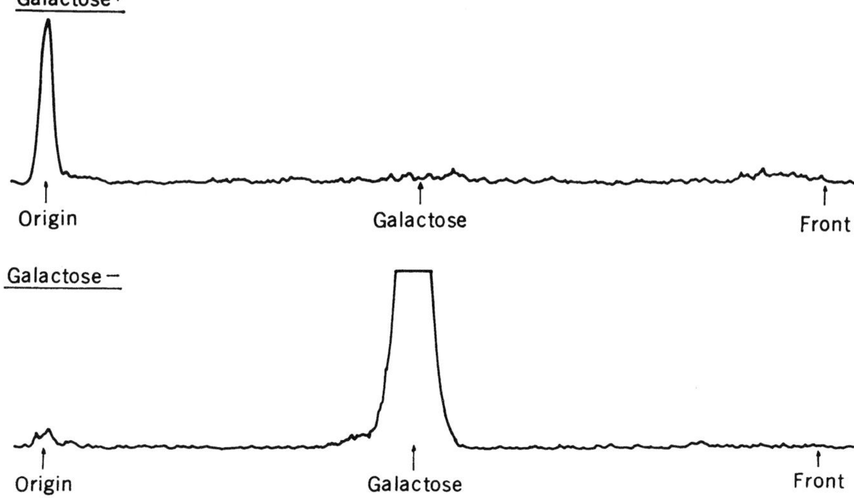

Fig. 8. Chromatrographic records showing lack of metabolism of galactose by the galactose⁻ mutant W4690. Mutant bacteria (bottom record) or wild-type bacteria, W3110 (top record), each at a concentration of 2×10^9 bacteria per milliliter in 1 milliliter of chemotaxis medium containing uniformly labeled C¹⁴-galactose (*14*) ($2 \times 10^{-4}M$; 6×10^5 counts per minute per milliliter), were incubated at 30°C for 2.5 hours. A 0.05-milliliter sample from each was then chromatographed in an ethylacetate, pyridine, H₂O (8:2:1) system for 12 hours. The top record shows that the wild-type bacteria consumed the galactose; radioactivity at the origin probably represents incorporation into cellular materials. The bacteria had been grown on mannose.

Metabolized Chemicals Do Not Block

3) *Chemicals attract bacteria even in the presence of a metabolizable chemical.* If bacteria detect metabolites of an attractant, or energy produced from it, then the addition of a metabolizable chemical should stop chemotaxis by flooding the cells with metabolites and energy. This was not found to be the case for either metabolizable or nonmetabolizable attractants.

Table 3 shows that the presence of the metabolizable chemicals pyruvate or succinate in the bacterial suspension and in the capillary did not block chemotaxis toward galactose, glucose, aspartate, or serine. Even the chemicals that are not metabolized—fucose, or galactose in the case of the galactose⁻ mutant, or glucose in the case of the glucose⁻ mutant—attracted bacteria perfectly well in the presence of pyruvate or succinate. Nor did glucose block chemotaxis toward aspartate or serine.

[It should be pointed out that no energy source has been added in any of the chemotaxis experiments reported in this article except for those of Table 3. Instead, the bacteria rely on an endogenous energy source (*23*). Stimulation by an added energy source in the case of the glucose⁻ mutant indicates that its endogenous energy source is inadequate, and this conclusion is supported by the observation that the endogenous uptake of oxygen is lower in this strain than it is in its glucose+ parent.]

Pyruvate, succinate, and glucose are in fact readily metabolized under these conditions, as shown by oxygen uptake measurements under exactly the conditions of the chemotaxis experiments. They are metabolized even in the presence of the attractants: the oxygen uptake on pyruvate or succinate at a concentration of $3 \times 10^{-3}M$ was not blocked by the addition of fucose ($10^{-2}M$; there was no inhibition), or by the addition of galactose in the case of the galactose⁻ mutant ($10^{-3}M$; there was no inhibition), or by the addition of glucose in the case of the glucose⁻ mutant ($10^{-4}M$; inhibition, 40 percent).

Structurally Related Attractants Compete

4) *Attractants that are closely related in structure compete with each other but not with structurally unre-*

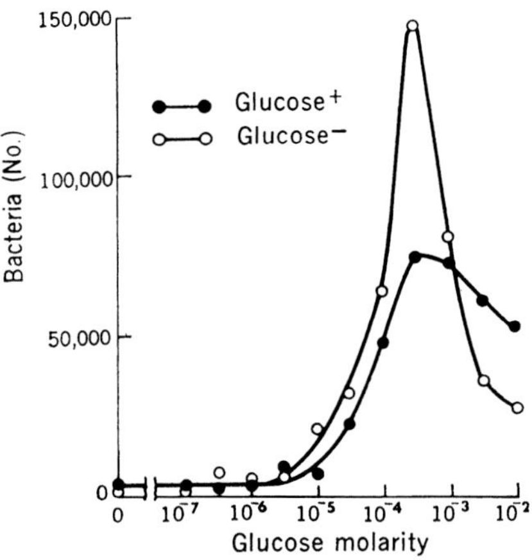

Fig. 9. Graph showing chemotaxis toward glucose by a glucose⁻ mutant, DF2000 (open circles), and its glucose⁺ parent, K10 (solid circles). Both strains of bacteria were grown on mannose. The experiment was carried out for 1 hour in the presence of succinate ($3 \times 10^{-3}M$) as energy source.

lated compounds. This finding supports the conclusion that it is the attractants themselves that are detected, and that there exists a variety of specific receptors.

In these experiments, one attractant at its optimum concentration is put into the capillary tube, and another attractant at a concentration of $0.01M$ is put into both the capillary and the bacterial suspension. If the two attractants use the same chemoreceptor, the response should be inhibited; if they do not, the response should not be affected. (Inhibition could result from other causes too, but failure to

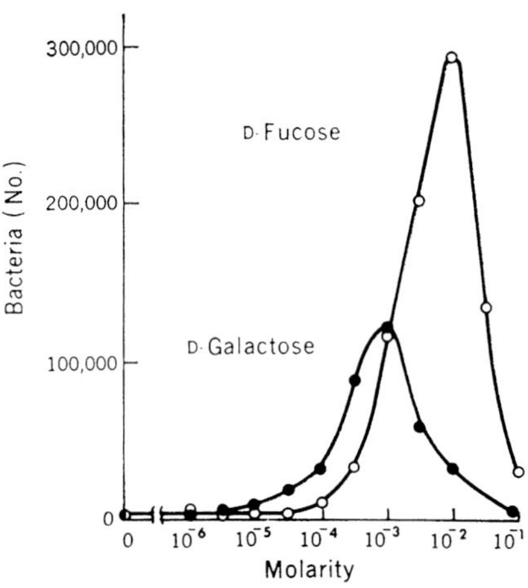

Fig. 10. Graph showing chemotaxis toward D-fucose (open circles) and D-galactose (solid circles). Wild-type bacteria (W3110), grown on D-galactose, were used in this experiment, which lasted 1 hour.

obtain inhibition is strong evidence that the attractants use different receptors.) Experiments of this nature were first carried out in the late 19th century (*26*). Some of our results follow (for a more detailed report, see *27* and *28*).

Chemotaxis toward fucose was completely inhibited by the presence of galactose, and in the reciprocal experiment there was nearly complete inhibition. This suggests that fucose and galactose use the same chemoreceptor (the "galactose receptor").

Glucose completely eliminated taxis toward galactose, but in the reciprocal experiment the inhibition was only about 60 to 70 percent, no matter how high the concentration of galactose was. This suggests that the receptor which detects galactose also detects glucose but that, in addition, there is another receptor that detects glucose but not galactose (the "glucose receptor").

Similar experiments show that ribose is detected by yet another receptor (the "ribose receptor"), which fails to detect either galactose or glucose.

Glucose did not block taxis toward serine or aspartate (Table 3), and neither did galactose, fucose, or ribose. In the reciprocal experiments, aspartate failed to inhibit taxis toward any of the sugars. (Serine slightly inhibits chemotaxis toward all other attractants, and this inhibition remains unexplained.) These results show that the receptors which detect the sugars are different from those which detect the amino acids.

Aspartate did not inhibit taxis toward serine, so evidently there are separate receptors for detecting these two amino acids (the "aspartate receptor" and the "serine receptor"). On the other hand, aspartate completely inhibited taxis toward glutamate, and complete inhibition was found in the reciprocal experiment, so it appears that aspartate and glutamate use the same receptor.

The various chemicals that are not attractants or that attract very weakly all failed to inhibit chemotaxis toward the attractants. (See, for example, the case of pyruvate and succinate in Table 3.)

Mutants Lacking Specific Taxes

5) *There are mutants which fail to carry out chemotaxis to certain attractants but are still able to metabolize them.* If there are chemoreceptors in bacteria and if they are specific, there

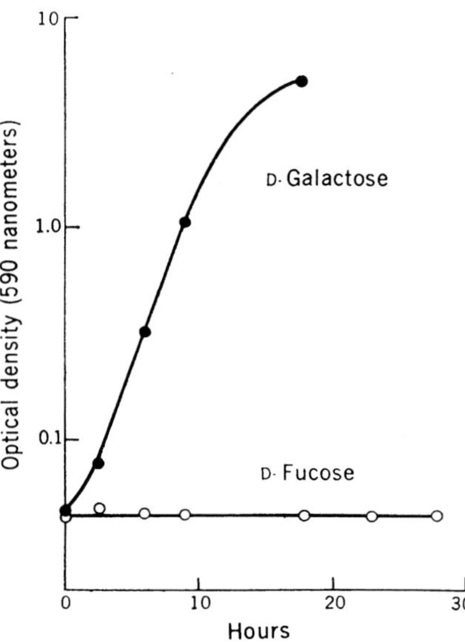

Fig. 11. Graph showing lack of growth on D-fucose by wild-type bacteria, W3110. Bacteria that had been grown on D-galactose were washed and inoculated into growth medium (*45*) containing D-fucose (open circles) or D-galactose (solid circles) at concentration of $0.05M$, and then shaken at 35°C.

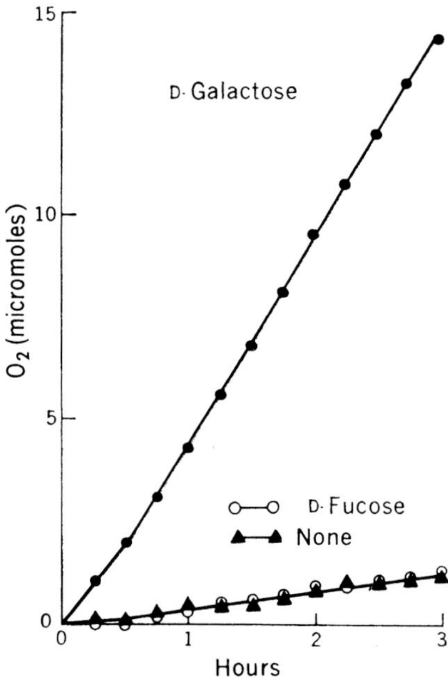

Fig. 12. Graph showing lack of oxidation of D-fucose. Wild-type bacteria (W3110), at a concentration of 8×10^8 bacteria per milliliter, were suspended in a Warburg flask containing 3 milliliters of chemotaxis medium to which D-fucose ($1 \times 10^{-2}M$) (open circles) or D-galactose ($1.5 \times 10^{-3}M$) (solid circles) was added, or to which no substrate was added (triangles), and uptake of oxygen was measured in a Warburg respirometer at 30°C. The results are expressed as micromoles of O₂ taken up per flask. The bacteria had been grown on D-galactose.

should be mutants that are defective in their response to some attractants but not to others, because of a defect in a single receptor. Such mutants of *Escherichia coli* have now been found by Hazelbauer and Mesibov (*24, 25*).

One mutant, defective in the "serine receptor," fails to be attracted to serine, except for a very weak response at the highest concentrations, and shows much-reduced taxis toward alanine, cysteine, and glycine. (These residual responses result from the "aspartate receptor.") The mutant is attracted normally to aspartate and glutamate, and to galactose, glucose, and ribose. It oxidizes and takes up L-serine at the same rate that its parent does.

Another mutant, lacking the "aspartate receptor," shows no chemotaxis toward aspartate and glutamate, and nearly normal taxis toward alanine, cysteine, glycine, serine, galactose, glucose, and ribose. The rate of oxidation and uptake of aspartate is the same for the mutant and its parent.

A third mutant, missing the "galactose receptor," is not attracted to galactose and fucose and is attracted to glucose at a higher-than-normal threshold ($3 \times 10^{-5}M$ instead of $4 \times 10^{-7}M$). (This response to glucose results from the "glucose receptor.") It is attracted normally to ribose, aspartate, and serine. At high concentrations of galactose, growth and oxidation are as fast for the mutant as for a strain, derived from the mutant, that has taxis toward galactose fully restored, but at low concentrations the rates are slower for the mutant. This slowness is caused by a defect in the uptake, rather than in the metabolism, of galactose.

The existence of these mutants argues for specific receptors and provides additional support for the idea that detection of the attractants is independent of their metabolism.

How Many Chemoreceptors?

To determine how many kinds of chemoreceptors there are, three approaches are being used. The first is to ask whether a given attractant is still effective when another attractant is present. The second is to try to isolate mutants defective in individual receptors. A third approach is to study the inducibility of specific taxes (presumably the inducibility of specific receptors) (*27*; see also *29*). For ex-

Table 4. Partial list of chemoreceptors in *Escherichia coli.** [Data from this article and from *24, 27,* and *28*]

Attractant	Threshold† molarity
Galactose receptor	
D-Galactose	4×10^{-7}
D-Glucose	4×10^{-7}
D-Fucose	3×10^{-5}
Glucose receptor	
D-Glucose	3×10^{-5}
Ribose receptor	
D-Ribose	3×10^{-7}
Aspartate receptor	
L-Aspartate	6×10^{-8}
L-Glutamate	1×10^{-5}
Serine receptor	
L-Serine	2×10^{-7}
L-Cysteine	5×10^{-6}
L-Alanine	5×10^{-5}
Glycine	5×10^{-5}

* A more complete description of the specificity of each receptor is in preparation (*24, 27, 28*). † "Threshold" is the concentration of attractant at which bacterial accumulation in the capillary tube first exceeds background accumulation. It is determined by plotting the log of the number of bacteria accumulated relative to the log of the attractant concentration. The threshold value depends on whether or not the chemical is taken up and used (see discussion of Fig. 13 in text). The chemicals listed here were all taken up and, except for D-fucose, were all utilizable; therefore the actual thresholds are considerably lower than those listed.

ample, taxis toward galactose and fucose is inducible by galactose (*30*).

The conclusion from results obtained so far (*24, 27, 28*) is that there are at least the five chemoreceptors shown in Table 4. There are probably no additional receptors for amino acids, since no amino acids besides those listed in Table 4 are strongly attractive (*28*). Among the monosaccharides, current work indicates that there is, in addition, a receptor specific for fructose. Among the disaccharides, research in progress indicates a receptor specific for maltose and another for trehalose. The attraction of bacteria to lactose probably results from chemotaxis toward the galactose (and possibly toward the glucose too) produced from the lactose; lactose itself is actually an extremely poor attractant (*27*). Oxygen is known to be an attractant for *Escherichia coli* (*3*), so there could be a receptor for it, but this question has not been investigated so far. A survey of possible other attractants or of repellents has not been completed.

It is conceivable that, besides chemoreceptors, at least some bacteria might have receptors specialized to detect light, gravity, or temperature, since all these stimuli are known to elicit tactic responses in some bacteria (*1*).

What is the nature of the chemoreceptors? One possibility is that they are the first enzymes in the metabolism of the chemicals. This possibility has been excluded in the case of galactose, because mutants that lack galactokinase still respond perfectly well. Another possibility is that the chemoreceptors are the permeases.

Role of Permeases

What role, if any, is played in chemotaxis by permeases and other components essential for transport of substances into the cell? To find out, mutants defective with respect to transport have been investigated from the standpoint of chemotaxis. It has been found in this study that the permeases and other transport-essential components that have been tested are not required for chemotaxis.

Figure 13 shows good attraction to galactose by an *Escherichia coli* mutant, 20SOK$^-$ (*20*), that is defective in the uptake of galactose (*20, 31*) to the extent of a 99.5 percent block (Fig. 14), owing to the absence of both galactose permease and methyl galactoside permease (*31*), and that, in addition, is unable to grow on, or to metabolize, galactose, as a result of a mutation in the gene for galactokinase. The threshold concentration for taxis toward galactose appears to be even lower (about $2 \times 10^{-9}M$) for the mutant than for strains that are wild-type with respect to galactose transport (compare Fig. 13 with Figs. 3 and 4); actually the thresholds are probably the same for the mutant and the wild-type bacteria, but the latter consume the galactose and in this way destroy the gradient.

Thus the two permeases most responsible (*31*) for transport of galactose in *Escherichia coli*—the galactose permease and the methyl galactoside permease—are not required for taxis toward galactose. A third permease that is in part responsible for transport of galactose, the thiomethyl galactoside permease II, is destroyed at 37°C (*32*), the temperature used for this experiment, and it was, moreover, not induced under the conditions of growth used. The lactose permease (thiomethyl galactoside permease I), normally capable of transporting galactose, does not transport galactose in this strain (*31*) and also was not induced under these conditions of growth. The small amount of galactose which does enter these mutant bacteria is known to be present in a phosphorylated form (*31*), so it is probably transported by the phosphorylation system described next. Mutants blocked in this system carry out chemotaxis toward galactose normally.

A two-enzyme system which catalyzes the phosphorylation of glucose and certain other sugars (*15, 16*) is required for the transport of these chemicals (*16, 33*). Enzyme I catalyzes the phosphorylation of a heat-stable protein by phosphoenolpyruvate; enzyme II then catalyzes the transfer of phosphate from the heat-stable protein to the sugar, and there are specific enzyme II's for different sugars. Three mutants of *Escherichia coli* that lack enzyme I activity [X17 and X19 (*34*) and W327 (*35*)] and a mutant [1101 (*36*)] defective in the heat-stable protein all showed normal chemotaxis toward glucose. An *E. coli* mutant (W1895-D1) defective in the enzyme II activity that phosphorylates glucose and α-methyl glucoside (*37*) was attracted normally to these chemicals. Thus neither enzyme I, nor the heat-stable protein, nor enzyme II activity is required for this chemotaxis.

Neither are the permeases and related transport systems sufficient for chemotaxis: many chemicals for which transport mechanisms exist fail to attract bacteria. Examples are the following amino acids for which active transport is known in *Escherichia coli* (*38*): L-glutamine, L-histidine, L-isoleucine, L-leucine, L-methionine, L-phenylalanine, L-tryptophan, L-tyrosine, and L-valine.

The results presented so far show that the chemoreceptors are not the enzymes that catalyze the metabolism of an attractant, nor are they the parts of the permeases or related systems of transport that have been tested. It remains possible that the chemoreceptors are components of the transport machinery which are intact in the transport mutants studied, or they may be new entities whose special function is the detection of chemicals during chemotaxis.

As discussed above, the *Escherichia coli* mutant that is not attracted to galactose (the "galactose taxis mutant") is defective in the uptake of galactose. This suggests that there may indeed be some component of the transport machinery that plays a role in chemoreception. Since the serine and aspartate taxis mutants show normal uptake of serine and aspartate, respectively, chemoreceptors (at least the serine and aspartate receptors) must contain a component that is not involved in transport.

The chemoreceptors appear to be located somewhere on the "outside" of the cell, since mutants which fail to transport the attractants still are attracted to them. However, this can be offered only as a suggestion, in view of the fact that some amount of attractant would be able to enter even these mutant cells.

Further efforts to identify the chemoreceptors, and attempts to isolate them, are in progress.

How Do Chemoreceptors Work?

The mechanism of chemoreception in bacteria is completely unknown. Somehow the gradient of the chemical affects the receptors (*39*), and this in turn causes a change that directs the flagella. This change could be in the cell membrane—a change in conformation of the membrane or a change in membrane potential. [In the protozoa such changes in potential have been observed prior to the reversal of cilia or to emission of light (*40*).] The change might be propagated along the membrane so as to reach all the flagella. In fact, the base of the flagellum is in close association with the cell membrane (*41*). The flagella could then respond by changing their orientation in some way to bring about an avoiding reaction (see *42*).

We have isolated and reported on 40 mutants which fail to carry out

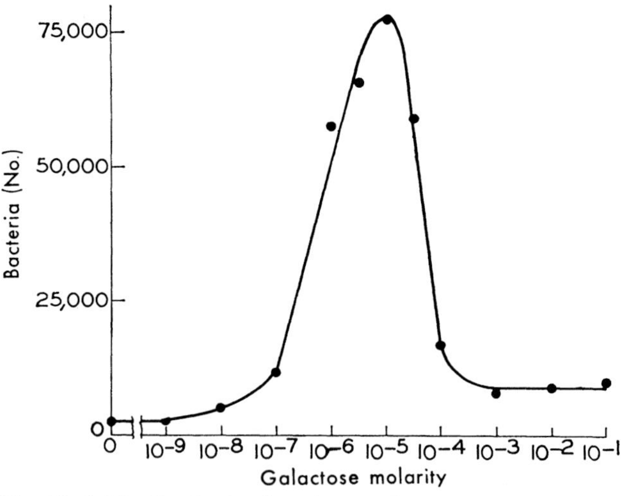

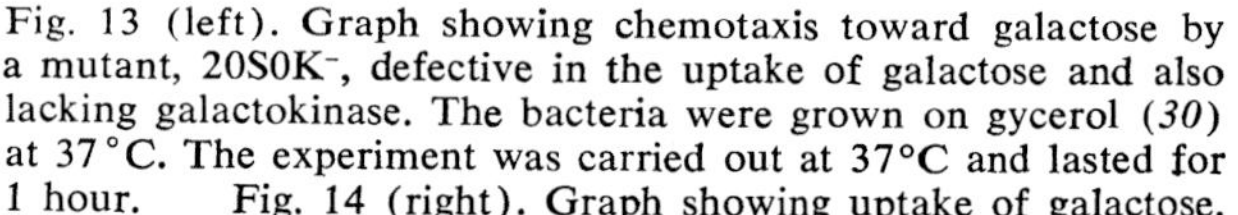

Fig. 13 (left). Graph showing chemotaxis toward galactose by a mutant, 20S0K⁻, defective in the uptake of galactose and also lacking galactokinase. The bacteria were grown on gycerol (*30*) at 37°C. The experiment was carried out at 37°C and lasted for 1 hour.   Fig. 14 (right). Graph showing uptake of galactose. About 5×10^8 bacteria were incubated in 1 milliliter of chemotaxis medium containing uniformly labeled C^{14}-galactose (*14*) ($10^{-6}M$; 7×10^4 counts per minute per milliliter), and at various times 0.05-milliliter samples were removed and washed on Millipore filters to measure the uptake of radioactivity. The permease⁺ strain is W3092 C, a mutant lacking galactokinase (*49*); the permease⁻ strain is 20S0K⁻, a mutant lacking galactokinase and defective in the uptake of galactose (*20*). Bacteria of both strains had been grown on glycerol. The results are expressed as picomoles of galactose taken up by bacteria in 1 milliliter of the incubation mixture.

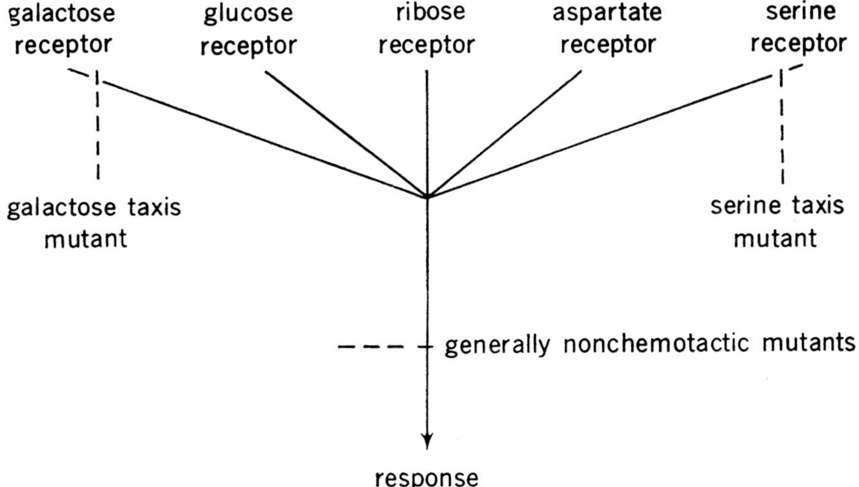

Fig. 15. The scheme for chemotaxis suggested in this article, showing location of defects in the various mutants.

chemotaxis toward any of the attractants—sugars, amino acids, or oxygen—though the bacteria are perfectly motile (43). Since it is unlikely that a single mutation would lead to a loss of all the kinds of chemoreceptors, these mutants are probably defective at some stage beyond the receptors, as shown diagrammatically in Fig. 15. The defect could be in a transmitting system through which information from all the receptors is channeled to the flagella, or in the responding mechanism itself. Genetic analyses of the mutants have shown that three genes are involved (43). Further studies of these mutants may lead to an understanding of the way in which the chemoreceptors direct the flagella.

Implications for Neurobiology and Behavioral Biology

The study of such stimulus-response systems in bacteria may have relevance for neurobiology and for behavioral biology of higher organisms. Possibly the chemoreceptors of bacteria are related to chemoreceptor sites in animal chemoreceptor cells, and perhaps knowledge of the way in which bacterial receptors function might lead to an increased understanding of the mechanism of smell and taste and other kinds of sensory reception (44). If there is an electrical signal that transmits information from the bacterial receptor to the flagellum, it might be similar to changes in membrane potential in higher organisms. The response of the flagellum to this signal may, in some ways, resemble the response of muscle to a nerve impulse.

The availability of behavioral mutants of bacteria—for example, mutants of the types reported here—together with the existence of a great body of knowledge about the genetics and biochemistry of *Escherichia coli*, should make the bacterial system a favorable one for studying simple forms of behavior and perhaps even some primitive kinds of "learning." From such studies might emerge a set of facts and concepts that can be applied to investigations of more complex phenomena in higher organisms.

Summary

Extensive metabolism of chemicals is neither required, nor sufficient, for attraction of bacteria to the chemicals. Instead, the bacteria detect the attractants themselves. The systems that carry out this detection are called "chemoreceptors." There are mutants that fail to be attracted to one particular chemical or to a group of closely related chemicals but still metabolize these chemicals normally. These mutants are regarded as being defective in specific chemoreceptors. Data obtained so far indicate that there are at least five different chemoreceptors in *Escherichia coli*. The chemoreceptors are not the enzymes that catalyze the metabolism of the attractants, nor are they the parts of the permeases and related transport systems that have been tested.

References and Notes

1. C. Weibull, *The Bacteria*, I. C. Gunsalus and R. Y. Stanier, Eds. (Academic Press, New York, 1960), vol. 1, p. 153.
2. J. Links, thesis, University of Leiden (1955) (in Dutch; summary in English); R. K. Clayton, in *Photophysiology*, A. C. Giese, Ed. (Academic Press, New York, 1964), vol. 2, p. 51. Another possibility was that bacteria respond to a physical property of the attractant solution—for example, the viscosity or osmotic strength. This is ruled out by the fact that the responses are highly specific: only certain sugars and certain amino acids are attractive. In addition, the attractants are active at low concentrations such as $10^{-6}M$.
3. J. Adler, *Science* 153, 708 (1966).
4. W. Pfeffer, *Untersuch. Botan. Inst. Tübingen* 1, 363 (1884); *ibid.* 2, 582 (1888).
5. A full account of the method will be given in J. Adler, "A quantitative method for studying chemotaxis," in preparation.
6. The capillary tubes are 1-microliter disposable micropipettes obtained from Drummond Scientific Company, Broomall, Pennsylvania. They are 3 centimeters long and have an internal diameter of 0.2 millimeter. One end was sealed in a flame. A solution was drawn into the capillary by passing the tube quickly through a flame several times and then immediately plunging the open end into the liquid. A chamber formed by laying a U-tube (bent from a 5-centimeter length of capillary tube, Kimax No. 34502, obtained from Owens-Illinois, Toledo, Ohio) between a microscope slide and a cover slip was filled with about 0.25 milliliter of a suspension of bacteria. The bacterial suspension (5×10^7 bacteria per milliliter) and the capillary both contained "chemotaxis medium": potassium phosphate buffer ($1 \times 10^{-2}M$) at pH 7.0; ethylenediaminetetraacetic acid ($1 \times 10^{-4}M$) to protect against inhibition of motility by heavy metal ions [J. Adler and B. Templeton, *J. Gen. Microbiol.* 46, 175 (1967)]; and L-methionine ($1 \times 10^{-6}M$), which is an absolute requirement for chemotaxis in those strains unable to synthesize it [J. Adler and M. M. Dahl, *J. Gen. Microbiol.* 46, 161 (1967)], and stimulates chemotaxis slightly even in wild-type strains. For all experiments reported in this article (chemotaxis experiments and others), the bacteria were grown at 35°C on the carbon and energy source indicated and then washed free of the growth medium, according to procedures described in J. Adler, *J. Bacteriol.* 92, 121 (1966).
7. The incubations were carried out on a slide warmer obtained from Fischer Scientific Company, Pittsburgh, Pa.
8. All sugars mentioned in this article (including fucose) have the D- configuration, and all amino acids mentioned have the L- configuration.
9. For a recent reference on aspartase, see K. Liess, D. Mecke, H. Holzer, *Biochem. Z.* 346, 244 (1966).
10. For a recent work on this enzyme, and a list of references, see L. Alföldi, I. Raskó, E. Kerekes, *J. Bacteriol.* 96, 1512 (1968).
11. Strain W4690, prepared by E. M. Lederberg, carries the mutations *gal* 1, 2, and 22. The content of enzymes for the metabolism of galactose was measured by H. E. Echols.
12. J. A. Shapiro, thesis, University of Cambridge (1967); ——— and S. Adhya. *Genetics*, in press.
13. D-Galactose, already "substantially glucose-free" (from Sigma Chemical Company, St. Louis, Mo.) was separated from any remaining glucose or other impurities by descending chromatography on Whatman No. 40 paper in an ethyl acetate, pyridine, H_2O (120:50:40) system for 12 hours [see I. Smith, *Chromatographic and Electrophoretic Techniques* (Interscience, New York, ed. 2, 1960), vol. 1, p. 248]. To remove any galactose, glucose, or other impurities, chemicals as follows were purified by descending chromatography on Whatman No. 40 paper. D-Fucose (Sigma Chemical Company) and α-methyl glucoside (Pfanstiehl Laboratories, Waukegan, Ill.) were chromatographed in an *n*-butanol, ethanol, H_2O (5:1:4) system for 25 hours [see E. L. Hirst and J. K. N. Jones, *Discussions Faraday Soc.* 7, 268 (1949)]; 2-deoxyglucose (Calbiochem, Los Angeles) was chromatographed in an isopropanol, H_2O (4:1) system for 27 hours [see I. Smith, *Chromatographic and Electrophoretic Techniques* (Interscience, New York,

ed. 2, 1960), vol. 1, p. 248]; and L-sorbose (Sigma Chemical Company) was chromatographed in an ethyl acetate, ethanol, H$_2$O (20:3:2) system for 90 hours (B. E. Butterworth, unpublished).

14. To remove an impurity oxidized by galactose-bacteria, uniformly labeled C^{14}-D-galactose (Mallinckrodt Chemical Works, Orlando, Fla.) was purified by descending chromatography on Whatman No. 40 paper in an *n*-butanol, pyridine, H$_2$O (63:35:48) system for 30 hours [see B. Rotman, A. K. Ganesan, R. Guzman, *J. Mol. Biol.* **36**, 247 (1968)]; it was then rechromatographed in an ethylacetate, pyridine, H$_2$O (8:2:1) system for 18 hours (B. E. Butterworth, unpublished). D-Fucose-6-^{3}H (Calbiochem) was purified by descending chromatography on Whatman No. 40 paper in an ethyl acetate, pyridine, H$_2$O (8:2:1) system for 8 hours, to remove a radioactive impurity.

15. W. Kundig, S. Ghosh, S. Roseman, *Proc. Nat. Acad. Sci. U.S.* **52**, 1067 (1964).

16. W. Kundig, F. D. Kundig, B. Anderson, S. Roseman, *J. Biol. Chem.* **241**, 3243 (1966).

17. D. Rogers and S. Yu, *J. Bacteriol.* **84**, 877 (1962); W. Kundig, F. D. Kundig, B. E. Anderson, S. Roseman, *Fed. Proc.* **24**, 658 (1965).

18. D. G. Fraenkel, *J. Bacteriol.* **95**, 1267 (1968).

19. Uniformly labeled C^{14}-D-glucose was obtained from Volk Radiochemical Company, Burbank, Calif.

20. G. Buttin, *J. Mol. Biol.* **7**, 164 (1963).

21. The procedure was the same as that described in the legend to Fig. 8, except that D-Fucose-6-^{3}H (purified as described above) was used ($5 \times 10^{-4}M$, 2×10^7 counts per minute per milliliter) and chromatography was carried out in an *n*-butanol, ethanol, H$_2$O (5:1:4) system for 24 hours [see E. L. Hirst and J. K. N. Jones, *Discussions Faraday Soc.* **7**, 268 (1949)]. Wild-type (W3110) bacteria were grown on galactose or on mannose, and galactose-bacteria (W4690), on mannose.

22. See, for example, H. Hagihira, T. H. Wilson, E. C. C. Lin, *Biochim. Biophys. Acta* **78**, 505 (1963); D. P. Kessler and H. V. Rickenberg, *ibid.* **90**, 609 (1964).

23. J. Adler and B. Templeton, *J. Gen. Microbiol.* **46**, 175 (1967).

24. G. L. Hazelbauer, R. E. Mesibov, J. Adler, *Proc. Nat. Acad. Sci. U.S.*, in press.

25. The serine taxis mutant was isolated from a strain (AW 405) that is wild-type with respect to chemotaxis, by repeatedly picking the center of the swarm that forms on a semisolid tryptone agar plate, according to the procedure described in J. B. Armstrong, J. Adler, M. M. Dahl, *J. Bacteriol.* **93**, 390 (1967). Several more such mutants were isolated similarly, but the tryptone was replaced by an amino acid mixture that had serine as the chief attractant. Likewise, the aspartate taxis mutant was isolated through use of an amino acid mixture that had aspartate as the chief attractant. The galactose taxis mutation was found in W3109 (*gal$_u$−*), a galactose- strain isolated by E. M. Lederberg [see *Microbiol. Genetics*, W. Hayes and R. C. Clowes, Eds. (Cambridge Univ. Press, Cambridge, England, 1960), p. 115)]. Strains able to metabolize galactose, derived from W3109 either by reversion or by transduction with the bacteriophage lambda dg, retain the galactose defect. The strain that has taxis toward galactose fully restored was isolated from such a strain that had acquired, by reversion, the ability to metabolize galactose.

26. W. Rothert, *Flora* **88**, 371 (1901).

27. J. Adler, G. L. Hazelbauer, M. M. Dahl, "Chemotaxis toward sugars by *Escherichia coli*," in preparation.

28. R. E. Mesibov and J. Adler, "Chemotaxis toward amino acids by *Escherichia coli*," in preparation.

29. R. W. Fleming and F. D. Williams, *Abstr. Amer. Soc. Microbiol.* (1968), p. 29.

30. Taxis toward galactose and fucose is much stronger when the bacteria have been grown on galactose than when they have been grown on glycerol. Growth on mannose, while not as effective as growth on galactose, gives very good taxis toward galactose and fucose. Galactose- bacteria were generally grown on mannose, but in a few cases (strain 20S0K−, for example) growth of galactose- bacteria on mannose resulted in very weak taxis toward galactose; in these cases growth on glycerol proved satisfactory.

31. B. Rotman, A. K. Ganesan, R. Guzman, *J. Mol. Biol.* **36**, 247 (1968).

32. L. S. Prestidge and A. B. Pardee, *Biochim. Biophys. Acta* **100**, 591 (1965).

33. R. D. Simoni, M. Levinthal, F. D. Kundig, W. Kundig, B. Anderson, P. E. Hartman, S. Roseman, *Proc. Nat. Acad. Sci. U.S.* **58**, 1963 (1967); H. R. Kaback, *J. Biol. Chem.* **243**, 3711 (1968).

34. Strains X17 and X19 were isolated by C. F. Fox from strain Hfr H. He found (personal communication) that extracts of the two mutants contain no detectable amounts of enzyme I (less than 0.1 percent of the amount in MO). The doubling time of MO, X17, and X19, on glucose, was found to be 1.0, 8.5, and 8.5 hours, respectively. Chemotaxis experiments were carried out with bacteria grown on lactate. In such mutant cells in chemotaxis medium (see *6*) the rate of accumulation of radioactivity from glucose was found to be inhibited by 88 percent relative to the parental strain MO.

35. Strain W327 is a glucose- mutant [M. Doudoroff, W. Z. Hassid, E. W. Putman, A. L. Potter, J. Lederberg, *J. Biol. Chem.* **179**, 921 (1949)] which fails to accumulate α-methyl glucoside [G. N. Cohen and J. Monod, *Bacteriol. Rev.* **21**, 169 (1957)]. Extracts of W327 lack detectable enzyme I activity (E. C. C. Lin and S. Tanaka, unpublished). The doubling time on glucose was found to be 15 hours. Chemotaxis experiments were carried out with bacteria grown on glycerol. In such mutant cells in chemotaxis medium the rate of accumulation of radioactivity from glucose was found to be inhibited by 91 percent relative to a wild-type strain W3110.

36. Strain 1101 has less than 2 percent of the heat-stable protein found in the parental strain (1100), according to C. F. Fox and G. Wilson, *Proc. Nat. Acad. Sci. U.S.* **59**, 988 (1968). The doubling time on glucose was found to be 1.2 hours for the parent and 8.5 hours for the mutant. Chemotaxis experiments were carried out with bacteria grown on lactate. In such mutant cells in chemotaxis medium the rate of accumulation of radioactivity from glucose was found to be inhibited by 78 percent relative to the wild-type strain W3110.

37. Strain W1895-D1 was isolated by D. P. Kessler as a mutant resistant to α-methyl glucoside; the growth of the parent, W1895, is inhibited by α-methyl glucoside. Cells of the mutant, unlike those of the parent, have been shown by Kessler (personal communication) not to accumulate α-methyl glucoside to any appreciable extent and not to phosphorylate the compound. Extracts of this mutant are defective in the enzyme II system for phosphorylation of glucose and α-methyl glucoside, according to C. F. Fox and G. Wilson [*Proc. Nat. Acad. Sci U.S.* **59**, 988 (1968)]. The doubling time on glucose was found to be 0.8 hour for the parent and 2.8 hours for the mutant. Chemotaxis experiments were carried out with bacteria grown on lactate. Accumulation of α-methyl glucoside by such mutant cells in chemotaxis medium was found to be inhibited by 70 percent relative to the parental strain.

38. For a review of the literature, see J. R. W. Piperno, thesis, University of Michigan (1966).

39. Does a bacterium compare the difference in the concentrations of the attractant at its two ends, or the difference between the concentration in its present location and in the place where it was a short time ago? The latter interpretation appears to be correct for phototaxis: Engelmann discovered that phototactic bacteria, swimming randomly, all back up—that is, carry out an avoiding reaction—when uniform illumination suddenly is dimmed (*1*). No comparable experiment has as yet been carried out for chemotaxis. The following phenomenon must be taken into account in investigating the mechanism of chemoreception. Pfeffer found (see *1*) that bacteria "suspended in 0.01% meat extract were attracted by a 0.05% meat extract solution but not by weaker solutions. When the bacteria were suspended in a 1% extract solution, a 5% solution was required in order to cause a tactic response. Thus the Weber law was found to hold within the concentration range tested, the relative threshold being 500%." Similar results have been obtained by R. E. Mesibov for *Escherichia coli* undergoing chemotaxis toward aspartate and serine.

40. Y. Naitoh, *Science* **154**, 660 (1966); H. Kinosita and A. Murakami, *Physiol. Rev.* **47**, 53 (1967); R. Eckert, *Science* **147**, 1140 (1965); ——, *ibid.*, p. 1142; Y. Naitoh and R. Eckert, *ibid.* **164**, 963 (1969).

41. For a review of the literature, see W. van Iterson, J. F. M. Hoeniger, E. N. van Zanten, *J. Cell Biol.* **31**, 585 (1966).

42. The avoiding reaction (or shock reaction) (see *1*) may be described as follows. A bacterium that happens to swim from a higher into a lower concentration of attractant tumbles for a fraction of a second, or jumps back, and then it goes off in a new direction; in some species that can swim equally well forward or backward, the organism just swims away backward. If the new direction takes the bacterium farther into the region of low concentration of attractant, the avoiding reaction is repeated, but if the bacterium encounters a higher concentration it continues to swim in the new direction. Thus the bacteria avoid low concentrations and accumulate in the region of higher concentration.

43. J. B. Armstrong, J. Adler, M. M. Dahl, *J. Bacteriol.* **93**, 390 (1967); J. B. Armstrong and J. Adler, *ibid.* **97**, 156 (1969); ——, *Genetics* **61**, 61 (1969).

44. In this connection, it is noteworthy that sensory receptor cells of animals generally contain, or are derived from, flagella (or, what is essentially the same thing, cilia); indeed, the suggestion has been made that animal receptors have evolved from single-celled, flagellated organisms [see J. A. Vinnikov, *Cold Spring Harbor Symp. Quant. Biol.* **30**, 293 (1965); R. M. Eakin, *ibid.*, p. 363].

45. J. Adler, *J. Bacteriol.* **92**, 121 (1966).

46. Essentially the same results were obtained when chemotaxis toward each of these chemicals was tested with bacteria grown on galactose (in the case of the sugar derivatives) or glycerol (in the case of the citric acid cycle compounds). In the case of glucose, bacteria grown on galactose were used for the experiment shown, since bacteria grown on glucose have poor motility (see *23*) and respond to glucose poorly, with a high threshold. In every case the bacteria were examined with the microscope for motility, and they were tested for their response to glucose or aspartate to make sure they were capable of carrying out chemotaxis. As in the case of bacteria grown on glucose, bacteria grown on gluconate or glucuronate had poor motility, and their response to glucose was poor.

47. The flasks were obtained from Kontes of Illinois, Franklin Park, Ill.

48. H. Jeffay and J. Alverez, *Anal. Chem.* **33**, 612 (1961).

49. H. C. P. Wu and H. M. Kalckar, *Proc. Nat. Acad. Sci. U.S.* **55**, 622 (1966); H. C. P. Wu, *J. Mol. Biol.* **24**, 213 (1967).

50. The research discussed was supported by a grant from the National Institutes of Health. I thank Margaret Dahl for having carried out many of the experiments described here.

Phylogenetic Structure of the Prokaryotic Domain: The Primary Kingdoms

C. R. WOESE AND G. E. FOX

The Woese and Fox paper opened a new era in biology by introducing the first outlines of a universal phylogeny. This was an early chapter in Woese's effort to determine a phylogeny of microorganisms by comparing ribosomal RNA (rRNA) sequences, at that stage of technology represented by catalogs of short oligonucleotide sequences. The prevailing notion of life's evolutionary diversity at the time was framed in the context of procaryote or eucaryote. Consequently, it was unexpected when the rRNA sequences from diverse organisms fell into three, not two, fundamentally distinct groups. There had to be three primary lines of evolutionary descent, phylogenetic "domains," now termed Archaea (formerly archaebacteria), Eubacteria, and Eucarya (eucaryotes).

The paper sparked publicity and controversy. The recognition of "archaebacteria" was heralded on the front page of the *New York Times* as discovery of a "third form of life." The concept of three primary domains touched off a flurry of refutations defending the procaryote-eucaryote or the five-kingdoms notions to account for biological organization. These familiar notions had never previously been tested, however, and the analysis of rRNA sequences proved them fundamentally incorrect. The shift in public and textbook treatment of the issue continues. Microbiologists, long without a way to relate microorganisms meaningfully, came to welcome the three-domains concept.

This paper led to a radical revision in our understanding of biological diversity and the course of evolution. The paper introduced new ways to think about ancient evolutionary history in the context of modern organisms. It led the way to a new approach to microbial identification, going beyond the anecdotes of physiological tests to the relative objectivity of numerical sequence comparisons. Coupled with molecular technology to analyze genes from organisms in the environment, the phylogenetic perspective is revising our understanding of the makeup of the microbial communities that dominate and maintain the biosphere. The sequence-based universal phylogenetic tree provides a metric for that otherwise nebulous concept, "biodiversity." This paper is a foundation of the modern era of microbiology.

NORMAN R. PACE

Reprinted from *Proceedings of the National Academy of Sciences USA* 74:5088–5090. Copyright © 1977, by permission of the authors.

Reprinted from
Proc. Natl. Acad. Sci. USA
Vol. 74, No. 11, pp. 5088–5090, November 1977
Evolution

Phylogenetic structure of the prokaryotic domain: The primary kingdoms

(archaebacteria/eubacteria/urkaryote/16S ribosomal RNA/molecular phylogeny)

CARL R. WOESE AND GEORGE E. FOX*

Department of Genetics and Development, University of Illinois, Urbana, Illinois 61801

Communicated by T. M. Sonneborn, August 18, 1977

ABSTRACT   A phylogenetic analysis based upon ribosomal RNA sequence characterization reveals that living systems represent one of three aboriginal lines of descent: (*i*) the eubacteria, comprising all typical bacteria; (*ii*) the archaebacteria, containing methanogenic bacteria; and (*iii*) the urkaryotes, now represented in the cytoplasmic component of eukaryotic cells.

The biologist has customarily structured his world in terms of certain basic dichotomies. Classically, what was not plant was animal. The discovery that bacteria, which initially had been considered plants, resembled both plants and animals less than plants and animals resembled one another led to a reformulation of the issue in terms of a yet more basic dichotomy, that of eukaryote versus prokaryote. The striking differences between eukaryotic and prokaryotic cells have now been documented in endless molecular detail. As a result, it is generally taken for granted that all extant life must be of these two basic types.

Thus, it appears that the biologist has solved the problem of the primary phylogenetic groupings. However, this is not the case. Dividing the living world into *Prokaryotae* and *Eukaryotae* has served, if anything, to obscure the problem of what extant groupings represent the various primeval branches from the common line of descent. The reason is that eukaryote/prokaryote is not primarily a phylogenetic distinction, although it is generally treated so. The eukaryotic cell is organized in a different and more complex way than is the prokaryote; this probably reflects the former's composite origin as a symbiotic collection of various simpler organisms (1–5). However striking, these organizational dissimilarities do not guarantee that eukaryote and prokaryote represent phylogenetic extremes.

The eukaryotic cell *per se* cannot be directly compared to the prokaryote. The composite nature of the eukaryotic cell makes it necessary that it first be conceptually reduced to its phylogenetically separate components, which arose from ancestors that were noncomposite and so individually are comparable to prokaryotes. In other words, the question of the primary phylogenetic groupings must be formulated solely in terms of relationships among "prokaryotes"—i.e., noncomposite entities. (Note that in this context there is no suggestion *a priori* that the living world is structured in a dichotomous way.)

The organizational differences between prokaryote and eukaryote and the composite nature of the latter indicate an important property of the evolutionary process: Evolution seems to progress in a "quantized" fashion. One level or domain of organization gives rise ultimately to a higher (more complex) one. What "prokaryote" and "eukaryote" actually represent are two such domains. Thus, although it is useful to define phylogenetic patterns within each domain, it is not meaningful to construct phylogenetic classifications between domains: Prokaryotic kingdoms are not comparable to eukaryotic ones. This should be recognized by an appropriate terminology. The highest phylogenetic unit in the prokaryotic domain we think should be called an "urkingdom"—or perhaps "primary kingdom." This would recognize the qualitative distinction between prokaryotic and eukaryotic kingdoms and emphasize that the former have primary evolutionary status.

The passage from one domain to a higher one then becomes a central problem. Initially one would like to know whether this is a frequent or a rare (unique) evolutionary event. It is traditionally assumed—without evidence—that the eukaryotic domain has arisen but once; all extant eukaryotes stem from a common ancestor, itself eukaryotic (2). A similar prejudice holds for the prokaryotic domain (2). [We elsewhere argue (6) that a hypothetical domain of lower complexity, that of "progenotes," may have preceded and given rise to the prokaryotes.] The present communication is a discussion of recent findings that relate to the urkingdom structure of the prokaryotic domain and the question of its unique as opposed to multiple origin.

Phylogenetic relationships cannot be reliably established in terms of noncomparable properties (7). A comparative approach that can measure degree of difference in comparable structures is required. An organism's genome seems to be the ultimate record of its evolutionary history (8). Thus, comparative analysis of molecular sequences has become a powerful approach to determining evolutionary relationships (9, 10).

To determine relationships covering the entire spectrum of extant living systems, one optimally needs a molecule of appropriately broad distribution. None of the readily characterized proteins fits this requirement. However, ribosomal RNA does. It is a component of all self-replicating systems; it is readily isolated; and its sequence changes but slowly with time—permitting the detection of relatedness among very distant species (11–13). To date, the primary structure of the 16S (18S) ribosomal RNA has been characterized in a moderately large and varied collection of organisms and organelles, and the general phylogenetic structure of the prokaryotic domain is beginning to emerge.

A comparative analysis of these data, summarized in Table 1, shows that the organisms clearly cluster into several primary kingdoms. The first of these contains all of the typical bacteria so far characterized, including the genera *Acetobacterium*, *Acinetobacter*, *Acholeplasma*, *Aeromonas*, *Alcaligenes*, *Anacystis*, *Aphanocapsa*, *Bacillus*, *Bdellovibrio*, *Chlorobium*, *Chromatium*, *Clostridium*, *Corynebacterium*, *Escherichia*, *Eubacterium*, *Lactobacillus*, *Leptospira*, *Micrococcus*, *Mycoplasma*, *Paracoccus*, *Photobacterium*, *Propionibacterium*,

---

* Present address: Department of Biophysical Sciences, University of Houston, Houston, TX 77004.

Molecular Biology and Physiology 441

Table 1.   Association coefficients (S_{AB}) between representative members of the three primary kingdoms

	1	2	3	4	5	6	7	8	9	10	11	12	13
1. *Saccharomyces cerevisiae,* 18S	—	0.29	0.33	0.05	0.06	0.08	0.09	0.11	0.08	0.11	0.11	0.08	0.08
2. *Lemna minor,* 18S	0.29	—	0.36	0.10	0.05	0.06	0.10	0.09	0.11	0.10	0.10	0.13	0.07
3. L cell, 18S	0.33	0.36	—	0.06	0.06	0.07	0.07	0.09	0.06	0.10	0.10	0.09	0.07
4. *Escherichia coli*	0.05	0.10	0.06	—	0.24	0.25	0.28	0.26	0.21	0.11	0.12	0.07	0.12
5. *Chlorobium vibrioforme*	0.06	0.05	0.06	0.24	—	0.22	0.22	0.20	0.19	0.06	0.07	0.06	0.09
6. *Bacillus firmus*	0.08	0.06	0.07	0.25	0.22	—	0.34	0.26	0.20	0.11	0.13	0.06	0.12
7. *Corynebacterium diphtheriae*	0.09	0.10	0.07	0.28	0.22	0.34	—	0.23	0.21	0.12	0.12	0.09	0.10
8. *Aphanocapsa* 6714	0.11	0.09	0.09	0.26	0.20	0.26	0.23	—	0.31	0.11	0.11	0.10	0.10
9. Chloroplast (*Lemna*)	0.08	0.11	0.06	0.21	0.19	0.20	0.21	0.31	—	0.14	0.12	0.10	0.12
10. *Methanobacterium thermoautotrophicum*	0.11	0.10	0.10	0.11	0.06	0.11	0.12	0.11	0.14	—	0.51	0.25	0.30
11. *M. ruminantium* strain M-1	0.11	0.10	0.10	0.12	0.07	0.13	0.12	0.11	0.12	0.51	—	0.25	0.24
12. *Methanobacterium* sp., Cariaco·isolate JR-1	0.08	0.13	0.09	0.07	0.06	0.06	0.09	0.10	0.10	0.25	0.25	—	0.32
13. *Methanosarcina barkeri*	0.08	0.07	0.07	0.12	0.09 ·	0.12	0.10	0.10	0.12	0.30	0.24	0.32	—

The 16S (18S) ribosomal RNA from the organisms (organelles) listed were digested with T1 RNase and the resulting digests were subjected to two-dimensional electrophoretic separation to produce an oligonucleotide fingerprint. The individual oligonucleotides on each fingerprint were then sequenced by established procedures (13, 14) to produce an oligonucleotide catalog characteristic of the given organism (3, 4, 13–17, 22, 23; unpublished data). Comparisons of all possible pairs of such catalogs defines a set of association coefficients (S_{AB}) given by: $S_{AB} = 2N_{AB}/(N_A + N_B)$, in which N_A, N_B, and N_{AB} are the total numbers of nucleotides in sequences of hexamers or larger in the catalog for organism A, in that for organism B, and in the interreaction of the two catalogs, respectively (13, 23).

Pseudomonas, Rhodopseudomonas, Rhodospirillum, Spirochaeta, Spiroplasma, Streptococcus, and *Vibrio* (refs. 13–17; unpublished data). The group has three major subdivisions, the blue-green bacteria and chloroplasts, the "Gram-positive" bacteria, and a broad "Gram-negative" subdivision (refs. 3, 4, 13–17; unpublished data). It is appropriate to call this urkingdom the *eubacteria.*

A second group is defined by the 18S rRNAs of the eukaryotic cytoplasm—animal, plant, fungal, and slime mold (unpublished data). It is uncertain what ancestral organism in the symbiosis that produced the eukaryotic cell this RNA represents. If there had been an "engulfing species" (1) in relation to which all the other organisms were endosymbionts, then it seems likely that 18S rRNA represents that species. This hypothetical group of organisms, in one sense the major ancestors of eukaryotic cells, might appropriately be called *urkaryotes.* Detailed study of anaerobic amoebae and the like (18), which seem not to contain mitochondria and in general are cytologically simpler than customary examples of eukaryotes, might help to resolve this question.

Eubacteria and urkaryotes correspond approximately to the conventional categories "prokaryote" and "eukaryote" when they are used in a phylogenetic sense. However, they do not constitute a dichotomy; they do not collectively exhaust the class of living systems. There exists a third kingdom which, to date, is represented solely by the methanogenic bacteria, a relatively unknown class of anaerobes that possess a unique metabolism based on the reduction of carbon dioxide to methane (19–21). *These "bacteria" appear to be no more related to typical bacteria than they are to eukaryotic cytoplasms.* Although the two divisions of this kingdom appear as remote from one another as blue-green algae are from other eubacteria, they nevertheless correspond to the *same* biochemical phenotype. The apparent antiquity of the methanogenic phenotype plus the fact that it seems well suited to the type of environment presumed to exist on earth 3–4 billion years ago lead us tentatively to name this urkingdom the *archaebacteria.* Whether or not other biochemically distinct phenotypes exist in this kingdom is clearly an important question upon which may turn our concept of the nature and ancestry of the first prokaryotes.

Table 1 shows the three urkingdoms to be equidistant from one another. Because the distances measured are actually proportional to numbers of mutations and not necessarily to time, it cannot be proven that the three lines of descent branched from the common ancestral line at about the same time. One of the three may represent a far earlier bifurcation than the other two, making there in effect only two urkingdoms. Of the three possible unequal branching patterns the case for which the initial bifurcation defines urkaryotes vs. all bacteria requires further comment because, as we have seen, there is a predilection to accept such a dichotomy.

The phenotype of the methanogens, although ostensibly "bacterial," on close scrutiny gives no indication of a specific phylogenetic resemblance to the eubacteria. For example, methanogens do have cell walls, but these do not contain peptidoglycan (24). The biochemistry of methane formation appears to involve totally unique coenzymes (23, 25, 26). The methanogen rRNAs are comparable in size to their eubacterial counterparts, but resemble the latter specifically in neither sequence (Table 1) nor in their pattern of base modification (23). The tRNAs from eubacteria and eukaryotes are characterized by a common modified sequence, TΨCG; methanogens modify this tRNA sequence in a quite different and unique way (23). It must be recognized that very little is known of the general biochemistry of the methanogens—and almost nothing is known regarding their molecular biology. Hence, although the above points are few in number, they represent most of what is now known. There is no reason at present to consider methanogens as any closer to eubacteria than to the "cytoplasmic component" of the eukaryote. Both in terms of rRNA sequence measurement and in terms of general phenotypic differences, then, the three groupings appear to be distinct urkingdoms.

If a third urkingdom exists, does this suggest that many more such will be found among yet to be characterized organisms? We think not, although the matter clearly requires an exhaustive search. As seen above, the number of species that can be classified as eubacteria is moderately large. To this list can be added *Spirillum* and *Desulfovibrio,* whose rRNAs appear typically eubacterial by nucleic acid hybridization measurements (27). Because the list is also phenotypically diverse, it seems unlikely that many, if any, of the yet uncharacterized

    Proc. Natl. Acad. Sci. USA 74 (1977)

prokaryotic groups will be shown to have coequal status with the present three. Conceivably the halophiles whose cell walls contain no peptidoglycan, are candidates for this distinction (28, 29).

Eukaryotic organelles, however, could be a different matter. There can be no doubt that the chloroplast is of specific eubacterial origin (3, 4). A question arises with the remaining organelles and structures. Mitochondria, for example, do not conform well to a "typically prokaryotic" phenotype, which has led some to conclude that they could not have arisen as endosymbionts (30). By using "prokaryote" in a phylogenetic sense, this formulation of the issue does not recognize a third alternative—that the organelle in question arose endosymbiotically from a separate line of descent whose phenotype is not "typically prokaryotic" (i.e., eubacterial). It is thus conceivable that some endosymbiotically formed structures represent still other major phylogenetic groups; some could even be the only extant representation thereof.

The question that remains to be answered is whether the common ancestor of all three major lines of descent was itself a prokaryote. If not, each urkingdom represents an independent evolution of the prokaryotic level of organization. Obviously, much more needs to be known about the general properties of all the urkingdoms before this matter can be definitely settled. At present we can point to two arguments suggesting that each urkingdom does represent a separate evolution of the prokaryotic level of organization.

The first argument concerns the stability of the general phenotypes. The general eubacterial phenotype has been stable for at least 3 billion years—i.e., the apparent age of blue-green algae (31). The methanogenic phenotype seems to be at least this old in that branchings within the two urkingdoms are comparably deep (see Table 1). The time available to form each phenotype (from their common ancestor) is then short by comparison, which seems paradoxical in that the two phenotypes are so fundamentally different. We think that this ostensible paradox implies that the common ancestor in this case was not a prokaryote. It was a far simpler entity; it probably did not evolve at the "slow" rate characteristic of prokaryotes; it did not possess many of the features possessed by prokaryotes, and so these evolved independently and differently in separate lines of descent.

The second argument concerns the quality of the differences in the three general phenotypes. It seems highly unlikely, for example, that differences in general patterns of base modification in rRNAs and tRNAs are related to the niches that organisms occupy. Rather, differences of this nature imply independent evolution of the properties in question. It has been argued elsewhere that features such as RNA base modification generally represent the final stage in the evolution of translation (32). If these features have evolved separately in two lines of descent, their common ancestor, lacking them, had a more rudimentary version of the translation mechanism and consequently, could not have been as complex as a prokaryote (6).

With the identification and characterization of the urkingdoms we are for the first time beginning to see the overall phylogenetic structure of the living world. It is not structured in a bipartite way along the lines of the organizationally dissimilar prokaryote and eukaryote. Rather, it is (at least) tripartite, comprising (*i*) the typical bacteria, (*ii*) the line of descent manifested in eukaryotic cytoplasms, and (*iii*) a little explored grouping, represented so far only by methanogenic bacteria.

The ideas expressed herein stem from research supported by the National Aeronautics and Space Administration and the National Science Foundation. We are grateful to a number of colleagues who have helped to generate the yet unpublished data that make these speculations possible: William Balch, Richard Blakemore, Linda Bonen, Tristan Dyer, Jane Gibson, Ramesh Gupta, Robert Hespell, Bobby Joe Lewis, Kenneth Luehrsen, Linda Magrum, Jack Maniloff, Norman Pace, Mitchel Sogin, Stephan Sogin, David Stahl, Ralph Tanner, Thomas Walker, Ralph Wolfe, and Lawrence Zablen. We thank Linda Magrum and David Nanney for suggesting the name "archaebacteria."

1. Stanier, R. Y. (1970) *Symp. Soc. Gen. Microbiol.* **20**, 1–38.
2. Margulis, L. (1970) *Origin of Eucaryotic Cells* (Yale University Press, New Haven).
3. Zablen, L. B., Kissel, M. S., Woese, C. R. & Buetow, D. E. (1975) *Proc. Natl. Acad. Sci. USA* **72**, 2418–2422.
4. Bonen, L. & Doolittle, W. F. (1975) *Proc. Natl. Acad. Sci. USA* **72**, 2310–2314.
5. Bonen, L., Cunningham, R. S., Gray, M. W. & Doolittle, W. F. (1977) *Nucleic Acid Res.* **4**, 663–671.
6. Woese, C. R. & Fox, G. E. (1977) *J. Mol. Evol.*, in press.
7. Sneath, P. H. A. & Sokal, R. R. (1973) *Numerical Taxonomy* (W. H. Freeman, San Francisco).
8. Zuckerkandl, E. & Pauling, L. (1965) *J. Theor. Biol.* **8**, 357–366.
9. Fitch, W. M. & Margoliash, E. (1967) *Science* **155**, 279–284.
10. Fitch, W. M. (1976) *J. Mol. Evol.* **8**, 13–40.
11. Sogin, S. J., Sogin, M. L. & Woese, C. R. (1972) *J. Mol. Evol.* **1**, 173–184.
12. Woese, C. R., Fox, G. E., Zablen, L., Uchida, T., Bonen, L., Pechman, K., Lewis, B. J. & Stahl, D. (1975) *Nature* **254**, 83–86.
13. Fox, G. E., Pechman, K. R. & Woese, C. R. (1977) *Int. J. Syst. Bacteriol.* **27**, 44–57.
14. Uchida, T., Bonen, L., Schaup, H. W., Lewis, B. J., Zablen, L. B. & Woese, C. R. (1974) *J. Mol. Evol.* **3**, 63–77.
15. Zablen, L. B. & Woese, C. R. (1975) *J. Mol. Evol.* **5**, 25–34.
16. Doolittle, W. F., Woese, C. R., Sogin, M. L., Bonen, L. & Stahl, D. (1975) *J. Mol. Evol.* **4**, 307–315.
17. Pechman, K. J., Lewis, B. J. & Woese, C. R. (1976) *Int. J. Syst. Bacteriol.* **26**, 305–310.
18. Bovee, E. C. & Jahn, T. L. (1973) in *The Biology of Amoeba*, ed. Jeon, K. W. (Academic Press, New York), p. 38.
19. Wolfe, R. S. (1972) *Adv. Microbiol. Phys.* **6**, 107–146.
20. Zeikus, J. G. (1977) *Bacteriol. Rev.* **41**, 514–541.
21. Zeikus, J. G. & Bowen, V. G. (1975) *Can. J. Microbiol.* **21**, 121–129.
22. Balch, W. E., Magrum, L. J., Fox, G. E., Wolfe, R. S. & Woese, C. R. (1977) *J. Mol. Evol.*, in press.
23. Fox, G. E., Magrum, L. J., Balch, W. E., Wolfe, R. S. & Woese, C. R. (1977) *Proc. Natl. Acad. Sci. USA*, **74**, 4537–4541.
24. Kandler, O. & Hippe, H. (1977) *Arch. Microbiol.* **113**, 57–60.
25. Taylor, C. D. & Wolfe, R. S. (1974) *J. Biol. Chem.* **249**, 4879–4885.
26. Cheeseman, P., Toms-Wood, A. & Wolfe, R. S. (1972) *J. Bacteriol.* **112**, 527–531.
27. Pace, B. & Campbell, L. L. (1971) *J. Bacteriol.* **107**, 543–547.
28. Brown, A. D. & Cho, K. Y. (1970) *J. Gen. Microbiol.* **62**, 267–270.
29. Reistad, R. (1972) *Arch. Mikrobiol.* **82**, 24–30.
30. Raff, R. A. & Mahler, H. R. (1973) *Science* **180**, 517–521.
31. Shopf, J. W. (1972) *Exobiology—Frontiers of Biology* (North Holland, Amsterdam), Vol. 23, pp. 16–61.
32. Woese, C. R. (1970) *Symp. Soc. Gen. Microbiol.* **20**, 39–54.

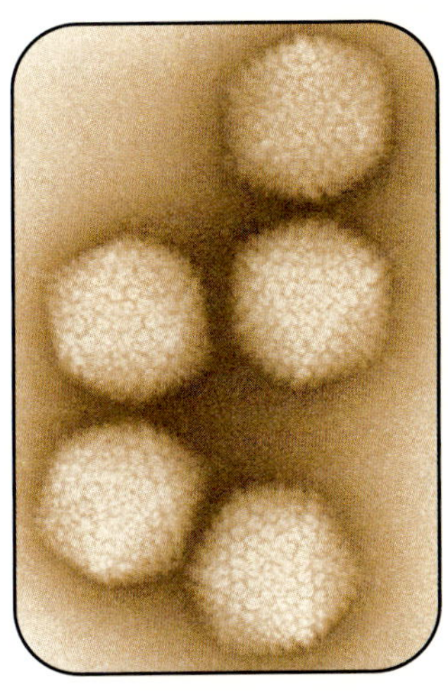

SECTION V
Virology

The landmark papers in 20th century virology were chosen by a committee of six: Robert E. Chanock, M.D. (virus-host interactions and pathogenesis); Kenneth N. Kreuzer, Ph.D. (bacteriophages); Bernard Moss, M.D., Ph.D. (DNA viruses); Peter K. Vogt, Ph.D. (tumor viruses); Milton B. Zaitlin, Ph.D. (plant viruses); and Wolfgang K. Joklik, D.Phil. (Committee Chair) (RNA viruses), who also wrote the introductions.

Choosing a selected number of influential papers in virology published in the 20th century proved to be a formidable undertaking. The problem was not that we, as committee members, disagreed as to what were or were not seminal papers of the front rank; the problem was that we found far too many of them. The background to our selections follows.

Dimitri Iwanowski, in 1892, discovered that an agent capable of causing tobacco mosaic disease was capable of passing through filters that retained bacteria, an observation replicated 6 years later by Friedrich Loeffler and Paul Frosch for the agent that causes foot-and-mouth disease in cattle and, in 1911, by Peyton Rous (*J. Exp. Med.* **13**:397–411, 1911) for the agent that causes malignant tumors (sarcomas) in chickens (Rous sarcoma virus). For the next 40 years, virologists focused on basic aspects of the virus-host interaction, using the techniques of epidemiology, immunology, and pathology. Plant viruses proved to be easy to obtain in large amounts, which permitted extensive chemical and physical studies. It was soon found that the plant viruses that were being studied consisted only of nucleic acid and protein, and when Wendell Stanley crystallized tobacco mosaic virus (TMV) in 1935 (*Science* **81**:644–645, 1935), the feat evoked great astonishment because it cut across preconceived ideas concerning the nature of "living" organisms: here agents capable of replicating in living cells behaved, under certain conditions, as typical macromolecules.

Bacteriophages were discovered independently during the first world war by Frederick Twort in England (*Lancet* **ii**:1241–1243, 1915) and Felix d'Hérelle in Canada (*C.R. Acad. Sci.* **165**:373–375, 1917). Early work with bacteriophages concentrated on their clinical application, but their ability to lyse bacteria was never matched by their activity in vivo, most probably because they were eliminated efficiently from the bloodstream. Soon after, the interaction between bacteriophages and their host cells began to be studied in terms of populations rather than single virus particles interacting with sin-

THE GROWTH OF BACTERIOPHAGE

By EMORY L. ELLIS AND MAX DELBRÜCK*

(From the William G. Kerckhoff Laboratories of the Biological Sciences, California Institute of Technology, Pasadena)

(Accepted for publication, September 7, 1938)

INTRODUCTION

Certain large protein molecules (viruses) possess the property of multiplying within living organisms. This process, which is at once so foreign to chemistry and so fundamental to biology, is exemplified in the multiplication of bacteriophage in the presence of susceptible bacteria.

Bacteriophage offers a number of advantages for the study of the multiplication process not available with viruses which multiply at the expense of more complex hosts. It can be stored indefinitely in the absence of a host without deterioration. Its concentration can be determined with fair accuracy by several methods, and even the individual particles can be counted by d'Herelle's method. It can be concentrated, purified, and generally handled like nucleoprotein, to which class of substances it apparently belongs (Schlesinger (1) and Northrop (2)). The host organism is easy to culture and in some cases can be grown in purely synthetic media, thus the conditions of growth of the host and of the phage can be controlled and varied in a quantitative and chemically well defined way.

Before the main problem, which is elucidation of the multiplication process, can be studied, certain information regarding the behavior of phage is needed. Above all, the "natural history" of bacteriophage, *i.e.* its growth under a well defined set of cultural conditions, is as yet insufficiently known, the only extensive quantitative work being that of Krueger and Northrop (3) on an anti-*staphylococcus* phage. The present work is a study of this problem, the growth of another phage (anti-*Escherichia coli* phage) under a standardized set of culture conditions.

* Fellow of The Rockefeller Foundation.

EXPERIMENTAL

Bacteria Culture.—Our host organism was a strain of *Escherichia coli*, which was kindly provided by Dr. C. C. Lindegren. Difco nutrient broth (pH 6.6–6.8) and nutrient agar were selected as culture media. These media were selected for the present work because of the complications which arise when synthetic media are used. We thus avoided the difficulties arising from the need for accessory growth factors.

Isolation, Culture, and Storage of Phage.—A bacteriophage active against this strain of *coli* was isolated in the usual way from fresh sewage filtrates. Its homogeneity was assured by five successive single plaque isolations. The properties of this phage remained constant throughout the work. The average plaque size on 1.5 per cent agar medium was 0.5 to 1.0 mm.

Phage was prepared by adding to 25 cc. of broth, 0.1 cc. of a 20 hour culture of bacteria, and 0.1 cc. of a previous phage preparation. After $3\frac{1}{2}$ hours at 37° the culture had become clear, and contained about 10^9 phage particles.

Such lysates even though stored in the ice box, decreased in phage concentration to about 20 per cent of their initial value in 1 day, and to about 2 per cent in a week, after which they remained constant. Part of this lost phage activity was found to be present in a small quantity of a precipitate which had sedimented during this storage period.

Therefore, lysates were always filtered through Jena sintered glass filters (5 on 3 grade) immediately after preparation. The phage concentration of these filtrates also decreased on storage, though more slowly, falling to 20 per cent in a week. However, 1:100 dilutions in distilled water of the fresh filtered lysates retained a constant assay value for several months, and these diluted preparations were used in the work reported here, except where otherwise specified.

This inactivation of our undiluted filtered phage suspensions on standing is probably a result of a combination of phage and specific phage inhibiting substances from the bacteria, as suggested by Burnett (4, 5). To test this hypothesis we prepared a polysaccharide fraction from agar cultures of these bacteria, according to a method reported by Heidelberger *et al.* (6). Aqueous solutions of this material, when mixed with phage suspensions, rapidly inactivated the phage.

Method of Assay.—We have used a modification of the plaque counting method of d'Herelle (7) throughout this work for the determination of phage concentrations. Although the plaque counting method has been reported unsatisfactory by various investigators, under our conditions it has proven to be entirely satisfactory.

Phage preparations suitably diluted in 18 hour broth cultures of bacteria to give a readily countable number of plaques (100 to 1000) were spread with a bent glass capillary over the surface of nutrient agar plates which had been dried by inverting on sterile filter paper overnight. The plates were then incubated 6 to 24 hours at 37°C. at which time the plaques were readily distinguishable. The 0.1 cc. used for spreading was completely soaked into the agar thus prepared in

2 to 3 minutes, thus giving no opportunity for the multiplication of phage in the liquid phase. Each step of each dilution was done with fresh sterile glassware. Tests of the amount of phage adhering to the glass spreaders showed that this quantity is negligible.

The time of contact between phage and bacteria in the final dilution before plating has no measurable influence on the plaque count, up to 5 minutes at 25°C. Even if phage alone is spread on the plate and allowed to soak in for 10 minutes, before seeding the plate with bacteria, only a small decrease in plaque count is apparent (about 20 per cent). This decrease we attribute to failure of some phage particles to come into contact with bacteria.

Under parallel conditions, the reproducibility of an assay is limited by the sampling error, which in this case is equal to the square root of the number of plaques (10 per cent for counts of 100; 3.2 per cent for counts of 1000). To test the effect of phage concentration on the number of plaques obtained, successive dilutions of a phage preparation were all plated, and the number of plaques enumerated. Over a 100-fold range of dilution, the plaque count was in linear proportion to the phage concentration. (See Fig. 1.)

Dreyer and Campbell-Renton (8) using a different anti-*coli* phage and an anti-*staphylococcus* phage, and a different technique found a complicated dependence of plaque count on dilution. Such a finding is incompatible with the concept that phage particles behave as single particles, *i.e.* without interaction, with respect to plaque formation. Our experiments showed no evidence of such a complicated behavior, and we ascribe it therefore to some secondary cause inherent in their procedure.

Bronfenbrenner and Korb (9) using a phage active against *B. dysenteriae* Shiga, and a different plating technique found that when the agar concentration was changed from 1 per cent to 2.5 per cent, the number of plaques was reduced to 1 per cent of its former value. They ascribed this to a change in the water supplied to the bacteria. With the technique which we have employed, variation of the agar concentration from 0.75 per cent to 3.0 per cent, had little influence on the number of plaques produced, though the size decreased noticeably with increasing agar concentration. (See Table I.)

Changes in the concentration of bacteria spread with the phage on the agar plates had no important influence on the number of plaques obtained. (See Table I.) The temperature at which plates were incubated had no significant effect on the number of plaques produced. (See Table I.)

In appraising the accuracy of this method, several points must be borne in mind. With our phage, our experiments confirm in the main the picture proposed by d'Herelle, according to which a phage particle grows in the following way: it becomes attached to a susceptible bacterium, multiplies upon or within it up to a critical time, when the newly formed phage particles are dispersed into the solution.

In the plaque counting method a single phage particle and an infected bacterium containing any number of phage particles will each give only one plaque.

This method therefore, does not give the number of phage particles but the number of loci within the solution at which one or more phage particles exist. These loci will hereafter be called "infective centers."   The linear relationship between phage concentration and plaque count (Fig. 1) does not prove that the number of plaques is equal to the number of infective centers, but only that it is proportional to this number.   We shall call the fraction of infective centers which

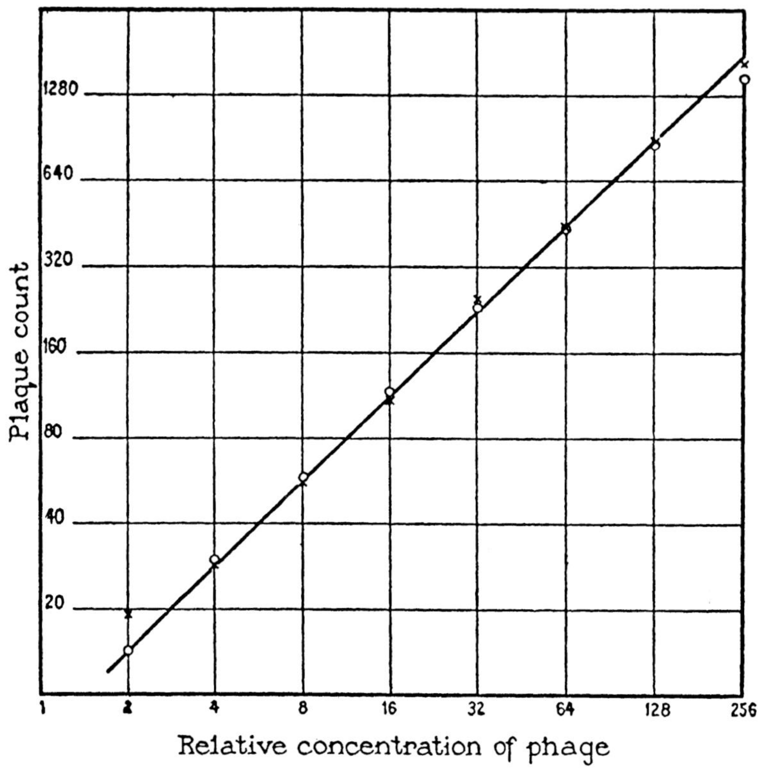

Fig. 1. Proportionality of the phage concentration to the plaque count.

Successive twofold dilutions of a phage preparation were plated in duplicate on nutrient agar; 0.1 cc. on each plate.   The plaque counts from two such series of dilutions are plotted against the relative phage concentration, both on a logarithmic scale.

produces plaques the "efficiency of plating."   With the concentrations of phage and bacteria which we have used this coefficient is essentially the fraction of infected bacteria in the suspension spread on the plate, which goes through to lysis under our cultural conditions on the agar medium.   After plating, the phage particles released by this lysis infect the surrounding bacteria, increasing only the size, and not the number of plaques.

TABLE I

Independence of Plaque Count on Plating Method

Agar concentration

Plates were prepared in which the agar strength varied, and all spread with 0.1 cc. of the same dilution of a phage preparation.   There is no significant difference in the numbers of plaques obtained.

Agar concentration, *per cent*	0.75	1.5	3.0
Plaque counts	394	373	424
	408	430	427
	376	443	455
	411	465	416
	373	404	469
Average	392	423	438
Plaque size, *mm*	2	0.5	0.2

Concentration of plating *coli*

A broth suspension of bacteria (10^9 bacteria/cc.) was prepared from a 24 hour agar slant and used at various dilutions, as the plating suspension for a single phage dilution. There are no significant differences in the plaque counts except at the highest dilution of the bacterial suspension, where the count is about 15 per cent lower.

Concentration	Plaque count
1	920
1/5	961
1/25	854
1/125	773

Temperature of plate incubation

Twelve plates were spread with 0.1 cc. of the same suspension of phage and bacteria, divided into three groups, and incubated at different temperatures.   There were no significant differences in the plaque counts obtained.

Temperature, °C	37	24	10
Plaque count	352	384	405
	343	405	377
	386	403	400
	422	479	406
Average	376	418	397

The experimental determination of the efficiency of plating is described in a later section (see p. 379). The coefficient varies from 0.3 to 0.5. This means that three to five out of every ten infected bacteria produce plaques. The fact that the efficiency of plating is relatively insensitive to variations in the temperature of plate incubation, density of plating *coli*, concentration of agar, etc. indicates that a definite fraction of the infected bacteria in the broth cultures do not readily go through to lysis when transferred to agar plates. For most experiments only the relative assay is significant; we have therefore, given the values derived directly from the plaque counts without taking into account the efficiency of plating, unless the contrary is stated.

Growth Measurements

The main features of the growth of this phage in broth cultures of the host are shown in Fig. 2. After a small initial increase (discussed below) the number of infective centers (individual phage particles, plus infected bacteria) in the suspension remains constant for a time, then rises sharply to a new value, after which it again remains constant. Later, a second sharp rise, not as clear-cut as the first, and finally a third rise occur. At this time visible lysis of the bacterial suspension takes place. A number of features of the growth process may be deduced from this and similar experiments, and this is the main concern of the present paper.

The Initial Rise

When a concentration of phage suitable for plating was added to a suspension of bacteria, and plated at once, a reproducible plaque count was obtained. If the suspension with added phage was allowed to stand 5 minutes at 37°C. (or 20 minutes at 25°C.) the number of plaques obtained on plating the suspension was found to be 1.6 times higher. This initial rise is not to be confused with the first "burst" which occurs later and increases the plaque count 70-fold. After the initial rise, the new value is readily duplicated and remains constant until the start of the first burst in the growth curve (30 minutes at 37° and 60 minutes at 25°).

This initial rise we attribute not to an increase in the number of infective centers, but to an increase in the probability of plaque formation (*i.e.* an increase in the efficiency of plating) by infected bacteria in a progressed state; that is, bacteria in which the phage particle has commenced to multiply. That this rise results from a change in the efficiency of plating and not from a quick increase in the number of infective centers is evident from the following experiment. Bacteria were grown for 24 hours at 25°C. on agar slants, then suspended in broth. Phage was added to this suspension and to a suspension of bacteria grown in the usual way, and the concentration of infective

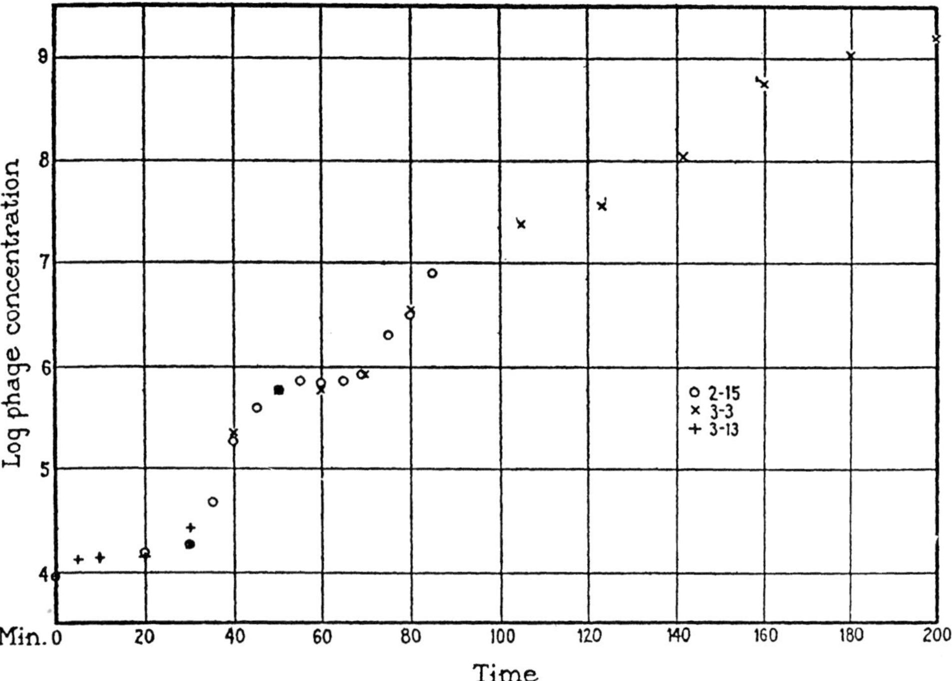

FIG. 2. Growth of phage in the presence of growing bacteria at 37°C.

A diluted phage preparation was mixed with a suspension of bacteria containing 2×10^8 organisms per cc., and diluted after 3 minutes 1 to 50 in broth. At this time about 70 per cent of the phage had become attached to bacteria. The total number of infective centers was determined at intervals on samples of this growth mixture. Three such experiments, done on different days, are plotted in this figure. The same curve was easily reproducible with all phage preparations stored under proper conditions.

centers was determined on both. The initial value was 1.6 times higher in the agar grown bacteria than in the control experiment, and remained constant until actual growth occurred. The initial rise was therefore absent in this case, clearly a result of an increase in the efficiency of plating. A sufficient number of experiments were performed with bacteria grown on agar to indicate that in other respects their behavior is similar to that of the bacteria grown in broth. The bacteria grown in this way on agar slants are in some way more susceptible to lysis than the broth cultured bacteria.

Adsorption

The first step in the growth of bacteriophage is its attachment to susceptible bacteria. The rate of this attachment can be readily measured by centrifuging the bacteria out of a suspension containing phage, at various times, and determining the amount of phage which remains unattached in the supernatant (*cf.* Krueger (10)).[1]

According to the picture of phage growth outlined above, phage cannot multiply except when attached to bacteria; therefore, the rate of attachment may, under certain conditions, limit the rate of growth. We wished to determine the rate of this adsorption so that it could be taken into account in the interpretation of growth experiments, or eliminated if possible, as a factor influencing the growth rate. Our growth curves show that there is no increase in the number of infective centers up to a critical time; we could therefore, make measurements of the adsorption on living bacteria suspended in broth, so long as the time allowed for attachment was less than the time to the start of the first burst in the growth curve. The adsorption proved to be so rapid that this time interval was ample to obtain adsorption of all but a few per cent of the free phage if the bacteria concentration was above 3×10^7. The number of bacteria remained constant; the lag phase in their growth was longer than the experimental period.

The rate of attachment was found to be first order with respect to the concentration of free phage (P_f) and first order with respect to the concentration of bacteria (B) over a wide range of concentrations, in agreement with the results reported by Krueger (10). That is, the concentration of free phage followed the equation

$$- \frac{d(P_f)}{dt} = k_a(P_f)(B)$$

[1] A very careful study of the adsorption of a *coli*-phage has also been made by Schlesinger (Schlesinger, M., *Z. Hyg. u. Infektionskrankh.*, 1932, **114**, 136, 149). Our results, which are less accurate and complete, agree qualitatively and quantitatively with the results of his detailed studies.

in which k_a was found to be 1.2 × 10⁻⁹ cm.³/min. at 15° and 1.9 × 10⁻⁹ cm.³/min. at 25°C.   These rate constants are about five times greater than those reported by Krueger (10).   With our ordinary 18 hour bacteria cultures (containing 2 × 10⁸ *B. coli*/cc.) we thus obtain 70 per cent attachment of phage in 3 minutes and 98 per cent in 10 minutes.   The adsorption follows the equation accurately until more than 90 per cent attachment has been accomplished, and then slows down somewhat, indicating either that not all the phage particles have the same affinity for the bacteria, or that equilibrium is being approached.   Other experiments not recorded here suggest that, if an equilibrium exists, it lies too far in favor of adsorption to be readily detected.   This equation expresses the rate of adsorption even when a tenfold excess of phage over bacteria is present, indicating that a single bacterium can accommodate a large number of phage particles on its surface, as found by several previous workers (5, 10).

Krueger (10) found a true equilibrium between free and adsorbed phage.   The absence of a detectable desorption in our case may result from the fixation of adsorbed phage by growth processes, since our conditions permitted growth, whereas Krueger's experiments were conducted at a temperature at which the phage could not grow.

Growth of Phage

Following adsorption of the phage particle on a susceptible bacterium, multiplication occurs, though this is not apparent as an increase in the number of plaques until the bacterium releases the resulting colony of phage particles into the solution.   Because the adsorption under proper conditions is so rapid and complete (as shown above) experiments could be devised in which only the influence of the processes following adsorption could be observed.

The details of these experiments were as follows: 0.1 cc. of a phage suspension of appropriate concentration was added to 0.9 cc. of an 18 hour broth bacterial culture, containing about 2 × 10⁸ *B. coli* / cc. After standing for a few minutes, 70 to 90 per cent of the phage was attached to the bacteria.   At this time, the mixture was diluted 50-fold in broth (previously adjusted to the required temperature) and incubated.   Samples were removed at regular intervals, and the concentration of infective centers determined.

The results of three experiments at 37°C. are plotted in Fig. 2, and confirm the suggestion of d'Herelle that phage multiplies under a spatial constraint, *i.e.* within or upon the bacterium, and is suddenly liberated in a burst. It is seen that after the initial rise (discussed above) the count of infective centers remains constant up to 30 minutes, and then rises about 70-fold above the initial value. The rise corresponds to the liberation of the phage particles which have multiplied in the initial constant period. This interpretation was verified by measurements of the free phage by centrifuging out the infected bacteria, and determining the number of phage particles in the supernatant liquid. The free phage concentration after adsorption was, of course, small compared to the total and remained constant up to the time of the first rise. It then rose steeply and became substantially equal to the total phage.

The number of bacteria lysed in this first burst is too small a fraction of the total bacteria used in these experiments to be measured as a change in turbidity; the ratio of uninfected bacteria to the total possible number of infected bacteria before the first burst is 400 to 1, the largest number of bacteria which can disappear in the first burst is therefore only 0.25 per cent of the total.

The phage particles liberated in the first burst are free to infect more bacteria. These phage particles then multiply within or on the newly infected bacteria; nevertheless, as before, the concentration of infective centers remains constant until these bacteria are lysed and release the phage which they contain into the medium. This gives the second burst which begins at about 70 minutes from the start of the experiment. Since the uninfected bacteria have been growing during this time, the bacteria lysed in the second burst amount to less than 5 per cent of the total bacteria present at this time. There is again therefore, no visible lysis.

This process is repeated, leading to a third rise of smaller magnitude starting at 120 minutes. At this time, inspection of the culture, which has until now been growing more turbid with the growth of the uninfected bacteria, shows a rapid lysis. The number of phage particles available at the end of the second rise was sufficient to infect the remainder of the bacteria.

These results are typical of a large number of such experiments, at

37°, all of which gave the 70-fold burst size, *i.e.* an average of 70 phage particles per infected bacterium, occurring quite accurately at the time shown, 30 minutes.   Indeed, one of the most striking features of these experiments was the constancy of the time interval from adsorption to the start of the first burst.   The magnitude of the rise (70-fold) was likewise readily reproducible by all phage preparations which had been stored under proper conditions to prevent deterioration (see above).

Multiple Infection

The adsorption measurements showed that a single bacterium can adsorb many phage particles.   The subsequent growth of phage in these "multiple infected" bacteria might conceivably lead to (*a*) an increase in burst size; (*b*) a burst at an earlier time, or (*c*) the same burst size at the same time, as if only one of the adsorbed particles had been effective, and the others inactivated.   In the presence of very great excesses of phage, Krueger and Northrop (3) and Northrop (2) report that visible lysis of the bacteria occurs in a very short time. It was possible therefore, that in our case, the latent period could be shortened by multiple infection.   To determine this point, we have made several experiments of which the following is an example.   0.8 cc. of a freshly prepared phage suspension containing 4×10^9 particles per cc. (assay corrected for efficiency of plating) was added to 0.2 cc. of bacterial suspension containing 4×10^9 bacteria per cc.   The ratio of phage to bacteria in this mixture was 4 to 1.   5 minutes were allowed for adsorption, and then the mixture was diluted 1 to 12,500 in broth, incubated at 25°, and the growth of the phage followed by plating at 20 minute intervals, with a control growth curve in which the phage to bacteria ratio was 1 to 10.   No significant difference was found either in the latent period or in the size of the burst.   The bacteria which had adsorbed several phage particles behaved as if only one of these particles was effective.

Effect of Temperature on Latent Period and Burst Size

A change in temperature might change either the latent period, *i.e.* the time of the burst, or change the size of the burst, or both.   In order to obtain more accurate estimates of the burst size it is desirable to minimize reinfection during the period of observation. This is obtained by diluting the phage-bacteria mixture (after initial contact to secure adsorption) to such an extent that the rate of adsorption then becomes extremely small. In this way, a single "cycle" of growth, (infection, growth, burst) was obtained as the following example

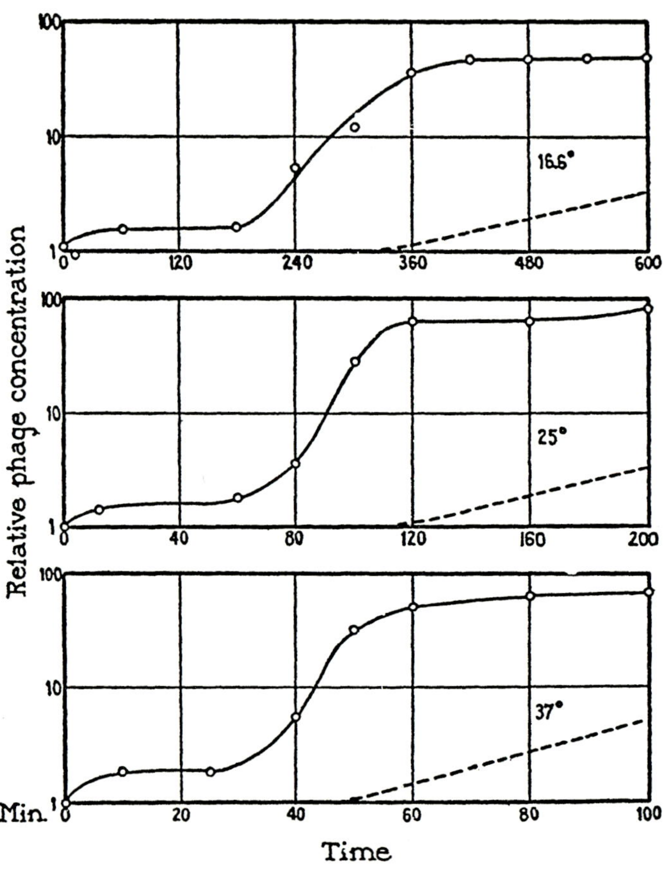

FIG. 3. One-step growth curves.

A suitable dilution of phage was mixed with a suspension of bacteria containing 2×10^8 organisms per cc. and allowed to stand at the indicated temperature for 10 minutes to obtain more than 90 per cent adsorption of the phage. This mixture was then diluted $1:10^4$ in broth, and incubated. It was again diluted $1:10$ at the start of the first rise to further decrease the rate of adsorption of the phage set free in the first step. The time scales are in the ratio $1:2:6$ for the temperatures 37, 25, and 16.6°C. Log P/P_0 is plotted, P_0 being the initial concentration of infective centers and P the concentration at time t. The broken line indicates the growth curve of the bacteria under the corresponding conditions.

shows. 0.1 cc. of phage of appropriate and known concentration was added to 0.9 cc. of an 18 hour culture and allowed to stand in this concentrated bacterial suspension for 10 minutes at the temperature of the experiment. This mixture was then diluted 1:10⁴ in broth and incubated at the temperature chosen. Samples of this diluted mixture were withdrawn at regular intervals and assayed. The results of three such experiments are plotted in Fig. 3. The rise corresponds to the average number of phage produced per burst, and its value can be appraised better in these experiments than in the complete growth curve previously given (Fig. 2) where there is probably some overlapping of the steps. In these experiments the rise is seen to be practically identical at the three temperatures, and equals about sixty particles per infected bacterium, but the time at which the rise occurred was 30 minutes at 37°, 60 minutes at 25°, and 180 minutes at 16.6°. This shows that the effect of temperature is solely on the latent period.

We have also made separate measurements of the rate of bacterial growth under the conditions of these experiments. They show that the average division period of the bacteria in their logarithmic growth phase varies in the same way with temperature, as the length of the latent period of phage growth. The figures are:

Temperature	Division period of B	Latent period of P growth
°C	min.	min.
16.6	About 120	180
25	42	60
37	21	30

There is a constant ratio (3/2) between the latent period of phage growth and the division period of the bacteria. This coincidence suggests a connection between the time required for division of a bacterium under optimum growth conditions, and the time from its infection by phage to its lysis.

Individual Phage Particle

The growth curves described above give averages only of large numbers of bursts. They can, however, also be studied individually, as was first done by Burnett (11).

If from a mixture containing many particles very small samples are withdrawn, containing each on the average only about one or less particles, then the fraction p_r of samples containing r particles is given by Poissons' (12) formula,

$$p_r = \frac{n^r e^{-n}}{r!} \tag{1}$$

where n is the average number of particles in a sample and e is the Napierian logarithm base. If the average number n is unknown, it can be evaluated from an experimental determination of any single

TABLE II

Distribution of Individual Particles among Small Samples

A suitably diluted phage preparation was added to 5 cc. of 18 hour bacteria culture and 0.1 cc. samples of this mixture were plated. The distribution of particles among the samples is that predicted by formula (1).

	p_r (experimental)	p_r (calculated)
0 plaques on 13 plates	0.394	0.441
1 plaque " 14 "	0.424	0.363
2 plaques " 5 "	0.151	0.148
3 " " 1 plate	0.033	0.040
4 " " 0 plates	0.000	0.008
27 " " 33 "	1.002	1.000

one of the p_r, for instance from a determination of p_0, the fraction of samples containing no particles:

$$n = -\ln p_0 \tag{2}$$

Let us now consider the following experiment. A small number of phage particles is added to a suspension containing bacteria in high concentration. Within a few minutes each phage particle has attached itself to a bacterium. The mixture is then diluted with a large volume of broth, in order to have the bacteria in low concentration so that after the first burst a long time elapses before reinfection, as in the one step growth curves. Samples (0.05 cc.) are removed from this mixture to separate small vials and incubated at the desired temperature. If these samples are plated separately (after adding a drop of bacterial suspension to each vial) before the occurrence of bursts, the fraction of the plates containing 0, 1, 2, *etc.* plaques is found to conform to formula (1) (see Table II). In this experiment we could also have inferred the average number of particles per sample, using formula (2), from the fraction of the plates showing no plaques (giving 0.93 per sample) instead of from the total number of plaques (27/33 = 0.82 per sample).

Experimental Measure of Efficiency of Plating

If the samples are incubated until the bursts have occurred, and then plated, the samples which had no particles will still show no plaques, those with one or more particles will show a large number, depending on the size of the burst, and on the efficiency of plating. In any case, if we wait until all bursts have occurred, only those samples which really contained no particle will show no plaques, quite independent of any inefficiency of plating. From this fraction of plates showing no plaques we can therefore evaluate the true number of particles originally present in the solution, and by comparison with the regular assay evaluate the efficiency of plating. In this way we have determined our efficiency of plating to be about 0.4. For instance, one such experiment gave no plaques on 23 out of 40 plates, and many plaques on each of the remaining plates. This gives $p_0 = \dfrac{23}{40}$ or 0.57 from which $n = 0.56$ particles per sample. A parallel assay of the stock phage used indicated 0.22 particles per sample; the plating efficiency was therefore $\dfrac{0.22}{0.56} = 0.39$. This plating efficiency remains fairly constant under our standard conditions for assay. The increase in probability of plaque formation which we suppose to take place following the infection of a bacterium by the phage particle, *i.e.* the initial rise, brings the plating efficiency up to 0.65.

The Burst Size

Single particle experiments such as that described above, revealed a great fluctuation in the magnitude of individual bursts, far larger than one would expect from the differences in size of the individual bacteria in a culture; indeed, they vary from a few particles to two hundred or more. Data from one such experiment are given in Table III.

We at first suspected that the fluctuation in burst size was connected with the time of the burst, in that early bursts were small and late bursts big, and the fluctuation was due to the experimental superposition of these.   However, measurement of a large number of bursts, plated at a time when only a small fraction of the bursts had occurred, showed the same large fluctuation.   We then suspected that the particles of a burst were not liberated simultaneously, but over an interval of time.   In this case one might expect a greater homogeneity

TABLE III

Fluctuation in Individual Burst Size

97.9 per cent of phage attached to bacteria in presence of excess bacteria (10 minutes), this mixture diluted, and samples incubated 200 minutes, then entire sample plated with added bacteria.

	Bursts
25 plates show 0 plaques	
1 plate shows 1 plaque	
14 plates show bursts	130
Average burst size, taking account of probable doubles = 48	58
	26
	123
	83
	9
	31
	5
	53
	48
	72
	45
	190
	9
Total..	882 plaques

in burst size, if measurements were made at a late time when they are at their maximum value.   This view also was found by experiment to be false.

The cause of the great fluctuation in burst size is therefore still obscure.

DISCUSSION

The results presented above show that the growth of this strain of phage is not uniform, but in bursts.   These bursts though of constant average size, under our conditions, vary widely in individual size.   A burst occurs after a definite latent period following the adsorption of the phage on susceptible bacteria, and visible lysis coincides only with the last step-wise rise in the growth curve when the phage particles outnumber the bacteria present.   It seemed reasonable to us to assume that the burst is identical with the lysis of the individual bacterium.

Krueger and Northrop (3), in their careful quantitative studies of an anti-*staphylococcus* phage came to an interpretation of their results which differs in some important respects from the above:

1. Their growth curves were smooth and gave no indication of steps; they concluded therefore that the production of phage is a continuous process.

2. In their case, the free phage during the logarithmic phase of a growth curve was an almost constant small fraction of the total phage. This led them to the view that there is an equilibrium between intracellular and extracellular phage.   With an improved technique, Krueger (10) found that the fraction of free phage decreased in proportion to the growth of the bacteria, in conformity with the assumption of an equilibrium between two phases.

3. Krueger and Northrop (3) found that visible lysis occurred when a critical ratio of total phage to bacteria had been attained, and they assumed that there was no lysis in the earlier period of phage growth.

To appreciate the nature of these differences it must be born in mind that their method of assay was essentially different from ours. They used, as a measure of the "activity" of the sample of phage assayed, the time required for it to lyse a test suspension of bacteria under standard conditions.   This time interval, according to the picture of the growth process given here, is the composite effect of a number of factors: the average time required for adsorption of free phage, its rate of growth in the infected bacteria, the time and size of burst, and the average time required for repetition of this process until the number of phage particles exceeds the number of bacteria and infects substantially all of them.   Then, after a time interval equal to the latent period, lysis occurs.

This lysis assay method tends to measure the total number of phage particles rather than the number of infective centers as the following considerations show. Let us take a sample of a growth mixture in which is suspended one infected bacterium containing fifty phage particles. If this sample is plated, it can show but a single plaque. However, if the sample is assayed by the lysis method, this single infective center soon sets free its fifty particles (or more, if multiplication is still proceeding) and the time required to attain lysis will approximate that for fifty free particles rather than that for a single particle.

Since the burst does not lead to an increase in the number of phage particles, but only to their dispersion into the solution, the lysis method cannot give any steps in the concentration of the *total* phage in a growth curve. On the other hand one might have expected a step-wise increase in the concentration of *free* phage. However, the adsorption rate of the phage used by Krueger (10) is so slow that the infection of the bacteria is spread over a time longer than the presumed latent period, and therefore the bursts would be similarly spread in time, smoothing out any steps which might otherwise appear. Moreover, their measurements were made at 30 minute intervals, which even in our case would have been insufficient to reveal the steps.

The ratio between intracellular and extracellular phage would be determined, according to this picture of phage growth, by the ratio of the average time of adsorption to the average latent period. The average time of adsorption would decrease as the bacteria increased, shifting the ratio of intracellular to extracellular phage in precisely the manner described by Krueger (10).

As we have indicated in the description of our growth curves, lysis of bacteria should become visible only at a late time. Infection of a large fraction of the bacteria is possible only after the free phage has attained a value comparable to the number of bacteria, and visible lysis should then set in after the lapse of a latent period. At this time the total phage (by activity assay) will be already large compared to the number of bacteria, in agreement with Krueger and Northrop's findings.

It appears therefore that while Krueger and Northrop's picture does not apply to our phage and bacteria, their results do not exclude for their phage the picture which we have adopted. It would be of fundamental importance if two phages behave in such a markedly different way.

SUMMARY

1. An anti-*Escherichia coli* phage has been isolated and its behavior studied.

2. A plaque counting method for this phage is described, and shown to give a number of plaques which is proportional to the phage concentration.   The number of plaques is shown to be independent of agar concentration, temperature of plate incubation, and concentration of the suspension of plating bacteria.

3. The efficiency of plating, *i.e.* the probability of plaque formation by a phage particle, depends somewhat on the culture of bacteria used for plating, and averages around 0.4.

4. Methods are described to avoid the inactivation of phage by substances in the fresh lysates.

5. The growth of phage can be divided into three periods: adsorption of the phage on the bacterium, growth upon or within the bacterium (latent period), and the release of the phage (burst).

6. The rate of adsorption of phage was found to be proportional to the concentration of phage and to the concentration of bacteria. The rate constant k_a is 1.2×10^{-9} cm.^{3}/min. at 15°C. and 1.9×10^{-9} cm.^{3}/min. at 25°.

7. The average latent period varies with the temperature in the same way as the division period of the bacteria.

8. The latent period before a burst of individual infected bacteria varies under constant conditions between a minimal value and about twice this value.

9. The average latent period and the average burst size are neither increased nor decreased by a fourfold infection of the bacteria with phage.

10. The average burst size is independent of the temperature, and is about 60 phage particles per bacterium.

11. The individual bursts vary in size from a few particles to about 200.   The same variability is found when the early bursts are measured separately, and when all the bursts are measured at a late time.

One of us (E. L. E.) wishes to acknowledge a grant in aid from Mrs. Seeley W. Mudd.   Acknowledgment is also made of the assistance of Mr. Dean Nichols during the preliminary phases of the work.

REFERENCES

1. Schlesinger, M., *Biochem. Z.*, Berlin, 1934, **273,** 306.
2. Northrop, J. H., *J. Gen. Physiol.*, 1938, **21,** 335.
3. Krueger, A. P., and Northrop, J. H., *J. Gen. Physiol.*, 1930, **14,** 223.
4. Burnett, F. M., *Brit. J. Exp. Path.*, 1927, **8,** 121.
5. Burnett, F. M., Keogh, E. V., and Lush, D., *Australian J. Exp. Biol. and Med. Sc.*, 1937, **15,** suppl. to part 3, p. 227.
6. Heidelberger, M., Kendall, F. E., and Scherp, H. W., *J. Exp. Med.*, 1936, **64,** 559.
7. d'Herelle, F., The bacteriophage and its behavior, Baltimore, The Williams & Wilkins Co., 1926.
8. Dreyer, C., and Campbell-Renton, M. L., *J. Path. and Bact.*, 1933, **36,** 399.
9. Bronfenbrenner, J. J., and Korb, C., *Proc. Soc. Exp. Biol. and Med.*, 1923, **21,** 315.
10. Krueger, A. P., *J. Gen. Physiol.*, 1931, **14,** 493.
11. Burnett, F. M., *Brit. J. Exp. Path.*, 1929, **10,** 109.
12. Poissons, S. D., Recherches sur la probabilité des jugements en matière criminelle et en matière civile, précédées des règles générales du calcul des probabilités, Paris, 1837.

Induction of Bacteriophage Lysis of an Entire Population of Lysogenic Bacteria

A. Lwoff, L. Siminovitch, and N. Kjeldgaard

At the time when this work was carried out, many of the basic principles of lysogeny had already been established. Thus it was known that lysogeny is an attribute of every bacterial cell; that lysogeny persists despite repeated passage in the presence of antiphage serum; that lysogenic phage adsorbs to the cells that produce it; and that artificial lysis of lysogenic bacteria liberates negligible amounts of phage. There was, however, no clear idea as to the nature of the lysogenic state. For example, A. D. Hershey and J. Bronfenbrenner wrote in 1948, "How virus is transmitted from cell to cell in lysogenic cultures seemingly refractory to lysis remains to be clarified. It must be concluded, however, that the phenomenon of lysogenesis, frequently cited as evidence for the spontaneous intracellular origin of virus, can equally well be explained as some sort of association between exogenous virus and incompletely susceptible bacteria" (*in Viral and Rickettsial Infections of Man*, Lippincott, Philadelphia, Pa., 1948).

The paper by Lwoff, Siminovitch, and Kjeldgaard is probably the most dramatic in revealing the true nature of lysogeny. While under normal conditions only a minute fraction of bacteria produce phage, they demonstrated that irradiation with UV light terminates the lysogenic state, which is then immediately followed by phage replication and lysis of the entire bacterial population. Clearly every cell possesses the phage genome, the expression of which is somehow repressed. This discovery opened up the field of lysogeny to detailed molecular analysis. The next step in this story was the characterization of the phage λ repressor and analysis of its function/mode of action. André Lwoff was awarded the Nobel Prize in 1965.

W. K. Joklik

Reprinted from *Comptes rendus de L'Academie des Sciences* 231:190–191, copyright © 1950, by permission of Elsevier Science Publishing.

Induction of Bacteriophage Lysis of an Entire Population of Lysogenic Bacteria[1]

by André Lwoff, Louis Siminovitch, and Niels Kjeldgaard
Presented by Robert Courrier

In a culture of lysogenic *Bacillus megaterium*, some bacteria multiply without liberating bacteriophages and perpetuate the lysogenic strain, while others produce bacteriophages which are liberated by lysis[2]. Under normal conditions of exponential growth, only a very small percentage of bacteria produce bacteriophages[2,3].

We have succeeded in inducing bacteriophage lysis of an entire population of lysogenic bacteria by means of irradiation by ultraviolet light.

B. megaterium is grown in a yeast extract medium.[2] During the exponential phase, when the number of bacteria has attained 34×10^6 per ml, a 2 mm layer of the culture is irradiated with ultraviolet light from a high pressure mercury vapor lamp, which delivers to the surface of the liquid an energy of 2000 ergs per mm^2 per minute of radiation of wave length 2537 Å. The culture is then shaken at 37°, and its optical density (O.D.) increases for about 80 minutes. The normal growth rate, that is to say the number of doublings per hour, is 3. Immediately after irradiation the growth rate is in the neighborhood of 1.5. At about the 80th minute, when the O.D. has increased by a factor of 3 to 4, bacterial lysis takes place: the culture clears in 40 to 50 minutes. The fraction of bacteria surviving is usually less than 10^{-4}. Lysis is accompanied by the liberation of about 70 to 150 bacteriophages per bacterium. Analogous results have been obtained with irradiation for 20, 30, 90, or 120 seconds.

A culture growing in synthetic medium and similarly irradiated from 1 to 60 seconds does not lyse. On the other hand, if a culture growing in yeast medium is centrifuged, resuspended in synthetic medium, and then irradiated for 5, 10, 20, or 30 seconds, bacteriophage lysis does take place. Finally if yeast extract is added to a culture growing in synthetic medium and the culture irradiated immediately for 10, 20, 30, or 60 seconds, no lysis takes place. But irradiation carried out 20 to 40 minutes after addition of the yeast extract (by which time the O.D. has increased by about 50%) induces bacteriophage lysis. UV irradiation thus induces production of bacteriophages only with bacteria that have grown for 20 to 40 minutes in a complex organic medium such as yeast extract.

Irradiation for 30 to 60 seconds of a nonlysogenic culture of *B. megaterium* growing in yeast extract does not appear to affect bacterial growth. The lysis of the lysogenic strain is thus probably not a direct effect of the UV irradiation.

It has been demonstrated previously that all bacteria of a lysogenic strain are capable of perpetuating the lysogenic character[2]. The experiments described here, which have been complemented by studies of bacteria irradiated and then isolated in microdrops, now demonstrate that under certain conditions, all the bacteria of a lysogenic population are capable of undergoing lysis with liberation of bacteriophages.

[1]Supported by a grant of the National Institutes of Health of the United States of America.

[2]A. Lwoff and A. Gutmann, *Ann. Inst. Pasteur*, **78**, p. 711–739, 1950.

[3]A. Lwoff, A. Siminovitch, and N. Kjeldgaard, *Comptes rendus*, **230**, p. 1219–1221, 1950; *Ann. Inst. Pasteur*, **79**, 815, 1950.

MICROBIOLOGIE. — *Induction de la lyse bactériophagique de la totalité d'une population microbienne lysogène* ([1]). Note de MM. Andrè Lwoff, Louis Siminovitch et Niels Kjeldgaard, présentée par M. Robert Courrier.

Dans une culture de *Bacillus megatherium* lysogène, certaines bactéries se multiplient sans libérer de bactériophages et perpétuent la souche lysogène alors que d'autres produisent des bactériophages qu'elles libèrent par lyse ([2]). Dans les conditions habituelles, durant la croissance exponentielle, seul un faible pourcentage de bactéries produit des bactériophages ([2]), ([3]).

Grâce à une irradiation ultraviolette effectuée dans certaines conditions, nous avons réussi à induire la lyse bactériophagique de la totalité d'une population bactérienne lysogène.

B. megatherium est cultivé dans un milieu à base d'extrait de levure ([2]). Durant la phase exponentielle, lorsque le nombre de bactéries atteint $34 \times 10^6/\text{cm}^3$ la culture est soumise, en couche de 2^{mm}, à l'action d'une lampe à vapeur de mercure à haute tension donnant à la surface du liquide une énergie de 2000 ergs/mm²/min pour la radiation 2537 Å. La culture est alors agitée à 37°. La densité optique (d. o.) augmente pendant 80 minutes environ. Le taux de croissance (nombre de doublements en 1 heure) normal est de 3. Il est voisin de 1,5 aussitôt après l'irradiation. Au environs de la 80ᵉ minute, alors que la d. o. a été multipliée par un facteur 3 à 4, on assiste à la lyse bactérienne : la culture s'éclaircit en 40 à 80 minutes. La proportion des bactéries survivantes est souvent inférieure à 10^{-4}. La lyse est accompagnée de la libération d'environ 70 à 150 bactériophages par bactérie. Des résultats analogues ont été obtenus avec des irradiations de 20, 30, 90 ou 120 secondes.

Les cultures en milieu synthétique, irradiées de 1 à 60 secondes dans ce même milieu, ne se lysent pas. En revanche, si une culture en milieu levuré est centrifugée et suspendue en milieu synthétique, puis irradiée 5, 10, 20 ou 30 secondes, elle subira une lyse bactériophagique. Enfin, si une culture en milieu synthétique est additionnée d'extrait de levure et irradiée immédiate-

---

([1]) Travail effectué avec l'aide d'une subvention du *National Institute of Health* des États-Unis d'Amérique.

([2]) A. Lwoff et A. Gutmann, *Ann. Inst. Pasteur*, **78**, 1950, p. 711-739.

([3]) A. Lwoff, L. Siminovitch et N. Kjeldgaard, *Comptes rendus*, **230**, 1950, p. 1219-1221; *Ann. Inst. Pasteur*, 1950 (à l'impression).

ment 10, 20, 30 ou 60 secondes, il n'y a pas de lyse. Mais l'irradiation effectuée après 20 ou 40 minutes (lorsque la d. o. a augmenté de 50 % environ), déclenche la lyse bactériophagique. Le rayonnement U. V. n'induit donc la production de bactériophages que chez des bactéries ayant vécu 20 à 40 minutes dans un milieu organique complexe, en l'espèce un extrait de levure.

L'irradiation de 30 ou 60 secondes en milieu levuré d'une culture de *B. megatherium* non lysogène ne modifie apparemment pas la croissance. La lyse de la souche lysogène n'est donc vraisemblablement pas l'effet direct du rayonnement U. V.

Il a été démontré précédemment que toutes les bactéries d'une souche lysogène sont capables de perpétuer le caractère lysogène ([2]). Les expériences décrites, qui ont été complétées par une étude des bactéries irradiées et isolées en microgouttes, démontrent maintenant que toutes les bactéries d'une population lysogène sont capables, dans des conditions déterminées, de se lyser avec libération de bactériophages.

(Extrait des *Comptes rendus des séances de l'Académie des Sciences*.
t. **231**, p. 190-191, séance du 10 juillet 1950.)

GAUTHIER-VILLARS, IMPRIMEUR-LIBRAIRE DES COMPTES RENDUS DES SÉANCES DE L'ACADÉMIE DES SCIENCES
137197-50   Paris. — Quai des Grands-Augustins, 55.

Independent Functions of Viral Protein and Nucleic Acid in Growth of Bacteriophage

A. D. HERSHEY AND M. CHASE

The results reported in this paper are another example of insights provided by the outstanding group of scientists at Cold Spring Harbor, headed by Max Delbrück, Salva Luria, Al Hershey, and Seymour Cohen, in the 1940s and 1950s. Little was known at that time about the nature and physical state of the "vegetative" form of infecting phage particles; all that was known was that, after adsorption, their infectivity was "eclipsed," and that after a roughly 10-min latent period infectious virus particles began to be generated during the "rise" period, which ended with host cell lysis and liberation of progeny phage. By labeling phage protein and DNA with different isotopes, Hershey and Chase showed, in an extremely beautiful and satisfying experiment, that the phage coat protein remains attached to the outer cell surface and that only the DNA gains access to the interior of the cell. This proved yet again, in a totally definitive manner, that it is DNA—and DNA alone—that carries genetic information. Al Hershey was awarded the Nobel Prize in 1969.

W. K. JOKLIK

Reproduced from *The Journal of General Physiology* 36:39–56, copyright © 1952, by permission of The Rockefeller University Press.

[Reprinted from THE JOURNAL OF GENERAL PHYSIOLOGY, September 20, 1952,
Vol. 36, No. 1, pp. 39–56]
Printed in U.S.A.

INDEPENDENT FUNCTIONS OF VIRAL PROTEIN AND NUCLEIC ACID IN GROWTH OF BACTERIOPHAGE*

By A. D. HERSHEY AND MARTHA CHASE

(*From the Department of Genetics, Carnegie Institution of Washington, Cold Spring Harbor, Long Island*)

(Received for publication, April 9, 1952)

The work of Doermann (1948), Doermann and Dissosway (1949), and Anderson and Doermann (1952) has shown that bacteriophages T2, T3, and T4 multiply in the bacterial cell in a non-infective form. The same is true of the phage carried by certain lysogenic bacteria (Lwoff and Gutmann, 1950). Little else is known about the vegetative phase of these viruses. The experiments reported in this paper show that one of the first steps in the growth of T2 is the release from its protein coat of the nucleic acid of the virus particle, after which the bulk of the sulfur-containing protein has no further function.

Materials and Methods.—Phage T2 means in this paper the variety called T2H (Hershey, 1946); T2h means one of the host range mutants of T2; UV-phage means phage irradiated with ultraviolet light from a germicidal lamp (General Electric Co.) to a fractional survival of 10^{-5}.

Sensitive bacteria means a strain (H) of *Escherichia coli* sensitive to T2 and its h mutant; resistant bacteria B/2 means a strain resistant to T2 but sensitive to its h mutant; resistant bacteria B/2h means a strain resistant to both. These bacteria do not adsorb the phages to which they are resistant.

"Salt-poor" broth contains per liter 10 gm. bacto-peptone, 1 gm. glucose, and 1 gm. NaCl. "Broth" contains, in addition, 3 gm. bacto-beef extract and 4 gm. NaCl.

Glycerol-lactate medium contains per liter 70 mM sodium lactate, 4 gm. glycerol, 5 gm. NaCl, 2 gm. KCl, 1 gm. NH$_4$Cl, 1 mM MgCl$_2$, 0.1 mM CaCl$_2$, 0.01 gm. gelatin, 10 mg. P (as orthophosphate), and 10 mg. S (as MgSO$_4$), at pH 7.0.

Adsorption medium contains per liter 4 gm. NaCl, 5 gm. K$_2$SO$_4$, 1.5 gm. KH$_2$PO$_4$, 3.0 gm. Na$_2$HPO$_4$, 1 mM MgSO$_4$, 0.1 mM CaCl$_2$, and 0.01 gm. gelatin, at pH 7.0.

Veronal buffer contains per liter 1 gm. sodium diethylbarbiturate, 3 mM MgSO$_4$, and 1 gm. gelatin, at pH 8.0.

The HCN referred to in this paper consists of molar sodium cyanide solution neutralized when needed with phosphoric acid.

---

* This investigation was supported in part by a research grant from the National Microbiological Institute of the National Institutes of Health, Public Health Service. Radioactive isotopes were supplied by the Oak Ridge National Laboratory on allocation from the Isotopes Division, United States Atomic Energy Commission.

Adsorption of isotope to bacteria was usually measured by mixing the sample in adsorption medium with bacteria from 18 hour broth cultures previously heated to 70°C. for 10 minutes and washed with adsorption medium. The mixtures were warmed for 5 minutes at 37°C., diluted with water, and centrifuged. Assays were made of both sediment and supernatant fractions.

Precipitation of isotope with antiserum was measured by mixing the sample in 0.5 per cent saline with about 10^{11} per ml. of non-radioactive phage and slightly more than the least quantity of antiphage serum (final dilution 1:160) that would cause visible precipitation. The mixture was centrifuged after 2 hours at 37°C.

Tests with DNase (desoxyribonuclease) were performed by warming samples diluted in veronal buffer for 15 minutes at 37°C. with 0.1 mg. per ml. of crystalline enzyme (Worthington Biochemical Laboratory).

Acid-soluble isotope was measured after the chilled sample had been precipitated with 5 per cent trichloroacetic acid in the presence of 1 mg./ml. of serum albumin, and centrifuged.

In all fractionations involving centrifugation, the sediments were not washed, and contained about 5 per cent of the supernatant. Both fractions were assayed.

Radioactivity was measured by means of an end-window Geiger counter, using dried samples sufficiently small to avoid losses by self-absorption. For absolute measurements, reference solutions of P^{32} obtained from the National Bureau of Standards, as well as a permanent simulated standard, were used. For absolute measurements of S^{35} we relied on the assays (± 20 per cent) furnished by the supplier of the isotope (Oak Ridge National Laboratory).

Glycerol-lactate medium was chosen to permit growth of bacteria without undesirable pH changes at low concentrations of phosphorus and sulfur, and proved useful also for certain experiments described in this paper. 18-hour cultures of sensitive bacteria grown in this medium contain about 2×10^9 cells per ml., which grow exponentially without lag or change in light-scattering per cell when subcultured in the same medium from either large or small seedings. The generation time is 1.5 hours at 37°C. The cells are smaller than those grown in broth. T2 shows a latent period of 22 to 25 minutes in this medium. The phage yield obtained by lysis with cyanide and UV-phage (described in context) is one per bacterium at 15 minutes and 16 per bacterium at 25 minutes. The final burst size in diluted cultures is 30 to 40 per bacterium, reached at 50 minutes. At 2×10^8 cells per ml., the culture lyses slowly, and yields 140 phage per bacterium. The growth of both bacteria and phage in this medium is as reproducible as that in broth.

For the preparation of radioactive phage, P^{32} of specific activity 0.5 mc./mg. or S^{35} of specific activity 8.0 mc./mg. was incorporated into glycerol-lactate medium, in which bacteria were allowed to grow at least 4 hours before seeding with phage. After infection with phage, the culture was aerated overnight, and the radioactive phage was isolated by three cycles of alternate slow (2000 G) and fast (12,000 G) centrifugation in adsorption medium. The suspensions were stored at a concentration not exceeding 4 μc./ml.

Preparations of this kind contain 1.0 to 3.0×10^{-12} μg. S and 2.5 to 3.5×10^{-11} μg. P per viable phage particle. Occasional preparations containing excessive amounts of sulfur can be improved by absorption with heat-killed bacteria that do not adsorb the phage. The radiochemical purity of the preparations is somewhat uncertain, owing to the possible presence of inactive phage particles and empty phage membranes. The presence in our preparations of sulfur (about 20 per cent) that is precipitated by antiphage serum (Table I) and either adsorbed by bacteria resistant to phage, or not adsorbed by bacteria sensitive to phage (Table VII), indicates contamination by membrane material. Contaminants of bacterial origin are probably negligible for present purposes as indicated by the data given in Table I. For proof that our principal findings reflect genuine properties of viable phage particles, we rely on some experiments with inactivated phage cited at the conclusion of this paper.

The Chemical Morphology of Resting Phage Particles.—Anderson (1949) found that bacteriophage T2 could be inactivated by suspending the particles in high concentrations of sodium chloride, and rapidly diluting the suspension with water. The inactivated phage was visible in electron micrographs as tadpole-shaped "ghosts." Since no inactivation occurred if the dilution was slow

TABLE I

Composition of Ghosts and Solution of Plasmolyzed Phage

Per cent of isotope	Whole phage labeled with		Plasmolyzed phage labeled with	
	P^{32}	S^{35}	P^{32}	S^{35}
Acid-soluble	—	—	1	—
Acid-soluble after treatment with DNase	1	1	80	1
Adsorbed to sensitive bacteria	85	90	2	90
Precipitated by antiphage	90	99	5	97

he attributed the inactivation to osmotic shock, and inferred that the particles possessed an osmotic membrane. Herriott (1951) found that osmotic shock released into solution the DNA (desoxypentose nucleic acid) of the phage particle, and that the ghosts could adsorb to bacteria and lyse them. He pointed out that this was a beginning toward the identification of viral functions with viral substances.

We have plasmolyzed isotopically labeled T2 by suspending the phage (10^{11} per ml.) in 3 M sodium chloride for 5 minutes at room temperature, and rapidly pouring into the suspension 40 volumes of distilled water. The plasmolyzed phage, containing not more than 2 per cent survivors, was then analyzed for phosphorus and sulfur in the several ways shown in Table I. The results confirm and extend previous findings as follows:—

1. Plasmolysis separates phage T2 into ghosts containing nearly all the sulfur and a solution containing nearly all the DNA of the intact particles.

2. The ghosts contain the principal antigens of the phage particle detectable by our antiserum. The DNA is released as the free acid, or possibly linked to sulfur-free, apparently non-antigenic substances.

3. The ghosts are specifically adsorbed to phage-susceptible bacteria; the DNA is not.

4. The ghosts represent protein coats that surround the DNA of the intact particles, react with antiserum, protect the DNA from DNase (desoxyribonuclease), and carry the organ of attachment to bacteria.

5. The effects noted are due to osmotic shock, because phage suspended in salt and diluted slowly is not inactivated, and its DNA is not exposed to DNase.

TABLE II

Sensitization of Phage DNA to DNase by Adsorption to Bacteria

Phage adsorbed to		Phage labeled with	Non-sedimentable isotope, *per cent*	
			After DNase	No DNase
Live bacteria.............................		S^{35}	2	1
" "		P^{32}	8	7
Bacteria heated before infection..............		S^{35}	15	11
" " " "		P^{32}	76	13
Bacteria heated after infection................		S^{35}	12	14
" " " "		P^{32}	66	23
Heated unadsorbed phage: acid-soluble P^{32}	70°.......	P^{32}	5	
	80°.......	P^{32}	13	
	90°......	P^{32}	81	
	100°.......	P^{32}	88	

Phage adsorbed to bacteria for 5 minutes at 37°C. in adsorption medium, followed by washing.

Bacteria heated for 10 minutes at 80°C. in adsorption medium (before infection) or in veronal buffer (after infection).

Unadsorbed phage heated in veronal buffer, treated with DNase, and precipitated with trichloroacetic acid.

All samples fractionated by centrifuging 10 minutes at 1300 *G*.

Sensitization of Phage DNA to DNase by Adsorption to Bacteria.—The structure of the resting phage particle described above suggests at once the possibility that multiplication of virus is preceded by the alteration or removal of the protective coats of the particles. This change might be expected to show itself as a sensitization of the phage DNA to DNase. The experiments described in Table II show that this happens. The results may be summarized as follows:—

1. Phage DNA becomes largely sensitive to DNase after adsorption to heat-killed bacteria.

2. The same is true of the DNA of phage adsorbed to live bacteria, and then

heated to 80°C. for 10 minutes, at which temperature unadsorbed phage is not sensitized to DNase.

3. The DNA of phage adsorbed to unheated bacteria is resistant to DNase, presumably because it is protected by cell structures impervious to the enzyme.

Graham and collaborators (personal communication) were the first to discover the sensitization of phage DNA to DNase by adsorption to heat-killed bacteria.

The DNA in infected cells is also made accessible to DNase by alternate freezing and thawing (followed by formaldehyde fixation to inactivate cellular enzymes), and to some extent by formaldehyde fixation alone, as illustrated by the following experiment.

Bacteria were grown in broth to 5×10^7 cells per ml., centrifuged, resuspended in adsorption medium, and infected with about two P^{32}-labeled phage per bacterium. After 5 minutes for adsorption, the suspension was diluted with water containing per liter 1.0 mM $MgSO_4$, 0.1 mM $CaCl_2$, and 10 mg. gelatin, and recentrifuged. The cells were resuspended in the fluid last mentioned at a concentration of 5×10^8 per ml. This suspension was frozen at $-15°C$. and thawed with a minimum of warming, three times in succession. Immediately after the third thawing, the cells were fixed by the addition of 0.5 per cent (v/v) of formalin (35 per cent HCHO). After 30 minutes at room temperature, the suspension was dialyzed free from formaldehyde and centrifuged at 2200 G for 15 minutes. Samples of P^{32}-labeled phage, frozen-thawed, fixed, and dialyzed, and of infected cells fixed only and dialyzed, were carried along as controls.

The analysis of these materials, given in Table III, shows that the effect of freezing and thawing is to make the intracellular DNA labile to DNase, without, however, causing much of it to leach out of the cells. Freezing and thawing and formaldehyde fixation have a negligible effect on unadsorbed phage, and formaldehyde fixation alone has only a mild effect on infected cells.

Both sensitization of the intracellular P^{32} to DNase, and its failure to leach out of the cells, are constant features of experiments of this type, independently of visible lysis. In the experiment just described, the frozen suspension cleared during the period of dialysis. Phase-contrast microscopy showed that the cells consisted largely of empty membranes, many apparently broken. In another experiment, samples of infected bacteria from a culture in salt-poor broth were repeatedly frozen and thawed at various times during the latent period of phage growth, fixed with formaldehyde, and then washed in the centrifuge. Clearing and microscopic lysis occurred only in suspensions frozen during the second half of the latent period, and occurred during the first or second thawing. In this case the lysed cells consisted wholly of intact cell membranes, appearing empty except for a few small, rather characteristic refractile bodies apparently attached to the cell walls. The behavior of intracellular P^{32} toward DNase, in either the lysed or unlysed cells, was not significantly different from that shown in Table III, and the content of P^{32} was only slightly less after lysis. The phage liberated during freezing and thawing was also titrated in this experiment. The lysis occurred without appreciable liberation of phage in suspensions frozen up to and including the 16th minute, and the 20 minute sample yielded only five per bacterium. Another sample of the culture formalinized at 30 minutes, and centrifuged without freezing, contained 66 per cent of the P^{32} in non-sedimentable form. The yield of extracellular phage at 30 minutes was 108 per bacterium, and the sedimented material consisted largely of formless debris but contained also many apparently intact cell membranes.

TABLE III

Sensitization of Intracellular Phage to DNase by Freezing, Thawing, and Fixation with Formaldehyde

	Unadsorbed phage frozen, thawed, fixed	Infected cells frozen, thawed, fixed	Infected cells fixed only
Low speed sediment fraction			
Total P^{32}	—	71	86
Acid-soluble	—	0	0.5
Acid-soluble after DNase	—	59	28
Low speed supernatant fraction			
Total P^{32}	—	29	14
Acid-soluble	1	0.8	0.4
Acid-soluble after DNase	11	21	5.5

The figures express per cent of total P^{32} in the original phage, or its adsorbed fraction.

We draw the following conclusions from the experiments in which cells infected with P^{32}-labeled phage are subjected to freezing and thawing.

1. Phage DNA becomes sensitive to DNAse after adsorption to bacteria in buffer under conditions in which no known growth process occurs (Benzer, 1952; Dulbecco, 1952).

2. The cell membrane can be made permeable to DNase under conditions that do not permit the escape of either the intracellular P^{32} or the bulk of the cell contents.

3. Even if the cells lyse as a result of freezing and thawing, permitting escape of other cell constituents, most of the P^{32} derived from phage remains inside the cell membranes, as do the mature phage progeny.

4. The intracellular P^{32} derived from phage is largely freed during spontaneous lysis accompanied by phage liberation.

We interpret these facts to mean that intracellular DNA derived from phage is not merely DNA in solution, but is part of an organized structure at all times during the latent period.

Liberation of DNA from Phage Particles by Adsorption to Bacterial Fragments.—The sensitization of phage DNA to specific depolymerase by adsorption to bacteria might mean that adsorption is followed by the ejection of the phage DNA from its protective coat. The following experiment shows that this is in fact what happens when phage attaches to fragmented bacterial cells.

TABLE IV

Release of DNA from Phage Adsorbed to Bacterial Debris

	Phage labeled with	
	S^{35}	P^{32}
Sediment fraction		
Surviving phage	16	22
Total isotope	87	55
Acid-soluble isotope	0	2
Acid-soluble after DNase	2	29
Supernatant fraction		
Surviving phage	5	5
Total isotope	13	45
Acid-soluble isotope	0.8	0.5
Acid-soluble after DNase	0.8	39

S^{35}- and P^{32}-labeled T2 were mixed with identical samples of bacterial debris in adsorption medium and warmed for 30 minutes at 37°C. The mixtures were then centrifuged for 15 minutes at 2200 G, and the sediment and supernatant fractions were analyzed separately. The results are expressed as per cent of input phage or isotope.

Bacterial debris was prepared by infecting cells in adsorption medium with four particles of T2 per bacterium, and transferring the cells to salt-poor broth at 37°C. The culture was aerated for 60 minutes, M/50 HCN was added, and incubation continued for 30 minutes longer. At this time the yield of extracellular phage was 400 particles per bacterium, which remained unadsorbed because of the low concentration of electrolytes. The debris from the lysed cells was washed by centrifugation at 1700 G, and resuspended in adsorption medium at a concentration equivalent to 3×10^9 lysed cells per ml. It consisted largely of collapsed and fragmented cell membranes. The adsorption of radioactive phage to this material is described in Table IV. The following facts should be noted.

1. The unadsorbed fraction contained only 5 per cent of the original phage particles in infective form, and only 13 per cent of the total sulfur. (Much of this sulfur must be the material that is not adsorbable to whole bacteria.)

2. About 80 per cent of the phage was inactivated. Most of the sulfur of this phage, as well as most of the surviving phage, was found in the sediment fraction.

3. The supernatant fraction contained 40 per cent of the total phage DNA (in a form labile to DNase) in addition to the DNA of the unadsorbed surviving phage. The labile DNA amounted to about half of the DNA of the inactivated phage particles, whose sulfur sedimented with the bacterial debris.

4. Most of the sedimentable DNA could be accounted for either as surviving phage, or as DNA labile to DNase, the latter amounting to about half the DNA of the inactivated particles.

Experiments of this kind are unsatisfactory in one respect: one cannot tell whether the liberated DNA represents all the DNA of some of the inactivated particles, or only part of it.

Similar results were obtained when bacteria (strain B) were lysed by large amounts of UV-killed phage T2 or T4 and then tested with P^{32}-labeled T2 and T4. The chief point of interest in this experiment is that bacterial debris saturated with UV-killed T2 adsorbs T4 better than T2, and debris saturated with T4 adsorbs T2 better than T4. As in the preceding experiment, some of the adsorbed phage was not inactivated and some of the DNA of the inactivated phage was not released from the debris.

These experiments show that some of the cell receptors for T2 are different from some of the cell receptors for T4, and that phage attaching to these specific receptors is inactivated by the same mechanism as phage attaching to unselected receptors. This mechanism is evidently an active one, and not merely the blocking of sites of attachment to bacteria.

Removal of Phage Coats from Infected Bacteria.—Anderson (1951) has obtained electron micrographs indicating that phage T2 attaches to bacteria by its tail. If this precarious attachment is preserved during the progress of the infection, and if the conclusions reached above are correct, it ought to be a simple matter to break the empty phage membranes off the infected bacteria, leaving the phage DNA inside the cells.

The following experiments show that this is readily accomplished by strong shearing forces applied to suspensions of infected cells, and further that infected cells from which 80 per cent of the sulfur of the parent virus has been removed remain capable of yielding phage progeny.

Broth-grown bacteria were infected with S^{35}- or P^{32}-labeled phage in adsorption medium, the unadsorbed material was removed by centrifugation, and the cells were resuspended in water containing per liter 1 mM $MgSO_4$, 0.1 mM $CaCl_2$, and 0.1 gm. gelatin. This suspension was spun in a Waring blendor (semimicro size) at 10,000 R.P.M. The suspension was cooled briefly in ice water at the end of each 60 second running period. Samples were removed at intervals, titrated (through antiphage serum) to measure the number of bacteria capable of yielding phage, and centrifuged to measure the proportion of isotope released from the cells.

The results of one experiment with each isotope are shown in Fig. 1. The data for S^{35} and survival of infected bacteria come from the same experiment, in which the ratio of added phage to bacteria was 0.28, and the concentrations

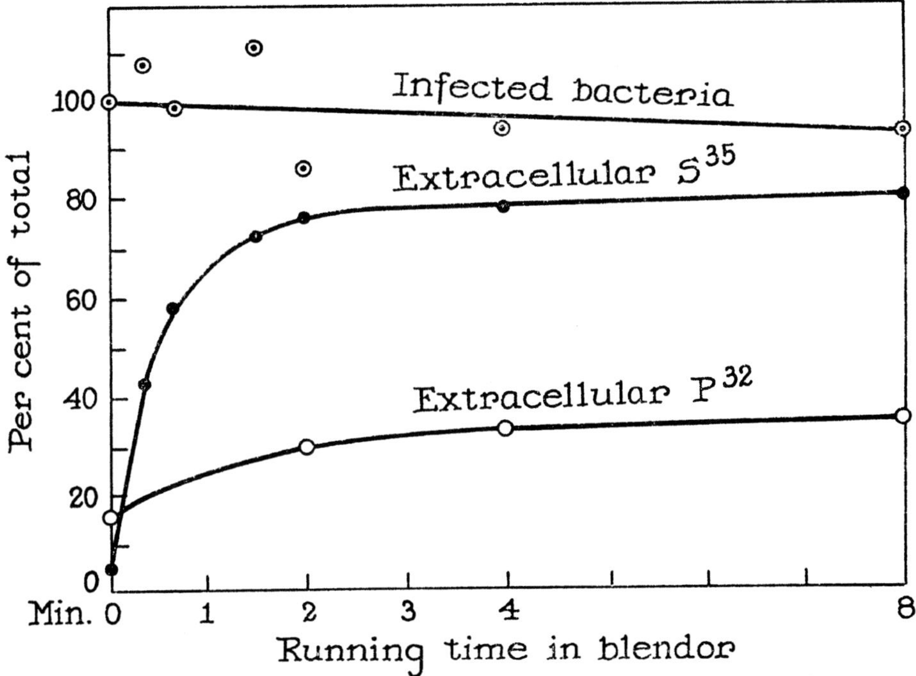

FIG. 1. Removal of S^{35} and P^{32} from bacteria infected with radioactive phage, and survival of the infected bacteria, during agitation in a Waring blendor.

of bacteria were 2.5×10^8 per ml. infected, and 9.7×10^8 per ml. total, by direct titration. The experiment with P^{32}-labeled phage was very similar. In connection with these results, it should be recalled that Anderson (1949) found that adsorption of phage to bacteria could be prevented by rapid stirring of the suspension.

At higher ratios of infection, considerable amounts of phage sulfur elute from the cells spontaneously under the conditions of these experiments, though the elution of P^{32} and the survival of infected cells are not affected by multiplicity of infection (Table V). This shows that there is a cooperative action among phage particles in producing alterations of the bacterial membrane which weaken the attachment of the phage. The cellular changes detected in this way may be related to those responsible for the release of bacterial components from infected bacteria (Prater, 1951; Price, 1952).

A variant of the preceding experiments was designed to test bacteria at a later stage in the growth of phage. For this purpose infected cells were aerated in broth for 5 or 15 minutes, fixed by the addition of 0.5 per cent (v/v) commercial formalin, centrifuged, resuspended in 0.1 per cent formalin in water, and subsequently handled as described above. The results were very similar to those already presented, except that the release of P^{32} from the cells was slightly less, and titrations of infected cells could not be made.

The S^{35}-labeled material detached from infected cells in the manner described possesses the following properties. It is sedimented at 12,000 G, though less completely than intact phage particles. It is completely precipitated by

TABLE V

Effect of Multiplicity of Infection on Elution of Phage Membranes from Infected Bacteria

Running time in blendor	Multiplicity of infection	P^{32}-labeled phage		S^{35}-labeled phage	
		Isotope eluted	Infected bacteria surviving	Isotope eluted	Infected bacteria surviving
min.		*per cent*	*per cent*	*per cent*	*per cent*
0	0.6	10	120	16	101
2.5	0.6	21	82	81	78
0	6.0	13	89	46	90
2.5	6.0	24	86	82	85

The infected bacteria were suspended at 10^9 cells per ml. in water containing per liter 1 mM MgSO$_4$, 0.1 mM CaCl$_2$, and 0.1 gm. gelatin. Samples were withdrawn for assay of extracellular isotope and infected bacteria before and after agitating the suspension. In either case the cells spent about 15 minutes at room temperature in the eluting fluid.

antiphage serum in the presence of whole phage carrier. 40 to 50 per cent of it readsorbs to sensitive bacteria, almost independently of bacterial concentration between 2×10^8 and 10^9 cells per ml., in 5 minutes at 37°C. The adsorption is not very specific: 10 to 25 per cent adsorbs to phage-resistant bacteria under the same conditions. The adsorption requires salt, and for this reason the efficient removal of S^{35} from infected bacteria can be accomplished only in a fluid poor in electrolytes.

The results of these experiments may be summarized as follows:—

1. 75 to 80 per cent of the phage sulfur can be stripped from infected cells by violent agitation of the suspension. At high multiplicity of infection, nearly 50 per cent elutes spontaneously. The properties of the S^{35}-labeled material show that it consists of more or less intact phage membranes, most of which have lost the ability to attach specifically to bacteria.

2. The release of sulfur is accompanied by the release of only 21 to 35 per

cent of the phage phosphorus, half of which is given up without any mechanical agitation.

3. The treatment does not cause any appreciable inactivation of intracellular phage.

4. These facts show that the bulk of the phage sulfur remains at the cell surface during infection, and takes no part in the multiplication of intracellular phage. The bulk of the phage DNA, on the other hand, enters the cell soon after adsorption of phage to bacteria.

Transfer of Sulfur and Phosphorus from Parental Phage to Progeny.—We have concluded above that the bulk of the sulfur-containing protein of the resting phage particle takes no part in the multiplication of phage, and in fact does not enter the cell. It follows that little or no sulfur should be transferred from parental phage to progeny. The experiments described below show that this expectation is correct, and that the maximal transfer is of the order 1 per cent

Bacteria were grown in glycerol-lactate medium overnight and subcultured in the same medium for 2 hours at 37°C. with aeration, the size of seeding being adjusted nephelometrically to yield 2×10^8 cells per ml. in the subculture. These bacteria were sedimented, resuspended in adsorption medium at a concentration of 10^9 cells per ml., and infected with S^{35}-labeled phage T2. After 5 minutes at 37°C., the suspension was diluted with 2 volumes of water and resedimented to remove unadsorbed phage (5 to 10 per cent by titer) and S^{35} (about 15 per cent). The cells were next suspended in glycerol-lactate medium at a concentration of 2×10^8 per ml. and aerated at 37°C. Growth of phage was terminated at the desired time by adding in rapid succession 0.02 mM HCN and 2×10^{11} UV-killed phage per ml. of culture. The cyanide stops the maturation of intracellular phage (Doermann, 1948), and the UV-killed phage minimizes losses of phage progeny by adsorption to bacterial debris, and promotes the lysis of bacteria (Maaløe and Watson, 1951). As mentioned in another connection, and also noted in these experiments, the lysing phage must be closely related to the phage undergoing multiplication (*e.g.*, T2H, its *h* mutant, or T2L, but not T4 or T6, in this instance) in order to prevent inactivation of progeny by adsorption to bacterial debris.

To obtain what we shall call the maximal yield of phage, the lysing phage was added 25 minutes after placing the infected cells in the culture medium, and the cyanide was added at the end of the 2nd hour. Under these conditions, lysis of infected cells occurs rather slowly.

Aeration was interrupted when the cyanide was added, and the cultures were left overnight at 37°C. The lysates were then fractionated by centrifugation into an initial low speed sediment (2500 *G* for 20 minutes), a high speed supernatant (12,000 *G* for 30 minutes), a second low speed sediment obtained by recentrifuging in adsorption medium the resuspended high speed sediment, and the clarified high speed sediment.

The distribution of S^{35} and phage among fractions obtained from three cultures of this kind is shown in Table VI. The results are typical (except for the excessively good recoveries of phage and S^{35}) of lysates in broth as well as lysates in glycerol-lactate medium.

The striking result of this experiment is that the distribution of S^{35} among the fractions is the same for early lysates that do not contain phage progeny, and later ones that do. This suggests that little or no S^{35} is contained in the mature phage progeny. Further fractionation by adsorption to bacteria confirms this suggestion.

Adsorption mixtures prepared for this purpose contained about 5×10^9 heat-killed bacteria (70°C. for 10 minutes) from 18 hour broth cultures, and

TABLE VI

Per Cent Distributions of Phage and S^{35} among Centrifugally Separated Fractions of Lysates after Infection with S^{35}-Labeled T2

Fraction	Lysis at $t = 0$ S^{35}	Lysis at $t = 10$ S^{35}	Maximal yield	
			S^{35}	Phage
1st low speed sediment....................	79	81	82	19
2nd " " "	2.4	2.1	2.8	14
High speed "	8.6	6.9	7.1	61
" " supernatant....................	10	10	7.5	7.0
Recovery....................	100	100	96	100

Infection with S^{35}-labeled T2, 0.8 particles per bacterium. Lysing phage UV-killed *h* mutant of T2. Phage yields per infected bacterium: <0.1 after lysis at $t = 0$; 0.12 at $t = 10$; maximal yield 29. Recovery of S^{35} means per cent of adsorbed input recovered in the four fractions; recovery of phage means per cent of total phage yield (by plaque count before fractionation) recovered by titration of fractions.

about 10^{11} phage (UV-killed lysing phage plus test phage), per ml. of adsorption medium. After warming to 37°C. for 5 minutes, the mixtures were diluted with 2 volumes of water, and centrifuged. Assays were made from supernatants and from unwashed resuspended sediments.

The results of tests of adsorption of S^{35} and phage to bacteria (H) adsorbing both T2 progeny and *h*-mutant lysing phage, to bacteria (B/2) adsorbing lysing phage only, and to bacteria (B/2*h*) adsorbing neither, are shown in Table VII, together with parallel tests of authentic S^{35}-labeled phage.

The adsorption tests show that the S^{35} present in the seed phage is adsorbed with the specificity of the phage, but that S^{35} present in lysates of bacteria infected with this phage shows a more complicated behavior. It is strongly adsorbed to bacteria adsorbing both progeny and lysing phage. It is weakly adsorbed to bacteria adsorbing neither. It is moderately well adsorbed to bacteria adsorbing lysing phage but not phage progeny. The latter test shows that the S^{35} is not contained in the phage progeny, and explains the fact that the S^{35} in early lysates not containing progeny behaves in the same way.

The specificity of the adsorption of S^{35}-labeled material contaminating the phage progeny is evidently due to the lysing phage, which is also adsorbed much more strongly to strain H than to B/2, as shown both by the visible reduction in Tyndall scattering (due to the lysing phage) in the supernatants of the test mixtures, and by independent measurements. This conclusion is further confirmed by the following facts.

TABLE VII

Adsorption Tests with Uniformly S^{35}-Labeled Phage and with Products of Their Growth in Non-Radioactive Medium

Adsorbing bacteria	Per cent adsorbed				
	Uniformly labeled S^{35} phage		Products of lysis at $t = 10$	Phage progeny (Maximal yield)	
	+ UV-h	No UV-h			
	S^{35}	S^{35}	S^{35}	S^{35}	Phage
Sensitive (H).....................	84	86	79	78	96
Resistant (B/2)..................	15	11	46	49	10
Resistant (B/2h).................	13	12	29	28	8

The uniformly labeled phage and the products of their growth are respectively the seed phage and the high speed sediment fractions from the experiment shown in Table VI.

The uniformly labeled phage is tested at a low ratio of phage to bacteria: +UV-h means with added UV-killed h mutant in equal concentration to that present in the other test materials.

The adsorption of phage is measured by plaque counts of supernatants, and also sediments in the case of the resistant bacteria, in the usual way.

1. If bacteria are infected with S^{35} phage, and then lysed near the midpoint of the latent period with cyanide alone (in salt-poor broth, to prevent readsorption of S^{35} to bacterial debris), the high speed sediment fraction contains S^{35} that is adsorbed weakly and non-specifically to bacteria.

2. If the lysing phage and the S^{35}-labeled infecting phage are the same (T2), or if the culture in salt-poor broth is allowed to lyse spontaneously (so that the yield of progeny is large), the S^{35} in the high speed sediment fraction is adsorbed with the specificity of the phage progeny (except for a weak non-specific adsorption). This is illustrated in Table VII by the adsorption to H and B/2h.

It should be noted that a phage progeny grown from S^{35}-labeled phage and containing a larger or smaller amount of contaminating radioactivity could not be distinguished by any known method from authentic S^{35}-labeled phage, except that a small amount of the contaminant could be removed by adsorption to bacteria resistant to the phage. In addition to the properties already mentioned, the contaminating S^{35} is completely precipitated with the phage by antiserum, and cannot be appreciably separated from the phage by further fractional sedimentation, at either high or low concentrations of electrolyte. On the other hand, the chemical contamination from this source would be very small in favorable circumstances, because the progeny of a single phage particle are numerous and the contaminant is evidently derived from the parents.

The properties of the S^{35}-labeled contaminant show that it consists of the remains of the coats of the parental phage particles, presumably identical with the material that can be removed from unlysed cells in the Waring blendor. The fact that it undergoes little chemical change is not surprising since it probably never enters the infected cell.

The properties described explain a mistaken preliminary report (Hershey *et al.*, 1951) of the transfer of S^{35} from parental to progeny phage.

It should be added that experiments identical to those shown in Tables VI and VII, but starting from phage labeled with P^{32}, show that phosphorus is transferred from parental to progeny phage to the extent of 30 per cent at yields of about 30 phage per infected bacterium, and that the P^{32} in prematurely lysed cultures is almost entirely non-sedimentable, becoming, in fact, acid-soluble on aging.

Similar measures of the transfer of P^{32} have been published by Putnam and Kozloff (1950) and others. Watson and Maaløe (1952) summarize this work, and report equal transfer (nearly 50 per cent) of phosphorus and adenine.

A Progeny of S^{35}-Labeled Phage Nearly Free from the Parental Label.—The following experiment shows clearly that the obligatory transfer of parental sulfur to offspring phage is less than 1 per cent, and probably considerably less. In this experiment, the phage yield from infected bacteria from which the S^{35}-labeled phage coats had been stripped in the Waring blendor was assayed directly for S^{35}.

Sensitive bacteria grown in broth were infected with five particles of S^{35}-labeled phage per bacterium, the high ratio of infection being necessary for purposes of assay. The infected bacteria were freed from unadsorbed phage and suspended in water containing per liter 1 mM $MgSO_4$, 0.1 mM $CaCl_2$, and 0.1 gm. gelatin. A sample of this suspension was agitated for 2.5 minutes in the Waring blendor, and centrifuged to remove the extracellular S^{35}. A second sample not run in the blendor was centrifuged at the same time. The cells from both samples were resuspended in warm salt-poor broth at a concentration of 10^8 bacteria per ml., and aerated for 80 minutes. The cultures were then lysed by the addition of 0.02 mM HCN, 2×10^{11} UV-killed T2, and 6 mg. NaCl per ml. of culture. The addition of salt at this point causes S^{35} that would otherwise be eluted (Hershey *et al.*, 1951) to remain attached to the bacterial debris. The lysates were fractionated and assayed as described previously, with the results shown in Table VIII.

The data show that stripping reduces more or less proportionately the S^{35}-content of all fractions. In particular, the S^{35}-content of the fraction containing most of the phage progeny is reduced from nearly 10 per cent to less than 1 per cent of the initially adsorbed isotope. This experiment shows that the bulk of the S^{35} appearing in all lysate fractions is derived from the remains of the coats of the parental phage particles.

Properties of Phage Inactivated by Formaldehyde.—Phage T2 warmed for 1 hour at 37°C. in adsorption medium containing 0.1 per cent (v/v) commercial formalin (35 per cent HCHO), and then dialyzed free from formalde-

TABLE VIII

Lysates of Bacteria Infected with S^{35}-Labeled T2 and Stripped in the Waring Blendor

Per cent of adsorbed S^{35} or of phage yield:	Cells stripped		Cells not stripped	
	S^{35}	Phage	S^{35}	Phage
Eluted in blendor fluid..........................	86	—	39	—
1st low-speed sediment........................	3.8	9.3	31	13
2nd " " • "	(0.2)	11	2.7	11
High-speed "	(0.7)	58	9.4	89
" " supernatant........................	(2.0)	1.1	(1.7)	1.6
Recovery...................................	93	79	84	115

All the input bacteria were recovered in assays of infected cells made during the latent period of both cultures. The phage yields were 270 (stripped cells) and 200 per bacterium, assayed before fractionation. Figures in parentheses were obtained from counting rates close to background.

hyde, shows a reduction in plaque titer by a factor 1000 or more. Inactivated phage of this kind possesses the following properties.

1. It is adsorbed to sensitive bacteria (as measured by either S^{35} or P^{32} labels), to the extent of about 70 per cent.

2. The adsorbed phage kills bacteria with an efficiency of about 35 per cent compared with the original phage stock.

3. The DNA of the inactive particles is resistant to DNase, but is made sensitive by osmotic shock.

4. The DNA of the inactive particles is not sensitized to DNase by adsorption to heat-killed bacteria, nor is it released into solution by adsorption to bacterial debris.

5. 70 per cent of the adsorbed phage DNA can be detached from infected cells spun in the Waring blendor. The detached DNA is almost entirely resistant to DNase.

These properties show that T2 inactivated by formaldehyde is largely incapable of injecting its DNA into the cells to which it attaches. Its behavior in the experiments outlined gives strong support to our interpretation of the corresponding experiments with active phage.

DISCUSSION

We have shown that when a particle of bacteriophage T2 attaches to a bacterial cell, most of the phage DNA enters the cell, and a residue containing at least 80 per cent of the sulfur-containing protein of the phage remains at the cell surface. This residue consists of the material forming the protective membrane of the resting phage particle, and it plays no further role in infection after the attachment of phage to bacterium.

These facts leave in question the possible function of the 20 per cent of sulfur-containing protein that may or may not enter the cell. We find that little or none of it is incorporated into the progeny of the infecting particle, and that at least part of it consists of additional material resembling the residue that can be shown to remain extracellular. Phosphorus and adenine (Watson and Maaløe, 1952) derived from the DNA of the infecting particle, on the other hand, are transferred to the phage progeny to a considerable and equal extent. We infer that sulfur-containing protein has no function in phage multiplication, and that DNA has some function.

It must be recalled that the following questions remain unanswered. (1) Does any sulfur-free phage material other than DNA enter the cell? (2) If so, is it transferred to the phage progeny? (3) Is the transfer of phosphorus (or hypothetical other substance) to progeny direct—that is, does it remain at all times in a form specifically identifiable as phage substance—or indirect?

Our experiments show clearly that a physical separation of the phage T2 into genetic and non-genetic parts is possible. A corresponding functional separation is seen in the partial independence of phenotype and genotype in the same phage (Novick and Szilard, 1951; Hershey *et al.*, 1951). The chemical identification of the genetic part must wait, however, until some of the questions asked above have been answered.

Two facts of significance for the immunologic method of attack on problems of viral growth should be emphasized here. First, the principal antigen of the infecting particles of phage T2 persists unchanged in infected cells. Second, it remains attached to the bacterial debris resulting from lysis of the cells. These possibilities seem to have been overlooked in a study by Rountree (1951) of viral antigens during the growth of phage T5.

SUMMARY

1. Osmotic shock disrupts particles of phage T2 into material containing nearly all the phage sulfur in a form precipitable by antiphage serum, and capable of specific adsorption to bacteria. It releases into solution nearly all

the phage DNA in a form not precipitable by antiserum and not adsorbable to bacteria. The sulfur-containing protein of the phage particle evidently makes up a membrane that protects the phage DNA from DNase, comprises the sole or principal antigenic material, and is responsible for attachment of the virus to bacteria.

2. Adsorption of T2 to heat-killed bacteria, and heating or alternate freezing and thawing of infected cells, sensitize the DNA of the adsorbed phage to DNase. These treatments have little or no sensitizing effect on unadsorbed phage. Neither heating nor freezing and thawing releases the phage DNA from infected cells, although other cell constituents can be extracted by these methods. These facts suggest that the phage DNA forms part of an organized intracellular structure throughout the period of phage growth.

3. Adsorption of phage T2 to bacterial debris causes part of the phage DNA to appear in solution, leaving the phage sulfur attached to the debris. Another part of the phage DNA, corresponding roughly to the remaining half of the DNA of the inactivated phage, remains attached to the debris but can be separated from it by DNase. Phage T4 behaves similarly, although the two phages can be shown to attach to different combining sites. The inactivation of phage by bacterial debris is evidently accompanied by the rupture of the viral membrane.

4. Suspensions of infected cells agitated in a Waring blendor release 75 per cent of the phage sulfur and only 15 per cent of the phage phosphorus to the solution as a result of the applied shearing force. The cells remain capable of yielding phage progeny.

5. The facts stated show that most of the phage sulfur remains at the cell surface and most of the phage DNA enters the cell on infection. Whether sulfur-free material other than DNA enters the cell has not been determined. The properties of the sulfur-containing residue identify it as essentially unchanged membranes of the phage particles. All types of evidence show that the passage of phage DNA into the cell occurs in non-nutrient medium under conditions in which other known steps in viral growth do not occur.

6. The phage progeny yielded by bacteria infected with phage labeled with radioactive sulfur contain less than 1 per cent of the parental radioactivity. The progeny of phage particles labeled with radioactive phosphorus contain 30 per cent or more of the parental phosphorus.

7. Phage inactivated by dilute formaldehyde is capable of adsorbing to bacteria, but does not release its DNA to the cell. This shows that the interaction between phage and bacterium resulting in release of the phage DNA from its protective membrane depends on labile components of the phage particle. By contrast, the components of the bacterium essential to this interaction are remarkably stable. The nature of the interaction is otherwise unknown.

8. The sulfur-containing protein of resting phage particles is confined to a

protective coat that is responsible for the adsorption to bacteria, and functions as an instrument for the injection of the phage DNA into the cell. This protein probably has no function in the growth of intracellular phage. The DNA has some function. Further chemical inferences should not be drawn from the experiments presented.

REFERENCES

Anderson, T. F., 1949, The reactions of bacterial viruses with their host cells, *Bot. Rev.*, **15,** 464.

Anderson, T. F., 1951, *Tr. New York Acad. Sc.*, **13,** 130.

Anderson, T. F., and Doermann, A. H., 1952, *J. Gen. Physiol.*, **35,** 657.

Benzer, S., 1952, *J. Bact.*, **63,** 59.

Doermann, A. H., 1948, *Carnegie Institution of Washington Yearbook, No. 47,* 176.

Doermann, A. H., and Dissosway, C., 1949, *Carnegie Institution of Washington Yearbook, No. 48,* 170.

Dulbecco, R., 1952, *J. Bact.*, **63,** 209.

Herriott, R. M., 1951, *J. Bact.*, **61,** 752.

Hershey, A. D., 1946, *Genetics*, **31,** 620.

Hershey, A. D., Roesel, C., Chase, M., and Forman, S., 1951, *Carnegie Institution of Washington Yearbook, No. 50,* 195.

Lwoff, A., and Gutmann, A., 1950, *Ann. Inst. Pasteur*, **78,** 711.

Maaløe, O., and Watson, J. D., 1951, *Proc. Nat. Acad. Sc.*, **37,** 507.

Novick, A., and Szilard, L., 1951, *Science*, **113,** 34.

Prater, C. D., 1951, Thesis, University of Pennsylvania.

Price, W. H., 1952, *J. Gen. Physiol.*, **35,** 409.

Putnam, F. W., and Kozloff, L.,1950, *J. Biol. Chem.*, **182,** 243.

Rountree, P. M., 1951, *Brit. J. Exp. Path.*, **32,** 341.

Watson, J. D., and Maaløe, O., 1952, *Acta path. et microbiol. scand.*, in press.

Production of Plaques in Monolayer Tissue Cultures by Single Particles of an Animal Virus

R. DULBECCO

For the first half century of animal virology, the major problem was lack of a simple method for quantitating infectious virus particles; the only method available at that time was some form or other of the serial-dilution end-point method in animals, all of which were both slow and expensive. Cloned cultured animal cells, which began to be available around 1950, provided Dulbecco with a new approach. He adapted the technique developed by Emory Ellis and Max Delbrück for assaying bacteriophage, that is, seeding serial dilutions of a given virus population onto a confluent lawn of host cells, to the measurement of Western equine encephalitis virus, and demonstrated that it also formed easily countable plaques in monolayers of chick embryo fibroblasts. The impact of this finding was enormous; animal virologists had been waiting for such a technique for decades. It was immediately found to be widely applicable to many types of cells and most viruses, gained quick acceptance, and is widely regarded as marking the beginning of molecular animal virology. Renato Dulbecco was awarded the Nobel Prize in 1975.

W. K. JOKLIK

Reprinted from *Proceedings of the National Academy of Sciences USA* 38:747–752. Copyright © 1952.

PRODUCTION OF PLAQUES IN MONOLAYER TISSUE CULTURES BY SINGLE PARTICLES OF AN ANIMAL VIRUS

By Renato Dulbecco

California Institute of Technology, Pasadena, California

Read before the Academy, April 29, 1952

Research on the growth characteristics and genetic properties of animal viruses has stood greatly in need of improved quantitative techniques, such as those used in the related field of bacteriophage studies.

The requirements for a quantitative virus technique are as follows: (1) The use of a uniform type of host cell; (2) an accurate assay technique; (3) the isolation of the progeny of a single virus particle; and (4) the separate isolation of each of the virus particles produced by a single infected cell. In bacterial virus work, production of plaques by single virus particles on a uniform bacterial layer fulfills the first three requirements. The fourth requirement is easily met, in the case of bacteriophages.

In this article we shall show that plaques can similarly be produced by animal viruses and that their properties fulfill the first three requirements.

We have found that the virus of Western Equine Encephalomyelitis,

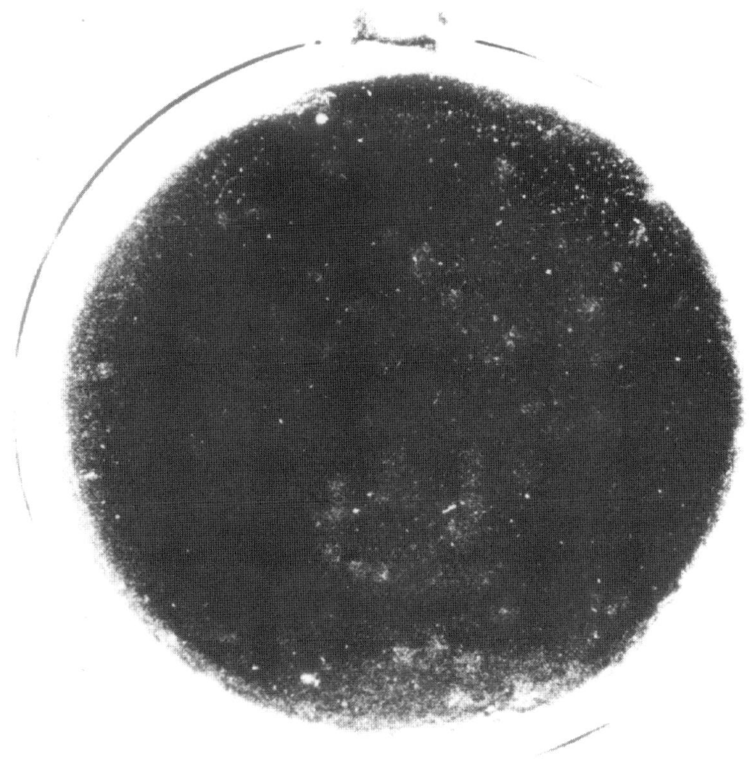

FIGURE 1

Plaques of Western Equine Encephalomyelitis virus on chicken fibroblasts. The plaques appear as round clear areas. The picture was taken against a distant dark background, sufficiently large to exactly cover the projection of the flask, surrounded by a uniformly brilliant light source, so as to detect the light scattered through a small angle by the cell debris in the plaques.

adapted to chicken embryo,[1] will produce plaques when it is grown on a monolayer of cells obtainable from chicken embryos by a modification of a technique devised by Shannon, Earle and Walts.[2] We use the following procedure: Under aseptic conditions, five 9-day-old chicken embryos are collected, decapitated, washed in Earle's saline solution (ES)[3] and pressed through a stainless steel, 24-mesh, wire cloth (supplied by Ludlow-Saylor

Wire Company, Los Angeles) which had been fitted at the bottom of a 50-ml. syringe. The resulting tissue pulp is collected in a 40-ml. centrifuge tube containing 15 ml. ES. After gravity sedimentation (10 minutes) the supernatant liquid containing cell debris and red blood cells is discarded and replaced with 15 ml. of a 0.5% solution of pure trypsin in ES. The tube is placed in a water bath at 37°C. for 10 minutes and then vigorously stirred by repeated pipetting with a 2-ml. automatic pipette. Owing to digestion of the connective fibers, a large fraction of the cells become free. The pieces of tissue which remain are eliminated by straining the suspension through the sieve unit described by Evans *et al.*[4] The cell suspension is then washed twice by centrifuging at 1000 r. p. m. for two minutes and resuspending the sediment in 15 ml. ES. The cell concentration in the final suspension is determined by counting in a hemocytometer.

An aliquot of the cell suspension containing 3×10^7 cells is introduced

TABLE 1

RELATIVE NUMBER OF PLAQUES PRODUCED BY WESTERN EQUINE ENCEPHALOMYELITIS VIRUS EXPOSED TO SPECIFIC ANTISERUM AND OTHER SUBSTANCES BEFORE INOCULATION INTO THE FLASKS

TIME OF CONTACT WITH THE VIRUS, MIN.	TREATMENT				
	DIST. WATER	ES	NORMAL HORSE SERUM 1:10 IN ES	HORSE SERUM HEATED AT 56°C. 30 MIN., 1:10 IN ES	HYPERIMMUNE HORSE SERUM HEATED AT 56°C. 30 MIN., 1:10 IN ES
0	1	1	1	1	1
7	. .	. .	. .	0.92	0.15
10	. .	. .	0.63	. .	. .
15	1.1	0.8	. .	0.74	0.085
20	. .	. .	0.25	. .	. .
22	. .	. .	. .	0.67	0.055
30	0.9	1	. .	. .	0.01

into a large Pyrex Carrel flask (80 mm. inside diameter) which had been acid cleaned by the procedure described by Earle.[5] The volume is brought up to 5 ml. with ES. There are added 5 ml. of horse serum and 1 ml. of 1:1 chicken embryo extract. The flask is hermetically closed with a rubber stopper. Approximately two such flasks can be prepared from one embryo.

The flasks are incubated in horizontal position at 37°C.

During the first few hours of incubation the cells settle out and stick to the bottom of the flask singly or in small clumps. They start to grow very soon and within 48 hours produce a continuous cell layer which covers the bottom of the flask. The layer is mostly one cell thick and is formed by rather uniform spindle cells and by a small number of macrophages. The spindle cells are similar in morphology and rate of growth to the fibroblasts obtained by culturing the connective tissue of the chicken embryo; for this reason we called them fibroblasts. Other cell types carried with the inoculum must obviously be present in the culture, but they are outnumbered by the fast-growing fibroblasts. A more uniform cell population can be obtained by using as inoculum a cell suspension obtained from the cell layer of one of these flasks by means of the described trypsin technique.

The relatively uniform cell layer is infected with the virus in the following way. The nutrient fluid is removed from the flask and the cell layer is washed with 5 ml. ES, which is then discarded. A virus sample of 0.5 ml. is introduced into the flask and uniformly spread on the cell layer by rocking. The flask is incubated for 30 minutes at 37°C. to allow attachment of of the virus to the cells, and the cell layer is then covered with a layer of

TABLE 2

NUMBER OF PLAQUES OBTAINED FROM STOCKS OF WESTERN EQUINE ENCEPHALOMYELITIS VIRUS AT DIFFERENT DILUTIONS

Experiments 1 and 2 were made with the same virus stock, experiments 3, 4 and 5 with a different stock. The virus consisted always of 20% extract of chicken embryo moribund after inoculation with the virus. Each flask was inoculated with 0.1 ml. of the diluted virus.

NUMBER OF THE EXPERIMENT	DILUTION OF THE VIRUS	NUMBER OF PLAQUES ON EACH FLASK
1	5×10^3	117; 180
	10^4	65; 87
2	10^4	59; 53
	3×10^4	28; 13; 18; 14; 19; 14; 18
3	10^3	79
	3×10^3	23
4	10^3	50
	4×10^3	12
	8×10^3	2
5	2×10^3	26
	6×10^3	10

5 ml. of melted agar containing nutrient fluid similar to that used for growing the cell layer. After solidification of the agar, the flasks are incubated for three days at 37°C. At this time they show round necrotic areas of 2 to 4 mm. in diameter, visible to the eye as bright areas against a dark background (Fig. 1). These areas are easily recognized under low-power magnification because the cells have been transformed into a granular debris that scatters the light. That these areas, which we call plaques, are produced by virus activity is shown by the fact that they are absent when the virus is absent and their number increases with the concentration of the virus. Prior treatment of the virus with specific antiserum reduces the number of plaques produced (table 1).

Each of these plaques is produced by one virus particle. This can be shown by determining the effect of dilution on the number of plaques produced by a given virus sample. The theoretical relation between the number of plaques and the virus concentration depends on the minimum number of virus particles required to produce a lesion. A linear dependence of the number of plaques on virus concentration can be obtained only if each plaque is produced by one virus particle. The results of several experiments show that this dependence is in fact linear, and so prove that one particle is sufficient to produce one plaque (table 2).

What fraction of the virus particles contained in a sample is able to produce plaques is not known. We have determined the relation between the number of particles able to produce a plaque under our conditions and the number of particles able to infect the chicken embryo by the standard inoculation procedure on the chorioallantoic membrane.[6] Two batches of 15 eggs each, incubated for 10 days, were inoculated with an amount of virus able to produce, respectively, 0.7 and 0.35 plaques on the average.

TABLE 3

COMPARISON OF THE TITER OF A SINGLE STOCK OF WESTERN EQUINE ENCEPHALOMYELITIS VIRUS BY PLAQUE COUNT AND EGG TITRATION

The plaque count titer is derived from Exp. 1, table 2 and equals 7.5×10^6/ml. The data of the egg titration are given here. Each egg was inoculated with 0.1 ml. of the virus dilution.

BATCH OF EGGS	VIRUS DILUTION	SURVIVING EGG FRACTION	$-\mathrm{LOG}_e$ SURVIVING EGG FRACTION	TITER, PER ML.
1	10^6	0.16	1.8	1.8×10^7
2	2×10^6	0.50	0.7	1.4×10^7

The eggs were then incubated for 30 hours at 37°C., then scored for live and dead. The live ones were further incubated for 24 hours to check their vitality; all survived. From the fraction of live eggs in each batch the average number of infecting virus particles with which each egg was inoculated was determined by assuming a Poisson distribution of the particles per egg. This assumption is legitimate in view of the proof previously given in this article that infection of the cells is produced by one virus particle. The data of one such experiment, reported in table 3, show that the titer was approximately twice the titer obtained by plaque count—a difference which is not significant because of the large standard error of egg titration. In another experiment the egg titration gave a titer somewhat lower than the plaque count.

We conclude from these experiments that there is a nearly one-to-one ratio between the number of plaques and the number of infective virus particles. This result on the one hand indicates that the plaque count is an efficient assay technique; on the other hand, it establishes a basic concept concerning animal virus action, namely, that infection of an embryo is produced by one virus particle.

The use of the plaque count as an assay technique has important practical advantages, as shown by the fact that the accuracy obtained with only one flask of this type is equal to that obtained with not less than one hundred chicken embryos in the currently used Reed and Muench technique.

We do not know whether many other viruses will exhibit this property of producing plaques under our conditions. So far, we have tried only one other virus, the Newcastle Disease virus,[7] and the results here were positive. The plaques formed by Newcastle Disease virus and those formed by Western Equine Encephalomyelitis virus are distinguishable under the microscope; the areas of cell destruction in the former are outlined by a halo of abnormally large cells or cell clumps of unknown nature, whereas in the latter there is seen only cell debris. The possibility of distinguishing the plaques produced by two different viruses on the same flask should be very valuable for the study of problems where two different viruses are involved, as in the phenomenon of interference.

Summary.—Plaque formation by single virus particles has been obtained with the virus of Western Equine Encephalomyelitis in a monolayer of chicken embryo fibroblasts grown *in vitro*. That a plaque is produced by a single virus particle is shown by the proportionality between number of plaques and virus concentration. The comparison between the number of plaques produced and the fraction of chicken embryos infected by the same virus sample indicates that nearly all virus particles able to infect the embryo produce a plaque; there is therefore a nearly 1:1 relation between infecting particles and plaques.

Plaques produced by this virus are distinguishable from the plaques produced by one other virus tested, the Newcastle Disease virus.

[1] Obtained from Dr. S. Lennette, California State Department of Public Health, Berkeley, Calif.

[2] Shannon, J. E., Earle, W. R., and Walts, H. K. , *J. Natl. Cancer Inst.* (in press).

[3] Earle, W. R., *J. Nat. Cancer Inst.* **4**, 165–212 (1943).

[4] Evans, V. J., Earle, W. R., Sanford, K. K., and Shannon, J. E., *Ibid.*, **11**, 907–927 (1951).

[5] Earle, W. R., *Ibid.*, **4**, 131–133 (1943).

[6] Beveridge, W. I. B., and Burnet, F. M., "The Cultivation of Viruses and Rickettsiae in the Chick Embryo," His Majesty's Stationery Office, London, 1946, p. 14.

[7] Strain B, obtained from F. B. Bang, Johns Hopkins Medical School, Baltimore, Md

Infectivity of Ribonucleic Acid from Tobacco Mosaic Virus

A. GIERER AND G. SCHRAMM

By the mid 1950s, there was no doubt that the genetic material of all living cells is DNA. It was also known, however, that many viruses—animal, plant, and bacterial—do not contain DNA, but RNA, and that therefore RNA must also be able to replicate, function as a repository for genetic information, and also express it. In fact, Gierer and Schramm went one step further: they showed that tobacco mosaic virus (TMV) RNA is infectious. That is, just as Al Hershey and Martha Chase had shown for DNA, Gierer and Schramm showed that TMV RNA is, by itself, functional genetic material and that it is also capable of gaining access to cells (albeit with fairly low efficiency) and of expressing the genetic information that it encodes. I well remember the furor that this demonstration caused. One older, distinguished biochemist colleague of mine told me at first that RNA must at least require the presence of a large peptide, then that of a small peptide, and in the end he was willing to settle for the presence of a handful of amino acids. All to no avail. The experiment with TMV RNA was quickly duplicated with the RNA genomes of several animal and other plant viruses and then also with the DNA genomes of several animal viruses. For DNA, obviously, the problem is trickier: it must first be transcribed into RNA, whereas the genomes of RNA viruses merely have to be translated. The major bottleneck, of course, is gaining access to the interior of host cells, accomplishing which is one of the primary functions of viral protein coats.

W. K. JOKLIK

Table 1. COMPARISON OF THE INFECTIVITY OF RIBONUCLEIC ACID AND TOBACCO MOSAIC VIRUS IN 0·1 M PHOSPHATE BUFFER

pH	Ribonucleic acid μgm./ml.	lesions	Tobacco mosaic virus μgm./ml.	lesions
6·1	10	153	0·09 0·8	95 445
7·3	10	815	0·27	1,048
7·3	1	524	0·05	795
7·5	10	998	0·27	685

number of lesions as does 0·2 μgm. tobacco mosaic virus. The infectivity of the ribonucleic acid preparation is thus about 2 per cent of that of the native virus.

The following experiments, the results of which are collected in Table 2, have been carried out to show that the infection is due to the nucleic acid rather than to contamination of the ribonucleic acid with native virus.

(*a*) In the ribonucleic acid solution protein was not detectable by chemical methods (Schuster, Schramm and Zillig, unpublished work) ; thus the amount must be less than 0·4 per cent of the ribonucleic acid content. By serological methods (complement fixation) it was shown that the ribonucleic acid contains less than 0·02 per cent of native tobacco mosaic virus protein.

(*b*) Treatment of both the ribonucleic acid and tobacco mosaic virus with normal rabbit serum (concentration 3×10^{-3}, applied for 10 min. at 4° C.) somewhat reduces the infectivity. There is no significant further reduction if the ribonucleic acid is treated with the same amount of tobacco mosaic virus antiserum, whereas with antiserum the infectivity of the virus itself is almost completely destroyed.

(*c*) Incubation of the stock solutions of ribonucleic acid (0·3 per cent) and tobacco mosaic virus (0·06 per cent) with 2 μgm. per ml. of ribonuclease at 4° C. for 10 min. reduces the activity of the ribonucleic acid to 0, whereas that of the virus remains almost unaffected.

(*d*) The sedimentation constant of the ribonucleic acid is 12–18 S, compared with 180 S for tobacco mosaic virus. We have centrifuged the stock solution of ribonucleic acid for 30 min. at 50,000 rev. per min. and found the supernatant liquid to be only a little less active than the original solution. If the solution of the virus is treated in the same manner, the activity of the supernatant liquid is very low.

(*e*) The ribonucleic acid is known to be unstable ; and, as would be expected, its infectivity is much reduced after 48 hr. at 20° C., whereas that of the virus is much less affected.

These experiments show that protein, if present at all, is only there in very small amounts and does not resemble closely the native protein of tobacco mosaic virus. We are thus led to conclude that the infectivity is due to the nucleic acid itself.

Infectivity of Ribonucleic Acid from Tobacco Mosaic Virus

IN their experiments with bacteriophages, Hershey and Chase[1] have shown that only the nucleic acid component plays a part in the intracellular multiplication. There are also indications that in simple viruses containing ribonucleic acid the nucleic acid plays a dominant part in the infection. Thus, experiments with tobacco mosaic virus have shown that the protein can be changed chemically without affecting the activity and the genetic properties[2] ; recently, it was even found[3] that part of the protein can be removed from tobacco mosaic virus without destroying the activity.

We have now obtained evidence that after complete removal of the protein, the ribonucleic acid itself is still infectious.

The protein was extracted from tobacco mosaic virus with phenol by a procedure elaborated by Schuster, Schramm and Zillig (to be published). After a short treatment at low temperatures, a preparation of ribonucleic acid is obtained which has a high molecular weight during the first few hours but depolymerizes in the course of time. Its physical properties will be described elsewhere.

A solution of 10 per cent tobacco mosaic virus in 0·02 M phosphate buffer of pH 7·3 is shaken for 8 min. at 5° C. with an equal amount of water-saturated phenol. The aqueous phase which contains the ribonucleic acid is separated by centrifugation, and the process of extraction with phenol is repeated at least twice for 2 min. The phenol is then extracted by ether from the aqueous phase. The whole procedure is carried out at 5° C. and takes about 50 min. ; it is followed immediately by testing the infectivity.

For that purpose, five to ten plants of *Nicotiana glutinosa* with five leaves each were inoculated with a diluted solution of the ribonucleic acid, and an equal number of plants with a standard solution of tobacco mosaic virus. The number of local lesions produced by ribonucleic acid and tobacco mosaic virus are compared in Table 1. It is found that 10 μgm. of ribonucleic acid produces about the same

Table 2. COMPARISON OF RIBONUCLEIC ACID (10 μgm./ml.) AND TOBACCO MOSAIC VIRUS (0·27 μgm./ml.) IN 0·1 M PHOSPHATE BUFFER OF pH 7·3
(Infectivity expressed as lesions per 30 leaves)

	Ribonucleic acid	Tobacco mosaic virus
Normal	488	629
With normal serum	180	117
With antiserum	145	0
With ribonuclease	0	473
After ultracentrifugation	367	31
After 48 hr. at 20° C.	2	130

The infectivity of the ribonucleic acid preparation is about 0·1 per cent of that of the same amount of ribonucleic acid contained in native tobacco mosaic virus. Whether this relatively low value is due to a large inactive fraction of the ribonucleic acid preparation, or to low efficiency of the mechanism of infection, has still to be determined.

Studies on the combination of the ribonucleic acid with proteins are being carried out and may elucidate the connexion between our findings and the reactivation experiments of Fraenkel-Conrat and Williams[4], of Lippincott and Commoner[5], and of Hart[6].

We are much indebted to Prof. H. Friedrich-Freksa for helpful discussions, to Mr. R. Engler and Dr. H. Schuster for their co-operation, and to Miss A. Kleih for assistance.

A detailed account of this work will be published in the *Zeitschrift für Naturforschung*.

A. GIERER
G. SCHRAMM

Max-Planck-Institut für Virusforschung,
    Tübingen.
      Feb. 10.

[1] Hershey, A. D., and Chase, M., *J. Gen. Physiol.*, **36**, 39 (1952).

[2] Schramm, G., and Müller, H., *Hoppe Seylers Z. physiol. Chem.*, **266**, 43 (1940); **274**, 267 (1942). Miller, G. L., and Stanley, W. M., *J. Biol. Chem.*, **141**, 905 (1941); **146**, 331 (1942). Harris, J. I., and Knight, C. A., *J. Biol. Chem.*, **214**, 215 (1955).

[3] Schramm, G., Schumacher, G., and Zillig, W., *Nature*, **175**, 549 (1955).

[4] Fraenkel-Conrat, H., and Williams, R. C., *Proc. U.S. Nat. Acad. Sci.*, **41**, 690 (1955).

[5] Lippincott, J. A., and Commoner, B., *Biochim. Biophys. Acta*, **19**, 198 (1956).

[6] Hart, R. G., *Nature*, **177**, 130 (1956).

Virus Interference. I. The Interferon

A. Isaacs and J. Lindenmann

For two decades or so before the discovery of interferon, studies of viral interference, either heterologous interference between different viruses or homologous interference between variously modified forms of a virus with active virus, were in vogue. These studies provided useful information: in particular, they strongly suggested that interference did not occur at the level of competition for virus receptors on cell surfaces, but within cells. The payoff came with the demonstration by Alick Isaacs and Jean Lindemann that exposure of chick embryo chorioallantoic membranes to heat-inactivated (but not too heat-inactivated: virus heated at 56°C caused the effect, but virus heated at 60°C did not) influenza virus generated a factor that, when added to fresh membranes, rendered them immune to influenza virus infection. The discovery of interferon was very important; it created a huge field of research. It turned out that there are three types of human interferon, and that they were the first representatives of intercellular messengers now known collectively as cytokines, lymphokines, chemokines, and so on, that regulate, control, and effect the interaction and reaction of cells with and to external stimuli.

W. K. Joklik

Reprinted by permission from *Proceedings of the Royal Society of London B* 147:258–267. Copyright © 1957, by The Royal Society.

Virus interference. I. The interferon

By A. Isaacs and J. Lindenmann*

National Institute for Medical Research, London

(*Communicated by C. H. Andrewes, F.R.S.—Received 7 March* 1957)

During a study of the interference produced by heat-inactivated influenza virus with the growth of live virus in fragments of chick chorio-allantoic membrane it was found that following incubation of heated virus with membrane a new factor was released. This factor, recognized by its ability to induce interference in fresh pieces of chorio-allantoic membrane, was called interferon. Following a lag phase interferon was first detected in the membranes after 3 h incubation and thereafter it was released into the surrounding fluid.

Introduction

One of the most useful situations for studying interference among animal viruses has been the interference produced by inactivated influenza viruses with the growth of live influenza virus in the chorio-allantoic membrane of the chick embryo. In this system, a number of variables have been measured, e.g. the effects of varying the dose of interfering and challenge virus or the time interval between the two inoculations, the effects of different methods of virus inactivation and the use of different virus strains (see review by Henle 1950). As a result of studies by different workers, it is generally agreed that interference cannot be explained by blockage of cell surface receptors. Fazekas de St Groth, Isaacs & Edney (1952) found that interference by influenza virus inactivated at 56°C took some hours until it was fully established, but it was difficult to decide by experiments in the intact chick embryo whether this time was required for the inactivated virus to be absorbed by the cells or for some further reactions to occur. We have studied this point with pieces of chorio-allantoic membrane suspended in buffered salt solution *in vitro* (Fulton & Armitage 1951; Tyrrell & Tamm 1955) a method which allows observation of fluid and cells seperately and manipulations which are not possible in the chick embryo. As a result, a number of new features of the interference reaction have emerged and these are described in this and the following paper.

Methods

Interfering virus

The Melbourne (1935) strain of influenza virus A was used as freshly harvested allantoic fluid. It was mixed with a 2 % sodium citrate solution in normal saline and borate buffer, pH 8·5, in the ratio 6 parts virus, 2 parts citrate-saline and 1 part borate buffer, and heated at 56°C for 1 h. This treatment abolishes the infectivity and enzymic activity of the virus while retaining its interfering activity (Isaacs & Edney 1950*a*). In the present experiments, it was shown by inoculating

---

* In receipt of a fellowship from the Swiss Academy of Medical Sciences.

[258]

eggs allantoically that heating had inactivated the virus; this test was included in most, but not in all of the present experiments. The heated virus is referred to as heated MEL.

Buffer

The buffer used to suspend and wash the pieces of chorio-allantoic membrane was that described by Earle (see Parker 1950).

Chorio-allantoic membrane

Pieces of chorio-allantoic membrane were removed from 10- or 11-day fertile hen's eggs by a technique similar to that described by Tamm, Folkers & Horsfall (1953). Six or seven pieces were taken from each egg and membranes were pooled from a group of eggs and randomized. In order to find the weight of a piece of membrane, ten pieces were selected at random, drained on filter paper and weighed. The average weight of a piece of membrane was found to be 20 mg.

Interference and challenge

An interference experiment was carried out in the following way: Six pieces of membrane were placed in $6 \times \frac{5}{8}$ in. test-tubes, and to each was added 1 ml. of test material. Six other tubes were similarly incubated with 1 ml. buffer. In each case 100 units of penicillin were added per ml. fluid. The tubes were stoppered and placed in a roller drum at 37°C (8 rev/h). After 24 h incubation the membranes were removed, washed in two changes of buffer and put in fresh tubes along with 1 ml. buffer in which MEL virus at a final dilution of 10^{-3} was incorporated. The MEL virus came from a stock of capillary tubes kept at -70°C and a single stock of virus lasted through almost all the experiments. The tubes were placed in the roller drum for a further 48 h at 37°C after which the fluids were titrated individually for their haemagglutinin content.

Haemagglutinin titrations

Two-fold dilutions (0·25 ml.) of test virus were made in normal saline using automatic pipettes and plastic plates. To each dilution 0·25 ml. of a 0·5 % suspension of chick red-blood cells was added, and the cells allowed to settle. That pattern which showed partial agglutination was taken as the end-point, and it was read by interpolation if necessary. One agglutinating dose (a.d.) is defined as the amount of virus present at the partial agglutination end-point. The readings in the tables are given as $\log_2$, i.e. tube number of a series of two-fold dilutions. Thus if the end-point of agglutination occurs at a 1/2 dilution the reading is taken as $\log_2 = 1$. If the end-point of agglutination occurs at a 1/4 dilution the reading is taken as $\log_2 = 2$. A 1/1 dilution end-point has a $\log_2 = 0$ and material which did not agglutinate red cells at a 1/1 dilution was given an arbitrary score of -1.

Assessment of results

A group of sixty titrations of control materials carried out in a single experiment was analyzed statistically. The logarithms of the haemagglutinin titres of individual fluids were found to occur in an approximately normal distribution. It was necessary, therefore, to measure geometric mean haemagglutinin titres when comparing two groups of materials. The mean ($\log_2$) titre of the sixty titrations was 6·78 and the variance 0·412 (standard deviation 0·642). If we took samples of six from this population we should expect that 99 % of the sample means would fall in the range 6·78 ± 0·67. Therefore, if the means of two samples, drawn at random, differ by more than 0·67 $\log_2$ it is likely that they are drawn from different populations. In practice, we have assumed that if an experimental group of six tubes showed a geometric mean titre of one $\log_2$ unit less than a group of six controls tested at the same time this indicates a slight but significant degree of interference; a difference of two $\log_2$ units was taken as showing definite interference.

One difficulty was that groups of controls tested on different days showed highly significant differences in titre. This finding appears to be due to the variable sensitivity of the red cells from different fowls to influenza viral haemagglutinin. In order to overcome this difficulty a control group was included in each experiment, and to facilitate comparison between different experiments the results for an experimental group are shown in the table as the proportion (expressed as a percentage) of the geometric mean haemagglutinin titre of the corresponding controls. Thus a difference of one $\log_2$ unit between experimental and control groups corresponds to a 50 % yield, a difference of two $\log_2$ units to a 25 % yield, etc. Small arithmetic differences in the percentage yield have therefore a much greater significance at low than at high percentage yields.

RESULTS

*Effect of varying the temperature and time of contact of heated virus
and cells on the degree of interference*

In preliminary experiments it was found that interference could be induced in pieces of chorio-allantoic membrane in the following way. Heat-inactivated MEL virus was added to the suspending fluid along with a piece of membrane, and this was then incubated in the roller drum for 24 h at 37°C; controls were incubated in buffer. The membranes were then washed, placed in fresh tubes with live MEL virus diluted 10^{-3} in 1 ml. of buffer, and incubated for a further 40 to 48 h at 37°C. Previous treatment of the membranes with heated MEL in this way caused pronounced interference with the growth of live MEL. In this system, 200 to 400 agglutinating doses of heated MEL almost completely suppressed haemagglutinin production by live virus and the effect of different doses of heated MEL can be seen from the information contained in table 4. Tyrrell & Tamm (1955) found that LEE virus heated at 56°C did not cause interference in a similar system; this difference may be due to virus strain variability.

It was soon found that the time interval between the application of interfering and challenge viruses had an influence on the degree of interference. In order to see what was the importance of the time interval, the following experiment was carried out. Pieces of membrane were mixed with a small dose of heated MEL virus (60 a.d.) in order to produce slight interference, and at varying time intervals, groups of membranes were removed, washed thoroughly in buffer, resuspended in buffer and further incubated either at 2 or 37°C. The total incubation time was 24 h and this was divided between a *primary* incubation period in contact with heated MEL at 37°C and a *secondary* incubation period of the washed membranes at 2 or 37°C. After the secondary incubation, the membranes were again washed and challenged with MEL virus. The results are shown in table 1. The results show that a primary period of 15 min contact between heated MEL and the membrane was sufficient to establish nearly as much interference as a primary period of 24 h at 37°C, provided the secondary incubation was carried out at 37°C. The findings suggest that the heated MEL virus is rapidly adsorbed to the cells, and thereafter that it does not act as an inert blocking agent. The fact that after 4 h *primary*

TABLE 1. EFFECT OF VARYING TIMES AND TEMPERATURES OF INCUBATION ON THE INTERFERING ACTIVITY OF HEATED MEL

interfering virus	primary incubation	secondary incubation	geometric mean HA titre $(\log_2)$	% of control titre
60 a.d. of heated MEL	15 min at 37°C	24 h at 37°C	3·9	15
60 a.d. of heated MEL	15 min at 37°C	24 h at 2°C	6·7	> 100
60 a.d. of heated MEL	1 h at 37°C	23 h at 37°C	4·0	33
60 a.d. of heated MEL	1 h at 37°C	23 h at 2°C	5·8	94
60 a.d. of heated MEL	4 h at 37°C	20 h at 37°C	2·3	10
60 a.d. of heated MEL	4 h at 37°C	20 h at 2°C	3·6	20
60 a.d. of heated MEL	24 h at 37°C	nil	1·5	3
buffer control	24 h at 37°C	nil	5·6	100
buffer control	24 h at 2°C	nil	5·9	100

incubation there is a slight difference in interference, depending on whether *secondary* incubation is carried out at 2 or 37°C, implies that some active metabolic process in the membrane requiring at least 4 h incubation at 37°C is necessary before interference is fully established. It is difficult to be sure how long this process takes since it might continue during the early stages of growth of the challenge virus, and 4 h is therefore only a minimal figure.

Stability of interfering activity of heated MEL

Incidental observations had pointed to some instability of the interfering activity of heated MEL during incubation at 37°C. An experiment illustrating this is shown in table 2. Table 2 shows that the interfering activity of heated MEL virus was reduced about ten-fold by incubating it for 24 h at 37°C before adding the membranes. This degree of instability is interesting since Paucker & Henle (1955) found that the infectivity of the PR 8 strain of influenza virus was inactivated at 37°C at the rate of about one $\log_{10}$ a day.

When an attempt was made to measure the amount of unabsorbed heated MEL virus after varying times of contact with chorio-allantoic membrane, a difficulty was soon encountered. Apparent rapid 'disappearance' of the haemagglutinin was found to be caused by combination of the virus with an inhibitor of agglutination released by the membrane into the surrounding fluid. This could be shown

262   A. Isaacs and J. Lindenmann

by incubating pieces of membrane with buffer at 37°C in the roller drum, when
after 2 h sufficient inhibitor of agglutination was released by the membrane into
the surrounding fluid to block agglutination by an equal volume of heated MEL
virus with an agglutinin titre of 100. A similar effect was noted by Schlesinger &
Karr (1956) and probably accounts for the difficulty described by Isaacs & Edney
(1950*b*) in removing the inhibitor in allantoic fluid by frequent washing of the
allantoic cavity.

TABLE 2. EFFECT OF INCUBATION AT 37°C FOR 24 H ON
THE INTERFERING ACTIVITY OF HEATED MEL

dose (a.d.) of interfering virus	incubation period before test	geometric mean HA titre ($\log_2$)	% of control titre
373	nil—control	0·8	2·3
124	nil—control	1·9	4
41	nil—control	3·9	20
1120	24 h at 37°C	2·3	5
373	24 h at 37°C	5·4	41

TABLE 3. EFFECT OF MEMBRANE INHIBITOR ON
INTERFERING ACTIVITY OF HEATED MEL

heated MEL suspended in	incubation period before test	geometric mean HA titre ($\log_2$)	% of control titre
buffer	nil	0·8	0·9
membrane inhibitor	nil	3·8	7
buffer	4 h at 37°C	2·3	2·4
membrane inhibitor	4 h at 37°C	5·5	22
control without heated MEL	nil	7·7	100

An experiment was carried out to see what effect this membrane inhibitor might
have on the interfering activity of heated MEL. A sample of inhibitor was prepared by incubating normal membranes in buffer. Heated MEL virus was then
tested for its interfering activity diluted in this membrane extract or in buffer as
a control; also, interfering activity was tested by adding the membranes to these
reagents at once, or after the two preparations of heated MEL had been incubated
for 4 h at 37°C. The results in table 3 show that the membrane extract had
a pronounced inhibitory effect on the interfering activity of heated MEL. Also, the
degree of instability of the interfering activity of heated MEL during 4 h incubation
at 37°C was not affected by the presence or absence of membrane inhibitor.

Residual interfering activity

These experiments have indicated that the interfering virus is rapidly taken up
by the cells, although interference in the cells takes some time to be established.
One would expect, too, that in these experiments little interfering activity would
remain in the fluid after 24 h contact between heated MEL and the membrane,
since any unabsorbed virus would lose interfering potency as a result of inactivation
at 37°C and combination with inhibitor. It was surprising, therefore, to find that

after 24 h incubation considerable interfering activity remained in the surrounding fluid. This can be seen from the experiment illustrated in table 4. In this experiment, different amounts of heated MEL were incubated with pieces of membrane for 24 h at 37°C. Membranes and fluids were then separated and the membranes were washed and challenged with MEL virus to assess the degree of *initial* interference. The amount of *residual* interference in the fluids was measured by adding fresh pieces of membrane to the fluids, incubating 24 h at 37°C and challenging with MEL virus in the same way. Clearly the degree of residual interfering activity is only slightly less than the amount of initial interfering activity.

TABLE 4. COMPARISON BETWEEN INITIAL INTERFERING ACTIVITY OF HEATED MEL AND RESIDUAL INTERFERING ACTIVITY AFTER INCUBATION WITH CHORIO-ALLANTOIC MEMBRANE FOR 24 H

interfering activity measured	dose (a.d.) of heated MEL initially present	geometric mean HA titre ($\log_2$)	% of control titre
initial	1120	< 0	< 1·3
initial	373	0·8	2·3
initial	124	1·9	4
initial	41	3·9	20
residual	1120	0·75	1·6
residual	373	3·6	12
residual	124	4·7	25
residual	41	5·7	50

In an effort to explain the results of the last experiment the possibility was considered that fresh interfering activity was produced by the membrane. This possibility was confirmed by the following experiment. Heated MEL virus was incubated with pieces of membrane for 2 h at 37°C. The membranes were then thoroughly washed and incubated in fresh buffer at 37°C. It was found that after some hours' incubation at 37°C fresh interfering activity could be detected in the incubating fluid. Some of the properties of this newly released interfering agent are described in the accompanying paper, but we can anticipate meanwhile by saying that the newly released interfering agent is a non-haemagglutinating macro-molecular particle which has many different properties from those of heated influenza virus. To distinguish it from the heated influenza virus we have called the newly released interfering agent 'interferon'. It was also found that the membranes which were liberating 'interferon' showed a diminished production of MEL virus on challenge, i.e. establishment of interference was accompanied by liberation of interferon.

Time of appearance of interferon in the membrane and release into the surrounding fluid

We next studied the appearance of interferon in the membranes and its liberation into the surrounding medium at different time-intervals after inoculating heated MEL virus. The experiment which is illustrated in figure 1 was carried out in the following way:

Pieces of chorio-allantoic membrane were mixed with a large dose of heated MEL (4000 agglutinating doses/membrane piece) and incubated in the roller drum for 3 h at 37°C. The membranes were then removed, washed thoroughly in buffer, resuspended in fresh test-tubes with 1 ml. of buffer/membrane piece and re-incubated at 37°C. The end of this time was taken as zero hour and at various time intervals thereafter groups of tubes were removed, and the fluids and membranes tested separately for interferon activity. The membranes were pooled in groups of six, ground in a Ten Broeck grinder, suspended in 6 ml. buffer, lightly centrifuged and the supernatant fluid tested by adding six fresh pieces of membrane. (In a control experiment it had been shown that after lightly centrifuging a membrane extract in this way all the interfering activity was present in the supernatant fluid.) The further test of interfering activity of fluids and membrane extracts was carried out by incubating test fluids with fresh pieces of membrane for 18 to 24 h at 37°C, washing the membranes and challenging with MEL virus. The yields of

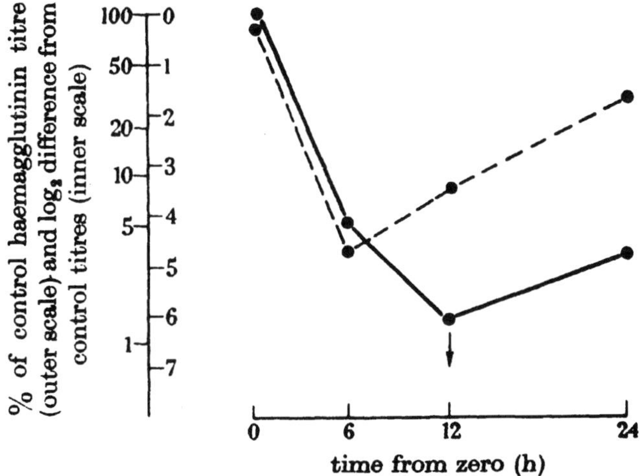

FIGURE 1. Presence of interferon in membranes and fluids.
—— fluids, – – – membrane extracts.

haemagglutinin were compared with those of control membranes incubated in buffer. In the experiment illustrated in figure 1, the samples harvested at 0 and 6 h had a different control group from the samples harvested at 12 and 24 h, and in each case the results in the experimental group were compared with their corresponding controls. The fluids and membrane extracts tested at 0 h showed no interfering activity, although the membranes had absorbed large amounts of heated MEL virus, and would themselves probably have been resistant to challenge as shown by control experiments. At 6 h there was a high degree of interferon activity in the membrane extracts and slightly less in the fluids. The greatest interfering activity in the fluids was found at 12 h, a finding which was confirmed in a second experiment, and by this time the activity in the membrane extracts had declined slightly. At 24 h there was little remaining activity in the membrane extracts and most of the interferon had been liberated into the medium.

This type of experiment was repeated with a number of variations. In one experiment an initial period of 2 h contact between heated MEL and membranes was allowed, and samples were taken at 0, 3. 6, 9 and 12 h. In this case, significant interfering activity was found in membranes and fluids at 3 h, but the potency of both was less than at 6 h. The only difference from the results shown in figure 1 was that in this experiment the zero samples of both membranes and fluids showed slight but significant interfering activity. This finding may have been due to small amounts of heated MEL remaining loosely attached to the membrane; since the same finding was noted when the heated MEL and membranes were left in contact for 24 h at 2°C. Under these conditions, Ishida & Ackermann (1956) have shown that influenza virus becomes loosely and reversibly attached to the chorio-allantoic membrane, irreversible attachment requiring incubation at 37°C. Preliminary incubation of heated MEL and membranes at 2°C followed by washing the membranes did not give satisfactory results and there was a low production of interferons.

In another experiment, the *differential* release of interferons into the fluid was studied. Heated MEL and membranes were kept 2 h in contact at 37°C and the membranes were then washed and incubated in buffer. At two-hourly intervals thereafter the membranes were removed, washed and incubated in fresh buffer; all the manipulations were carried out with warm reagents and at 37°C. The yields of interferon during the different time intervals were then measured and the results are shown in table 5.

TABLE 5. DIFFERENTIAL YIELD OF INTERFERONS AT INTERVALS
AFTER INOCULATING HEATED MEL

differential sample	geometric mean HA titre ($\log_2$)	% of control titre
0–2 h	6·7	81
2–4 h	5·3	29
4–6 h	5·2	27
6–8 h	5·7	36
8–10 h	5·8	40
control	6·9	100
10–24 h	7·0	81
control	7·2	100

The maximal liberation of interferon occurred between the second and sixth hours. The yield of interferon was much smaller at each interval, largely due to the fact that the total yield was divided among so many samples. (In the accompanying paper it is shown that interferon prepared as described here has little interfering activity at a dilution of 1 in 10.) In this experiment little interferon was liberated after 10 h, but in another experiment slight activity was found in a 10 to 24 h sample.

Relationship between interfering activity of the virus and interferon production

The amount of interferon produced depends on the amount of heated MEL used, as indicated by the results shown in table 4. The following experiment showed that interferon production is also dependent on the possession of interfering activity by the heated MEL. It is based on the fact that MEL virus heated for 1 h at 56°C had strong interfering activity, whereas the same virus heated for 1 h at 60°C had no significant interfering activity (Isaacs & Edney 1950a).

Aliquots of a preparation of MEL virus in citrate-borate buffer were heated for 1 h at 56 and 60°C and incubated with pieces of membrane for 2 h at 37°C. The membranes were then washed and incubated in buffer for a further 20 h at 37°C. Thereafter, the membranes were removed and tested for their ability to support the multiplication of MEL virus, while the fluids were tested for their interferon content. The results are shown in table 6.

TABLE 6. RELATIONSHIP OF INTERFERING ACTIVITY AND INTERFERON PRODUCTION BY MEL HEATED AT 56 AND 60°C

MEL heated at	material tested	geometric mean HA titre ($\log_2$)	% of control titre
56°C for 1 h	membranes for interference	0·88	1·1
60°C for 1 h	membranes for interference	6·4	53
56°C for 1 h	interferons liberated	2·6	4
60°C for 1 h	interferons liberated	7·1	100

The virus which had been heated at 56°C induced significant interference in the membranes which also liberated interferon. The virus which had been heated at 60°C caused a barely detectable degree of interference and no interferon was liberated.

Attempted passage of interferon in series

The results obtained have not made it clear whether interferon is part of the heated MEL virus which is liberated by the membrane, or whether it is newly synthesized in the membrane. In either event it was interesting to test the possibility that interferon might be able to replicate in series. In order to test this, advantage was taken of the fact that interferon exerts its activity if it is left in contact with the membranes for 4 h at 37°C, and the membranes are then washed and incubated in buffer for a further 20 h at 37°C before challenge. Under these conditions no new interferon activity could be detected in ground membranes or in the medium after a single passage or after two serial passages. Controls showed that the interferon grown in eggs or tissue cultures did not produce live virus.

DISCUSSION

In earlier studies on virus interference attempts were sometimes made to explain the phenomenon as due to an inert blocking action of the interfering virus preventing the challenge virus from entering the cells. This view has been contested by those who could find no evidence that the interfering virus prevented uptake of challenge virus, and in the present studies it was found that in order to establish interference more than 4 h incubation at 37°C was required. Tyrrell & Tamm (1955) found that incubation at 37°C was necessary for the establishment of interference in a similar system, but using virus 'inactivated' at 22 or 37°C. However, it is known that virus incubated at 22 or 37°C retains some infectivity and is able to undergo a modified cycle of virus multiplication resulting in the production of virus resembling incomplete virus (Henle 1953; Horsfall 1954). In contrast, virus heated at 56°C for 1 h has hitherto shown no evidence of infectivity, or of the ability to produce virus haemagglutinin or soluble antigens (Isaacs & Fulton 1953). The present results suggest, nevertheless, that interference shows some of the characters we might expect of an abortive attempt at a single cycle of virus multiplication. A second finding which supports this idea (in addition to the fact that interference requires some metabolic activity on the part of the membrane) is that the interfering action of heated MEL is inactivated during incubation at 37°C to approximately the same extent as is the infectivity of unheated virus. But the best support for the idea arises if we consider the interferon provisionally as an abortive product of virus multiplication. This suggestion is made mainly as a guide to further experimentation until the interferon is better characterized chemically and serologically.

The analogies between interferon and virus production are as follows: Little or no interferon activity could be detected in membranes or fluids shortly after inoculating heated MEL; this is analogous to the so-called 'eclipse period', when only a small fraction of the inoculated influenza virus can be recovered from infected cells (Hoyle 1948; Henle 1949). Secondly, the times at which interferon activity can be detected in membranes and fluids correspond very well with those at which new virus antigens appear after inoculating live influenza virus. The fact that interferon is able to inhibit influenza virus growth, but is unable to replicate in series, suggests similarities to the single cycle of viral haemagglutinin production caused by infection with incomplete virus (Burnet, Lind & Stevens 1955; Paucker & Henle 1955).

The experiment which showed that virus heated at 60°C for 1 h had no interfering action and did not lead to the production of interferon, suggests a close relationship between the two phenomena. However, there is insufficient evidence to postulate yet that the influenza viral interference phenomenon is due directly to interferon production. It is also not yet known whether interferon is simply liberated from the heated MEL or is newly synthesized in the membrane. In favour of its being newly synthesized is an observation that in order to obtain good yields of interferon, adequate oxygenation of the membrane is necessary.

REFERENCES

Burnet, F. M., Lind, P. E. & Stevens, K. M. 1955 *Aust. J. Exp. Biol. Med. Sci.* **33**, 127.
Fazekas, de St Groth, S., Isaacs, A. & Edney, M. 1952 *Nature, Lond.* **170**, 573.
Fulton, F. & Armitage, P. 1951 *J. Hyg., Camb.* **49**, 247.
Henle, W. 1949 *J. Exp. Med.* **90**, 1.
Henle, W. 1950 *J. Immunol.* **64**, 203.
Henle, W. 1953 *Cold Spr. Harb. Symp. Quant. Biol.* **18**, 35.
Horsfall, F. L. Jr. 1954 *J. Exp. Med.* **100**, 135.
Hoyle, L. 1948 *Brit. J. Exp. Path.* **29**, 390.
Isaacs, A. & Edney, M. 1950*a* *Aust. J. Exp. Biol. Med. Sci.* **28**, 219.
Isaacs, A. & Edney, M. 1950*b* *Aust. J. Exp. Biol. Med. Sci.* **28**, 231.
Isaacs, A. & Fulton, F. 1953 *J. Gen. Microbiol.* **9**, 132.
Ishida, N. & Ackermann, W. W. 1956 *J. Exp. Med.* **104**, 501.
Parker, R. C. 1950 *Methods of tissue culture.* London and Toronto: Cassell and Co.
Paucker, K. & Henle, W. 1955 *J. Exp. Med.* **101**, 493.
Schlesinger, R. W. & Karr, H. V. 1956 *J. Exp. Med.* **103**, 309.
Tamm, I., Folkers, K. & Horsfall, F. L. Jr. 1953 *J. Exp. Med.* **98**, 229.
Tyrrell, D. A. J. & Tamm, I. 1955 *J. Immunol.* **75**, 43.

RNA-Dependent DNA Polymerase in Virions of RNA Tumour Viruses

D. BALTIMORE

Reprinted by permission from *Nature* 226:1209–1211. Copyright © 1970. Macmillan Magazines Ltd.

RNA-Dependent DNA Polymerase in Virions of Rous Sarcoma Virus

H. M. TEMIN AND S. MIZUTANI

Reprinted by permission from *Nature* 226:1211–1213. Copyright © 1970. Macmillan Magazines Ltd.

The discovery of the reverse transcriptase by Baltimore and by Temin and Mizutani was an extremely satisfying one because it had been anticipated for some time. Thus it was known that inhibition of DNA synthesis, as well as that of RNA synthesis, prevents infection and transformation by retroviruses; there was also good evidence that cells transformed by retroviruses possess DNA capable of hybridizing with retrovirus RNA. It seemed reasonable to postulate, therefore, that the RNA in infecting retrovirus particles serves as the template for the synthesis of DNA which remains in infected cells and serves as the template for the synthesis of the RNA in progeny retrovirus particles. As for the source of the enzyme synthesizing the retrovirus-specific DNA, there were two possibilities. First, the RNA resulting from the uncoating of infecting retrovirus particles might be translated into RNA-dependent DNA polymerase, which would then transcribe it—a tricky sequence of events because under conditions of a single virus particle infecting a cell, which would be the normal situation, the same RNA molecule would first have to be translated into at least several enzyme molecules and then serve as the template for DNA synthesis. The second possibility is that the enzyme might be formed in infected cells, translated from progeny RNA molecules, and then be incorporated into the progeny retrovirus particles. Ample precedent for this type of situation was already at hand: it was known that poxvirus, reovirus, and vesicular stomatitis virus particles contain DNA-dependent RNA polymerase, double-stranded RNA-dependent RNA polymerase, and single-stranded RNA-dependent RNA polymerase, respectively. Further, it turned out that retrovirus particles are not the only particles in nature that contain an enzyme capable of transcribing RNA into DNA; there are also the particles collectively known as retrotransposons. These are encoded by sequences in cellular genomes known as retroelements and include intracisternal A-type particles and VL30 elements in vertebrate cells, numerous families of retroelements in *Drosophila*-like *copia* and *gypsy*, and the Ty elements in yeast. As for the reason why retroelements, which may have originated in an RNA world, continue to exist, that may lie in their ability to generate that most essential prerequisite of evolution, namely genetic diversity. Both David Baltimore and Howard Temin were awarded the Nobel Prize in 1975.

W. K. JOKLIK

Viral RNA-dependent DNA Polymerase

Two independent groups of investigators have found evidence of an enzyme in virions of RNA tumour viruses which synthesizes DNA from an RNA template. This discovery, if upheld, will have important implications not only for carcinogenesis by RNA viruses but also for the general understanding of genetic transcription: apparently the classical process of information transfer from DNA to RNA can be inverted.

RNA-dependent DNA Polymerase in Virions of RNA Tumour Viruses

DNA seems to have a critical role in the multiplication and transforming ability of RNA tumour viruses[1]. Infection and transformation by these viruses can be prevented by inhibitors of DNA synthesis added during the first 8–12 h after exposure of cells to the virus[1-4]. The necessary DNA synthesis seems to involve the production of DNA which is genetically specific for the infecting virus[5,6], although hybridization studies intended to demonstrate virus-specific DNA have been inconclusive[1]. Also, the formation of virions by the RNA tumour viruses is sensitive to actinomycin D and therefore seems to involve DNA-dependent RNA synthesis[1-4,7]. One model which explains these data postulates the transfer of the information of the infecting RNA to a DNA copy which then serves as template for the synthesis of viral RNA[1,2,7]. This model requires a unique enzyme, an RNA-dependent DNA polymerase.

No enzyme which synthesizes DNA from an RNA template has been found in any type of cell. Unless such an enzyme exists in uninfected cells, the RNA tumour viruses must either induce its synthesis soon after infection or carry the enzyme into the cell as part of the virion. Precedents exist for the occurrence of nucleotide polymerases in the virions of animal viruses. Vaccinia[8,9]— a DNA virus, Reo[10,11]—a double-stranded RNA virus, and vesicular stomatitis virus (VSV)[12]—a single-stranded RNA virus, have all been shown to contain RNA polymerases. This study demonstrates that an RNA-dependent DNA polymerase is present in the virions of two RNA tumour viruses: Rauscher mouse leukaemia virus (R-MLV) and Rous sarcoma virus. Temin[13] has also identified this activity in Rous sarcoma virus.

Incorporation of Radioactivity from ³H-TTP by R-MLV

A preparation of purified R-MLV was incubated in conditions of DNA polymerase assay. The preparation incorporated radioactivity from ³H-TTP into an acid-insoluble product (Table 1). The reaction required Mg^{2+}, although Mn^{2+} could partially substitute and each of the four deoxyribonucleoside triphosphates was necessary for activity. The reaction was stimulated strongly by dithiothreitol and weakly by NaCl (Table 1). The kinetics of incorporation of radioactivity from ³H-TTP by R-MLV are shown in Fig. 1, curve 1. The reaction rate accelerates for about 1 h and then declines. This time-course may indicate the occurrence of a slow activation of the polymerase in the reaction mixture. The activity is approximately proportional to the amount of added virus.

For other viruses which have nucleotide polymerases in their virions, there is little or no activity demonstrable unless the virions are activated by heat, proteolytic enzymes or detergents[8-12]. None of these treatments increased the activity of the R-MLV DNA polymerase. In fact, incubation at 50° C for 10 min totally inactivated the R-MLV enzyme as did inclusion of trypsin (50 μg/ml.) in the reaction mixture. Addition of as little as 0.01 per cent 'Triton N-101' (a non-ionic detergent) also markedly depressed activity.

Table 1. PROPERTIES OF THE RAUSCHER MOUSE LEUKAEMIA VIRUS DNA POLYMERASE

Reaction system	pmoles ³H-TMP incorporated in 45 min
Complete	3.31
Without magnesium acetate	0.04
Without magnesium acetate + 6 mM $MnCl_2$	1.59
Without dithiothreitol	0.38
Without NaCl	2.18
Without dATP	< 0.10
Without dCTP	0.12
Without dGTP	< 0.10

A preparation of R-MLV was provided by the Viral Resources Program of the National Cancer Institute. The virus had been purified from the plasma of infected Swiss mice by differential centrifugation. The preparation had a titre of $10^{4.88}$ spleen enlarging doses (50 per cent end point) per ml. Before use the preparation was centrifuged at $105,000g$ for 30 min and the pellet was suspended in 0.137 M NaCl-0.003 M KCl-0.01 M phosphate buffer (pH 7.4)-0.6 mM EDTA (PBS–EDTA) at 1/20 of the initial volume. The concentrated virus suspension contained 3.1 mg/ml. of protein. The assay mixture contained, in 0.1 ml., 5 μmoles Tris-HCl (pH 8.3) at 37° C, 0.6 μmole magnesium acetate, 6 μmoles NaCl, 2 μmoles dithiothreitol, 0.08 μmole each of dATP, dCTP and dGTP, 0.001 μmole [³H-*methyl*]-TTP (708 c.p.m. per pmole) (New England Nuclear) and 15 μg viral protein. The reaction mixture was incubated for 45 min at 37° C. The acid-insoluble radioactivity in the sample was then determined by addition of sodium pyrophosphate, carrier yeast RNA and trichloroacetic acid followed by filtration through a membrane filter and counting in a scintillation spectrometer, all as previously described[12]. The radioactivity of an unincubated sample was subtracted from each value (less than 7 per cent of the incorporation in the complete reaction mixture).

Characterization of the Product

The nature of the reaction product was investigated by determining its sensitivity to various treatments. The product could be rendered acid-soluble by either pancreatic deoxyribonuclease or micrococcal nuclease but was unaffected by pancreatic ribonuclease or by alkaline hydrolysis (Table 2). The product therefore has the properties of DNA. If 50 μg/ml. of deoxyribonuclease was added to a reaction mixture there was no loss of acid-insoluble product. The product is therefore protected from the enzyme, probably by the envelope of the virion, although merely diluting the reaction mixture into 10 mM $MgCl_2$ enables the product to be digested by deoxyribonuclease (Table 2).

Table 2. CHARACTERIZATION OF THE POLYMERASE PRODUCT

Expt.	Treatment	Acid-insoluble radioactivity	Percentage undigested product
1	Untreated	1,425	(100)
	20 μg deoxyribonuclease	125	9
	20 μg micrococcal nuclease	69	5
	20 μg ribonuclease	1,361	96
2	Untreated	1,644	(100)
	NaOH hydrolysed	1,684	100

For experiment 1, 93 μg of viral protein was incubated for 2 h in a reaction mixture twice the size of that described in Table 1, with ³H-TTP having a specific activity of 1,133 c.p.m. per pmole. A 50 μl. portion of the reaction mixture was diluted to 5 ml. with 10 mM $MgCl_2$ and 0·5 ml. aliquots were incubated for 1·5 h at 37° C with the indicated enzymes. (The sample with micrococcal nuclease also contained 5 mM $CaCl_2$.) The samples were then chilled, precipitated with trichloroacetic acid and radioactivity was counted. For experiment 2, two standard reaction mixtures were incubated for 45 min at 37° C, then to one sample was added 0·1 ml. of 1 M NaOH and it was boiled for 5 min. It was then chilled and both samples were precipitated with trichloroacetic acid and counted. In a separate experiment (unpublished) it was shown that the alkaline hydrolysis conditions would completely degrade the RNA product of the VSV virion polymerase.

Localization of the Enzyme and its Template

To investigate whether the DNA polymerase and its template were associated with the virions, a R-MLV suspension was centrifuged to equilibrium in a 15–50 per cent sucrose gradient and fractions of the gradient were assayed for DNA polymerase activity. Most of the activity was

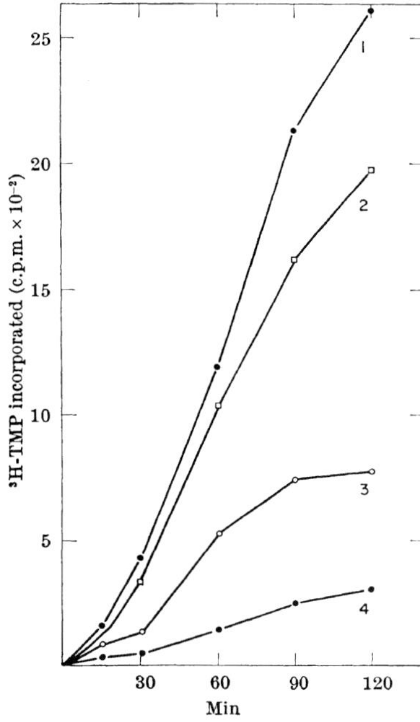

Fig. 1. Incorporation of radioactivity from ³H-TTP by the R-MLV DNA polymerase in the presence and absence of ribonuclease. A 1·5-fold standard reaction mixture was prepared with 30 μg of viral protein and ³H–TTP (specific activity 950 c.p.m. per pmole). At various times, 20 μl. aliquots were added to 0·5 ml. of non-radioactive 0·1 M sodium pyrophosphate and acid insoluble radioactivity was determined[12]. For the preincubated samples, 0·06 ml. of H_2O and 0·01 ml. of R-MLV (30 μg of protein) were incubated with or without 10 μg of pancreatic ribonuclease at 22° C for 20 min, chilled and brought to 0·15 ml. with a concentrated mixture of the components of the assay system. Curve 1, no treatment; curve 2, preincubated; curve 3, 10 μg ribonuclease added to the reaction mixture; curve 4, preincubated with 10 μg ribonuclease.

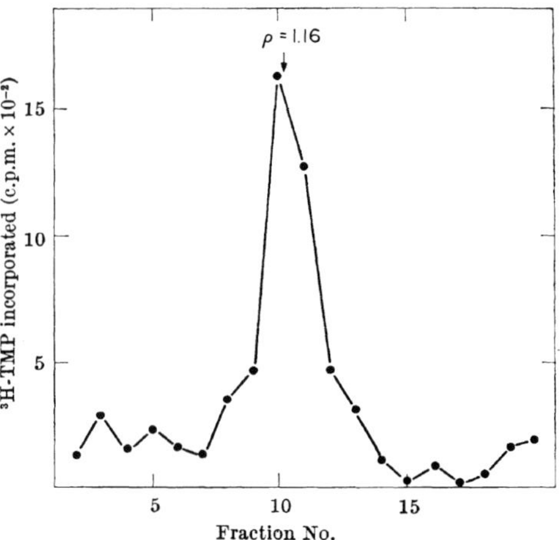

Fig. 2. Localization of DNA polymerase activity in R-MLV by isopycnic centrifugation. A preparation of R-MLV containing 150 μg of protein in 50 μl. was layered over a linear 5·2 ml. gradient of 15–50 per cent sucrose in PBS–EDTA. After centrifugation for 2 h at 60,000 r.p.m. in the Spinco 'SW65' rotor, 0·27 ml. fractions of the gradient were collected and 0·1 ml. portions of each fraction were incubated for 60 min in a standard reaction mixture. The acid-precipitable radioactivity was then collected and counted. The density of each fraction was determined from its refractive index. The arrow indicates the position of a sharp, visible band of light-scattering material which occurred at a density of 1·16.

found at the position of the visible band of virions (Fig. 2). The density at this band was 1·16 g/cm³, in agreement with the known density of the virions[14]. The polymerase and its template therefore seem to be constituents of the virion.

The Template is RNA

Virions of the RNA tumour viruses contain RNA but no DNA[15,16]. The template for the virion DNA polymerase is therefore probably the viral RNA. To substantiate further that RNA is the template, the effect of ribonuclease on the reaction was investigated. When 50 μg/ml. of pancreatic ribonuclease was included in the reaction mixture, there was a 50 per cent inhibition of activity during the first hour and more than 80 per cent inhibition during the second hour of incubation (Fig. 1, curve 3). If the virions were preincubated with the enzyme in water at 22° C and the components of the reaction mixture were then added, an earlier and more extensive inhibition was evident (Fig. 1, curve 4). Preincubation in water without ribonuclease caused only a slight inactivation of the virion polymerase activity (Fig. 1, curve 2). Increasing the concentration of ribonuclease during preincubation could inhibit more than 95 per cent of the DNA polymerase activity (Table 3). To ensure that the inhibition by ribonuclease was attributable to the enzymic activity of the added protein, two other basic proteins were preincubated with the virions. Only ribonuclease was able to inhibit the reaction (Table 3). These experiments substantiate the idea that RNA is the template for the reaction. Hybridization experiments are in progress to determine if the DNA is complementary in base sequence to the viral RNA.

Ability of the Enzyme to Incorporate Ribonucleotides

The deoxyribonucleotide incorporation measured in these experiments could be the result of an RNA polymerase activity in the virion which can polymerize deoxyribonucleotides when they are provided in the reaction mixture. The VSV RNA polymerase and the R-MLV DNA polymerase were therefore compared. The VSV RNA polymerase incorporated only ribonucleotides. At its pH optimum of 7·3 (my unpublished observation), in the presence of the four common ribonucleoside triphosphates, the enzyme incorporated ^{3}H-GMP extensively[12]. At this pH, however, in the presence of the four deoxyribonucleoside triphosphates, no ^{3}H-TMP incorporation was demonstrable (Table 4). Furthermore, replacement of even a single ribonucleotide by its homologous deoxyribonucleotide led to no detectable synthesis (my unpublished observation). At pH 8·3, the optimum for the R-MLV DNA polymerase, the VSV polymerase catalysed much less ribonucleotide incorporation and no significant deoxyribonucleotide incorporation could be detected.

Table 3. EFFECT OF RIBONUCLEASE ON THE DNA POLYMERASE ACTIVITY OF RAUSCHER MOUSE LEUKAEMIA VIRUS

Conditions	pmoles ^{3}H-TMP incorporation
No preincubation	2·50
Preincubated with no addition	2·20
Preincubated with 20 μg/ml. ribonuclease	0·69
Preincubated with 50 μg/ml. ribonuclease	0·31
Preincubated with 200 μg/ml. ribonuclease	0·08
Preincubated with no addition	3·69
Preincubated with 50 μg/ml. ribonuclease	0·52
Preincubated with 50 μg/ml. lysozyme	3·67
Preincubated with 50 μg/ml. cytochrome c	3·97

In experiment 1, for the preincubation, 15 μg of viral protein in 5 μl. of solution was added to 45 μl. of water at 4° C containing the indicated amounts of enzyme. After incubation for 30 min at 22° C, the samples were chilled and 50 μl. of a 2-fold concentrated standard reaction mixture was added. The samples were then incubated at 37° C for 45 min and acid-insoluble radioactivity was measured. In experiment 2, the same procedure was followed, except that the preincubation was for 20 min at 22° C and the 37° C incubation was for 60 min.

Table 4. COMPARISON OF NUCLEOTIDE INCORPORATION BY VESICULAR STOMATITIS VIRUS AND RAUSCHER MOUSE LEUKAEMIA VIRUS

Precursor	pH	Incorporation in 45 min (pmoles) Vesicular stomatitis virus	Mouse leukaemia virus
[3]H-TTP	8·3	< 0·01	2·3
[3]H-TTP (omit dATP)	8·3	N.D.	0·06
[3]H-TTP (omit dATP; plus ATP)	8·3	N.D.	0·08
[3]H-GTP	8·3	0·43	< 0·03
[3]H-GTP	7·3	3·7	< 0·03

When ^{3}H-TTP was the precursor, standard reaction conditions were used (see Table 1). When ^{3}H-GTP was the precursor, the reaction mixture contained, in 0·1 ml., 5 μmoles Tris-HCl (pH as indicated), 0·6 μmoles magnesium acetate, 0·3 μmoles mercaptoethanol, 9 μmoles NaCl, 0·08 μmole each of ATP, CTP, UTP; and 0·001 μmole ^{3}H-GTP (1,040 c.p.m. per pmole). All VSV assays included 0·1 per cent 'Triton N-101' (ref. 12) and 2–5 μg of viral protein. The R-MLV assays contained 15 μg of viral protein.

The R-MLV polymerase incorporated only deoxyribonucleotides. At pH 8·3, ^{3}H-TMP incorporation was readily demonstrable but replacement of dATP by ATP completely prevented synthesis (Table 4). Furthermore, no significant incorporation of ^{3}H-GMP could be detected in the presence of the four ribonucleotides. At pH 7·3, the R-MLV polymerase was also inactive with ribonucleotides. The polymerase in the R-MLV virions is therefore highly specific for deoxyribonucleotides.

DNA Polymerase in Rous Sarcoma Virus

A preparation of the Prague strain of Rous sarcoma virus was assayed for DNA polymerase activity (Table 5). Incorporation of radioactivity from ^{3}H-TTP was demonstrable and the activity was severely reduced by omission of either Mg^{2+} or dATP from the reaction mixture. RNA-dependent DNA polymerase is therefore probably a constituent of all RNA tumour viruses.

These experiments indicate that the virions of Rauscher mouse leukaemia virus and Rous sarcoma virus contain a DNA polymerase. The inhibition of its activity by ribonuclease suggests that the enzyme is an RNA-dependent DNA polymerase. It seems probable that all RNA tumour viruses have such an activity. The existence of this enzyme strongly supports the earlier suggestions[1-7] that

genetically specific DNA synthesis is an early event in the replication cycle of the RNA tumour viruses and that DNA is the template for viral RNA synthesis. Whether the viral DNA ("provirus")[2] is integrated into the host genome or remains as a free template for RNA synthesis will require further study. It will also be necessary to determine whether the host DNA-dependent RNA polymerase or a virus-specific enzyme catalyses the synthesis of viral RNA from the DNA.

Table 5. PROPERTIES OF THE ROUS SARCOMA VIRUS DNA POLYMERASE

Reaction system	pmoles ^{3}H-TMP incorporated in 120 min
Complete	2·06
Without magnesium acetate	0·12
Without dATP	0·19

A preparation of the Prague strain (sub-group C) of Rous sarcoma virus[16] having a titre of 5×10^7 focus forming units per ml. was provided by Dr Peter Vogt. The virus was purified from tissue culture fluid by differential centrifugation. Before use the preparation was centrifuged and the pellet dissolved in 1/10 of the initial volume as described for the R-MLV preparation. For each assay 15 μl. of the concentrated Rous sarcoma virus preparation was assayed in a standard reaction mixture by incubation for 2 h. An unincubated control sample had radioactivity corresponding to 0·14 pmole which was subtracted from the experimental values.

I thank Drs G. Todaro, F. Rauscher and R. Holdenreid for their assistance in providing the mouse leukaemia virus. This work was supported by grants from the US Public Health Service and the American Cancer Society and was carried out during the tenure of an American Society Faculty Research Award.

DAVID BALTIMORE

Department of Biology,
Massachusetts Institute of Technology,
Cambridge,
Massachusetts 02139.

Received June 2, 1970.

[1] Green, M., *Ann. Rev. Biochem.*, **39** (1970, in the press).
[2] Temin, H. M., *Virology*, **23**, 486 (1964).
[3] Bader, J. P., *Virology*, **22**, 462 (1964).
[4] Vigier, P., and Golde, A., *Virology*, **23**, 511 (1964).
[5] Duesberg, P. H., and Vogt, P. K., *Proc. US Nat. Acad. Sci.*, **64**, 939 (1969).
[6] Temin, H. M., in *Biology of Large RNA Viruses* (edit. by Barry, R., and Mahy, B.) (Academic Press, London, 1970).
[7] Temin, H. M., *Virology*, **20**, 577 (1963).
[8] Kates, J. R., and McAuslan, B. R., *Proc. US Nat. Acad. Sci.*, **58**, 134 (1967).
[9] Munyon, W., Paoletti, E., and Grace, J. T. J., *Proc. US Nat. Acad. Sci.*, **58**, 2280 (1967).
[10] Shatkin, A. J., and Sipe, J. D., *Proc. US Nat. Acad. Sci.*, **61**, 1462 (1968).
[11] Borsa, J., and Graham, A. F., *Biochem. Biophys. Res. Commun.*, **33**, 895 (1968).
[12] Baltimore, D., Huang, A. S., and Stampfer, M., *Proc. US Nat. Acad. Sci.*, **66** (1970, in the press).
[13] Temin, H. M., and Mizutani, S., *Nature*, **226**, 1211 (1970) (following article).
[14] O'Conner, T. E., Rauscher, F. J., and Zeigel, R. F., *Science*, **144**, 1144 (1964).
[15] Crawford, L. V., and Crawford, E. M., *Virology*, **13**, 227 (1961).
[16] Duesberg, P., and Robinson, W. S., *Proc. US Nat. Acad. Sci.*, **55**, 219 (1966).
[17] Duff, R. G., and Vogt, P. K., *Virology*, **39**, 18 (1969).

(*Reprinted from Nature*, Vol. 226, No. 5252, pp. 1211–1213,
June 27, 1970)

RNA-dependent DNA Polymerase in Virions of Rous Sarcoma Virus

INFECTION of sensitive cells by RNA sarcoma viruses requires the synthesis of new DNA different from that synthesized in the *S*-phase of the cell cycle (refs. 1, 2 and D. Boettiger and H. M. T. (*Nature*, in the press)); production of RNA tumour viruses is sensitive to actinomycin D[3,4]; and cells transformed by RNA tumour viruses have new DNA which hybridizes with viral RNA[5,6]. These are the basic observations essential to the DNA provirus hypothesis—replication of RNA tumour viruses takes place through a DNA intermediate, not through an RNA intermediate as does the replication of other RNA viruses[7].

Formation of the provirus is normal in stationary chicken cells exposed to Rous sarcoma virus (RSV), even in the presence of 0·5 µg/ml. cycloheximide (our unpublished results). This finding, together with the discovery of polymerases in virions of vaccinia virus and of reovirus[8-11], suggested that an enzyme that would synthesize DNA from an RNA template might be present in virions of RSV. We now report data supporting the existence of such an enzyme, and we learn that David Baltimore has independently discovered a similar enzyme in virions of Rauscher leukaemia virus[12].

The sources of virus and methods of concentration have been described[13]. All preparations were carried out in sterile conditions. Concentrated virus was placed on a layer of 15 per cent sucrose and centrifuged at 25,000 r.p.m. for 1 h in the 'SW 25.1' rotor of the Spinco ultracentrifuge on to a cushion of 60 per cent sucrose. The virus band was collected from the interphase and further purified by equilibrium sucrose density gradient centrifugation[14]. Virus further purified by sucrose velocity density gradient centrifugation gave the same results.

The polymerase assay consisted of 0·125 µmoles each of dATP, dCTP, and dGTP (Calbiochem) (in 0·02 M Tris-HCl buffer at pH 8·0, containing 0·33 Mm EDTA and 1·7 mM 2-mercaptoethanol); 1·25 µmoles of $MgCl_2$ and 2·5 µmoles of KCl; 2·5 µg phosphoenolpyruvate (Calbiochem); 10 µg pyruvate kinase (Calbiochem); 2·5 µCi of ^{3}H-TTP (Schwarz) (12 Ci/mmole); and 0·025 ml. of enzyme (10^8 focus forming units of disrupted Schmidt-Ruppin virus, $A_{280\,nm} = 0·30$) in a total volume of 0·125 ml. Incubation was at 40° C for 1 h. 0·025 ml. of the reaction mixture was withdrawn and assayed for acid-insoluble counts by the method of Furlong[15].

Table 1. ACTIVATION OF ENZYME

System	³H-TTP incorporated (d.p.m.)
No virions	0
Non-disrupted virions	255
Virions disrupted with 'Nonidet'	
At 0° + DTT	6,730
At 0° − DTT	4,420
At 40° + DTT	5,000
At 40° − DTT	425

Purified virions untreated or incubated for 5 min at 0° C or 40° C with 0·25 per cent 'Nonidet P–40' (Shell Chemical Co.) with 0 or 1 per cent dithiothreitol (DTT) (Sigma) were assayed in the standard polymerase assay.

To observe full activity of the enzyme, it was necessary to treat the virions with a non-ionic detergent (Tables 1 and 4). If the treatment was at 40° C the presence of dithiothreitol (DTT) was necessary to recover activity. In most preparations of virions, however, there was some activity, 5–20 per cent of the disrupted virions, in the absence of detergent treatment, which probably represents disrupted virions in the preparation. It is known that virions of RNA tumour viruses are easily disrupted[16,17], so that the activity is probably present in the nucleoid of the virion.

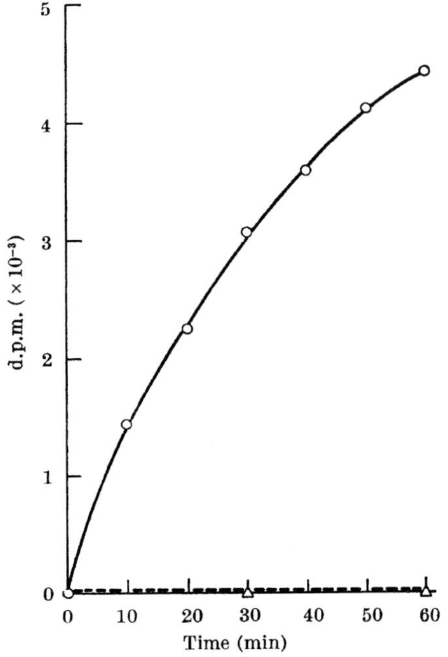

Fig. 1. Kinetics of incorporation. Virus treated with 'Nonidet' and dithiothreitol at 0° C and incubated at 37° C (○—○) or 80° C (△ - - - △) for 10 min was assayed in a standard polymerase assay.

2

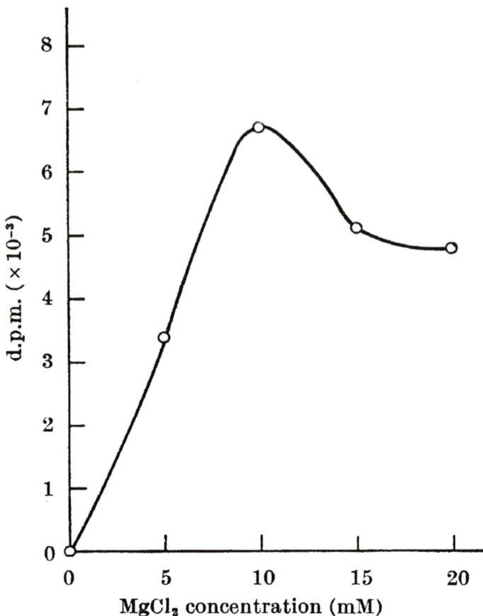

Fig. 2. MgCl$_2$ requirement. Virus treated with 'Nonidet' and dithio-threitol at 0° C was incubated in the standard polymerase assay with different concentrations of MgCl$_2$.

The kinetics of incorporation with disrupted virions are shown in Fig. 1. Incorporation is rapid for 1 h. Other experiments show that incorporation continues at about the same rate for the second hour. Preheating disrupted virus at 80° C prevents any incorporation, and so does pretreatment of disrupted virus with crystalline trypsin.

Fig. 2 demonstrates that there is an absolute require-ment for MgCl$_2$, 10 mM being the optimum concentration. The data in Table 2 show that MnCl$_2$ can substitute for MgCl$_2$ in the polymerase assay, but CaCl$_2$ cannot. Other experiments show that a monovalent cation is not required for activity, although 20 mM KCl causes a 15 per cent stimulation. Higher concentrations of KCl are inhibitory: 60 per cent inhibition was observed at 80 mM.

Table 2. REQUIREMENTS FOR ENZYME ACTIVITY

System	^{3}H-TTP incor-porated (d.p.m.)
Complete	5,675
Without MgCl$_2$	186
Without MgCl$_2$, with MnCl$_2$	5,570
Without MgCl$_2$, with CaCl$_2$	18
Without dATP	897
Without dCTP	1,780
Without dGTP	2,190

Virus treated with 'Nonidet' and dithiothreitol at 0° C was incubated in the standard polymerase assay with the substitutions listed.

3

Table 3. RNA DEPENDENCE OF POLYMERASE ACTIVITY

Treatment	³H-TTP incorporated (d.p.m.)
Non-treated disrupted virions	9,110
Disrupted virions preincubated with ribonuclease A (50 μg/ml.) at 20° C for 1 h	2,650
Disrupted virions preincubated with ribonuclease A (1 mg/ml.) at 0° C for 1 h	137
Disrupted virions preincubated with lysozyme (50 μg/ml.) at 0° C for 1 h	9,650

Disrupted virions were incubated with ribonuclease A (Worthington) which was heated at 80° C for 10 min, or with lysozyme at the indicated concentration in the specified conditions, and a standard polymerase assay was performed.

When the amount of disrupted virions present in the polymerase assay was varied, the amount of incorporation varied with second-order kinetics. When incubation was carried out at different temperatures, a broad optimum between 40° C and 50° C was found. (The high temperature of this optimum may relate to the fact that the normal host of the virus is the chicken.) When incubation was carried out at different pHs, a broad optimum at pH 8–9·5 was found.

Table 2 demonstrates that all four deoxyribonucleoside triphosphates are required for full activity, but some activity was present when only three deoxyribonucleoside triphosphates were added and 10–20 per cent of full activity was still present with only two deoxyribonucleoside triphosphates. The activity in the presence of three deoxyribonucleoside triphosphates is probably the result of the presence of deoxyribonucleoside triphosphates in the virion. Other host components are known to be incorporated in the virion of RNA tumour viruses[18,19].

The data in Table 3 demonstrate that incorporation of thymidine triphosphate was more than 99 per cent abolished if the virions were pretreated at 0° with 1 mg ribonuclease per ml. Treatment with 50 μg/ml. ribo-

Table 4. SOURCE OF POLYMERASE

Source	³H-TTP incorporated (d.p.m.)
Virions of SRV	1,410
Disrupted virions of SRV	5,675
Virions of AMV	1,875
Disrupted virions of AMV	12,850
Disrupted pellet from supernatant of uninfected cells	0

Virions of Schmidt-Ruppin virus (SRV) were prepared as before (experiment of Table 2). Virions of avian myeloblastosis virus (AMV) and a pellet from uninfected cells were prepared by differential centrifugation. All disrupted preparations were treated with 'Nonidet' and dithiothreitol at 0° C and assayed in a standard polymerase assay. The material used per tube was originally from 45 ml. of culture fluid for SRV, 20 ml. for AMV, and 20 ml. for uninfected cells.

4

Table 5. NATURE OF PRODUCT

| | Residual acid-insoluble ^{3}H-TTP (d.p.m.) | |
Treatment	Experiment A	Experiment B
Buffer	10,200	8,350
Deoxyribonuclease	697	1,520
Ribonuclease	10,900	7,200
KOH	—	8,250

A standard polymerase assay was performed with 'Nonidet' treated virions. The product was incubated in buffer or 0·3 M KOH at 37° C for 20 h or with (A) 1 mg/ml. or (B) 50 μg/ml. of deoxyribonuclease I (Worthington), or with 1 mg/ml. of ribonuclease A (Worthington) for 1 h at 37° C, and portions were removed and tested for acid-insoluble counts.

nuclease at 20° C did not prevent all incorporation of thymidine triphosphate, which suggests that the RNA of the virion may be masked by protein. (Lysozyme was added as a control for non-specific binding of ribonuclease to DNA.) Because the ribonuclease was heated for 10 min at 80° C or 100° C before use to destroy deoxyribonuclease it seems that intact RNA is necessary for incorporation of thymidine triphosphate.

To determine whether the enzyme is present in supernatants of normal cells or in RNA leukaemia viruses, the experiment of Table 4 was performed. Normal cell supernatant did not contain activity even after treatment with 'Nonidet'. Virions of avian myeloblastosis virus (AMV) contained activity that was increased ten-fold by treatment with 'Nonidet'.

The nature of the product of the polymerase assay was investigated by treating portions with deoxyribonuclease, ribonuclease or KOH. About 80 per cent of the product was made acid soluble by treatment with deoxyribonuclease, and the product was resistant to ribonuclease and KOH (Table 5).

To determine if the polymerase might also make RNA, disrupted virions were incubated with the four ribonucleoside triphosphates, including ^{3}H-UTP (Schwarz, 3·2 Ci/mmole). With either $MgCl_2$ or $MnCl_2$ in the incubation mixture, no incorporation was detected. In a parallel incubation with deoxyribonucleoside triphosphates, 12,200 d.p.m. of ^{3}H-TTP was incorporated.

These results demonstrate that there is a new polymerase inside the virions of RNA tumour viruses. It is not present in supernatants of normal cells but is present in virions of avian sarcoma and leukaemia RNA tumour viruses. The polymerase seems to catalyse the incorporation of deoxyribonucleoside triphosphates into DNA from an RNA template. Work is being performed to characterize further the reaction and the product. If the present results and Baltimore's results[12] with Rauscher leukaemia virus are upheld, they will constitute strong evidence that the DNA provirus hypothesis is correct and that RNA tumour viruses have a DNA genome when they are in

5

cells and an RNA genome when they are in virions. This result would have strong implications for theories of viral carcinogenesis and, possibly, for theories of information transfer in other biological systems[20].

This work was supported by a US Public Health Service research grant from the National Cancer Institute. H. M. T. holds a research career development award from the National Cancer Institute.

HOWARD M. TEMIN
SATOSHI MIZUTANI

McArdle Laboratory for Cancer Research,
University of Wisconsin,
Madison,
Wisconsin 53706.

Received June 15, 1970.

[1] Temin, H. M., *Cancer Res.*, **28**, 1835 (1968).
[2] Murray, R. K., and Temin, H. M., *Intern. J. Cancer*, **5**, 320 (1970).
[3] Temin, H. M., *Virology*, **20**, 577 (1963).
[4] Baluda, M. B., and Nayak, D. P., *J. Virol.*, **4**, 554 (1969).
[5] Temin, H. M., *Proc. US Nat. Acad. Sci.*, **52**, 323 (1964).
[6] Baluda, M. B., and Nayak, D. P., in *Biology of Large RNA Viruses* (edit. by Barry, R., and Mahy, B.) (Academic Press, London, 1970).
[7] Temin, H. M., *Nat. Cancer Inst. Monog.*, **17**, 557 (1964).
[8] Kates, J. R., and McAuslan, B. R., *Proc. US Nat. Acad. Sci.*, **57**, 314 (1967).
[9] Munyon, W., Paoletti, E., and Grace, J. T., *Proc. US Nat. Acad. Sci.*, **58**, 2280 (1967).
[10] Borsa, J., and Graham, A. F., *Biochem. Biophys. Res. Commun.*, **33**, 895 (1968).
[11] Shatkin, A. J., and Sipe, J. D., *Proc. US Nat. Acad. Sci.*, **61**, 1462 (1968).
[12] Baltimore, D., *Nature*, **226**, 1209 (1970) (preceding article).
[13] Altaner, C., and Temin, H. M., *Virology*, **40**, 118 (1970).
[14] Robinson, W. S., Pitkanen, A., and Rubin, H., *Proc. US Nat. Acad. Sci.*, **54**, 137 (1965).
[15] Furlong, N. B., *Meth. Cancer Res.*, **3**, 27 (1967).
[16] Vogt, P. K., *Adv. Virus. Res.*, **11**, 293 (1965).
[17] Bauer, H., and Schafer, W., *Virology*, **29**, 494 (1966).
[18] Bauer, H., *Z. Naturforsch.*, 21b, 453 (1966).
[19] Erikson, R. L., *Virology*, **37**, 124 (1969).
[20] Temin, H. M., *Persp. Biol. Med.*, **5**, 320 (1970).

Biochemical Method for Inserting New Genetic Information into DNA of Simian Virus 40: Circular SV40 DNA Molecules Containing Lambda Phage Genes and the Galactose Operon of Escherichia coli

D. A. Jackson, R. H. Symons, and P. Berg

Like the year 1940, the year 1970 occupies a unique position in the history of virology. In 1940, the conceptualization of the one-step growth cycle by Emory Ellis and Max Delbrück; the discovery of hemagglutination, which provided a rapid and simple means of quantitating virus particles; and the invention of the electron microscope and the ultracentrifuge, which provided the means for studying viruses by physical techniques, converged to initiate the era of molecular virology, molecular biology, molecular cell biology, and molecular genetics. In 1970, the discovery of restriction endonucleases by Ham Smith and K. W. Willcox (*J. Mol. Biol.* **51**:379–391, 1970) provided the means for localizing any gene and of joining any DNA sequence to any other; the discovery of the reverse transcriptase by David Baltimore and by Howard Temin and Satoshi Mizutani extended this technology to RNA by permitting the transcription of RNA into DNA. These two discoveries initiated the age of genetic engineering.

Very soon scientists realized that these discoveries would enable them to insert new or modified genetic information into the genomes of living creatures. In order to achieve this, appropriate vectors were required. The vectors of choice for introducing DNA into cells are viruses; and since the new genetic information was to be introduced into cellular genomes, the virus had to be one whose genome was inserted into the cellular genome. The first choices here were retroviruses and papovaviruses. In this paper Jackson, Symons, and Berg chose simian virus 40 (SV40) and devised techniques for inserting foreign genetic information into its genome via use of appropriate restriction endonucleases. This paper is one of the cornerstones on which all subsequent "vectorology" is based, which is the reason why it was selected. Paul Berg was awarded the Nobel Prize in 1980.

W. K. Joklik

Reprinted from *Proceedings of the National Academy of Sciences USA* 69:2904–2909. Copyright © 1972, by permission of the authors.

Proc. Nat. Acad. Sci. USA
Vol. 69, No. 10, pp. 2904–2909, October 1972

Biochemical Method for Inserting New Genetic Information into DNA of Simian Virus 40: Circular SV40 DNA Molecules Containing Lambda Phage Genes and the Galactose Operon of *Escherichia coli*

(molecular hybrids/DNA joining/viral transformation/genetic transfer)

DAVID A. JACKSON*, ROBERT H. SYMONS†, AND PAUL BERG

Department of Biochemistry, Stanford University Medical Center, Stanford, California 94305

Contributed by Paul Berg, July 31, 1972

ABSTRACT   We have developed methods for covalently joining duplex DNA molecules to one another and have used these techniques to construct circular dimers of SV40 DNA and to insert a DNA segment containing lambda phage genes and the galactose operon of *E. coli* into SV40 DNA. The method involves: (*a*) converting circular SV40 DNA to a linear form, (*b*) adding single-stranded homodeoxypolymeric extensions of defined composition and length to the 3′ ends of one of the DNA strands with the enzyme terminal deoxynucleotidyl transferase (*c*) adding complementary homodeoxypolymeric extensions to the other DNA strand, (*d*) annealing the two DNA molecules to form a circular duplex structure, and (*e*) filling the gaps and sealing nicks in this structure with *E. coli* DNA polymerase and DNA ligase to form a covalently closed-circular DNA molecule.

Our goal is to develop a method by which new, functionally defined segments of genetic information can be introduced into mammalian cells. It is known that the DNA of the transforming virus SV40 can enter into a stable, heritable, and presumably covalent association with the genomes of various mammalian cells (1, 2). Since purified SV40 DNA can also transform cells (although with reduced efficiency), it seemed possible that SV40 DNA molecules, into which a segment of functionally defined, nonviral DNA had been covalently integrated, could serve as vectors to transport and stabilize these nonviral DNA sequences in the cell genome. Accordingly, we have developed biochemical techniques that are generally applicable for joining covalently any two DNA molecules.‡ Using these techniques, we have constructed circular dimers of SV40 DNA; moreover, a DNA segment containing λ phage genes and the galactose operon of *Escherichia coli* has been covalently integrated into the circular SV40 DNA molecule. Such hybrid DNA molecules and others like them can be tested for their capacity to transduce foreign DNA sequences into mammalian cells, and can be used to determine whether these new nonviral genes can be expressed in a novel environment.

---

* Present address: Department of Microbiology, University of Michigan Medical Center, Ann Arbor, Mich. 48104.
† Present address: Department of Biochemistry, University of Adelaide, Adelaide, South Australia, 5001 Australia.
‡ Drs. Peter Lobban and A. D. Kaiser of this department have performed experiments similar to ours and have obtained similar results using bacteriophage P22 DNA (Lobban, P. and Kaiser, A. D., in preparation).

MATERIALS AND METHODS

DNA. (*a*) Covalently closed-circular duplex SV40 DNA [SV40(I)] (labeled with [³H]dT, 5 × 10⁴ cpm/µg), free from SV40 linear or oligomeric molecules [but containing 3–5% of nicked double-stranded circles—SV40(II)] was purified from SV40-infected CV-1 cells (Jackson, D., & Berg, P., in preparation). (*b*) Closed-circular duplex λ*dvgal* DNA labeled with [³H]dT (2.5 × 10⁴ cpm/µg), was isolated from an *E. coli* strain containing this DNA as an autonomously replicating plasmid (see ref. 3) by equilibrium sedimentation in CsCl–ethidium bromide gradients (4) after lysis of the cells with detergent. A more detailed characterization of this DNA will be published later. Present information indicates that the λ*dvgal* (λ*dv-120*) DNA is a circular dimer containing tandem duplications of a sequence of several λ phage genes (including C_I, O, and P) joined to the entire galactose operon of *E. coli* (Berg, D., Mertz, J., & Jackson, D., in preparation). DNA concentrations are given as molecular concentrations.

Enzymes. The circular SV40 and λ*dvgal* DNA molecules were cleaved with the bacterial restriction endonuclease RI (Yoshimori and Boyer, unpublished; the enzyme was generously made available to us by these workers). Phage λ-exonuclease (given to us by Peter Lobban) was prepared according to Little *et al.* (5), calf-thymus deoxynucleotidyl terminal transferase (terminal transferase), prepared according to Kato *et al.* (6), was generously sent to us by F. N. Hayes; *E. coli* DNA polymerase I Fraction VII (7) was a gift of Douglas Brutlag; and *E. coli* DNA ligase (8) and exonuclease III (9) were kindly supplied by Paul Modrich.

Substrates. [α-³²P]deoxynucleoside triphosphates (specific activities 5–10 Ci/µmol) were synthesized by the method of Symons (10). All other reagents were obtained from commercial sources.

Centrifugations. Alkaline sucrose gradients were formed by diffusion from equal volumes of 5, 10, 15, and 20% sucrose solutions with 2 mM EDTA containing, respectively, 0.2, 0.4, 0.6, and 0.8 M NaOH, and 0.8, 0.6, 0.4, 0.2 M NaCl. 100-µl samples were run on 3.8-ml gradients in a Beckman SW56 Ti rotor in a Beckman L2-65B ultracentrifuge at 4° and 55,000 rpm for the indicated times. 2- to 10-drop fractions were collected onto 2.5-cm diameter Whatman 3MM discs, dried without washing, and counted in PPO–dimethyl POPOP–toluene scintillator in a Nuclear Chicago Mark II scintillation spectrometer. An overlap of 0.4% of ^{32}P into the ^{3}H channel was not corrected for.

CsCl–ethidium bromide equilibrium centrifugation was performed in a Beckman Type 50 rotor at 4° and 37,000 rpm for 48 hr. SV40 DNA in 10 mM Tris·HCl (pH 8.1)–1 mM Na EDTA–10 mM NaCl was adjusted to 1.566 g/ml of CsCl and 350 μg/ml of ethidium bromide. 30-Drop fractions were collected and aliquots were precipitated on Whatman GF/C filters with cold 2 N HCl; the filters were washed and counted.

Electron Microscopy. DNA was spread for electron microscopy by the aqueous method of Davis *et al.* (11) and photographed in a Phillips EM 300. Projections of the molecules were traced on paper and measured with a Keuffel and Esser map measurer. Plaque-purified SV40(II) DNA was used as an internal length standard.

Conversion of SV40(I) DNA to Unit Length Linear DNA [SV40(L$_{RI}$)] with R$_I$ Endonuclease. [^{3}H]SV40(I) DNA (18.7 nM) in 100 mM Tris·HCl buffer (pH 7.5)–10 mM MgCl$_2$–2 mM 2-mercaptoethanol was incubated for 30 min at 37° with an amount of R$_I$ previously determined to convert 1.5 times this amount of SV40(I) to linear molecules [SV40(L$_{RI}$)]; Na EDTA (30 mM) was added to stop the reaction, and the DNA was precipitated in 67% ethanol.

Removal of 5′-Terminal Regions from SV40(L$_{RI}$) with λ Exonuclease. [^{3}H]SV40(L$_{RI}$) (15 nM) in 67 mM K-glycinate (pH 9.5), 4 mM MgCl$_2$, 0.1 mM EDTA was incubated at 0° with λ-exonuclease (20 μg/ml) to yield [^{3}H]SV40(L$_{RI}$exo) DNA. Release of [^{3}H]dTMP was measured by chromatographing aliquots of the reaction on polyethyleneimine thin-layer sheets (Brinkmann) in 0.6 M NH$_4$HCO$_3$ and counting the dTMP spot and the origin (undegraded DNA).

Addition of Homopolymeric Extensions to SV40(L$_{RI}$exo) with Terminal Transferase. [^{3}H]SV40(L$_{RI}$exo) (50 nM) in 100 mM K-cacodylate (pH 7.0), 8 mM MgCl$_2$, 2 mM 2-mercaptoethanol, 150 μg/ml of bovine serum albumin, [α-^{32}P]dNTP (0.2 mM for dATP, 0.4 mM for dTTP) was incubated with terminal transferase (30–60 μg/ml) at 37°. Addition of [^{32}P]dNMP residues to SV40 DNA was measured by spotting aliquots of the reaction mixture on DEAE-paper discs (Whatman DE-81), washing each disc by suction with 50 ml (each) of 0.3 M NH$_4$-formate (pH 7.8) and 0.25 M NH$_4$HCO$_3$, and then with 20 ml of ethanol. To determine the proportion of SV40 linear DNA molecules that had acquired at least one "functional" (dA)$_n$ tail, we measured the amount of SV40 DNA (^{3}H counts) that could be bound to a Whatman GF/C filter (2.4-cm diameter) to which 150 μg of polyuridylic acid had been fixed (13). 15-μl Aliquots of the reaction mixture were mixed with 5 ml of 0.70 M NaCl–0.07 M Na citrate (pH 7.0)–2% Sarkosyl, and filtered at room temperature through the poly(U) filters, at a flow rate of 3–5 ml/min. Each filter was washed by rapid suction with 50 ml of the same buffer at 0°, dried, and counted. Control experiments showed that 98–100% of [^{3}H]oligo(dA)$_{125}$ bound to the filters under these conditions. When the ratio of [^{32}P]dNMP to [^{3}H]DNA reached the value equivalent to the desired length of the extension, the reaction was stopped with EDTA (30 mM) and 2% Sarkosyl. The [^{3}H]SV40(L$_{RI}$exo)–[^{32}P]dA or –dT DNA was purified by neutral sucrose gradient zone sedimentation to remove unincorporated dNTP, as well as any traces of SV40(I) or SV40(II) DNA.

Formation of Hydrogen-Bonded Circular DNA Molecules. [^{32}P]dA and -dT DNAs were mixed at concentrations of 0.15 nM each in 0.1 M NaCl–10 mM Tris·HCl (pH 8.1)–1 mM EDTA. The mixture was kept at 51° for 30 min, then cooled slowly to room temperature.

Formation of Covalently Closed-Circular DNA Molecules. After annealing of the DNA, a mixture of the enzymes, substrates, and cofactors needed for closure was added to the DNA solution and the mixture was incubated at 20° for 3–5 hr. The final concentrations in the reaction mixture were: 20 mM Tris·HCl (pH 8.1), 1 mM EDTA, 6 mM MgCl$_2$, 50 μg/ml bovine-serum albumin, 10 mM NH$_4$Cl, 80 mM NaCl, 0.052 mM DPN, 0.08 mM (each), dATP, dGTP, dCTP, and dTTP, (0.4 μg/ml) *E. coli* DNA polymerase I, (15 units/ml) *E. coli* ligase, and (0.4 unit/ml) *E. coli* exonuclease III.

RESULTS

General approach

Fig. 1 outlines the general approach used to generate circular, covalently-closed DNA molecules from two separate DNAs. Since, in the present case, the units to be joined are themselves circular, the first step requires conversion of the circular structures to linear duplexes. This could be achieved by a double-strand scission at random locations (see *Discussion*) or, as we describe in this paper, at a unique site with R$_I$ restriction endonuclease. Relatively short (50–100 nucleotides) poly(dA) or poly(dT) extensions are added on the 3′-hydroxyl termini of the linear duplexes with terminal transferase; prior

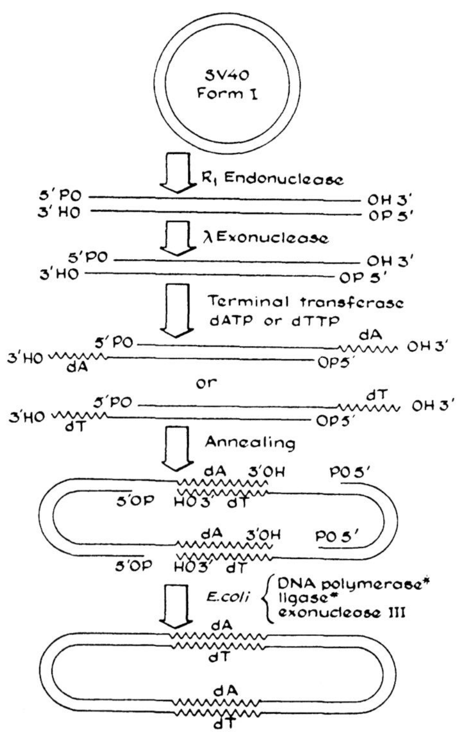

FIG. 1. General protocol for producing covalently closed SV40 dimer circles from SV40(I) DNA.

* The four deoxynucleoside triphosphates and DPN are also present for the DNA polymerase and ligase reactions, respectively.

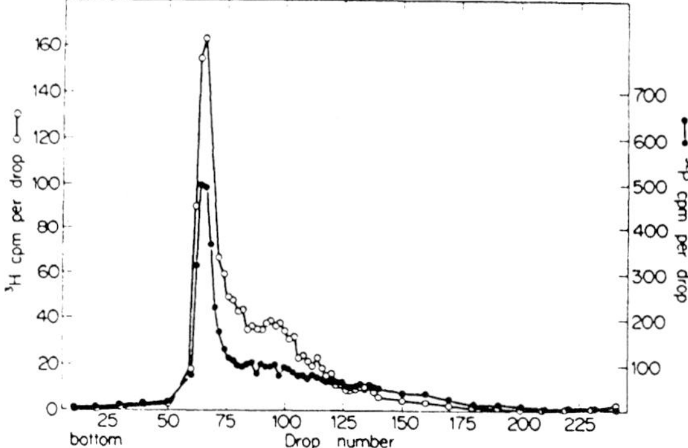

FIG. 2. Alkaline sucrose gradient sedimentation of [³H]SV40-(L$_{RI}$exo)–[³²P](dA)$_{80}$ DNA. 0.16 μg of DNA was centrifuged for 6.0 hr.

removal of a short sequence (30–50 nucleotides) from the 5′-phosphoryl termini by digestion with λ exonuclease facilitates the terminal transferase reaction. Linear duplexes containing (dA)$_n$ extensions are annealed to the DNA to be joined containing (dT)$_n$ extensions at relatively low concentrations. The circular structure formed contains the two DNAs, held together by two hydrogen-bonded homopolymeric regions (Fig. 1). Repair of the four gaps is mediated by *E. coli* DNA polymerase with the four deoxynucleosidetriphosphates, and covalent closure of the ring structure is effected by *E. coli* DNA ligase; *E. coli* exonuclease III removes 3′-phosphoryl residues at any nicks inadvertently introduced during the manipulations (nicks with 3′-phosphoryl ends cannot be sealed by ligase).

Principal steps in the procedure

Circular SV40 DNA Can Be Opened to Linear Duplexes by R$_I$ Endonuclease. Digestion of SV40(I) DNA with excess R$_I$ endonuclease yields a product that sediments at 14.5 S in neutral sucrose gradients and appears as a linear duplex with the same contour length as SV40(II) DNA when examined by electron microscopy [(18); Jackson and Berg, in preparation; see Table 1]. The point of cleavage is at a unique site on the SV40 DNA, and few if any single-strand breaks are introduced elsewhere in the molecule (18); moreover, the termini at each end are 5′-phosphoryl, 3′-hydroxyl (Mertz, J., Davis, R., in preparation). Digestion of plaque-purified SV40 DNA under our conditions yields about 87% linear molecules, 10% nicked circles, and 3% residual supercoiled circles.

Addition of Oligo(dA) or -(dT) Extensions to the 3′-Hydroxyl Termini of SV40 (L$_{RI}$). Terminal transferase has been used to generate deoxyhomopolymeric extensions on the 3′-hydroxyl termini of DNA (7); once the chain is initiated, chain propagation is statistical in that each chain grows at about the same rate (12). Although the length of the extensions can be controlled by variation of either the time of incubation or the amount of substrate, we have varied the time of incubation to minimize spurious nicking of the DNA by trace amounts of endonuclease activity in the enzyme preparation; we have so far been unable to remove or selectively inhibit these nucleases (Jackson and Berg, in preparation). Incubation of SV40(L$_{RI}$) with terminal transferase and

either dATP or dTTP resulted in appreciable addition of mononucleotidyl units to the DNA. But, for example, after addition of 100 residues of dA per end, only a small proportion of the modified SV40 DNA would bind to filter discs containing poly(U) (13). This result indicated that initiation of terminal nucleotidyl addition was infrequent with SV40(L$_{RI}$), but that once initiated those termini served as preferential primers for extensive homopolymer synthesis.

Lobban and Kaiser (unpublished) found that P22 phage DNA became a better primer for homopolymer synthesis after incubation of the DNA with λ exonuclease. This enzyme removes, successively, deoxymononucleotides from 5′-phosphoryl termini of double-stranded DNA (15), thereby rendering the 3′-hydroxyl termini single-stranded. We confirmed their finding with SV40(L$_{RI}$) DNA; after removal of 30–50

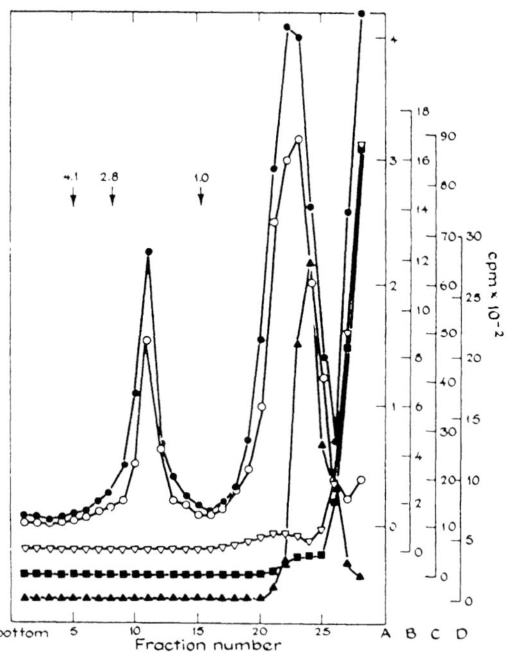

FIG. 3. Alkaline sucrose gradient sedimentation of [³H]-SV40(L$_{RI}$exo)–[³²P](dA)$_{80}$ and —(dT)$_{80}$ DNA incubated 4 hr with and without addition of *E. coli* DNA polymerase I (*P*), ligase (*L*), and exonuclease III (*III*). Conditions are described in *Methods*. 8-Drop fractions were collected. Samples *A*, *C*, and *D* were centrifuged for 60 min, sample *B* for 90 min. Line *A*, dA-ended, plus dT-ended SV40 linears, plus (P+L+III) (³²P, ●; ³H, ○); line *B*, dT-ended, SV40 omitted, plus (P+L+III) (³²P, ▼); line *C*, dA-ended SV40 omitted, plus (P+L+III) (³²P, ■); line *D*, dA-ended plus dT-ended SV40 linears, without (P+L+III) (³²P, ▲). ³H profiles are not shown for lines *B*, *C*, and *D*, but all show that the SV40 DNA sediments in its normal monomeric position. The ³²P and ³H profiles in line *A* are shifted to a faster-sedimenting position with respect to the ³²P profile in line *D* because SV40 strands are covalently linked to one another through (dA)$_{80}$ or (dT)$_{80}$ bridges in most of the molecules, whether or not covalently closed-circles are formed. Very little ³²P remains associated with the SV40 DNA in lines *B* and *C* because tails that remain single-stranded are degraded to 5′-mononucleotides by the 3′- to 5′-exonuclease activity of *E. coli* DNA polymerase I (7).

The *arrows* indicate the position in the gradient of different size supercoiled marker DNAs; the *number* is the multiple of SV40 DNA molecular size (1.0).

nucleotides per 5′-end (see *Methods*), the number of SV40(L$_{RI}$) molecules that could be bound to poly(U) filters after incubation with terminal transferase and dATP increased 5- to 6-fold. Even after separation of the strands of the SV40(L$_{RI}$exo)-dA, a substantial proportion of the ³H-label in the DNA was still bound by the poly(U) filter, indicating that both 3′-hydroxy termini in the duplex DNA can serve as primers.

The weight-average length of the homopolymer extensions was 50–100 residues per end. Zone sedimentation of [³H]-SV40(L$_{RI}$exo)-[³²P](dA)$_{80}$ (this particular preparation, which is described in *Methods*, had on the average, 80 dA residues per end) in an alkaline sucrose gradient showed that (*i*) 60–70% of the SV40 DNA strands are intact, (*ii*) the [³²P](dA)$_{80}$ is covalently attached to the [³H]SV40 DNA, and (*iii*) the distribution of oligo(dA) chain lengths attached to the SV40 DNA is narrow, indicating that the deviation from the calculated mean length of 80 is small (Fig. 2). SV40(L$_{RI}$exo), having (dT)$_{80}$ extensions, was prepared with [³²P]dTTP and gave essentially the same results when analyzed as described above.

Hydrogen-Bonded Circular Molecules Are Formed by Annealing SV40(L$_{RI}$exo)-(dA)$_{80}$ and SV40(L$_{RI}$exo)-(dT)$_{80}$ Together. When SV40(L$_{RI}$exo)-(dA)$_{80}$ and SV40(L$_{RI}$exo)-(dT)$_{80}$ were annealed together, 30–60% of the molecules seen by electron microscopy were circular dimers; linear monomers, linear dimers, and more complex branched forms were also seen. If SV40(L$_{RI}$exo)-(dA)$_{80}$ or -(dT)$_{80}$ alone was annealed, no circles were found. Centrifugation of annealed preparations in neutral sucrose gradients showed that the bulk of the SV40 DNA sedimented faster than modified unit-length linears (as would be expected for circular and linear dimers, as well as for higher oligomers). Sedimentation in alkaline gradients, however, showed only unit-length single strands containing the oligonucleotide tails (as seen in Fig. 2).

Covalently Closed-Circular DNA Molecules Are Formed by Incubation of Hydrogen-Bonded Complexes with DNA Polymerase, Ligase, and Exonuclease III. The hydrogen-bonded complexes described above can be sealed by incubation with the *E. coli* enzymes DNA polymerase I, ligase, and exonuclease III, plus their substrates and cofactors. Zone sedimentation in alkaline sucrose gradients (Fig. 3) shows that 20% of the

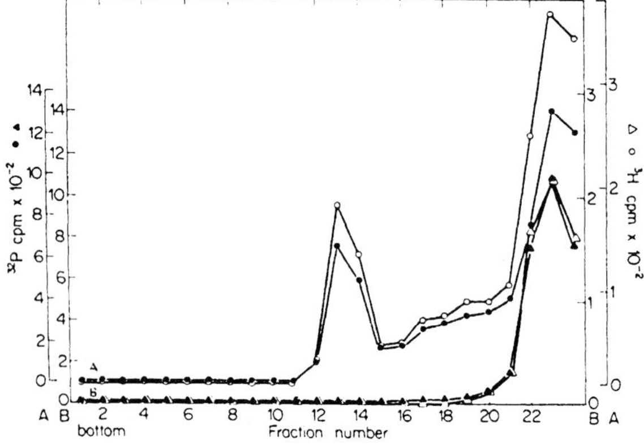

FIG. 4. CsCl–ethidium bromide equilibrium centrifugation of the products analyzed in Fig. 4. Line *A*, dA-ended, plus dT-ended SV40 linears, plus (P+L+III) (³²P, ●; ³H, ○); line *B*, the same mixture without (P+L+III) (³²P, ▲; ³H, △).

TABLE 1. *Relative lengths of SV40 and λdvgal-120 DNA molecules*

DNA species	Length ± standard deviation in SV40 units*	Number of molecules in sample
SV40(II)	1.00	224
SV40(L$_{RI}$)†	1.00 ± 0.03	108
(SV40–dA dT)$_2$	2.06 ± 0.19	23
λdvgal-120(I)	4.09 ± 0.14	65
λdvgal-120(L$_{RI}$)	2.00 ± 0.04	163
λdvgal-SV40	2.95 ± 0.04	76
λdv-1	2.78 ± 0.05	13

* The contour length of plaque-purified SV40(II) DNA is defined as 1.00 unit.

† Data supplied by J. Morrow.

input ³²P label derived from the oligo(dA) and -(dT) tails sediments with the ³H label present in the SV40 DNA, in the position expected of a covalently closed-circular SV40 dimer (70–75 S). About the same amount of labeled DNA bands in a CsCl–ethidium bromide gradient at a buoyant density characteristic of covalently closed-circular DNA (Fig. 4).

DNA isolated from the heavy band of the CsCl–ethidium bromide gradient contains primarily circular molecules, with a contour length twice that of SV40(II) DNA (Table 1) when viewed by electron microscopy. No covalently closed DNA is formed if either one of the linear precursors is omitted from the annealing step or if the enzymes are left out of the closure reaction. We conclude, therefore, that two unit-length linear SV40 molecules have been joined to form a covalently closed-circular dimer.

Covalent closure of the hydrogen-bonded SV40 DNA dimers is dependent on Mg²⁺, all four deoxynucleoside triphosphates, *E. coli* DNA polymerase I, and ligase, and is inhibited by 98% if exonuclease III is omitted (Lobban and Kaiser first observed the need for exonuclease III in the joining of P22 molecules; we confirmed their finding with this system). Exonuclease III is probably needed to remove 3′-phosphate groups from 3′-phosphoryl, 5′-hydroxyl nicks introduced by the endonuclease contaminating the terminal transferase preparation. 3′-phosphoryl groups are potent inhibitors of *E. coli* DNA polymerase I (14) and termini having 5′-hydroxyl groups cannot be sealed by *E. coli* ligase (8). The 5′-hydroxyl group can be removed and replaced by a 5′-phosphoryl group by the 5′- to 3′-exonuclease activity of *E. coli* DNA polymerase I (7).

Preparation of the Galactose Operon for Insertion into SV40 DNA. The galactose operon of *E. coli* was obtained from a λdvgal DNA; λdvgal is a covalently closed, supercoiled DNA molecule four times as long as SV40(II) DNA (Table 1). After complete digestion of λdvgal DNA with the R$_I$ endonuclease, linear molecules two times the length of SV40(II) DNA are virtually the exclusive product (Table 1). This population has a unimodal length distribution by electron microscopy and appears to be homogeneous by ultracentrifugal criteria (Jackson and Berg, in preparation). The R$_I$ endonuclease seems, therefore, to cut λdvgal circular DNA into two equal length linear molecules. Since one R$_I$ endonuclease cleavage per λdv monomeric unit occurs in the closely related λdv-204 (Jackson and Berg, in preparation), it is likely that λdvgal is cleaved at the

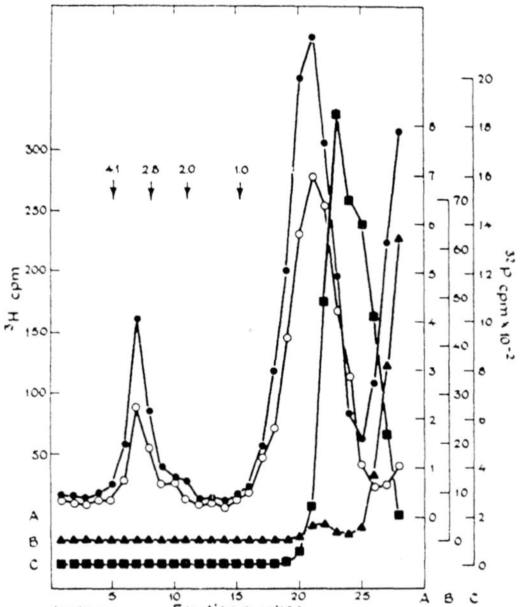

FIG. 5. Alkaline sucrose gradient sedimentation of annealed [³H]SV40(L$_{RI}$exo)–[³²P](dA)$_{80}$ and [³H]λ*dvgal–120*) (L$_{RI}$exo)–[³²P](dT)$_{80}$ incubated for 3 hr with and without (P+L+III). Centrifugation was for 60 min. Line *A*, dA-ended SV40, plus dT-ended λ*dvgal–120* linears, plus (P+L+III)(³²P, ●; ³H, ○); line *B*, dT-ended λ*dvgal–120* linears, plus dT-ended SV40 linears, plus (P+L+III) (³²P, ▲); line *C*,dA-ended SV40 linears, plus dT-ended λ*dvgal–120* linears, without (P+L+III) (³²P, ■).

The *arrows* indicate the position in the gradient of supercoiled marker DNAs having the indicated multiple of SV40 DNA molecular size.

same sites and, therefore, that each linear piece contains an intact galactose operon.

The purified λ*dvgal* (L$_{RI}$) DNA was prepared for joining to SV40 DNA by treatment with λ-exonuclease, followed by terminal transferase and [³²P]dTTP, as described for SV40-(L$_{RI}$).

Formation of Covalently Closed-Circular DNA Molecules Containing both SV40 and λdvgal DNA. Annealing of [³H]-SV40(L$_{RI}$exo)–[³²P](dA)$_{80}$ with [³H]λ*dvgal*(L$_{RI}$exo)–[³²P]-(dT)$_{80}$, followed by incubation with the enzymes, substrates, and cofactors needed for closure, produced a species of DNA (in about 15% yield) that sedimented rapidly in alkaline sucrose gradients (Fig. 5) and that formed a band in a CsCl-ethidium bromide gradient at the position expected for covalently closed DNA (Fig. 6). The putative λ*dvgal*–SV40 circular DNA sediments just ahead of λ*dv*-1, a supercoiled circular DNA marker [2.8 times the length of SV40(II)DNA], and behind λ*dvgal* supercoiled circles [4.1 times SV40(II)DNA] in the alkaline sucrose gradient. Electron microscopic measurements of the DNA recovered from the dense band of the CsCl-ethidium bromide gradient showed a mean contour length for the major species of 2.95 ± 0.04 times that of SV40(II) DNA (Table 1). Each of these measurements supports the conclusion that the newly formed, covalently closed-circular DNA contains one SV40 DNA segment and one λ*dvgal* DNA monomeric segment.

Omission of the enzymes from the reaction mixture prevents λ*dvgal*–SV40 DNA formation (Figs. 5 and 6). No covalently closed product is detectable (Fig. 5) if λ*dvgal* and SV40 linear molecules with identical, rather than complementary, tails are annealed and incubated with the enzymes. This result demonstrates directly that the formation of covalently closed DNA depends on complementarity of the homopolymeric tails.

We conclude from the experiments described above that λ*dvgal* DNA containing the intact galactose operon from *E. coli*, together with some phage λ genes, has been covalently inserted into an SV40 genome. These molecules should be useful for testing whether these bacterial genes can be introduced into a mammalian cell genome and whether they can be expressed there.

DISCUSSION

The methods described in this report for the covalent joining of two SV40 molecules and for the insertion of a segment of DNA containing the galactose operon of *E. coli* into SV40 are general and offer an approach for covalently joining any two DNA molecules together. With the exception of the fortuitous property of the R$_I$ endonuclease, which creates convenient linear DNA precursors, none of the techniques used depends upon any unique property of SV40 and/or the λ*dvgal* DNA. By the use of known enzymes and only minor modifications of the methods described here, it should be possible to join DNA molecules even if they have the wrong combination of hydroxyl and phosphoryl groups at their termini. By judicious use of generally available enzymes, even DNA duplexes with protruding 5′- or 3′-ends can be modified to become suitable substrates for the joining reaction.

One important feature of this method, which is different from some other techniques that can be used to join unrelated DNA molecules to one another (16, 19), is that here the joining is directed by the homopolymeric tails on the DNA. In our protocol, molecule A and molecule B can only be joined to each other; all AA and BB intermolecular joinings and all A and B intramolecular joinings (circularizations) are prevented. The yield of the desired product is thus increased, and subsequent purification problems are greatly reduced.

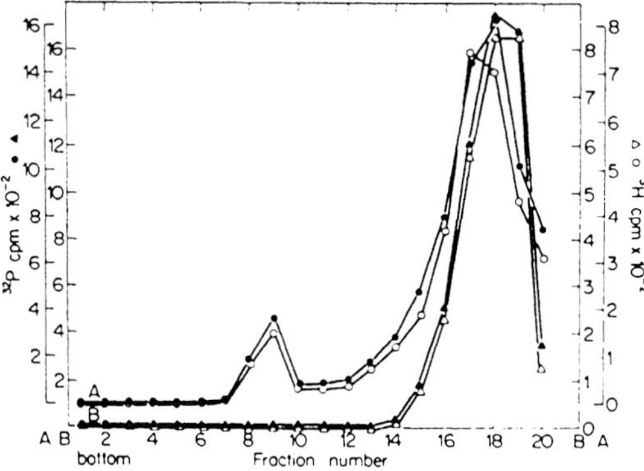

FIG. 6. CsCl–ethidium bromide equilibrium centrifugations of joined [³H]SV40(L$_{RI}$exo)–[³²P](dA)$_{80}$ and [³H]λ*dvgal–120*(L$_{RI}$exo)–[³²P](dT)$_{80}$ DNA. The samples were those referred to in Fig. 5. Line *A*, dA-ended SV40 linears, plus dT-ended λ*dvgal–120* linears, plus (P+L+III) (³²P, ●; ³H, ○); line *B*, the same mixture without (P+L+III) (³²P, ▲; ³H, △).

For some purposes, however, it may be desirable to insert λ*dvgal* or other DNA molecules at other specific, or even random, locations in the SV40 genome. Other specific placements could be accomplished if other endonucleases could be found that cleave the SV40 circular DNA specifically. Since pancreatic DNase in the presence of Mn^{2+} produces randomly located, double-strand scissions (17) of SV40 circular DNA (Jackson and Berg, in preparation), it should be possible to insert a DNA segment at a large number of positions in the SV40 genome.

Although the λ*dvgal* DNA segment is integrated at the same location in each SV40 DNA molecule, it should be emphasized that the orientation of the two DNA segments to each other is probably not identical. This follows from the fact that each of the two strands of a duplex can be joined to *either* of the two strands of the other duplex $\left(e.g., \begin{smallmatrix} W & W \\ C & C \end{smallmatrix} \text{ or } \begin{smallmatrix} W & C \\ C & W \end{smallmatrix} \right)$§. What possible consequences this fact has on the genetic expression of these segments remains to be seen.

We have no information concerning the biological activities of the SV40 dimer or the λ*dvgal*–SV40 DNAs, but appropriate experiments are in progress. It is clear, however, that the location of the R_I break in the SV40 genome will be crucial in determining the biological potential of these molecules; preliminary evidence suggests that the break occurs in the late genes of SV40 (Morrow, Kelly, Berg, and Lewis, in preparation).

A further feature of these molecules that may bear on their usefulness is the $(dA \cdot dT)_n$ tracts that join the two DNA segments. They could be helpful (as a physical or genetic marker) or a hindrance (by making the molecule more sensitive to degradation) for their potential use as a transducer.

The λ*dvgal*–SV40 DNA produced in these experiments is, in effect, a trivalent biological reagent. It contains the genetic information to code for most of the functions of SV40, all of the functions of the *E. coli* galactose operon, and those functions of the λ bacteriophage required for autonomous replication of circular DNA molecules in *E. coli*. Each of these

---

§ The symbols W and C refer to one or the other complementary strands of a DNA duplex, and the "connectors" indicate how the strands can be joined in the closed-circular duplex.

sets of functions has a wide range of potential uses in studying the molecular biology of SV40 and the mammalian cells with which this virus interacts.

We are grateful to Peter Lobban for many helpful discussions. D. A. J. was a Basic Science Fellow of the National Cystic Fibrosis Research Foundation; R. H. S. was on study leave from the Department of Biochemistry, University of Adelaide, Australia and was supported in part by a grant from the USPHS. This research was supported by Grant GM-13235 from the USPHS and Grant VC-23A from the American Cancer Society.

1. Sambrook, J., Westphal, H., Srinivasan, P. R. & Dulbecco, R. (1968) *Proc. Nat. Acad. Sci. USA* **60**, 1288–1295.
2. Dulbecco, R. (1969) *Science* **166**, 962–968.
3. Matsubara, K., & Kaiser, A. D. (1968) *Cold Spring Harbor Symp. Quant. Biol.* **33**, 27–34.
4. Radloff, R., Bauer, W., Vinograd, J. (1967) *Proc. Nat. Acad. Sci. USA* **57**, 1514–1521.
5. Little, J. W., Lehman, I. R. & Kaiser, A. D. (1967) *J. Biol. Chem.* **242**, 672–678.
6. Kato, K., Goncalves, J. M., Houts, G. E., & Bollum, F. J. (1967) *J. Biol. Chem.* **242**, 2780–2789.
7. Jovin, T. M., Englund, P. T. & Kornberg, A. (1969) *J. Biol. Chem.* **244**, 2996–3008.
8. Olivera, B. M., Hall, Z. W., Anraku, Y., Chien, J. R. & Lehman, I. R. (1968) *Cold Spring Harbor Symp. Quant. Biol.* **33**, 27–34.
9. Richardson, C. C., Lehman, I. R. & Kornberg, A. (1964) *J. Biol Chem.* **239**, 251–258.
10. Symons, R. H. (1969) *Biochim. Biophys. Acta* **190**, 548–550.
11. Davis, R., Simon, M. & Davidson, N. (1971) in *Methods in Enzymology*, eds. Grossman, L. & Moldave, K. (Academic Press, New York), Vol. 21, pp. 413–428.
12. Chang, L. M. S. & Bollum, F. J. (1971) *Biochemistry* **10**, 536–542.
13. Sheldon, R., Jurale, C. & Kates, J. (1972) *Proc. Nat. Acad. Sci. USA* **69**, 417–421.
14. Richardson, C. C., Schildkraut, C. L. & Kornberg, A. (1963) *Cold Spring Harbor Symp. Quant. Biol.* **28**, 9–19.
15. Little, J. W. (1967) *J. Biol. Chem.* **242**, 679–686.
16. Sgaramella, V., van de Sande, J. H. & Khorana, H. G. (1970) *Proc. Nat. Acad. Sci. USA* **67**, 1468–1475.
17. Melgar, E. & Goldthwait, D. A. (1968) *J. Biol. Chem.* **243**, 4409–4416.
18. Morrow, J. F. & Berg, P. (1972) *Proc. Nat. Acad. Sci. USA* **69**, in press.
19. Sgaramella, V. & Lobban, P. (1972) *Nature*, in press.

A Map of Simian Virus 40 Transcription Sites Expressed in Productively Infected Cells

G. Khoury, M. A. Martin, T. N. H. Lee, K. J. Danna, and D. Nathans

All that was known in the early 1970s about viral transcription programs was that there was early mRNA and, once progeny genomes were available, late mRNA. Except for some early mRNA species that were also transcribed during the late period, these were nonoverlapping populations, as revealed by hybridization competition analysis. These insights were gained first from work with bacteriophage and then, in the mid 1960s, with poxviruses, followed by adenoviruses and herpesviruses. The discovery of restriction endonucleases provided the opportunity for a far more definitive characterization of transcription programs; now viral genomes could be split into any number of fragments with precisely known locations, and mRNA populations at any stage of the replication cycle could be tested for ability to hybridize with any of them. The leading player in this field was Dan Nathans, as demonstrated in the beautiful analysis presented in this paper. This type of analysis has now been duplicated an infinite number of times in this or related forms. Dan Nathans was awarded the Nobel Prize in 1978.

W. K. Joklik

Reprinted from *Journal of Molecular Biology* 78:377–389. Copyright © 1973, by permission of the publisher, Academic Press.

J. Mol. Biol. (1973) **78**, 377–389

A Map of Simian Virus 40 Transcription Sites Expressed in Productively Infected Cells

George Khoury, Malcolm A. Martin

Laboratory of Biology of Viruses
National Institute of Allergy and Infectious Diseases
National Institutes of Health
Bethesda, Md 20014, U.S.A.

Theresa N. H. Lee, Kathleen J. Danna and Daniel Nathans

Department of Microbiology
The John Hopkins University School of Medicine
Baltimore, Md 21205, U.S.A.

(*Received 31 January 1973, and in revised form 12 April 1973*)

The topographical location of "early" and "late" sequences on the physical map of the simian virus 40 genome was determined by reacting unlabeled RNA from monkey cells productively infected with simian virus 40 with the separated strands of each of the 11 DNA fragments formed by digesting supercoiled virus DNA with the restriction endonuclease of *Hemophilus influenzae*. Stable species of early RNA are complementary to the minus strands of the contiguous fragments A, H, I and B, while the late RNA is transcribed predominantly from the plus strands of fragments A, C, D, E, K, F, J, G and B, which also form a continuous set on the physical map. This result is in agreement with previous findings, which indicated that transcription before DNA synthesis is confined to the minus DNA strand, while late transcription includes sequences located predominantly on the plus strand.

The direction of DNA transcription on the plus and minus template strands was ascertained by first preparing a population of linear simian virus 40 DNA molecules oriented by cleavage with the *Escherichia coli* R_1 restriction endonuclease and then digesting them with *E. coli* exonuclease III. DNA reassociation studies using the separated strands of these "half-molecules" and specific simian virus 40 DNA fragments indicated that transcription proceeds from A → H → I → B on the minus DNA strand, and in the opposite direction on the plus DNA strand.

1. Introduction

Transcription of the simian virus 40 genome in productively-infected cells occurs in two recognizable stages. Before the onset of virus DNA synthesis a portion of the genome is transcribed into stable "early" RNA (Aloni *et al.*, 1968; Oda & Dulbecco, 1968; Sauer & Kidwai, 1968; Carp *et al.*, 1969). Following the onset of virus DNA synthesis, early RNA synthesis continues and is accompanied by the appearance of a new species of RNA; together, these RNA species are complementary to the equivalent of one strand of SV40† DNA (Aloni *et al.*, 1968; Martin & Axelrod, 1969). Recently it has been found that early SV40 RNA is transcribed from 30 to 40% of one strand of the virus DNA (designated the minus strand), while the specifically late SV40 RNA (i.e. RNA found only after the onset of virus DNA synthesis) is transcribed from 60 to 70% of the plus strand (Khoury & Martin, 1972; Lindstrom & Dulbecco, 1972; Khoury *et al.*, 1972; Sambrook *et al.*, 1972). Although these results have clearly identified the early and late template strands and have indicated what fraction of each strand is transcribed into stable virus RNA, they do not indicate the topology of early and late regions of the SV40 genome.

To localize the early and late template regions we have made use of specific fragments of SV40 DNA generated by bacterial restriction endonucleases. Of most use have been the 11 fragments produced by *Hemophilus influenzae* restriction endonuclease (Danna & Nathans, 1971), which vary in size from 22·5% to 4% of the length of the SV40 genome. The order of these fragments in the SV40 DNA molecule is now known (Danna & Nathans, 1972), as is their relation to the cleavage sites of other restriction enzymes, thus providing a physical map of the SV40 genome (Danna *et al.*, 1973). We have used these enzyme cleavage products to determine the regions of the plus and minus strands of SV40 DNA that are transcribed into stable RNA in productively infected cells, and also to determine the direction of early and late transcription. In a subsequent communication, we shall present a similar analysis of SV40 transcription in virus transformed cells.

2. Materials and Methods

(a) *Preparation of virus and virus DNA*

Small plaque SV40 (from strain 776, obtained from K. Takemoto) was plaque purified and a stock prepared in the BSC-1 line of African green monkey kidney cells, as described previously (Danna & Nathans, 1971). ^{32}P-labeled SV40 DNA I was obtained from virus purified by isopycnic centrifugation in CsCl as described previously (Yoshiike, 1968; Trilling & Axelrod, 1970). The specific activities of the labeled DNA preparations varied from 1 to 4×10^5 cts/min per μg. ^{14}C-labeled SV40 DNA I was extracted from infected monkey cells by differential salt precipitation (Hirt, 1967) and purified by equilibrium density centrifugation in CsCl/ethidium bromide (Radloff *et al.*, 1967) followed by sedimentation through a neutral sucrose gradient; the specific activity was 3×10^4 cts/min per μg.

(b) *Digestion of SV40 DNA with* H. influenzae *restriction endonuclease*

To obtain specific fragments of SV40 DNA, [^{32}P] or [^{14}C]DNA I was incubated with *H. influenzae* restriction endonuclease in a mixture containing 25 to 70 μg of DNA, 0·03 unit of enzyme (Smith & Wilcox, 1970), 50 mM-NaCl, and 6 mM each of $MgCl_2$ and Tris·HCl (pH 7·5), in a volume of 0·10 ml, as described previously (Danna & Nathans, 1972). After incubation at 37°C for 2 h, the reaction was complete as determined by electrophoresis and autoradiography of a portion of the reaction mixture (see below).

(c) *Isolation of DNA fragments*

The 11 fragments of SV40 DNA present in an *H. influenzae* enzyme digest (fragments A to K) were separated by electrophoresis in 4% polyacrylamide slab gels measuring 15 cm $\times$ 40 cm $\times$ 0·16 cm, as described elsewhere (Danna & Nathans, 1972). ^{32}P-labeled

† Abbreviation used: SV40, simian virus 40.

fragments were detected by autoradiography of the wet gel slab and DNA was eluted from segments of gel containing individual fragments with 0·1 times SSC (SSC is 0·15 M-NaCl, 0·015 M-sodium citrate). To localize ^{14}C-labeled fragments, the ^{14}C-labeled digest was mixed with a small amount of ^{32}P-labeled digest and the above procedure followed. The ^{32}P was then allowed to decay to undetectable levels before use of the ^{14}C-labeled fragments. Individual ^{14}C-labeled fragments were then tested for purity by electrophoresis and autoradiography of dried gel slabs; all fragments were pure by this criterion.

(d) *Cleavage of SV40 DNA with* E. coli R_I *restriction endonuclease*

E. coli R_I restriction endonuclease (Yoshimori, 1971) generously supplied by H. Boyer, was incubated with ^{32}P-labeled SV40 DNA I as previously described (Danna *et al.*, 1973). The product had a sedimentation rate of 14·5 S in a 5% to 20% neutral sucrose gradient, corresponding to full length linear molecules (Morrow & Berg, 1972).

(e) *Exonuclease III digestion of* R_I *linear DNA*

The ^{32}P-labeled linear DNA product of the *E. coli* R_I cleavage was incubated at 35°C with 10 units of *E. coli* exonuclease III (kindly provided by H. Gerry) in a reaction mixture containing 5 μg of DNA, 5 mM-MgCl$_2$, 5 mM-β-mercaptoethanol, 50 μg bovine serum albumin/ml, and 50 mM-Tris·HCl (pH 7·4) in a volume of 0·21 ml. Portions of 5 μl of the reaction mixture were removed at various times and the percentage of DNA soluble in cold 5% trichloroacetic acid determined. The values observed were $< 0·1\%$, 24%, 45% and 49% at 0, 20, 40 and 60 min, respectively. The reaction was stopped at 60 min by addition of an equal vol. of water-saturated phenol and the aqueous layer plus two 0·2-ml washes of the phenol layer were extracted 5 times with ether. The ether was removed by a stream of nitrogen and the solution was then dialyzed against 0·05 M-NaCl. The exonuclease-treated DNA sedimented in an alkaline sucrose gradient as a broad band with an S value of 13·5 (expected value for half-length strands is 11·5 S), suggesting some variability in the extent of exonuclease digestion of individual molecules. Since exonuclease III digests duplex DNA from the 3′ end, this preparation will be referred to as 5′ "half molecules".

(f) *Preparation of RNA*

"Early lytic" RNA was prepared from BSC-1 cells 24 to 36 h after infection with SV40 (multiplicity of 40 to 80 plaque-forming units/cell) in the presence of 20 μg arabinosyl cytosine/ml as described previously (Khoury *et al.*, 1972). Late lytic RNA was purified from BSC-1 cells 40 to 60 h after infection using the same multiplicity of virus. SV40 complementary RNA was prepared *in vitro* with *E. coli* RNA polymerase and SV40 DNA I as described previously (Khoury & Martin, 1972).

(g) *Separation and purification of SV40 DNA strands*

The ^{32}P-labeled fragments of SV40 DNA, obtained by digestion with the *H. influenzae* restriction endonuclease, were dialyzed against 10 mM-Tris·HCl (pH 7·5), and 10 mM-NaCl. Reaction mixtures containing approximately 0·05 to 0·15 μg of a specific, heat-denatured DNA fragment, 10 μg of complementary RNA, 10 mM-NaCl, 10 mM-Tris·HCl in 1·5 ml were incubated at 60° for 1 h. The plus and minus strands of each of the 11 fragments were then separated on hydroxyapatite and were purified by methods described previously (Khoury *et al.*, 1972).

(h) *DNA-RNA hybridization*

Small amounts (1 to 5 ng) of the plus or minus strand of each specific SV40 DNA fragment were incubated with early (0·2 to 0·5 mg) or late (0·1 to 0·3 mg) lytic RNA in 1·0 M-NaCl, 80 mM-phosphate buffer and 0·2% sodium dodecyl sulfate in 0·15 to 0·25 ml at 68°C for 18 to 24 h. DNA–RNA hybrid formation was monitored on hydroxyapatite (Kohne, 1969). DNA–RNA hybrids formed with fragments A or B, and lytic RNA, were also examined by treatment with the single-strand specific nuclease, S$_1$, as has been described in a previous publication (Khoury *et al.*, 1973). The reaction mixtures for S$_1$ nuclease analysis were diluted to 3 ml to give a final concentration of 300 mM-NaCl, 10^{-2} mM-ZnSO$_4$, 30 mM-sodium acetate buffer (pH 4·0) and 30 μg denatured salmon sperm DNA/ml, and divided into 3 equal portions, each containing about 350 cts/min of ^{32}P-labeled DNA. 2 of the samples were treated with an excess of the S$_1$ enzyme and the third served as a control. The fraction of the DNA resistant to the single strand specific enzyme was determined from the average of the nuclease resistant radioactivity in the 2 treated samples compared to the untreated sample.

(i) *DNA–DNA hybridization*

DNA–DNA hybridization was carried out with ^{14}C-labeled denatured fragments of SV40 DNA and the separated strands of the ^{32}P-labeled 5′ half-molecules produced by sequential cleavage of DNA I with *E. coli* R$_I$ endonuclease and exonuclease III. The strands of 5′ half molecules were separated as described above for the separation of the strands of DNA fragments. These separated strands will be referred to as the plus and minus 5′ half strands. For hybridization, approximately 1 to 4 ng of either the plus or minus 5′ half strand was incubated with approximately 20 to 40 ng of a specific heat-denatured ^{14}C-labeled DNA fragment in 1·2 M-NaCl, 100 mM-phosphate buffer, 0·3% sodium dodecyl sulfate for 24 h at 68°C. The extent of DNA reassociation was then analyzed on hydroxyapatite (Britten & Kohne, 1968).

(j) *Detection of radioactivity*

After the addition of 0·2 mg of carrier yeast RNA, samples were adjusted to 5% trichloroacetic acid and collected on nitrocellulose membrane filters (B-6, Carl Schleicher and Schuell Co., Keene, N. H.). The filters were washed with 10 ml of cold 0·01 N-HCl, dried, and counted in a toluene-based scintillation fluid. A background of 20 cts/min was subtracted from each result.

3. Results

(a) *Separation of fragment strands*

The plus and minus strands of each of the 11 fragments (A to K), produced by digestion of SV40 DNA I with the *H. influenzae* restriction endonuclease, were separated by incubating denatured fragments with an excess of SV40 complementary RNA followed by chromatography on hydroxyapatite, as described previously for unfractionated virus DNA (Khoury *et al.*, 1972). Less than 4% of the purified separated strands of each fragment "self-associated" when incubated under the conditions used for DNA–RNA hybridization. The strand which reacted with SV40 complementary RNA has been designated the minus strand, and the other, the plus strand.

(b) *Hybridization of the plus and minus strands of DNA fragments with RNA from SV40-infected BSC-1 cells*

As shown previously, 30 to 40% of the minus strand of SV40 DNA is expressed early in a productive infection and 60 to 70% of the plus strand is expressed late in infection (Lindstrom & Dulbecco, 1972; Khoury *et al.*, 1972; Sambrook *et al.*, 1972). In order to determine which of the 11 DNA fragments contained the sequences complementary to early RNA and to late RNA, the individual strands of each DNA fragment were incubated with early or late lytic RNA. The percentage of each [^{32}P] DNA fragment which hybridized to added RNA was then determined by hydroxyapatite chromatography. The results of these experiments, shown in Table 1, indicated a complex pattern of transcription.

TABLE 1

Hybridization of early or late lytic RNA to the separated strands of unique SV40 DNA fragments

DNA fragment	[^{32}P]DNA in hybrid molecules (%)			
	(−) DNA strand		(+) DNA strand	
	Early RNA	Late RNA	Early RNA	Late RNA
A	87	91	9	83
B	87	85	12	87
C	9	46	15	90
D	5	14	19	88
E	8	12	16	90
F	3	7	21	91
G	10	14	16	91
H	39	51	6	35
I	23	30	7	46
J	10	10	14	85
K	5	7	12	87

RNA from SV40-infected BSC-1 cells (early, 0·2 to 0·5 mg; late, 0·1 to 0·3 mg) was hybridized under the conditions described in Materials and Methods to the plus or minus strand of each SV40 DNA fragment produced by cleavage with the *H. influenzae* restriction endonuclease. Samples containing 250 to 400 cts/min were analyzed by hydroxyapatite chromatography to determine the percentage of [^{32}P]DNA in hybrid molecules. In the absence of RNA, this value was less than 4% for either strand of the 11 fragments. Similarly, low levels of hybridization (<2%) were found when uninfected monkey cell RNA was incubated with either strand of a control fragment (E).

Focusing first on the reaction between early RNA and the minus strands of DNA fragments, which are derived from the early template strand (column 1 of Table 1) we see that only fragments A, B, H and I hybridized with early RNA to a significant extent. At the level of RNA used, fragments A and B were essentially saturated, whereas H and I were not. Therefore, we conclude that at least a portion of the SV40 genome corresponding to each of the A, B, H and I fragments is transcribed into early SV40 RNA, and that stable "early" transcripts from segments A and B are probably more abundant than those from H and I.

Considering next the reaction between late RNA and the plus strands of the DNA fragments, which are derived from the specifically late template strand (column 4 of Table 1), we see that all fragments hybridized with late RNA. At the level of RNA used, fragments A, B, C, D, E, F, G, J and K were essentially saturated, whereas fragments H and I were not. We conclude from these results that at least a portion of the SV40 genome corresponding to each of the fragments is transcribed into late RNA and that stable late transcripts from A, B, C, D, E, F, G, J and K are probably more abundant than those from H and I.

The reaction between the minus strands of DNA fragments and late RNA (column 2 of Table 1) shows the expected hybridization with fragments A, B, H and I since early SV40 RNA species are known to be present during the late phase of productive infection (Aloni *et al.*, 1968; Oda & Dulbecco, 1968; Sauer & Kidwai, 1968; Carp *et al.*, 1969). In addition, the minus strand of fragment C is unique in that it hybridizes with late SV40 RNA but not with early RNA. The possible significance of this last result will be considered in the Discussion.

Also shown in Table 1 (column 3) is the reaction between early RNA and the plus strands of DNA fragments. As shown in the Table, the plus strands of several fragments showed a low level of hybridization with early RNA. The significance of the low levels of reaction between plus (late template) strands of some of the other fragments and early RNA is not clear (see Discussion).

(c) *Hybridization of fragments H and I with different concentrations of RNA*

The minus strands of fragments A and B and the plus strands of all fragments except H and I were saturated by the amounts of late lytic RNA used in this study. The low levels of hybridization of both the plus and minus strands of fragments H and I with late RNA (Table 1) suggested that stable transcripts from these regions of SV40 DNA may be less abundant than transcripts from other regions of the genome. To explore this possibility further, the hybridization reaction with these fragments was done with different concentrations of late RNA. As shown in Table 2, the extent

TABLE 2

Effect of RNA concentration on hybridization experiments with
DNA fragments H and I

DNA fragment	Strand	RNA concentration† (mg/ml)	Duplex molecules (%)
H	+	2·4	16
		3·2	35
H	−	2·4	39
		3·2	51
I	+	2·4	27
		3·2	46
I	−	2·4	24
		3·2	30
J	+	1·6	85
J	−	1·6	10
K	+	1·2	87
K	−	1·2	7

Under conditions described in Materials and Methods, 1 to 4 ng (approx. 300 cts/min) of either strand of a DNA fragment was incubated with the indicated concentrations of late lytic RNA. The percentage of radioactivity in duplex molecules was analyzed by hydroxyapatite chromatography.

† Total cellular RNA obtained late in the productive cycle of SV40 infection of BSC-1 cells.

of hybridization of each strand of fragment H and of fragment I was dependent on the concentration of late RNA. This was the case even at RNA levels which were at least two to three times higher than that needed to saturate the plus strand of the other DNA fragments, including fragments J and K which are smaller than H and I. These data, together with those presented in Table 1, thus suggest that stable transcripts of both strands of fragments H and I are present in lower relative concentrations than RNA species transcribed from SV40 DNA sequences corresponding to the other nine DNA fragments. Alternatively, the results obtained with fragments H and I could point to a previously unsuspected heterogeneity of these fragments (see Discussion).

(d) *Estimation of the proportion of each strand of fragments A and B which hybridize with lytic RNA*

The results presented in Table 1 indicate that more than 85% of the plus and minus strands of fragments A and B reacted with late lytic RNA. In this experiment, the DNA–RNA reaction was monitored by hydroxyapatite chromatography which fails to discriminate between complete and partial DNA–RNA hybrids. For this reason we decided to use the S_1 single-strand specific nuclease assay to determine the fraction of the minus and plus strands which form DNA–RNA hybrids. Since early SV40 RNA sequences are represented in late lytic RNA, and since there is more SV40-specific RNA late after infection, late lytic RNA was used in these studies to insure that the DNA–RNA reactions had reached saturation.

The separated strands of ^{32}P-labeled fragments A and B were incubated with saturating amounts of late lytic RNA and then exposed to the S_1 nuclease as described in Materials and Methods. The proportion of either DNA strand of these fragments which formed hybrid structures with RNA was determined as the percentage of [^{32}P]DNA resistant to the S_1 enzyme. This value provided an estimate of the extent of transcription on each strand of fragment A or B (Table 3). Approximately 33% of the plus strand and 60% of the minus strand of fragment A appear to be transcribed, while 51% of the plus strand and 40% of the minus strand of fragment B are expressed. Since each of these sets adds up to nearly 100%, one interpretation of these results is that the minus and plus strand templates do not overlap.

TABLE 3

Extent of hybridization of lytic RNA to a specific DNA fragment

DNA fragment	Strand	Resistance of hybrid to S_1 enzyme (%)
A	+	33
A	−	60
B	+	51
B	−	40

Approximately 5 ng of the plus or minus strand of fragments A and B were incubated with 0·8 to 1·0 mg of late lytic RNA in 0·75 ml as described in Materials and Methods. After 24 h incubation, the sample was diluted to 3 ml and divided into thirds; 2 of these were treated with an excess of S_1 enzyme, while the third served as a control. The percentage resistance to S_1 nuclease is the average of the radioactive counts in the 2 treated samples compared to the control. The values in this Table represent the average of 2 of these experiments and differed by less than 10%

(e) *Map positions of early and late template regions*

Danna *et al.* (1973) have constructed a physical map of the SV40 genome based on the sites of cleavage of SV40 DNA by restriction endonuclease from *H. influenzae*, *H. parainfluenzae*, and *E. coli* (R_I) (Fig. 1). We can therefore relate the results of hybridization of DNA fragments with SV40 lytic RNA to the map position of individual fragments.

Early RNA hybridized to the minus strands of fragments A, H, I and B, which form a contiguous group in the map (Fig. 1). Since only 60% of the minus strand of A and

40% of the minus strand of B are expressed, the early template region appears to extend from near the middle of A to near the middle of B, i.e. about 30% of the length of the virus genome. This interpretation assumes that the DNA sequences transcribed into stable early RNA are entirely contiguous. Late lytic RNA, on the other hand, hybridized almost completely with the plus strands of fragments A, C, D, E, K, F, J, G and B. As seen in the cleavage map shown in Figure 1, these fragments also form a contiguous group. Since about 33% of the plus strand of A and about 51% of the plus strand of B are expressed, the late template region forms a segment encompassing about 70% of the length of the virus genome (Fig. 1). As noted above, very high concentrations of late lytic RNA hybridize with the plus strands of fragments H and I but the reaction is incomplete. This suggests that a minor species of late lytic RNA is transcribed from parts of the sequences corresponding to fragments H and I (see Discussion).

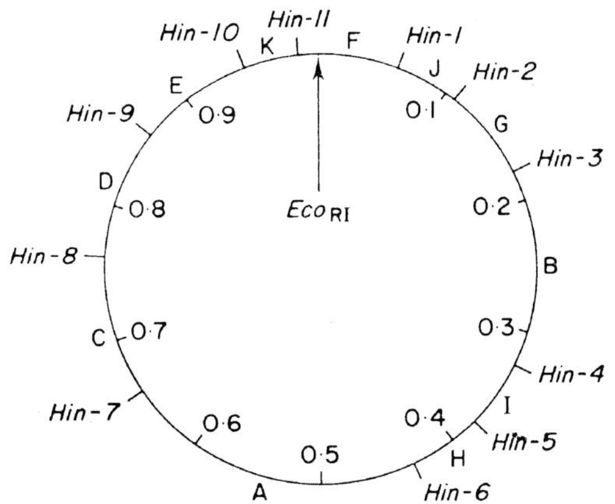

Fig. 1. A cleavage map of the SV40 genome (Danna *et al.*, 1973). The *E. coli* R_I endonuclease site (*Eco* R_I) is the zero point and map distances are given as fractional length of SV40 DNA in the direction F → J → G‒ ‒ ‒ ‒. *Hin-1, 2, 3*, etc. refer to the *H. influenzae* endonuclease cleavage sites. The map positions of these are: *Hin-1*, 0·060; *Hin-2*, 0·105; *Hin-3*, 0·175; *Hin-4*, 0·325; *Hin-5*, 0·375; *Hin-6*, 0·430; *Hin-7*, 0·655; *Hin-8*, 0·760; *Hin-9*, 0·860; *Hin-10*, 0·945; *Hin-11*, 0·985.

(f) *Direction of transcription*

Although the results just presented indicate the position of early and late template regions on the SV40 map, they do not indicate the direction of transcription of the early and late genes, i.e. clockwise or counter-clockwise transcription as the map is drawn. To determine the direction of transcription, it is sufficient to establish the 5′ to 3′ direction of the minus and plus strands of SV40 DNA relative to the cleavage map, since RNA is transcribed in the 3′ to 5′ direction of the template strand, and the minus and plus strands are known to be templates for early and late RNA, respectively.

In order to relate the 5′ to 3′ direction of the minus and plus strands of SV40 DNA to the cleavage map, we have carried out the experiment diagrammed in Figure 2. ^{32}P-labeled SV40 DNA I was first cleaved with the *E. coli* R_I restriction endonuclease to obtain unique, full length linear molecules. Since the R_I enzyme cleaves within fragment F (Danna *et al.*, 1973), these molecules have the map order F_1 J G B I H A

C D E K F$_2$ (see Fig. 2). In the next step, the linear molecules were digested with *E. coli* exonuclease III, which removed the 3′ halves of each strand, leaving 5′ half molecules. The individual minus and plus 5′ half strands were then isolated by annealing with SV40 complementary RNA. Finally, we determined which 5′ half strand (plus or minus) contains fragment J and fragment G sequences, and which contains fragment K and fragment E sequences by annealing each half strand with denatured ^{14}C-labeled fragments J, G, K or E. The sequences corresponding to these four fragments are nearest to sequence F, which is present at each end of the R$_I$ linear molecules (Fig. 2).

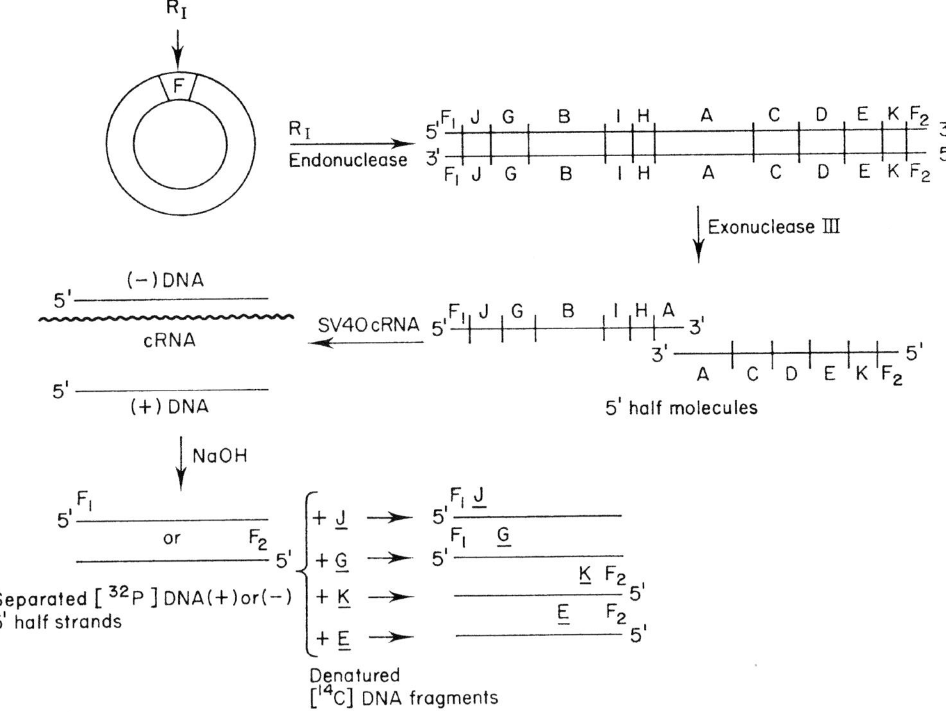

FIG. 2. Scheme for determining the 5′→3′ orientation of SV40 DNA strands. See the text for a description of each step. F$_1$ and F$_2$ are the parts of fragment F resulting from cleavage by the R$_1$ restriction enzyme. cRNA, complementary RNA.

The results of duplicate experiments are presented in Table 4. Plus or minus ^{32}P-labeled 5′ half strands were incubated with a tenfold excess of a single, denatured, ^{14}C-labeled DNA fragment (G, J, E or K) under annealing conditions, and the fraction of [^{32}P]DNA which formed duplexes with the [^{14}C]DNA was determined by hydroxyapatite chromatography. The results indicate that fragments E and K hybridize preferentially with the *plus* 5′ half strand, while fragments G and J hybridize preferentially with the *minus* 5′ half strand of SV40 DNA. In each case, more than 85% of the ^{14}C-labeled DNA fragments reassociated. Theoretically one would expect that nearly all of the ^{32}P-labeled DNA would anneal in the presence of an excess of the appropriate [^{14}C]DNA fragment if the half-strands had remained intact. However, when the separated ^{32}P-labeled 5′ half strands used in this experiment were examined in alkaline sucrose, they sedimented in the 5·4 to 10·7 S region of the gradient which corresponds to a single-stranded molecule 1·7 to 6·1 × 10^5 daltons in size, thus explaining why no more than 40% of the [^{32}P]DNA reassociated. The data, nevertheless, are

TABLE 4

Hybridization of the separated strands of ^{32}P*-labeled SV40 DNA half-molecules with specific* ^{14}C*-labeled SV40 DNA fragments*

Denatured [^{14}C]DNA fragment	[^{32}P]DNA in hybrid molecules (%)	
	(+) DNA strand	(−) DNA strand
G	6	40
J	11	35
E	40	20
K	30	15

Hybridization reactions were done as described in Materials and Methods. The percentage of [^{32}P]DNA in hybrid molecules was analyzed on hydroxyapatite as previously described. Values represent the average of 2 experiments and differed by less than 10%.

internally consistent, i.e. adjacent fragments E and K preferentially bind to one 5′ half strand (plus), while fragments G and J bind preferentially to the other (minus) 5′ half strand. Since the minus strand is the template for early SV40 RNA, we can conclude that the orientation of the early template strand is 5′ F J G B I H A . . . . . 3′, i.e. 5′—3′ clockwise on the cleavage map shown in Figure 1, and the orientation of the plus strand (late template strand) is 5′ F K E D C A . . . . 3′, i.e. 5′—3′ counterclockwise on the cleavage map. It follows, therefore, that transcription of early genes proceeds from A to B counter-clockwise on the minus DNA strand and transcription of late genes proceeds from A to B in a clockwise direction on the plus DNA strand.

4. Discussion

The major conclusions of this study are presented in the map shown in Figure 3. We have confirmed previous results indicating that early in productive infection, transcription of the SV40 genome occurs on the minus strand template. Later in infection, additional transcription occurs on the plus strand template. The early region of SV40 DNA includes segments of fragments A, B, H and I, while the late region of the virus genome encompasses polynucleotide sequences present in fragments C, D, E, K, F, J, G and parts of A and B. In addition, there is evidence for late transcription of at least part of the sequences corresponding to the plus strands of fragments H and I. These transcripts, however, are at a much reduced level compared to those from the other parts of the genome. Within fragments A and B, which have relatively abundant RNA species complementary to each strand, nearly all of the stable virus-specific RNA appears to be asymmetrically transcribed, since in each case we were able to account for over 90% of the transcribing activity by summing the hybridization to each DNA strand. Although we cannot make any statement about the precise location of transcribed regions on the plus and minus strands of fragments A and B, the simplest explanation of this result is that the parts of A and B nearest H and I are transcribed from the minus strand template and the parts nearest C and G, respectively, are transcribed from the plus strand template. The early region would then encompass sequences present between 0·26 to 0·57 map unit and the late region would comprise the rest of the genome (Fig. 3).

The direction of transcription of early and late SV40 RNA has been established by

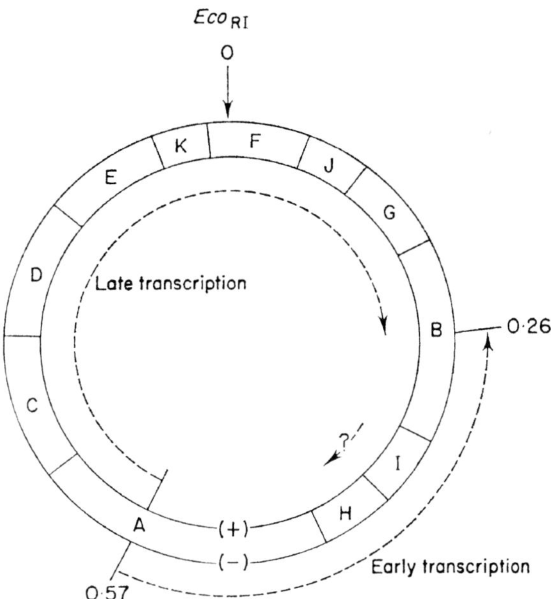

Fig. 3. Diagrammatic representation of transcription, early and late in the SV40 lytic cycle as it relates to the physical map of the viral genome.

determining the 5′ to 3′ orientation of the minus and plus strands of SV40 DNA, which serve as the early and late templates, respectively. As indicated in Figure 3, early transcription occurs in a counter-clockwise direction from A to B and late transcription occurs in a clockwise direction from A to B. On the assumption that the abundant species of stable SV40 RNA that we have used in these studies is the immediate transcription product, we can localize initiation sites for both early and late transcription near the middle of fragment A (at 0·57 map unit) and termination sites for both early and late transcription near the middle of fragment B (at 0·26 map unit). Alternatively, if large regions of the DNA are symmetrically transcribed (Aloni, 1972) with subsequent degradation of specific RNA sequences, the true initiation and termination sites may be at different locations. Furthermore, we have not excluded the possibility that other initiation and termination sites may be present within both the early and late regions. In this regard, experiments carried out *in vitro* with *E. coli* RNA polymerase and an SV40 DNA I template suggest that transcription with this enzyme begins outside fragment A (Westphal *et al.*, 1973; Zain, Dhar, Weissman, Lebowitz & Lewis, unpublished observations). It will be of considerable interest to determine what special nucleotide sequences may exist at the transition sites within fragments A and B, and whether these sites are related to the specific cleavage of SV40 component I DNA by the single-strand specific nuclease S_1 (J. Morrow & P. Berg, personal communication).

The map position for the early region of the SV40 genome shown in Figure 3 is entirely consistent with observations made with adeno-SV40 hybrid viruses by Kelly & Lewis (personal communication) who found that the entire early region of SV40 is included in the DNA of a hybrid virus containing 43% of the SV40 genome. They have localized within the early segment those regions which are required for expression of U antigen, T antigen, and the SV40 tumor-specific transplantation antigen. Recently

26

Patch *et al.* (1972) found that the SV40 DNA segment of $AD2^+ND_1$ contains parts of both early and late information and, therefore, has a probable transcriptional initiation or termination site. A comparison of the map of the SV40 segment of the adenovirus hybrid with the cleavage map of the SV40 genome (Morrow & Berg, 1972; T. Kelly, personal communication; P. Lebowitz, personal communication) suggests that this site corresponds to the termination point within fragment B. Therefore, it is likely that transcription of SV40 DNA sequences in $AD2^+ND_1$ infected cells begins within the adenovirus portion of the DNA.

The incomplete hybridization of *both* the minus and plus strands of fragments H and I with the RNA from infected monkey cells (Table 1) suggests that transcription of SV40 DNA may be more complex than initially thought. As pointed out above, fragments H and I, which are physically located in the middle of the early region of the viral genome, were the only two that failed to react completely with amounts of lytic RNA known to be saturating for the other nine *H. influenzae* DNA fragments. Although heterogeneity of these two fragments could explain this finding, there is at present no other evidence for heterogeneity (Danna *et al.*, 1973). The results presented in Table 2 suggest that the relative abundance of RNA complementary to fragment H or I DNA is much lower than the other, stable lytic RNA species encountered thus far. It is not clear at present whether the low concentrations of virus specific RNA complementary to DNA fragments H and I reflect a diminished rate of synthesis or the degradation of specific polyribonucleotide sequences transcribed from this region of the virus genome. Either of these possibilities would reduce the concentrations of specific RNA sequences and explain the results presented in Table 2. The reaction of a significant proportion of the *plus* strands of fragments H and I with late lytic RNA (Tables 1 and 2) suggests the existence of late DNA sequences (or early information on the plus DNA strand) within the early segment of the virus DNA. As such, this class of RNA may have an important role with respect to the regulation of SV40 DNA transcription.

Low levels of reaction (6 to 21%) were also observed between early lytic RNA and the plus strands of the 11 SV40 DNA fragments (Table 1). Previous reports have indicated that early RNA is complementary to the minus strand of virus DNA (Lindstrom & Dulbecco, 1972; Khoury *et al.*, 1972; Sambrook *et al.*, 1972). The early lytic RNA used in the experiment shown in Table 1 was prepared from monkey cells infected in the presence of arabinosyl cytosine. Incomplete inhibition of SV40 DNA synthesis at a time during productive infection (24 to 36 h) when late RNA is known to be present may account for the observed low levels of reaction between early RNA and the plus strands of the various fragments. Alternatively, this result could reflect the incomplete post-transcriptional degradation of RNA complementary to the plus strand of SV40 DNA. Several reports have appeared recently which suggest that integration of virus genetic information into chromosomal DNA may occur during productive infection by SV40 (Lavi & Winocour, 1972; Tai *et al.*, 1972; Hirai & Defendi, 1972). Whether the RNA transcribed from such integrated SV40 DNA (Jaenisch, 1972; Rozenblatt & Winocour, 1972) is related to those RNA sequences that hybridize incompletely with the DNA fragments remains to be determined.

The partial hybridization of late, but not early RNA, to the minus strand of fragment C (Table 1) is worthy of comment. Since fragment C contains the origin of DNA replication (Nathans & Danna, 1972; Danna & Nathans, 1972), it is possible that this RNA represents a primer for initiation of DNA replication (Wickner *et al.*, 1972). Since SV40 DNA replication proceeds bidirectionally from fragment C (Danna &

Nathans, 1972; Fareed *et al.*, 1972), one might expect primers for each strand at the initiation site.

It has been shown that the pattern of transcription in SV40 transformed cells is significantly different from that observed during lytic infection (Khoury *et al.*, 1973). The use of SV40 DNA fragments produced by restriction endonucleases should permit a topographic analysis of these differences.

This work was supported by grants from the National Institutes of Health and the Whitehall Foundation.

REFERENCES

Aloni, Y. (1972). *Proc. Nat. Acad. Sci.*, U.S.A. **69**, 2404–2409.
Aloni, Y., Winocour, E. & Sachs, L. (1968). *J. Mol. Biol.* **31**, 415–429.
Britten, R. J. & Kohne, D. E. (1968). *Science,* **161**, 529–540.
Carp, R. I., Sauer, G. & Sokol, F. (1969). *Virology,* **37**, 214–226.
Danna, K. J. & Nathans, D. (1971). *Proc. Nat. Acad. Sci.,* U.S.A. **68**, 2913–2917.
Danna, K. J. & Nathans, D. (1972). *Proc. Nat. Acad. Sci.,* U.S.A. **69**, 3097–3100.
Danna, K. J., Sack, G. & Nathans, D. (1973). *J. Mol. Biol.* **78**, 363–376.
Fareed, G. C., Garon, C. F. & Salzman, N. P. (1972). *J. Virol.* **10**, 484–491.
Hirai, K. & Defendi, V. (1972). *J. Virol.* **9**, 705–707.
Hirt, B. (1967). *J. Mol. Biol.* **26**, 365–369.
Jaenisch, R. (1972). *Nature New Biol.* **235**, 46–47.
Khoury, G. & Martin, M. A. (1972). *Nature New Biol.* **238**, 4–6.
Khoury, G., Byrne, J. C. & Martin, M. A. (1972). *Proc. Nat. Acad. Sci.,* U.S.A. **69**, 1925–1928.
Khoury, G., Byrne, J. C., Takemoto, K. K. & Martin, M. A. (1973). *J. Virol.* **11**, 54–60.
Kohne, D. E. (1969). *Carnegie Inst. of Wash. Yearbook,* **67**, 310–320.
Lavi, S. & Winocour, E. (1972). *J. Virol.* **9**, 309–316.
Lindstrom, D. M. & Dulbecco, R. (1972). *Proc. Nat. Acad. Sci.,* U.S.A. **69**, 1517–1520.
Martin, M. A. & Axelrod, D. (1969). *Proc. Nat. Acad. Sci.,* U.S.A. **64**, 1203–1210.
Morrow, J. F. & Berg, P. (1972). *Proc. Nat. Acad. Sci.,* U.S.A. **69**, 3365–3369.
Nathans, D. & Danna, K. J. (1972). *Nature New Biol.* **236**, 200–202.
Oda, K. & Dulbecco, R. (1968). *Proc. Nat. Acad. Sci.,* U.S.A. **60**, 525–532.
Patch, C., Lewis, A. M. & Levine, A. S. (1972). *Proc. Nat. Acad. Sci.,* U.S.A. **69**, 3375–3379.
Radloff, R., Bauer, W. & Vinograd, J. (1967). *Proc. Nat. Acad. Sci.,* U.S.A. **57**, 1514–1521.
Rozenblatt, S. & Winocour, E. (1972). *Virology,* **50**, 558–566.
Sambrook, J., Sharp, P. A. & Keller, W. (1972). *J. Mol. Biol.* **70**, 57–71.
Sauer, G. & Kidwai, J. R. (1968). *Proc. Nat. Acad. Sci.,* U.S.A. **61**, 1256–1263.
Smith, H. O. & Wilcox, K. (1970). *J. Mol. Biol.* **51**, 379–391.
Tai, H. T., Smith, A. C., Sharp, P. A. & Vinograd, J. (1972). *J. Virol.* **9**, 317–325.
Trilling, D. M. & Axelrod, D. (1970). *Science,* **168**, 268–271.
Westphal, H., Delius, H. & Mulder, C. (1973). In *Lepetit Colloquia on Biology and Medicine,* Vol. **4**, North Holland Publishing Co., Amsterdam and London.
Wickner, W., Brutlag, D., Schekman, R. & Kornberg, A. (1972). *Proc. Nat. Acad. Sci.,* U.S.A. **69**, 965–969.
Yoshiike, K. (1968). *Virology,* **34**, 391–401.
Yoshimori, R. N. (1971). Ph.D. Thesis, University of California, San Francisco Medical Center.

Methylated Nucleotides Block 5'-Terminus of Vaccinia Virus Messenger RNA

C. M. WEI AND B. MOSS

Reprinted from *Proceedings of the National Academy of Sciences USA* 72:318–322. Copyright © 1975, by permission of the authors.

Reovirus Messenger RNA Contains a Methylated, Blocked 5'-Terminal Structure: $m^7G(5')ppp(5')G^mpCp-$

Y. FURUICHI, M. MORGAN, S. MUTHUKRISHNAN, AND A. J. SHATKIN

Reprinted from *Proceedings of the National Academy of Sciences USA* 72:362–367. Copyright © 1975, by permission of the authors.

All the previous papers cited and discussed in this section were seminal in the truest sense of the word: tours de force, opening up entire fields of research. The three discoveries that we cite now, this one (Wei and Moss, and Furuichi et al.) and the next two (Stehelin et al.; Berget et al. and Chow et al., see below), share in addition the attribute of being totally unexpected. The first is the simultaneous discovery by Wei and Moss and by Furuichi et al. that the messenger RNAs (mRNAs) of a DNA virus, vaccinia virus, and of a double-stranded RNA virus, reovirus, possess at their 5' termini the structure m^7GpppG^mpCp-. This structure is novel in three respects: it is a terminal base-methylated nucleotide linked to a ribose-methylated nucleotide through a 5'-5' pyrophosphate bond. This was later found to be a widespread property: although not all mRNAs possess this cap, the vast majority, including cellular mRNAs, do. Further, it was subsequently found that although uncapped mRNAs are not completely inactive, in most cases capping greatly increases the efficiency with which mRNAs are translated. This is another example of a major discovery, affecting every cell, that was made first with viruses, the advantage of viruses being the fact that they possess genomes that are three to four logs smaller than cellular genomes.

W. K. JOKLIK

Reprinted from
Proc. Nat. Acad. Sci. USA
Vol. 72, No. 1, pp. 318–322, January 1975

Methylated Nucleotides Block 5′-Terminus of Vaccinia Virus Messenger RNA

(RNA methyl transferase/7-methylguanosine/2′-*O*-methylguanosine/2′-*O*-methyladenosine)

CHA MER WEI AND BERNARD MOSS

Laboratory of Biology of Viruses, National Institute of Allergy and Infectious Diseases, National Institutes of Health, Bethesda, Maryland 20014

Communicated by David Baltimore, November 8, 1974

ABSTRACT   **Studies on the nature and location of the methylated nucleotides in mRNA synthesized *in vitro* by vaccinia virus particles revealed an unusual 5′-terminal structure. Evidence that the pyrophosphate group is blocked by 7-methylguanosine and that both 2′-*O*-methyladenosine and 2′-*O*-methylguanosine occupy penultimate positions was presented. According to this model, the 5′-termini of vaccinia virus mRNAs are:**

7MeG⁵′ppp⁵′GMepNp and 7MeG⁵′ppp⁵′AMepNp.

The presence of a low number of methylated nucleotides is a common, although recently discovered, characteristic of mRNAs synthesized *in vivo* by eukaryotic cells (1, 2) and *in vitro* by virion-associated enzymes of cytoplasmic polyhedrosis virus of insects (3), vaccinia virus (4), and reovirus (5). In the case of mRNA made by the double-stranded RNA viruses, cytoplasmic polyhedrosis virus (3) and reovirus (5), evidence for the location of a single 2′-*O*-methylribonucleotide at or near the 5′-terminus was presented. Other studies had suggested a terminal 2′-*O*-substituted ribonucleotide in the + strands of the genomes of cytoplasmic polyhedrosis virus (6) and reovirus (7). Experiments with mRNA synthesized by vaccinia, a double-stranded DNA virus, indicated the presence of either 5′-terminal methylated nucleotides or an internal sequence of adjacent methylated nucleotides (4). Further studies with the latter RNA now reveal an unusual structure consisting of a terminal base-methylated nucleotide linked to a ribosemethylated nucleotide through a 5′–5′ pyrophosphate bond.

MATERIALS AND METHODS

Preparation of Methylated Vaccinia Virus mRNA. Procedures for the purification of vaccinia virus and synthesis *in vitro* of methylated mRNA by virus particles have been described (4).

Chemical and Enzymatic Treatment of RNA. Alkali digestion and DEAE–cellulose chromatography in 7 M urea at pH 7.6 were carried out as described (4). ³H was measured by liquid scintillation counting, and ³²P by either the latter method or from Čerenkov radiation. Nucleotides were desalted by readsorption to and elution from a DEAE–cellulose column equilibrated with ammonium carbonate. Procedures for depurination and chromatography on Dowex 50(H⁺) were adapted from Culp and Brown (8). Periodate oxidation followed by β-elimination with aniline was performed as described by Fraenkel-Conrat and Steinschneider (9). Digestion with alkaline phosphatase (0.1 mg/ml) and/or snake venom phosphodiesterase (0.2 mg/ml) was in 50 mM Tris·HCl (pH 8.5) and 5 mM MgCl₂ for 2 hr at 37°.

Thin-layer Electrophoresis and Chromatography. Cellulose-coated sheets (Eastman, 20 × 20 cm) were used for both electrophoresis and chromatography. Electrophoresis was performed under the following conditions: (A) 50 mM sodium formate (pH 3.5), 1000 V, 75 min; (B) 0.2 M sodium borate (pH 9), 800 V, 110 min; (C) 1 M formic acid, 800 V, 70 min. Chromatography solvents that were used included: (A) ethyl acetate–isopropanol–7.5 M NH₄OH–*n*-butanol (3:2:2:1); (B) isopropanol–concentrated HCl–H₂O (68:17.6:14.3); (C) isopropanol–H₂O–NH₄OH (7:2:1); (D) methanol–concentrated HCl–H₂O (7:2:1).

Enzymes, Chemicals, and Isotopes. Alkaline phosphatase and snake venom phosphodiesterase were purchased from Worthington Biochemical Corp.; 2′-*O*-methylguanosine, synthesized by R. Robins, was a gift of M. Sporn; other 2′-*O*-methylribonucleosides were obtained from P-L Biochemicals and Ash Stevens, Inc.; base methylated compounds were from Sigma Chemical Co. [β,γ-³²P]GTP (5.5 Ci/mmol) and *S*-adenosyl[*methyl*-³H]methionine (8.5 Ci/mmol) were products of ICN and New England Nuclear Corp., respectively.

RESULTS

Location of Methylated Nucleotides at the 5′-Terminus of Vaccinia mRNA. mRNA synthesized *in vitro* by vaccinia virus particles in the presence of *S*-adenosylmethionine contained about 2.3 methyl groups per 1000 nucleotides (4). After alkali digestion, the major methyl-labeled material chromatographed on DEAE-cellulose with a net charge of slightly more than −5. This charge suggested either a 5′-terminal methylated nucleotide containing a pyrophosphate group or an internal alkali-resistant sequence of ribosemethylated nucleotides (4). Attempts were made to label the pyrophosphate group with [β,γ-³²P]ribonucleoside triphosphates and to simultaneously label the nucleosides with *S*-adenosyl[*methyl*-³H]methionine. When [β,γ-³²P]GTP was used and the RNA then subjected to alkali hydrolysis and DEAE–cellulose chromatography, the major peaks of ³H and ³²P eluted together with a net charge of slightly more than −5 (Fig. 1B, peak II). A minor ³H peak (Fig. 1B, peak I) had a net charge between −4 and −5 and was overlapped by a broader peak of ³²P-labeled material. By contrast, after alkali hydrolysis of RNA made in the absence of *S*-adenosylmethionine, very little ³²P-labeled material chromatographed with a net charge of −5 (Fig. 1A).

Pyrophosphate Group at the 5′-Terminus of Vaccinia mRNA Is Blocked. Experiments in the preceding section were consistent with a structure of the type ppNMepNp- at the 5′-

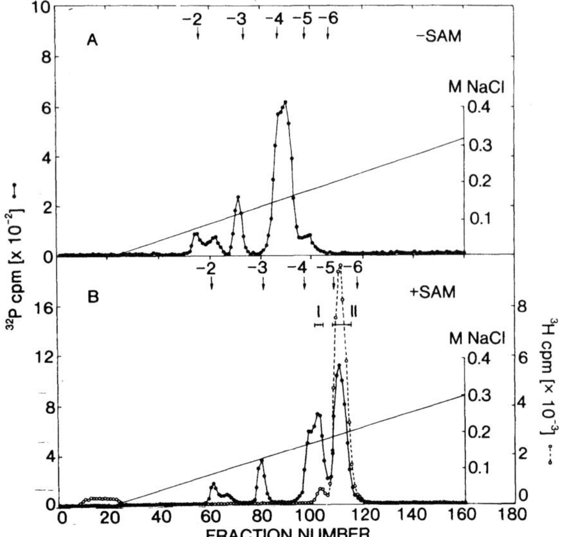

Fig. 1. DEAE–cellulose column chromatography of alkali hydrolysate of $[\beta,\gamma\text{-}^{32}\text{P}]$GTP-labeled vaccinia mRNA synthesized *in vitro* in the absence and presence of *S*-adenosyl[*methyl*-^{3}H]methionine. The reaction mixture (3 ml) containing 50 mM Tris·HCl (pH 8.5), 10 mM dithiothreitol, 5 mM MgCl$_2$, 2.5 mM each of ATP, CTP, and UTP, 0.1 mM $[\beta,\gamma\text{-}^{32}\text{P}]$GTP (1 Ci/mmol), 0.05% Nonidet P-40 detergent, 0.52 mg of vaccinia virus in the absence or presence of 1.68 μM *S*-adenosyl[*methyl*-^{3}H]methionine (8.5 Ci/mmol) was incubated at 37° for 30 min. Unincorporated radioactivity was removed by three cycles of trichloroacetic acid precipitation, and the RNA was digested with KOH and chromatographed on a DEAE–cellulose column (4). The arrows indicate the absorbance peaks of marker oligonucleotides, and the numbers above them represent their net charges.

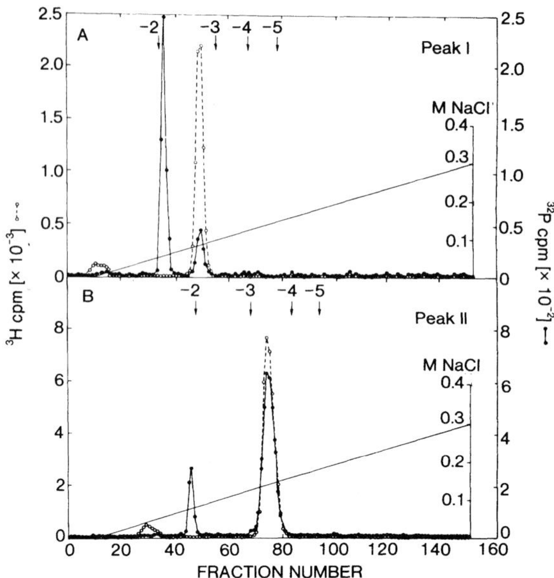

Fig. 2. DEAE–cellulose column chromatography of alkaline phosphatase-treated material in peaks I and II of Fig. IB. (A) Phosphatase-treated peak I material; (B) phosphatase-treated peak II material.

terminus of vaccinia mRNA. However, the 5′-terminal phosphates were now found to be resistant to alkaline phosphatase, which suggested the presence of a blocking group. Treatment of intact [*methyl*-^{3}H]RNA with this enzyme had no effect on the net charge of the labeled products obtained by alkali hydrolysis. Furthermore, when isolated material from peak II, which had a net charge of slightly more than −5, was treated with alkaline phosphatase and rechromatographed on DEAE–cellulose, most of the ^{3}H and ^{32}P still eluted together but now with a net charge between −3 and −4 (Fig. 2B). This 2 charge reduction was consistent with the removal of only an unlabeled 3′-phosphate. Methyl-labeled material in peak I, which originally had a charge between −4 and −5, also lost 2 charges when treated with alkaline phosphatase (Fig. 2A). However, only a fraction of the ^{32}P co-chromatographed with the ^{3}H and the remainder, presumably formed from unmethylated material that overlapped peak I, was converted to ^{32}P$_i$ with a charge of −2. The −4 peak derived from RNA made in the absence of *S*-adenosylmethionine (Fig. 1A) was entirely converted to ^{32}P$_i$ (not shown).

Isolation of Methylated Nucleotides. Peak II material that had been treated with alkaline phosphatase to remove the 3′-phosphate and purified by DEAE–cellulose chromatography as in Fig. 2B was completely resistant to spleen phosphodiesterase, a 5′-exonuclease. However, after venom phosphodiesterase treatment, three [^{3}H]*methyl*-labeled nucleotides were detected on electrophoresis: one migrated with marker

AMP; another with marker GMP; and the third, which was the only one also labeled with ^{32}P derived from $[\beta,\gamma\text{-}^{32}\text{P}]$GTP, migrated ahead of all the marker mononucleotides (Fig. 3). The latter nucleotide was eluted from the thin-layer sheet and shown to chromatograph on a DEAE–cellulose column as a nucleoside diphosphate (Fig. 4). A methylated nucleoside diphosphate and methylated derivatives of AMP and GMP could result from venom phosphodiesterase cleavage of MeN$^{5'}$pp/p$^{5'}$AMe/pN and MeN$^{5'}$pp/p$^{5'}$GMe/pN at the positions indicated by the slashes. The derivatives of AMP and GMP were thought to be methylated on the ribose because of previous evidence for 2′-*O*-methyl groups (4). The additional product of venom phosphodiesterase digestion, pN, could not be detected since it was unlabeled. An alternative model such as MeN$^{5'}$ ppp$^{5'}$ GMepAMepN seemed un-

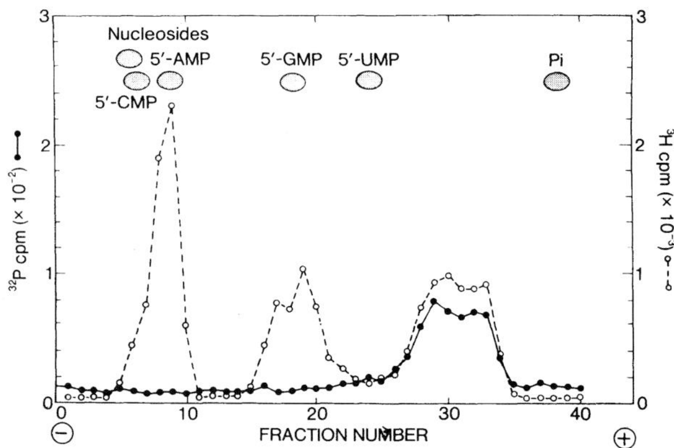

Fig. 3. Thin-layer electrophoresis of nucleotides produced by venom phosphodiesterase digestion of alkaline phosphatase-treated material in peak II. The −3 peak of Fig. 2B was isolated, desalted, treated with phosphodiesterase, and analyzed by electrophoresis in system A.

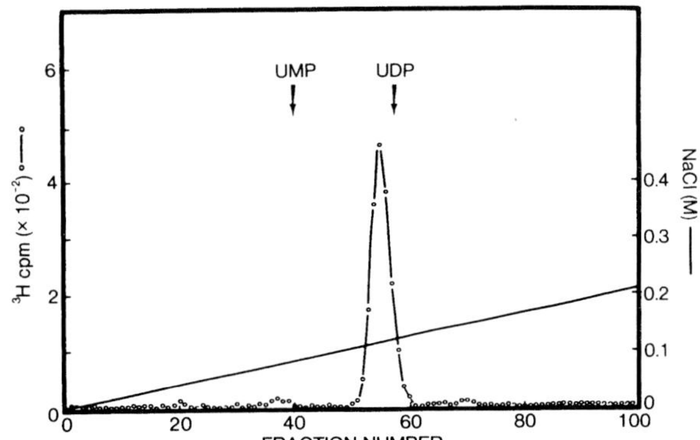

Fig. 4. DEAE–cellulose column chromatography of nucleoside diphosphate. The rapidly migrating nucleotide from Fig. 3 was chromatographed with UMP and UDP markers. ^{32}P was not indicated because of the lower levels of radioactivity.

likely because of its greater net charge and for additional reasons to be discussed.

Isolation of Methylated Nucleosides. Although venom phosphodiesterase digestion alone produced nucleotides, combined digestion with venom phosphodiesterase and alkaline phosphatase resulted in the formation of nucleosides. A single methylated nucleoside, designated MeN, was derived from peak I (Fig. 5A), whereas MeN plus two additional nucleosides were derived from peak II (Fig. 5B). The latter two nucleosides were identified as 2′-O-methyladenosine and 2′-O-methylguanosine by chromatography with authentic markers as indicated in Fig. 5B. MeN did not chromatograph with any of the 2′-O-methylribonucleoside markers, suggesting that it might be base-methylated. Since ribonucleosides with free *cis*-glycols complex with borate and migrate on electrophoresis toward the anode at pH 9 whereas 2′-O-methylribonucleosides migrate toward the cathode (10), this procedure provided a useful discriminatory test. As anticipated, MeN migrated toward the anode and the other two nucleosides migrated toward the cathode.

Additional experiments were done to correlate the results of this section with those of the previous one. The expected 2′-O-methylribonucleosides were produced by alkaline phosphatase treatment of the methylated nucleoside monophosphates isolated as described in Fig. 3. These results, as well as the demonstration that MeN was derived from the methylated nucleoside diphosphate, are documented in Fig. 5C, D, and E. Another finding, that the ratio of 2′-O-methylguanosine to 2′-O-methyladenosine derived from peak II was not an integral number but that MeN was always equal to the sum of the 2′-O-methylribonucleosides, was consistent with the model favored in the previous section and incompatible with the alternative model. Isolation of only MeN from peak I material suggested an analogous structure MeN$^{5′}$ppp$^{5′}$Np for the minor component.

Removal of the Blocking Group from mRNA. The terminal location and the free 2′,3′-OH groups on MeN suggested that it should be possible to remove the latter nucleoside from vaccinia mRNA by periodate oxidation and β-elimination. As predicted, after such treatment the phosphates at the 5′-terminus of the RNA were susceptible to alkaline phosphatase and a major component with a net charge of slightly

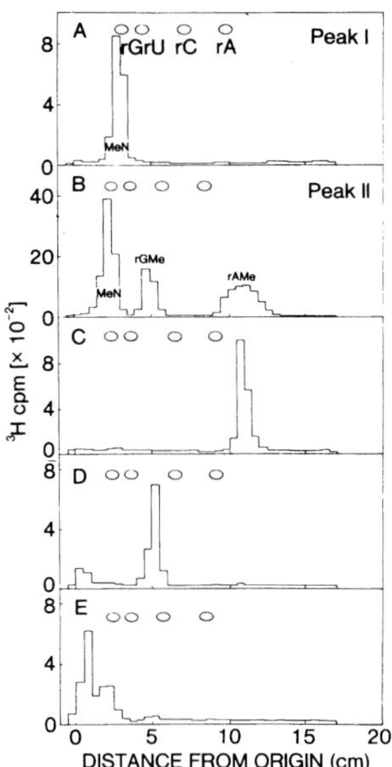

Fig. 5. Thin-layer chromatography of nucleosides produced by venom phosphodiesterase and alkaline phosphatase. Material of (A) peak I and (B) peak II of Fig. 1B was treated with phosphodiesterase and phosphatase. (C) Material in the peak corresponding to AMP, (D) in the peak corresponding to GMP, and (E) in the rapidly migrating peak of Fig. 3, was treated with phosphatase. Chromatography solvent A was used. The positions of the four internal ribonucleoside markers are indicated in each segment. The positions of 2′-O-methylguanosine (rGMe) and 2′-O-methyladenosine (rAMe) in relation to the radioactively labeled peaks were determined in a separate experiment but is indicated in (B).

more than −3, corresponding to GMepNp and AMepNp, was obtained by subsequent alkali hydrolysis and DEAE–cellulose chromatography (Fig. 6B). Control methylated RNA that had not undergone treatment to remove the blocking group was unaffected by alkaline phosphatase since following alkali hydrolysis the major product still had a charge of slightly more than −5 (Fig. 6A).

To prove that MeN was removed from the 5′-terminus of the RNA by periodate oxidation and β-elimination, the material with a net charge of −3 (Fig. 6B) was digested with a combination of alkaline phosphatase and venom phosphodiesterase and analyzed by thin-layer chromatography. 2′-O-Methyladenosine and 2′-O-methylguanosine were identified, whereas MeN was virtually absent (Fig. 7).

Identification of MeN. Our results thus far suggested that MeN was a base-methylated nucleoside. Accordingly, methylated vaccinia RNA was depurinated and analyzed by chromatography on a Dowex 50 (H$^+$) column. Approximately half of the methyl-labeled material, derived from the 2′-O-methylribonucleosides, did not adsorb and the remainder eluted soon after the guanine marker, suggesting that it was a methylguanine derivative (Fig. 8A). By contrast, when the material from the −3 peak (Fig. 6B) was depurinated, a methylated purine was not detected (Fig. 8B).

The guanine derivative was identified as 7-methylguanine by chromatography in several systems (B, C, and D), one of which is shown in Fig. 9A. In addition, marker 7-methylguanosine migrated as MeN in the alkaline chromatography

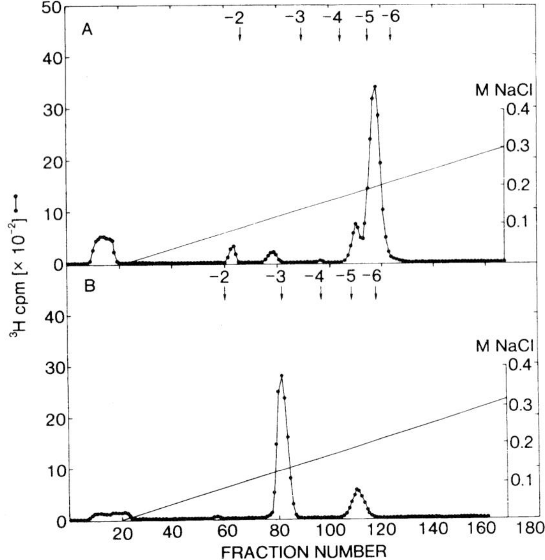

FIG. 6.   DEAE–cellulose column chromatography of alkali hydrolysate of phosphatase-treated RNA. (A) Control, [³H]-methyl-labeled vaccinia mRNA was treated with phosphatase and then hydrolyzed with 0.3 M KOH. (B) Prior to phosphatase treatment and alkali hydrolysis, [³H]methyl-labeled RNA was subjected to periodate oxidation and aniline cleavage.

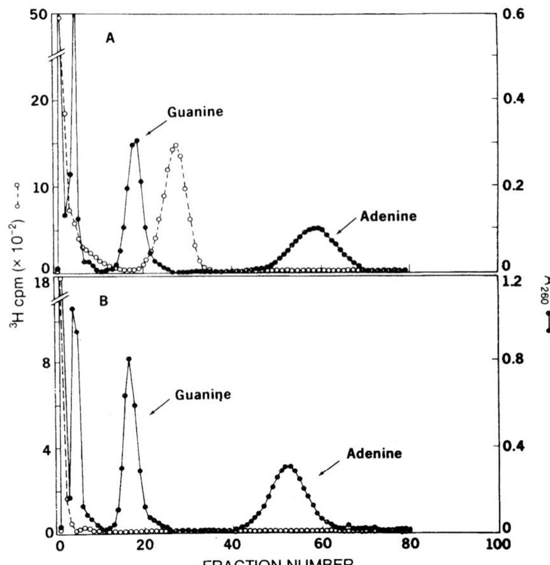

FIG. 8. Column chromatography of products of depurination reaction. [³H]Methyl-labeled RNA was mixed with 1 mg of yeast tRNA and heated in 1 M HCl at 100° for 30 min and chromatographed on a Dowex 50 (H⁺) column. (A) From intact RNA. (B) From −3 charge peak of Fig. 6B.

system used in Fig. 5. Actually 7-methylguanosine undergoes ring scission at neutral or alkaline pH values (11) and MeN was isolated predominantly in the ring-opened form as shown in Fig. 9B. This alteration may explain why, in a previous section, venom phosphodiesterase did not also cleave in the following manner: 7MeG⁵′p/pp⁵′MeG/pN.

DISCUSSION

We propose that the 5′-terminus of vaccinia mRNA, synthesized *in vitro* by purified vaccinia virus particles in the presence of *S*-adenosylmethionine, is blocked and has the unusual structure:

7MeG⁵′ppp⁵′GMepNp- and 7MeG⁵′ppp⁵′AMepNp-.

In addition, since vaccinia mRNA is on the order of 1000 nucleotides long (12) and there are 2.3 methyl groups per 1000 nucleotides (4), the majority of mRNA molecules must contain such a methylated terminus.

Most of our analyses were carried out on 7MeG⁵′ppp⁵′-GMepNp and 7MeG⁵′ppp⁵′AMepNp obtained by alkali hydrolysis of RNA labeled with [β,γ-³²P]GTP and *S*-adenosyl[*methyl*-³H]methionine. The 3′-terminal phosphate was removed with alkaline phosphatase and the remaining product was digested with venom phosphodiesterase to yield ³²P- and ³H-labeled pp7MeG, ³H-labeled pAMe and pGMe, and unlabeled pN. Formation of these products provided evidence for three phosphates in the pyrophosphate bridge. The number of phosphates had been a stumbling block since the

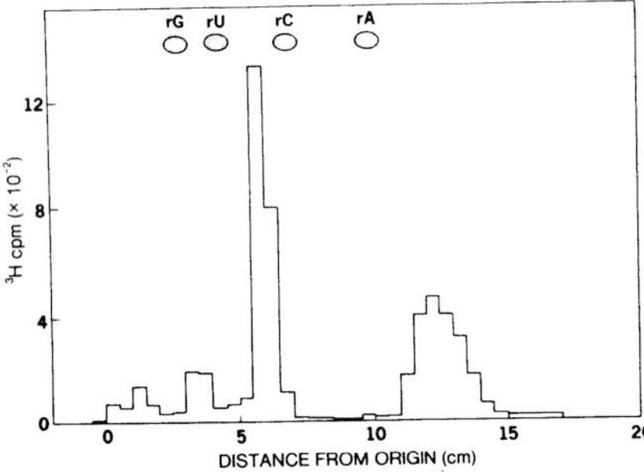

FIG. 7.   Thin-layer chromatography of digest of material in −3 charge peak of Fig. 6B. The material in the −3 peak was isolated, desalted, and digested with a mixture of phosphatase and phosphodiesterase. Chromatography solvent A was used.

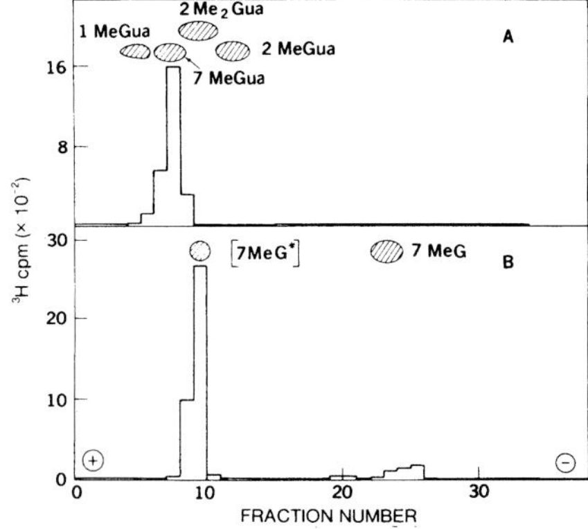

FIG. 9. Identification of 7-methylguanine and ring-open form of 7-methylguanosine. (A) Methylated vaccinia mRNA was depurinated and chromatographed in solvent B with methylated guanine standards. (B) Methylated RNA was digested with RNase A and T1 followed by a combination of alkaline phosphatase and venom phosphodiesterase, and MeN was then isolated by chromatography as in Fig. 5. Purified MeN, marker 7-methylguanosine (7MeG), and the ring-opened form [7MeG*], obtained by treatment of the latter with Tris·HCl (pH 8.5) or 0.3 M KOH, were analyzed by electrophoresis in system C.

net charge of the alkali hydrolysis fragments, as determined by DEAE–cellulose chromatography, was slightly more than −5 whereas a charge of −6 was predicted for this structure. Further digestion of the nucleotides with alkaline phosphatase produced 7-methylguanosine (ring-opened form), 2′-O-methyladenosine, and 2′-O-methylguanosine. Intact 7-methylguanine was isolated by depurination of RNA, indicating that the ring-opened form was an artifact of isolation. In another sequence of experiments, 7-methylguanosine was removed from RNA by periodate oxidation and β-elimination, leaving the unblocked terminal structures pppGMepNp- and pppAMepNp-, which were then susceptible to alkaline phosphatase digestion. A minor alkali hydrolysis product, 7MeG⁵′ppp⁵′Np, was derived from undermethylated RNA since it was not detected when saturating concentrations of S-adenosylmethionine were used for synthesis.

The terminal modification of vaccinia mRNA appears to occur post-transcriptionally since S-adenosylmethionine had no effect on the rate of mRNA synthesis *in vitro* (4) and, moreover, enzymes capable of this modification have been isolated and are being studied in our laboratory.

The possibility that 5′-termini of the double-stranded RNAs of cytoplasmic polyhedrosis virus and reovirus are blocked is under investigation (5, 7). Sequences quite similar to those of vaccinia mRNA have been proposed for the 5′-termini of certain low molecular weight, nonpolyadenylated RNA species of unknown function from nuclei of Novikoff hepatoma cells (13). One of these sequences is 2,2,7Me₃G⁵′-pp⁵′AMepUMep. Although Perry and Kelley (1) did not determine the location of the methylated nucleotides of L cell mRNA, a portion of the alkali digest eluted from DEAE–cellulose at 0.3 M NaCl, which suggested either an attached pyrophosphate group or an alkali-resistant internal sequence. More recently, the sequences 7MeG⁵′ppp⁵′NMep and 7-MeG⁵′ppp⁵′NMepNMep were found at the 5′-termini of HeLa cell mRNAs (C. M. Wei, A. Gershowitz, and B. Moss, submitted for publication). At present there is no evidence

for the function of modified 5′-termini, although the resistance of the blocked 5′-terminal oligonucleotide of vaccinia mRNA to spleen phosphodiesterase, a 5′-exonuclease, suggests one possibility.

Note Added in Proof. Additional evidence for the presence of three phosphates in the pyrophosphate bridge was obtained by digesting vaccinia mRNA with nuclease P₁ to yield 7Me-G⁵′ppp⁵′GMe and 7MeG⁵′ppp⁵′AMe. 7MeG was then removed by periodate oxidation and β-elimination, and pppGMe and pppAMe were identified by DEAE–cellulose chromatography.

We thank Michael Sporn for a sample of 2′-O-methylguanosine, Michael Cashel, Scott Martin, Marcia Ensinger, and David Baltimore for helpful discussions, and Aaron Shatkin for manuscripts of refs. 5 and 7 prior to publication.

1. Perry, R. P. & Kelley, D. E. (1974) *Cell* **1**, 37–42.
2. Desrosiers, R., Frederici, K. & Rottman, F. (1974) *Fed. Proc.* **33**, 1432.
3. Furuichi, Y. (1974) *Nucl. Acids Res.* **1**, 809–822.
4. Wei, C. M. & Moss, B. (1974) *Proc. Nat. Acad. Sci., USA* **71**, 3014–3018.
5. Shatkin, A. J. (1974) *Proc. Nat. Acad. Sci., USA* **71**, 3204–3207.
6. Miura, K., Watanabe, K. & Sugiura, M. (1974) *J. Mol. Biol.* **86**, 31–48.
7. Miura, K., Watanabe, K., Sugiura, M. & Shatkin, A. J. (1974) *Proc. Nat. Acad. Sci., USA* **71**, 3979–3983.
8. Culp, L. & Brown, G. M. (1968), *Arch. Biochem. Biophys.* **124**, 483–492.
9. Fraenkel-Conrat, H. & Steinschneider, A. (1967) in *Methods in Enzymology*, eds. Grossman, L. & Moldave, K. (Academic Press, New York), Vol. 12, Sect. B, pp. 243–246.
10. Khym, J. X. (1967) in *Methods in Enzymology*, eds. Grossman, L. & Moldave, K. (Academic Press, New York), Vol. 12, Sect. A, pp. 93–101.
11. Hall, R. (1971) *The Modified Nucleosides in Nucleic Acids* (Columbia University Press, New York).
12. Kates, J. & Beeson, J. (1970) *J. Mol. Biol.* **50**, 1–18.
13. Ro-Choi, T. S., Reddy, R., Choi, Y. C., Raj, N. B. & Hennings, D. (1974) *Fed. Proc.* **33**, 1832.

Reprinted from
Proc. Nat. Acad. Sci. USA
Vol. 72, No. 1, pp. 362–366, January 1975

Reovirus Messenger RNA Contains a Methylated, Blocked 5′-Terminal Structure: $m^7G(5′)ppp(5′)G^mpCp$-

(methylation/5′-sequence)

Y. FURUICHI, M. MORGAN, S. MUTHUKRISHNAN, AND A. J. SHATKIN

Department of Cell Biology, Roche Institute of Molecular Biology, Nutley, New Jersey 07110

Communicated by Edward L. Tatum, November 1, 1974

ABSTRACT   Reovirus mRNA synthesized *in vitro* by the virus-associated RNA polymerase in the presence of *S*-adenosylmethionine contains blocked, methylated 5′-termini with the structure, $m^7G(5′)ppp(5′)G^mpCp$. The functional significance and possible mechanism of formation of this novel 5′–5′ terminal nucleotide linkage are discussed.

Many animal viruses contain RNA polymerases that synthesize mRNA by transcribing the viral genome (1). The mRNAs of several viruses were recently found to be methylated when synthesized in the presence of the methyl donor, *S*-adenosylmethionine (SAdoMet). These include purified cytoplasmic polyhedrosis virus (CPV) (2) and reovirus (3) (two double-stranded RNA viruses), vesicular stomatitis virus [which contains a single-stranded RNA (4)], and the double-stranded DNA-containing vaccinia virus (5). Methylated nucleotides have been observed not only in mRNAs from animal viruses with different types of genomes, but also in mammalian cell mRNA (6, 7). The sites of methylation in each case are at or near the 5′-termini of the mRNA, and initiation of transcription by CPV is coupled to methylation (2). In addition to a role in transcription and possible post-transcriptional processing (8), methylation of the 5′-termini of mRNA is also likely to have an important influence on the regulation of protein synthesis. Consequently, it was of interest to analyze in detail the 5′-terminal sequence of a methylated mRNA. We selected reovirus mRNA for analysis because it can be synthesized in large quantities *in vitro* and is translated with fidelity in cell-free extracts*.

MATERIALS AND METHODS

Reovirus type 3 Dearing strain was grown in mouse L cells and purified as described (9). Viral mRNA synthesized *in vitro* by chymotrypsin-treated purified reovirus was extracted with phenol, passed through Sephadex G-100, and collected by alcohol precipitation*. For *Penicillium* nuclease digestion, the RNA was dissolved in 0.3 ml of H_2O, adjusted to 10 mM sodium acetate buffer (pH 6.0), and incubated with 100 μg of enzyme. RNA was treated with alkaline phosphatase (20 units/ml) in 50 mM Tris·HCl buffer (pH 8.0). Nucleotide pyrophosphatase was used at 0.05 μg/ml in 20 mM Tris·HCl buffer (pH 7.5) containing 1 mM Mg^{2+}. All incubations were at 37° for 30 min. Samples were spotted on Whatman no. 3MM paper and analyzed by electrophoresis at 2600 V for 40 min in pyridine acetate buffer (pH 3.5). Authentic compounds were

---

Abbreviations: SAdoMet, *S*-adenosylmethionine; CPV, cytoplasmic polyhedrosis virus. In this paper, $m^7G(5′)ppp(5′)G^mpCp$ is represented by $7mG5′ppp5′GmpCp$.

* Both, G. W., Lavi, S. & Shatkin, A. J., *Cell*, in press.

located under ultraviolet light. The dried paper was cut into 1-cm strips for determination of radioactivity in scintillation fluid.

Alkaline phosphatase, nucleotide pyrophosphatase, and RNase T2 were purchased from Worthington Biochemical Corp., Sigma Chemical Co., and Sankyo Co., respectively. *Penicillium* nuclease (10) was kindly provided by Dr. K. -I. Miura. *S*-Adenosyl-L-[*methyl*-³H]methionine (8.5 Ci/mmol) and [α-³²P]GTP (5.6 Ci/mmol) were obtained from New England Nuclear; [¹⁴C]CTP (44.4 mCi/mmol) and [³H]GTP (13 Ci/mmol) were from Schwarz/Mann; [β,γ-³²P]GTP (3.5–9 Ci/mmol) and *S*-adenosyl-L-[*2*-³H]methionine (2.1 Ci/mmol) were from ICN and Amersham, respectively.

RESULTS

Phosphatase-Resistant 5′-Terminal Phosphates in Methylated RNA. Reovirus mRNA synthesized *in vitro* in the absence of a methyl donor with [β,γ-³²P]GTP as the radioactive precursor contains 5′-terminal ³²ppGp (11–13). Little or no incorporation of ³²P into mRNA was obtained with [γ-³²P]GTP. The γ-phosphate in each case is removed by a virion-associated phosphohydrolase (13–15). Similarly, methylated mRNA made in the presence of SAdoMet is ³²P-labeled only with [β,γ-³²P]GTP (Table 1). However, in contrast to the 5′-terminal ³²ppGp in nonmethylated mRNA, which is converted to ³²Pᵢ by incubation with alkaline phosphatase (13), the 5′-terminal ³²P in methylated mRNA was 54% resistant to removal by phosphatase digestion (Table 1). All three size classes of mRNA separated by velocity sedimentation in a glycerol density gradient (3) contained similar proportions of molecules with phosphatase-resistant 5′-termini (data not shown). Consistent with a methyl group donor function of SAdoMet, mRNA synthesized in the presence of SAdo[2-³H]Met was not ³H-labeled (Table 1).

Structure of the Blocked 5′-Termini. In order to study the nature of the phosphate-protecting group, mRNA was synthesized in the presence of three radioactive precursors: [β,γ-³²P]GTP to label the 5′ ends, [¹⁴C]CTP as a measure of total RNA formation, and SAdo[*methyl*-³H]Met to identify the methylated residues. The radioactive mRNA was digested with *Penicillium* nuclease which cleaves phosphodiester linkages in nucleic acid to yield 5′-mononucleotides (10). The digestion products were analyzed by paper electrophoresis (Fig. 1a). A single ¹⁴C-labeled component (peak I) was obtained in the position of authentic pC. All of the [³H]methyl-labeled material migrated as a peak (II) between pA and pG. It also contained 5′-terminal ³²P, and the molar ratio of ³H to ³²P was 2.1. Another fraction of the ³²P migrated with ppG

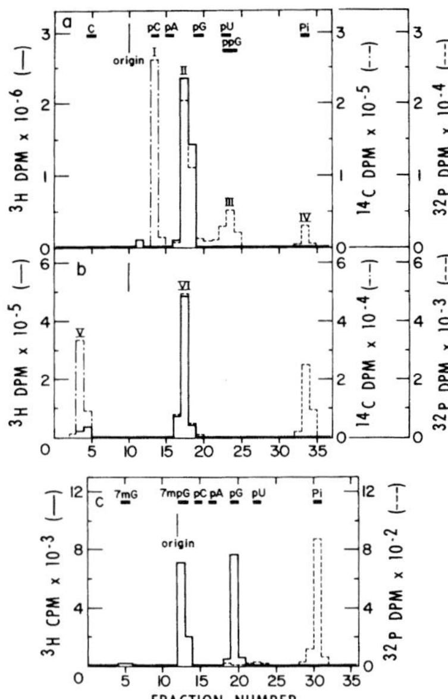

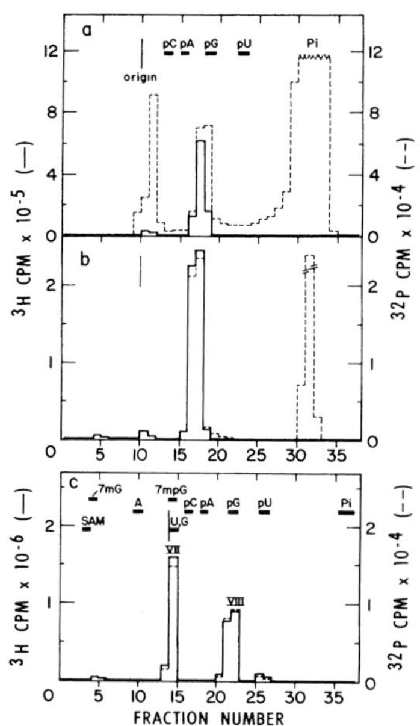

Fig. 1. Electrophoretic analyses of enzymatic digests of mRNA. mRNA was synthesized under standard conditions with [β,γ-^{32}P]GTP, *S*Ado[*methyl*-^{3}H]Met, and [^{14}C]CTP as the radioactive precursors. The purified products were digested with (a) *Penicillium* nuclease or (b) *Penicillium* nuclease followed by alkaline phosphatase, and analyzed by paper electrophoresis. (c) Peak VI material was eluted with 0.1 ml of H$_2$O, digested with nucleotide pyrophosphatase, and analyzed again.

Fig. 2. Analyses of digests of mRNA made in the presence of [α-^{32}P]GTP and *S*Ado[*methyl*-^{3}H]Met. RNA was digested with (a) *Penicillium* nuclease followed by alkaline phosphatase and analyzed by electrophoresis. (b) Material in fractions 17–19 was eluted with H$_2$O, redigested with phosphatase, and again analyzed. (c) Fractions 17 and 18 were eluted, treated with nucleotide pyrophosphatase, and analyzed again by electrophoresis.

(peak III), which presumably is derived from unblocked 5′-terminal ^{32}ppG. Peak IV contains ^{32}P$_i$, possibly due to contaminating phosphatase.

Digestion with alkaline phosphatase after *Penicillium* nuclease treatment converted the ^{14}C-labeled pC (peak I) to cytidine (Fig. 1b, peak V). However, the migration of the ^{3}H-labeled material (II) was not altered, and its 5′-terminal phosphates were retained (VI). The ^{32}ppG (III) was quantitatively converted to P$_i$.

The *Penicillium* nuclease- and phosphatase-resistant 5′-terminal structure (VI) was hydrolyzed with nucleotide pyrophosphatase and analyzed again (Fig. 1c). All the ^{32}P radioactivity was converted to P$_i$, but the [^{3}H]methyl radioactivity was equally divided between nucleotides migrating with authentic 7-methyl-pG and pG. These results, together with the previous report (3) that reovirus mRNA contains alkali-resistant 2′-*O*-methyl-GpC, tentatively establish the blocked 5′-terminal structure in peak II as 7mGp^{32}ppGm.

Origin of the Phosphates in the 5′-Terminal Sequence, *7mGpppGm.* The formation of blocked 5′-termini in reovirus mRNA occurs by an unknown series of reactions, but synthesis of a 5′-sequence with the structure 7mGpppGm presumably would require the incorporation of both α- and β-position phosphates from [^{32}P]GTP. In order to test this scheme, mRNA was synthesized in the presence of *S*Ado[*methyl*-^{3}H]Met as described in Fig. 1, with the exception that [α-^{32}P]GTP replaced [β,γ-^{32}P]GTP. The mRNA was digested with *Penicillium* nuclease and alkaline phosphatase and analyzed by electrophoresis (Fig. 2a). A single ^{3}H-labeled component that also contained 2.4% of the total ^{32}P was observed in the position of presumptive 7mGpppGm (peak VI of Fig. 1b). The remaining ^{32}P was present predominantly as P$_i$, with smaller amounts migrating near the origin probably corresponding to residual, undigested material. The ^{3}H, ^{32}P-labeled material was eluted, redigested with alkaline phosphatase to remove residual [^{32}P] nucleotides, and repurified

TABLE 1. *Incorporation of radioactive precursors into reovirus mRNA*

Exp.	Precursors	Incorporation (pmol)			% ^{32}P phosphatase-resistant
		^{3}H	^{14}C	^{32}P	
1	*S*AdoMet, [^{14}C]CTP, [γ-^{32}P]GTP	—	5,200	0	—
2	*S*Ado[*methyl*-^{3}H]Met, [^{14}C]CTP, [β,γ-^{32}P]GTP	80	29,630	71	54
3	*S*Ado[*2*-^{3}H]Met, [^{14}C]CTP	0	11,616	—	—

Reovirus mRNA was synthesized under standard conditions (3,*) with chymotrypsin-treated virus and the indicated precursors, including 0.01 mM *S*AdoMet in each case. The ^{32}P-labeled product was digested with *Penicillium* nuclease and alkaline phosphatase (*Materials and Methods*) and analyzed by paper electrophoresis (Fig. 1). The percent resistance was estimated on the basis of the radioactivity in regions corresponding to P$_i$ versus peak II. The ratio of ^{3}H to ^{32}P in peak II material was 2.1.

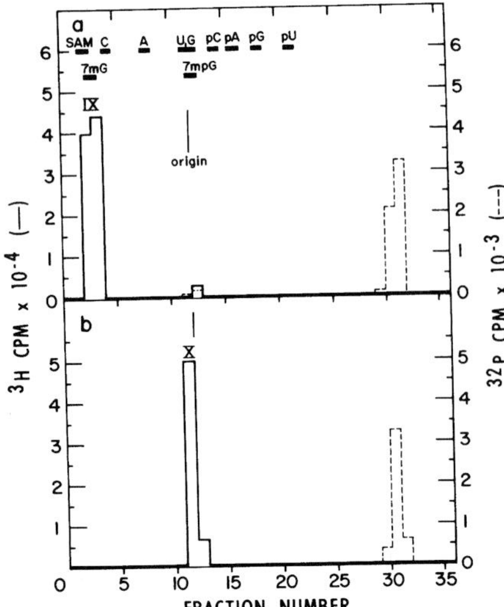

FIG. 3. Identification of ^{3}H, ^{32}P-labeled 7mpG and 2′-*O*-mpG from mRNA synthesized with [α-^{32}P]GTP and SAdo[*methyl*-^{3}H]Met. Material of peaks VII (a) and VIII (b) was eluted, digested with alkaline phosphatase, and analyzed by paper electrophoresis.

by electrophoresis (Fig. 2b). Fractions 15–20, containing the presumptive 7mG^{32}pp^{32}pGm, were eluted and digested with pyrophosphatase. The resulting mononucleotides were separated (Fig. 2c). Two [^{3}H]methyl, ^{32}P-labeled nucleotides were resolved in the positions of 7mpG and pG. Furthermore, the molar ratio of ^{3}H to ^{32}P was one in each peak. Thus when [α-^{32}P]GTP and SAdo[*methyl*-^{3}H]Met were used to label the 5′-terminal structure in reovirus mRNA, the two phosphates in 5′ linkage to the two modified guanosine residues were ^{32}P-labeled. In contrast, use of [β,γ-^{32}P]GTP and SAdo[*methyl*-^{3}H]Met as mRNA precursors did not result in ^{32}P-labeled mononucleotides, indicating that the 5′-terminal structure is 7mGpppGm.

Identification of 7mpG as the 5′-Terminal Blocking Group. 7-Methyl guanosine and 7mpG have unusual electrophoretic properties due to the extra positive charge at N^7. The 5′-mononucleotide was neutral at pH 3.5 (peak VII, Fig. 2c). This double-labeled 7mpG was eluted, treated with alkaline phosphatase, and again analyzed. Positively charged 7-[*methyl*-^{3}H]guanosine (peak IX) and ^{32}P$_i$ were obtained (Fig. 3a). In contrast, the double-labeled 2′-*O*-methyl-guanylic acid, which migrates in the position of pG (peak VIII, Fig. 2c), yielded 2′-*O*-[*methyl*-^{3}H]guanosine (peak X) of neutral charge and ^{32}P$_i$ after digestion with phosphatase (Fig. 3b).

Treatment of 7mpG at alkaline pH results in its partial conversion to the ring-opened structure, 2-amino-4-hydroxy-5-(*N*-methyl) carboxamide-6-ribosylamino-pyrimidine-5′-monophosphate with loss of the N^7 positive charge (16). As shown in Fig. 4a, treatment of double-labeled 7mpG (peak VII, Fig. 2c), and authentic 7mpG in 10 M NH$_4$OH at 37° for 1 hr converted them from neutral to negatively charged structures. Phosphatase treatment after alkali conversion altered the mobility of both the radioactive sample and authentic marker as predicted; the ^{3}H-label migrated as a neutral component [presumably 2-amino-4-hydroxy-5-(*N*-methyl) carboxamide-6-ribosylaminopyrimidine)] and the phosphate was released

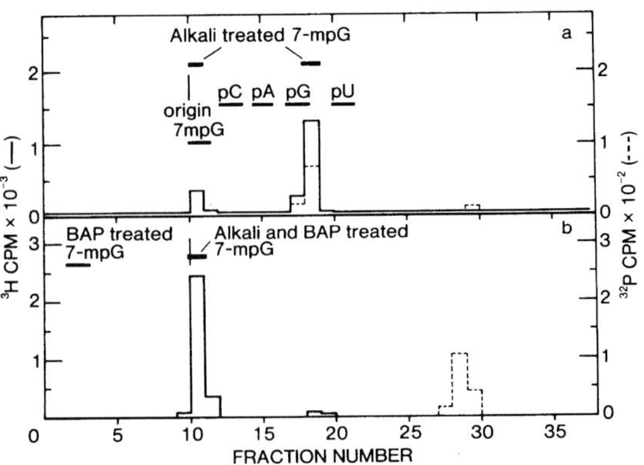

FIG. 4. Effect of alkali treatment on migration of 7mpG. (a) Peak VII material and authentic 7mpG (200 μg) were incubated for 1 hr at 37° in 10 M NH$_4$OH, evaporated to dryness, dissolved in H$_2$O, and analyzed by paper electrophoresis. For (b), the evaporated sample was dissolved in 10 mM Tris·HCl buffer (pH 8.0), and digested with alkaline phosphatase (BAP) before electrophoresis.

as ^{32}P$_i$ (Fig. 4b). The same type of experiment was done with the 5′-terminal structure, 7mGpppGm, derived by *Penicillium* nuclease digestion of mRNA labeled with SAdo[*methyl*-^{3}H]-Met and [β,γ-^{32}P]GTP. After alkaline conversion, the 5′-terminal structure migrated more rapidly toward the anode (Fig. 5a and b). Some degradation apparently also occurred during exposure to strong alkali, and a small fraction (15%) of the ^{3}H-label appeared in a neutral compound (Fig. 5b). Alkaline phosphatase digestion of the alkali-treated, double-labeled 5′-terminal structure also yielded a small amount (18%) of ^{32}P$_i$, but most of the radioactivity was retained in a component migrating faster than the untreated 7mGpppGm (Fig. 5c). It presumably corresponds to 2-amino-4-hydroxy-5-(*N*-methyl)carboxamide-6-ribosylaminopyrimidine-5′-triphospho-5′-(2′-*O*-methyl) guanosine. Alkali conversion of the 5′-terminal, methylated structure obtained by RNase T2 digestion (17) to a more negative component was also demonstrated by chromatography on DEAE–cellulose in 7 M urea (data not shown).

Identification of 2′-O-Methylguanylic Acid. The methylated residues in reovirus mRNA 5′-termini were further identified on the basis of their R_F values in several chromatographic solvent systems. mRNA synthesized with [^{3}H]SAdoMet and [α-^{32}P]GTP as the radioactive precursors were digested with *Penicillium* nuclease and alkaline phosphatase, and the separated 5′-termini were treated with pyrophosphatase as in Fig. 2. The migration of the two resulting double-labeled nucleotides, peaks VII and VIII in Fig. 2c, were compared to several authentic samples both before and after phosphatase treatment. The R_F values confirm that peaks VII and VIII correspond to 7mpG and 2′-*O*-mpG, respectively (Table 2). The corresponding methylated nucleosides (peaks IX and X) were also identified by chromatography on Bio-Rad Aminex A-5 cation exchange resin and DEAE–borate after digestion of reovirus mRNA with a mixture of venom phosphodiesterase, pancreatic RNase, and alkaline phosphatase (F. Rottman, personal communication).

5′–5′ Inverted Linkage of the Terminal Nucleotides. The incorporation of ^{32}P from [α-^{32}P]5′-GTP into 5′-7mGp in

TABLE 2. *Identification of methylated compounds by paper chromatography*

	Solvent system*			
Compound	A	B	C	D
Peak VII	0.40†	0.12	0.10	0.54
Peak IX	0.65	0.13	0.30	0.54
Peak VIII	0.34	0.16	0.11	0.60
Peak X	0.65	0.21	0.50	0.59
7mpG	0.40	0.12	0.10	0.54
7mG	0.65	0.13	0.30	0.54
2′-O-mpG	0.34	0.16	—	—
2′-O-mG	0.65	0.21	0.50	0.59

* A: isobutyric acid–0.5 M NH₄OH (10:6 v/v); B: *n*-butanol–HCl–ethanol (50:50:20 v/v); C: *n*-butanol–acetic acid–H₂O (3:2:1 v/v); D: isopropanol–HCl–H₂O (68:17.6:14.4 v/v).

† R_F values. —, not done.

reovirus mRNA indicates that the terminal sequence includes a 5′–5′ inverted linkage: 7mG5′ppp5′Gmp. A nucleotide linkage of this type contains 2′,3′-hydroxyls that are available for periodate oxidation and reduction with [³H]borohydride. To test for a 5′–5′-terminal linkage, reovirus mRNA was synthesized with [α-³²P]GTP as radioactive precursor. After purification*, the RNA was oxidized with periodate, reduced with [³H]borohydride, and re-purified by gel filtration (27). After treatment with *Penicillium* nuclease and alkaline phosphatase, the digest was analyzed by paper electrophoresis. Equal amounts of ³H were present in the positions of C [derived from 3′-termini (27)] and 7mGpppmG, and the latter was also ³²P-labeled (see Fig. 1b, peaks V and VI, respectively). The 7mGpppGm was eluted, digested with nucleotide pyrophosphatase and alkaline phosphatase, and again analyzed by paper electrophoresis. As shown in Fig. 6, this [³H]borohydride-reduced, enzyme-digested compound yielded ³²Pᵢ and a ³H-labeled component that migrated with authentic 7mG trialcohol (7mG′). A small fraction of the ³H remained at the origin, as expected for the ring-opened structure derived from 7mG (see Fig. 4b). Thus, the 5′-ends of reovirus mRNA are 7mG5′ppp5′GmpCp in which the 2′-3′-hydroxyls of 7mG are free.

DISCUSSION

Reovirus mRNA synthesized by the virion polymerase in the presence of *S*-adenosylmethionine contains a novel 5′-terminal structure: 7mG5′ppp5′Gmp. Messenger RNAs made *in vitro* by the polymerase present in purified CPV† and vaccinia virus (Urushibara, Furuichi, Nishimura, and Miura, unpublished results) recently were also found to contain blocked, methylated 5′-sequences, and the structure was identified as 7mGpppAmp in CPV mRNA. This type of blocked 5′-terminus appears not to be limited to virion transcriptase products. Several low molecular weight RNAs isolated from Novikoff hepatoma cell nuclei contain 5′-terminal m₃²·²·⁷-G⁴′pp⁴′Amp (18). Mouse L cell mRNA also is methylated (6, 7), and analyses of the methylated sequences in L cell (ref. 8 and Perry, Kelly, Friderici, and Rottman, unpublished results) and BSC-1 monkey cell mRNA (Lavi and Shatkin, un-

† Furuichi, Y. & Miura, K.-I., *Nature*, in press.

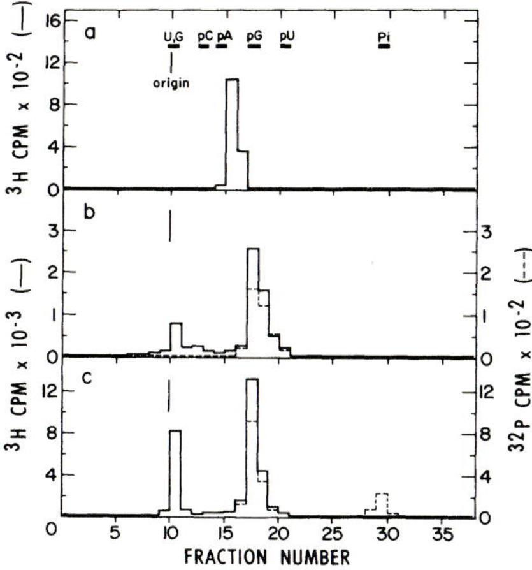

FIG. 5. Effect of alkali treatment on migration of 7mGpppGm. (a) RNA synthesized in the presence of SAdo[*methyl*-³H]Met was digested with *Penicillium* nuclease and analyzed by electrophoresis. The 5′-terminal 7mGpppGm contained all the radioactivity and migrated as a single component between pA and pG in the position of peak II. (b) Peak II material was eluted and an aliquot was treated with 10 M NH₄OH for 1 hr at 37° before electrophoresis. (c) A portion of the material in (b) after alkali treatment was evaporated to dryness, dissolved in 10 mM Tris·-HCl buffer (pH 8.0), digested with alkaline phosphatase, and again analyzed.

published results) are consistent with the presence of blocked 5′-termini.

Identification of a blocking nucleotide covalently linked in a 5′–5′ configuration to the initiating nucleotide at the 5′-

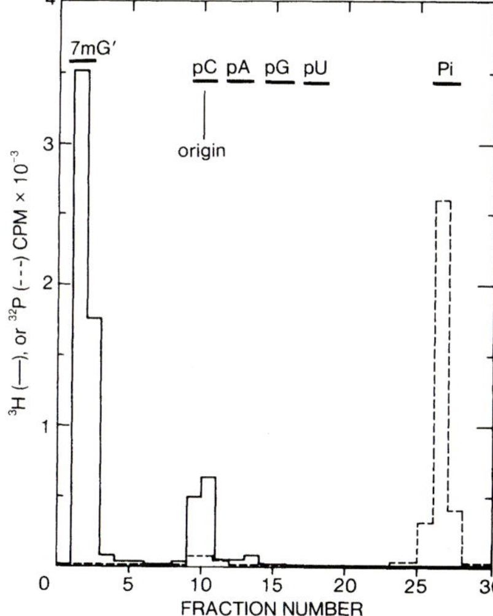

FIG. 6. Characterization of 5′-linkage of 7 mG. mRNA was synthesized with [α-³²P]GTP and unlabeled SAdoMet. The RNA was oxidized with sodium periodate and reduced with 20 mCi of potassium [³H]borohydride (27). Carrier L cell ribosomal RNA (2 mg) was added, and the RNA was purified free of genome RNA by precipitation twice in 2 M LiCl at 0° followed by gel filtration in Sephadex G-100. After alcohol precipitation, the ³H, ³²P-labeled RNA was digested with *Penicillium* nuclease and analyzed by electrophoresis. The material corresponding to peak II was eluted, digested with nucleotide pyrophosphatase and alkaline phosphatase, and again analyzed.

end of mRNA has several important implications for structural studies. For example, RNA with this 5′-sequence would contain a 3′-hydroxyl available for polyadenylation at both ends of the molecule. In addition, the 5′-ends would remain resistant to ^{32}P-labeling by the polynucleotide kinase technique (19), even after phosphatase treatment. However, because the 2′–3′-hydroxyls of the 5′-terminal nucleoside are not involved in internucleotide phosphate bonds, they would be oxidized by exposure to periodate. Treatment of the oxidized RNA with β-aniline would eliminate the 5′-terminal nucleoside, rendering the terminus available for kinase labeling after removal of the unblocked phosphates by phosphatase digestion. This series of reactions was previously shown to be required for efficient kinase labeling of the 5′-ends of CPV (20) and reovirus genome RNA (21), and analysis of uniformly ^{32}P-labeled reovirus double-stranded RNA confirmed that the strand corresponding in sequence to the viral mRNA contains blocked 5′-termini with the structure, 7mGpppGmp (Chow, Furuichi, Muthukrishnan, Morgan, and Shatkin, unpublished results). Attempts to use polynucleotide kinase to label other mammalian cell mRNAs, i.e., hemoglobin mRNA (R. Williamson, personal communication) have also been unsuccessful, possibly due to the presence of blocked 5′-termini. It will be of interest to reexamine the nature of the 5′-termini of mRNAs in the light of this possibility.

The mechanism of formation of the blocked 5′-ends in viral mRNA is unknown, but several observations indicate that it occurs via a coupled reaction involving methylation and addition of a nucleoside-5′-monophosphate to the initiating 5′-nucleotide. The 5′-sequence of reovirus mRNA made in the absence of *S*AdoMet is ppGpCp. The terminal phosphate is removed by a virion phosphohydrolase, an activity that has also been found in CPV (22) and vaccinia virus (23, 24). The same unblocked, unmethylated 5′-sequence was present in a fraction of the reovirus mRNA molecules synthesized in the presence of *S*AdoMet. Presumably, this RNA was made by particles that lack methylase activity. In no instance was 7mGpppGpCp or ppGmpCp observed, consistent with a coupling between methylation and blocking of the 5′- termini of reovirus mRNA.

The biological role of blocked, methylated 5-terminal structures in animal cell and virus mRNAs remains to be determined, but their appearance in a wide variety of mRNAs is indicative of an important function. Methylated residues in cellular heterogeneous nuclear RNA may be recognized as cleavage sites by processing enzymes that are responsible for mRNA formation (8). The 5′-terminal protecting group could be added by a condensation reaction during or after cleavage (8). This same type of mechanism may be used in virus-infected and/or -transformed cells for generating virus-specified mRNAs. Reovirus (and CPV) mRNAs appear to be synthesized by the virion-associated polymerase by end-to-end transcription of the genome template RNA segments (25–27). All species of reovirus mRNA synthesized in the absence of *S*AdoMet contain initiating 5′-ends, ppGp, and thus are probably not formed by post-transcriptional cleavage of a larger precursor RNA. However, in the presence of *S*AdoMet, the appearance of initiated 5′-termini in blocked, methylated mRNA could also be due to synthesis of a complete transcript of the 10 reovirus genome segments followed by specific methylation and cleavage. This mechanism is made unlikely, however, by the observation that in the 5′-sequence of CPV mRNA, the middle phosphate is derived from the β-phosphate of ATP† but not GTP (Furuichi, unpublished results). Therefore, the blocked, methylated 5′-structures probably also have a role in mRNA function, possibly as protection against exonucleolytic attack or in translational control at the level of ribosome binding and/or initiation of polypeptide synthesis.

We thank Alba LaFiandra for valuable assistance and Fritz Rottman for helpful discussion.

1. Shatkin, A. J. (1974) *Annu. Rev. Biochem.* **43**, 643–665.
2. Furuichi, Y. (1974) *Nucl. Acid. Res.* **1**, 809–822.
3. Shatkin, A. J. (1974) *Proc. Nat. Acad. Sci. USA* **71**, 3204–3207.
4. Rhodes, D. P., Moyer, S. A. & Banerjee, A. K. (1974) *Cell* **3**, 327–333.
5. Wei, C. W. & Moss, B. (1974) *Proc. Nat. Acad. Sci. USA* **71**, 3014–3018.
6. Perry, R. P. & Kelley, D. E. (1974) *Cell* **1**, 37–42.
7. Desrosiers, R., Friderici, K. & Rottman, F. (1974) *Proc. Nat. Acad. Sci. USA* **71**, 3971–3975.
8. Rottman, F., Shatkin, A. J. & Perry, R. P. (1974) *Cell* **3**, 197–199.
9. Shatkin, A. J. & LaFiandra, A. (1972) *J. Virol.* **10**, 698–706.
10. Fujimoto, M., Kuninaka, A. & Yoshino, H. (1969) *Agr. Biol. Chem.* **33**, 1517–1518.
11. Nichols, J. L., Hay, A. J. & Joklik, W. K. (1972) *Nature New Biol.* **235**, 105–107.
12. Levin, D. H., Acs, G. & Silverstein, S. C. (1970) *Nature* **227**, 603–604.
13. Banerjee, A. K., Ward, R. & Shatkin, A. J. (1971) *Nature New Biol.* **230**, 169–172.
14. Borsa, J., Grover, J. & Chapman, J. D. (1970) *J. Virol.* **6**, 295–302.
15. Kapuler, A. M., Mendelsohn, N., Klett, H. & Acs, G. (1970) *Nature* **225**, 1209–1213.
16. Hains, J. A., Reese, C. B. & Todd, A. R. (1967) *J. Chem. Soc.*, 5281.
17. Hiramaru, M., Uchida, T. & Egami, F. (1966) *Anal. Biochem.*, **17**, 135–142.
18. Ro-Choi, T. S., Reddy, R., Choi, Y. C., Kaj, N. B. & Henning, D. (1974) *Fed. Proc.* **33**, 1548.
19. Richardson, C. C. (1965) *Proc. Nat. Acad. Sci. USA* **54**, 158–165.
20. Miura, K. I., Watanabe, K. & Sugiura, M. (1974) *J. Mol. Biol.* **86**, 31–48.
21. Miura, K., Watanabe, K., Sugiura, M. & Shatkin, A. J. (1974) *Proc. Nat. Acad. Sci. USA* **71**, 3979–3983.
22. Storer, G. B., Shepherd, M. G. & Kalmakoff, J. (1973) *Intervirology* **2**, 87–94.
23. Gold, P. & Dales, S. (1968) *Proc. Nat. Acad. Sci. USA* **60**, 845–852.
24. Munyon, W., Paoletti, E., Ospina, J. & Grace, J. T. (1968) *J. Virol.* **2**, 167–172.
25. Shimotohno, K. & Miura, K.-I. (1974) *J. Mol. Biol.* **86**, 21–30.
26. Schonberg, M., Silverstein, S. C., Levin, D. H. & Acs, G. (1971) *Proc. Nat. Acad. Sci. USA* **68**, 505–508.
27. Banerjee, A. K., Ward, R. & Shatkin, A. J. (1971) *Nature New Biol.* **232**, 114–115.

DNA Related to the Transforming Gene(s) of Avian Sarcoma Viruses Is Present in Normal Avian DNA

D. STEHELIN, H. E. VARMUS, J. M. BISHOP, AND P. K. VOGT

At the time when this work was carried out, it was known that neoplastic transformation of chick embryo fibroblasts by avian sarcoma virus (ASV) is due to a single gene; although several attempts had been made to determine whether DNA homologous to sequences in ASV RNA exists in the genome of normal chick cells, the results were inconclusive.

The implications of the results reported in this paper, which were soon expanded to show definitively that the oncogene of ASV, namely *src*, is a gene pirated from cells and modified so as to escape regulation by cellular control mechanisms, were enormous. Soon a variety of other retroviral oncogenes were also found to be modified cellular genes; and when the functions of these genes were examined and found to be cytokines and their receptors, protein kinases, G proteins, and nuclear proteins, including transcription factors, it was realized that here were the components of a cascade pathway for transmitting signals resulting from the binding of intercellular messengers to receptors on cell surfaces all the way to components of the nuclear mechanisms that control gene expression. Thus the vast field of intracellular signal transduction derives from the observation that retroviral oncogenes are modified pirated cellular genes. Both Mike Bishop and Harold Varmus were awarded the Nobel Prize in 1989.

W. K. JOKLIK

DNA related to the transforming gene(s) of avian sarcoma viruses is present in normal avian DNA

INFECTION of fibroblasts by avian sarcoma virus (ASV) leads to neoplastic transformation of the host cell. Genetic analyses have implicated specific viral genes in the transforming process[1-4], and recent results suggest that a single viral gene is responsible[4]. Normal chicken cells contain DNA homologous to part of the ASV genome[5-8]; moreover, embryonic fibroblasts from certain strains of chickens can produce low titres of infectious type C viruses either spontaneously[9] or in response to various inducing agents[10]. None of the viruses obtained from normal chicken cells, however, can transform fibroblasts, and results with molecular hybridisation indicate that the nucleotide sequences responsible for transformation by ASV are not part of the genetic complement of the normal cell[11]. We demonstrate here that the DNA of normal chicken cells contains nucleotide sequences closely related to at least a portion of the transforming gene(s) of ASV; in addition, we have found that similar sequences are widely distributed among DNA of avian species and that they have diverged roughly according to phylogenetic distances among the species. Our data are relevant to current hypotheses of the origin of the genomes of RNA tumour viruses[12] and the potential role of these genomes in oncogenesis[13].

We have prepared radioactive DNA ($cDNA_{sarc}$) complementary to nucleotide sequences which represent most or all of the viral gene(s) required for transformation of fibroblasts by ASV[14]. Our procedure to isolate $cDNA_{sarc}$ exploited the existence of deletion mutants of ASV which lack 10–20% of the viral genome (transformation defective, or td viruses)[11,15-17]; results of genetic analyses indicate that the deleted nucleotide sequences include part or all of the gene(s) responsible for oncogenesis and cellular transformation[4,14]. In our procedure the genome of the Prague-C strain (Pr-C) of ASV was transcribed into complementary DNA by endogenous RNA-directed DNA polymerase activity; we then used molecular hybridisation to select DNA specific for the region missing from the genome of the td deletion mutants. The preparation of $cDNA_{sarc}$ used was a virtually uniform transcript from about 16% of the Pr-C ASV genome[14], a region equivalent in size to the entire deletion in the strain of td virus used in our experiments[11,15-17]. Since the unit genome of ASV contains about 10,000 nucleotides[18,19], the genetic complexity of $cDNA_{sarc}$ is about 1,600 nucleotides, sufficient to represent an entire cistron. Nucleotide sequences homologous to $cDNA_{sarc}$ seem to be ubiquitous in the genomes of ASVs, but are not present in the genomes of avian leukosis viruses (including the endogenous chicken virus, RAV-0) or sarcoma-leukosis viruses from other species[14].

DNAs from several avian species (chicken, turkey, quail, duck and emu) contain nucleotide sequences which can anneal with $cDNA_{sarc}$ (Fig. 1 and Table 1). In contrast, we detected no homology between $cDNA_{sarc}$ and DNA from mammals (mouse and calf thymus; Table 1). The kinetics of the reactions between $cDNA_{sarc}$ and DNAs from chicken, quail and duck (Fig. 1a, c and d) were similar to the kinetics for the reassociation of unique nucleotide sequences in avian DNAs (C_0t about 1,000 mol s^{-1})[20]; thus, nucleotide sequences homologous to $cDNA_{sarc}$ are present as single (or a few) copies in each haploid complement of the avian DNAs tested. This conclusion was substantiated by measuring in a single reaction

Table 1 Homology between cDNA$_{sarc}$ and normal DNAs

Assay	Hybridisation conditions		Extent of reaction between cDNA$_{sarc}$ and DNA from						
	[Na⁺]	Temperature	Chicken	Quail	Turkey	Duck	Emu	Mouse	Calf
S1	0.9 M	68°	52%	46%	48%	45%	24%	<2%	<2%
HAP	0.9 M	68°					36%		<5%
HAP	1.5 M	59°					54%		

DNA was extracted from 10–11-d-old embryos of chickens, ducks and quails, 3-d-old mice (strain RIII), livers of adult turkeys, liver and heart of a 22-d-old emu, and calf thymus. Reaction mixtures contining denatured DNA (8 mg ml⁻¹) and ³H-cDNA$_{sarc}$ (0.32 ng ml⁻¹, 7,000 c.p.m. ml⁻¹) in a final volume of 0.3 ml were incubated at either 59 or 68 °C for 48 h. Samples incubated at 59 °C were in 1.5 M NaCl (final C_0t = 40,000), those incubated at 68 °C were in 0.9 M NaCl (final C_0t = 32,000); all reactions also continued 0.001 M EDTA–0.02 M Tris-HCl, *p*H 7.4. Duplex formation was measured by either hydrolysis with S1 nuclease[28] (in 0.3 M NaCl at 50 °C) or fractionation on hydroxyapatite (HAP) (samples adsorbed in 0.14 M sodium phosphate at 50 °C).

mixture the rates of reassociation between chicken DNA and both cDNA$_{sarc}$ (labelled with ³H) and unique nucleotide sequences purified from chicken DNA (labelled with ¹⁴C) (Fig. 1*b*); the rates were similar for both labelled DNAs.

The extent of duplex formation with cDNA$_{sarc}$ varied with the amount of chicken DNA used in the reactions

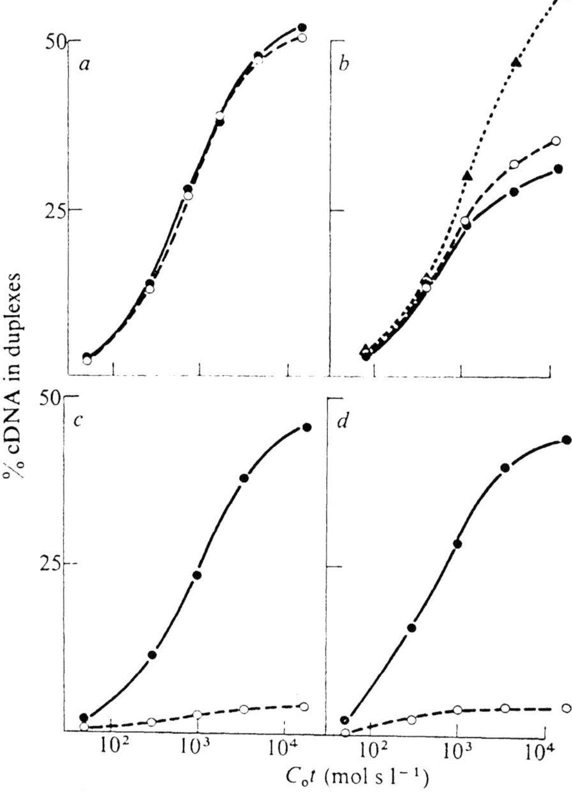

Fig. 1 Annealing of cDNA$_{sarc}$ to normal avian DNAs. DNA was prepared from 10–11-d-old embryos[7], denatured and incubated with ³H-cDNA$_{sarc}$ and ³²P-cDNA$_{B77}$ in 0.6 M NaCl–0.001 EDTA–0.02 M Tris-HCl, *p*H 7.4 at 68 °C for 0.1 to 32 h. Samples (0.1–0.2 ml) were withdrawn at different times and the extent of duplex formation with cDNA$_{sarc}$ and cDNA$_{B77}$ was measured by hydrolysis with the single-strand specific nuclease, S1 (ref. 28). The preparation of cDNA$_{sarc}$ and cDNA$_{B77}$ has been described elsewhere[14] and is outlined in the text. Unique sequence chicken DNA was prepared by reannealing denatured ¹⁴C-chicken DNA to a C_0t of 500; unique sequence DNA was eluted from a hydroxyapatite column at 60 °C in 0.14 M phosphate buffer[20]. *a*, Chicken DNA (9 mg ml⁻¹) with ³H-cDNA$_{sarc}$ (0.5 ng ml⁻¹, 10,000 c.p.m. ml⁻¹) (●) and ³²P-cDNA$_{B77}$ (0.3 ng ml⁻¹, 10,000 c.p.m. ml⁻¹) (○). *b*, Chicken DNA (5 mg ml⁻¹) with ³H-cDNA$_{sarc}$ (0.45 ng ml⁻¹, 9,000 cpm ml⁻¹) (●); ³²P-cDNA$_{B77}$ (0.5 ng ml⁻¹, 6,000 c.p.m. ml⁻¹) (○), and ¹⁴C-chicken unique sequence DNA (▲). *c*, Quail DNA (9 mg ml⁻¹) with ³H-cDNA$_{sarc}$ and ³²P-cDNA$_{B77}$ as in *a*. *d*, Duck DNA (9 mg ml⁻¹) with ³H-cDNA$_{sarc}$ and ³²P-cDNA$_{B77}$ as in *a*.

(Fig. 1*a* and *b*), as expected in reactions where identical labelled and unlabelled DNA strands are competing for unlabelled complementary strands, and the reactions were incomplete (about 50%) at the highest value of C_0t (Fig. 1*a*, *c* and *d* and Table 1). Nevertheless, we believe that most or all of the nucleotide sequences of cDNA$_{sarc}$ are present in the DNA of quail since cDNA$_{sarc}$ anneals completely with RNA from certain quail cells (unpublished results of ourselves and C. Moscovici). Moreover, the rates and extents of annealing between cDNA$_{sarc}$ and DNA from chicken, duck and turkey are approximately the same as the rate and extent of annealing between cDNA$_{sarc}$ and quail DNA (Table 1 and Fig. 1). Thus, it is likely that most or all of the nucleotide sequences of cDNA$_{sarc}$ are present in the DNA of all these birds.

The extent of the reaction between cDNA$_{sarc}$ and DNA from emu, a relatively primitive Australian bird, was limited (24%) when tested by hydrolysis with a single strand-specific nuclease (Table 1); the extent of the reaction was greater when analysed on hydroxyapatite, a less stringent procedure than the nuclease test (Table 1), and was further augmented when the annealings were performed in conditions which facilitate pairing of partially matched nucleotide sequences (1.5 M NaCl, 59 °C, ref. 21) (Table 1). These date indicate that the nucleotide sequences homologous to cDNA$_{sarc}$ in emu DNA are substantially diverged from the homologous sequences in the other avian DNAs; we have obtained further data to sustain this conclusion by analysing the thermal stability of the duplexes formed between cDNA$_{sarc}$ and various avian DNAs (see below, Table 2 and Fig. 2).

Avian DNAs were also tested with ³²P-labelled single-stranded DNA, complementary to the RNA genome of the B77 strain of ASV (cDNA$_{B77}$), synthesised with detergent-disrupted virions, and purified as described previously[14]. The virus used to prepare cDNA$_{B77}$ consisted mainly of td variants (about 90% of the particles; unpublished observations). Consequently, DNA synthesised with the virus was deficient in nucleotide sequences homologous to cDNA$_{sarc}$ and served principally to detect other portions of the ASV genome. A substantial fraction (about 50%) of cDNA$_{B77}$ reacted with normal chicken DNA, but there was little or no reaction with quail, duck, turkey and emu DNAs (Fig. 1 and unpublished data); these results conform to previous reports[20,22,23]. We conclude that the DNAs from widely divergent avian species all contain nucleotide sequences which are at least partially related to transforming gene(s) of ASV, whereas only chicken DNA has appreciable homology with the remainder of the ASV genome.

The relatedness of DNA sequences homologous to cDNA$_{sarc}$ in different avian species was analysed by denaturing duplexes formed between cDNA$_{sarc}$ and normal avian DNAs (Fig. 2 and Table 2). In addition, we denatured duplexes between cDNA$_{sarc}$ and DNA from XC cells (rat cells transformed by Pr-C ASV) to test completely matched duplexes containing cDNA$_{sarc}$ sequences. The mammalian DNAs we have tested (calf and mouse) contain no nucleotide sequences homologous to cDNA$_{sarc}$ before in-

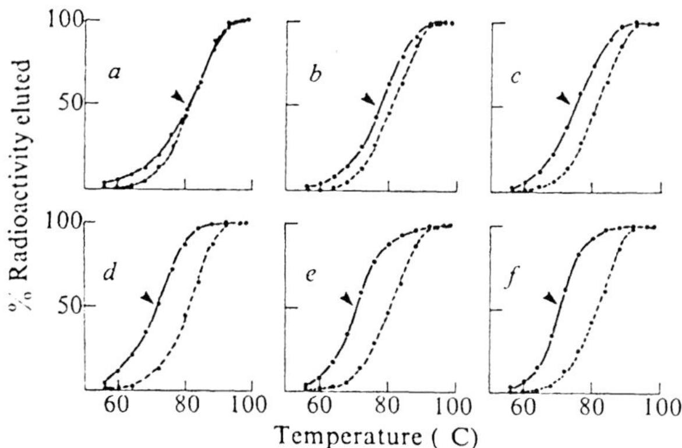

Fig. 2 Thermal denaturation of duplexes formed between cDNA$_{sarc}$ and cellular DNAs. DNA was extracted from XC cells grown in culture, from 10–11-d-old embryos of chickens, ducks and quails, from the livers of adult turkeys, and from liver and heart of a 22-d-old emu. Denatured DNAs (0.5 mg) were incubated with ^{3}H-cDNA$_{sarc}$ (0.25 ng, 5,000 c.p.m) in 0.1 ml of 0.9 M NaCl–0.001 M EDTA–0.02 M Tris-HCl, pH 7.4, for 100 h at 68 °C, final $C_0t = 40,000$. In a separate reaction, ^{32}P-cDNA$_{B77}$ (2.5 ng, 100,000 c.p.m.) was annealed with 7 mg of denatured chicken DNA (final $C_0t = 80,000$). Samples (0.4 mg of chicken DNA and 6,000 c.p.m. of cDNA$_{B77}$) of this reaction were added to the DNAs which had been annealed with cDNA$_{sarc}$. The mixtures were then passed through a column of hydroxyapatite (3 ml packed vol) in 0.12 M sodium phosphate, pH 6.8, at 56 °C; in these conditions, stable duplexes bind to the hydroxyapatite, whereas single-stranded DNA does not. The columns were then washed continuously with 0.12 M sodium phosphate (0.4 ml min^{-1}) while the temperature of the column was raised in increments of 4 °C every 10 min. Fractions (4 ml) were collected and analysed for acid-precipitable radioactivity. ●——●, Duplexes with cDNA$_{sarc}$; ●– – – –●, duplexes with cDNA$_{B77}$ and chicken DNA (internal standard). *a*, Provirus (XC); *b*, chicken; *c*, quail; *d*, turkey; *e*, duck; *f*, emu.

fection by ASV; consequently, the DNA homologous to cDNA$_{sarc}$ in rat cells after infection presumably represents nucleotide sequences contained entirely in recently integrated provirus for ASV rather than related sequences present in the DNA of uninfected cells. Denaturation was standardised internally by adding duplexes formed in a separate annealing reaction between normal chicken DNA and ^{32}P-labelled cDNA$_{B77}$; these duplexes denatured with a T_m of 81 ± 1 °C.

Duplexes between cDNA$_{sarc}$ and DNA from XC cells had the highest T_m (81 °C; Table 2), as expected for completely matched complementary nucleotide sequences. Duplexes with chicken DNA were slightly less stable ($T_m = 77$ °C); the nucleotide sequences of cDNA$_{sarc}$ in the ASV genome are apparently diverged from the homologous sequences in the avian host. The reduction in the T_m is consistent with about 3% mismatching of bases[24]. The stability of duplexes with other avian DNAs ($T_m = 70–74$ °C) decreased roughly in accord with phylogenetic distances among the species

tested (Table 2); the T_mS suggest 5–8% mismatching of base pairs[24]. We made no effort in these tests to obtain maximum duplex formation with cDNA$_{sarc}$, and the reactions with quail, turkey, duck and emu were relatively limited in extent (15–37%). In other experiments where at least 50% of cDNA$_{sarc}$ was annealed into duplexes with chicken, quail and duck DNAs, we observed T_mS similar to those given in Table 2. cDNA$_{sarc}$ seems to represent nucleotide sequences which arose and diverged during the course of avian speciation.

We have shown previously that the nucleotide sequences of cDNA$_{sarc}$ include genetic information required for transformation of fibroblasts by ASV[14]; the data reported here indicate that similar or partially related information is present in the genome of the normal chicken cell and widely distributed among the avian species. We suggest that part or all of the transforming gene(s) of ASV was derived from the chicken genome or a species closely related to chicken, either by a process akin to transduction[25] or by other events, including recombination, which are alleged to have generated type C viruses from normal cellular genes[12]. If the entire ASV genome originated from cellular genes[12], then nucleotide sequences in the transforming gene(s) have been conserved relative to other viral genes, because the nucleotide sequences of cDNA$_{sarc}$ are the only portion of the ASV genome which we can detect in DNA from birds other than chickens. We cannot exclude, however, the possibility that viral genes other than those represented by cDNA$_{sarc}$ have been introduced into the germ line of chickens after speciation. The sequences homologous to cDNA$_{sarc}$ in the genome of ASV are slightly diverged from the analogous sequences in chicken genome; this could be the consequence of either the process which generated viral genes from cellular genes[12] or mutations during the course of repeated viral propagation.

Others have reported evidence concerning the genetic divergence during possible spread of unidentified portions of type C viral genomes during evolution[26,27] and the apparent transduction of cellular nucleotide sequences by type C viruses[25], but our study provides the first data on the origins of a genetically identified set of viral nucleotide sequences. It should be possible to carry out similar studies for at least one other gene of ASV (the gene coding for the type-specific glycoprotein), using techniques similar to those reported here.

Neiman and his colleagues have reported that the hybridisation of ASV 70S RNA to normal chicken DNA was competed completely by RNA from non-transforming viruses[11]. They therefore concluded that the genes responsible for transformation by ASV were present in the avian genome only after viral infection. We cannot explain the discrepancy between their results and ours.

We anticipate that cellular DNA homologous to cDNA$_{sarc}$ serves some function which accounts for its conservation during avian speciation. The nucleotide sequences which anneal with cDNA$_{sarc}$ are part of the

Table 2 Thermal stabilities of duplexes between cDNA$_{sarc}$ and normal DNAs

DNA	%cDNA$_{sarc}$ in duplexes	T_m	ΔT_m	Phylogenetic distance from chicken (Myr)
Provirus (XC)	56	81	0	
Chicken	52	77	− 4	0
Quail	37	74	− 7	35–40
Turkey	30	72	− 9	40
Duck	16	71	−10	80
Emu	15	70	−11	100

Denatured DNAs (0.5 mg) were annealed with ^{3}H-cDNA$_{sarc}$ (0.25 ng, 5,000 c.p.m.), adsorbed to hydroxyapatite in 0.12 M sodium phosphate at 56 °C, and denatured with a thermal gradient, all as described for Fig. 2. Duplexes between ^{32}P-cDNA$_{B77}$ and normal chicken DNA, included in each analysis as an internal standard, denatured with $T_m = 8 1 \pm 1$ °C. The estimates of phylogenetic distance, deduced from fossil records and antigenic relationships among proteins[29], were provided by Professor Allan Wilson (personal communication).

unique fraction of cellular DNA and could represent either structural or regulatory genes. But the function of those sequences is unknown. We are testing the possibilities that they are involved in the normal regulation of cell growth and development or in the transformation of cell behaviour by physical, chemical or viral agents.

We thank William E. Meeker of Sacramento Zoo for assistance in obtaining the emu, Karen Smith and Jean Jackson for technical assistance, and H. Temin and L. Levintow for editorial advice. This work was supported by grants from the US Public Health Services and the American Cancer Society, and a contract within the Virus Cancer Program of the National Cancer Institute. D.S. was supported by CNRS and H.E.V. is a recipient of a research career development award from the National Cancer Institute.

D. Stehelin*
H. E. Varmus
J. M. Bishop

*Department of Microbiology,
University of California,
San Francisco, California 94143*

P. K. Vogt

*Department of Microbiology,
University of California,
Los Angeles, California 90033*

Received November 10, 1975; accepted January 2, 1976.

*Present address: IRSC BP8, 94800–Villejuif, France.

1 Martin, G. S., *Nature*, 227, 1021–1023 (1970).
2 Kawai, S., and Hanafusa, H., *Virology*, 46, 470–479 (1971).
3 Bader, J. P., *J. Virol.*, 10, 267–276 (1972).
4 Wyke, J. A., Bell, J. G., and Beamand, J. A., *Cold Spring Harb. Symp. quant. Biol.*, 39, 897–905 (1974).
5 Baluda, M. A., *Proc. natn. Acad. Sci. U.S.A.*, 69, 576–580 (1972).
6 Rosenthal, P. N., Robinson, H. L., Robinson, W. S., Hanafusa, T., and Hanafusa, H., *Proc. natn. Acad. Sci. U.S.A.*, 68, 2336–2340 (1971).
7 Varmus, H. E., Weiss, R. A., Friis, R. R., Levinson, W., and Bishop, J. M., *Proc. natn. Acad. Sci. U.S.A.*, 69, 20–24 (1972).
8 Wright, S. E., and Neiman, P. E., *Biochemistry*, 13, 1549–1554 (1974).
9 Vogt, P. K., and Friis, R. R., *Virology*, 43, 223–234 (1971).
10 Weiss, R. A., Friis, R. R., Katz, E., and Vogt, P. K., *Virology*, 46, 920–938. (1971).
11 Neiman, P. E., Wright, S. E., McMillin, C., and MacDonnell, D., *J. Virol.*, 13, 837–846 (1974).
12 Temin, H. M., *Cancer Res.*, 34, 2835–2841 (1974).
13 Huebner, R. J., and Todaro, G. J., *Proc. natn. Acad. Sci. U.S.A.*, 64, 1087–1094 (1969).
14 Stehein, D., Guntaka, R. V., Varmus, H. E., and Bishop, J. M., *J. molec. Biol.* (in the press).
15 Duesberg, P. H., and Vogt, P. K., *Proc. natn. Acad. Sci. U.S.A.*, 67, 1673–1680 (1970).
16 Lai, M. M. C., Duesberg, P. H., Horst, J., and Vogt, P. K., *Proc. natn. Acad. Sci. U.S.A.*, 70, 2266–2270 (1973).
17 Duesberg, P. H., and Vogt, P. K., *J. Virol.*, 12, 594–599 (1973).
18 Beemon, K., Duesberg, P., and Vogt, P. K., *Proc. natn. Acad. Sci. U.S.A.*, 71, 4254–4258 (1974).
19 Billeter, M. A., Parsons, J. T., and Coffin, J. M., *Proc. natn. Acad. Sci. U.S.A.*, 71, 3560–3564 (1974).
20 Varmus, H. E., Heasley, S., and Bishop, J. M., *J. Virol.*, 14, 895–903 (1974).
21 Rice, N., and Paul, P., *Yb. Carnegie Instn. Wash.*, 71, 262–264 (1972).
22 Tereba, A., Skoog, L., and Vogt, P. K., *Virology*, 65, 524–534 (1975).
23 Kang, C. Y., and Temin, H. M., *J. Virol.*, 14, 1179–1188 (1974).
24 Ullman, J. S., and McCarthy, B. J., *Biochim. biophys. Acta*, 294, 405–415 (1973).
25 Scolnick, E. M., Rands, E., Williams, D., and Parks, W. P., *J. Virol.*, 12, 458–463 (1973).
26 Benveniste, R. E., and Todaro, G. J., *Proc. natn. Acad. Sci. U.S.A.*, 71, 4513–4518 (1974).
27 Benveniste, R. E., and Todaro, G. J., *Nature*, 252, 456–459 (1974).
28 Leong, J. A., *et al.*, *J. Virol.*, 9, 891–902 (1972).
29 Praeger, E. M., Brush, A. H., Nolan, R. A., Nakaniski, M., and Wilson, A. C., *J. molec. Evol.*, 3, 243 (1974).

Spliced Segments at the 5' Terminus of Adenovirus 2 Late mRNA

S. M. BERGET, C. MOORE, AND P. A. SHARP

Reprinted from *Proceedings of the National Academy of Sciences USA* 74:3171–3175. Copyright © 1977, by permission of the authors.

An Amazing Sequence Arrangement at the 5' Ends of Adenovirus 2 Messenger RNA

L. T. CHOW, R. E. GELINAS, T. R. BROKER, AND R. J. ROBERTS

Reprinted from *Cell* 12:1–8. Copyright © 1977, by permission of Cell Press

Again, a completely unexpected discovery. In contrast to the situation in procaryotes, little, if anything, was known in the mid 1970s concerning the signals that specify where transcription initiates and terminates on eucaryotic genes. There had long been evidence, inferential at first and gradually more and more definitive, that a very significant fraction of newly synthesized RNA does not reach the cytoplasm. This could be interpreted as transcription proceeding in the form of large transcripts that are subsequently cleaved to yield functional mRNA; in fact, Lennart Phillipson and coworkers had shown, in 1971, that long polyadenylated transcripts formed in the nucleus appear subsequently as smaller polyadenylated molecules associated with polysomes in the cytoplasm. Defining the relation between large transcripts and much smaller mRNAs was not, however, a problem that could be solved by examining cellular RNA transcription; and both the Sharp and the Roberts laboratories chose the adenovirus system to analyze it in detail. Both used variations of electron microscopic techniques to visualize heteroduplexes of late viral mRNA and adenovirus DNA and were able to demonstrate that late adenovirus mRNAs consist of coding sequences preceded by a leader sequence that is derived from three widely separated regions in the viral genome. The sequences between these regions (the intervening sequences or introns) were postulated and subsequently indeed shown to be present in the original transcripts and spliced out by a highly complicated mechanism, the exact nature of which is still not entirely clear. It was soon demonstrated that this situation is not peculiar to adenovirus, but that it applies to cellular mRNAs as well, many of which are derived from very long transcripts that contain multiple introns separated by thousands of base pairs. The realization that coding sequences are often separated in genomes by long noncoding sequences has drastically changed our ideas concerning the evolution and organization of genetic material. Both Phil Sharp and Rich Roberts were awarded the Nobel Prize in 1993.

W. K. JOKLIK

Proc. Natl. Acad. Sci. USA
Vol. 74, No. 8, pp. 3171–3175, August 1977
Biochemistry

Spliced segments at the 5′ terminus of adenovirus 2 late mRNA*

(adenovirus 2 mRNA processing/5′ tails on mRNAs/electron microscopy of mRNA·DNA hybrids)

Susan M. Berget, Claire Moore, and Phillip A. Sharp

Center for Cancer Research and Department of Biology, Massachusetts Institute of Technology, Cambridge, Massachusetts 02139

Communicated by David Baltimore, May 9, 1977

ABSTRACT   An mRNA fraction coding for hexon polypeptide, the major virion structural protein, was purified by gel electrophoresis from extracts of adenovirus 2-infected cells late in the lytic cycle. The mRNA sequences in this fraction were mapped between 51.7 and 61.3 units on the genome by visualizing RNA·DNA hybrids in the electron microscope. When hybrids of hexon mRNA and single-stranded restriction endonuclease cleavage fragments of viral DNA were visualized in the electron microscope, branched forms were observed in which 160 nucleotides of RNA from the 5′ terminus were not hydrogen bonded to the single-stranded DNA. DNA sequences complementary to the RNA sequences in each 5′ tail were found by electron microscopy to be located at 17, 20, and 27 units on the same strand as that coding for the body of the hexon mRNA. Thus, four segments of viral RNA may be joined together during the synthesis of mature hexon mRNA. A model is presented for adenovirus late mRNA synthesis that involves multiple splicing during maturation of a larger precursor nuclear RNA.

Most eukaryotic mRNAs bear modifications at both termini; their 3′ termini have a tract of poly(A) that ranges in length from 30 to 200 bases (1–4), while their 5′ termini are typically capped with a methylated guanine joined through a 5′-5′ pyrophosphate linkage to a second nucleotide methylated at its 2′ position (5, 6). Both types of modifications of eukaryotic mRNA are known to occur after transcription.

All adenovirus mRNAs are thought to contain poly(A) tracts at their 3′ termini (7) and be capped with a methylated guanine (8, 9). Specific restriction endonuclease cleavage fragments of adenovirus 2 (Ad2) DNA have permitted the mapping of regions of the genome expressed as mRNA and viral proteins during different stages of the lytic cycle (10–12). Little is known about the molecular mechanisms of viral mRNA synthesis. An important aspect of late mRNA synthesis is thought to be the processing and selection of viral mRNAs from the nucleus (13, 14). We have purified a late Ad2 hexon mRNA and found evidence providing some insight into the mechanism of synthesis of this mRNA.

MATERIALS AND METHODS

Isolation of Ad2 DNA and RNA. Polyribosomal RNA was prepared from Ad2-infected cells 32 hr after infection as described by Flint and Sharp (14, 15) and selected by chromatography on poly(U)-Sephadex (16).

R-Loop Mapping. The R-loop hybridization mixture was essentially that of Thomas *et al.* (17) and contained 70% (vol/vol) formamide [Matheson, Coleman, and Bell, 99%, further purified as described by Duesberg and Vogt (18)]; 0.20 M Tris·HCl, pH 7.91; 0.50 M NaCl; 0.01 M EDTA; Ad2 DNA at 10 μg/ml; and purified hexon mRNA at 1–10 μg/ml. This mixture was incubated at 52.5° for 2–3 hr and spread on a hy-

pophase of water with internal length standards of DNA from bacteriophage φX174, 5375 bases (19).

Hybridization to Single-Stranded Ad2 DNA. Hybridizations of either polyribosomal poly(A) or purified hexon mRNA with restriction endonuclease fragments of Ad2 DNA were carried out in reaction mixtures of 80% formamide; 0.40 M NaCl; 0.04 M 2-(*N*-morpholino)ethanesulfonic acid (Mes), pH 6.2; 0.01 M EDTA; DNA at 10 μg/ml; and hexon mRNA at 1.0–10 μg/ml (20). The sample was incubated at 57–60° for 2–3 hr.

RESULTS

Adenovirus late mRNAs begin to appear on polyribosomes about 13 hr after infection and continue to accumulate in the cell throughout the lytic cycle (21). Thus, to fractionate the most abundant late mRNAs, polyribosomes were prepared from cells 32 hr after infection with Ad2 and poly(A)-containing mRNA was selected by chromatography on poly(U)-Sephadex columns. These mRNAs were then resolved into different molecular weight fractions by electrophoresis in 2.4–4.0% linear gradient polyacrylamide gels containing a uniform concentration of 7 M urea. After staining with ethidium bromide, distinct fluorescent bands were present in gels containing mRNA from virus-infected cells that were not found in gels containing identically prepared HeLa cell mRNA (Fig. 1A). These virus-specific RNAs were selectively labeled when [^{32}P]phosphate was added to infected cells 24 hr after infection and the same mRNA fractions were prepared (Fig. 1B). RNA from the predominant ethidium bromide-staining band migrating 1.5 times faster than 28S rRNA in Fig. 1A and marked with a large arrow has been shown to code for the hexon polypeptide by *in vitro* translation (S. M. Berget, B. E. Roberts, and P. A. Sharp, data not shown). Furthermore, this RNA has been mapped by the R-loop technique (see below) to a region of the genome known to code for hexon (12) and is complementary to the *r* strand of the viral DNA (11). This mRNA species, therefore, will be referred to in the following sections as hexon mRNA.

R-Loop Mapping of Hexon RNA. The R-loop technique developed by White and Hogness (22) and Thomas *et al.* (17) was used to position purified hexon mRNA on the viral genome. RNA eluted from a gel similar to that shown in Fig. 1A was incubated, as described in *Materials and Methods*, with either total Ad2 DNA or restriction endonuclease fragments. Of the 43 total Ad2 DNA molecules scored as containing R-loops, 41 were observed to have a single region of hybrid, while two molecules contain a second R-loop, apparently in the region of the genome coding for the 100K polypeptide (12).

Fig. 2 shows two examples of R-loops resulting from hybridization of hexon mRNA to fragments generated by the cleavage of Ad2 DNA with the *Eco*RI restriction endonuclease

Abbreviation: Ad2, adenovirus 2.
* *We dedicate this work to the memory of Jerome Vinograd, a man who loved science.*

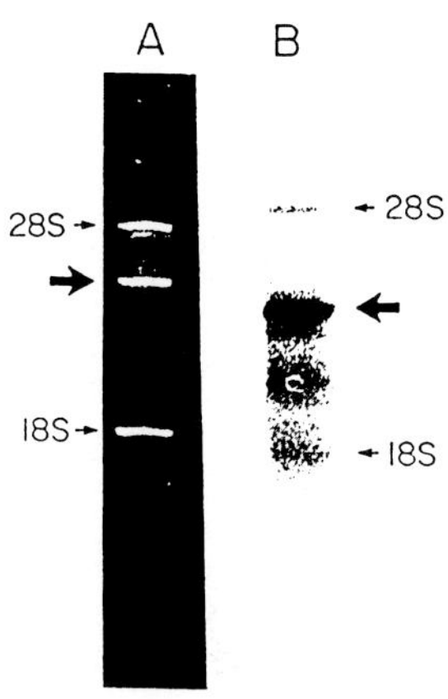

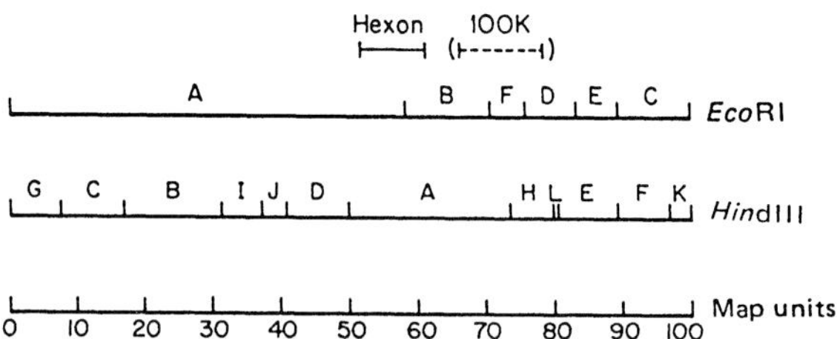

FIG. 3. Restriction map of Ad2 DNA. The vertical lines indicate the positions of cleavage for either the *Eco*RI or *Hin*dIII restriction endonucleases. The positions of genes coding for the hexon and 100,000 molecular weight (100K) polypeptides are from Lewis *et al.* (12).

FIG. 1. Polyacrylamide gel electrophoresis of Ad2 mRNA. Polyribosomal poly(A)-containing RNA was prepared 32 hr after infection as described in *Materials and Methods* from either Ad2-infected cells (*A*) or infected cells to which [^{32}P]phosphate was added 24 hr after infection (*B*). Approximately 25 μg of this RNA was resolved by electrophoresis on 2.4–4.0% polyacrylamide gels containing 7 M urea for 16 hr at 100 V. The gel was either stained with 0.50 μg/ml of ethidium bromide and photographed (*A*) or autoradiographed (*B*).

(see Fig. 3 for fragment location). Hexon RNA spans the junction between the *Eco*RI A and B fragments, thus creating branches at one end of each fragment; one strand of the branch is single-stranded, the other is double-stranded and terminated in a ball of single-stranded RNA. Comparison of the lengths of the two hybrids with an internal standard of double-stranded φX174 DNA maps the 5′ end of the RNA at 51.7 ± 0.5 units (uncertainties are as indicated in ref. 19) of the genome and the 3′ end at 61.3 ± 0.5 units, in close agreement with other estimations based on both R-loop mapping with total cytoplasmic RNA (23, 24) and the viral polypeptide mapping of Lewis *et al.* (12).

When hexon mRNA was incubated under conditions to form R-loops with Ad2 DNA that had been cleaved with the *Hin*dIII restriction endonuclease, R-loops of the type shown in Fig. 4

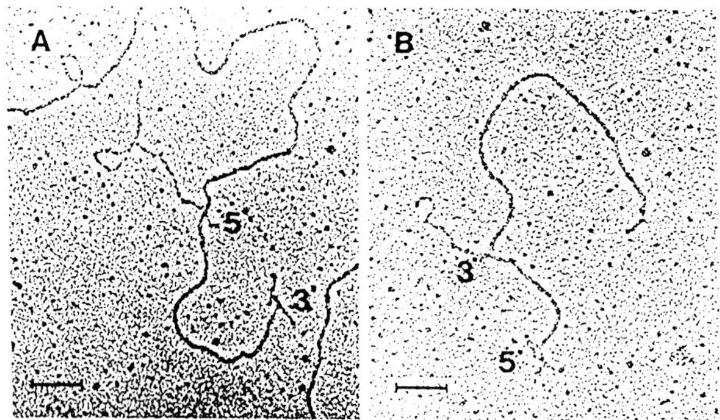

FIG. 2. R-Loops of purified hexon mRNA and *Eco*RI cleavage products of Ad2 DNA. Total Ad2 DNA cleaved by *Eco*RI and hexon mRNA were incubated and then spread to visualize R-loops as described in the *text*. Two examples of the 60 R-loop structures photographed and measured are shown; hybrid structures are observed at the ends of the *Eco*RI A fragment (*A*) and the *Eco*RI B fragment (*B*). The junction of these two fragments maps at 58.5 units on the Ad2 genome. The strand specificity of the mRNA sequences from this region (11) and the *Eco*RI cleavage map of Ad2 DNA were used to assign the 5′ and 3′ polarity of the RNA forming these R-loops. Bars represent 0.1 μm.

A, B, and *C* were observed. The mRNA is totally included within the *Hin*dIII A fragment, forming R-loops terminating at positions 6.3 ± 0.1% and 50.2 ± 0.8% from the end of the fragment positioned at 50.1–73.6 map units on the Ad2 genome. Because hexon mRNA is known to be transcribed in the rightward direction (11), the map coordinates of hexon mRNA established from the R-loops to *Eco*RI fragments indicate that the 5′ end of hexon mRNA is 6.3% from the end of the *Hin*dIII A fragment. Small single-stranded "tails" were visible at both ends of the R-loop; such tails appeared on 88% of the 5′ ends (all polarities given with respect to the mRNA orientation) and on 75% of the 3′ ends of those R-loops with termini that mapped within two standard deviations of the mean position of the termini expected for full-length hybrids. Although the 3′ tail might in part be attributed to the poly(A) tracts of these mRNAs, the tail of the 5′ end of the mRNA was not expected and prompted further investigation.

Hybridization of Hexon RNA to Single-Stranded DNA. A possible interpretation of single-stranded tail-like structures at the ends of R-loops would be that branch migration creating DNA·DNA duplex had occurred, displacing the ends of the RNA (25). To examine this possibility, hybrids were formed with hexon mRNA and totally single-stranded *Hin*dIII A DNA. In such hybrids there would be no competing DNA·DNA renaturation to displace RNA·DNA hybrids. Such RNA·DNA hybrids were formed by incubating hexon mRNA with denatured *Hin*dIII A DNA at high formamide concentration (80%) and at 57° (20). After an appropriate incubation the hybrids were spread and examined under the electron microscope. As expected, little or no duplex *Hin*dIII A fragment was observed. Fig. 4 *D, E,* and *F* shows the types of hybrids that were observed with hexon mRNA and *Hin*dIII A DNA. Double-stranded hybrid segments were terminated with clearly visible tails at both the 5′ and 3′ ends of the mRNA molecules. Of those molecules having full-length hybrid, 90% had 5′ tails and 64% had 3′ tails. There may have been a bias in favor of selecting full-length hybrids with tails for screening because it was difficult to accurately position the ends of a duplex region that did not terminate in a forked structure. However, approximately 20% of the total *Hin*dIII A strands displayed a forked structure at the 5′-end position of the hexon mRNA. This result strongly suggests that the sequences in the tails are not complementary to the adjacent DNA sequences.

Histograms comparing the lengths of the hybrid regions observed in the R-loop technique to those produced by hybridization of hexon mRNA with denatured *Hin*dIII A DNA are shown in Fig. 5 *A* and *B*. When lengths are calculated from the hatched area of the histogram (those molecules containing full-length hexon RNA) the hybrid length is 3330 ± 290 base pairs for R-loop hybrids and 3540 ± 240 base pairs for hybrids formed with single-stranded *Hin*dIII A DNA. This agrees very

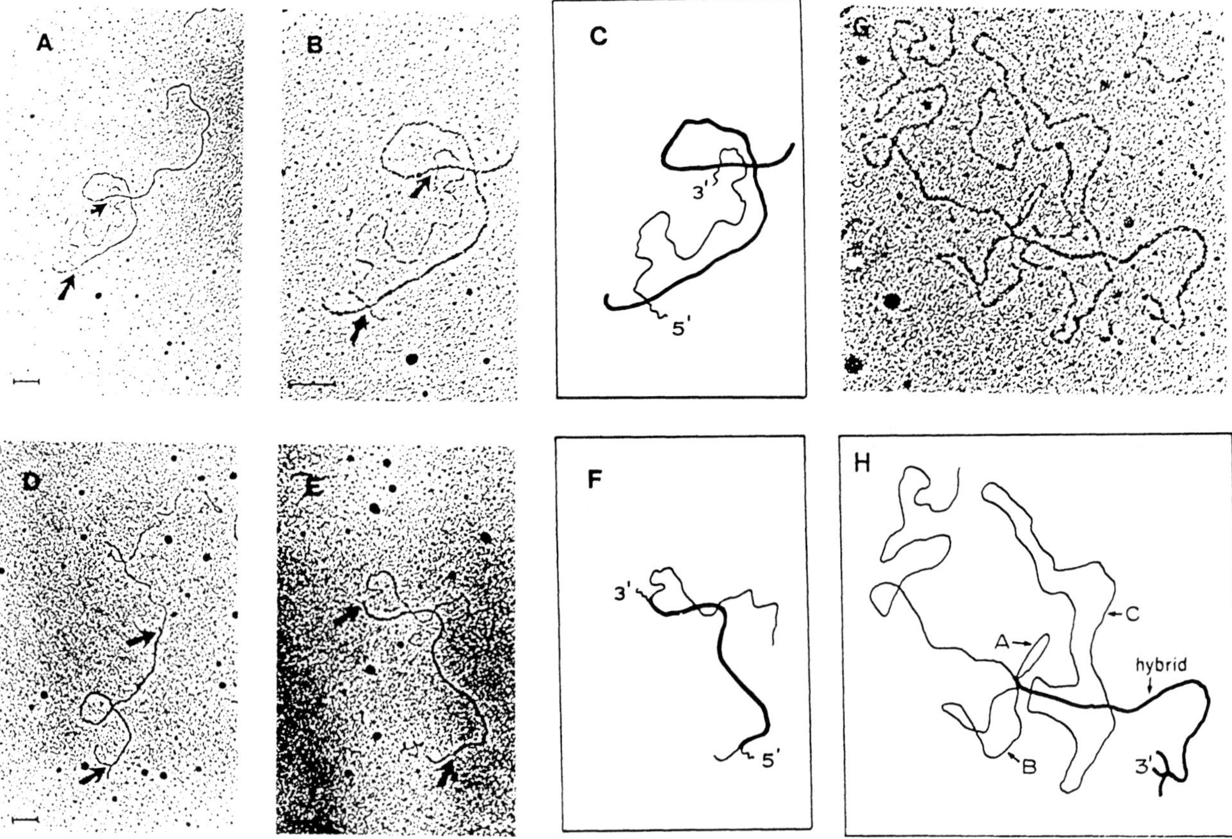

FIG. 4.   Electron micrographs of hybrids of hexon mRNA and fragments of Ad2 DNA. An example of an R-loop hybrid observed after incubation of hexon mRNA and duplex *Hin*dIII A fragment DNA is shown in A and B and is diagrammed schematically in C. Similarly, two examples of hybrids of hexon mRNA and single-stranded *Hin*dIII A fragment are shown in D and E. A schematic of the hybrid structure shown in E is given in F. The single-stranded RNA at the end of the hybrid region is represented by a wave-like line. The hybrids of hexon mRNA and single-stranded *Hin*dIII A fragment DNA shown in D and c were mounted from an 80% formamide solution. In A, B, D, and E the positions of the RNA tails at the 5′ and 3′ ends of the hybrids are denoted by arrows. An example of a hybrid between single-stranded *Eco*RI A DNA and hexon RNA is shown in G and diagrammed in H. The hybrid region is indicated by a heavy line; loops A, B, and C (single-stranded unhybridized DNA) are joined by hybrid regions resulting from annealing of upstream DNA sequences to the 5′ tail of hexon mRNA. Bars on micrographs represent 0.1 um.

well with the length of the RNA itself, 3510 ± 180 bases, as determined by visualization in the electron microscope following spreading by the urea/formamide technique (Fig. 5C) (26).

The measured lengths of 5′ and 3′ tails on the two types of hybrids are similar; the 5′ tails measure 170 ± 40 bases on R-loops and 160 ± 50 bases on hybrids with single-stranded DNA (Fig. 5 D and E); and the 3′ tail measures 150 ± 60 nucleotides on R-loops and 110 ± 40 nucleotides on hybrids with single-stranded DNA (Fig. 5 F and G). These contour lengths may be an underestimate because they were calculated assuming that single-stranded RNA chains were fully extended under the formamide spreading conditions employed. Those molecules having hybrids but no tails are also scored on the histograms in Fig. 5. In both techniques, those molecules having less than a full-length hybrid region were always missing at least one tail, suggesting that neither method artifactually generates such structures. Fig. 5 also contains two histograms showing the position of the 5′ tail with respect to the 50.1 unit end of the Ad2 *Hin*dIII A fragment (H and I); the 5′ tail appears to begin 480 ± 40 base pairs from the end of R-loop molecules and 410 ± 50 bases from an end of the hybrid molecules formed with the single-stranded *Hin*dIII A fragment.

The presence of branched structures suggested that the 5′ end of hexon mRNA may not be complementary to the adjacent region of DNA. Several alternate explanations remained to be eliminated by control experiments. The first involved the possibility that the branched structure was due to an unusually (A+T)-rich set of DNA sequences at this position which would be melted at the high formamide–high temperature spreading conditions employed. To eliminate this possibility, hybridization mixtures of hexon RNA and the purified denatured *Hin*dIII A fragment were diluted 70-fold into either 50% or 40% formamide solution and prepared for electron microscopy; under these conditions melting of even highly (A+T)-rich complementary sequences should not be observed. However, hybrid structures still contained 5′ and 3′ tails at the same frequency as those scored at the higher formamide concentrations (histogram not shown).

Another possibility was that the tails might arise from palindromic sequences at the ends of the mRNA molecules which were more stable as RNA·RNA hybrids and thus would not form hybrids with the complementary DNA. If such palindromic sequences were present in the Ad2 *Hin*dIII A fragment at this position, they should be visible as hairpins on single-stranded *Hin*dIII A DNA spread under low formamide

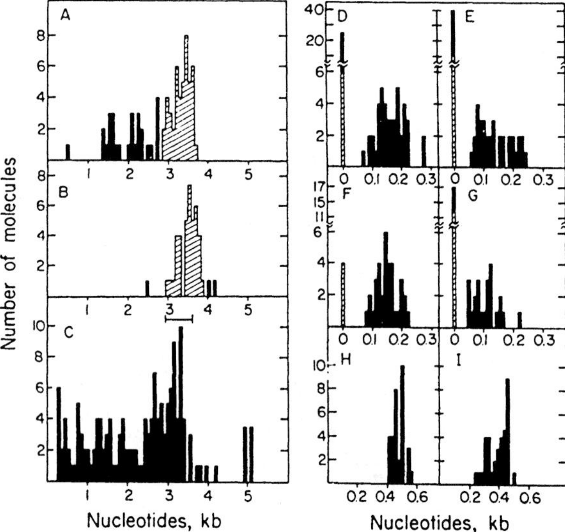

FIG. 5.   Histograms of contour lengths of various parts of hexon mRNA and *Hin*dIII A DNA hybrids. kb is kilobases, 1000 bases or base pairs. *A* is a histogram of the contour length of the RNA·DNA duplex in R-loops of hexon mRNA and duplex *Hin*dIII A DNA. Similarly, *B* is a histogram of the contour length of RNA·DNA duplex in hybrids of hexon mRNA and single-stranded *Hin*dIII A DNA. An example of each of these types of hybrids is shown in *A* and *D*, respectively, of Fig. 4. In both *A* and *B* the hatched area represents hybrids found by intact RNA chains and these molecules were used in calculating average lengths of RNA·DNA duplex. *C* is the contour length histogram of the hexon mRNA as it is eluted from the gel. The molecules bracketed by the bar were assumed to be intact. *D* and *E* are the contour length histograms of the 5′ and 3′ single-stranded RNA tails, respectively, for the R-loop hybrids scored in *A*. Similar histograms for the 5′ and 3′ single-stranded RNA tails of the hybrids formed with hexon mRNA and single-stranded *Hin*dIII A DNA, respectively, are shown in *F* and *G*, respectively. A histogram of the duplex contour length between the end of the *Hin*dIII A DNA and the beginning of the R-loop hybrid is given in *H*. The histogram for the equivalent contour length between the end of the single-stranded *Hin*dIII A DNA and the beginning of the RNA·DNA hybrid region is shown in *I*. The hatched and solid areas scored on the 0 nucleotide position in *D* through *G* represent molecules falling in the hatched and solid areas, respectively, of *A* and *B*.

conditions. Therefore, single-stranded *Hin*dIII A DNA fragment was spread from a 50% formamide solution and visualized with the electron microscope. A histogram of the position of all hairpin structures relative to the nearest end of the single-stranded DNA segment was constructed for 51 molecules. No hairpin structures were observed at the map position of either end of the hexon mRNA (data not shown).

To ensure that the tails were linked to the hybrid by ribonuclease-sensitive bonds, hybrids were spread for visualization after treatment with pancreatic RNase under conditions where hybrid structures should be resistant to degradation. After RNase treatment, no 5′ or 3′ tails were observed, though the appropriate RNA·DNA hybrid length was seen when spread from an 80% formamide solution (data not shown).

If the sequences in the 5′ tail of hexon mRNA are transcribed upstream from the same template strand as the other 95% of the RNA sequences, then a hybrid of this mRNA and single-stranded *Eco*RI A DNA should form single-stranded DNA loops at the 5′ terminus of the duplex part of the hybrid. The single-stranded DNA forming the loop would correspond to the viral DNA sequences between the two regions of the template strand that were transcribed and joined during synthesis of the

hexon mRNA. This experiment was performed and RNA·DNA hybrids of *Eco*RI A DNA and hexon mRNA were selected for scoring which, as expected, had a duplex region on one terminus terminating in a collapsed ball of single-stranded RNA. The R-loop data described in Fig. 1 predict that 70.9% of the hexon mRNA adjacent to the 5′ terminus should form hybrids with single-stranded *Eco*RI A DNA; the remaining 29.1% would be collapsed under these spreading conditions. An example of the RNA·DNA hybrids observed is shown in Fig. 4*G* and schematically in Fig. 4*H*. Data from 24 such structures were averaged for the following discussion. At the left end of the structure there is a single-stranded segment 5770 ± 390 bases (16.8% of the genome) in length followed by three deletion type loops of single-stranded DNA originating within 200 bases of the 5′ terminus of the RNA·DNA hybrid segment. The RNA·DNA hybrid has the expected contour length of 2710 ± 320 bases (7.74%). The order and contour length of the three loops from the left end of the genome to the right are: loop A, 1010 ± 130 (2.90%); loop B, 2350 ± 130 (6.70%); and loop C, 8060 ± 830 (23.0%). Loops A and B are separated by 80 ± 20 bases and loops B and C by 110 ± 10 bases. The simplest interpretation of this structure is that the 5′ tail of the hexon mRNA is composed of sequences transcribed from three different regions of the same strand of the viral DNA. The map positions of these three regions are 16.8 ± 1.1, 19.8 ± 1.1, and 26.9 ± 1.1. The segment of RNA·DNA hybrid at the 5′ terminus of the hexon mRNA creating loop A is too short to be distinguished in our electron micrographs. A comparison of the sum of the lengths of the duplex segments separating the three loops, 190 ± 30 bases, with the measured length of the 5′ tail on the hexon mRNA, 160 ± 50 bases, suggests that this region may be quite short. However, this segment would probably have to be at least 15 bases long to be stable under the denaturing conditions used for spreading these samples. Fifty single-stranded *Eco*RI A DNA molecules that contained one or more loops were scored from the same grid; no loop structures were observed that corresponded to the loops seen in the hexon mRNA/*Eco*RI A hybrids.

DISCUSSION

The most abundant viral mRNA found on the polyribosomes of cells 32 hr after infection with adenovirus 2 maps by the R-loop technique in the region of the genome that codes for the hexon polypeptide (12). When R-loops between this mRNA and the *Hin*dIII A fragment were examined in the electron microscope, almost all molecules containing an intact mRNA had single-stranded RNA tails of 160 nucleotides at their 5′ ends. To test whether this single-stranded 5′-end RNA tail was produced by branch migration forming homologous DNA·DNA base pairs, hybrids were formed between the purified mRNA and denatured *Hin*dIII A fragment DNA. A forked structure was observed at the 5′ end of this mRNA in almost all hybrids formed by the annealing of an intact mRNA chain to single-stranded DNA. This forked structure was observed under a variety of different conditions of mounting for visualization in the electron microscope and strongly suggests that a segment of the 5′ end of the mRNA is not complementary to the adjacent viral DNA sequences. The RNA sequences in each 5′ tail are apparently transcribed from the *r* strand of Ad2 DNA upstream from those coding for the body of the hexon mRNA. The structure of hexon mRNA and single-stranded *Eco*RI A DNA hybrids (see Fig. 4 *G* and *H*) suggests that RNA sequences of unknown length from 16.8 ± 1.1%, of 80 ± 20 bases from 19.8 ± 1.1%, and 110 ± 10 bases from 26.9 ± 1.1% are joined in the 5′ tail of hexon mRNA.

When total poly(A)-containing polyribosomal RNA was hybridized to denatured *Hin*dIII A under the same conditions, a second mRNA mapping in the region of the genome coding for the 100K polypeptide (12) was observed to have a similar forked structure at its 5′ terminus. Thus, a common short sequence of RNA might be attached to several late mRNAs. This is consistent with the observation of R. Gelinas, D. Klessig, and R. Roberts (personal communication) that a single T1 ribonuclease oligonucleotide containing a capped structure is found on total viral mRNAs isolated during the late stage of infection.

The three short segments forming the 5′ tail of hexon mRNA are probably spliced to the body of this mRNA during posttranscriptional processing. During the late stage of the lytic cycle the *r* strand of Ad2 is transcribed into long transcripts that originate in the left third of the genome and terminate near the right end (27–30). The region of the genome coding for the body of the hexon mRNA and the sequences in these three short RNA segments in the 5′ tail of this mRNA are probably included in this long transcript. Thus, a plausible model for the synthesis of the mature hexon mRNA would be the intramolecular joining of these short segments to the body of the hexon mRNA during the processing of a nuclear precursor to generate the mature mRNA. This would result in the maturation of one mRNA species from each longer precursor and would explain the large abundance of accumulated viral RNA sequences in the nucleus of cells during the late stage of the lytic cycle and the selective transport of certain viral RNA sequences to the cytoplasm (14). It is interesting to speculate on how general such a model for the processing of eukaryotic mRNAs could be. Assuming that eukaryotic mRNA sequences are adjacent to the 3′ terminus of heterogeneous nuclear RNA, this mechanism would certainly explain the observations by Perry and Kelley (31) that the 5′-terminal cap 1 structures of heterogeneous nuclear RNA from mouse cells are conserved during the processing of these sequences to cytoplasmic mRNAs, though the lengths of the RNA chains differ by a factor of 4 between these RNA fractions.

The role of the spliced RNA segment at the 5′ end of adenovirus late mRNA is subject to speculation. This RNA segment could be involved in the selection of certain viral RNA sequences for transport to the cell cytoplasm or could be responsible for the preferential translation of viral mRNA during the late stage of infection. Because the capped 5′ terminus of eukaryotic mRNA is thought to be directly involved in the initiation of translation of mRNA, an involvement of these sequences in the control of translation would be expected.

We would like to thank Arnold J. Berk, Timothy J. Harrison, Daniel Donoghue, and David Baltimore for comments on the manuscript, and Ms. Margarita Siafaca for typing the manuscript. We gratefully acknowledge the suggestion by David Baltimore that we map the RNA sequences in the 5′ tail by electron microscopy of RNA·DNA hybrids. This work was supported by an American Cancer Society Grant and career development support (VC-151A) to P.A.S., a Cancer Center Core Grant (CA-14051), and a National Institutes of Health Postdoctoral Fellowship to S.M.B. (CA02391-01).

1. Kates, J. (1970) *Cold Spring Harbor Symp. Quant. Biol.* **35**, 743–752.
2. Edmonds, M., Vaughan, M. H., Jr. & Nakazoto, H. (1971) *Proc. Natl. Acad. Sci. USA* **68**, 1336–1340.
3. Lee, S. Y., Mendecki, J. & Brawerman, G. (1971) *Proc. Natl. Acad. Sci. USA* **68**, 1331–1335.
4. Darnell, J. E., Jr., Wall, R. & Tushinski, R. J. (1971) *Proc. Natl. Acad. Sci. USA* **68**, 1321–1325.
5. Furuichi, Y., Morgan, M., Muthukrishnan, S. & Shatkin, A. J. (1975) *Proc. Natl. Acad. Sci. USA* **72**, 362–366.
6. Wei, C. M. & Moss, B. (1974) *Proc. Natl. Acad. Sci. USA* **71**, 3014–3018.
7. Philipson, L., Wall, R., Glickman, G. & Darnell, J. E. (1971) *Proc. Natl. Acad. Sci. USA* **68**, 2806–2809.
8. Hashimoto, S. & Green, M. (1976) *J. Virol.* **20**, 425–435.
9. Moss, B. & Koczot, F. (1976) *J. Virol.* **17**, 385–392.
10. Sharp, P. A. & Flint, S. J. (1976) *Current Topics in Microbiology and Immunology* **74**, 137–158.
11. Sharp, P. A., Gallimore, P. H. & Flint, S. J. (1974) *Cold Spring Harbor Symp. Quant. Biol.* **34**, 457–474.
12. Lewis, J., Atkins, J. F., Anderson, C., Baum, P. R. & Gesteland, R. F. (1975) *Proc. Natl. Acad. Sci. USA* **72**, 1344–1348.
13. Bachenheimer, S. & Darnell, J. E. (1975) *Proc. Natl. Acad. Sci. USA* **72**, 4445–4449.
14. Flint, S. J. & Sharp, P. A. (1976) *J. Mol. Biol.* **106**, 749–771.
15. Flint, S. J., Gallimore, P. H. & Sharp, P. A. (1975) *J. Mol. Biol.* **96**, 47–68.
16. Lindberg, U., Persson, T. & Philipson, L. (1972) *J. Virol.* **10**, 909–919.
17. Thomas, M., White, R. L. & Davis, R. W. (1976) *Proc. Natl. Acad. Sci. USA* **73**, 2294–2298.
18. Duesberg, P. H. & Vogt, P. K. (1973) *J. Virol.* **12**, 594–599.
19. Davis, R. W., Simon, M. & Davidson, N. (1971) in *Methods in Enzymology*, eds. Grossman, L. & Moldave, K. (Academic Press, New York), Vol. 21, pp. 413–428.
20. Casey, J. & Davidson, N. (1977) *Nucleic Acid Res.* **4**, 1539–1552.
21. Green, M. (1970) *Annu. Rev. Biochem.* **39**, 701–756.
22. White, R. L. & Hogness, D. S. (1977) *Cell* **10**, 177–192.
23. Westphal, H., Meyer, J. & Maizel, J. (1976) *Proc. Natl. Acad. Sci. USA* **73**, 2069–2071.
24. Chow, L. T., Roberts, J. M., Lewis, J. B. & Broker, T. M. (1977) *Cell*, in press.
25. Lee, C. S., Davis, R. W. & Davidson, N. (1970) *J. Mol. Biol.* **48**, 1–22.
26. Robberson, D., Aloni, Y., Attardi, G. & Davidson, N. (1971) *J. Mol. Biol.* **60**, 473–484.
27. Parsons, J. T. & Green, M. (1971) *Virology* **45**, 154–162.
28. Wall, R., Philipson, L. & Darnell, J. E. (1972) *Virology* **50**, 27–34.
29. McGuire, P. M., Swart, C. & Hodge, L. D. (1972) *Proc. Natl. Acad. Sci. USA* **69**, 1578–1582.
30. Goldberg, S., Weber, J. & Darnell, J. E. (1977) *Cell* **10**, 617–622.
31. Perry, R. P. & Kelley, D. E. (1976) *Cell* **8**, 433–442.

Cell, Vol. 12, 1–8, September 1977, Copyright © 1977 by MIT

An Amazing Sequence Arrangement at the 5′ Ends of Adenovirus 2 Messenger RNA

Louise T. Chow, Richard E. Gelinas, Thomas R. Broker and Richard J. Roberts
Cold Spring Harbor Laboratory
Cold Spring Harbor, New York 11724

Summary

The 5′ terminal sequences of several adenovirus 2 (Ad2) mRNAs, isolated late in infection, are complementary to sequences within the Ad2 genome which are remote from the DNA from which the main coding sequence of each mRNA is transcribed. This has been observed by forming RNA displacement loops (R loops) between Ad2 DNA and unfractionated polysomal RNA from infected cells. The 5′ terminal sequences of mRNAs in R loops, variously located between positions 36 and 92, form complex secondary hybrids with single-stranded DNA from restriction endonuclease fragments containing sequences to the left of position 36 on the Ad2 genome. The structures visualized in the electron microscope show that short sequences coded at map positions 16.6, 19.6 and 26.6 on the R strand are joined to form a leader sequence of 150–200 nucleotides at the 5′ end of many late mRNAs. A late mRNA which maps to the left of position 16.6 shows a different pattern of second site hybridization. It contains sequences from 4.9–6.0 linked directly to those from 9.6–10.9. These findings imply a new mechanism for the biosynthesis of Ad2 mRNA in mammalian cells.

Introduction

In contrast to the detailed knowledge of the mechanics of transcription in procaryotic cells (Losick and Chamberlin, 1976), little is known about this process in eucaryotic cells. Several possible schemes exist: one, analogous to the bacterial system, requires independent promoters for each mRNA; a second postulates the production of long primary transcripts in the nucleus which are subsequently cleaved to yield individual mRNAs (Darnell, Jelinek and Molloy, 1973); and a third invokes the use of RNA primers coded at one region on the genome but acting at some other region(s) and becoming elongated into mRNAs (Dickson and Robertson, 1976). Experiments to test these hypotheses directly have been hampered by the complexity of the eucaryotic genome. We have chosen to study these processes in a simpler system — lytic infection of human cells by adenovirus 2 (Ad2).

Ad2 DNA is transcribed by RNA polymerase II (Price and Penman, 1972; Wallace and Kates, 1972), and its transcription shows features charac-

teristic of that of the host genome (Lewin, 1975a, 1975b). For example, long polyadenylated transcripts appear in the nucleus, but only a small percentage of this nuclear RNA appears as polyadenylated mRNA on cytoplasmic polysomes (Philipson et al., 1971). These mRNAs are "capped" at their 5′ ends (Moss and Koczot, 1976; Sommer et al., 1976). Gelinas and Roberts (1977) found that most Ad2 mRNAs isolated at late times during infection contain the same "capped" 11 nucleotide sequence at their 5′ ends. This sequence was sensitive to ribonuclease cleavage in mRNA:DNA hybrids (Gelinas and Roberts, 1977; Klessig, 1977) and led to the suggestion that this 5′ terminal sequence might not be coded immediately adjacent to the main body of the mRNA.

Thomas, White and Davis (1976) have shown that individual RNA molecules can be displayed as RNA displacement loops (R loops) in the electron microscope, and map coordinates have been obtained for many Ad2 mRNAs (Meyer et al., 1977; Chow et al., 1977). In the present studies, we have used mRNAs visualized in such R loops to examine more closely the sequences present at the 5′ end of late Ad2 mRNAs.

Results

R loops were formed between Ad2 DNA and polysomal RNA isolated 22 hr after Ad2 infection. The 5′ ends of the mRNA should form single-stranded projections if they are not coded immediately adjacent to the rest of the mRNA, and so might be visualized by hybridization to a single-stranded DNA fragment containing their complement. We therefore prepared a set of restriction endonuclease fragments of the Ad2 genome, separated their strands by agarose gel electrophoresis (Hayward, 1972; Sharp, Gallimore and Flint, 1974) and added each single strand in turn as a third hybridization component after the preparation of the R loops. Since R loops were formed from a mixed population of late mRNAs, many different species were examined simultaneously. By using a restriction endonuclease fragment as the single-stranded probe, complicated structures which might arise from hybridization of the probe to the single-stranded DNA segment of the R loop were limited to one region of the genome. Figure 1a shows the results of such an experiment using the slow strand of Hind III-B (map position 17.0–31.5) as the single-stranded DNA probe. The probe hybridized with the 5′ end of hexon mRNA in the R loop but not with the displaced DNA strand. It adopted a looped configuration, indicating that sequences from the 5′ end of the mRNA were complementary to two separate regions within the probe. The 5′ ends of other

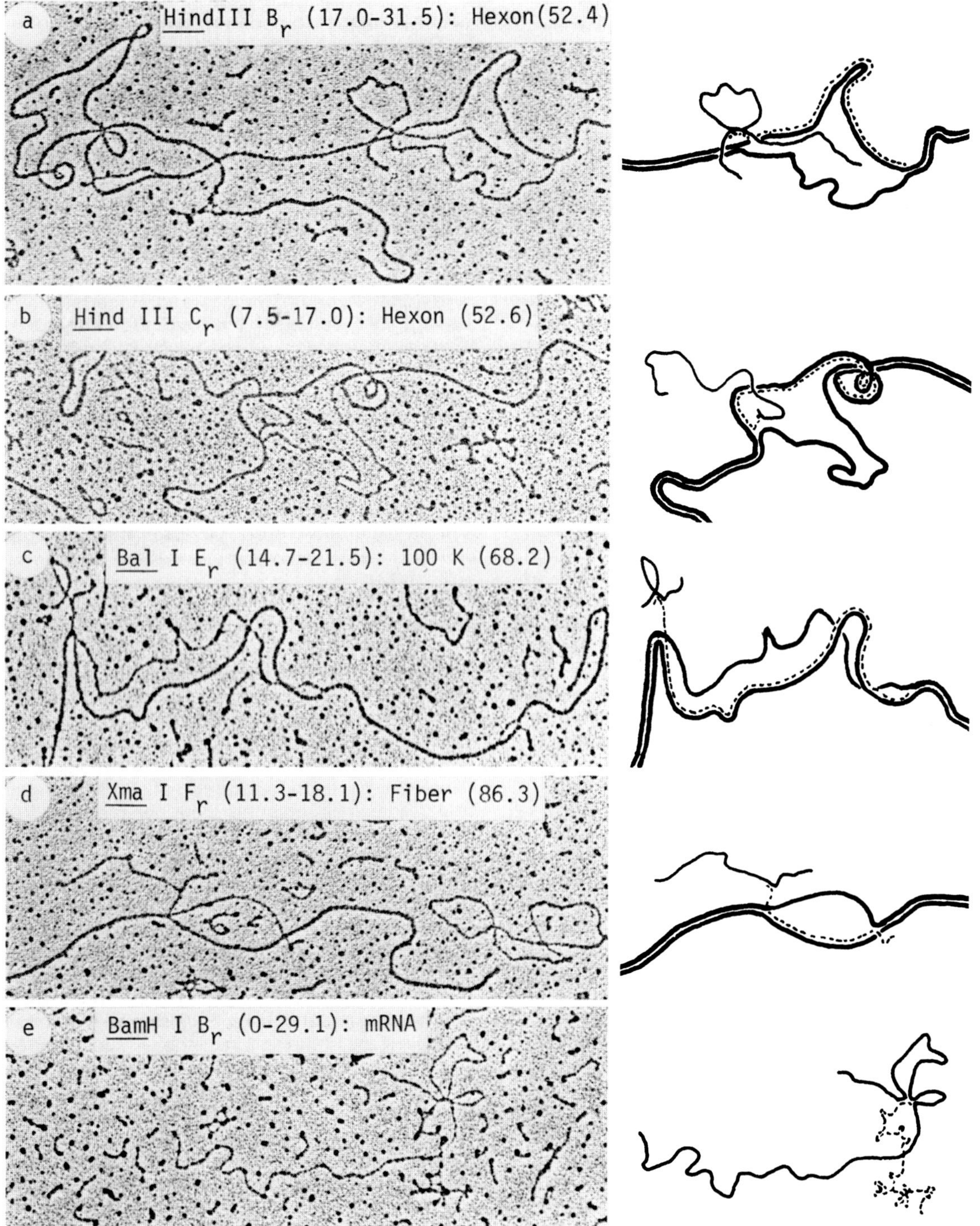

a HindIII B_r (17.0-31.5): Hexon(52.4)
b Hind III C_r (7.5-17.0): Hexon (52.6)
c Bal I E_r (14.7-21.5): 100 K (68.2)
d Xma I F_r (11.3-18.1): Fiber (86.3)
e BamH I B_r (0-29.1): mRNA

Table 1. Map Coordinates of Labeled 5′ Termini of Late Ad2 mRNAs

mRNA Assignment	Previous Map Coordinates[a]	Map Coordinates of 5′ Label (Mean ± Standard Deviation)	Number R Loop Molecules Labeled with Restriction Endonuclease Fragments				
			Bam HI-B	Hind III-B	Bal I-E	Xma I-F	Total[b]
Core	−	36.6 ± 0.6	2	1	0	2	5
Penton	38.8	38.9 ± 1.0	1	3	3	0	7
Core	45.4	45.0 ± 1.0	1	6	1	1	9
pVI	49.9	49.8 ± 0.5	3	4	2	2	11
Hexon	51.9	52.2 ± 0.8	28	20	4	4	56
100K	67.9	67.9 ± 0.4	3	3	2	3	11
pVIII	74.6	74.1 ± 0.5	5	2	0	2	9
Fiber	86.3	86.4 ± 0.5	29	26	5	2	62
Totals			72	65	17	16	170

[a] Chow et al. (1977).
[b] Not included in the table are two molecules, each labeled at 66 and 71, that could be alternative 5′ ends for 100K and pVIII, respectively.

mRNAs also show identical two-site hybridization with the slow strand of Hind III-B and are compiled in Table 1. Length measurements place the contact points between the Hind III-B single strand and the mRNA at approximately 900 ± 60 nucleotides (42 measurements) from the end of the short arm and 1800 ± 120 nucleotides (42 measurements) from the end of the long arm. The distance between the two contact points on the DNA (the loop) was about 2400 ± 90 nucleotides (49 measurements). To orient these two arms, determine the strandedness and obtain accurate map positions for the points of hybridization, we used the separated strands of an overlapping fragment Bam HI-B (map position 0–29.1) in a similar experiment. The results are shown in Figure 2a, in which the slow strand of Bam HI-B is hybridized to the 5′ ends of both fiber and hexon mRNAs. In these cases and in others reported in Table 1, more complicated structures were observed. Three contact points between the single-stranded DNA probe and the 5′ end of the mRNA are now evident, and the Bam HI-B fragment is held into two loops. Length measurements give values of 5800 ± 180 nucleotides (39 measurements) for the long arm, 950 ± 100 nucleotides (58 measurements) for the short arm, 2400 ± 130 nucleotides (48 measurements) for the large loop and 1000 ± 100 nucleotides (47 measurements) for the small loop. Comparison with the hybridization sites on the Hind III-B strand suggests map positions of

16.6, 19.6 and 29.6 for the three segments of Ad2 DNA which hybridize to the 5′ ends of mRNA. Examination of these structures revealed that the contact point closest to the main portion of the mRNA was on the long arm of the Hind III-B fragment and on the short arm of the Bam HI-B fragment. Thus the 3′ end of the leader sequence is at 26.6 and the 5′ end is at 16.6. Because the mRNAs labeled by these probes are transcribed from the R strand from left to right, and because nucleic acids form anti-parallel base pairs, we conclude that these probes are from the R strand. Weingartner et al. (1976) have also shown that the slow strand of Bam HI-B is the R strand. All the mRNA species labeled are transcribed from the R strand (Sharp, Gallimore and Flint, 1974; Pettersson, Tibbetts and Philipson, 1976).

The 5′ terminal leader sequence of an mRNA in an R loop occasionally formed an intramolecular structure by hybridizing to its complementary DNA at coordinates 19.6 or 26.6 within the same DNA molecule. One example involving hexon mRNA is shown in Figure 2b. Such interaction constrains the intervening DNA, which often assumes a supercoiled configuration during spreading for electron microscopy.

To ensure that our interpretation of these structures was correct, we performed a number of control experiments. In separate hybridizations, single strands from restriction fragments encompassing

Figure 1. Hybridization of Rightward-Transcribed Strands (r) of Restriction Fragments to the Common 5′ Leader Sequences of Late Ad2 mRNA (e) or mRNAs in R Loops on Ad2 DNA (a–d)

(a) represents Hind III-B_r annealed to mRNA for hexon; (b) represents Hind III-C_r annealed to mRNA for hexon; (c) represents Bal I-E_r annealed to mRNA for the 100K protein; (d) represents Xma I-F_r annealed to mRNA for fiber; (e) represents Bam HI-B_r annealed to free mRNA. Map coordinates covered by each restriction fragment and locations of the hybridization are given in parentheses. Illustrative tracings are provided. (——) Ad2 DNA; (——) restriction fragments; (----) mRNA. In (c) and (d), most of the RNA "bridge" between the R loop and the restriction fragment is due to branch migration of the mRNA. The remaining portion is due to the unhybridized leader sequence.

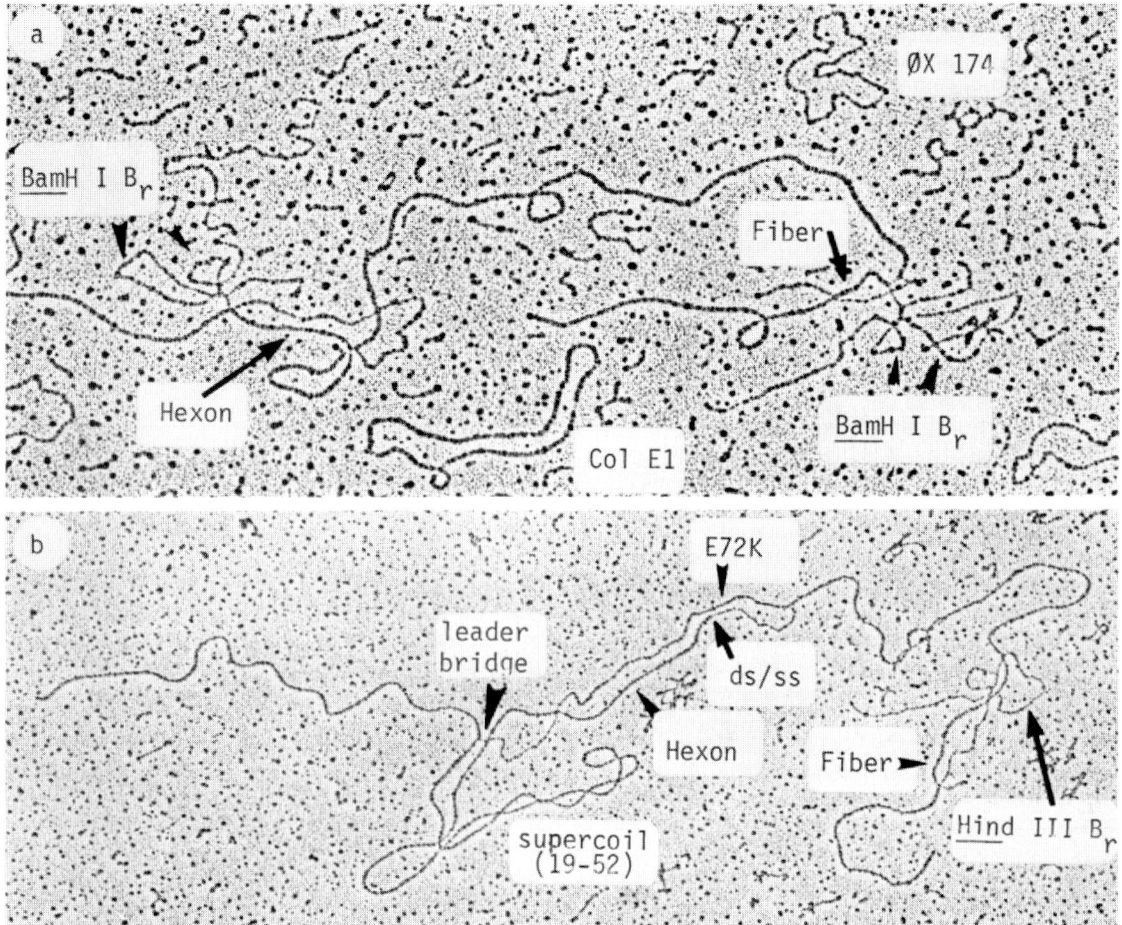

Figure 2. Multiple Site Hybridization of the 5′ Leader Sequences of the Hexon and Fiber mRNAs in R Loops

(a) Hybridization with the R strand (r) of the Bam HI-B restriction fragment. Arrows point to the DNA:RNA hybrids. Arrowheads point to the large and small loops formed in Bam HI-B$_r$ DNA due to the hybridization. An additional 100 nucleotides at the 5′ ends of the hexon R loops have been displaced by branch migration. The spreading force during the preparation of grids has denatured about 200 nucleotides at the 5′ end of the fiber mRNA/DNA hybrid. (b) The leader of the fiber mRNA in an R loop was labeled by an added R strand of the Hind III-B$_r$ fragment. The leader on the hexon mRNA was labeled by intramolecular hybridization to complementary DNA at coordinate 19 on the same molecule. The intervening DNA segment was constrained, and it formed tertiary superhelical twists when solvent conditions were changed during preparation of the sample for electron microscopy. The hexon RNA formed a convergent R loop with the mRNA for the E72K protein hybridized to the opposite (L) DNA strand.

the entire Ad2 genome were used as probes, and a summary of these data is shown in Figure 3. Only the slow strands of Hind III-B (Figure 1a, two contacts at 19.6 and 26.6), Hind III-C (Figure 1b, one contact at 16.6), Bam HI-B (Figure 2a, three contacts at 16.6, 19.6 and 26.6), Bal I-E (Figure 1c, two contacts at 16.6 and 19.6) and Xma I-F (Figure 1d, one contact at 16.6) showed consistent hybridization to RNA branches at the 5′ ends of R loops. In particular, it should be noted that the fast strands of these five fragments did not interact with any of the R loops. When the slow strands of Bam HI-B or Bal I-E were incubated alone and spread under identical conditions, no loops of the same size or with the same coordinates as those formed in the presence of mRNA were detected. If polysomal RNA was present during the incubation of the slow strand, but not the fast strand, of Bam HI-B, however, loops of the type shown in Figure 1e (identical to those seen at the 5′ ends of many late mRNAs in R loops) were frequently observed and were associated with collapsed RNA. The possibility that sequences at map positions 16.6, 19.6 and 26.6 were reiterated on the Ad2 genome was tested by isolating small fragments of the genome containing these sequences, labeling them to high specific

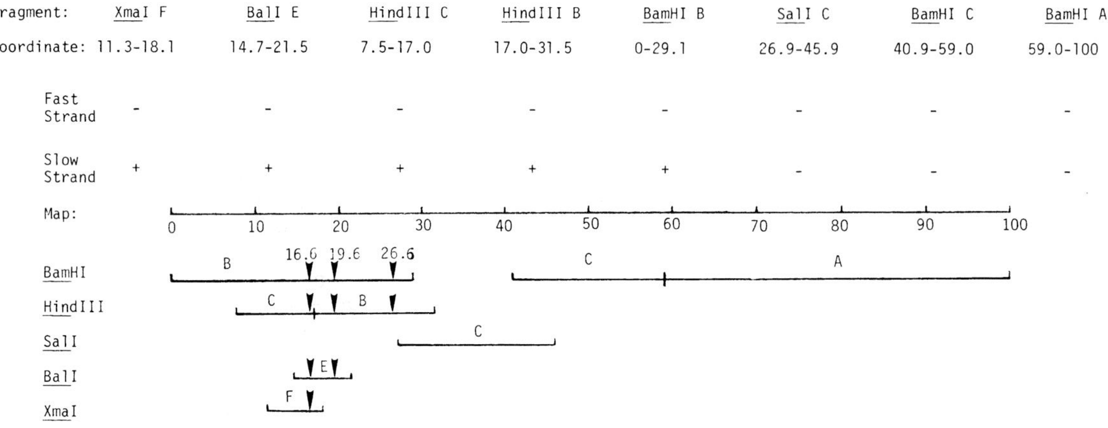

Figure 3. Hybridization of Separated Strands of Ad2 DNA Restriction Fragments to the 5′ Leader Sequence of Late Polysomal mRNAs.
Strands from restriction fragments spanning the entire Ad2 genome were used in separate experiments to label the R loops. All (+) slow strands come from the R strand, as discussed in Results. (+) indicates consistent hybridization of the strands to the leader sequences of the mRNAs in R loops; (−) indicates negative results. Arrowheads point to the locations of hybridization on the R strands of the fragments. The map coordinates of the restriction fragments are obtained from C. Mulder and R. Greene for Bam HI and Xma I (unpublished observations), from R. J. Roberts and J. Sambrook for Hind III; from J. R. Arrand and R. J. Roberts for Sal I (unpublished observations); and from R. E. Gelinas and R. J. Roberts for Bal I (unpublished observations).

activity in vitro by nick translation and using them as hybridization probes against fragments of the Ad2 genome immobilized on nitrocellulose filters (Southern, 1975). In each case, as shown in Figure 4, the fragments rehybridized only to that region of the genome from which they were derived and failed to hybridize to any other sequences on the Ad2 genome.

Hybrids between any one component of the leader sequences of the mRNA in an R loop and single-stranded DNA probes are stable in 70% formamide, 0.4 M NaCl, 0.1 M HEPES at 30°C, and yet there is only a hint of a duplex at positions 26.6 and 19.6 when Bam HI-B and Hind III-B are used, or at position 19.6 in the Bal I-E fragment. The duplex regions were measured to be 50–100 nucleotides at each of these two positions, and we believe it is improbable that more than a total of 200 nucleotides are involved at all three contact points.

The results described above refer to transcripts located to the right of position 36. Several other late mRNAs are known to map to the left of this coordinate. One of these, coding for polypeptide IVa₂ (map position 14.9–11.2), is transcribed from the L strand, and a second, coding for virion-associated component IX (map position 9.7–11.0), is transcribed from the R strand (Chow et al., 1977; U. Pettersson and M. B. Mathews, manuscript submitted). Both have been visualized in R loops, but neither showed secondary hybridization with any of the fragments used in this study. Some of the R loops formed by a polysomal RNA which contains sequences from coordinates 9.6 (± 0.2)–10.9 (± 0.2) (24 measurements each), however, have an unusual structure. Sequences from the 5′ end of this RNA form a second R loop with a noncontiguous region of the Ad2 genome located between coordinates 4.9 (± 0.3)–6.0 (± 0.2) (Figures 5a and 5b). As a result, the intervening double-stranded DNA was held into a third loop, and a short bridge of displaced RNA between the two R loops is clearly visible. This structure has frequently been observed in molecules containing a convergent R loop formed between IVa₂ mRNA and this new RNA (Figure 5c). Because a strand switch at 11.0 can be seen in this structure, as has been observed earlier (Chow et al., 1977), the new RNA species can be assigned to the R strand.

Discussion

The results presented in this paper show that sequences present at three separated sites (16.6, 19.6, 26.6) on the R strand of the Ad2 genome are complementary to a continuous sequence at the 5′ end of late Ad2 mRNAs that are transcribed from the R strand and map to the right of position 36. Since these sequences are available for hybridization when mRNA is displayed in R loops and are not reiterated elsewhere in the Ad2 genome, we conclude that they are not coded at a site immediately adjacent to the main portion of the mRNAs. Biochemical evidence by Gelinas and Roberts (1977) and Klessig (1977) has been presented to support this idea. Since it seems improbable to us that the sequence present at the 5′ end of many of these late Ad2 mRNAs is actually coded by the host genome and is only complementary to these three

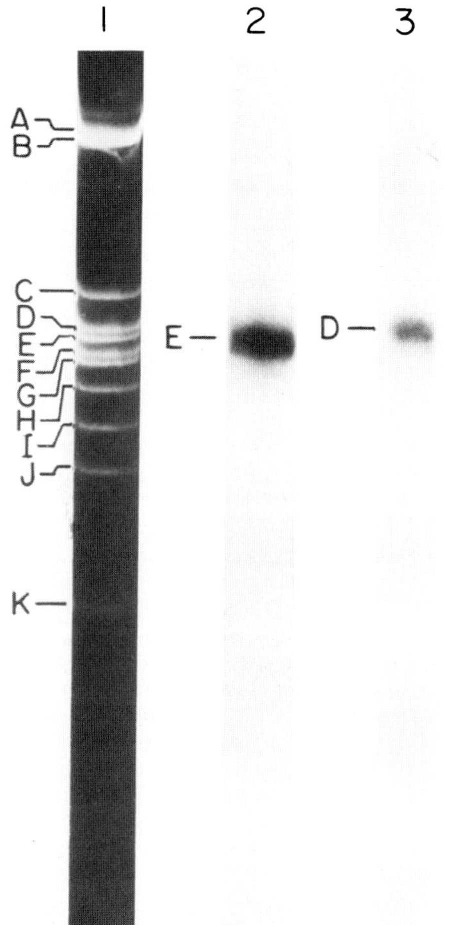

Figure 4. Hybridization of Bal I-E DNA (14.7–21.5) and Bal I-D DNA (21.5–28.5) to Bal I Fragments of Ad2 DNA

Bal I-E and -D fragments of Ad2 DNA were isolated after two cycles of purification by agarose gel electrophoresis, labeled with ^{32}P by nick translation and used as hybridization probes against all Bal I fragments of Ad2 DNA bound to nitrocellulose membranes. Slot 1 represents 2.0 μg of Bal I fragments of Ad2 DNA fractionated on a 1.4% agarose gel and stained with ethidium bromide. The same amount of DNA was present in slots 2 and 3. Slot 2 represents ^{32}P–Bal I-E DNA (10⁶ dpm; about 20 μg) hybridized to Bal I fragments of Ad2 DNA. Slot 3 represents ^{32}P–Bal I-D DNA (10⁶ dpm; about 20 μg) hybridized to Bal I fragments of Ad2 DNA. The minimal length of sequence homology which can be detected by this method has not been determined.

Ad2 sequences by chance, we believe that these sequences are probably transcribed from positions 16.6, 19.6 and 26.6 on the R strand of the Ad2 genome, and that their juxtaposition is an inherent feature of Ad2 mRNA biosynthesis.

Two mRNAs (for polypeptides IVa₂ and IX) mapping to the left of position 30 seem to have a different sequence arrangement at their 5′ ends. Particularly surprising is the finding that a polysomal RNA

containing sequences from coordinate 9.6–10.9, the coding region for component IX, has an additional sequence at its 5′ end which is complementary to a noncontinguous segment from 4.9–6.0. This RNA may be related to early transcripts for E15K, which map between 5.0–11.0 or between 5.0–6.4 (Chow et al., 1977), and also to the component IX mRNA, which maps between 9.7–11.0 (Chow et al., 1977; U. Pettersson and M. B. Mathews, manuscript submitted). The absence of the tripartite leader and the occurrence of this new mRNA would account for the hybridizational and translational data reported for mRNAs originating from this region of the genome (Lewis, Anderson and Atkins, 1977).

These observations, together with the results presented in the accompanying papers on late Ad2 mRNA (Klessig, 1977; Lewis, Anderson and Atkins, 1977) and on Ad2-SV40 mRNA (Dunn and Hassell, 1977) are not directly consistent with any mechanism previously suggested for the biosynthesis of mRNA in eucaryotic cells. They imply that an alternate scheme must exist for Ad2 mRNAs, and perhaps for eucaryotic mRNA in general. One such mechanism is outlined in the accompanying paper by Klessig (1977). The experiments described herein provide a convenient method to map accurately the 5′ termini of Ad2 mRNAs, and have confirmed many of the previous assignments (Chow et al., 1977) and established new ones. We have recently learned of similar experiments by Berget, Moore and Sharp (1977) who used electron microscopy to examine hybrids between purified hexon mRNA and single strands of DNA. They observed that the 5′ terminal mRNA sequence appeared as a single-stranded tail, which was complementary to three noncontiguous regions of the Ad2 genome with map coordinates essentially identical to those reported here.

Experimental Procedures

Restriction Endonucleases
Bal I (Gelinas et al., 1977) and Xma I (Endow and Roberts, 1977) were purified as described. Bam HI, Sal I and Xma I were purified by unpublished procedures of P. A. Myers and R. J. Roberts. In all cases, DNA was digested at 37°C in 6 mM Tris–HCl (pH 7.9), 6 mM MgCl₂ and 6 mM 2–mercaptoethanol.

Isolation of Viral DNA and RNA
DNA was prepared from Ad2 virions grown on HeLa or KB cells in suspension cultures as described by Pettersson and Sambrook (1973) and Pettersson et al. (1973). Fragments of the Ad2 genome were produced by digestion with the restriction endonucleases Bal I, Bam H-I, Hind III, Sal I and Xma I. DNA fragments were fractionated by agarose slab gel electrophoresis (Sugden et al., 1975) and recovered from the agarose by chromatography on hydroxylapatite (Lewis et al., 1975), or by homogenization and diffusion followed by phenol extraction. Ad2 mRNA was a gift from Dr. J. B. Lewis. Polysomes were isolated 22 hr after Ad2 infection of KB cells by the method of Schreier and Staehelin (1973), and the RNA was recovered by the method of Anderson et al. (1974).

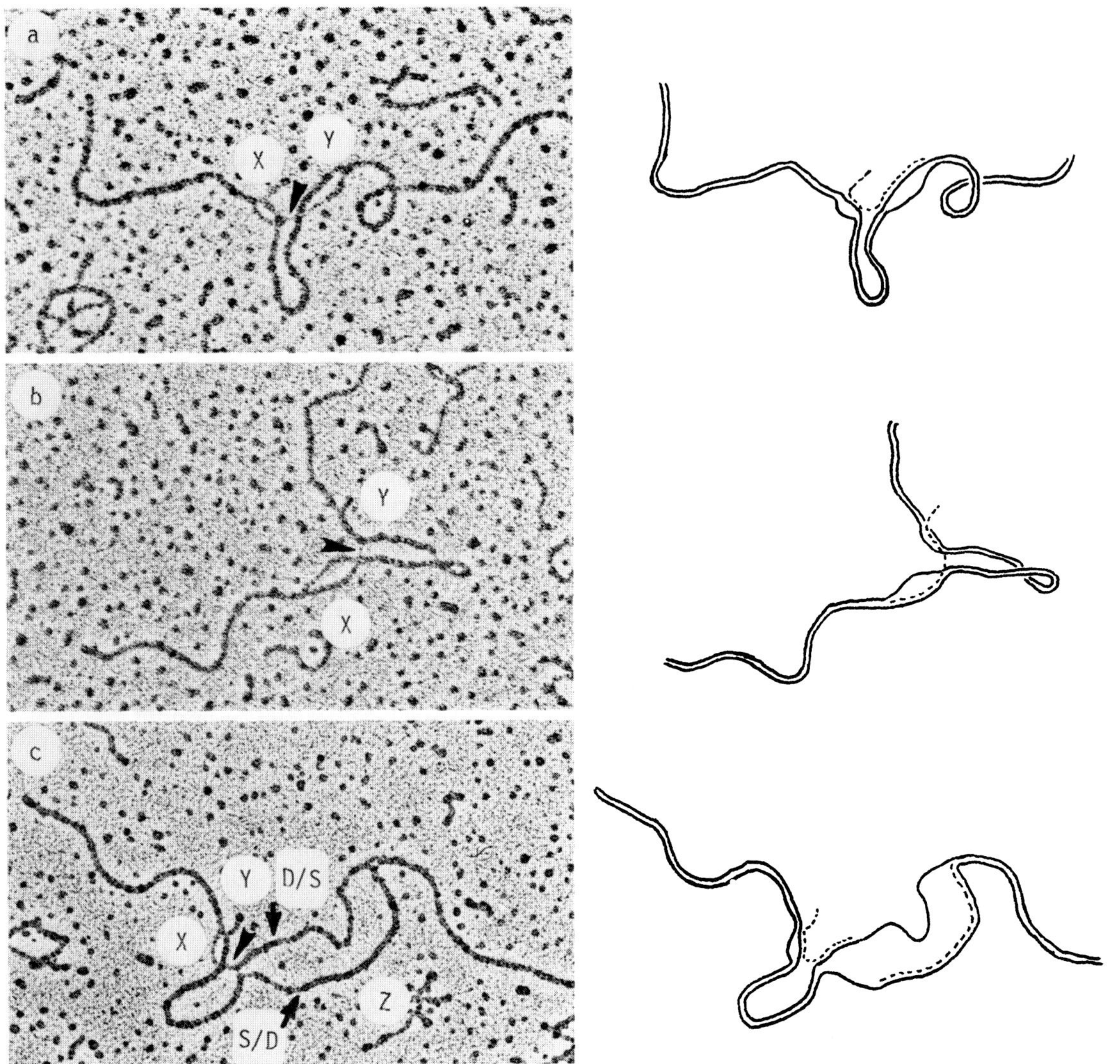

Figure 5. R Loops Formed between Ad2 DNA and a Polysomal RNA Containing Sequences from Map Coordinates 4.9–6.0 (X) and 9.6–10.9 (Y)

(a) Sequence Y, with the same coordinates as the mRNA for peptide IX, is totally contained in an R loop, whereas sequence X, possibly coding for the 15K protein, is only partially contained in an R loop with its 5′ end displaced as a tail. (b) Y is present in a collapsed and partially displaced R loop, whereas X is totally contained in an R loop. (c) X is in a partial R loop. Y is in a convergent R loop with the mRNA (Z), tentatively assigned to peptide IVa$_2$, on the opposite DNA strand. The RNA bridge (indicated by arrowheads) between X and Y is visible because of some RNA displacement by the intervening DNA segment. D/S and S/D indicate the double-strand/single-strand junctions in the convergent R loop.

Strand Separation of Endonuclease Fragments

Purified restriction fragments were denatured in 0.25 M NaOH and subjected to electrophoresis on 1.4% agarose slab gels containing the Tris-phosphate-EDTA buffer described by Hayward (1972), but at half the stated ionic strength. After electrophoresis, bands of single strands were located by staining with ethidium bromide. Single-stranded DNA was recovered by homogenizing the gel in several volumes of 0.01 M Tris–HCl (pH 7.9), 0.001 M EDTA and allowing the DNA to diffuse out for several hours. The aqueous supernatant was extracted first with phenol and then with chloroform. E. coli rRNA was added as carrier, and the single-stranded DNA was recovered by ethanol precipitation.

Filter Hybridizations

Bal I-E (14.7–21.5) and Bal I-D (21.5–28.5) DNAs were labeled in vitro by nick translation (Kelly et al., 1970) as described by Maniatis et al. (1975) and were used as probes to challenge Bal I fragments of Ad2 DNA adsorbed to nitrocellulose membranes by the method of Southern (1975).

Electron Microscopy

R loops were formed on intact Ad2 DNA at 51.5°C for 14–16 hr as described previously (Chow et al., 1977). Aliquots were diluted with an equal volume of the same buffer-formamide mixture containing purified, separated strands of Ad2 restriction fragments.

580   Microbiology: A Centenary Perspective

Cell
8

The concentration of the single strands was 5–10 μg/ml. The solution was returned to the water bath and cooled to 42 or 30°C over a period of 3–5 hr. Electron microscope grid preparation and data processing have been described by Chow et al. (1977). Single-stranded φX174 (5375 bases) and double-stranded φX174 RF or Col E1 DNA (6300 base pairs) were included as internal length standards.

Acknowledgments

We thank Dr. J. B. Lewis for a gift of late polysomal RNA; J. Bonventre, P. A. Myers and J. Scott for technical assistance; and M. Moschitta for secretarial help. This work was supported by a grant from the National Cancer Institute.

Received June 9, 1977; revised July 5, 1977.

Note Added in Proof

The mRNA for IVa_2 (14.9–11.2, L strand) also has a short single component leader present at its 5′ end. There is only a short deletion of the RNA sequences between the leader and the coding sequences, which is visible in Figure 5c as a small loop.

References

Anderson, C. W., Lewis, J. B., Atkins, J. F. and Gesteland, R. F. (1974). Proc. Nat. Acad. Sci. USA 71, 2756–2760.

Berget, S. M., Moore, C. and Sharp, P. A. (1977). Proc. Nat. Acad. Sci. USA, in press.

Chow, L., Roberts, J. M., Lewis, J. B. and Broker, T. R. (1977). Cell 11, 819–836.

Darnell, J. E., Jelinek, W. R. and Molloy, G. R. (1973). Science 181, 1215–1221.

Dickson, E. and Robertson, H. D. (1976). Cancer Res. 36, 3387–3393.

Dunn, A. R. and Hassell, J. A. (1977). Cell 12, 23–36.

Endow, S. A. and Roberts, R. J. (1977). J. Mol. Biol. 112, 521–529.

Gelinas, R. E. and Roberts, R. J. (1977). Cell 11, 533–544.

Gelinas, R. E., Myers, P. A., Weiss, G. H., Murray, K. and Roberts, R. J. (1977). J. Mol. Biol., in press.

Hayward, G. S. (1972). Virology 49, 342–344.

Kelly, R. B., Cozzarelli, N. R., Deutscher, M. P., Lehman, I. R. and Kornberg, A. (1970). J. Biol. Chem. 245, 39–45.

Klessig, D. F. (1977). Cell 12, 9–21.

Lewin, B. (1975a). Cell 4, 11–20.

Lewin, B. (1975b). Cell 4, 77–93.

Lewis, J. B., Anderson, C. W. and Atkins, J. F. (1977). Cell 12, 37–44.

Lewis, J. B., Atkins, J., Anderson, C., Baum, P. and Gesteland, R. (1975). Proc. Nat. Acad. Sci. USA 72, 1344–1348.

Losick, R. and Chamberlin, M. (1976). RNA Polymerase (Cold Spring Harbor, New York: Cold Spring Harbor Laboratory Press), p. 899.

Maniatis, T., Jeffrey, A. and Kleid, D. G. (1975). Proc. Nat. Acad. Sci. USA 72, 1184–1188.

Meyer, J., Neuwald, P. D., Lai, S. P., Maizel, J. V., Jr. and Westphal, H. (1977). J. Virol. 21, 1010–1018.

Moss, B. and Koczot, F. (1976). J. Virol. 17, 385–392.

Pettersson, U. and Sambrook, J. (1973). J. Mol. Biol. 73, 125–130.

Pettersson, U., Tibbetts, C. and Philipson, L. (1976). J. Mol. Biol. 101, 479–501.

Pettersson, U., Mulder, C., Delius, H. and Sharp, P. A. (1973).

Proc. Nat. Acad. Sci. 70, 200–204.

Philipson, L., Wall, R., Glickman, G. and Darnell, J. E. (1971). Proc. Nat. Acad. Sci. USA 68, 2806–2809.

Price, R. and Penman, S. (1972). J. Virol. 9, 621–626.

Schreier, M. H. and Staehelin, T. (1973). J. Mol. Biol. 73, 329–349.

Sharp, P. A., Gallimore, P. H. and Flint, S. J. (1974). Cold Spring Harbor Symp. Quant. Biol. 39, 457–474.

Sommer, S., Salditt-Georgieff, M., Bachenheimer, S., Darnell, J. E., Furuichi, Y., Morgan, M. and Shatkin, A. J. (1976). Nucl. Acids Res. 3, 749–765.

Southern, E. M. (1975). J. Mol. Biol. 98, 503–517.

Sugden, B., DeTroy, B., Roberts, R. J. and Sambrook, J. (1975). Anal. Biochem. 68, 36–46.

Thomas, M., White, R. L. and Davis, R. W. (1976). Proc. Nat. Acad. Sci. USA 73, 2294–2298.

Wallace, R. D. and Kates, J. (1972). J. Virol. 9, 627–635.

Weingartner, B., Winnacker, E-L., Tolun, A. and Pettersson, U. (1976). Cell 9, 259–268.